Glossary of Notation

Numerical Analysis

third edition

Richard L. Burden
Youngstown State University

J. Douglas Faires
Youngstown State University

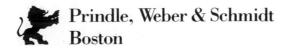 Prindle, Weber & Schmidt
Boston

PWS PUBLISHERS

Prindle, Weber & Schmidt •♧• Duxbury Press •♠• PWS Engineering •⚲• Breton Publishers •⚙
20 Park Plaza • Boston, Massachusetts 02116

PWS Publishers is a division of Wadsworth, Inc.

Library of Congress Cataloging in Publication Data

Burden, Richard L.
    Numerical analysis.

    Bibliography: p.
    Includes index.
    1. Numerical analysis.   I. Faires, J. Douglas.
II. Title.
QA297.B84   1985   519.4   84-16073
ISBN 0-87150-857-5

ISBN 0-87150-857-5

Printed in the United States of America.

10 9 8 7 6 5 4   — 88 87 86

Text and cover design by Elise Kaiser. Cover art "Reaction" 1984 by Masaaki Noda; a silk screen print used with the permission of the artist. Production handled by Unicorn Production Services, Inc. Text composition by H. Charlesworth & Co. Ltd. Covers printed by New England Book Components. Text printed and bound by Halliday Lithograph.

Contents

Preface

About the Text

The material in this text was developed over a period of years for a sequence of courses in the theory and application of numerical methods. The students for whom this sequence was designed are primarily junior-level science and engineering majors who have completed at least the first year of the college calculus sequence. In addition, these students have a knowledge of some high-level programming language such as BASIC, FORTRAN, or Pascal. Although a knowledge of the fundamentals of matrix algebra and differential equations is also useful, adequate introductory material on these topics is presented in the text; so these topics are not prerequisites.

Sufficient material is included to serve as a basis for a full year of study, but we expect that most readers will use the text for only a single-term course. In such a course the students learn to recognize the type of problems that require numerical techniques for their solution, see some examples of the error propagation that can occur when numerical methods are applied, and have the opportunity to accurately approximate the solution to some problems that cannot be solved exactly. This permits the remainder of the text to be used as a reference for the type of problems that will occur in future work. Either the full-year or single-course treatment is consistent with the philosophy of the text: *to give an introduction to numerical methods; to tell how, why, and when they can be expected to work; and to provide a firm basis for future study.*

Virtually every concept in the text is illustrated by an example, and this edition contains nearly 2000 class-tested exercises. The exercises range from elementary applications of the algorithms to generalizations and extensions of the theory. In addition, a large number of applied problems are presented from diverse areas of engineering and the physical, biological, and social sciences. The applications chosen are concise and demonstrate how numerical methods can be, and are applied in "real-life" situations.

Changes in the Third Edition

This edition contains an improved treatment of many of the topics in the previous edition, as well as a few additions:
· Chapter 1 has been expanded to include more discussion on round-off error, with new examples to demonstrate when problems commonly occur.

- Theoretical material in Chapter 2 on solving single nonlinear equations has been condensed and simplified.
- Müller's method has been added in Chapter 2 as the general purpose method for approximating the roots of polynomials.
- More emphasis has been placed on the divided–difference techniques in Chapter 3 by adding two new algorithms: one for Newton's interpolatory divided–difference method, the other using divided differences to simplify Hermite interpolation.
- Chapter 4 more heavily emphasizes the importance of the Richardson extrapolation technique.
- In Chapter 4 the section on Gaussian quadrature has been modified to eliminate much of the overlap with the Chapter 7 orthogonal polynomial material.
- Algorithms have been added in Chapter 5 for solving systems of differential equations and stiff differential equations.
- The Chapter 5 sections on stability and stiffness have been rewritten and expanded.
- The Chapter 5 introductory material on multistep methods is now separated from the treatment of variable step–size predictor–corrector techniques.
- Chapter 6 has been reorganized to make it easier to delete the material normally covered in a linear algebra course.
- In Chapter 6 the section on pivoting techniques has been expanded to more fully describe this important topic.
- In Chapter 7, the discussion of orthogonal polynomials has been streamlined, and treatment of the fast Fourier transform has been substantially expanded.
- The first section of Chapter 8 has been simplified and the material in the latter sections reordered.
- A section on the method of steepest descent has been added to the end of Chapter 9 to provide a means to determine initial approximations for the Newton and Broyden methods.
- In Chapter 10 the section on finite difference methods for solving boundary–value problems has been split to separate the linear and nonlinear cases.
- The method of characteristics has been deleted from Chapter 11, as reviewers suggested it is a procedure infrequently used.

Algorithms

As in the previous edition, a detailed structured algorithm without program listing is given for each significant method presented in the text. The algorithms are in a form that can be coded by students with even limited programming experience. Actual programs are not included because, in our experience, this encourages some students to generate results without fully understanding the method involved. However, a FORTRAN listing of programs for all the algorithms is available from the publisher for instructors using the book. The programs are contained in an instructor's manual that also includes the answers to the exerecises not given in the back of the book. Most of the results in the answer section, and examples, were generated using these programs run in single precision Structured WATFIV on an Amdah 470 V/6 computer at Youngstown State University.

Although the algorithms will lead to correct programs for the examples and exercises in the text, it must be emphasized that there has been no attempt to write general-purpose software. In particular, the algorithms have not always been listed in the form that leads to the most efficient program in terms of either time or storage requirements. When a conflict occurred between writing a form leading to an extremely efficient program and writing a slightly different one illustrating the important features of the method, the latter path was invariably taken.

It was our intention in this edition to provide detailed descriptions of the standard software routines that are available for numerical methods. Recently, however, a book has appeared by John Rice [108] that is dedicated to this task. This book, *Numerical Methods, Software, and Analysis: IMSL Reference Edition*, presents complete information on most of the appropriate subroutines available from IMSL (International Mathematical and Statistical Library) as well as a discussion of the ACM algorithms and relevant software available from other sources. It now seems more appropriate to recommend the reader to Professor Rice's book, if the library routines will be needed on a regular basis, than to summarize this information in our text.

Suggested Course Outlines

The text has been designed to allow instructors flexibility in the choice of topics as well as the level of theoretical rigor and emphasis on applications. In line with these aims, detailed references have been provided for all results that are not demonstrated in the text, as well as for the applications that are used to indicate the practical importance of the methods. The references have been specifically chosen on the basis of being the most generally available sources in most college libraries.

A flow chart follows that indicates chapter prerequisites. The only deviations from this chart are described in footnotes at the bottom of the first pages of Sections 3.6 and 4.7. Most of the possible sequences that can be generated from this chart have been successfully taught by the authors at Youngstown State University.

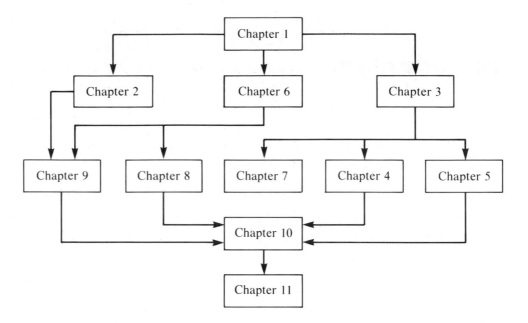

Acknowledgments

The authors would like to express their appreciation to the many individuals who made suggestions for improvement on the previous editions of this book. These include our past reviewers, the reviewers of this version: G. S. Gill, Brigham Young University; William Siegmann, Rensselaer Polytechnic Institute; Jozef Brody, Concordia University; Alain Lan, Ryerson Polytechnic; W. Gesing, University of Toronto; Charles Chen, University of Alabama, Huntsville; G. J. Kurowski, University of California, Davis; Richard E. Goodrick, California State University, Hayward; R. A. Mathon, University of Toronto; Heath K. Riggs, University of Vermont; Roger C. McCann, Mississippi State University; Laxmi N. Gupta, Rochester Institute of Technology; Nicholas D. Goodman, SUNY, Buffalo; and Ralph Lee, University of Missouri, Rolla, as well as the many students and faculty who dropped us a line or made a suggestion in a class or at a meeting. In particular, we want to thank Don Maxwell, a graduate student at Youngstown State, who checked the bibliography and helped us prepare the book for production.

Introduction

The subject of numerical analysis is concerned with devising methods for approximating, in an efficient manner, the solutions to mathematically expressed problems. The efficiency of the method depends both upon the accuracy required of the method and the ease with which it can be implemented. In a practical situation, the mathematical problem is derived from a physical phenomenon where some simplifying assumptions have been made to allow the mathematical representation to develop. Generally a relaxation on the physical assumptions leads to a more appropriate mathematical model, but at the same time one that is more difficult or impossible to solve explicitly. Since the mathematical problem ordinarily does not solve the physical problem exactly in any case, it is often more appropriate to find an approximate solution to a more complicated mathematical model of a physical problem than to find an exact solution of a simplified model. To obtain such an approximation a method called an *algorithm* is devised. The algorithm consists of a sequence of algebraic and logical operations that produces the approximation to the mathematical problem, and, it is hoped, to the physical problem as well, within a prescribed tolerance or accuracy.

Since the efficiency of a method depends upon its ease of implementation, the choice of the appropriate method for approximating the solution to a problem is influenced significantly by changes in calculator and computer technology. Twenty-five years ago, before the widespread use of digital computing equipment, methods requiring a large amount of computational effort could not be reasonably applied. Since that time, however, the advances in computing equipment have made some of these methods increasingly attractive. At present, the limiting factor generally involves the amount of computer storage requirements of the method, although the cost factor associated with a large amount of computation time is, of course, also important. The availability of personal computers and low-cost programmable calculators is also an influencing factor in the choice of an approximation method, since these can be used to solve many relatively simple problems.

The basic ideas that underlie most current numerical techniques have been known for some time, as have the methods used in predicting bounds for the maximum error that can be produced in an application of the methods. It is of primary interest, then, to determine the way in which these methods have developed and how their error can be estimated, since variations of these techniques will undoubtedly be used to develop and apply numerical procedures in the future, irrespective of the technology.

The methods we discuss in this text include those that are commonly used at the present time and those on which improvements will most likely be based in the near future.

1

Mathematical Preliminaries

In beginning chemistry courses, students are confronted with a relationship known as the *ideal gas law*,

$$PV = NRT,$$

which relates the pressure P, volume V, temperature T, and number of moles N of an "ideal" gas. The R in this equation is a constant depending only on the measurement system being used.

Suppose two experiments are conducted to test this law, using the same gas in each case. In the first experiment,

$$P = 1.0 \text{ atmosphere,} \qquad V = 0.10 \text{ cubic meter,}$$
$$N = 0.0042 \text{ mole,} \qquad R = 0.082.$$

Using the ideal gas law, we predict the temperature of the gas to be

$$T = \frac{PV}{NR} = \frac{(1.0)(0.10)}{(0.082)(0.0042)} = 290° \text{ Kelvin or } 17° \text{ Celsius.}$$

When we measure the temperature of the gas, we find that the true temperature is 15° Celsius.

The experiment is then repeated, using the same values of R and N, but increasing the pressure by a factor of four while reducing the volume by the same factor. Since the product PV remains the same, the predicted temperature would still be 290° Kelvin (or 17° Celsius), but now we find that the actual temperature of the gas is 32° Celsius.

Clearly the ideal gas law is suspect when an error of this magnitude is obtained. Before concluding that the law is invalid in this situation, however, we should examine our data to determine whether the error could be attributed to the experimental results. If so, it would be of interest to determine how much more accurate our experimental results would need to be to ensure that an error of this magnitude could not occur.

Analysis of the error involved in calculations is an important topic in numerical analysis and will be introduced in Section 1.2.

This chapter contains a short review of those topics from elementary single-variable calculus that will repeatedly be needed in later chapters, together with an introduction to the terminology used in discussing convergence, error analysis, and the machine representation of numbers.

1.1 Review of Calculus

Fundamental to the study of calculus are the concepts of **limit** and **continuity** of a function.

DEFINITION 1.1   Let f be a function defined on a set X of real numbers; f is said to have the **limit** L at x_0, written $\lim_{x \to x_0} f(x) = L$, if, given any real number $\varepsilon > 0$, there exists a real number $\delta > 0$ such that $|f(x) - L| < \varepsilon$, whenever $x \in X$ and $0 < |x - x_0| < \delta$. (*See Fig. 1.1.*)

FIGURE 1.1

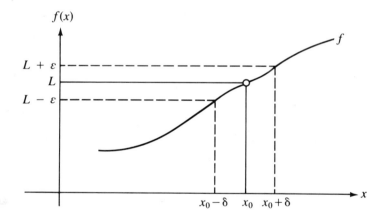

DEFINITION 1.2   Let f be a function defined on a set X of real numbers and $x_0 \in X$; f is said to be **continuous** at x_0 if $\lim_{x \to x_0} f(x) = f(x_0)$. The function f is said to be continuous on X if it is continuous at each number in X; $C(X)$ denotes the set of all functions continuous on X. When X is an interval of the real line, the parentheses in this notation will be omitted. For example, the set of all functions continuous on the closed interval $[a, b]$ will be denoted $C[a, b]$.

In a similar manner, the **limit of a sequence** of real or complex numbers can be defined.

DEFINITION 1.3   Let $\{x_n\}_{n=1}^{\infty}$ be an infinite sequence of real or complex numbers. The sequence is said to **converge** to a number x (called the limit) if, for any $\varepsilon > 0$, there exists a positive integer $N(\varepsilon)$ such that $n > N(\varepsilon)$ implies $|x_n - x| < \varepsilon$. The notation $\lim_{n \to \infty} x_n = x$, or $x_n \to x$ as $n \to \infty$, means that the sequence $\{x_n\}_{n=1}^{\infty}$ converges to x.

The following theorem relates the concepts of convergence and continuity.

THEOREM 1.4   If f is a function defined on a set X of real numbers and $x_0 \in X$, then the following are equivalent:

a) f is continuous at x_0;
b) if $\{x_n\}_{n=1}^{\infty}$ is any sequence in X converging to x_0, then

$$\lim_{n \to \infty} f(x_n) = f(x_0).$$

DEFINITION 1.5   If f is a function defined in an open interval containing x_0, f is said to be **differentiable** at x_0 if

$$\lim_{x \to x_0} \frac{f(x) - f(x_0)}{x - x_0}$$

exists. When this limit exists it is denoted by $f'(x_0)$ and is called the **derivative** of f at x_0. A function that has a derivative at each number in a set X is said to be **differentiable** on X.

THEOREM 1.6   If the function f is differentiable at x_0, then f is continuous at x_0.

The set of all functions that have n continuous derivatives on X is denoted $C^n(X)$, and the set of functions that have derivatives of all orders at each number in X is denoted $C^\infty(X)$. Polynomial, rational, trigonometric, exponential, and logarithmic functions are in class $C^\infty(X)$, where X consists of all numbers at which the functions are defined. When X is an interval of the real line, we will again omit the parentheses in this notation.

The next theorems are of fundamental importance in deriving methods for error estimation. The proofs of these theorems and the other unreferenced results in this section can be found in any standard calculus text.

THEOREM 1.7   (Rolle's Theorem)   Suppose $f \in C[a, b]$ and f is differentiable on (a, b). If $f(a) = f(b) = 0$, then a number c, $a < c < b$, exists with $f'(c) = 0$. (*See Fig. 1.2.*)

FIGURE 1.2

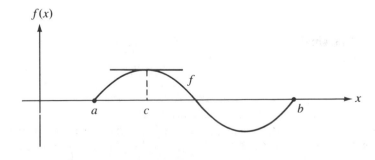

THEOREM 1.8　　(Mean Value Theorem)　If $f \in C[a, b]$ and f is differentiable on (a, b), then a number $c, a < c < b$, exists such that

$$f'(c) = \frac{f(b) - f(a)}{b - a}. \quad \text{(See Fig. 1.3.)}$$

FIGURE 1.3

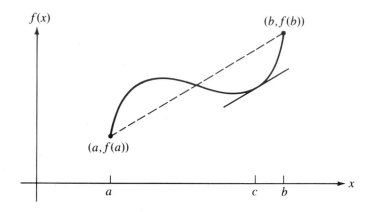

THEOREM 1.9　　(Extreme Value Theorem)　If $f \in C[a, b]$, then $c_1, c_2 \in [a, b]$ exist with $f(c_1) \leq f(x) \leq f(c_2)$ for each $x \in [a, b]$. If, in addition, f is differentiable on (a, b), then the numbers c_1 and c_2 occur either at endpoints of $[a, b]$ or where f' is zero.

Two other results will be needed in our study of numerical methods. The first is a generalization of the usual Mean Value Theorem for Integrals.

THEOREM 1.10　　(Weighted Mean Value Theorem for Integrals)　If $f \in C[a, b]$, g is integrable on $[a, b]$ and $g(x)$ does not change sign on $[a, b]$, then there exists a number $c, a < c < b$, such that:

$$\int_a^b f(x)g(x)\,dx = f(c) \int_a^b g(x)\,dx.$$

When $g(x) \equiv 1$, this theorem gives what is called the **average value** of the function over the interval $[a, b]$. (*See Fig. 1.4.*) The proof of Theorem 1.10 is not generally given in a basic calculus course, but can be found in any standard advanced calculus text (see, for example, Fulks [50], page 162).

The other theorem we will need that is not generally presented in a basic calculus course is derived by applying Rolle's Theorem (Theorem 1.7) successively to $f, f', \ldots$, and finally to $f^{(n-1)}$.

FIGURE 1.4

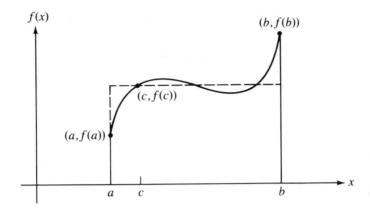

THEOREM 1.11   (Generalized Rolle's Theorem)   Let $f \in C[a, b]$ be n times differentiable on (a, b). If f vanishes at the $n + 1$ distinct numbers $x_0, \ldots, x_n$ in $[a, b]$, then a number c in (a, b) exists with $f^{(n)}(c) = 0$.

The next theorem presented is the Intermediate Value Theorem. Although its statement is intuitively clear, the proof is beyond the scope of the usual calculus course. The proof can be found in most advanced calculus texts (see, for example, Fulks [50], page 67).

THEOREM 1.12   (Intermediate Value Theorem)   If $f \in C[a, b]$ and K is any number between $f(a)$ and $f(b)$, then there exists c in (a, b) for which $f(c) = K$. (*See Fig. 1.5.*)

FIGURE 1.5

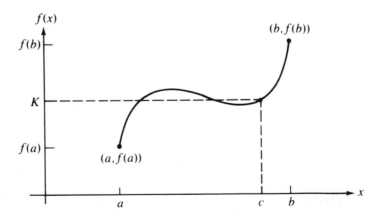

EXAMPLE 1   To show that $x^5 - 2x^3 + 3x^2 - 1 = 0$ has a solution on the interval $[0, 1]$, consider the function $f(x) = x^5 - 2x^3 + 3x^2 - 1$. Clearly f is continuous on $[0, 1]$ and $f(0) = -1$ while $f(1) = 1$. Since $f(0) < 0 < f(1)$, the Intermediate Value Theorem implies that there is a number x, with $0 < x < 1$, for which $x^5 - 2x^3 + 3x^2 - 1 = 0$. ■

As seen in Example 1, the Intermediate Value Theorem is important as an aid to determine when solutions to certain problems exist. It does not, however, give a means for finding these solutions. This topic will be discussed more thoroughly in Chapter 2.

The final theorem in this review from calculus describes the development of the Taylor polynomials. The importance of the Taylor polynomials to the study of numerical analysis cannot be overemphasized, and the following result will be used repeatedly.

THEOREM 1.13

(Taylor's Theorem)   Suppose $f \in C^n[a, b]$ and $f^{(n+1)}$ exists on $[a, b]$. Let $x_0 \in [a, b]$. For every $x \in [a, b]$, there exists $\xi(x)$ between x_0 and x with

$$f(x) = P_n(x) + R_n(x),$$

where

$$P_n(x) = f(x_0) + f'(x_0)(x - x_0) + \frac{f''(x_0)}{2!}(x - x_0)^2 + \cdots + \frac{f^{(n)}(x_0)}{n!}(x - x_0)^n$$

$$= \sum_{k=0}^{n} \frac{f^{(k)}(x_0)}{k!}(x - x_0)^k,$$

and

$$R_n(x) = \frac{f^{(n+1)}(\xi(x))}{(n+1)!}(x - x_0)^{n+1}.$$

Here $P_n(x)$ is called the **nth-degree Taylor polynomial** for f about x_0 and $R_n(x)$ is called the **remainder term** (or **truncation error**) associated with $P_n(x)$. The infinite series obtained by taking the limit of $P_n(x)$ as $n \to \infty$ is called the **Taylor series** for f about x_0. In case $x_0 = 0$, the Taylor polynomial is often called a **Maclaurin polynomial** and the Taylor series is called a **Maclaurin series**.

The term **truncation error** generally refers to the error involved in using a truncated or finite summation to approximate the sum of an infinite series. This terminology will be reintroduced in subsequent chapters.

EXAMPLE 2

Let $f(x) = \cos x$. Since $f \in C^\infty(R)$, Theorem 1.13 can be applied for any $n > 0$. For $n = 2$ and $x_0 = 0$, Theorem 1.13 gives

$$\cos x = 1 - \tfrac{1}{2}x^2 + \tfrac{1}{6}x^3 \sin \xi(x),$$

where $\xi(x)$ is a number between 0 and x.

With $x = 0.001$, the Taylor polynomial and remainder term is

$$\cos 0.001 = 1 - \tfrac{1}{2}(0.001)^2 + \tfrac{1}{6}(0.001)^3 \sin \xi(x)$$

$$= 0.9999995 + (0.16\bar{6}) \cdot 10^{-9} \sin \xi(x),$$

where $0 < \xi(x) < 0.001$. (The bar over the last digit in 0.166 is used to indicate that this digit repeats indefinitely.)

Since $|\sin \xi(x)| < 1$, 0.9999995 can be used as an approximation to $\cos 0.001$ with assurance of at least nine decimal-place accuracy. Using standard tables, it can be found that

$$\cos 0.001 = 0.999999500000042;$$

so there is actually 13-decimal-place accuracy.

If, in this example, the third-degree Taylor polynomial had been used with $x_0 = 0$, then

$$\cos x = 1 - \tfrac{1}{2}x^2 + \tfrac{1}{24}x^4 \cos \xi(x),$$

where $0 < \xi(x) < 0.001$, since $f'''(0) = 0$. The approximating polynomial remains the same, and the approximation would still be 0.9999995, but 13-decimal-place accuracy would be expected since

$$\left| \tfrac{1}{24}x^4 \cos \xi(x) \right| \leq \tfrac{1}{24}(0.001)^4(1) \approx 4.2 \times 10^{-14}.$$

This corresponds more closely to the actual accuracy obtained. ∎

Exercise Set 1.1

1. Let $f(x) = 1 - e^x + (e-1)\sin((\pi/2)x)$. Show that $f'(x)$ is zero at least once in $[0, 1]$.

2. Show that the equation $x^3 = e^x \sin x$ must have at least one solution in $[1, 4]$.

3. Show that the equation $x = 3^{-x}$ has a solution in $[0, 1]$.

4. Let $f(x) = (x-1)\tan x + x \sin \pi x$. Show that $f'(x) = 0$ for some x in $[0, 1]$.

5. Let $f(x) = x \sin \pi x - (x-2)\ln x$. Show that $f'(x) = 0$ for some x in $[1, 2]$.

6. Let $f(x) = (x-2)\sin x \ln(x+2)$. Show that $f''(x) = 0$ for some x in $[-1, 3]$.

7. Use the Intermediate Value Theorem and Rolle's Theorem to show that the graph of $f(x) = x^3 + 2x + k$ crosses the x-axis exactly once, regardless of the value of the constant k.

8. Suppose $f \in C[a, b]$ and $f'(x)$ exists on (a, b). Show that if $f'(x) \neq 0$ for all x in (a, b), then there can exist at most one number p in $[a, b]$ with $f(p) = 0$.

9. Find the Taylor polynomial of degree four for f expanded about $x_0 = 0$ if $f(x) = e^x \cos x$. Use this polynomial to approximate $f(\pi/16)$, and find a bound for the error in this approximation.

10. Find the Taylor polynomial of degree four for f expanded about $x_0 = 1$ if $f(x) = e^{x^2}$. Use this polynomial to approximate $f(1.1)$, and find a bound for the error in this approximation.

11. Let $f(x) = e^{-x}$. Find the third-degree Taylor polynomial for f expanded about $x_0 = 1$, and approximate $e^{-0.99}$ using the Taylor polynomial. How many decimal places of accuracy are expected?

12. Using Taylor's Theorem with $n = 2$ and $x_0 = 0$, find an approximation for $\sin 0.001$. Do you expect the accuracy in this problem to be that obtained for $\cos 0.001$ in Example 2?

13. Use a Taylor polynomial about $\pi/4$ to approximate $\cos 42°$ to an accuracy of 10^{-6}.

14. Use the third-degree Taylor polynomial about 1 to approximate $\ln 1.1$ and find the maximum error for this approximation.

15. Use a Taylor polynomial for the function $f(x) = \ln x$ about e to find an approximation to $\ln 3$ that is accurate to within 10^{-4}.

16. A Maclaurin polynomial for e^x is used to give the approximation 2.5 to e. The error bound in this approximation is established to be $E = 1/6$. Find a bound for the error in E.

17. Let $f(x) = |x|$ for $x \in R$. Show that f is differentiable at any $x \neq 0$, but that f is not differentiable at $x = 0$.

18. Use the Mean Value Theorem to show that $|\sin a - \sin b| \leq |a - b|$ and deduce from this result that $|\sin a + \sin b| \leq |a + b|$.

19. Determine whether the Mean Value Theorem applies in the following situations. If so, find a number c satisfying the conclusion of this theorem; if not, show that no such number exists.

   a) $f(x) = x^{2/3}$, $[a, b] = [-1, 8]$.   b) $f(x) = x^{2/3}$, $[a, b] = [0, 8]$.

   c) $f(x) = |x|$, $[a, b] = [-1, 1]$.   d) $f(x) = |x|$, $[a, b] = [0, 1]$.

20. Give an example of a function that is continuous on $[0, 1]$ and differentiable except at $x = 1/2$, but for which the conclusion of the Mean Value Theorem does not hold.

21. A function $f: [a, b] \to R$ is said to satisfy a Lipschitz condition with Lipschitz constant L on $[a, b]$ if, for every $x, y \in [a, b]$,

$$|f(x) - f(y)| \leq L|x - y|.$$

   a) Show that if f satisfies a Lipschitz condition with Lipschitz constant L on an interval $[a, b]$, then $f \in C[a, b]$.

   b) Show that if f has a derivative that is bounded on $[a, b]$ by L, then f satisfies a Lipschitz condition with Lipschitz constant L on $[a, b]$.

   c) Give an example of a function that is continuous on a closed interval but does not satisfy a Lipschitz condition on the interval.

1.2 Round-Off Errors and Computer Arithmetic

When a calculator or digital computer is used to perform numerical calculations, an unavoidable error, called **round-off error**, must be considered. This error arises because the arithmetic performed in a machine involves numbers with only a finite number of digits, with the result that many calculations are performed with approximate representations of the actual numbers. In a typical computer, only a relatively small subset of the real number system is used for the representation of all real numbers. This subset contains only rational numbers, both positive and negative, and stores a fractional part, called the **mantissa**, together with an exponential part, called the **characteristic**. For example, a single-precision floating-point number used in the IBM 370 or 3000 series consists of a 1-binary-digit (**bit**) sign indicator, a 7-bit exponent with a base of 16, and a 24-bit mantissa. Since 24 binary digits correspond to between 6 and 7 decimal digits, we can assume that this number has at least 6 decimal digits of precision for the floating-point number system. The exponent of 7 binary digits gives a range of 0 to 127, but because of an exponential bias the range is actually -64 to $+63$, that is, 64 is automatically subtracted from the listed exponent.

Consider for example, the machine number

0	1000010	10110011000001000000000

The leftmost bit is a zero, which indicates that the number is positive. The next seven bits, 1000010, are equivalent to the decimal number

$$1 \cdot 2^6 + 0 \cdot 2^5 + 0 \cdot 2^4 + 0 \cdot 2^3 + 0 \cdot 2^2 + 1 \cdot 2^1 + 0 \cdot 2^0 = 66$$

and are used to describe the characteristic. The final 24 bits indicate that the mantissa is

$$1 \cdot (1/2)^1 + 1 \cdot (1/2)^3 + 1 \cdot (1/2)^4 + 1 \cdot (1/2)^7 + 1 \cdot (1/2)^8 + 1 \cdot (1/2)^{14}.$$

As a consequence, this machine number precisely represents the decimal number

$$+ [(1/2)^1 + (1/2)^3 + (1/2)^4 + (1/2)^7 + (1/2)^8 + (1/2)^{14}]16^{66-64} = 179.015625.$$

However, the next smallest machine number is

0	1000010	10110011000001111111111	$= 179.0156097412109375,$

while the next largest machine number is

0	1000010	10110011000010000000001	$= 179.0156402587890625.$

This means that our original machine number must represent not only 179.015625, but many real numbers that are between this number and its nearest machine number neighbors. To be precise, the original machine number is used to represent any real number in the interval

$$[179.01561737060546875, 179.01563262939453125).$$

To ensure uniqueness of representation and obtain all the available precision of the system, at least one of the four leftmost bits of the mantissa of a machine number is required to be a one. Consequently, 15×2^{28} numbers of the form

$$\pm 0.d_1 d_2 \cdots d_{24} \times 16^{e_1 e_2 \cdots e_7},$$

are used by this system to represent all real numbers. This requirement also implies that the smallest normalized, positive machine number that can be represented is

0	0000000	00010000000000000000000	$= 16^{-65} \approx 10^{-78}$

while the largest is

| 0 | 1111111 | 111111111111111111111111111 | $\approx 16^{63} \approx 10^{76}.$ |

Numbers occurring in calculations that have a magnitude of less than 16^{-65} result in what is called **underflow**, and are often set to zero, while numbers greater than 16^{63} result in an **overflow** condition and cause the computations to halt.

The use of binary digits tends to conceal the computational difficulties that occur when a finite collection of machine numbers is used to represent all the real numbers. To explain the problems that can arise, we will now assume, for simplicity, that machine numbers are represented in the normalized decimal form

(1.1) $$\pm 0.d_1 d_2 \cdots d_k \times 10^n, \qquad 1 \le d_1 \le 9, \quad 0 \le d_i \le 9,$$

for each $i = 2, \ldots, k$, where, from what we have just discussed, the IBM machines have approximately $k = 6$ and $-77 \le n \le 76$. Numbers of this form will be called *decimal machine numbers*.

It is useful to consider the representation of an arbitrary real number in the **floating-point form** (1.1). Any positive real number y can be normalized to achieve the form

$$y = 0.d_1 d_2 \cdots d_k d_{k+1} d_{k+2} \cdots \times 10^n,$$

if we assume y is within the numerical range of the machine. The floating-point form (1.1), denoted by $fl(y)$, is obtained by terminating the mantissa of y at k decimal digits. There are two ways of performing this termination. One method is to simply chop off the digits $d_{k+1} d_{k+2} \cdots$ to obtain

$$fl(y) = 0.d_1 d_2 \cdots d_k \times 10^n.$$

This method is quite accurately called **chopping** the number. The other method is to add $5 \times 10^{n-(k+1)}$ to y and then chop to obtain

$$fl(y) = 0.\delta_1 \delta_2 \cdots \delta_k \times 10^n.$$

The latter method is often referred to as **rounding** the number. In this method, if $d_{k+1} \ge 5$, we add one to d_k to obtain $fl(y)$; that is, we round up. If $d_{k+1} < 5$, we merely chop off all but the first k digits; so we round down.

EXAMPLE 1   The number π has an infinite decimal expansion

$$\pi = 3.14159265 \ldots = 0.314159265 \ldots \times 10^1.$$

Suppose that $k = 5$ and that chopping is employed. Then the floating-point form of π is

$$fl(\pi) = 0.31415 \times 10^1 = 3.1415.$$

Since the sixth digit of the decimal expansion of π is a nine, the floating-point form of π using five-digit rounding is

$$fl(\pi) = (0.31415 + 0.00001) \times 10^1 = 3.1416. \qquad \blacksquare$$

The error that results from replacing a number with its floating-point form is called **round-off error** (regardless of whether the rounding or chopping method is used). The following definition specifies two methods for measuring approximation errors. These methods will be used throughout the text.

DEFINITION 1.14   If p^* is an approximation to p, the **absolute error** is given by $|p - p^*|$, and the **relative error** is given by $|p - p^*|/|p|$, provided that $p \neq 0$.

Consider the absolute and relative errors in representing p by p^* in the following example.

EXAMPLE 2   a)   If $p = 0.3000 \times 10^1$ and $p^* = 0.3100 \times 10^1$, the absolute error is 0.1 and the relative error is $0.333\overline{3} \times 10^{-1}$.

b)   If $p = 0.3000 \times 10^{-3}$ and $p^* = 0.3100 \times 10^{-3}$, the absolute error is 0.1×10^{-4} and the relative error is $0.333\overline{3} \times 10^{-1}$.

c)   If $p = 0.3000 \times 10^4$ and $p^* = 0.3100 \times 10^4$, the absolute error is 0.1×10^3 and the relative error is $0.333\overline{3} \times 10^{-1}$.

This example shows that the same relative error, $0.333\overline{3} \times 10^{-1}$, occurs for widely varying absolute errors. Consequently, as a measure of accuracy, the absolute error may be misleading and the relative error more meaningful. ∎

Returning to the machine representation of numbers we see that the floating-point representation $fl(y)$ for the number y has the relative error

$$\left| \frac{y - fl(y)}{y} \right|.$$

If k decimal digits and chopping are used for the machine representation of

$$y = 0.d_1 d_2 \cdots d_k d_{k+1} \cdots \times 10^n,$$

then

$$\left| \frac{y - fl(y)}{y} \right| = \left| \frac{0.d_1 d_2 \cdots d_k d_{k+1} \cdots \times 10^n - 0.d_1 d_2 \cdots d_k \times 10^n}{0.d_1 d_2 \cdots \times 10^n} \right|$$

$$= \left| \frac{0.d_{k+1} d_{k+2} \cdots \times 10^{n-k}}{0.d_1 d_2 \cdots \times 10^n} \right| = \left| \frac{0.d_{k+1} d_{k+2} \cdots}{0.d_1 d_2 \cdots} \right| \times 10^{-k}.$$

Since $d_1 \neq 0$, the minimal value of the denominator is 0.1. The numerator is bounded by 1. As a consequence,

$$\left| \frac{y - fl(y)}{y} \right| \leq \frac{1}{0.1} \times 10^{-k} = 10^{-k+1}$$

In a similar manner, a bound for the relative error when using k-digit rounding arithmetic is $0.5 \times 10^{-k+1}$. (See Exercise 14.)

Note that the bounds for the relative error using k-digit arithmetic are independent of the number being represented. This is due to the manner in which the machine numbers are distributed along the real line. Because of the exponential form of the characteristic, the same number of decimal machine numbers are used to represent each of the intervals $[0.1, 1]$, $[1, 10]$, and $[10, 100]$. In fact, within the limits of the machine, the number of decimal machine numbers in $[10^n, 10^{n+1}]$ is constant for all integers n. In a binary representation system, the situation is the same, except that the base of the characteristic is 16 rather than 10.

The common use of rounding arithmetic in computing devices leads to the following definition.

DEFINITION 1.15   The number p^* is said to approximate p to t **significant digits** (or figures) if t is the largest nonnegative integer for which

$$\frac{|p - p^*|}{|p|} < 5 \times 10^{-t}.$$

The reason for using relative error in the definition is to obtain a continuous concept. Consider the numbers 1000, 5000, 9990, and 10,000. For p^* to approximate 1000 to four significant figures by this definition, p^* has to satisfy

$$\left| \frac{p^* - 1000}{1000} \right| < 5 \times 10^{-4}.$$

This requires that $999.5 < p^* < 1000.5$ and agrees with the intuitive definition of significant digits. Considering this same problem for the numbers 5000 and 9990, p^* would need to satisfy $4997.5 < p^* < 5002.5$ and $9985.005 < p^* < 9994.995$, respectively, to be accurate to four significant digits. This may not agree with the intuitive idea of significant digits. Note, however, that, for p^* to be a four-significant-digit approximation to 10,000, p^* must satisfy $9995 < p^* < 10005$, which again agrees with intuition. The following table illustrates the continuous nature of this concept by listing, for the various values of p, the least upper bound of $|p - p^*|$, denoted $\max|p - p^*|$, when p^* agrees with p to four significant figures.

p	0.1	0.5	100	1000	5000	9990	10000
$\max|p - p^*|$	0.00005	0.00025	0.05	0.5	2.5	4.995	5.0

In addition to inaccurate representation of numbers, the arithmetic performed in a computer is not exact. The arithmetic generally involves manipulating binary digits by various shifting or logical operations. Since the actual mechanics of these operations are not pertinent to this presentation, we shall devise our own approximation to computer arithmetic. Although our arithmetic will not give the exact picture, it should suffice to explain the problems that occur. (For an explanation of the manipulations actually involved, the reader is urged to consult more technically oriented computer science texts, such as Mano, [87], *Computer System Architecture*.)

Assume the floating-point representations $fl(x)$ and $fl(y)$ are given for the real numbers x and y and that the symbols $\oplus$, $\ominus$, $\otimes$, $\oslash$ represent the machine addition, subtraction, multiplication, and division operations, respectively. We will assume a k-digit arithmetic given by

$$x \oplus y = fl(fl(x) + fl(y)),$$
$$x \ominus y = fl(fl(x) - fl(y)),$$
$$x \otimes y = fl(fl(x) \times fl(y)),$$
$$x \oslash y = fl(fl(x) \div fl(y)).$$

This idealized arithmetic corresponds to performing exact arithmetic on the floating-point representations of x and y and then converting the exact result to its floating-point representation.

EXAMPLE 3   Suppose that $x = \frac{1}{3}$, $y = \frac{5}{7}$ and that five-digit chopping is used for arithmetic calculations involving x and y. Table 1.1 lists the values of these computer-type operations on $fl(x) = 0.33333 \times 10^0$ and $fl(y) = 0.71428 \times 10^0$.

TABLE 1.1

Operation	Result	Actual value	Absolute error	Relative error
$x \oplus y$	0.10476×10^1	22/21	0.190×10^{-4}	0.182×10^{-4}
$y \ominus x$	0.38095×10^0	8/21	0.238×10^{-5}	0.625×10^{-5}
$x \otimes y$	0.23809×10^0	5/21	0.524×10^{-5}	0.220×10^{-4}
$y \oslash x$	0.21428×10^1	15/7	0.571×10^{-4}	0.267×10^{-4}

Since the maximum relative error for these operations is 0.267×10^{-4}, the arithmetic produces satisfactory five-digit results. Suppose, however, that we also have $u = 0.714251$, $v = 98765.9$, and $w = 0.111111 \times 10^{-4}$ so that $fl(u) = 0.71425 \times 10^0$, $fl(v) = 0.98765 \times 10^5$, and $fl(w) = 0.11111 \times 10^{-4}$. These numbers were chosen to illustrate some problems that can arise with finite-digit arithmetic. In Table 1.2, $y \ominus u$ results in a small absolute error but a large relative error. The subsequent division by the small number w or multiplication by the large number v magnifies the absolute error without modifying the relative error. The addition of the large and small numbers u and v produces large absolute error but not large relative error. ∎

TABLE 1.2

Operation	Result	Actual value	Absolute error	Relative error
$y \ominus u$	0.30000×10^{-4}	0.34714×10^{-4}	0.471×10^{-5}	0.136
$(y \ominus u) \oslash w$	0.27000×10^1	0.31243×10^1	0.424	0.136
$(y \ominus u) \otimes v$	0.29629×10^1	0.34285×10^1	0.465	0.136
$u \oplus v$	0.98765×10^5	0.98766×10^5	0.161×10^1	0.163×10^{-4}

The computations in Example 3 illustrate that, although calculations involving finite-digit arithmetic generally lead to acceptable results, significant errors can occur in some circumstances. One of the most common of these involves the division of a finite-digit result by a number with small magnitude (or equivalently, multiplying by a relatively large number). Suppose that the actual number r has the finite-digit approximation $r + \varepsilon$, where ε denotes the round-off error. Dividing by δ results in a new round-off error ε/δ. The significance of ε/δ depends on the magnitude of δ. If δ is relatively large, the new round-off error will be small; but if the magnitude of δ is small, the round-off error ε/δ will be large.

Another calculation that customarily produces unacceptable round-off error involves the subtraction of nearly equal numbers. Suppose that the two numbers x and y have the k-digit representations

$$fl(x) = 0.d_1 d_2 \cdots d_p \alpha_{p+1} \alpha_{p+2} \cdots \alpha_k \times 10^n$$

and

$$fl(y) = 0.d_1 d_2 \cdots d_p \beta_{p+1} \beta_{p+2} \cdots \beta_k \times 10^n.$$

Then the floating-point form of $x - y$ will be

$$fl(fl(x) - fl(y)) = 0.\gamma_{p+1} \gamma_{p+2} \cdots \gamma_k \times 10^{n-p}$$

for some constants $\gamma_{p+1}, \gamma_{p+2}, \ldots, \gamma_k$. As a consequence, the number used to represent $x - y$ will have only $k - p$ digits of significance. (In most calculation devices, $x - y$ would actually be assigned k digits, but the last p might simply be randomly assigned.) Any further calculations involving $x - y$ retain the problem of having only $k - p$ digits of significance, since any chain of calculations would not be expected to be more accurate than its weakest portion.

The loss of significant digits due to round-off error can often be avoided by a reformulation of the problem, as shown in the next examples.

EXAMPLE 4   The quadratic formula states that the roots of $ax^2 + bx + c = 0$, when $a \neq 0$, are

$$x_1 = \frac{-b + \sqrt{b^2 - 4ac}}{2a} \qquad \text{and} \qquad x_2 = \frac{-b - \sqrt{b^2 - 4ac}}{2a}.$$

Consider

$$x^2 + 62.10x + 1 = 0,$$

a quadratic equation with roots approximately

$$x_1 = -0.01610723 \qquad \text{and} \qquad x_2 = -62.08390.$$

In this equation b^2 is much larger than $4ac$, so the numerator in the calculation for x_1 involves the subtraction of nearly equal numbers. Suppose we perform the calculations for x_1 using four-digit rounding arithmetic. First we have

$$\sqrt{b^2 - 4ac} = \sqrt{(62.10)^2 - 4.000}$$
$$= \sqrt{3856. - 4.000}$$
$$= \sqrt{3852.} = 62.06.$$

So

$$fl(x_1) = \frac{-b + \sqrt{b^2 - 4ac}}{2a} = \frac{-62.10 + 62.06}{2.000} = \frac{-0.04000}{2.000} = -0.02000,$$

a rather poor approximation to $x_1 = -0.01611$. On the other hand, the calculation for x_2 involves the *addition* of the nearly equal numbers $-b$ and $-\sqrt{b^2 - 4ac}$ and presents no problem:

$$fl(x_2) = \frac{-b - \sqrt{b^2 - 4ac}}{2a} = \frac{-62.10 - 62.06}{2.000} = \frac{-124.2}{2.000} = -62.10,$$

an accurate approximation to $x_2 = -62.08$.

To obtain a more accurate four-digit rounding approximation for x_1, we change the form of the quadratic formula by "rationalizing the numerator":

$$x_1 = \frac{-b + \sqrt{b^2 - 4ac}}{2a} \left(\frac{-b - \sqrt{b^2 - 4ac}}{-b - \sqrt{b^2 - 4ac}} \right) = \frac{b^2 - (b^2 - 4ac)}{2a(-b - \sqrt{b^2 - 4ac})}$$

which simplifies to

(1.2)
$$x_1 = \frac{-2c}{b + \sqrt{b^2 - 4ac}}.$$

Using (1.2) gives the accurate result

$$fl(x_1) = \frac{-2.000}{62.10 + 62.06} = \frac{-2.000}{124.2} = -0.0161.$$

The "rationalization" technique can also be applied to give the alternate form for x_2

(1.3)
$$x_2 = \frac{-2c}{b - \sqrt{b^2 - 4ac}}.$$

This would be the form to use if b were a negative number. In our problem, however, the use of this formula results in not only the subtraction of nearly equal numbers, but also the division by the small result of this subtraction. The inaccuracy that this produces is dramatic

$$fl(x_2) = \frac{-2c}{b - \sqrt{b^2 - 4ac}} = \frac{-2.000}{62.10 - 62.06} = \frac{-2.000}{0.04000} = -50.00. \qquad \blacksquare$$

EXAMPLE 5   Evaluate $f(x) = x^3 - 6x^2 + 3x - 0.149$ at $x = 4.71$ using three-digit arithmetic.

Table 1.3 gives the intermediate results in the calculations. Note that the three-digit chopping values simply retain the leading three digits, with no rounding involved, and differ significantly from the three-digit rounding values.

TABLE 1.3

	x	x^2	x^3	$6x^2$	$3x$
Exact	4.71	22.1841	104.487111	133.1046	14.13
Three-digit (chopping)	4.71	22.1	104.	132.	14.1
Three-digit (rounding)	4.71	22.2	105.	133.	14.1

Exact:

$$f(4.71) = 104.487111 - 133.1046 + 14.13 - 0.149$$
$$= -14.636489;$$

Three-digit (chopping):

$$f(4.71) = 104. - 132. + 14.1 - 0.149$$
$$= -14.0;$$

Three-digit (rounding):

$$f(4.71) = 105. - 133. + 14.1 - 0.149$$
$$= -14.0.$$

The relative error for both the three-digit methods is

$$\left| \frac{-14.636489 + 14.0}{-14.636489} \right| \approx 0.04.$$

As an alternative approach, $f(x)$ could be written in a *nested* manner as

$$f(x) = x^3 - 6x^2 + 3x - 0.149 = ((x - 6)x + 3)x - 0.149.$$

This gives $x(x^2 - 6x + 3)$

Three-digit (chopping): $f(4.71) = ((4.71 - 6)4.71 + 3)4.71 - 0.149 = -14.5,$

and a three-digit rounding answer of -14.6. The new relative errors are

Three-digit (chopping): $\left| \dfrac{-14.636489 + 14.5}{-14.636489} \right| \approx 0.0093$

Three-digit (rounding): $\left| \dfrac{-14.636489 + 14.6}{-14.636489} \right| \approx 0.0025.$

Although both three-digit chopping approximations are correct to two significant digits the relative error has been reduced to about one-fourth that of the original procedure. In the case of the three-digit rounding procedure, an additional significant digit has been obtained.   ■

Polynomials should always be expressed in nested form before performing an evaluation, since this form minimizes the number of required arithmetic calculations. The decreased error in Example 5 is due to the fact that the number of computations has been reduced from four multiplications and three additions to two multiplications and three additions. Evidently one way to reduce this error is to reduce the number of error-producing computations.

Exercise Set 1.2

1. Find bounds for x, using standard tables and Definition 1.15, if x is a four-significant-digit approximation to

   a) π b) e

2. Suppose that p^* approximates p to three significant digits. Find the interval in which p^* must lie if p is

   a) 150   b) 900

   c) 1500   d) 90

3. Consider the following values of p and p^*. What is (i) the absolute error, (ii) the relative error in approximating p^* by p?

   a) $p = \pi, \quad p^* = 3.1$ b) $p = \dfrac{1}{3}, \quad p^* = 0.333$

   c) $p = \dfrac{\pi}{1000}, \quad p^* = 0.0031$ d) $p = \dfrac{100}{3}, \quad p^* = 33.3$

4. To how many significant digits does p^* approximate p in Exercise 3?

5. Perform the following computations (i) exactly, (ii) using three-digit chopping arithmetic, (iii) using three-digit rounding arithmetic. Then determine any loss in significant digits, assuming that the given numbers are exact.

   a) $14.1 + 0.0981$ b) $0.0218 \times 179.$

   c) $(164. + 0.913) - (143. + 21.0)$ d) $(164. - 143.) + (0.913 - 21.0)$

6. Perform the following computations (i) exactly, (ii) using three-digit chopping arithmetic, (iii) using three-digit rounding arithmetic.

   a) $\dfrac{4}{5} + \dfrac{1}{3}$ b) $\dfrac{4}{5} \cdot \dfrac{1}{3}$

   c) $\left(\dfrac{1}{3} - \dfrac{3}{11}\right) + \dfrac{3}{20}$ d) $\left(\dfrac{1}{3} + \dfrac{3}{11}\right) - \dfrac{3}{20}$

7. Convert the following machine numbers into decimal numbers

   a) | 0 | 0111101 | 1001000010000000000000000 |

   b) | 1 | 0111101 | 1001000010000000000000000 |

   c) | 0 | 1000010 | 1001000010000000000000000 |

   d) | 1 | 1000010 | 1001000010000000000000000 |

8. Find the next largest and next smallest machine numbers in decimal form to the numbers given in Exercise 7.

9. Determine the exact decimal equivalent for 2^{-24}.

10. Suppose two points (x_0, y_0) and (x_1, y_1) are on a straight line. Two formulas are available to find the x-intercept of the line:

$$x = \frac{x_0 y_1 - x_1 y_0}{y_1 - y_0} \quad \text{and} \quad x = x_0 - \frac{(x_1 - x_0)y_0}{y_1 - y_0}$$

   a) Show that both formulas are algebraically correct.

   b) Using the data $(x_0, y_0) = (1.31, 3.24)$ and $(x_1, y_1) = (1.93, 4.76)$ and three-digit rounding arithmetic, compute the x-intercept both ways. Which method is better and why?

11. The Taylor polynomial of degree n for $f(x) = e^x$ is $\sum_{i=0}^{n} (x^i/i!)$. Use the Taylor polynomial of degree 9 to find an approximation to e^{-5} by:

   a) $\quad e^{-5} \approx \sum_{i=0}^{9} \frac{(-5)^i}{i!} = \sum_{i=0}^{9} \frac{(-1)^i 5^i}{i!}$ b) $\quad e^{-5} \approx \frac{1}{e^5} \approx \frac{1}{\sum_{i=0}^{9} (5^i/i!)}$

   An approximate value of e^{-5} correct to three digits is 6.74×10^{-3}. Which formula, (a) or (b), gives the most accuracy, and why?

12. Consider the equations:

   (1) $\qquad\qquad\qquad\qquad\qquad 31.69x + 14.31y = 45.00,$

   (2) $\qquad\qquad\qquad\qquad\qquad 13.11x + 5.89y = 19.00.$

   The unique solution to this set of equations is $x = 7.2$ and $y = -12.8$. A method often presented in elementary algebra courses for solving problems of this type is to multiply Eq. (1) by the coefficient of x in Eq. (2), multiply Eq. (2) by the coefficient of x in Eq. (1), and then subtract the resulting equations. For this problem, we would obtain

   (3) $\qquad [(13.11)(14.31) - (31.69)(5.89)]y = (13.11)(45.00) - (31.69)(19.00).$

   a) Perform this operation using four-digit chopping arithmetic, and use your result to find four-digit values of y and x.

   b) Explain why the answers obtained in (a) differ significantly from the actual values for x and y.

13. A rectangular parallelepiped has sides 3 cm (centimeters), 4 cm, and 5 cm, measured only to the nearest centimeter. What are the best upper and lower bounds for the volume of this parallelpiped? What are the best upper and lower bounds for the surface area?

14. Suppose that $fl(y)$ is a k-digit rounding approximation to y. Show that

$$\left| \frac{y - fl(y)}{y} \right| \leq 0.5 \times 10^{-k+1}$$

   [*Hint:* If $d_{k+1} < 5$, then $fl(y) = 0.d_1 d_2 \ldots d_k \times 10^n$. If $d_{k+1} \geq 5$, then $fl(y) = 0.d_1 d_2 \ldots d_k \times 10^n + 10^{n-k}$.]

1.3 Algorithms and Convergence

The examples in Section 1.2 demonstrate ways that machine calculations involving approximations can result in the growth of rounding errors. Throughout this text we will be examining approximation procedures involving sequences of calculations. These procedures are called **algorithms**. To be precise, an algorithm is a procedure that describes, in an unambiguous manner, a finite sequence of steps to be performed in a specified order. The object of the algorithm generally is to implement a numerical procedure to solve a problem or approximate a solution to the problem.

As a vehicle for describing algorithms a **pseudocode** is used. This pseudocode specifies the form of the input to be supplied and the form of the desired output. Not all numerical procedures give satisfactory output for arbitrarily chosen input. As a consequence, a stopping technique independent of the numerical technique is incorporated into each algorithm so that infinite loops are unlikely to occur.

The steps in the algorithms have been arranged so that there should be minimal difficulty translating pseudocode into ALGOL, BASIC, FORTRAN, Pascal, PL/1, or any other programming language that is suitable for scientific applications. Looping techniques in the algorithms are either counter controlled, for example,

For $i = 1, 2, ..., n$
Set $x_i = a_i + i \cdot h$

or condition controlled, such as

While $i < N$ do Steps 3–6.

Two punctuation symbols are used in the algorithms: the period (.) indicates the termination of a step, while the semicolon (;) separates tasks within a step. Indentation is used to indicate that groups of statements are to be treated as a single entity.

To allow for conditional execution, we use the standard

If ... then

or

If ... then
else

constructions.

The algorithms are liberally laced with comments written in italics and contained within parentheses to distinguish them from the algorithmic statements.

EXAMPLE 1   An algorithm to compute

$$\sum_{i=1}^{N} x_i = x_1 + x_2 + \cdots + x_N,$$

where N and the numbers $x_1, x_2, ..., x_N$ are given, is described by the following:

INPUT $N, x_1, x_2, \ldots, x_N$.

OUTPUT $SUM = \sum_{i=1}^{N} x_i$.

Step 1   Set $SUM = 0$.

Step 2   For $i = 1, 2, \ldots, N$ do
            set $SUM = SUM + x_i$.

Step 3   OUTPUT (SUM).
            STOP.   ■

EXAMPLE 2   The Taylor polynomial $P_N(x)$ for $f(x) = \ln x$ expanded about $x_0 = 1$ is

$$P_N(x) = \sum_{i=1}^{N} \frac{(-1)^{i+1}}{i} (x-1)^i.$$

The value of ln 1.5 to eight decimal places is 0.40546511. Suppose we want to compute the minimal value of N required for

$$|\ln 1.5 - P_N(1.5)| < 10^{-5},$$

without using the Taylor polynomial truncation error formula. An algorithm to solve this problem is:

INPUT   value x, tolerance TOL, maximum number of iterations M.

OUTPUT   degree N of the polynomial or a message of failure.

Step 1   Set $N = 1$;

        $y = x - 1$;

        $VALUE = \ln(x)$;

        $SUM = 0$;

        $POWER = 1$;

        $SIGN = -1$.

Step 2   While $N \le M$ do Steps 3–5.

      Step 3   Set $POWER = POWER * y$;

           $SIGN = -SIGN$;

           $SUM = SUM + SIGN * POWER/N$.

      Step 4   If $|SUM - VALUE| < TOL$ then

           OUTPUT (N);

           STOP.

      Step 5   Set $N = N + 1$;

Step 6   OUTPUT ('Method Failed');
            STOP.

The input for our problem would be $x = 1.5$, $TOL = 10^{-5}$, and perhaps $M = 15$. This choice of M would constitute an upper bound for the number of calculations we would be willing to have performed, recognizing that the algorithm would be likely to fail if this bound were exceeded.

Whether the output would be a value for N or the failure message depends on the precision of the computational device being used.   ∎

We are primarily interested in choosing methods that will produce dependably accurate results. One criterion we will impose on an algorithm whenever possible is that small changes in the initial data produce correspondingly small changes in the final results. An algorithm that satisfies this property is called **stable**; it is **unstable** when this criterion is not fulfilled. Some algorithms will be stable for certain choices of initial data but not for all choices. We will attempt to characterize the stability properties of algorithms whenever possible.

To consider further the subject of rounding error growth and its connection to algorithm stability, suppose that an error ε is introduced at some stage in the calculations and that the error after n subsequent operations is denoted by E_n. The two cases that arise most often in practice are defined as follows.

DEFINITION 1.16   Suppose that E_n represents the growth of an error after n subsequent operations. If $|E_n| \approx Cn\varepsilon$, where C is a constant independent of n, the growth of error is said to be **linear**. If $|E_n| \approx k^n \varepsilon$, for some $k > 1$, the growth of error is **exponential**.

Linear growth of error is usually unavoidable and, when C and ε are small, the results are generally acceptable. Exponential growth of error should be avoided, since the term k^n becomes large for even relatively small values of n. This leads to unacceptable inaccuracies, regardless of the size of ε. As a consequence, an algorithm that exhibits linear growth of error is stable, while an algorithm exhibiting exponential-error growth is unstable. (*See Fig. 1.6.*)

FIGURE 1.6

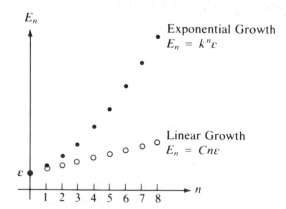

Exponential Growth
$E_n = k^n \varepsilon$

Linear Growth
$E_n = Cn\varepsilon$

EXAMPLE 3

The sequence $p_n = (1/3)^n$, $n > 0$, could be generated recursively by letting $p_0 = 1$ and defining $p_n = (1/3)p_{n-1}$, whenever $n > 1$. If the sequence is generated in this manner using five-digit rounding arithmetic, the results are

$$0.10000 \times 10^1, \quad 0.33333 \times 10^0, \quad 0.11111 \times 10^0,$$
$$0.37036 \times 10^{-1}, \quad 0.12345 \times 10^{-1}, \quad \dots .$$

The rounding error introduced by replacing $1/3$ by 0.33333 produces an error of only $(0.33333)^n \times 10^{-5}$ in the nth term of the sequence. This method of generating the sequence is clearly stable.

Another way to generate the sequence is to define $p_0 = 1$, $p_1 = 1/3$, and compute, for each $n \geq 2$,

(1.4) $$p_n = (10/3)p_{n-1} - p_{n-2}.$$

Table 1.4 lists the exact and five-digit rounding results using this formula. This method quite clearly is unstable.

TABLE 1.4

n	Computed p_n	Correct value p_n
0	0.10000×10^0	0.10000×10^0
1	0.33333×10^0	0.33333×10^0
2	0.11110×10^0	0.11111×10^0
3	0.37000×10^{-1}	0.37037×10^{-1}
4	0.12230×10^{-1}	0.12346×10^{-1}
5	0.37660×10^{-2}	0.41152×10^{-2}
6	0.32300×10^{-3}	0.13717×10^{-2}
7	-0.26893×10^{-2}	0.45725×10^{-3}
8	-0.92872×10^{-2}	0.15242×10^{-3}

Note that formula (1.4) will be satisfied whenever p_n is of the form

$$p_n = C_1(1/3)^n + C_2 3^n$$

for any pair of constants C_1 and C_2. To verify this,

$$\frac{10}{3}p_{n-1} - p_{n-2} = \frac{10}{3}\left[C_1\left(\frac{1}{3}\right)^{n-1} + C_2 3^{n-1} \right] - \left[C_1\left(\frac{1}{3}\right)^{n-2} + C_2 3^{n-2} \right]$$

$$= C_1\left[\frac{10}{3}\left(\frac{1}{3}\right)^{n-1} - \left(\frac{1}{3}\right)^{n-2} \right] + C_2\left[\frac{10}{3}3^{n-1} - 3^{n-2} \right]$$

$$= C_1\left(\frac{1}{3}\right)^n + C_2 3^n = p_n.$$

To have $p_0 = 1$ and $p_1 = 1/3$ in Eq. (1.4), the constants C_1 and C_2 must be chosen as $C_1 = 1$ and $C_2 = 0$. However, in the five-digit approximation, the first two terms are $p_0 = 0.10000 \times 10^1$ and $p_1 = 0.33333 \times 10^0$, which requires a modification of these constants to $C_1 = 0.10000 \times 10^1$ and $C_2 = -0.12500 \times 10^{-5}$. This small change in C_2

results in a rounding error of $3^n(-0.12500 \times 10^{-5})$ in producing p_n. As a consequence, an exponential growth of error results, which is reflected in the extreme inaccuracies found in the entries of Table 1.4.   ∎

To reduce the effects of rounding error, we can use large-order digit arithmetic such as the double- or multiple-precision option available on most digital computers. A disadvantage in using double-precision arithmetic is that it takes a great deal more computer time. Also, the serious growth of rounding error will not be eliminated entirely, but only postponed if a large number of subsequent computations are performed.

Another approach to estimating rounding error is to use interval arithmetic, that is, to retain the largest and smallest possible values at each step, so that, in the end, we obtain an interval which contains the true value. Unfortunately, the true answer may be near the extremes of the interval, and we may have to find a very small interval for reasonable implementation.

It is also possible to study error from a statistical standpoint. This study, however, involves considerable analysis and as such is beyond the scope of this text. Henrici [62], pages 305–309, presents a discussion of a statistical approach to estimate accumulated round-off error.

Since iterative techniques involving sequences are often used, this section will conclude with a brief discussion of some terminology that is used to describe the rate at which convergence occurs when employing a numerical technique. In general, we would like the technique to converge as rapidly as possible. To compare the convergence rules of various methods, we use the following definition.

DEFINITION 1.17   Suppose $\{\alpha_n\}_{n=1}^{\infty}$ is a sequence that converges to a number α. We say that $\{\alpha_n\}_{n=1}^{\infty}$ converges to α with **rate of convergence** $O(\beta_n)$, where $\{\beta_n\}_{n=1}^{\infty}$ is another sequence with $\beta_n \neq 0$ for each n, if

$$\frac{|\alpha_n - \alpha|}{|\beta_n|} \leq K \qquad \text{for sufficiently large } n,$$

and K is a constant independent of n. This case is often indicated by writing $\alpha_n = \alpha + O(\beta_n)$ or $\alpha_n \to \alpha$ with rate of convergence $O(\beta_n)$.

EXAMPLE 4   Suppose that the sequences $\{\alpha_n\}$ and $\{\hat{\alpha}_n\}$ are described by $\alpha_n = (\sin n)/n$ and $\hat{\alpha}_n = (n+3)/n^3$ for each integer $n \geq 1$. Although $\lim_{n \to \infty} \alpha_n = \lim_{n \to \infty} \hat{\alpha}_n = 0$, the sequence $\{\hat{\alpha}_n\}$ converges to this limit much faster than the sequence $\{\alpha_n\}$.

If we let $\beta_n = 1/n$ and $\hat{\beta}_n = 1/n^2$ for each n we see that

$$\left| \frac{(\sin n)/n - 0}{(1/n)} \right| = |\sin n| \leq 1$$

and

$$\left| \frac{(n+3)/n^3 - 0}{(1/n^2)} \right| = \left| \frac{n+3}{n} \right| \leq 4$$

so $$\alpha_n = 0 + O\left(\frac{1}{n}\right) \qquad \text{while } \hat{\alpha} = 0 + O\left(\frac{1}{n^2}\right).$$

This implies that the rate of convergence of $\{\alpha_n\}$ to zero is similar to the convergence of $\{1/n\}$ to zero, while $\{\hat{\alpha}_n\}$ converges to zero in a manner similar to the more rapidly convergent sequence $\{1/n^2\}$. ■

This concept generalizes to functions as follows:

DEFINITION 1.18 If $\lim_{x \to 0} F(x) = L$, the convergence is said to be $O(G(x))$ if there exists a number $K > 0$, independent of x, for which

$$\frac{|F(x) - L|}{|G(x)|} \le K \qquad \text{for sufficiently small } x > 0.$$

This situation is often indicated by writing $F(x) = L + O(G(x))$ or $F(x) \to L$ with rate of convergence $O(G(x))$.

EXAMPLE 5 In Example 2 of Section 1.1 we found that by using a Taylor polynomial of degree three,

$$\cos x = 1 - \frac{1}{2}x^2 + \frac{1}{24}x^4 \cos \xi(x)$$

for some number $\xi(x)$ between zero and x.
  Consequently,

$$\cos x + \frac{1}{2}x^2 = 1 + \frac{1}{24}x^4 \cos \xi(x).$$

This implies that

$$\cos x + \frac{1}{2}x^2 = 1 + O(x^4)$$

since

$$\frac{\left|\left(\cos x + \frac{1}{2}x^2\right) - 1\right|}{x^4} = \left|\frac{1}{24}\cos \xi(x)\right| \le \frac{1}{24}$$

The implication is that $\cos x + (1/2)x^2$ converges to its limit, 1, at roughly the same rate that x^4 converges to zero. ■

Exercise Set 1.3

1. All calculus students know that

$$\lim_{h \to 0} \frac{\sin h}{h} = 1,$$

   but show that

$$\frac{\sin h}{h} = 1 + O(h^2).$$

2. $\lim\limits_{h \to 0} \dfrac{1 - \cos h}{h} = 0$ at what rate of convergence?

3. The sequence of numbers $1, \dfrac{1}{3}, \dfrac{1}{9}, \dfrac{1}{27}, \ldots$ considered in Example 3 can be generated by either of the two recurrence relations

   i) $p_0 = 1,\ p_1 = 1/3,\ p_n = \dfrac{5}{6}p_{n-1} - \dfrac{1}{6}p_{n-2},\ n = 2, 3, \ldots$

   ii) $q_0 = 1,\ q_1 = 1/3,\ q_n = \dfrac{5}{3}q_{n-1} - \dfrac{4}{9}q_{n-2},\ n = 2, 3, \ldots$

   a) Compute p_n for $n = 2, \ldots, 8$, using five-digit rounding arithmetic and compare the results to those obtained in Example 3.

   b) Compute q_n for $n = 2, \ldots, 8$, using five-digit rounding arithmetic and compare the results to those obtained in Example 3 and in part (a).

   c) Show whether procedure (i) is stable.

   d) Show whether procedure (ii) is stable.

4. a) Use three-digit chopping arithmetic to compute the sum $\sum_{i=1}^{10} (1/i^2)$ first by $\frac{1}{1} + \frac{1}{4} + \cdots + \frac{1}{100}$ and then by $\frac{1}{100} + \frac{1}{81} + \cdots + \frac{1}{1}$. Which method is most accurate and why?

   b) Write an algorithm to sum the finite series $\sum\limits_{i=1}^{N} x_i$ in reverse order.

5. Construct an algorithm that has as input an integer $n \geq 1$, $n + 1$ points $x_0, x_1, \ldots, x_n$, and a point x and which produces as output the product $P = (x - x_0)(x - x_1) \cdots (x - x_n)$.

6. Example 4 of Section 1.2 gives alternate formulas for the roots x_1 and x_2 of $ax^2 + bx + c = 0$. Construct an algorithm with input a, b, c and output x_1, x_2 that computes the roots x_1 and x_2 (which may be equal or even complex conjugates) so as to employ the best formula for each.

7. Assume that

$$\frac{1 - 2x}{1 - x + x^2} + \frac{2x - 4x^3}{1 - x^2 + x^4} + \frac{4x^3 - 8x^7}{1 - x^4 + x^8} + \cdots = \frac{1 + 2x}{1 + x + x^2}$$

   for $x < 1$. If $x = 0.25$, write and execute an algorithm that determines the number of terms needed on the left side of the equation so that the left side differs from the right side by less than 10^{-6}.

8. Construct an algorithm that has as input a number x, $0 < x < \pi/4$, a tolerance TOL, and maximum number of iterations M to approximate $\sin x$ using a Maclaurin series. The approximation is to be based on

$$\left| \sin x - \sum_{k=0}^{n} \frac{f^{(k)}(0)}{k!} x^k \right| \le \frac{|f^{(n+1)}(\xi)| x^{n+1}}{(n+1)!} < TOL$$

where $f(x) = \sin x$ and $n \le M$. The algorithm should either output the approximation to $\sin x$ or a message that the procedure failed.

Solutions of Equations in One Variable

The growth of large populations can be modeled over short periods of time by assuming that the population grows continuously with time at a rate proportional to the number of individuals present at that time. If we let $N(t)$ denote the number of individuals at time t and λ denote the constant birth rate of the population, the population satisfies the differential equation

$$\frac{dN(t)}{dt} = \lambda N(t).$$

The solution to this equation is $N(t) = N_0 e^{\lambda t}$, where N_0 denotes the initial population.

This model is valid only when the population is isolated with no immigration from outside the community. If immigration is permitted at a constant rate v, the differential equation governing the situation becomes

$$\frac{dN(t)}{dt} = \lambda N(t) + v,$$

whose solution is

$$N(t) = N_0 e^{\lambda t} + \frac{v}{\lambda}(e^{\lambda t} - 1).$$

Suppose a certain population contains one million individuals initially, that 435,000 individuals immigrate into the community in the first year, and that 1,564,000 individuals are present at the end of one year. To determine the birth rate of this population necessitates solving for λ in the equation

$$1,564,000 = 1,000,000 e^{\lambda} + \frac{435,000}{\lambda}(e^{\lambda} - 1).$$

The numerical methods discussed in this chapter are used to find approximations to solutions of equations of this type, when the exact solutions cannot be obtained by algebraic methods. The solution to this particular problem is considered in Exercise 19 of Section 2.3.

In this chapter we will discuss one of the most basic problems in numerical analysis. The problem is called a **root-finding problem** and consists of finding values of the variable x that satisfy the equation $f(x) = 0$, for a given function f. A solution to this problem is called a **zero** of f or a **root** of $f(x) = 0$. This is one of the oldest numerical approximation problems and yet research is continuing actively in this area at the present time. The procedures we will discuss range from the classical Newton–Raphson method, basically developed by Isaac Newton over 300 years ago, to methods that were first published quite recently.

2.1 The Bisection Algorithm

The first technique, based on the Intermediate Value Theorem (Theorem 1.12), is called the **bisection algorithm**, or **binary-search method**. Suppose a continuous function f, defined on the interval $[a, b]$, is given, with $f(a)$ and $f(b)$ of opposite sign. Then by Theorem 1.12, there exists p, $a < p < b$, for which $f(p) = 0$. Although the procedure will work for the case when $f(a)$ and $f(b)$ have opposite signs and there is more than one root in the interval $[a, b]$, it will be assumed for simplicity that the root in this interval is unique. The method calls for a repeated halving of subintervals of $[a, b]$ and, at each step, locating the "half" containing p. To begin, set $a_1 = a$ and $b_1 = b$, and let p_1 be the midpoint of $[a, b]$; that is,

$$p_1 = \tfrac{1}{2}(a_1 + b_1).$$

If $f(p_1) = 0$, then $p = p_1$; if not, then $f(p_1)$ has the same sign as either $f(a_1)$ or $f(b_1)$. If $f(p_1)$ and $f(a_1)$ have the same sign, then $p \in (p_1, b_1)$, and we set $a_2 = p_1$ and $b_2 = b_1$. If $f(p_1)$ and $f(b_1)$ are of the same sign, then $p \in (a_1, p_1)$, and we set $a_2 = a_1$ and $b_2 = p_1$. Now we reapply the process to the interval $[a_2, b_2]$. This produces the following algorithm. (*See Fig. 2.1.*)

FIGURE 2.1

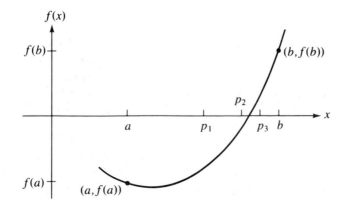

Bisection Algorithm 2.1

To find a solution to $f(x) = 0$ given the continuous function f on the interval $[a, b]$ where $f(a)$ and $f(b)$ have opposite signs:

INPUT endpoints a, b; tolerance TOL; maximum number of iterations N_0.

OUTPUT approximate solution p or message of failure.

Step 1   Set $i = 1$.

Step 2   While $i \leq N_0$ do Steps 3–6.

Step 3   Set $p = a + (b - a)/2$.   (*Compute p_i.*)

Step 4   If $f(p) = 0$ or $(b - a)/2 < TOL$ then
OUTPUT (p);   (*Procedure completed successfully.*)
STOP.

Step 5   Set $i = i + 1$.

Step 6   If $f(a)f(p) > 0$ then set $a = p$ (*Compute a_i, b_i.*)
else set $b = p$.

Step 7   OUTPUT ('Method failed after N_0 iterations, $N_0 = $ ', N_0);
(*Procedure completed unsuccessfully.*)
STOP.

We briefly mention some other stopping procedures that could also be applied in Step 4 of Algorithm 2.1, all of which apply to any iterative technique considered in this chapter. Select a tolerance $\varepsilon > 0$ and generate $p_1, \ldots, p_N$ until one of the following conditions is met:

$$(2.1) \qquad\qquad |p_N - p_{N-1}| < \varepsilon,$$

$$(2.2) \qquad\qquad \frac{|p_N - p_{N-1}|}{|p_N|} < \varepsilon, \qquad p_N \neq 0, \text{ or}$$

$$(2.3) \qquad\qquad |f(p_N)| < \varepsilon.$$

Unfortunately, difficulties can arise using any of these stopping criteria. For example, there exist sequences $\{p_n\}$ with the property that the differences $p_n - p_{n-1}$ converge to zero while the sequence itself diverges. (See Exercise 13.) It is also possible for $f(p_n)$ to be close to zero while p_n differs significantly from p. (See Exercise 14.) Without additional knowledge about f or p, inequality (2.2) is the best stopping criterion to apply because it tests relative error.

When using a computer to generate the approximations, it is also good practice to add a condition that will set an upper bound on the number of iterations performed. This eliminates the possibility of the machine being put into an infinite loop, a possibility that can arise when the sequence diverges (and also when the program is incorrectly coded). This is easily done by setting an initial bound N_0 and requiring the procedure to terminate if $i \geq N_0$, as was done in Step 2 of Algorithm 2.1.

Note that in order to start the bisection algorithm, an interval $[a, b]$ must be found with $f(a) \cdot f(b) < 0$. At each step of the bisection algorithm, the length of the interval known to contain a zero of f is reduced by a factor of two; hence it is advantageous to choose the initial interval $[a, b]$ as small as possible. For example, if $f(x) = 2x^3 - x^2 + x - 1$,

$$f(-4) \cdot f(4) < 0 \qquad \text{and} \qquad f(0) \cdot f(1) < 0,$$

so the bisection algorithm could be used on either of the intervals $[-4, 4]$ or $[0, 1]$. Starting the bisection algorithm on $[0, 1]$ instead of $[-4, 4]$ will reduce the number of iterations required to achieve any specified accuracy by three.

To illustrate the bisection algorithm, consider the following example. The iteration in this example is terminated when $|p_{n-1} - p_n|/|p_n| < 10^{-4}$.

EXAMPLE 1   The function $f(x) = x^3 + 4x^2 - 10$ has a root in $[1, 2]$ since $f(1) = -5$ and $f(2) = 14$. It is easily seen that there is only one root in $[1, 2]$. The bisection algorithm gives the values in Table 2.1.

TABLE 2.1

n	a_n	b_n	p_n	$f(p_n)$
1	1.0	2.0	1.5	2.375
2	1.0	1.5	1.25	-1.79687
3	1.25	1.5	1.375	0.16211
4	1.25	1.375	1.3125	-0.84839
5	1.3125	1.375	1.34375	-0.35098
6	1.34375	1.375	1.359375	-0.09641
7	1.359375	1.375	1.3671875	0.03236
8	1.359375	1.3671875	1.36328125	-0.03215
9	1.36328125	1.3671875	1.365234375	0.000072
10	1.36328125	1.365234375	1.364257813	-0.01605
11	1.364257813	1.365234375	1.364746094	-0.00799
12	1.364746094	1.365234375	1.364990235	-0.00396
13	1.364990235	1.365234375	1.365112305	-0.00194

After 13 iterations, we can see that $p_{13} = 1.365112305$ approximates the root p with an error

$$|p - p_{13}| < |b_{14} - a_{14}| = |1.365234375 - 1.365112305|$$
$$= 0.000122070$$

and since $|a_{14}| < |p|$,

$$\frac{|p - p_{13}|}{|p|} < \frac{|b_{14} - a_{14}|}{|a_{14}|} \leq 9.0 \times 10^{-5},$$

the approximation is correct to at least four significant digits. The correct value of p, to nine decimal places, is $p = 1.365230013$. It is interesting to note that p_9 is closer to p than is the final approximation p_{13}, but there is no way of determining this exactly unless the true answer is known.   ∎

    The bisection algorithm, though conceptually clear, has significant drawbacks. It is very slow in converging (that is, N may become quite large before $|p - p_N|$ is sufficiently small) and, moreover, a good intermediate approximation may be inadvertently discarded. However, the method has the important property that it will always converge to a solution and for that reason is often used as a "starter" for the more efficient methods presented later in this chapter.

THEOREM 2.1
Let $f \in C[a, b]$ and suppose $f(a) \cdot f(b) < 0$. The bisection procedure (Algorithm 2.1) generates a sequence $\{p_n\}$ approximating p with the property

(2.4)
$$|p_n - p| \leq \frac{b - a}{2^n}, \qquad n \geq 1.$$

PROOF
For each $n \geq 1$, we have

$$b_n - a_n = \frac{1}{2^{n-1}}(b - a) \qquad \text{and} \qquad p \in (a_n, b_n).$$

Since $p_n = \frac{1}{2}(a_n + b_n)$, for all $n \geq 1$, it follows that

$$|p_n - p| \leq \tfrac{1}{2}(b_n - a_n) = 2^{-n}(b - a).$$ □

    According to Definition 1.17 (p. 23), inequality (2.4) implies that $\{p_n\}_{n=1}^{\infty}$ converges to p and is bounded by a sequence that converges to zero with $O(2^{-n})$ rate of convergence. It is important to realize that theorems such as this give only bounds for errors in approximation. For example, this bound applied to the problem in Example 1 ensures only that

$$|p - p_9| \leq \frac{2 - 1}{2^9} \approx 2 \times 10^{-3}.$$

The actual error is much smaller than this

$$|p - p_9| = |1.365230013 - 1.365234375| \approx 4.4 \times 10^{-6}.$$

EXAMPLE 2
Determine approximately how many iterations are necessary to solve $f(x) = x^3 + 4x^2 - 10 = 0$ with an accuracy of $\varepsilon = 10^{-5}$ for $a_1 = 1$ and $b_1 = 2$. This requires finding an integer N that will satisfy:

$$|p_N - p| \leq 2^{-N}(b - a) = 2^{-N} < 10^{-5}.$$

    To determine N we will use logarithms. Although logarithms to any base would suffice, we will use base 10 logarithms since the tolerance is given as a power of 10.

Since $2^{-N} < 10^{-5}$ implies that $\log_{10} 2^{-N} < \log_{10} 10^{-5} = -5$,

$$-N \log_{10} 2 < -5 \qquad \text{or} \qquad N > \frac{5}{\log_{10} 2} \approx 16.6.$$

It would appear to require 17 iterations to obtain an approximation accurate to 10^{-5}. With $\varepsilon = 10^{-3}$, $N \geq 10$ iterations are required and the value of $p_9 = 1.36523475$ is accurate to within 10^{-4}. It is important to note that these techniques give only a *bound* for the number of iterations necessary, and in many cases this bound is much larger than the actual number required. ■

Exercise Set 2.1

1. Show that $f(x) = x^3 - x - 1$ has exactly one zero in the interval $[1, 2]$. Approximate the zero to within 10^{-2} using the Bisection Algorithm.

2. Use the Bisection Algorithm to find solutions accurate to within 10^{-2} for $x^4 - 2x^3 - 4x^2 + 4x + 4 = 0$ on

   a) $[-2, 0]$ b) $[0, 2]$

   c) $[1, 2]$

3. Use the Bisection Algorithm to find a solution accurate to within 10^{-2} for $x = \tan x$ on $[4, 4.5]$.

4. Use the Bisection Algorithm to find all solutions of $x^3 - 7x^2 + 14x - 6 = 0$ to an accuracy of 10^{-3}.

5. Use the Bisection Algorithm to find solutions accurate to 10^{-5} for the following problems:

   a) $x - 2^{-x} = 0$ for $0 \leq x \leq 1$,

   b) $e^x + 2^{-x} + 2 \cos x - 6 = 0$ for $1 \leq x \leq 2$,

   c) $e^x - x^2 + 3x - 2 = 0$ for $0 \leq x \leq 1$.

6. a) Use the Bisection Algorithm to find a solution, accurate to within 10^{-2} for $x + 0.5 + 2 \cos \pi x = 0$ on $[0.5, 1.5]$.

   b) Change Step 6 in the Bisection Algorithm to

   "If $f(b)f(p) > 0$ then set $b = p$ else set $a = p$."

   Use this new algorithm to find a solution accurate to within 10^{-2} for $x + 0.5 + 2 \cos \pi x = 0$.

   c) Discuss any discrepancy between parts (a) and (b).

7. Find an approximation to $\sqrt[3]{25}$ correct to within 10^{-4} using the Bisection Algorithm. [*Hint*: Consider $f(x) = x^3 - 25$.]

8. Find an approximation to $\sqrt{3}$ correct to within 10^{-4} using the Bisection Algorithm.

9. Use Theorem 2.1 to find a bound for the number of iterations needed to achieve an approximation with accuracy 10^{-4} to the solution of $x^3 - x - 1 = 0$ lying in the interval $[1, 2]$. Find an approximation to the root with this degree of accuracy.

10. Use Theorem 2.1 to find a bound for the number of iterations needed to achieve an approximation with accuracy 10^{-3} to the solution of $x^3 + x - 4 = 0$ lying in the interval $[1, 4]$. Find an approximation to the root with this degree of accuracy.

11. Apply the Bisection Algorithm to the equation

$$\frac{4x - 7}{(x - 2)^2} = 0,$$

using the intervals $[1.2, 2.2]$ and $[1.5, 2.5]$. Explain the results graphically.

12. Find all the zeros of $f(x) = x^2 + 10 \cos x$ accurate to within 10^{-4}. [*Hint:* Consider the graph of f.]

13. Let $\{p_n\}$ be the sequence defined by $p_n = \sum_{k=1}^{n} (1/k)$. Show that $\lim_{n \to \infty} (p_n - p_{n-1}) = 0$, but that $\{p_n\}$ diverges.

14. Let $f(x) = (x - 1)^{10}$, $p = 1$, and $p_n = 1 + 1/n$. Show that $|f(p_n)| < 10^{-3}$ whenever $n > 1$, but that $|p - p_n| < 10^{-3}$ requires that $n > 1000$.

2.2 Fixed-Point Iteration

In this section we consider methods for determining the solution to an equation that is expressed, for some function g, in the form

$$g(x) = x.$$

A solution to such an equation is said to be a **fixed point** of the function g.

If a fixed point could be found for any given g, then every root-finding problem could also be solved. For example, the root-finding problem $f(x) = 0$ has solutions that correspond precisely to the fixed points of $g(x) = x$ when $g(x) = x - f(x)$. The first task, then, is to decide when a function will have a fixed point and how the fixed points can be determined. (In numerical analysis "determined" generally means approximated to a sufficient degree of accuracy.)

EXAMPLE 1
a) The function $g(x) = x$, $0 \le x \le 1$, has a fixed point at each x in $[0, 1]$.
b) The function $g(x) = x - \sin \pi x$ has exactly two fixed points in $[0, 1]$, $x = 0$ and $x = 1$. (*See Fig. 2.2.*) ■

The following theorem gives sufficient conditions for the existence and uniqueness of a fixed point.

THEOREM 2.2   If $g \in C[a, b]$ and $g(x) \in [a, b]$ for all $x \in [a, b]$, then g has a fixed point in $[a, b]$.

FIGURE 2.2

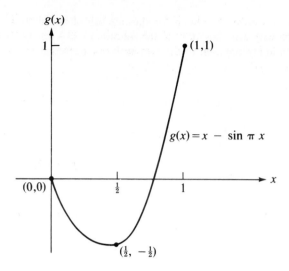

Further, suppose $g'(x)$ exists on (a, b) and

(2.5) $$|g'(x)| \le k < 1 \qquad \text{for all } x \in (a, b).$$

Then g has a unique fixed point p in $[a, b]$. (*See Fig. 2.3.*)

FIGURE 2.3

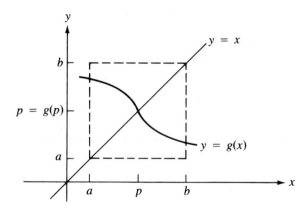

PROOF

If $g(a) = a$ or $g(b) = b$, the existence of a fixed point is obvious. Suppose not, then it must be true that $g(a) > a$ and $g(b) < b$. Define $h(x) = g(x) - x$; h is continuous on $[a, b]$, and, moreover,

$$h(a) = g(a) - a > 0, \qquad h(b) = g(b) - b < 0.$$

The Intermediate Value Theorem implies that there exists $p \in (a, b)$ for which $h(p) = 0$. Thus, $g(p) - p = 0$ and p is a fixed point of g.

Suppose in addition that inequality (2.5) holds and that p and q are both fixed points in $[a, b]$ with $p \ne q$. By the Mean Value Theorem (Theorem 1.8), a number ξ exists between p and q, and hence in $[a, b]$, with

$$|p - q| = |g(p) - g(q)| = |g'(\xi)||p - q| \le k|p - q| < |p - q|,$$

which is a contradiction. This contradiction must come from the only supposition, $p \ne q$. Hence, $p = q$ and the fixed point in $[a, b]$ is unique. $\qquad \square$

EXAMPLE 2   a) Let $g(x) = (x^2 - 1)/3$ on $[-1, 1]$. Using the Extreme Value Theorem (Theorem 1.9), it is easy to show that the absolute minimum of g occurs at $x = 0$ and is $g(0) = -\frac{1}{3}$. Similarly, the absolute maximum of g occurs at $x = \pm 1$ and has the value $g(\pm 1) = 0$. Moreover, g is continuous and

$$|g'(x)| = \left|\frac{2x}{3}\right| \le \frac{2}{3} \qquad \text{for all } x \in [-1, 1],$$

so g satisfies the hypotheses of Theorem 2.2 and has a unique fixed point in $[-1, 1]$.

In this example, the unique fixed point p in the interval $[-1, 1]$ can be determined exactly. If

$$p = g(p) = \frac{p^2 - 1}{3}, \qquad \text{then } p^2 - 3p - 1 = 0,$$

which, by the quadratic formula, implies that

$$p = \frac{3 - \sqrt{13}}{2}.$$

Note that g also has a unique fixed point $p = (3 + \sqrt{13})/2$ for the interval $[3, 4]$. However, $g(4) = 5$ and $g'(4) = \frac{8}{3} > 1$; so g does not satisfy the hypotheses of Theorem 2.2. This shows that the hypotheses of Theorem 2.2 are sufficient to guarantee a unique fixed point, but are not necessary.

   b) Let $g(x) = 3^{-x}$. Since $g'(x) = -3^{-x} \ln 3 < 0$ on $[0, 1]$, the function g is decreasing on $[0, 1]$. Hence, $g(1) = \frac{1}{3} \le g(x) \le 1 = g(0)$ for $0 \le x \le 1$. Thus for $x \in [0, 1]$, $g(x) \in [0, 1]$. Therefore, g has a fixed point in $[0, 1]$. Since,

$$g'(0) = -\ln 3 = -1.098612289,$$

$|g'(x)| \not\le 1$ on $[0, 1]$ and Theorem 2.2 cannot be used to determine uniqueness. However, g is decreasing so it is clear that the fixed point must be unique. (*See Fig. 2.4*). ∎

FIGURE 2.4

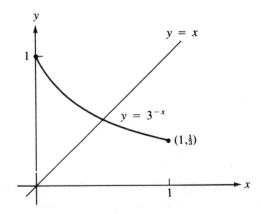

To approximate the fixed point of a function g, we choose an initial approximation p_0 and generate the sequence $\{p_n\}_{n=0}^{\infty}$ by letting $p_n = g(p_{n-1})$ for each $n \geq 1$. If the sequence converges to p and g is continuous, then, by Theorem 1.4

$$p = \lim_{n \to \infty} p_n = \lim_{n \to \infty} g(p_{n-1}) = g\left(\lim_{n \to \infty} p_{n-1}\right) = g(p),$$

and a solution to $x = g(x)$ is obtained. This technique is called the **fixed-point iterative technique**, or **functional iteration**. The procedure is detailed in Algorithm 2.2 and described in Fig. 2.5.

FIGURE 2.5

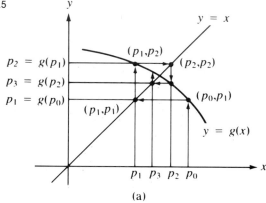

(a)

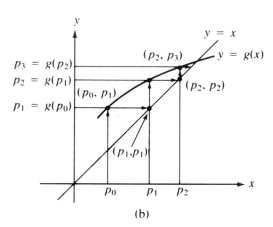

(b)

Fixed-Point Algorithm 2.2

To find a solution to $p = g(p)$ given an initial approximation p_0:

INPUT initial approximation p_0; tolerance TOL; maximum number of iterations N_0.

OUTPUT approximate solution p or message of failure.

Step 1   Set $i = 1$.

Step 2   While $i \leq N_0$ do Steps 3–6.

   Step 3   Set $p = g(p_0)$.   (*Compute p_i.*)

   Step 4   If $|p - p_0| < TOL$ then
             OUTPUT (p);   (*Procedure completed successfully.*)
             STOP.

   Step 5   Set $i = i + 1$.

   Step 6   Set $p_0 = p$.   (*Update p_0.*)

Step 7   OUTPUT ('Method failed after N_0 iterations, $N_0 = $', N_0);
          (*Procedure completed unsuccessfully.*)
          STOP.

To illustrate the technique of functional iteration, consider the following example.

EXAMPLE 3   The equation $x^3 + 4x^2 - 10 = 0$ has a unique root in $[1, 2]$. There are many ways to change the equation to the form $x = g(x)$ by simple algebraic manipulation. For example, to obtain the function g described in (c) we could manipulate the equation $x^3 + 4x^2 - 10 = 0$ as follows:

$$4x^2 = 10 - x^3 \qquad \text{so} \qquad x^2 = \tfrac{1}{4}(10 - x^3)$$

and

$$x = \pm\tfrac{1}{2}(10 - x^3)^{1/2}.$$

To obtain a positive solution $g_3(x)$ is chosen as shown. It is not important to derive these functions, but it should be verified that the fixed point of each is actually a solution to the original equation.

a) $x = g_1(x) = x - x^3 - 4x^2 + 10,$ b) $x = g_2(x) = \left(\dfrac{10}{x} - 4x\right)^{1/2},$

c) $x = g_3(x) = \tfrac{1}{2}(10 - x^3)^{1/2},$ d) $x = g_4(x) = \left(\dfrac{10}{4+x}\right)^{1/2},$

e) $x = g_5(x) = x - \dfrac{x^3 + 4x^2 - 10}{3x^2 + 8x}.$

With $p_0 = 1.5$, Table 2.2 lists the results of the fixed-point iteration method for all five choices of g.

TABLE 2.2

n/p_n	(a)	(b)	(c)	(d)	(e)
0	1.5	1.5	1.5	1.5	1.5
1	-0.875	0.8165	1.286953768	1.348399725	1.373333333
2	6.732	2.9969	1.402540804	1.367376372	1.365262015
3	-469.7	$(-8.65)^{1/2}$	1.345458374	1.364957015	1.365230014
4	1.03×10^8		1.375170253	1.365264748	1.365230013
5			1.360094193	1.365225594	
6			1.367846968	1.365230576	
7			1.363887004	1.365229942	
8			1.365916734	1.365230022	
9			1.364878217	1.365230012	
10			1.365410062	1.365230014	
15			1.365223680	1.365230013	
20			1.365230236		
25			1.365230006		
30			1.365230013		

The actual root is 1.365230013, as was noted in Example 1 of Section 2.1. Comparing the results to the bisection algorithm given in that example, it can be seen that excellent results have been obtained for choices (c), (d), and (e), whereas the bisection technique requires 27 iterations for such accuracy. It is interesting to note that choice (a) led to divergence and (b) became undefined because it involved the square root of a negative number.   ∎

The previous example illustrates the need for a procedure which will guarantee that the function g converges to a solution of $x = g(x)$ and will also choose g in a manner to make this convergence as rapid as possible. The following theorem is the first step in determining such a procedure.

THEOREM 2.3 Let $g \in C[a, b]$ and suppose that $g(x) \in [a, b]$ for all $x \in [a, b]$. Further, let g' exist on (a, b) with

(2.6) $$|g'(x)| \leq k < 1 \qquad \text{for all } x \in (a, b).$$

If p_0 is any number in $[a, b]$, then the sequence defined by

$$p_n = g(p_{n-1}), \qquad n \geq 1,$$

will converge to the unique fixed point p in $[a, b]$.

PROOF By Theorem 2.2, a unique fixed point exists in $[a, b]$. Since g maps $[a, b]$ into itself, the sequence $\{p_n\}_{n=0}^{\infty}$ is defined for all $n \geq 0$ and $p_n \in [a, b]$ for all n. Using inequality (2.6) and the Mean Value Theorem,

(2.7) $$|p_n - p| = |g(p_{n-1}) - g(p)| \leq |g'(\xi)||p_{n-1} - p| \leq k|p_{n-1} - p|,$$

where $\xi \in (a, b)$. Applying inequality (2.7) inductively gives:

(2.8) $$|p_n - p| \leq k|p_{n-1} - p| \leq k^2|p_{n-2} - p| \leq \cdots \leq k^n|p_0 - p|.$$

Since $k < 1$,

$$\lim_{n \to \infty} |p_n - p| \leq \lim_{n \to \infty} k^n|p_0 - p| = 0$$

and $\{p_n\}_{n=0}^{\infty}$ converges to p.   □

COROLLARY 2.4 If g satisfies the hypotheses of Theorem 2.3, a bound for the error involved in using p_n to approximate p is given by

(2.9) $$|p_n - p| \leq k^n \max\{p_0 - a, b - p_0\} \qquad \text{for all } n \geq 1.$$

PROOF From inequality (2.8),

$$|p_n - p| \leq k^n|p_0 - p| \leq k^n \max\{p_0 - a, b - p_0\},$$

since $p \in [a, b]$.   □

COROLLARY 2.5   If g satisfies the hypotheses of Theorem 2.3, then

(2.10) $$|p_n - p| \le \frac{k^n}{1-k}|p_0 - p_1| \qquad \text{for all } n \ge 1.$$

PROOF   For $n \ge 1$, the procedure used in the proof of Theorem 2.3 implies that

$$|p_{n+1} - p_n| = |g(p_n) - g(p_{n-1})| \le k|p_n - p_{n-1}| \le \cdots \le k^n|p_1 - p_0|.$$

Thus, for $m > n \ge 1$,

$$
\begin{aligned}
|p_m - p_n| &= |p_m - p_{m-1} + p_{m-1} - \cdots + p_{n+1} - p_n| \\
&\le |p_m - p_{m-1}| + |p_{m-1} - p_{m-2}| + \cdots + |p_{n+1} - p_n| \\
&\le k^{m-1}|p_1 - p_0| + k^{m-2}|p_1 - p_0| + \cdots + k^n|p_1 - p_0| \\
&= k^n(1 + k + k^2 + \cdots + k^{m-n-1})|p_1 - p_0|.
\end{aligned}
$$

By Theorem 2.3, $\lim_{m \to \infty} p_m = p$, so

$$|p - p_n| = \lim_{m \to \infty} |p_m - p_n| \le k^n|p_1 - p_0| \sum_{i=0}^{\infty} k^i = \frac{k^n}{1-k}|p_1 - p_0|. \qquad \square$$

Both corollaries relate the rate of convergence to the bound k on the first derivative. It is clear that the rate of convergence depends on the factor $k^n/(1-k)$, and that the smaller k can be made, the faster the convergence. The convergence may be very slow if k is close to 1. The fixed-point methods in Example 3 will be reconsidered in light of the results described in Theorem 2.3.

EXAMPLE 4
a) When $g_1(x) = x - x^3 - 4x^2 + 10$, $g_1'(x) = 1 - 3x^2 - 8x$. There is no interval $[a, b]$ containing p for which $|g_1'(x)| < 1$. Although Theorem 2.3 does not guarantee that the method must fail for this choice of g, there is no reason to suspect convergence.

b) With $g_2(x) = [(10/x) - 4x]^{1/2}$, we can see that g_2 does not map $[1, 2]$ into $[1, 2]$ and the sequence $\{p_n\}_{n=0}^{\infty}$ is not defined with $p_0 = 1.5$. Moreover, there is no interval containing p such that

$$|g_2'(x)| < 1, \qquad \text{since } |g_2'(p)| \approx 3.4.$$

c) For the function $g_3(x) = \frac{1}{2}(10 - x^3)^{1/2}$,

$$g_3'(x) = -\frac{3}{4}x^2(10 - x^3)^{-1/2} < 0 \qquad \text{on } [1, 2],$$

so g is strictly decreasing on $[1, 2]$. However, $|g_3'(2)| \approx 2.12$, so inequality (2.6) does not hold on $[1, 2]$. A closer examination of the sequence $\{p_n\}_{n=0}^{\infty}$ starting with $p_0 = 1.5$ will show that it suffices to consider the interval $[1, 1.5]$ instead of $[1, 2]$. On this interval it is still true that $g_3'(x) < 0$ and g is strictly decreasing, but, additionally,

$$1 < 1.28 \approx g_3(1.5) \le g_3(x) \le g_3(1) = 1.5$$

for all $x \in [1, 1.5]$. This shows that g_3 maps the interval $[1, 1.5]$ into itself. Since it is also true that $|g_3'(x)| \le |g_3'(1.5)| \approx 0.66$ on this interval, Theorem 2.3 confirms the convergence of which we were already aware.

d) For $g_4(x) = \left(\dfrac{10}{4+x}\right)^{1/2}$,

$$|g_4'(x)| = \left|\frac{-5}{\sqrt{10}(4+x)^{3/2}}\right| \le \frac{5}{\sqrt{10}(5)^{3/2}} < 0.15 \qquad \text{for all } x \in [1, 2].$$

The bound on the magnitude of $g_4'(x)$ is much smaller than the bound on the magnitude of $g_3'(x)$, which explains the reason for the more rapid convergence using g_4. The other part of Example 3 could be handled in a similar manner.

∎

Exercise Set 2.2

1. a) Show that each of the following functions has a fixed point at p precisely when $f(p) = 0$, where $f(x) = x^4 + 2x^2 - x - 3$.

   i) $g_1(x) = [3 + x - 2x^2]^{1/4}$

   ii) $g_2(x) = \left[\dfrac{x + 3 - x^4}{2}\right]^{1/2}$

   iii) $g_3(x) = \left[\dfrac{x + 3}{x^2 + 2}\right]^{1/2}$

   iv) $g_4(x) = \dfrac{3x^4 + 2x^2 + 3}{4x^3 + 4x - 1}$

   b) Perform four iterations, if possible, on each of the functions g defined in part (a). Let $p_0 = 1$ and $p_{n+1} = g(p_n)$ for $n = 0, 1, 2, 3$.

   c) Which function do you think gives the best approximation to the solution?

2. Use a fixed-point iteration method to determine a solution accurate to within 10^{-2} for $2 \sin \pi x + x = 0$ on $[1, 2]$. Use $p_0 = 1$.

3. Use Theorem 2.2 to show that $g(x) = 2^{-x}$ has a unique fixed point on $[\frac{1}{3}, 1]$. Use fixed-point iteration to find an approximation to the fixed point accurate to within 10^{-4}. Use Corollary 2.4 or 2.5 to estimate the number of iterations required to achieve 10^{-4} accuracy, and compare this theoretical estimate to the number actually needed.

4. Use Theorem 2.2 to show that $g(x) = \pi + 0.5 \sin x$ has a unique fixed point on $[0, 2\pi]$. Use fixed-point iteration to find an approximation to the fixed point that is accurate to within 10^{-2}. Use Corollary 2.4 or 2.5 to estimate the number of iterations required to achieve 10^{-2} accuracy, and compare this theoretical estimate to the number actually needed.

5. Solve $x^3 - x - 1 = 0$ for the root in $[1, 2]$, using fixed-point iteration. Obtain an approximation to the root accurate to within 10^{-2}.

6. Use a fixed-point iteration procedure to find an approximation to $\sqrt{3}$ that is accurate to within 10^{-4}. Compare your result and the number of iterations required with the answer obtained in Exercise 8 of Section 2.1.

7. Use a fixed-point iteration procedure to find an approximation to $\sqrt[3]{25}$ that is accurate to within 10^{-4}. Compare your result and the number of iterations required with the answer obtained in Exercise 7 of Section 2.1.

8. For each of the following equations, determine an interval $[a, b]$ on which fixed-point iteration will converge. Estimate the number of iterations necessary to obtain approximations accurate to within 10^{-5}, and perform the calculations.

   a) $x = \dfrac{2 - e^x + x^2}{3}$

   b) $x = \sqrt{\dfrac{1}{3}e^x}$

   c) $x = 5^{-x}$

   d) $x = 6^{-x}$

   e) $x = 1.75 + \dfrac{4x - 7}{x - 2}$

   f) $x = \dfrac{5}{x^2} + 2$

9. For each of the following equations, determine a function g and an interval $[a, b]$ on which fixed-point iteration will converge to a positive solution of the equation.

   a) $3x^2 - e^x = 0$

   b) $x - \cos x = 0.$

   Find the solutions to within 10^{-5}.

10. Find all the zeros of $f(x) = x^2 + 10 \cos x$ by using the fixed-point iteration method for an appropriate iteration function g. Find the zeros accurate to within 10^{-4} and compare the number of required iterations to the number required in Exercise 12, Section 2.1.

11. Use a fixed-point iteration method to determine a solution accurate to within 10^{-4} for $x = \tan x,\ 4 \leq x \leq 5$.

12. Find a function g defined on $[0, 1]$ that satisfies none of the hypotheses of Theorem 2.2, but still has a unique fixed point on $[0, 1]$.

13. Prove that the sequence defined by

$$x_n = \frac{1}{2}\left(x_{n-1} + \frac{2}{x_{n-1}}\right), \qquad \text{for } n \geq 1.$$

converges to $\sqrt{2}$ for any $x_0 > 0$.

14. a) Show that Theorem 2.2 is true if inequality (2.5) is replaced by $g'(x) \leq k < 1$ for all $x \in (a, b)$.

    b) Show that Theorem 2.3 may not hold if inequality (2.6) is replaced by the hypothesis in part (a).

15. Replace the assumption in inequality (2.6) of Theorem 2.3 with "g satisfies a Lipschitz condition on the interval $[a, b]$ with Lipschitz constant $L < 1$." (See Exercise 21, Section 1.1.) Show that the conclusions of this theorem are still valid.

16. Suppose that g is continuously differentiable on some interval (c, d) which contains the fixed point p of g. Show that if $|g'(p)| < 1$, then there exists a $\delta > 0$ such that the fixed-point iteration converges for any initial approximation p_0, whenever $|p_0 - p| \leq \delta$.

2.3 The Newton–Raphson Method

The **Newton–Raphson** (or simply **Newton's**) method is one of the most powerful and well-known numerical methods for solving a root-finding problem $f(x) = 0$. There are at least three common ways of introducing Newton's method. The most common approach is by considering the technique graphically (see Exercise 9). Another possibility is to derive Newton's method as a simple technique to obtain faster convergence than offered by many other types of functional iteration (see Section 2.4). The third means of introducing Newton's method, which will be discussed below, is an intuitive approach based on the Taylor polynomial defined in Theorem 1.13.

Suppose that the function f is twice continuously differentiable on the interval $[a, b]$; that is, $f \in C^2[a, b]$. Let $\bar{x} \in [a, b]$ be an approximation to p such that $f'(\bar{x}) \neq 0$ and $|\bar{x} - p|$ is "small." Consider the first-degree Taylor polynomial for $f(x)$, expanded about $\bar{x}$,

$$(2.11) \qquad f(x) = f(\bar{x}) + (x - \bar{x})f'(\bar{x}) + \frac{(x - \bar{x})^2}{2} f''(\xi(x)),$$

where $\xi(x)$ lies between x and $\bar{x}$. Since $f(p) = 0$, Eq. (2.11), with $x = p$, gives

$$(2.12) \qquad 0 = f(\bar{x}) + (p - \bar{x})f'(\bar{x}) + \frac{(p - \bar{x})^2}{2} f''(\xi(p)).$$

Newton's method is derived by assuming that the term involving $(p - \bar{x})^2$ is negligible and that

$$(2.13) \qquad 0 \approx f(\bar{x}) + (p - \bar{x})f'(\bar{x}).$$

Solving for p in this equation gives:

$$(2.14) \qquad p \approx \bar{x} - \frac{f(\bar{x})}{f'(\bar{x})},$$

which should be a better approximation to p than is $\bar{x}$. This sets the stage for the Newton–Raphson method, which involves generating the sequence $\{p_n\}$ defined by

$$(2.15) \qquad p_n = p_{n-1} - \frac{f(p_{n-1})}{f'(p_{n-1})}, \qquad n \geq 1.$$

Figure 2.6 illustrates graphically how the approximations are obtained using successive tangents. (See also Exercise 9.)

Newton–Raphson Algorithm 2.3

To find a solution to $f(x) = 0$ given an initial approximation p_0:

INPUT initial approximation p_0; tolerance *TOL*; maximum number of iterations N_0.

OUTPUT approximate solution p or message of failure.

Step 1   Set $i = 1$.

Step 2 While $i \le N_0$ do Steps 3–6.

    Step 3 Set $p = p_0 - f(p_0)/f'(p_0)$.   (*Compute p_i.*)

    Step 4 If $|p - p_0| < TOL$ then
          OUTPUT (p);   (*Procedure completed successfully.*)
          STOP.

    Step 5 Set $i = i + 1$.

    Step 6 Set $p_0 = p$.   (*Update p_0.*)

Step 7 OUTPUT ('Method failed after N_0 iterations, $N_0 =$ ', N_0);
    (*Procedure completed unsuccessfully.*)
    STOP.

FIGURE 2.6

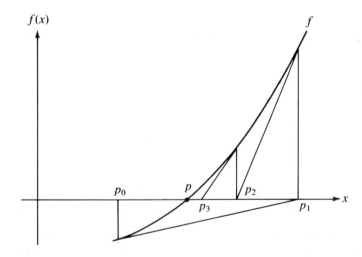

The stopping-technique inequalities (2.1), (2.2), and (2.3) are applicable to Newton's method; that is: select a tolerance $\varepsilon > 0$ and construct $p_1, \ldots, p_N$ until

$$(2.16) \qquad\qquad |p_N - p_{N-1}| < \varepsilon,$$

$$(2.17) \qquad\qquad \frac{|p_N - p_{N-1}|}{|p_N|} < \varepsilon, \qquad p_N \neq 0,$$

or

$$(2.18) \qquad\qquad |f(p_N)| < \varepsilon.$$

A form of inequality (2.16) is used in Step 4 of Algorithm 2.3. Note that inequality (2.18) may not give much information about the actual error $|p_N - p|$. (See Exercise 14 Section 2.1.)

Newton's method is a functional iteration technique $p_n = g(p_{n-1})$, $n \geq 1$ for which

$$g(p_{n-1}) = p_{n-1} - \frac{f(p_{n-1})}{f'(p_{n-1})}, \qquad n \geq 1.$$

It is clear from this equation that Newton's method cannot be continued if $f'(p_{n-1}) = 0$ for some n. We will see that the method is most effective when f' is bounded away from zero near the fixed point p.

EXAMPLE 1 a) Suppose a solution to the equation $x = \cos x$ is needed. Let $f(x) = \cos x - x$. Then

$$f\left(\frac{\pi}{2}\right) = -\frac{\pi}{2} < 0 < 1 = f(0),$$

and, by the Intermediate Value Theorem (Theorem 1.12), there exists a zero of f in $[0, \pi/2]$. The graphs of the equations $y = x$ and $y = \cos x$ appear in Fig. 2.7; their intersection is the fixed point of $g(x) = \cos x$.

FIGURE 2.7

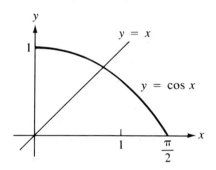

From the graph it is clear that $f(x) = 0$ has a unique solution in $[0, \pi/2]$. Since $f'(x) = -\sin x - 1$, Newton's method has the form

$$p_n = p_{n-1} - \frac{(\cos p_{n-1} - p_{n-1})}{(-\sin p_{n-1} - 1)}, \qquad n \geq 1,$$

where p_0 is yet to be selected. For some problems it suffices to let p_0 be arbitrary, while for others it is important to select a good initial approximation. For the problem under consideration, the graph in Fig. 2.7 suggests $p_0 = \pi/4$ as an initial approximation. With $p_0 = \pi/4$, the approximations in Table 2.3 are generated. An excellent approximation is obtained with $n = 3$.

b) To obtain the unique solution to $x^3 + 4x^2 - 10 = 0$ on the interval $[1, 2]$ by Newton's method, generate the sequence $\{p_n\}_{n=1}^{\infty}$ by

$$p_n = p_{n-1} - \frac{p_{n-1}^3 + 4p_{n-1}^2 - 10}{3p_{n-1}^2 + 8p_{n-1}}, \qquad n \geq 1.$$

Selecting $p_0 = 1.5$ produces the results of Example 3(e) of Section 2.2, in which $p_3 = 1.36523001$ is correct to the eighth decimal place. ∎

TABLE 2.3

n	p_n	n	p_n
0	0.7853981635	3	0.7390851332
1	0.7395361337	4	0.7390851332
2	0.7390851781	5	0.7390851332

The Taylor series derivation of Newton's method points out the importance of an accurate initial approximation. The crucial assumption, going from Eq. (2.12) to Eq. (2.13), is that the term involving $(p - \bar{x})^2$ can be deleted. This will clearly be a false assumption unless $\bar{x}$ is a good approximation to p. In particular, if p_0 is not sufficiently close to the actual root, Newton's method may not converge to the root. This, however, is not always the case. (Exercises 11 and 13 illustrate some of the possibilities that can occur.)

The following convergence theorem for Newton's method illustrates the theoretical importance of the choice of p_0.

THEOREM 2.6   Let $f \in C^2[a, b]$. If $p \in [a, b]$ is such that $f(p) = 0$ and $f'(p) \neq 0$, then there exists $\delta > 0$ such that Newton's method generates a sequence $\{p_n\}_{n=1}^\infty$ converging to p for any initial approximation $p_0 \in [p - \delta, p + \delta]$.

PROOF   The proof will be based on analyzing Newton's method as a functional iteration scheme $p_n = g(p_{n-1})$, for $n \geq 1$, with

$$g(x) = x - \frac{f(x)}{f'(x)}.$$

The object is to find, for a value k in $(0, 1)$, an interval $[p - \delta, p + \delta]$ such that g maps the interval $[p - \delta, p + \delta]$ into itself and $|g'(x)| \leq k < 1$ for $x \in [p - \delta, p + \delta]$.

Since $f'(p) \neq 0$ and f' is continuous, there exists $\delta_1 > 0$ such that $f'(x) \neq 0$ for $x \in [p - \delta_1, p + \delta_1] \subset [a, b]$. Thus, g is defined and continuous on $[p - \delta_1, p + \delta_1]$. Also,

$$g'(x) = 1 - \frac{f'(x)f'(x) - f(x)f''(x)}{[f'(x)]^2} = \frac{f(x)f''(x)}{[f'(x)]^2}$$

for $x \in [p - \delta_1, p + \delta_1]$; and since $f \in C^2[a, b], g \in C^1[p - \delta_1, p + \delta_1]$. By assumption, $f(p) = 0$, so

(2.19)
$$g'(p) = \frac{f(p)f''(p)}{[f'(p)]^2} = 0.$$

Since g' is continuous, Eq. (2.19) implies that there exists a δ with $0 < \delta < \delta_1$ and

$$|g'(x)| \leq k \qquad \text{for } x \in [p - \delta, p + \delta].$$

It remains to show that $g:[p - \delta, p + \delta] \to [p - \delta, p + \delta]$. If $x \in [p - \delta, p + \delta]$, the Mean Value Theorem implies that, for some number ξ between x and p, $|g(x) - g(p)| = |g'(\xi)||x - p|$. So

$$|g(x) - p| = |g(x) - g(p)| = |g'(\xi)||x - p| \leq k|x - p| < |x - p|.$$

Since $x \in [p - \delta, p + \delta]$, it follows that $|x - p| < \delta$ and that $|g(x) - p| < \delta$. This implies $g:[p - \delta, p + \delta] \to [p - \delta, p + \delta]$.

All the hypotheses of Theorem 2.3 (p. 38) are now satisfied for $g(x) = x - f(x)/f'(x)$, so the sequence $\{p_n\}_{n=1}^{\infty}$ defined by

$$p_n = g(p_{n-1}) \qquad \text{for } n = 1, 2, 3, \ldots$$

converges to p for any $p_0 \in [p - \delta, p + \delta]$.   □

Theorem 2.6 states that, under reasonable assumptions, Newton's method will always converge provided a sufficiently accurate initial approximation is chosen. It also implies that the constant k that bounds the derivative of g, and, consequently, indicates the speed of convergence of the method, decreases as the procedure continues.

Newton's method is an extremely powerful technique, but it has a major difficulty: the need to know the value of the derivative of f at each approximation. Frequently $f'(x)$ is far more difficult and needs more arithmetic operations to calculate than $f(x)$. As a simple example, let $f(x) = x^2 3^x \cos 2x$, then $f'(x) = 2x3^x \cos 2x + x^2 3x(\cos 2x)\ln 3 - 2x^2 3^x \sin 2x$, which is extremely tiresome to evaluate, particularly if the method is being performed on a nonprogrammable calculator.

To circumvent the problem of the derivative evaluation in Newton's method, we derive a slight variation. By definition,

$$f'(p_{n-1}) = \lim_{x \to p_{n-1}} \frac{f(x) - f(p_{n-1})}{x - p_{n-1}}.$$

Letting $x = p_{n-2}$,

$$f'(p_{n-1}) \approx \frac{f(p_{n-2}) - f(p_{n-1})}{p_{n-2} - p_{n-1}} = \frac{f(p_{n-1}) - f(p_{n-2})}{p_{n-1} - p_{n-2}}.$$

FIGURE 2.8

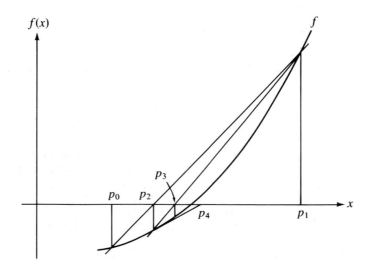

Using this approximation for $f'(p_{n-1})$ in Newton's formula gives

(2.20)
$$p_n = p_{n-1} - \frac{f(p_{n-1})(p_{n-1} - p_{n-2})}{f(p_{n-1}) - f(p_{n-2})}.$$

The technique using this formula is called the **Secant method** and is presented in Algorithm 2.4. (*See Fig. 2.8.*)

Secant Algorithm 2.4

To find a solution to $f(x) = 0$ given initial approximations p_0 and p_1:

INPUT initial approximations p_0, p_1; tolerance *TOL*; maximum number of iterations N_0.

OUTPUT approximate solution p or message of failure.

Step 1   Set $i = 2$;
  $q_0 = f(p_0)$;
  $q_1 = f(p_1)$.

Step 2   While $i \leq N_0$ do Steps 3–6.

  Step 3   Set $p = p_1 - q_1(p_1 - p_0)/(q_1 - q_0)$.   (*Compute p_i.*)

  Step 4   If $|p - p_1| < TOL$ then
       OUTPUT (p);   (*Procedure completed successfully.*)
       STOP.

  Step 5   Set $i = i + 1$.

  Step 6   Set $p_0 = p_1$;   (*Update p_0, q_0, p_1, q_1.*)
       $q_0 = q_1$;
       $p_1 = p$;
       $q_1 = f(p)$.

Step 7   OUTPUT ('Method failed after N_0 iterations, $N_0 = $', N_0);
  (*Procedure completed unsuccessfully.*)
  STOP.

The next example involves a problem we considered in Example 1(a), where we used Newton's method with $p_0 = \pi/4$.

EXAMPLE 2   Find a zero of $f(x) = \cos x - x$, using the Secant method. In Example 1 we used an initial approximation of $p_0 = \pi/4$. Here we need two initial approximations. Table 2.4 lists the calculations with $p_0 = 0.5$, $p_1 = \pi/4$, and the formula

$$p_n = p_{n-1} - \frac{(p_{n-1} - p_{n-2})(\cos p_{n-1} - p_{n-1})}{(\cos p_{n-1} - p_{n-1}) - (\cos p_{n-2} - p_{n-2})} \qquad \text{for } n \geq 2,$$

from Algorithm 2.4.   ∎

TABLE 2.4

n	p_n
0	0.5
1	0.7853981635
2	0.7363841390
3	0.7390581394
4	0.7390851492
5	0.7390851334

By comparing the results here with those in Example 1, we can see that p_5 is accurate to the tenth decimal place. It is interesting to note that the convergence of the Secant method is slightly slower in this example than that of Newton's method, which obtained this degree of accuracy with p_3. This result is generally true, as will be shown in Section 2.4. (See Exercise 16 of Section 2.4.)

The Secant method (or Newton's method) is often used to refine an answer obtained by another technique, such as the Bisection method. Since these methods require a good first approximation, but generally give rapid convergence, they serve this purpose well.

Exercise Set 2.3

1. Approximate to within 10^{-4} the roots of the following equations in the given intervals by using Newton's method.

   a) $x^3 - 2x^2 - 5 = 0$, $[1, 4]$ b) $x^3 + 3x^2 - 1 = 0$, $[-4, 0]$

   c) $x - \cos x = 0$, $[0, \pi/2]$ d) $x - 0.8 - 0.2 \sin x = 0$, $[0, \pi/2]$

2. Repeat Exercise 1 using the Secant method.

3. Find an approximate root of $x^3 - x - 1 = 0$ in $[1, 2]$ with 10^{-5} accuracy, first by Newton's method and then by the Secant method.

4. Use Newton's method to approximate the solutions of the following equations to within 10^{-5}.

   a) $x = \dfrac{2 - e^x + x^2}{3}$ b) $3x^2 - e^x = 0$

   c) $e^x + 2^{-x} + 2 \cos x - 6 = 0$ d) $x^2 + 10 \cos x = 0$

5. Repeat Exercise 4 using the Secant method.

6. Use Newton's method to approximate, to within 10^{-4}, the value of x that produces the point on the graph of $y = x^2$ that is closest to $(1, 0)$. (*Hint:*   Minimize $[d(x)]^2$, where $d(x)$ represents the distance from (x, x^2) to $(1, 0)$).

7. Use Newton's method to approximate, to within 10^{-4}, the value of x that produces the point on the graph of $y = 1/x$ that is closest to $(2, 1)$.

8. Solve $4 \cos x = e^x$ with accuracy 10^{-4}, by using:

a)   Newton's method with $p_0 = 1$.

b)   The Secant method with $p_0 = \pi/4$ and $p_1 = \pi/2$.

9. The following describes Newton's method graphically: Suppose that $f'(x)$ exists on $[a, b]$ and that $f'(x) \neq 0$ on $[a, b]$. Further, suppose there exists one $p \in [a, b]$ such that $f(p) = 0$. Let $p_0 \in [a, b]$ be arbitrary. Let p_1 be the point at which the tangent line to f at $(p_0, f(p_0))$ crosses the x-axis. For each $n \geq 1$, let p_n be the x-intercept of the line tangent to f at $(p_{n-1}, f(p_{n-1}))$. Derive the formula describing this method.

10. Describe the Secant method geometrically. [*Hint*:   Refer to Exercise 9 and the definition of the secant line.]

11. Use Newton's method to solve the equation

$$0 = \left(\sin x - \frac{x}{2} \right)^2 \qquad \text{with } p_0 = \frac{\pi}{2}.$$

Iterate using Newton's method until an accuracy of 10^{-5} is obtained for the approximate root with $f(x) = (\sin x - x/2)^2$. Do the results seem unusual for Newton's method? Also, solve the equation with $p_0 = 5\pi$ and $p_0 = 10\pi$.

12. a)   Compute an approximation to $\sqrt{3}$ accurate to within 10^{-4} using Newton's method with $p_0 = 2$.

b)   Repeat part (a), using the Secant method, and then compare your results with those obtained in Exercise 6 of Section 2.2 and Exercise 8 of Section 2.1.

13. The function $f(x) = (4x - 7)/(x - 2)$ has a zero at $p = 1.75$. Use Newton's method with the following initial approximations.

a) $p_0 = 1.625$ b) $p_0 = 1.875$

c) $p_0 = 1.5$ d) $p_0 = 1.95$

e) $p_0 = 3$ f) $p_0 = 7$

Explain the results graphically.

14. For what values of p_0 and p_1 can the Secant method be used to solve the following equation?

$$f(x) = \frac{4x - 7}{x - 2} = 0.$$

15. The iteration equation for the Secant method can also be written as:

$$p_n = \frac{f(p_{n-1})p_{n-2} - f(p_{n-2})p_{n-1}}{f(p_{n-1}) - f(p_{n-2})}.$$

Can you explain why, in general, this iteration equation is inferior to the one given in Algorithm 2.4?

16. **Method of False Position (or *Regula Falsi*)**   This is another method for finding a root of the equation $f(x) = 0$ lying in the interval $[a, b]$. The method is similar to the Bisection technique in that intervals $[a_i, b_i]$ are generated bracketing a root, and the method is similar to the Secant method in the manner of obtaining new approximate intervals. Assuming that the interval $[a_i, b_i]$ contains a root of $f(x) = 0$, compute the value of the x-intercept of the line joining the points $(a_i, f(a_i))$ and $(b_i, f(b_i))$ and label this point p_i. If $f(p_i)f(a_i) < 0$, define $a_{i+1} = a_i$ and $b_{i+1} = p_i$ otherwise, define $a_{i+1} = p_i$ and $b_{i+1} = b_i$. Construct an algorithm, similar to Algorithms 2.1 and 2.4, which describes the Method of False Position.

17. Repeat Exercise 4 using the Method of False Position described in Exercise 16.

18. Use the algorithm constructed in Exercise 16 to find an approximate root of $\cos x - x = 0$ lying on the interval $[0.5, \pi/4]$.

19. Use Newton's method to find an approximation for λ, accurate to within 10^{-4}, for the population equation

$$1,564,000 = 1,000,000e^\lambda + \frac{435,000}{\lambda}(e^\lambda - 1),$$

discussed in the introduction to this chapter. Use this value to predict the population at the end of the second year, assuming that the immigration rate during this year remains at 435,000 individuals per year.

20. The function described by $f(x) = \ln(x^2 + 1) - e^{0.4x} \cos \pi x$ has an infinite number of zeros.

    a) Use Newton's method to determine, within 10^{-6}, the only negative zero.

    b) Use Newton's method to determine, within 10^{-6}, the four smallest positive zeros.

    c) Determine a reasonable starting approximation to determine the nth smallest positive zero of f. (*Hint:* Sketch a rough graph of f).

    d) Use part (c) to determine, within 10^{-6}, the 25th smallest positive zero of f.

21. The sum of two numbers is 20. If each number is added to its square root, the product of the two sums equals 155.55. Determine the two numbers to within 10^{-4}.

22. A drug administered to a patient produces a concentration in the blood stream given by $c(t) = Ate^{-t/3} \frac{\text{mg}}{\text{ml}}$, t hours after A units have been injected. The maximum safe concentration is $1 \frac{\text{mg}}{\text{ml}}$.

    a) What amount should be injected to reach this maximum safe concentration and when does this maximum occur?

    b) An additional amount of this drug is to be administered to the patient after the concentration falls to $0.25 \frac{\text{mg}}{\text{ml}}$. Determine, to the nearest minute, when this second injection should be given.

    c) Assuming that the concentration from consecutive injections is additive and that 75% of the amount originally injected is administered in the second injection, when is it time for the third injection?

23. The accumulated value of a savings account based on regular periodic payments can be determined from the *annuity due equation,*

$$A = \frac{P}{i}[(1+i)^n - 1].$$

In this equation A is the amount in the account, P is the amount regularly deposited, and i is the rate of interest per period for the n deposit periods.

   An engineer would like to have a savings account valued at \$75,000 upon retirement in 20 years and can afford to put \$150 per month toward this goal. What is the minimal interest rate at which this amount can be deposited, assuming that the interest is compounded quarterly? What is the minimal interest rate if the interest is compounded daily? (Assume a 360-day year, the common banking practice.)

24. Problems relating the amount of money required to pay off a mortgage over a fixed period of time involve a formula,

$$A = \frac{P}{i}[1 - (1+i)^{-n}],$$

known as an *ordinary annuity equation.* In this equation A is the amount of the mortgage, P is the amount of each payment, and i is the interest rate per period for the n payment periods.

   Suppose that a 30-year home mortgage in the amount of \$50,000 is needed and that the borrower can afford house payments of at most \$450 per month. What is the maximal interest rate the borrower can afford to pay?

25. The logistic population growth model is described by an equation of the form

$$P(t) = \frac{P_L}{1 - ce^{-kt}},$$

where P_L, c, and k are constants and $P(t)$ is the population at time t. P_L represents the limiting value of the population since $\lim_{t \to \infty} P(t) = P_L$, provided that $k > 0$. Use the census data for the years 1950, 1960, and 1970 listed in the table on page 78 to determine the constants P_L, c, and k for a logistic growth model. Use the logistic model to predict the population of the United States in 1980 and in 2000, assuming $t = 0$ at 1950. Compare the 1980 prediction to the actual value.

26. The Gompertz population growth model is described by

$$P(t) = P_L e^{-ce^{-kt}}$$

where P_L, c, and k are constants and $P(t)$ is the population at time t. Repeat Exercise 25 using the Gompertz growth model in place of the logistic model.

27. In a paper entitled "Holdup and Axial Mixing in Bubble Columns Containing Screen Cylinders" [26], B. H. Chen computes the gas holdup in a gas–liquid bubble column by first approximating the quantity

(1) $$2 \sum_{n=1}^{\infty} \frac{S_n \sin S_n}{S_n^2 + M^2 + 2M} \exp\left[M - \frac{S_n^2 + M^2}{2M} \cdot \frac{t}{\theta} \right],$$

where t and M are physical parameters and the S_n's are the smallest values (in magnitude) satisfying

(2) $$S_n \tan\left(\frac{S_n}{2}\right) = M, \quad \text{when } n \text{ is odd, and}$$

(3) $$S_n \cot\left(\frac{S_n}{2}\right) = -M, \quad \text{when } n \text{ is even.}$$

a) Assuming $M = 3.7$, find S_1, S_2, S_3, and S_4.

b) Use the results in part (a) to approximate the sum in Eq. (1), when $t = 0$.

28. In the design of terrain vehicles, it is necessary to consider the failure of the vehicle when attempting to negotiate two types of obstacles. One type of failure is called Hang-Up Failure (HUF) and typically occurs when the vehicle attempts to cross an obstacle that causes the bottom of the vehicle to touch the ground (or the obstacle). The other type of failure is called Nose-In Failure (NIF) and commonly occurs when the vehicle descends into a ditch and its nose touches the ground.

The following figure, adapted from Bekker [7], shows the components associated with the NIF of a vehicle. In that reference it is shown that the maximum angle α, which can be negotiated by a vehicle when β is the maximum angle at which HUF does *not* occur, satisfies the equation

$$A \sin \alpha \cos \alpha + B \sin^2 \alpha - C \cos \alpha - E \sin \alpha = 0,$$

where

$$A = l \sin \beta_1,$$
$$B = l \cos \beta_1,$$
$$C = (h + 0.5D)\sin \beta_1 - 0.5D \tan \beta_1,$$
$$E = (h + 0.5D)\cos \beta_1 - 0.5D.$$

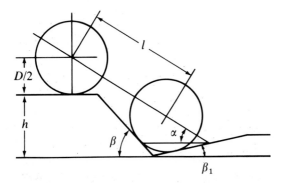

a) It is stated that when $l = 89$ in., $h = 49$ in., $D = 55$ in., and $\beta_1 = 11.5°$, angle α is approximately $33°$. Verify this result.

b) Find α for the situation when l, h, and β_1 are the same as in part (a) but $D = 30$ in.

29. To find approximations for the tension T and angle of inclination from the horizontal ϕ at a particular point on a cable or pipeline run underwater, a number of equations of the form

(1) $$\phi = \tan^{-1}\left[\frac{T_0 \sin \phi_0 - F}{T_0 \cos \phi_0 - G}\right], \quad \text{and}$$

(2) $$T = [(T_0 \sin \phi_0 - F)^2 + (T_0 \cos \phi_0 - G)^2]^{1/2}$$

must be solved for ϕ and T (see Wang [138]). The functions F and G in (1) and (2) both involve ϕ and have the form

$$F = \frac{s}{2}[-f(\phi_0)\sin \phi_0 - f(\phi)\sin \phi + g(\phi_0)\cos \phi_0 + g(\phi)\cos \phi] - ws$$

and $$G = -\frac{s}{2}[f(\phi_0)\cos \phi_0 + f(\phi)\cos \phi + g(\phi_0)\sin \phi_0 + g(\phi)\sin \phi],$$

for given tangential and normal hydrodynamic loading functions f and g.

Suppose that $\phi_0 = \pi/2$, $s = 0.1$, $T_0 = 2$, $w = 1$, and the loading functions f and g are given by

$$f(\phi) = 0.02 \cos \phi \quad \text{and} \quad g(\phi) = 0.98 \sin^2 \phi + 0.02 \sin \phi.$$

a) Use the Secant method to find an approximation to ϕ starting with both initial approximations close to $\pi/2$. What type of accuracy for ϕ is sufficient?

b) Use the Method of False Position to find this approximation. (See Exercise 16.)

30. Player A will shutout (win by a score of 21-0) a player B in a game of racquetball with probability

$$P = \frac{1+p}{2}\left(\frac{p}{1-p+p^2}\right)^{21},$$

where p denotes the probability A will win any specific rally (independent of the server). (See Keller [76], p. 267.) Determine, to within 10^{-3}, the minimal value of p that will ensure that A will shutout B in at least half the matches they play.

2.4 Error Analysis for Iterative Methods

This section is devoted to investigating the order of convergence of functional iteration schemes and, as a means of obtaining rapid convergence, rediscovering Newton's method. We will also consider ways of accelerating the convergence of Newton's method in special circumstances. Before methods for accelerating convergence can be presented, it is necessary to define a procedure for measuring the speed of convergence.

DEFINITION 2.7   Suppose $\{p_n\}_{n=0}^{\infty}$ is a sequence that converges to p and that $e_n = p_n - p$ for each $n \geq 0$. If positive constants λ and α exist with

$$\lim_{n \to \infty} \frac{|p_{n+1} - p|}{|p_n - p|^{\alpha}} = \lim_{n \to \infty} \frac{|e_{n+1}|}{|e_n|^{\alpha}} = \lambda,$$

then $\{p_n\}_{n=0}^{\infty}$ is said to **converge to p of order α,** with asymptotic error constant λ.

An iterative technique for solving a problem of the form $x = g(x)$ is said to be of order α if, whenever the method gives convergence for a sequence $\{p_n\}_{n=0}^{\infty}$, where $p_n = g(p_{n-1})$ for $n \geq 1$, then the sequence $\{p_n\}_{n=0}^{\infty}$ converges to the solution of order α.

In general, a sequence with a large order of convergence will converge more rapidly than a sequence with a lower order. The asymptotic constant will affect the speed of convergence, but it is not as important as the order. Two cases of order will be given special attention.

1. If $\alpha = 1$, the method is called **linear.**
2. If $\alpha = 2$, the method is called **quadratic.**

EXAMPLE 1   Suppose we want to find an approximate solution to $g(x) = x$, using the fixed-point iteration scheme $p_n = g(p_{n-1})$, for all $n \geq 1$.

Assume that g maps the interval $[a, b]$ into itself and that a positive number k exists with $|g'(x)| \leq k < 1$ for all $x \in [a, b]$. Theorem 2.3 (p. 38) implies that g has a unique fixed point $p \in [a, b]$ and if $p_0 \in [a, b]$, the fixed-point sequence $\{p_n\}_{n=0}^{\infty}$ converges to p. The convergence will be shown to be linear, provided that $g'(p) \neq 0$. If n is any positive integer, then

$$e_{n+1} = p_{n+1} - p = g(p_n) - g(p) = g'(\xi_n)(p_n - p) = g'(\xi_n)e_n,$$

where ξ_n is between p_n and p. Since $\{p_n\}_{n=0}^{\infty}$ converges to p, $\{\xi_n\}_{n=0}^{\infty}$ also converges to p. Assuming that g' is continuous on $[a, b]$, we have

$$\lim_{n \to \infty} g'(\xi_n) = g'(p).$$

Thus,

$$\lim_{n \to \infty} \frac{e_{n+1}}{e_n} = \lim_{n \to \infty} g'(\xi_n) = g'(p) \quad \text{and} \quad \lim_{n \to \infty} \frac{|e_{n+1}|}{|e_n|} = |g'(p)|.$$

Hence, fixed-point iteration exhibits linear convergence if $g'(p) \neq 0$. Higher-order convergence can occur only when $g'(p) = 0$.   ∎

The next example compares a linearly convergent method to one that is quadratically convergent and demonstrates why we will be trying to find higher-order convergent methods.

EXAMPLE 2   Suppose we have two convergent iterative schemes described by

(2.21) $$\lim_{n \to \infty} \frac{|e_{n+1}|}{|e_n|} = 0.75, \quad \text{a linear method,}$$

and

(2.22) $$\lim_{n \to \infty} \frac{|\tilde{e}_{n+1}|}{|\tilde{e}_n|^2} = 0.75, \quad \text{a quadratic method.}$$

Suppose also that, for simplicity,

$$\frac{|e_{n+1}|}{|e_n|} \approx 0.75 \qquad \text{and} \qquad \frac{|\tilde{e}_{n+1}|}{|\tilde{e}_n|^2} \approx 0.75.$$

For the linearly convergent scheme, this means that

$$|e_n| \approx 0.75|e_{n-1}| \approx (0.75)^2|e_{n-2}| \approx \cdots \approx (0.75)^n|e_0|,$$

while the quadratically convergent procedure has

$$|\tilde{e}_n| \approx 0.75|\tilde{e}_{n-1}|^2 \approx 0.75[0.75|\tilde{e}_{n-2}|^2]^2 = (0.75)^3|\tilde{e}_{n-2}|^4$$

$$\approx (0.75)^3[(0.75)|\tilde{e}_{n-3}|^2]^4 = (0.75)^7|\tilde{e}_{n-3}|^8 \approx \cdots \approx (0.75)^{2^{n+1}-1}|\tilde{e}_0|^{2^{n+1}}.$$

To compare the speed of convergence, assume that $|e_0| = |\tilde{e}_0| = 0.5$ and use the estimates to determine the minimal value of n needed to obtain an error not exceeding 10^{-8}. For the linear method, this implies that n should be such that

$$|e_n| = (0.75)^n(0.5) \le 10^{-8}, \qquad \text{that is, } n \ge \frac{\log_{10} 2 - 8}{\log_{10} 0.75} \approx 62.$$

For the quadratic convergent method

$$|\tilde{e}_n| = (0.75)^{2^{n+1}-1}(0.5)^{2^{n+1}} = (0.75)^{-1}(0.375)^{2^{n+1}} \le 10^{-8}$$

implies that $\qquad\qquad 2^{n+1}\log_{10} 0.375 \le \log_{10} 0.75 - 8.$

Thus, $\qquad\qquad 2^{n+1} \ge \dfrac{\log_{10} 0.75 - 8}{\log_{10} 0.375} \approx 19, \qquad \text{so } n \ge 4.$

In this circumstance, the quadratically convergent method requiring only 4 iterations is vastly superior to the linear method requiring 62.   ∎

Having discussed the desirability of quadratic convergence, the next step is to determine and characterize quadratic functional iteration schemes.

THEOREM 2.8 Let p be a solution of $x = g(x)$. Suppose that $g'(p) = 0$ and g'' is continuous in an open interval containing p. Then there exists a $\delta > 0$ such that, for $p_0 \in [p - \delta, p + \delta]$, the sequence defined by $p_n = g(p_{n-1})$, for all $n \ge 1$, is quadratically convergent.

PROOF Choose $\delta > 0$ such that on the interval $[p - \delta, p + \delta]$, $|g'(x)| \le k < 1$ and g'' is continuous. Since $|g'(x)| \le k < 1$, it follows that the terms of the sequence $\{p_n\}_{n=0}^{\infty}$ are contained in $[p - \delta, p + \delta]$. Expanding $g(x)$ in a linear Taylor polynomial for $x \in [p - \delta, p + \delta]$ gives

$$g(x) = g(p) + g'(p)(x - p) + \frac{g''(\xi)}{2}(x - p)^2,$$

where ξ lies between x and p. Using the hypotheses $g(p) = p$ and $g'(p) = 0$, we infer that:

$$g(x) = p + \frac{g''(\xi)}{2}(x - p)^2.$$

In particular, when $x = p_n$ for some n,

$$p_{n+1} = g(p_n) = p + \frac{g''(\xi_n)}{2}(p_n - p)^2$$

with ξ_n between p_n and p. Thus,

(2.23)
$$p_{n+1} - p = e_{n+1} = \frac{g''(\xi_n)}{2} e_n^2.$$

Since $|g'(x)| \leq k < 1$ on $[p - \delta, p + \delta]$, and g maps $[p - \delta, p + \delta]$ into itself, it follows from Theorem 2.3 (p. 38) that $\{p_n\}_{n=0}^{\infty}$ converges to p. Since ξ_n is between p and p_n for each n, $\{\xi_n\}_{n=0}^{\infty}$ converges to p also, and

$$\lim_{n \to \infty} \frac{|e_{n+1}|}{|e_n|^2} = \frac{|g''(p)|}{2}.$$

This implies that the sequence $\{p_n\}_{n=0}^{\infty}$ is quadratically convergent.   □

To use Theorem 2.8 to solve an equation of the form $f(x) = 0$, suppose the equation $f(x) = 0$ has a solution p for which $f'(p) \neq 0$. Consider a fixed-point scheme

$$p_n = g(p_{n-1}), \qquad n \geq 1,$$

with g of the form

$$g(x) = x - \phi(x) f(x),$$

where ϕ is an arbitrary function to be chosen later.

If $\phi(x)$ is bounded, then $g(p) = p$, and, for the iterative procedure derived from g to be quadratically convergent, it suffices to have $g'(p) = 0$. But

$$g'(x) = 1 - \phi'(x) f(x) - f'(x) \phi(x) \qquad \text{and} \qquad g'(p) = 1 - f'(p) \phi(p).$$

Consequently, $g'(p) = 0$ if and only if $\phi(p) = 1/f'(p)$.

In particular, quadratic convergence holds for the scheme

$$p_n = g(p_{n-1}) = p_{n-1} - \frac{f(p_{n-1})}{f'(p)}$$

under suitable conditions on f. However, since p, and consequently $f'(p)$, are generally unknown, a reasonable approach is to let $\phi(x) = 1/f'(x)$, so that $\phi(p) = 1/f'(p)$. The procedure then becomes

$$p_n = g(p_{n-1}) = p_{n-1} - \frac{f(p_{n-1})}{f'(p_{n-1})},$$

which can be recognized as Newton's method.

In the preceding discussion the restriction was made that $f'(p) \neq 0$, where p is the solution to $f(x) = 0$. From the definition of Newton's method, it is clear that difficulties might occur if $f'(p_n)$ goes to zero simultaneously with $f(p_n)$. In particular, this method and the Secant method will generally give problems if $f'(p) = 0$ when $f(p) = 0$. To examine these difficulties in more detail we make the following definition.

DEFINITION 2.9   A solution p of $f(x) = 0$ is said to be a **zero of multiplicity m** of f if $f(x)$ can be written as $f(x) = (x - p)^m q(x)$, for $x \neq p$, where $\lim_{x \to p} q(x) \neq 0$.

In essence, $q(x)$ represents that portion of $f(x)$ that does not contribute to the zero of f. The following result gives a means to easily identify zeros of a function that have multiplicity one. Such zeros are often called **simple**. A generalization of this theorem is considered in Exercise 14.

THEOREM 2.10 $f \in C'[a, b]$ has a simple zero at p in (a, b) if and only if $f(p) = 0$, but $f'(p) \neq 0$.

PROOF   If f has a simple zero at p, then $f(p) = 0$ and $f(x) = (x - p)q(x)$, where $\lim_{x \to p} q(x) \neq 0$. Since $f \in C'[a, b]$,

$$f'(p) = \lim_{x \to p} f'(x) = \lim_{x \to p} [q(x) + (x-p)q'(x)] = \lim_{x \to p} q(x) \neq 0.$$

Conversely, if $f(p) = 0$, but $f'(p) \neq 0$, expand f in a Taylor series of degree zero about p. Then

$$f(x) = f(p) + f'(\xi(x))(x - p) = (x - p)f'(\xi(x)),$$

where $\xi(x)$ is between x and p. Since $f \in C'[a, b]$,

$$\lim_{x \to p} f'(\xi(x)) = f'\left(\lim_{x \to p} \xi(x) \right) = f'(p) \neq 0.$$

Letting $q = f' \circ \xi$ gives $f(x) = (x - p)q(x)$, where $\lim_{x \to p} q(x) \neq 0$. Thus f has a simple root at p.   □

The result in Theorem 2.10 implies that an interval about p exists such that Newton's method converges quadratically to p for any initial approximation, provided that p is a simple root. The following example shows that quadratic convergence need not occur if the root is not simple.

EXAMPLE 3   The function described by $f(x) = e^x - x - 1$ has a root of multiplicity two at $p = 0$. (*See Fig. 2.9.*) To show this consider the function

$$q(x) = \frac{e^x - x - 1}{x^2}.$$

Using L'Hôpital's rule,

$$\lim_{x \to 0} \frac{e^x - x - 1}{x^2} = \lim_{x \to 0} \frac{e^x - 1}{2x} = \lim_{x \to 0} \frac{e^x}{2} = \frac{1}{2} \neq 0.$$

Thus $f(x) = (x - 0)^2 \dfrac{e^x - x - 1}{x^2}$,   where $\lim_{x \to 0} \dfrac{e^x - x - 1}{x^2} \neq 0$,

and the multiplicity of the root at $p = 0$ is two. (Using the result of Exercise 14 we could show this fact by noting that $f(0) = 0$, $f'(0) = 0$, but $f''(0) \neq 0$.)

The terms generated by Newton's method applied to f with $p_0 = 1$ are shown in Table 2.5. The sequence is clearly not quadratically convergent to zero.   ■

FIGURE 2.9

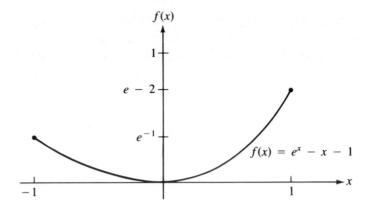

TABLE 2.5

n	p_n	n	p_n
0	1.0	9	2.7750×10^{-3}
1	0.58198	10	1.3881×10^{-3}
2	0.31906	11	6.9411×10^{-4}
3	0.16800	12	3.4703×10^{-4}
4	0.08635	13	1.7416×10^{-4}
5	0.04380	14	8.8041×10^{-5}
6	0.02206	15	4.2610×10^{-5}
7	0.01107	16	1.9142×10^{-5}
8	0.005545		

One method of handling the problem of multiple roots is to define a function $\mu(x)$ by

$$\mu(x) = \frac{f(x)}{f'(x)}.$$

If p is a root of multiplicity $m \geq 1$ and $f(x) = (x - p)^m q(x)$, then

$$\mu(x) = \frac{(x - p)^m q(x)}{m(x - p)^{m-1} q(x) + (x - p)^m q'(x)} = \frac{(x - p)q(x)}{mq(x) + (x - p)q'(x)}$$

also has a root at p, but of multiplicity one. Newton's method can then be applied to the function μ to give

$$g(x) = x - \frac{\mu(x)}{\mu'(x)} = x - \frac{f(x)/f'(x)}{\{[f'(x)]^2 - [f(x)][f''(x)]\}/[f'(x)]^2}$$

or

(2.24)
$$g(x) = x - \frac{f(x)f'(x)}{[f'(x)]^2 - [f(x)][f''(x)]}.$$

If g has the required continuity conditions, functional iteration applied to g will be quadratically convergent regardless of the multiplicity of the root. Theoretically, the only drawback to this method is the additional calculation of $f''(x)$ and the more laborious procedure of calculating the iterates. In practice, however, the presence of a multiple root can cause severe round-off problems.

EXAMPLE 4 a) In Example 3 of Section 2.2 we solved $f(x) = x^3 + 4x^2 - 10 = 0$ for the root $p = 1.36523001$. To compare convergence for a root of multiplicity one by Newton's method and the modified Newton's method listed in Eq. (2.24), let

i) $$p_n = p_{n-1} - \frac{p_{n-1}^3 + 4p_{n-1}^2 - 10}{3p_{n-1}^2 + 8p_{n-1}}, \qquad \text{from Newton's method}$$

and, from Eq. (2.24),

ii) $$p_n = p_{n-1} - \frac{(p_{n-1}^3 + 4p_{n-1}^2 - 10)(3p_{n-1}^2 + 8p_{n-1})}{(3p_{n-1}^2 + 8p_{n-1})^2 - (p_{n-1}^3 + 4p_{n-1}^2 - 10)(6p_{n-1} + 8)}.$$

With $p_0 = 1.5$, the first three iterates for (i) and (ii) are as follows.

	(i)	(ii)
p_1	1.37333333	1.35689898
p_2	1.36526201	1.36519585
p_3	1.36523001	1.36523001

b) For an illustration of the situation that occurs at a multiple root, consider the equation $f(x) = x^4 - 4x^2 + 4 = 0$, which has a root of multiplicity two at $x = \sqrt{2}$. Using both Newton's method and the modification in Eq. (2.24) produces, after simplification, the sequences with terms

i) $$p_n = p_{n-1} - \frac{(p_{n-1}^2 - 2)}{4p_{n-1}}, \qquad \text{from Newton's method,}$$

and

ii) $$p_n = p_{n-1} - \frac{(p_{n-1}^2 - 2)p_{n-1}}{(p_{n-1}^2 + 2)}, \qquad \text{from Eq. (2.24).}$$

With $p_0 = 1.5$, the first three iterates for (i) and (ii), are as follows:

	(i)	(ii)
p_1	1.458333333	1.411764706
p_2	1.436607143	1.414211438
p_3	1.425497619	1.414213562

The actual answer correct to 10^{-9} is the value listed for p_3 in (ii). To obtain this accuracy by the standard Newton–Raphson method requires 20 iterations.

∎

Exercise Set 2.4

1. Use Eq. (2.24) and functional iteration to find a root, accurate within 10^{-5}, of $f(x) = (\sin x - (x/2))^2 = 0$, starting with $p_0 = \pi/2$. Compare your results with those obtained in Exercise 11 in Section 2.3.

2. Use the modified Newton-Raphson method described in Eq. (2.24) to find an approximation to a root of

$$f(x) = x^2 + 2xe^x + e^{2x} = 0$$

   starting with $p_0 = 0$ and performing 10 iterations. If you are using a computer, redo the calculations using extended precision.

3. Consider the function $f(x) = e^{6x} + 3(\ln 2)^2 e^{2x} - \ln 8e^{4x} - (\ln 2)^3$. Use Eq. (2.24) with $p_0 = 0$ to determine the root of $f(x) = 0$. Generate terms until $|p_{n+1} - p_n| < 0.0002$.

4. Repeat Exercise 3 with the constants in $f(x)$ replaced by their four-digit approximations, that is, with $f(x) = e^{6x} + 1.441e^{2x} - 2.079e^{4x} - 0.3330$. Compare the solutions to the results in Exercise 3.

5. Show that the following sequences $\{p_n\}$ converge linearly to $p = 0$. Find how many terms must be generated before $|p_n - p| \le 5 \times 10^{-2}$.

   a) $p_n = \dfrac{1}{n}, \quad n \ge 1$ b) $p_n = \dfrac{1}{n^2}, \quad n \ge 1$

6. Show that the sequence defined by $p_n = 1/n^k$, $n \ge 1$, for any positive integer k, converges linearly to $p = 0$. For each pair of integers k and m, determine a number N for which $1/N^k < 10^{-m}$.

7. Show that the sequence $p_n = 10^{-2^n}$ converges quadratically to zero.

8. Suppose $\alpha > 1$. Construct a sequence that converges to zero of order α.

9. Suppose p is a root of multiplicity m of $f(x) = 0$ where f''' is continuous on an open interval containing p. Use Theorem 2.8 to show that functional iteration using

$$g(x) = x - \frac{mf(x)}{f'(x)}$$

   gives quadratic convergence.

10. Repeat Exercise 2 using the method of Exercise 9 with $m = 1, 2, 3, 4$.

11. Repeat Exercise 1 using the method of Exercise 9 with $m = 1, 2, 3, 4$.

12. Repeat Exercises 3 and 4 using the method of Exercise 9 with $m = 1, 2, 3$, and 4.

13. Show that the Bisection algorithm 2.1 gives a sequence with an error bound that converges linearly to zero.

14. Suppose that f has m continuous derivatives. Show that f has a zero of multiplicity m at p if and only if

$$0 = f(p) = f'(p) = \cdots = f^{(m-1)}(p), \qquad \text{but } f^{(m)}(p) \neq 0.$$

15. Show that the iterative method to solve $f(x) = 0$, given by

$$p_n = p_{n-1} - \frac{f(p_{n-1})}{f'(p_{n-1})} - \frac{f''(p_{n-1})}{2f'(p_{n-1})} \left[\frac{f(p_{n-1})}{f'(p_{n-1})} \right]^2, \qquad n = 1, 2, 3, \ldots,$$

will generally yield cubic ($\alpha = 3$) convergence. Using the analysis of Example 2, compare quadratic and cubic convergence.

16. It can be shown, see for example Dahlquist and Björck [31], p. 228–229, that if $\{p_n\}_{n=0}^{\infty}$ are convergent Secant method approximations to p, the solution to $f(x) = 0$, then a constant C exists with $|p_{n+1} - p| \approx C|p_n - p||p_{n-1} - p|$, for sufficiently large values of n. Assume $\{p_n\}$ converges to p of order α and show that $\alpha = (1 + \sqrt{5})/2$. This implies that the order of convergence of the Secant method is approximately 1.62.

17. The concept stated in Definition 2.7 could have been defined to include sequences that do not converge. A sequence $\{p_n\}_{n=0}^{\infty}$ is sometimes said to be of order α, with respect to a number p, if

$$\lim_{n \to \infty} \frac{|p_{n+1} - p|}{|p_n - p|^{\alpha}} = \lambda > 0, \qquad \text{for some } \lambda.$$

a) Suppose that $\{p_n\}_{n=0}^{\infty}$ is of order α, with respect to p, for $\alpha = 1$.

   i) Show that if $\lambda < 1$, then $\{p_n\}_{n=0}^{\infty}$ converges to p.

   ii) Show that if $\lambda > 1$, then $\{p_n\}_{n=0}^{\infty}$ does not converge to p.

   iii) What can be concluded when $\lambda = 1$?

b) Suppose that $\{p_n\}_{n=0}^{\infty}$ is of order α, with respect to p, for $\alpha > 1$. Show that if $|\lambda(p - p_0)| < 1$, then $\{p_n\}_{n=0}^{\infty}$ converges to p.

2.5 Accelerating Convergence

In this section we consider a technique, called Aitken's Δ^2 method, that can be used to accelerate the convergence of *any* sequence that is linearly convergent, regardless of its origin.

Assume that $\{p_n\}_{n=0}^{\infty}$ is a linearly convergent sequence with limit p; that is, for $e_n = p_n - p$,

$$\lim_{n \to \infty} \frac{|e_{n+1}|}{|e_n|} = \lambda \qquad \text{and} \qquad 0 < \lambda < 1.$$

To investigate the construction of a sequence $\{\hat{p}_n\}_{n=0}^{\infty}$ that converges more rapidly to p, suppose n is sufficiently large that the ratio can be used to approximate the limit. If we assume also that the e_n's have the same sign, then

$$e_{n+1} \approx \lambda e_n \text{ and } e_{n+2} \approx e_{n+1}.$$

So
$$p_{n+2} \approx e_{n+2} + p \approx \lambda e_{n+1} + p, \qquad \text{or}$$

(2.25)
$$p_{n+2} \approx \lambda(p_{n+1} - p) + p.$$

Replacing $(n + 1)$ by n in Eq. (2.25) gives

(2.26)
$$p_{n+1} \approx \lambda(p_n - p) + p;$$

and solving equations (2.25) and (2.26) for p while eliminating λ leads to

$$\begin{aligned}
p &\approx \frac{p_{n+2}p_n - p_{n+1}^2}{p_{n+2} - 2p_{n+1} + p_n} \\
&\approx \frac{p_n^2 + p_n p_{n+2} + 2p_n p_{n+1} - 2p_n p_{n+1} - p_n^2 - p_{n+1}^2}{p_{n+2} - 2p_{n+1} + p_n} \\
&\approx \frac{(p_n^2 + p_n p_{n+2} - 2p_n p_{n+1}) - (p_n^2 - 2p_n p_{n+1} + p_{n+1}^2)}{p_{n+2} - 2p_{n+1} + p_n} \\
&\approx p_n - \frac{(p_{n+1} - p_n)^2}{p_{n+2} - 2p_{n+1} + p_n}.
\end{aligned}$$

Aitken's Δ^2 method is based on the assumption that the sequence $\{\hat{p}_n\}_{n=0}^{\infty}$, defined by

(2.27)
$$\hat{p}_n = p_n - \frac{(p_{n+1} - p_n)^2}{p_{n+2} - 2p_{n+1} + p_n},$$

converges more rapidly to p than the original sequence $\{p_n\}_{n=0}^{\infty}$.

EXAMPLE 1   The sequence $\{p_n\}_{n=1}^{\infty}$, where $p_n = \cos(1/n)$, converges linearly to $p = 1$. The first few terms of the sequences $\{p_n\}_{n=1}^{\infty}$ and $\{\hat{p}_n\}_{n=1}^{\infty}$ are given in Table 2.6.

TABLE 2.6

n	p_n	$\hat{p}_n$
1	0.54030	0.96178
2	0.87758	0.98213
3	0.94496	0.98979
4	0.96891	0.99342
5	0.98007	0.99541
6	0.98614	
7	0.98981	

It certainly appears that $\{\hat{p}_n\}_{n=1}^{\infty}$ converges more rapidly to $p = 1$ than $\{p_n\}_{n=1}^{\infty}$. ∎

The Δ notation associated with this technique has its origin in the following definition.

DEFINITION 2.11   Given the sequence $\{p_n\}_{n=0}^{\infty}$, define the **forward difference** Δp_n by

$$\Delta p_n = p_{n+1} - p_n \qquad \text{for } n \geq 0.$$

Higher powers $\Delta^k p_n$ are defined recursively by

$$\Delta^k p_n = \Delta^{k-1}(\Delta p_n) \qquad \text{for } k \geq 2.$$

Because of the definition,

$$
\begin{aligned}
\Delta^2 p_n &= \Delta(p_{n+1} - p_n) \\
&= \Delta p_{n+1} - \Delta p_n \\
&= (p_{n+2} - p_{n+1}) - (p_{n+1} - p_n) \\
&= p_{n+2} - 2p_{n+1} + p_n.
\end{aligned}
$$

Thus, the formula for $\hat{p}_n$ given in Eq. (2.27) can be written as

$$(2.28) \qquad \hat{p}_n = p_n - \frac{(\Delta p_n)^2}{\Delta^2 p_n} \qquad \text{for all } n \geq 0.$$

To this point in our discussion of Aitken's Δ^2 method, we have stated that the sequence $\{\hat{p}_n\}_{n=0}^{\infty}$ converges to p more rapidly than does the original sequence $\{p_n\}_{n=0}^{\infty}$, but we have not said exactly what is meant by the term "more rapid" convergence. Theorem 2.12 explains and justifies this terminology. The proof of this theorem is considered in Exercise 8.

THEOREM 2.12   Let p_n be any sequence converging linearly to the limit p with $e_n = p_n - p \neq 0$ for all $n \geq 0$. Then the sequence $\{\hat{p}_n\}_{n=0}^{\infty}$ converges to p faster than $\{p_n\}_{n=0}^{\infty}$ in the sense that

$$\lim_{n \to \infty} \frac{\hat{p}_n - p}{p_n - p} = 0.$$

By applying Aitken's Δ^2 method to a linearly convergent sequence obtained from fixed-point iteration, we can accelerate the convergence to quadratic. This procedure is known as Steffensen's method and differs slightly from applying Aitken's Δ^2 method directly to the linearly convergent fixed-point iteration sequence. The direct procedure would construct in order

$$p_0, \ p_1 = g(p_0), \ p_2 = g(p_1), \ \hat{p}_0 = \{\Delta^2\}p_0, \ p_3 = g(p_2), \ \hat{p}_1 = \{\Delta^2\}p_1, \ \ldots,$$

where $\{\Delta^2\}$ is used to indicate that the Aitken's Δ^2 technique is employed. Steffensen's method constructs the same first four terms, p_0, p_1, p_2, and $\hat{p}_0$. However, at this step it is assumed that $\hat{p}_0$ is a better approximation to p than is p_2 and applies fixed-point iteration to $\hat{p}_0$ instead of to p_2. Using the notation in Algorithm 2.5, the sequence generated is

$$p_0^{(0)}, \; p_1^{(0)} = g(p_0^{(0)}), \; p_2^{(0)} = g(p_1^{(0)}), \; p_0^{(1)} = \{\Delta^2\} p_0^{(0)}, \; p_1^{(1)} = g(p_0^{(1)}), \; \dots$$

Every third term is generated using the Δ^2 technique; the others use fixed-point iteration on the previous term.

Steffensen's Algorithm 2.5

To find a solution to $p = g(p)$ given an initial approximation p_0:

INPUT initial approximation p_0; tolerance TOL; maximum number of iterations N_0.

OUTPUT approximate solution p or message of failure.

Step 1   Set $i = 1$.

Step 2   While $i \le N_0$ do Steps 3–6.

    Step 3   Set $p_1 = g(p_0)$;   (*Compute $p_1^{(i-1)}$.*)
                  $p_2 = g(p_1)$;   (*Compute $p_2^{(i-1)}$.*)
                  $p = p_0 - (p_1 - p_0)^2 / (p_2 - 2p_1 + p_0)$.   (*Compute $p_0^{(i)}$.*)

    Step 4   If $|p - p_0| < TOL$ then
                  OUTPUT (p);   (*Procedure completed successfully.*)
                  STOP.

    Step 5   Set $i = i + 1$.

    Step 6   Set $p_0 = p$.   (*Update p_0.*)

Step 7   OUTPUT ('Method failed after N_0 iterations, $N_0 = $', N_0);
          (*Procedure completed unsuccessfully.*)
          STOP.

Note that $\Delta^2 p_n$ may be zero. If that should happen, we terminate the sequence and select $p_2^{(n-1)}$ as the approximate answer, since otherwise this would introduce a zero in the denominator of the next iterate.

EXAMPLE 2   To solve $x^3 + 4x^2 - 10 = 0$ using Steffensen's method, let $x^3 + 4x^2 = 10$ and solve for x by dividing by $x + 4$. Thus, if

$$g(x) = \left(\frac{10}{x + 4} \right)^{1/2},$$

then $x = g(x)$ implies $x^3 + 4x^2 - 10 = 0$.

Using $p_0 = 1.5$, Steffensen's procedure gives

k	$p_0^{(k)}$	$p_1^{(k)}$	$p_2^{(k)}$
0	1.5	1.348399725	1.367376372
1	1.365265224	1.365225534	1.365230583
2	1.365230013		

The iterate $p_0^{(2)} = 1.365230013$ is accurate to the ninth decimal place. In this example, Steffensen's method gave about the same rate of convergence as Newton's method (see Example 4 in Section 2.4).   ■

From Example 2, it appears that Steffensen's method gives quadratic convergence without evaluating a derivative. Theorem 2.13 verifies that this is the case. The proof of this theorem can be found in Henrici [62], pages 90–92, or Isaacson and Keller [67], pages 103–107.

THEOREM 2.13   Suppose that $x = g(x)$ has the solution p with $g'(p) \neq 1$. If there exists a $\delta > 0$ such that $g \in C^3[p - \delta, p + \delta]$, then Steffensen's method gives quadratic convergence for any $p_0 \in [p - \delta, p + \delta]$.

The weakness in Steffensen's method occurs because of the necessity that $g'(p) \neq 1$, a condition that is equivalent to requiring that the multiplicity of the zero p be one for the corresponding root-finding problem $f(x) = 0$. As a consequence, Steffensen's method cannot be expected to accelerate to quadratic the linear convergence that generally results when Newton's method is used to approximate a zero of multiplicity greater than one.

Exercise Set 2.5

1. Solve $x^3 - x - 1 = 0$ for the root in $[1, 2]$ to an accuracy of 10^{-4} using Steffensen's method and the results of Exercise 5 of Section 2.2.

2. Solve $x - 2^{-x} = 0$ for the root in $[0, 1]$ to an accuracy of 10^{-4} using Steffensen's method, and compare to the results of Exercise 3 of Section 2.2.

3. Use Steffensen's method with $p_0 = 2$ to compute an approximation to $\sqrt{3}$ accurate to within 10^{-4}. Compare this result with those obtained in Exercise 12 of Section 2.3, Exercise 6, of Section 2.2, and Exercise 8 of Section 2.1.

4. Approximate the solutions to within 10^{-5} of the following equations, using Steffensen's method.

    a) $x = \dfrac{2 - e^x + x^2}{3}$, where g is the function in Exercise 8(a) of Section 2.2.

b) $3x^2 - e^x = 0$, where g is the function in Exercise 9(a) of Section 2.2.

c) $x - \cos x = 0$, where g is the function in Exercise 9(b) of Section 2.2.

5.   For the following linearly convergent sequences $\{p_n\}$, use Aitken's Δ^2 method to generate a sequence $\{\hat{p}_n\}$ until $|\hat{p}_n - p| \leq 5 \times 10^{-2}$.

a) $p_n = \dfrac{1}{n}, \quad n \geq 1$ b) $p_n = \dfrac{1}{n^2}, \quad n \geq 1$

6.   A sequence $\{p_n\}$ is said to be **superlinearly convergent** to p if

$$\lim_{n \to \infty} \frac{p_{n+1} - p}{p_n - p} = 0.$$

a)   Show that if $p_n \to p$ of order α for $\alpha > 1$, then $\{p_n\}$ is superlinearly convergent to p.

b)   Find a sequence $\{p_n\}$ that is superlinearly convergent to zero, but does not converge to zero of order α for any $\alpha > 1$.

7.   Suppose that $\{p_n\}$ is superlinearly convergent to p. Show that

$$\lim_{n \to \infty} \frac{|p_{n+1} - p_n|}{|p_n - p|} = 1.$$

8.   Prove Theorem 2.12 [*Hint*: Let $e_n = (p_{n+1} - p)/(p_n - p) - \lambda$ and express $(\hat{p}_n - p_n)/(p_n - \lambda)$ in terms of e_n, e_{n-1}, and λ].

2.6 Zeros of Polynomials and Müller's Method

A function of the form

$$P(x) = a_n x^n + a_{n-1} x^{n-1} + \cdots + a_1 x + a_0,$$

where the a_i's, called the **coefficients** of P, are constants and $a_n \neq 0$, is called a **polynomial of degree n**. The zero function, $P(x) = 0$ for all values of x, is considered a polynomial but is assigned no degree.

THEOREM 2.14   (Fundamental Theorem of Algebra)   If P is a polynomial of degree $n \geq 1$, then $P(x) = 0$ has at least one (possibly complex) root.

Although Theorem 2.14 is basic to any study of elementary functions, the usual proof requires techniques from the study of complex-function theory. The reader is referred to Saff and Snider [112], page 155, for the culmination of a systematic development of the topics needed to prove Theorem 2.14.

An important consequence of Theorem 2.14 is Corollary 2.15.

COROLLARY 2.15   If $P(x) = a_n x^n + a_{n-1} x^{n-1} + \cdots + a_1 x + a_0$ is a polynomial of degree $n \geq 1$, then there exist unique constants $x_1, x_2, \ldots, x_k$, possibly complex, and positive integers, $m_1, m_2,$ $\ldots, m_k$, such that $\sum_{i=1}^{k} m_i = n$ and

$$P(x) = a_n(x - x_1)^{m_1}(x - x_2)^{m_2} \cdots (x - x_k)^{m_k}.$$

Corollary 2.15 states that the zeros of a polynomial are unique and that, if each zero x_i is counted as many times as its multiplicity m_i, then a polynomial of degree n has exactly n zeros. Its proof is considered in Exercise 5(a).

The following corollary of the Fundamental Theorem of Algebra will be used often in this section and in later chapters. The proof of this result is considered in Exercise 5(b).

COROLLARY 2.16   Let P and Q be polynomials of degree at most n. If $x_1, x_2, \ldots, x_k$, $k > n$, are distinct numbers with $P(x_i) = Q(x_i)$ for $i = 1, 2, \ldots, k$, then $P(x) = Q(x)$ for all values of x.

To use the Newton–Raphson procedure to locate approximate zeros of a polynomial P, it is necessary to evaluate P and its derivative at specified values. Since both P and its derivative are polynomials, computational efficiency requires that the evaluation of these functions be done in the nested manner discussed in Section 1.2. Horner's method described in Theorem 2.17 incorporates this nesting technique and as a consequence requires only n multiplications and n additions to evaluate an arbitrary nth degree polynomial.

THEOREM 2.17   (Horner's Method)   Let

$$P(x) = a_n x^n + a_{n-1} x^{n-1} + \cdots + a_1 x + a_0 \qquad \text{and} \qquad b_n = a_n.$$

If

$$b_k = a_k + b_{k+1} x_0 \qquad \text{for } k = n-1, n-2, \ldots, 1, 0,$$

then $b_0 = P(x_0)$. Moreover, if

$$Q(x) = b_n x^{n-1} + b_{n-1} x^{n-2} + \cdots + b_2 x + b_1,$$

then

(2.29) $$P(x) = (x - x_0)Q(x) + b_0.$$

PROOF   By the definition of $Q(x)$,

$$\begin{aligned}
(x - x_0)Q(x) + b_0 &= (x - x_0)(b_n x^{n-1} + \cdots + b_2 x + b_1) + b_0 \\
&= (b_n x^n + b_{n-1} x^{n-1} + \cdots + b_2 x^2 + b_1 x) \\
&\quad - (b_n x_0 x^{n-1} + \cdots + b_2 x_0 x + b_1 x_0) + b_0 \\
&= b_n x^n + (b_{n-1} - b_n x_0)x^{n-1} + \cdots + (b_1 - b_2 x_0)x + (b_0 - b_1 x_0).
\end{aligned}$$

By the hypotheses of the theorem, $b_n = a_n$ and $b_k - b_{k+1} x_0 = a_k$, so

$$(x - x_0)Q(x) + b_0 = P(x) \qquad \text{and} \qquad b_0 = P(x_0). \qquad \square$$

EXAMPLE 1   Evaluate $P(x) = 2x^4 - 3x^2 + 3x - 4$ at $x_0 = -2$ using Horner's method. Using Theorem 2.17,

$$b_4 = 2, \qquad b_3 = 2(-2) + 0 = -4,$$

$$b_2 = (-4)(-2) - 3 = 5, \qquad b_1 = 5(-2) + 3 = -7$$

and finally,

$$P(-2) = b_0 = (-7)(-2) - 4 = 10.$$

Moreover, Theorem 2.17 gives

$$P(x) = (x + 2)(2x^3 - 4x^2 + 5x - 7) + 10. \qquad \blacksquare$$

When we use hand calculation in Horner's method, we first construct a table, which suggests the "synthetic division" name often applied to the technique. For the problem in the preceding example, the table would appear as:

	$\begin{pmatrix} \text{Coefficient} \\ \text{of } x^4 \end{pmatrix}$	$\begin{pmatrix} \text{Coefficient} \\ \text{of } x^3 \end{pmatrix}$	$\begin{pmatrix} \text{Coefficient} \\ \text{of } x^2 \end{pmatrix}$	$\begin{pmatrix} \text{Coefficient} \\ \text{of } x \end{pmatrix}$	$\begin{pmatrix} \text{Constant} \\ \text{term} \end{pmatrix}$
$x_0 = -2$	$a_4 = 2$	$a_3 = 0$	$a_2 = -3$	$a_1 = 3$	$a_0 = -4$
		$b_4 x_0 = -4$	$b_3 x_0 = 8$	$b_2 x_0 = -10$	$b_1 x_0 = 14$
	$b_4 = 2$	$b_3 = -4$	$b_2 = 5$	$b_1 = -7$	$b_0 = 10$

An additional advantage of using the Horner's (or synthetic-division) procedure is that, since

$$P(x) = (x - x_0)Q(x) + b_0,$$

where $$Q(x) = b_n x^{n-1} + b_{n-1} x^{n-2} + \cdots + b_2 x + b_1,$$

differentiating with respect to x gives

$$P'(x) = Q(x) + (x - x_0)Q'(x) \qquad \text{and}$$

(2.30) $$P'(x_0) = Q(x_0).$$

Therefore, when the Newton–Raphson method is being used to find an approximate zero of a polynomial P, both P and P' can be evaluated in the same manner. The following algorithm computes $P(x_0)$ and $P'(x_0)$ using Horner's method.

Horner's Algorithm 2.6

To evaluate the polynomial

$$P(x) = a_n x^n + a_{n-1} x^{n-1} + \cdots + a_1 x + a_0$$

and its derivative at x_0:

INPUT degree n; coefficients $a_0, a_1, \ldots, a_n; x_0.$

OUTPUT $y = P(x_0)$; $z = P'(x_0)$.

Step 1 Set $y = a_n$; (*Compute b_n for P.*)
    $z = a_n$. (*Compute b_{n-1} for Q.*)

Step 2 For $j = n - 1, n - 2, \ldots, 1$
    set $y = x_0 y + a_j$; (*Compute b_j for P.*)
    $z = x_0 z + y$. (*Compute b_{j-1} for Q.*)

Step 3 Set $y = x_0 y + a_0$. (*Compute b_0 for P.*)

Step 4 OUTPUT (y, z);
    STOP.

EXAMPLE 2   Find an approximation to one of the zeros of

$$P(x) = 2x^4 - 3x^2 + 3x - 4$$

by using the Newton–Raphson procedure and synthetic division to evaluate $P(x_n)$ and $P'(x_n)$ for each iterate x_n. Using $x_0 = -2$ as an initial approximation, we obtained $P(-2)$ in Example 1 by:

$$
\begin{array}{r|rrrrr}
x_0 = -2 & 2 & 0 & -3 & 3 & -4 \\
& & -4 & 8 & -10 & 14 \\
\hline
& 2 & -4 & 5 & -7 & 10 = P(-2)
\end{array}
$$

Using Theorem 2.17 and Eq. (2.30),

$$Q(x) = 2x^3 - 4x^2 + 5x - 7 \quad \text{and} \quad P'(-2) = Q(-2);$$

so $P'(-2)$ can be found by evaluating $Q(-2)$ in a similar manner:

$$
\begin{array}{r|rrrr}
x_0 = -2 & 2 & -4 & 5 & -7 \\
& & -4 & 16 & -42 \\
\hline
& 2 & -8 & 21 & -49 = Q(-2) = P'(-2)
\end{array}
$$

and $\qquad x_1 = x_0 - \dfrac{P(x_0)}{P'(x_0)} = -2 - \dfrac{10}{-49} \approx -1.796.$

Repeating the procedure to find x_2,

$$
\begin{array}{r|rrrrr}
-1.796 & 2 & 0 & -3 & 3 & -4 \\
& & -3.592 & 6.451 & -6.197 & 5.742 \\
\hline
& 2 & -3.592 & 3.451 & -3.197 & 1.742 = P(x_1) \\
& & -3.592 & 12.902 & -29.368 & \\
\hline
& 2 & -7.184 & 16.353 & -32.565 = Q(x_1) = P'(x_1)
\end{array}
$$

So $P(-1.796) = 1.742$, $P'(-1.796) = -32.565$, and

$$x_2 = -1.796 - \frac{1.742}{-32.565} \approx -1.7425.$$

An actual zero to five decimal places is -1.73896.   ∎

Note that the polynomial denoted Q depends on the approximation being used and changes from iterate to iterate.

If the Nth iterate, x_N, in the Newton–Raphson procedure is an approximate zero of P, then

$$P(x) = (x - x_N)Q(x) + b_0 = (x - x_N)Q(x) + P(x_N) \approx (x - x_N)Q(x);$$

so $x - x_N$ is an approximate factor of $P(x)$. Letting $\hat{x}_1 = x_N$ be the approximate zero of P and $Q_1(x)$ the approximate factor,

$$P(x) \approx (x - \hat{x}_1)Q_1(x),$$

we can find a second approximate zero of P by applying the Newton–Raphson procedure to $Q_1(x)$. If P is an nth-degree polynomial with n real zeros, this procedure applied repeatedly will eventually result in $(n - 2)$ approximate zeros of P and an approximate quadratic factor $Q_{n-2}(x)$. At this stage, $Q_{n-2}(x) = 0$ can be solved by the quadratic formula to find the last two approximate zeros of P. Although this method can be used to find approximate zeros of many polynomials, it depends on repeated use of approximations and on occasion can lead to very inaccurate approximations.

The procedure just described is called **deflation**. The accuracy difficulty with deflation is due to the fact that, when we obtain the approximate zeros of P, the Newton–Raphson procedure is used on the reduced polynomial Q_k, i.e., the polynomial having the property that

$$P(x) \approx (x - \hat{x}_1)(x - \hat{x}_2) \cdots (x - \hat{x}_k)Q_k(x).$$

An approximate zero $\hat{x}_{k+1}$ of Q_k will generally not approximate a root of $P(x) = 0$ as well as a root of $Q_k(x) = 0$. The inaccuracy will usually increase as k increases. One way to eliminate this difficulty is to use the reduced equations, that is, the approximate factors of the original polynomial P, to find approximations, $\hat{x}_2, \hat{x}_3, \ldots, \hat{x}_k$, to the zeros of P and then improve these approximations by applying the Newton–Raphson procedure to the original polynomial P.

It has been noted previously that the success of Newton's method often depends on obtaining a good initial approximation. The basic idea for finding approximate zeros of P is as follows: evaluate P at points x_i for $i = 1, 2, \ldots, k$. If $P(x_i)P(x_j) < 0$, then P has a zero between x_i and x_j. The problem becomes a matter of choosing the x_i's so that the chance of missing a change of sign is minimized, while keeping the number of x_i's reasonably small.

To illustrate the possible difficulty of this problem, consider the polynomial

$$P(x) = 16x^4 - 40x^3 + 5x^2 + 20x + 6.$$

If x_i is any integer, it can be shown that $P(x_i) > 0$; hence, evaluating $P(x)$ at this infinite number of x_i's would not locate any intervals (x_i, x_j) containing zeros of P. However, P does have real zeros. It happened that this particular choice of the x_i's is inappropriate for this polynomial because of the closeness of the roots.

Another problem with applying Newton's method to polynomials concerns the possibility of the polynomial having complex roots even when all the coefficients are real numbers. If the initial approximation using Newton's method is a real number, all subsequent approximations will also be real numbers.

One way to overcome this difficulty is to begin with a nonreal initial approximation and do all the computations using complex arithmetic. An alternative approach has its basis in the following theorem.

THEOREM 2.18   If $z = a + bi$ is a complex zero of multiplicity m of the polynomial P, then $\bar{z} = a - bi$ is also a zero of multiplicity m of the polynomial P and $(x^2 - 2ax + a^2 + b^2)^m$ is a factor of P.

A synthetic division involving quadratic polynomials can be devised to approximately factor the polynomial so that one term will be a quadratic polynomial whose complex roots are approximations to the roots of the original polynomial. This technique was described in some detail in our previous edition.

Instead of proceeding along these lines we will now consider a method first presented by D. E. Müller [90] in 1956. This technique can be used for any root-finding problem, but it is particularly useful for approximating the roots of polynomials.

Müller's method is a generalization of the Secant method. The Secant method begins with two initial approximations x_0 and x_1 and determines the next approximation x_2 as the intersection of the x-axis with the line through $(x_0, f(x_0))$ and $(x_1, f(x_1))$. (*See Figure 2.10(a)*.) Müller's method uses three initial approximations x_0, x_1, and x_2 and determines the next approximation x_3 by considering the intersection of the x-axis with the parabola through $(x_0, f(x_0))$, $(x_1, f(x_1))$, and $(x_2, f(x_2))$. (*See Figure 2.10(b)*.)

FIGURE 2.10

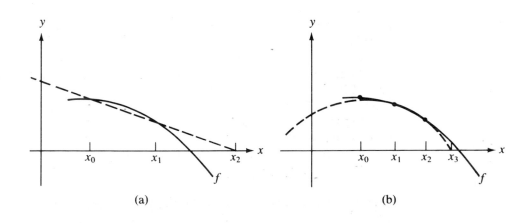

(a)   (b)

The derivation of Müller's method begins by considering the quadratic polynomial

$$P(x) = a(x - x_2)^2 + b(x - x_2) + c$$

that passes through $(x_0, f(x_0))$, $(x_1, f(x_1))$, and $(x_2, f(x_2))$. The constants a, b, and c can be determined from the conditions

$$f(x_0) = a(x_0 - x_2)^2 + b(x_0 - x_2) + c,$$
$$f(x_1) = a(x_1 - x_2)^2 + b(x_1 - x_2) + c,$$

and

$$f(x_2) = a \cdot 0^2 + b \cdot 0 + c$$

to be

(2.31)

$$c = f(x_2),$$

$$b = \frac{(x_0 - x_2)^2[f(x_1) - f(x_2)] - (x_1 - x_2)^2[f(x_0) - f(x_2)]}{(x_0 - x_2)(x_1 - x_2)(x_0 - x_1)}$$

and

$$a = \frac{(x_1 - x_2)[f(x_0) - f(x_2)] - (x_0 - x_2)[f(x_1) - f(x_2)]}{(x_0 - x_2)(x_1 - x_2)(x_0 - x_1)}.$$

To determine x_3, the root of P, we apply the quadratic formula to P. Because of round-off error problems caused by the subtraction of nearly equal numbers, however, we apply the formula in the manner prescribed in Example 4 of Section 1.2:

(2.32)

$$x_3 - x_2 = \frac{-2c}{b \pm \sqrt{b^2 - 4ac}}.$$

This gives two possibilities for x_3 depending on the sign preceding the radical term in Eq. (2.32). In Müller's method, the sign is chosen to agree with the sign of b. Chosen in this manner, the denominator will be the largest in magnitude and will result in x_3 being selected as the closest root of P to x_2. Thus

$$x_3 = x_2 - \frac{2c}{b + \text{sign}\,(b)\sqrt{b^2 - 4ac}}$$

where a, b, and c are given in Eq. (2.31).

Once x_3 is determined, the procedure is reinitialized using x_1, x_2, and x_3 in place of x_0, x_1, and x_2 to determine the next approximation x_4. The method continues until a satisfactory conclusion is obtained. Since the method involves at each step the radical $\sqrt{b^2 - 4ac}$, the method will approximate complex roots when it is appropriate to do so.

Müller's Algorithm 2.7

To find a solution to $f(x) = 0$ given three approximations x_0, x_1, and x_2:

INPUT x_0, x_1, x_2; tolerance TOL; maximum number of iterations N_0.

OUTPUT approximate solution p or message of failure.

Step 1   Set $h_1 = x_1 - x_0$;
$\qquad h_2 = x_2 - x_1$;
$\qquad \delta_1 = (f(x_1) - f(x_0))/h_1$;
$\qquad \delta_2 = (f(x_2) - f(x_1))/h_2$;
$\qquad d = (\delta_2 - \delta_1)/(h_2 + h_1)$;
$\qquad i = 2$.

Step 2   While $i \leq N_0$ do Steps 3–7.

Step 3 $b = \delta_2 + h_2 d$;
$\qquad D = (b^2 - 4f(x_2)d)^{1/2}$. (*Note: May be complex arithmetic.*)

Step 4   If $|b - D| < |b + D|$ then set $E = b + d$
$\qquad\qquad\qquad\qquad\qquad$ else set $E = b - d$.

Step 5   Set $h = -2f(x_2)/E$;
$\qquad p = x_2 + h$.

Step 6   If $|h| < TOL$ then
$\qquad$ OUTPUT (p);   (*Procedure completed successfully.*)
$\qquad$ STOP.

Step 7   Set $x_0 = x_1$;   (*Prepare for next iteration.*)
$\qquad x_1 = x_2$;
$\qquad x_2 = p$;
$\qquad h_1 = x_1 - x_0$;
$\qquad h_2 = x_2 - x_1$;
$\qquad \delta_1 = (f(x_1) - f(x_0))/h_1$;
$\qquad \delta_2 = (f(x_2) - f(x_1))/h_2$;
$\qquad d = (\delta_2 - \delta_1)/(h_2 + h_1)$;
$\qquad i = i + 1$.

Step 8   OUTPUT ('Method failed after N_0 iterations, $N_0 = $ ', N_0);
$\qquad$ (*Procedure completed unsuccessfully.*)
$\qquad$ STOP.

EXAMPLE 3   Consider the polynomial $P(x) = 16x^4 - 40x^3 + 5x^2 + 20x + 6$. Using Algorithm 2.7 with $TOL = 10^{-5}$ and different values of x_0, x_1, and x_2 produces the results in Table 2.7. The actual values for the roots of the equation are 1.241677, 1.970446, -0.356062 $\pm 0.162758i$ which shows that the approximations from Müller's method are excellent.   ∎

Example 3 illustrates that Müller's method can approximate the roots of polynomials with a variety of starting values. In fact, the importance of Müller's method is that the technique generally will converge to the root of a polynomial for any initial approximation choice. Problems can be constructed for which convergence will not occur for certain choices of initial approximations. For example, if x_i, x_{i+1},

TABLE 2.7

$x_0 = 0.5$, $x_1 = -0.5$, $x_2 = 0$		
i	x_i	$f(x_i)$
3	$-0.555556 + 0.598352i$	$-29.4007 - 3.89872i$
4	$-0.435450 + 0.102101i$	$1.33223 - 1.19309i$
5	$-0.390631 + 0.141852i$	$0.375057 - 0.670164i$
6	$-0.357699 + 0.169926i$	$-0.146746 - 0.00744629i$
7	$-0.356051 + 0.162856i$	$-0.183868 \times 10^{-2} + 0.539780 \times 10^{-3}i$
8	$-0.356062 + 0.162758i$	$0.286102 \times 10^{-5} + 0.953674 \times 10^{-6}i$

$x_0 = 0.5$, $x_1 = 1.0$, $x_2 = 1.5$		
i	x_i	$f(x_i)$
3	1.28785	-1.37624
4	1.23746	0.126941
5	1.24160	0.219440×10^{-2}
6	1.24168	0.257492×10^{-4}
7	1.24168	0.257492×10^{-4}

$x_0 = 2.5$, $x_1 = 2.0$, $x_2 = 2.25$		
i	x_i	$f(x_i)$
3	1.96059	-0.611255
4	1.97056	0.748825×10^{-2}
5	1.97044	-0.295639×10^{-4}
6	1.97044	-0.259639×10^{-4}

and x_{i+2} for some i happen to have the property that $f(x_i) = f(x_{i+1}) = f(x_{i+2})$, the quadratic equation will reduce to a nonzero constant function and never intersect the x-axis. This is not usually the case, however, and general-purpose software using Müller's method, for example, the ZANLYT subroutine in the IMSL, only requests one initial approximation per root and will even supply this approximation as an option.

Müller's method is not quite as efficient as Newton's method; its order of convergence near a root is approximately $\alpha = 1.84$ compared to the quadratic, $\alpha = 2$, of Newton's method, but it is better than the Secant method, whose order is approximately $\alpha = 1.62$.

Deflation is generally used with Müller's method once an approximate root has been determined. After an approximation to the root of the deflated equation has been determined, use either Müller's method or Newton's method in the original polynomial with this root as the initial approximation. This will ensure that the root being approximated is a solution to the true equation, not to the deflated equation.

Other high-order methods are available for determining the roots of polynomials. If this topic is of particular interest, we recommend that consideration be given to Laguerre's method, which gives cubic convergence and also approximates complex roots (see Householder [65] pages 176–179 for a complete discussion), the

Jenkins–Traub method (see Jenkins and Traub [70], and Brent's method (see Brent [13]), which is based on the bisection and regula-falsi methods. Another method of interest, called Cauchy's method, is similar to Müller's method, but it avoids the failure problem of Müller's method when $f(x_i) = f(x_{i+1}) = f(x_{i+2})$, for some i. For interesting discussion of this method, as well as more detail on Müller's method, we recommend Young and Gregory [146], Sections 4.10, 4.11, and 5.4.

Exercise Set 2.6

1. Find approximations to within 10^{-4} to all the real zeros of the following polynomials using Newton's method and deflation.

    a) $P(x) = x^3 - 2x^2 - 5$ (see Exercise 1(a) of Section 2.3).

    b) $P(x) = x^3 + 3x^2 - 1$ (see Exercise 1(b) of Section 2.3).

    c) $P(x) = x^3 - x - 1$ (see Exercise 3 of Section 2.3).

    d) $P(x) = x^4 + 2x^2 - x - 3$ (see Exercise 1 of Section 2.2).

2. Find approximations to within 10^{-5} to all of the zeros of each of the following polynomials by first finding the real zeros and then reducing to polynomials of lower degree to determine any complex zeros.

    a) $P(x) = x^4 + 5x^3 - 9x^2 - 85x - 136$

    b) $P(x) = x^4 - 2x^3 - 12x^2 + 16x - 40$

    c) $P(x) = x^4 + x^3 + 3x^2 + 2x + 2$

    d) $P(x) = x^5 + 11x^4 - 21x^3 - 10x^2 - 21x - 5$

    e) $P(x) = 16x^4 + 88x^3 + 159x^2 + 76x - 240$

    f) $P(x) = x^4 - 4x^2 - 3x + 5$

    g) $P(x) = x^4 - 2x^3 - 4x^2 + 4x + 4$ (See Exercise 2 of Section 2.1.)

    h) $P(x) = x^3 - 7x^2 + 14x - 6$ (See Exercise 4 of Section 2.1.)

3. Repeat Exercise 1 using Müller's method.

4. Repeat Exercise 2 using Müller's method.

5. a) Prove Corollary 2.15.   b) Prove Corollary 2.16.

6. Prove Theorem 2.18. (*Hint:*   First consider the case $m = 1$ and note what happens to the constants if $a - bi$ is not a zero of $P(x)$.)

7. Prove the following theorem.

    Theorem   Let $P(x) = a_n x^n + a_{n-1} x^{n-1} + \cdots + a_1 x + a_0$ be a polynomial of degree n and let x_0 be a positive real number with $P(x_0) > 0$. If $Q(x)$ satisfies

    $$P(x) = (x - x_0)Q(x) + P(x_0)$$
    $$= (x - x_0)(b_n x^{n-1} + \cdots + b_2 x + b_1) + P(x_0),$$

and $b_i > 0$ for $i = 1, 2, \ldots, n$, then all zeros of P are less than or equal to x_0.

8. $P(x) = 10x^3 - 8.3x^2 + 2.295x - 0.21141 = 0$ has a root at $x = 0.29$. Use Newton's method with an initial approximation $x_0 = 0.28$ to attempt to find this root. What happens?

9. Use Newton's method to find, within 10^{-3}, the zeros and critical points of the following functions. Use this information to sketch the graph of f.

    a) $f(x) = x^3 - 9x^2 + 12$ b) $f(x) = x^4 - 2x^3 - 5x^2 + 12x - 5$

10. A can in the shape of a right circular cylinder is to be constructed to contain 1000 cm^3. The circular top and bottom of the can must have a radius 0.25 cm more than the radius of the can so that the excess can be used to form a seal with the side. The sheet of material being formed into the side of the can must also be 0.25 cm longer than the circumference of the can so that a seal can be formed. Find, to within 10^{-4}, the minimal amount of material needed to construct the can.

11. Two ladders crisscross an alley. Each ladder reaches from the base of one wall to some point on the opposite wall. The ladders cross at a height H above the pavement. Given that the lengths of the ladders are $x_1 = 20$ feet and $x_2 = 30$ feet and that $H = 8$ feet, find W.

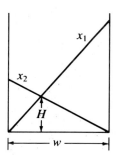

12. In describing a mathematical model for machine-tool chatter [97], the authors need the roots of a polynomial equation of the form

$$P_{2n}(B) = 1 - \phi_1 B - \phi_2 B^2 - \cdots - \phi_{2n} B^{2n} = 0,$$

for various values of n and collections of constants $\phi_1, \phi_2, \ldots, \phi_{2n}$. The value of n depends on the amount of experimental data being used, and approximations for the constants ϕ_i, $i = 1, 2, \ldots, 2n$ are obtained using this data and a linear least squares technique, a topic discussed in Section 7.1.

    a) In an experimental situation involving a Gisholt Turret Lathe, it was found that, for $n = 2$, reasonable approximations for the ϕ_i's are:

    $$\phi_1 = 1.8310, \qquad \phi_2 = -0.5218,$$

    $$\phi_3 = -0.4754, \qquad \phi_4 = 0.1595.$$

    Use these values to find the zeros of P_4.

    b) In the same experiment, with $n = 3$, values of ϕ_i were found to be:

    $$\phi_1 = 1.742, \qquad \phi_2 = -0.0385, \qquad \phi_3 = -0.8133,$$

$$\phi_4 = -0.1061, \qquad \phi_5 = 0.2019, \qquad \phi_6 = 0.0383.$$

Use these values to find the zeros of P_6.

13. The *Legendre* polynomials can be generated recursively by $P_0(x) = 1$, $P_1(x) = x$ and
$$P_{n+2}(x) = \frac{2n+3}{n+2} x P_{n+1}(x) - \frac{n+1}{n+2} P_n(x), n \geq 0.$$ These polynomials and their roots will be considered in Sections 4.7 and 7.2. In Table 4.12 on page 188 a table lists the zeros of P_2, P_3, P_4, and P_5.

a) Determine these polynomials and verify that the values in the table are correct.

b) Determine P_6 and approximate the zeros of this polynomial to within 10^{-6}.

14. The *Chebyshev* polynomials can be generated recursively by $T_0(x) = 1$, $T_1(x) = x$ and $T_{n+2}(x) = 2x T_{n+1}(x) - T_n(x), n \geq 0$. These polynomials and their roots will be considered in Sections 7.2 and 7.3.

a) Determine T_2, T_3, T_4, and T_5.

b) Approximate, to within 10^{-6}, the zeros of T_3, T_4, and T_5.

c) Show that the results in part (b) are consistent with the result in Eq. (7.16) of Theorem 7.8.

15. The *Laguerre* polynomials can be generated recursively by $L_0(x) = 1$, $L_1(x) = 1 - x$ and $L_{n+2}(x) = (2n + 3 - x)L_{n+1}(x) - (n + 1)^2 L_n(x), n \geq 0$.

a) Determine L_2, L_3, L_4, and L_5.

b) Approximate, to within 10^{-4}, the zeros of L_3, L_4, and L_5.

16. The *Hermite* polynomials can be generated recursively by $H_0(x) = 1$, $H_1(x) = 2x$, and $H_{n+2}(x) = 2x H_{n+1}(x) - 2(n + 1)H_n(x), n \geq 0$.

a) Determine H_2, H_3, H_4, and H_5.

b) Approximate, to within 10^{-4}, the zeros of H_3, H_4, and H_5.

3

Interpolation and Polynomial Approximation

A census of the population of the United States is taken every 10 years. The following table lists the population, in thousands of people, from 1930 to 1980.

Year	1930	1940	1950	1960	1970	1980
Population (in thousands)	123,203	131,669	150,697	179,323	203,212	226,505

In reviewing this data, we might ask whether it could be used to reasonably estimate the population, say, in 1965 or even in the year 2000. Some predictions of this type can be obtained by using a function that fits the given data. This is a topic called *interpolation* and is the subject of this chapter.

One of the most useful and well-known classes of functions mapping the set of real numbers into itself is the class of **algebraic polynomials**, i.e., the set of functions of the form

$$P_n(x) = a_0 + a_1 x + \cdots + a_n x^n,$$

where n is a nonnegative integer and $a_0, \ldots, a_n$ are real constants. One major reason for their importance is that they uniformly approximate continuous functions; that is, given any function, defined and continuous on a closed interval, there exists a polynomial that is as "close" to the given function as desired. This result is expressed more precisely in the following theorem. (*See Fig. 3.1.*)

FIGURE 3.1

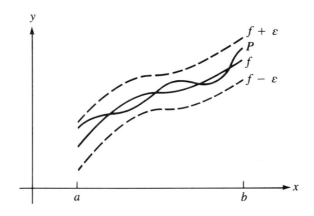

THEOREM 3.1

(Weierstrass Approximation Theorem)   If f is defined and continuous on $[a, b]$ and $\varepsilon > 0$ is given, then there exists a polynomial P, defined on $[a, b]$, with the property that

$$|f(x) - P(x)| < \varepsilon \qquad \text{for all } x \in [a, b].$$

The proof of this theorem can be found in any elementary text on real analysis (see, for example, Bartle [6], pages 165–172).

Other important reasons for considering the class of polynomials in the approximation of functions are that the derivative and indefinite integral of any polynomial are easy to determine and the result is again a polynomial. For these reasons, the class of polynomials is often used for approximating other functions that are known or assumed to be continuous.

3.1 The Taylor Polynomials

The Weierstrass Theorem is very important from a theoretical standpoint, but it cannot be used effectively for computational purposes. In this section we consider the problem of finding a polynomial of a specific degree that is "close" to a given function at a point. A polynomial P agrees with a function f at the number x_0 precisely when

$P(x_0) = f(x_0)$. The polynomial also has the same "direction" as the function f at $(x_0, f(x_0))$ if $P'(x_0) = f'(x_0)$. In a similar manner, the polynomial of nth degree that best approximates the function f near x_0 has as many derivatives as possible at x_0 in agreement with those of f. This is precisely the condition satisfied by the nth degree Taylor polynomial for the function f at x_0:

$$P_n(x) = f(x_0) + f'(x_0)(x - x_0) + f''(x_0)\frac{(x - x_0)^2}{2!} + \cdots + f^{(n)}(x_0)\frac{(x - x_0)^n}{n!},$$

a function with a known remainder, or error term,

$$P_n(x) - f(x) = R_n(x) = \frac{f^{(n+1)}(\xi(x))}{(n+1)!}(x - x_0)^{n+1},$$

for some number $\xi(x)$ between x and x_0.

EXAMPLE 1

a) Calculate the third-degree Taylor polynomial about $x_0 = 0$ for $f(x) = (1 + x)^{1/2}$.
b) Use the polynomial in part (a) to approximate $\sqrt{1.1}$, and find a bound for the error involved.
c) Use the polynomial in part (a) to approximate $\int_0^{0.1} (1 + x)^{1/2}\, dx$, and find a bound for the error of this approximation.

Performing the necessary differentiation yields

$$f(x) = (1 + x)^{1/2}, \quad \text{so } f(0) = 1,$$
$$f'(x) = \tfrac{1}{2}(1 + x)^{-1/2}, \quad \text{so } f'(0) = \tfrac{1}{2},$$
$$f''(x) = -\tfrac{1}{4}(1 + x)^{-3/2}, \quad \text{so } f''(0) = -\tfrac{1}{4},$$
$$f'''(x) = \tfrac{3}{8}(1 + x)^{-5/2}, \quad \text{so } f'''(0) = \tfrac{3}{8},$$
$$f^{(iv)}(x) = -\tfrac{15}{16}(1 + x)^{-7/2}, \quad \text{so } f^{(iv)}(\xi) = -\tfrac{15}{16}(1 + \xi)^{-7/2},$$

where ξ is between zero and x. From Taylor's theorem,

(3.1)
$$P_3(x) = f(0) + f'(0)x + \frac{f''(0)}{2!}x^2 + \frac{f'''(0)}{3!}x^3$$

$$= 1 + \frac{1}{2}x - \frac{1}{4}\cdot\frac{1}{2!}x^2 + \frac{3}{8}\cdot\frac{1}{3!}x^3 = 1 + \frac{1}{2}x - \frac{1}{8}x^2 + \frac{1}{16}x^3,$$

is the third-degree Taylor polynomial requested in part (a).

To answer part (b),

$$\sqrt{1.1} = f(0.1) \approx P_3(0.1) = 1 + \tfrac{1}{2}(0.1) - \tfrac{1}{8}(0.1)^2 + \tfrac{1}{16}(0.1)^3 = 1.0488125.$$

The error is given by $R_3(0.1)$, and a bound is derived as follows:

$$|R_3(0.1)| = \frac{\left|-\tfrac{15}{16}(1 + \xi)^{-7/2}\right|}{4!}(0.1)^4$$

$$\leq \frac{15}{(16)(24)}(0.1)^4 \max_{\xi \in [0, 0.1]} (1 + \xi)^{-7/2} = \frac{0.0005}{128}(1) \leq 3.91 \times 10^{-6}.$$

Since the true value of $\sqrt{1.1}$ is 1.0488088, the actual error is about 3.7×10^{-6}.

The computations for part (c) proceed as follows:

$$\int_0^{0.1} (1 + x)^{1/2} \, dx \approx \int_0^{0.1} P_3(x) \, dx$$

$$= \int_0^{0.1} \left(1 + \frac{x}{2} - \frac{x^2}{8} + \frac{x^3}{16} \right) dx$$

$$= \left[x + \frac{x^2}{4} - \frac{x^3}{24} + \frac{x^4}{64} \right]_0^{0.1} = 0.1024598958,$$

with a remainder given by $\int_0^{0.1} R_3(x) \, dx$. Using techniques similar to those used in part (b),

$$\left| \int_0^{0.1} R_3(x) \, dx \right| = \frac{15}{(16)4!} \int_0^{0.1} (1 + \xi)^{-7/2} x^4 \, dx$$

$$\leq \frac{5}{128} \int_0^{0.1} x^4 \, dx = \frac{5}{128} \cdot \frac{x^5}{5} \bigg]_0^{0.1} \leq 7.82 \times 10^{-8}.$$

Note that the actual remainder term $\int_0^{0.1} R_3(x) \, dx$ is negative, so the integral cannot exceed the approximation. Hence,

$$\int_0^{0.1} P_3(x) \, dx - \int_0^{0.1} |R_3(x)| \, dx \leq \int_0^{0.1} (1 + x)^{1/2} \, dx \leq \int_0^{0.1} P_3(x) \, dx$$

or $$0.1024598176 \leq \int_0^{0.1} (1 + x)^{1/2} \leq 0.1024598958.$$

Since the actual value of $\int_0^{0.1} (1 + x)^{1/2} \, dx$ is 0.102459822, the true error is about 7.4×10^{-8}.   ■

One difficulty associated with approximation by Taylor series is illustrated in the following example.

EXAMPLE 2   The following table lists the values of the third-degree Taylor polynomial (3.1) for the function $f(x) = \sqrt{1 + x}$ considered in the previous example, and the error associated with using this polynomial for various values of x.

x	0.1	0.5	1	2	10
$P_3(x)$	1.048813	1.2266	1.438	2.00	56.00
$f(x)$	1.048809	1.2247	1.414	1.73	3.32
$|P_3(x) - f(x)|$	0.000004	0.0019	0.024	0.27	52.68

Although in some instances better approximations can be obtained if slightly higher-degree Taylor polynomials are used, this is not always the case. Consider, as an extreme example, the problem of using Taylor polynomials of various degrees for $f(x) = 1/x$ expanded about $x_0 = 1$, to approximate $f(3) = \frac{1}{3}$. Since $f(x) = x^{-1}$, $f'(x) = -x^{-2}$, $f''(x) = (-1)^2 2 \cdot x^{-3}$, and in general, $f^{(n)}(x) = (-1)^n n! x^{-n-1}$, the Taylor polynomial for $n \geq 1$ is

$$P_n(x) = \sum_{k=0}^{n} \frac{f^{(k)}(1)}{k!}(x-1)^k = \sum_{k=0}^{n} (-1)^k (x-1)^k.$$

To approximate $f(3) = \frac{1}{3}$ by $P_n(3)$ for increasing values of n, we obtain the following:

n	0	1	2	3	4	5	6	7
$P_n(3)$	1	-1	3	-5	11	-21	43	-85

The reason this approximating technique fails is that the error term,

$$R_n(x) = \frac{(-1)^{n+1}(x-1)^{n+1}}{\xi^{n+2}} \qquad \text{where } 1 < \xi < x,$$

grows in absolute value as n increases. (*See Fig. 3.2.*) The growth in error results because $x = 3$ is not "near enough" to $x_0 = 1$. For a more complete discussion of the difficulties associated with power series, consult a text on analytic function theory, for example, Saff and Snider [112]. ∎

FIGURE 3.2

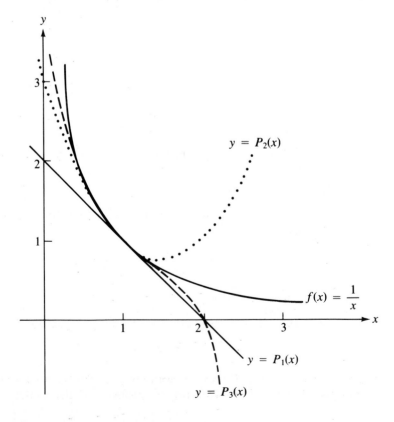

Since the Taylor polynomials have the property that all the information used in the approximation is concentrated at one point, that is, at x_0, the type of difficulty that occurs in Example 2 is quite common. This generally limits the use of approximation by Taylor polynomials to the situation where approximations are needed at points very close to x_0. For ordinary computational purposes, it is more efficient to use methods that include information at various points, and it is the construction of this type of polynomial which will be considered in the remainder of this chapter. The primary use of Taylor polynomials in numerical analysis is not for approximation purposes, but for the derivation of numerical techniques.

Exercise Set 3.1

1. Find the Taylor polynomial of degree 2 for $f(x) = x^2 - 3$ expanded about:

   a) $x_0 = 1$.   b) $x_0 = 0$.

2. Obtain the third-degree Taylor polynomial for $f(x) = (1 + x)^{-2}$ about $x_0 = 0$, and use this polynomial to approximate $f(0.05)$. Find an error bound for this approximation, and compare your result to the exact value of $f(0.05)$.

3. Use the Taylor polynomial in Exercise 2 to approximate $\int_0^{0.05} (1 + x)^{-2} \, dx$. Find an error bound for this approximation, and compare your result with the actual value for this integral.

4. Find the smallest integer n necessary to approximate $f(x) = 1/x$ at $x = 1.25$ with accuracy 10^{-8}, using the Taylor polynomial of degree n about $x_0 = 1$.

5. Use the error term of a Taylor polynomial to estimate the error involved in using $\sin x \approx x$ to approximate $\sin 1°$.

6. Let $f(x) = \ln(1 + x)$. Find the fourth-degree Taylor polynomial for f expanded about $x_0 = 0$, and use it to approximate $\ln(1.1)$. Find a bound for the error in this approximation.

7. Let $F(x) = \int_0^x (1 + t)^{-1} \, dt$. Using the third-degree Taylor polynomial of $f(x) = (1 + x)^{-1}$ expanded about $x_0 = 0$, approximate $F(0.1)$. Compare your results to those obtained in Exercise 6.

Bernstein Polynomials   Given a function f defined on $[0, 1]$, the Bernstein polynomial of degree n for f is given by

$$B_n(x) = \sum_{k=0}^{n} \binom{n}{k} f\left(\frac{k}{n}\right) x^k (1 - x)^{n-k}$$

where $\binom{n}{k}$ denotes $\dfrac{n!}{k!(n-k)!}$.

It can be shown that, if f is continuous on $[0, 1]$ and $x_0 \in [0, 1]$, then

$$\lim_{n \to \infty} B_n(x_0) = f(x_0).$$

The polynomials can be used in a constructive proof of the Weierstrass Theorem (see Bartle [6]). Exercises 8–10 refer to the Bernstein polynomials.

8. Find $B_3(x)$ for the functions

   a) $f(x) = x$, $x \in [0, 1]$.   b) $f(x) = 1$, $x \in [0, 1]$.

9. Show that for each $k \leq n$,

$$\binom{n-1}{k-1} = \left(\frac{k}{n}\right)\binom{n}{k}.$$

10. Use Exercise 9 and the fact that

$$1 = \sum_{k=0}^{n} \binom{n}{k} x^k (1-x)^{n-k} \qquad \text{for each } n,$$

to show that, for $f(x) = x^2$,

$$B_n(x) = \left(\frac{n-1}{n}\right) x^2 + \frac{1}{n} x.$$

11. Using Exercise 10, estimate the value of n necessary for $|B_n(x) - x^2| \leq 10^{-6}$ to hold for all x in $[0, 1]$.

3.2 Interpolation and the Lagrange Polynomial

The previous section discussed approximating polynomials that agree with a given function and some of its derivatives at a single point. These polynomials are useful over small intervals for functions whose derivatives exist and are easily evaluated, but this is clearly not always the case. Consequently, the Taylor polynomial is often of little use, and alternative methods of approximation must be sought. The material in this section is concerned with finding approximating polynomials that can be determined simply by specifying certain points on the plane through which they must pass.

Consider the problem of determining a polynomial of degree 1 that passes through the distinct points (x_0, y_0) and (x_1, y_1). This problem is the same as approximating a function f for which $f(x_0) = y_0$ and $f(x_1) = y_1$, by means of a first-degree polynomial interpolating, or agreeing with, the values of f at the given points.

Consider the polynomial

$$P(x) = \frac{(x - x_1)}{(x_0 - x_1)} y_0 + \frac{(x - x_0)}{(x_1 - x_0)} y_1.$$

When $x = x_0$,

$$P(x_0) = 1 \cdot y_0 + 0 \cdot y_1 = y_0 = f(x_0)$$

and when $x = x_1$,

$$P(x_1) = 0 \cdot y_0 + 1 \cdot y_1 = y_1 = f(x_1),$$

so P has the required properties.

The technique used to construct P is the method of "interpolation" often used in trigonometric or logarithmic tables. What may not be obvious is that P is the only polynomial of degree 1 or less with the interpolating property. This result, however, follows immediately from Corollary 2.16, page 67.

To generalize the concept of linear interpolation, consider the construction of a polynomial of degree at most n that passes through the $n+1$ points $(x_0, f(x_0))$, $(x_1, f(x_1))$, ..., $(x_n, f(x_n))$. (*See Fig. 3.3.*) The linear polynomial passing through $(x_0, f(x_0))$ and $(x_1, f(x_1))$ is constructed by using the quotients

$$L_0(x) = \frac{(x-x_1)}{(x_0-x_1)} \quad \text{and} \quad L_1(x) = \frac{(x-x_0)}{(x_1-x_0)}.$$

When $x=x_0$, $L_0(x_0)=1$ while $L_1(x_0)=0$. When $x=x_1$, $L_0(x_1)=0$ while $L_1(x_1)=1$.

FIGURE 3.3

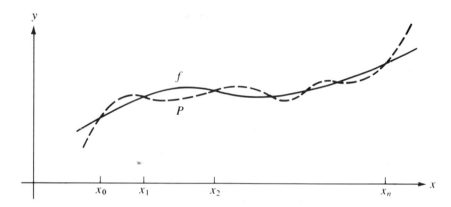

For the general case we need to construct, for each $k=0, 1, ..., n$, a quotient $L_{n,k}(x)$ with the property that $L_{n,k}(x_i)=0$ when $i \neq k$ and $L_{n,k}(x_k)=1$. To satisfy $L_{n,k}(x_i)=0$ for each $i \neq k$ requires that the numerator of $L_{n,k}$ contain the term

(3.2) $$(x-x_0)(x-x_1)\cdots(x-x_{k-1})(x-x_{k+1})\cdots(x-x_n).$$

To satisfy $L_{n,k}(x_k)=1$, the denominator of L_k must be equal to (3.2) when $x=x_k$. Thus,

$$L_{n,k}(x) = \frac{(x-x_0)\cdots(x-x_{k-1})(x-x_{k+1})\cdots(x-x_n)}{(x_k-x_0)\cdots(x_k-x_{k-1})(x_k-x_{k+1})\cdots(x_k-x_n)} = \prod_{\substack{i=0 \\ i \neq k}}^{n} \frac{(x-x_i)}{(x_k-x_i)}.$$

A sketch of the graph of $L_{n,k}$ is shown in Fig. 3.4.

FIGURE 3.4

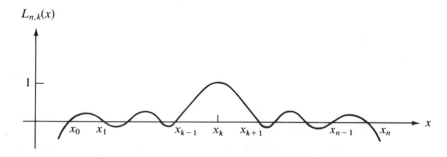

The interpolating polynomial is easily described now that the form of $L_{n,k}$ is known. This polynomial is called the **Lagrange interpolating polynomial** and defined in the following theorem.

THEOREM 3.2   If $x_0, x_1, \ldots, x_n$ are $(n+1)$ distinct numbers and f is a function whose values are given at these numbers, then there exists a unique polynomial P of degree at most n with the property that

$$f(x_k) = P(x_k) \qquad \text{for each } k = 0, 1, \ldots, n.$$

This polynomial is given by

$$(3.3) \qquad P(x) = f(x_0)L_{n,0}(x) + \cdots + f(x_n)L_{n,n}(x) = \sum_{k=0}^{n} f(x_k)L_{n,k}(x),$$

where

$$(3.4) \qquad L_{n,k}(x) = \frac{(x-x_0)(x-x_1)\cdots(x-x_{k-1})(x-x_{k+1})\cdots(x-x_n)}{(x_k-x_0)(x_k-x_1)\cdots(x_k-x_{k-1})(x_k-x_{k+1})\cdots(x_k-x_n)}$$

$$= \prod_{\substack{i=0 \\ i \neq k}}^{n} \frac{(x-x_i)}{(x_k-x_i)} \qquad \text{for each } k = 0, 1, \ldots, n.$$

We will write $L_{n,k}(x)$ simply as $L_k(x)$ when there should be no confusion as to its degree.

EXAMPLE 1   Using the numbers, or nodes, $x_0 = 2$, $x_1 = 2.5$, and $x_2 = 4$ to find the second-degree interpolating polynomial for $f(x) = 1/x$ requires that we first determine the coefficient polynomials L_0, L_1, and L_2:

$$L_0(x) = \frac{(x-2.5)(x-4)}{(2-2.5)(2-4)} = x^2 - 6.5x + 10,$$

$$L_1(x) = \frac{(x-2)(x-4)}{(2.5-2)(2.5-4)} = \tfrac{1}{3}(-4x^2 + 24x - 32),$$

and

$$L_2(x) = \frac{(x-2)(x-2.5)}{(4-2)(4-2.5)} = \tfrac{1}{3}(x^2 - 4.5x + 5).$$

Since $f(x_0) = f(2) = 0.5$, $f(x_1) = f(2.5) = 0.4$, and $f(x_2) = f(4) = 0.25$,

$$P(x) = \sum_{k=0}^{2} f(x_k)L_k(x)$$

$$= 0.5(x^2 - 6.5x + 10) + \frac{0.4}{3}(-4x^2 + 24x - 32)$$

$$+ \frac{0.25}{3}(x^2 - 4.5x + 5)$$

$$= 0.05x^2 - 0.425x + 1.15$$

$$= (0.05x - 0.425)x + 1.15.$$

An approximation to $f(3) = \frac{1}{3}$ is

$$f(3) \approx P(3) = 0.325.$$

Compare this example to Example 2 of Section 3.1 where no Taylor polynomial (expanded about $x_0 = 1$) could be used to reasonably approximate $f(3) = \frac{1}{3}$. (*See Fig. 3.5.*) ∎

FIGURE 3.5

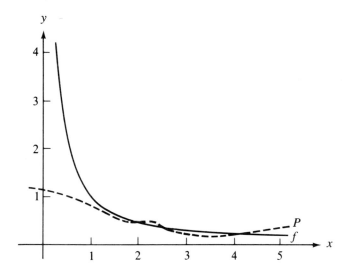

The next step is to calculate a remainder term or bound for the error involved in approximating a function by an interpolating polynomial. This is done in the following theorem.

THEOREM 3.3

If $x_0, x_1, \ldots, x_n$ are distinct numbers in the interval $[a, b]$ and if $f \in C^{n+1}[a, b]$, then, for each x in $[a, b]$, a number $\xi(x)$ in (a, b) exists with

(3.5)
$$f(x) = P(x) + \frac{f^{(n+1)}(\xi(x))}{(n+1)!}(x - x_0)(x - x_1) \cdots (x - x_n),$$

where P is the interpolating polynomial given in Eq. (3.3).

PROOF

Note first that, if $x = x_k$ for $k = 0, 1, \ldots, n$, then $f(x_k) = P(x_k)$ and choosing $\xi(x_k)$ arbitrarily in (a, b) yields Eq. (3.5). If $x \neq x_k$ for any $k = 0, 1, \ldots, n$, define the function g for t in $[a, b]$ by

$$g(t) = f(t) - P(t) - [f(x) - P(x)] \frac{(t - x_0)(t - x_1) \cdots (t - x_n)}{(x - x_0)(x - x_1) \cdots (x - x_n)}$$

$$= f(t) - P(t) - [f(x) - P(x)] \prod_{i=0}^{n} \frac{(t - x_i)}{(x - x_i)}.$$

Since $f \in C^{n+1}[a, b]$, $P \in C^{\infty}[a, b]$, and $x \neq x_k$ for any k, it follows that $g \in C^{n+1}[a, b]$. For $t = x_k$

$$g(x_k) = f(x_k) - P(x_k) - [f(x) - P(x)] \prod_{i=0}^{n} \frac{(x_k - x_i)}{(x - x_i)}$$

$$= 0 - [f(x) - P(x)] \cdot 0$$

$$= 0.$$

Moreover, $$g(x) = f(x) - P(x) - [f(x) - P(x)] \prod_{i=0}^{n} \frac{(x - x_i)}{(x - x_i)}$$

$$= f(x) - P(x) - [f(x) - P(x)]$$

$$= 0.$$

Thus, $g \in C^{n+1}[a, b]$ and g vanishes at the $n + 2$ distinct numbers $x, x_0, x_1, \ldots, x_n$. By the Generalized Rolle's Theorem (Theorem 1.11, p. 5), there exists $\xi \equiv \xi(x)$ in (a, b) for which $g^{(n+1)}(\xi) = 0$. Evaluating $g^{(n+1)}$ at ξ gives

(3.6) $$0 = g^{(n+1)}(\xi) = f^{(n+1)}(\xi) - P^{(n+1)}(\xi) - [f(x) - P(x)] \frac{d^{n+1}}{dt^{n+1}} \left(\prod_{i=0}^{n} \frac{(t - x_i)}{(x - x_i)} \right) \Bigg|_{t=\xi}.$$

Since P is a polynomial of degree at most n, the $(n + 1)$st derivative, $P^{(n+1)}$, must be identically zero. Also, $\prod_{i=0}^{n} [(t - x_i)/(x - x_i)]$ is a polynomial of degree $(n + 1)$, so

$$\prod_{i=0}^{n} \frac{(t - x_i)}{(x - x_i)} = \left(\frac{1}{\prod_{i=0}^{n} (x - x_i)} \right) t^{n+1} + \text{(lower-degree terms in } t),$$

and $$\frac{d^{n+1}}{dt^{n+1}} \prod_{i=0}^{n} \frac{(t - x_i)}{(x - x_i)} = \frac{(n + 1)!}{\prod_{i=0}^{n} (x - x_i)}.$$

Equation (3.6) now becomes

$$0 = f^{(n+1)}(\xi) - 0 - [f(x) - P(x)] \frac{(n + 1)!}{\prod_{i=0}^{n} (x - x_i)}$$

and, upon solving for $f(x)$,

(3.7) $$f(x) = P(x) + \frac{f^{(n+1)}(\xi)}{(n + 1)!} \prod_{i=0}^{n} (x - x_i). \qquad \square$$

The error formula derived in Theorem 3.3 is an important theoretical result because Lagrange polynomials are used extensively for deriving numerical differentiation and integration methods. Error bounds for these techniques are obtained from the Lagrange error formula.

Note that the error form for the Lagrange polynomial is quite similar to that for the Taylor polynomial. The Taylor polynomial of degree n about x_0 concentrates all the known information at x_0 and has an error term of the form

$$\frac{f^{(n+1)}(\xi(x))}{(n + 1)!} (x - x_0)^{n+1}.$$

The Lagrange polynomial of degree n uses information at the distinct numbers $x_0, x_1, \ldots, x_n$ and, in place of $(x - x_0)^n$, its error formula uses a product of the $n + 1$ terms $(x - x_0), (x - x_1), \ldots, (x - x_n)$:

$$\frac{f^{(n+1)}(\xi(x))}{(n+1)!}(x-x_0)(x-x_1)\cdots(x-x_n).$$

Although the error formula in Eq. (3.5) is an important theoretical result, its practical use is restricted to those functions whose derivatives have known bounds. (This is often the case for tabulated functions such as the trigonometric or logarithmic functions.)

EXAMPLE 2   Suppose a table is to be prepared for the function $f(x) = e^x$, $0 \le x \le 1$. Assume the number of decimal places to be given per entry will be d; for example, if $d = 5$, then $f(1) = 2.71828$, and that the difference between adjacent x-values, the step size, is h.

a) Assuming $d \ge 6$, what should h be for linear interpolation (i.e., the Lagrange polynomial of degree 1) to give an absolute error of at most 10^{-6}?

b) If $d < 6$, what should h be for linear interpolation to give an absolute error of at most 10^{-6}?

Let $x \in [0, 1]$ and suppose j is such that $x_j \le x \le x_{j+1}$. From Eq. (3.5), the error in linear interpolation is

$$|f(x) - P(x)| = \left|\frac{f^{(2)}(\xi)}{2!}(x-x_j)(x-x_{j+1})\right| = \frac{|f^{(2)}(\xi)|}{2}|(x-x_j)||(x-x_{j+1})|.$$

Since the step size is h, it follows that $x_j = jh$, $x_{j+1} = (j+1)h$, and

$$|f(x) - P(x)| \le \frac{|f^{(2)}(\xi)|}{2!}|(x-jh)(x-(j+1)h)|.$$

Hence

$$|f(x) - P(x)| \le \tfrac{1}{2} \max_{\xi \in [0,1]} |f^{(2)}(\xi)| \max_{x_j \le x \le x_{j+1}} |(x-jh)(x-(j+1)h)|$$

$$= \tfrac{1}{2} \max_{\xi \in [0,1]} e^{\xi} \max_{x_j \le x \le x_{j+1}} |(x-jh)(x-(j+1)h)|$$

$$\le \tfrac{1}{2}e \max_{x_j \le x \le x_{j+1}} |(x-jh)(x-(j+1)h)|.$$

By considering $g(x) = (x-jh)(x-(j+1)h)$ for $jh \le x \le (j+1)h$ and using techniques of calculus (see Exercise 14), it can be shown that

(3.8) $$\max_{x_j \le x \le x_{j+1}} |g(x)| = |g((j+\tfrac{1}{2})h)| = \frac{h^2}{4}.$$

Consequently, the error in linear interpolation is bounded by

$$|f(x) - P(x)| \le \frac{eh^2}{8}.$$

To answer part (a), it is sufficient for h to be chosen so that

$$\frac{eh^2}{8} \le 10^{-6}, \quad h^2 \le \frac{8}{e} \cdot 10^{-6}, \quad h^2 < 2.944 \times 10^{-6}, \quad \text{or} \quad h < 1.72 \times 10^{-3}.$$

Letting $h = 0.001$ would be one logical choice for the step size.

The reason we distinguish between parts (a) and (b) is to emphasize that, if the tables are accurate only to the fifth decimal place, it is impossible to obtain accurate values to the sixth place via interpolation. The errors in rounding to the fifth decimal place will remain, so there is no answer for part (b). ∎

The last example of this section illustrates interpolation techniques for a situation when the error portion of Eq. (3.5) cannot be used. This example also serves to illustrate that we should look for a more efficient way to obtain an approximation via interpolation.

EXAMPLE 3   Table 3.1 lists values of a function (the Bessel function of the first kind of order zero) at various points. The approximations to $f(1.5)$ obtained by various Lagrange polynomials will be compared.

TABLE 3.1

x	$f(x)$
1.0	0.7651977
1.3	0.6200860
1.6	0.4554022
1.9	0.2818186
2.2	0.1103623

Since 1.5 is between 1.3 and 1.6, the linear polynomial will use $x_0 = 1.3$ and $x_1 = 1.6$. The value of the interpolating polynomial at 1.5 is given by

$$P_1(1.5) = \frac{(1.5 - 1.6)}{(1.3 - 1.6)}(0.6200860) + \frac{(1.5 - 1.3)}{(1.6 - 1.3)}(0.4554022) = 0.5102968.$$

Two polynomials of degree two could reasonably be used, one by letting $x_0 = 1.3$, $x_1 = 1.6$, and $x_2 = 1.9$, which gives

$$P_2(1.5) = \frac{(1.5 - 1.6)(1.5 - 1.9)}{(1.3 - 1.6)(1.3 - 1.9)}(0.6200860) + \frac{(1.5 - 1.3)(1.5 - 1.9)}{(1.6 - 1.3)(1.6 - 1.9)}(0.4554022)$$

$$+ \frac{(1.5 - 1.3)(1.5 - 1.6)}{(1.9 - 1.3)(1.9 - 1.6)}(0.2818186)$$

$$= 0.5112857$$

and the other by letting $x_0 = 1.0$, $x_1 = 1.3$, and $x_2 = 1.6$, in which case

$$\hat{P}_2(1.5) = 0.5124715.$$

In the third-degree case there are also two choices for the polynomial. One is with $x_0 = 1.3$, $x_1 = 1.6$, $x_2 = 1.9$, and $x_3 = 2.2$, which gives

$$P_3(1.5) = 0.5118302.$$

The other is obtained by letting $x_0 = 1.0$, $x_1 = 1.3$, $x_2 = 1.6$, and $x_3 = 1.9$, giving

$$\hat{P}_3(1.5) = 0.5118127.$$

The fourth-degree Lagrange polynomial uses all the entries in the table. With $x_0 = 1.0$, $x_1 = 1.3$, $x_2 = 1.6$, $x_3 = 1.9$, and $x_4 = 2.2$, it can be shown that

$$P_4(1.5) = 0.5118200.$$

Since $P_3(1.5)$, $\hat{P}_3(1.5)$, and $P_4(1.5)$ all agree to within 2×10^{-5} units, we expect $P_4(1.5)$ to be the most accurate approximation and to be correct to within 2×10^{-5} units.

The actual value of $f(1.5)$ is known to be 0.5118277, so the true accuracies of the approximations are as follows:

$$|P_1(1.5) - f(1.5)| \approx 1.53 \times 10^{-3},$$
$$|P_2(1.5) - f(1.5)| \approx 5.42 \times 10^{-4},$$
$$|\hat{P}_2(1.5) - f(1.5)| \approx 6.44 \times 10^{-4},$$
$$|P_3(1.5) - f(1.5)| \approx 2.5 \times 10^{-6},$$
$$|\hat{P}_3(1.5) - f(1.5)| \approx 1.50 \times 10^{-5},$$
$$|P_4(1.5) - f(1.5)| \approx 7.7 \times 10^{-6}.$$

P_3 is the most accurate approximation. However, with no knowledge of the actual value of $f(1.5)$, P_4 would be accepted as the best approximation. Note that the error or remainder term derived in Theorem 3.3 cannot be applied here, since no knowledge of the fourth derivative of f is available. Unfortunately, this is generally the case. ■

Exercise Set 3.2

1. Use appropriate Lagrange interpolating polynomials of degree one, two, three, and four to approximate

    a) $f(2.5)$ if $f(2.0) = 0.5103757$, $f(2.2) = 0.5207843$, $f(2.4) = 0.5104147$, $f(2.6) = 0.4813306$, $f(2.8) = 0.4359160$.

    b) $f(0)$ if $f(-0.3) = -0.20431$, $f(-0.1) = -0.08993$, $f(0.1) = 0.11007$, $f(0.3) = 0.39569$, $f(0.5) = 0.79845$.

    c) $f(1.25)$ if $f(1.0) = 0.24255$, $f(1.1) = 0.48603$, $f(1.2) = 0.86160$, $f(1.3) = 1.59751$, $f(1.4) = 3.76155$.

    d) $f(0.5)$ if $f(0.2) = 0.9798652$, $f(0.4) = 0.9177710$, $f(0.6) = 0.8080348$, $f(0.8) = 0.6386093$, $f(1.0) = 0.3843735$.

    e) $f(0.2)$ if $f(0.1) = 1.2314028$, $f(0.3) = 1.9121188$, $f(0.4) = 2.3855409$, $f(0.5) = 2.9682818$, $f(0.6) = 3.6801169$.

2. Use the following values to construct a Lagrange polynomial of degree two or less. Find an approximation to sin 0.34 and use Eq. (3.5) to determine an error bound for the approximation.

$$\sin 0.30 = 0.29552 \qquad \sin 0.32 = 0.31457 \qquad \sin 0.35 = 0.34290$$

3. Add the value $\sin 0.33 = 0.32404$ to the data in Exercise 2 and construct a Lagrange polynomial of degree three or less. Approximate sin 0.34 and find a bound for the error.

4. Let $f(x) = 3xe^x - 2e^x$. Approximate $f(1.03)$ using the interpolating polynomial of degree two or less, using $x_0 = 1$, $x_1 = 1.05$, and $x_2 = 1.07$. Compare the actual error to the error bound obtained from Eq. (3.5).

5. Use the following values to construct a third-degree Lagrange polynomial approximation to $f(1.09)$. The function being approximated is $f(x) = \log_{10} \tan x$. Use this knowledge to find a bound for the error in the approximation.

$$f(1.00) = 0.1924 \qquad f(1.05) = 0.2414$$

$$f(1.10) = 0.2933 \qquad f(1.15) = 0.3492$$

6. Use the Lagrange interpolating polynomial of degree three or less to approximate cos 0.750 using the values below. Find an error bound using Eq. (3.5).

$$\cos 0.698 = 0.7661 \qquad \cos 0.768 = 0.7193$$

$$\cos 0.733 = 0.7432 \qquad \cos 0.803 = 0.6946$$

The actual value of cos 0.750 is 0.7317 (to four decimal places). If there is a discrepancy between the actual error and your error bound, explain why this occurred.

7. Use the following values to construct a fourth-degree Lagrange polynomial approximation to $f(1.25)$. The function being approximated is $f(x) = e^{x^2 - 1}$. Use this knowledge to find a bound for the error in the approximation.

$$f(1.0) = 1.00000 \qquad f(1.2) = 1.55271 \qquad f(1.4) = 2.61170$$

$$f(1.1) = 1.23368 \qquad f(1.3) = 1.99372$$

8. Let $f(x) = (4x - 7)/(x - 2)$ and $x_0 = 1.7$, $x_1 = 1.8$, $x_2 = 1.9$, and $x_3 = 2.1$.

    a) Approximate $f(1.75)$ using the interpolating polynomial of degree at most two on the nodes $x_0, x_1,$ and x_2.

    b) Approximate $f(1.75)$ and $f(2.00)$ using the interpolating polynomial on $x_0, x_1, x_2,$ and x_3.

    c) Can an error bound from Eq. (3.5) be applied to part (a) or (b)? What does the bound give as an error estimate?

9. Let $f(x) = e^x$, $0 \le x \le 2$. Using the values given, perform the following computation:

    a) Approximate $f(0.25)$ using linear interpolation with $x_0 = 0$ and $x_1 = 0.5$.

    b) Approximate $f(0.75)$ using linear interpolation with $x_0 = 0.5$ and $x_1 = 1$.

    c) Approximate $f(0.25)$ and $f(0.75)$ by using the second-degree interpolating polynomial with $x_0 = 0$, $x_1 = 1$, and $x_2 = 2$.

d)   Which approximations are better? Why?

x	0.0	0.5	1.0	2.0
$f(x)$	1.00000	1.64872	2.71828	7.38906

10.   Suppose it is desired to construct six-place tables for the common or base-10 logarithm function from $x = 1$ to $x = 10$ in such a way that linear interpolation is accurate to the sixth decimal place. Determine the largest possible step size for this table.

11.   Show that

$$\sum_{k=0}^{n} L_k(x) = 1 \qquad \text{for all } x.$$

(*Hint*:   Consider $f(x) \equiv 1$.)

12.   Let $\omega(x) = \prod_{k=0}^{n} (x - x_k)$. Show that the interpolating polynomial of degree n on $x_0, \ldots, x_n$ can be written as

$$P(x) = \omega(x) \sum_{k=0}^{n} \frac{f(x_k)}{(x - x_k)\omega'(x_k)}.$$

13.   Prove Theorem 1.13, page 6. [*Hint*:   Let

$$g(t) = f(t) - P(t) - [f(x) - P(x)] \cdot \frac{(t - x_0)^n}{(x - x_0)^n},$$

where P is the nth-degree Taylor polynomial, and use Theorem 1.11, page 5.]

14.   Show that

$$\max_{x_j \le x \le x_{j+1}} |g(x)| = \frac{h^2}{4}$$

where $g(x) = (x - jh)(x - (j+1)h)$. This establishes the result in Eq. (3.8).

15.   In the introduction to this chapter, the following table was given, listing the population of the United States from 1930 to 1980.

Year	1930	1940	1950	1960	1970	1980
Population (in thousands)	123,203	131,669	150,697	179,323	203,212	226,505

Find the Lagrange polynomial of degree 5 fitting this data, and use this polynomial to estimate the population in the years 1920, 1965, and 2000. The population in 1920 was approximately 105,711,000. How accurate do you think your 1965 and 2000 figures are?

3.3 Iterated Interpolation

One difficulty that arises in using the method of Section 3.2 is that since the error term given by Theorem 3.3 is difficult to work with, the degree of the polynomial needed for the desired accuracy is generally not known until the computations have been completed. The usual practice is to compute the results given from various polynomials until appropriate agreement is obtained. This was done, for instance, in Example 3. In this example, the work done in calculating the approximation by the second-degree polynomial does not lessen the work needed to calculate the third-degree approximation; nor is the fourth-degree approximation easier to obtain once the third-degree approximation is known. The purpose of this section is to derive these approximating polynomials in a manner that uses the previous calculations to the greatest advantage.

DEFINITION 3.4   Let f be a function defined at $x_0, x_1, x_2, \ldots, x_n$, and suppose that $m_1, m_2, \ldots, m_k$ are k distinct integers with $0 \le m_i \le n$ for each i. The Lagrange polynomial of degree less than k that agrees with f at the k points $x_{m_1}, x_{m_2}, \ldots, x_{m_k}$ is denoted by $P_{m_1, m_2, \ldots, m_k}$.

EXAMPLE 1   If $x_0 = 1$, $x_1 = 2$, $x_2 = 3$, $x_3 = 4$, $x_4 = 6$, and $f(x) = x^3$, then $P_{1,2,4}$ is the polynomial that agrees with f at $x_1 = 2$, $x_2 = 3$, and $x_4 = 6$; that is,

$$P_{1,2,4}(x) = \frac{(x-3)(x-6)}{(2-3)(2-6)}(8) + \frac{(x-2)(x-6)}{(3-2)(3-6)}(27) + \frac{(x-2)(x-3)}{(6-2)(6-3)}(216). \qquad \blacksquare$$

The next result describes a method for recursively generating Lagrange polynomial approximations.

THEOREM 3.5   Let f be defined at $x_0, x_1, \ldots, x_k$ and x_j, x_i be two distinct numbers in this set. If

$$(3.9) \qquad P(x) = \frac{(x - x_j)P_{0,1,\ldots,j-1,j+1,\ldots,k}(x) - (x - x_i)P_{0,1,\ldots,i-1,i+1,\ldots,k}(x)}{(x_i - x_j)},$$

then P is the Lagrange polynomial of degree less than or equal to k which interpolates f at the $k + 1$ points $x_0, x_1, \ldots, x_k$.

PROOF   For ease of notation, let $Q \equiv P_{0,1,\ldots,i-1,i+1,\ldots,k}$ and $\hat{Q} \equiv P_{0,1,\ldots,j-1,j+1,\ldots,k}$. Q and $\hat{Q}$ are polynomials of degree $k - 1$ or less; hence, P must be of degree at most k. If $0 \le r \le k$ and $r \ne i, j$, then $Q(x_r) = \hat{Q}(x_r) = f(x_r)$, so

$$P(x_r) = \frac{(x_r - x_j)\hat{Q}(x_r) - (x_r - x_i)Q(x_r)}{x_i - x_j} = \frac{(x_i - x_j)}{(x_i - x_j)}f(x_r) = f(x_r).$$

Moreover, $$P(x_i) = \frac{(x_i - x_j)\hat{Q}(x_i) - (x_i - x_i)Q(x_i)}{x_i - x_j} = \frac{(x_i - x_j)}{(x_i - x_j)}f(x_i) = f(x_i),$$

and similarly $P(x_j) = f(x_j)$. But, by definition, $P_{0,1,\ldots,k}$ is the unique polynomial of degree at most k which agrees with f at $x_0, x_1, \ldots, x_k$. Thus, $P \equiv P_{0,1,\ldots,k}$. $\square$

EXAMPLE 2   In Example 3 of Section 3.2, values of various Lagrange polynomials at $x = 1.5$ were obtained, using the data in Table 3.2.

In this example we calculate the approximation of $f(1.5)$ using Theorem 3.5. If $x_0 = 1.0$, $x_1 = 1.3$, $x_2 = 1.6$, $x_3 = 1.9$, $x_4 = 2.2$, the notation of Definition 3.4 implies

TABLE 3.2

	$n_{,}0$	$n_{,}1$	
	x	$f(x)$	
0	1.0	0.7651977	
1	1.3	0.6200860	
2	1.6	0.4554022	
3	1.9	0.2818186	
4	2.2	0.1103623	

that $f(1.0) = P_0$, $f(1.3) = P_1$, $f(1.6) = P_2$, $f(1.9) = P_3$, and $f(2.2) = P_4$; so these are the five polynomials of degree zero (constants) that approximate $f(1.5)$.

Calculating $P_{0,1}(1.5)$ yields

$$P_{0,1}(1.5) = \frac{(1.5 - 1.0)P_1 - (1.5 - 1.3)P_0}{(1.3 - 1.0)}$$

$$= \frac{0.5(0.6200860) - 0.2(0.7651977)}{0.3} = 0.5233449.$$

Similarly,

$$P_{1,2}(1.5) = \frac{(1.5 - 1.3)(0.4554022) - (1.5 - 1.6)(0.6200860)}{(1.6 - 1.3)} = 0.5102968,$$

$$P_{2,3}(1.5) = 0.5132634, \quad \text{and} \quad P_{3,4}(1.5) = 0.5104270.$$

These give the approximations using first-degree polynomials. $P_{1,2}$ is expected to be the best approximation since 1.5 is between $x_1 = 1.3$ and $x_2 = 1.6$.

The approximations using second-degree polynomials are given by

$$P_{0,1,2}(1.5) = \frac{(1.5 - 1.0)(0.5102968) - (1.5 - 1.6)(0.5233449)}{(1.6 - 1.0)} = 0.5124715,$$

$$P_{1,2,3}(1.5) = 0.5112857, \quad \text{and} \quad P_{2,3,4}(1.5) = 0.5137361.$$

The higher-degree approximations are generated in a similar manner and are listed in Table 3.4 in the form shown in Table 3.3.

TABLE 3.3

x_0	P_0				
x_1	P_1	$P_{0,1}$			
x_2	P_2	$P_{1,2}$	$P_{0,1,2}$		
x_3	P_3	$P_{2,3}$	$P_{1,2,3}$	$P_{0,1,2,3}$	
x_4	P_4	$P_{3,4}$	$P_{2,3,4}$	$P_{1,2,3,4}$	$P_{0,1,2,3,4}$

TABLE 3.4

1.0	0.7651977				
1.3	0.6200860	0.5233449			
1.6	0.4554022	0.5102968	0.5124715		
1.9	0.2818186	0.5132634	0.5112857	0.5118127	
2.2	0.1103623	0.5104270	0.5137361	0.5118302	0.5118200

Suppose that at this point it is decided that the latest approximation $P_{0,1,2,3,4}$ is not as accurate as desired. Another node x_5 could be selected, another row added to the table;

$$x_5 \quad P_5 \quad P_{4,5} \quad P_{3,4,5} \quad P_{2,3,4,5} \quad P_{1,2,3,4,5} \quad P_{0,1,2,3,4,5},$$

and $P_{0,1,2,3,4}$, $P_{1,2,3,4,5}$, and $P_{0,1,2,3,4,5}$ could be compared to determine further accuracy.

In the example we have been considering, the value of the Bessel function of the first kind of order zero at 2.5 is -0.0483838. Using this to construct the new row gives:

2.5 -0.0483838 0.4807699   0.5301984   0.5119070   0.5118430   0.5118277.

The new entry is correct to six decimal places.   ■

The procedure just outlined is called **Neville's method**. The notation used in Table 3.3 is cumbersome because of the number of subscripts used to represent the entries. Note, however, that as an array is being constructed, only two subscripts are actually needed. Proceeding down the table corresponds to using consecutive points x_i with larger i, and proceeding to the right corresponds to increasing the degree of the interpolating polynomial. Since the points appear consecutively in each entry, we need to describe only a starting point and the number of additional points used in constructing the approximation.

Let $Q_{i,j}$, $i \geq j$, denote the interpolating polynomial of degree j on the $(j+1)$ numbers $x_{i-j}, x_{i-j+1}, \ldots, x_{i-1}, x_i$. To compute

$$Q_{i,j} = P_{i-j,i-j+1,\ldots,i-1,i}$$

by Neville's method use

$$Q_{i,j-1} = P_{i-j+1,\ldots,i-1,i} \quad \text{and} \quad Q_{i-1,j-1} = P_{i-j,i-j+1,\ldots,i-1}$$

in Eq. (3.9) to obtain:

$$Q_{i,j}(x) = \frac{(x - x_{i-j})Q_{i,j-1}(x) - (x - x_i)Q_{i-1,j-1}(x)}{x_i - x_{i-j}}$$

for each $j = 1, 2, 3, \ldots$, and $i = j, j+1, \ldots$. Additionally, let $Q_{i,0} = f(x_i)$ for each i. This notation for Neville's method provides the array in Table 3.5. This table and Table 3.3

TABLE 3.5

x_0	$Q_{0,0}$				
x_1	$Q_{1,0}$	$Q_{1,1}$			
x_2	$Q_{2,0}$	$Q_{2,1}$	$Q_{2,2}$		
x_3	$Q_{3,0}$	$Q_{3,1}$	$Q_{3,2}$	$Q_{3,3}$	
x_4	$Q_{4,0}$	$Q_{4,1}$	$Q_{4,2}$	$Q_{4,3}$	$Q_{4,4}$

involving the P's are the same, except for the notation, but Table 3.5 is much easier to set up for computer utilization.

The following algorithm constructs the entries in Table 3.5 by rows.

Neville's Iterated Interpolation Algorithm 3.1

To evaluate the interpolating polynomial P on the $(n + 1)$ distinct numbers $x_0, \ldots, x_n$ at the number x for the function f:

INPUT numbers $x_0, x_1, \ldots, x_n$; values $f(x_0), f(x_1), \ldots, f(x_n)$ as the first column $Q_{0,0}$, $Q_{1,0}, \ldots, Q_{n,0}$ of Q.

OUTPUT the table Q with $P(x) = Q_{n,n}$.

Step 1   For $i = 1, 2, \ldots, n$
         for $j = 1, 2, \ldots, i$

$$\text{set } Q_{i,j} = \frac{(x - x_{i-j})Q_{i,j-1} - (x - x_i)Q_{i-1,j-1}}{x_i - x_{i-j}}.$$

Step 2   OUTPUT (Q);
         STOP.

The algorithm can be modified to allow for the addition of new interpolating nodes. For example, the inequality

$$|Q_{i,i} - Q_{i-1,i-1}| < \varepsilon,$$

could be used as a stopping criterion, where ε is a prescribed error tolerance. If the inequality is true, $Q_{i,i}$ is a reasonable approximation to $f(x)$. If the inequality is false, a new interpolation point x_{i+1} is added.

Exercise Set 3.3

1. Use Neville's method to obtain the approximations for Exercise 1 of Section 3.2.

2. Approximate $\sqrt{3}$ using Neville's method with the function $f(x) = 3^x$ and the values $x_0 = -2$, $x_1 = -1$, $x_2 = 0$, $x_3 = 1$, and $x_4 = 2$.

3. Use Neville's method to approximate $f(-0.78)$ for the function $f(x) = x^2 e^x \cos x$ using $x_0 = -1.0$, $x_1 = -0.9$, $x_2 = -0.8$, $x_3 = -0.7$, and $x_4 = -0.6$.

4. Use Neville's method to approximate $f(1.09)$ for the data given in Exercise 5 of Section 3.2.

5. Use Neville's method to approximate $f(1.25)$ for the data given in Exercise 7 of Section 3.2.

6. a)   Use Neville's method to approximate $f(1.03)$ with $P_{0,1,2}$ for the function $f(x) = 3xe^x - e^{2x}$ using $x_0 = 1$, $x_1 = 1.05$, and $x_2 = 1.07$.

b) Suppose the approximation of (a) is not sufficiently accurate. Compute $P_{0,1,2,3}$ where $x_3 = 1.04$.

7. Repeat Exercise 6 using four-digit arithmetic. Do you think Neville's method is sensitive to rounding errors?

8. **Aitken's method**, which is similar to Neville's method, constructs the following table of interpolating values:

x_0	P_0				
x_1	P_1	$P_{0,1}$			
x_2	P_2	$P_{0,2}$	$P_{0,1,2}$		
x_3	P_3	$P_{0,3}$	$P_{0,1,3}$	$P_{0,1,2,3}$	
x_4	P_4	$P_{0,4}$	$P_{0,1,4}$	$P_{0,1,2,4}$	$P_{0,1,2,3,4}$
$\vdots$	$\vdots$	$\vdots$	$\vdots$	$\vdots$	$\vdots$

To compute each new value, use the value at the top of the preceding column with the value in the same row, preceding column; for example,

$$P_{0,1,3}(x) = \frac{(x - x_3)P_{0,1}(x) - (x - x_1)P_{0,3}(x)}{x_1 - x_3}.$$

Using the same notation $Q_{i,j}$ as in Neville's method, construct the algorithm to compute $Q_{i,j}$ for Aitken's method.

9. Approximate $\sqrt{3}$ using Aitken's method on the function $f(x) = 3^x$ for the nodes $x_0 = -2$, $x_1 = -1$, $x_2 = 0$, $x_3 = 1$, $x_4 = 2$. Compare your result with the result of Exercise 2. Which method do you prefer?

10. Repeat Exercise 4 using Aitken's method instead of Neville's method. Compare your approximations to the exact value $f(1.09) = \log_{10} \tan(1.09)$.

11. Repeat Exercise 5 using Aitken's method instead of Neville's method. Compare your approximations to the exact value $f(1.25) = e^{(1.25)^2 - 1}$.

12. Construct a sequence of interpolating values, y_n, to $f(1 + \sqrt{10})$, where $f(x) = (1 + x^2)^{-1}$ for $-5 \leq x \leq 5$, as follows: For each $n = 1, 2, \ldots, 10$, let $h = 10/n$ and $y_n = P_n(1 + \sqrt{10})$, where $P_n(x)$ is the interpolating polynomial for $f(x)$ at the nodes $x_0^{(n)}, x_1^{(n)}, \ldots, x_n^{(n)}$ and $x_j^{(n)} = -5 + jh$ for each $j = 0, 1, 2, \ldots, n$. Does the sequence $\{y_n\}$ seem to converge to $f(1 + \sqrt{10})$?

Inverse Interpolation   Suppose $f \in C^1[a, b]$, $f'(x) \neq 0$ on $[a, b]$ and f has one zero p in $[a, b]$. Let $x_0, \ldots, x_n$ be $n + 1$ distinct numbers in $[a, b]$ with $f(x_k) = y_k$ for each $k = 0, 1, \ldots, n$. To approximate p, construct the interpolating polynomial of degree n on the nodes $y_0, \ldots, y_n$ for f^{-1}. Since $y_k = f(x_k)$ and $0 = f(p)$, it follows that $f^{-1}(y_k) = x_k$ and $p = f^{-1}(0)$. Using iterated interpolation to approximate $f^{-1}(0)$ is called *iterated inverse interpolation*.

13. Use iterated inverse interpolation to find an approximation to the solution of $x - e^{-x} = 0$, using the data

x	0.3	0.4	0.5	0.6
e^{-x}	0.740818	0.670320	0.606531	0.548812

14. Construct an algorithm that can be used for inverse interpolation.

3.4 Divided Differences

The iterated interpolation techniques discussed in the previous section are useful for determining the values of successively higher-degree interpolating polynomials at a particular point. Each of the entries in the interpolation table, however, depends on the point being evaluated, so the table cannot be used to give an explicit representation for the interpolating polynomial.

Methods for determining the explicit representation of an interpolating polynomial from tabulated data are known as **divided-difference methods.** These methods were more widely used for computational purposes before digital computing equipment became readily available. However, the methods can be used to derive techniques for approximating the derivatives and integrals of functions, as well as for approximating the solutions to differential equations.

Our treatment of divided-difference methods will be brief since the results in this section will not be used extensively in subsequent material. Most classical texts on the subject of numerical analysis have extensive treatments of divided-difference methods. If a more comprehensive treatment is needed, the text by Hildebrand [64] is a particularly good reference.

Suppose that P_n is the Lagrange polynomial of degree at most n that agrees with the function f at the distinct numbers $x_0, x_1, \ldots, x_n$. The divided differences of f with respect to $x_0, x_1, \ldots, x_n$ can be derived by showing that P_n has the representation

$$(3.10) \qquad P_n(x) = a_0 + a_1(x - x_0) + a_2(x - x_0)(x - x_1) + \cdots$$
$$+ a_n(x - x_0)(x - x_1) \cdots (x - x_{n-1})$$

for appropriate constants $a_0, a_1, \ldots, a_n$.

To determine the first of these constants, a_0, note that if $P_n(x)$ can be written in the form of Eq. (3.10), then evaluating P_n at x_0 leaves only the constant term a_0; that is, $a_0 = P_n(x_0) = f(x_0)$.

Similarly, when P_n is evaluated at x_1, the only nonzero terms in the evaluation of $P_n(x_1)$ are the constant and linear terms,

$$f(x_0) + a_1(x_1 - x_0) = P_n(x_1) = f(x_1);$$

so

$$(3.11) \qquad a_1 = \frac{f(x_1) - f(x_0)}{x_1 - x_0}.$$

At this stage we introduce what is known as the **divided-difference notation**. The zeroth divided difference of the function f, with respect to x_i, is denoted $f[x_i]$ and is simply the evaluation of f at x_i,

$$f[x_i] = f(x_i).$$

The remaining divided differences are defined inductively; the first divided difference of f with respect to x_i and x_{i+1} is denoted $f[x_i, x_{i+1}]$ and defined as

$$(3.12) \qquad f[x_i, x_{i+1}] = \frac{f[x_{i+1}] - f[x_i]}{x_{i+1} - x_i}.$$

When the $(k-1)$st divided differences

$$f[x_i, x_{i+1}, x_{i+2}, \ldots, x_{i+k-1}] \qquad \text{and} \qquad f[x_{i+1}, x_{i+2}, \ldots, x_{i+k-1}, x_{i+k}]$$

have both been determined, the kth divided difference relative to $x_i, x_{i+1}, x_{i+2}, \ldots, x_{i+k}$ is given by

$$(3.13) \quad f[x_i, x_{i+1}, \ldots, x_{i+k-1}, x_{i+k}] = \frac{f[x_{i+1}, x_{i+2}, \ldots, x_{i+k}] - f[x_i, x_{i+1}, \ldots, x_{i+k-1}]}{x_{i+k} - x_i}.$$

With this notation, Eq. (3.11) can be re-expressed as $a_1 = f[x_0, x_1]$ and the interpolating polynomial in Eq. (3.10) is:

$$P_n(x) = f[x_0] + f[x_0, x_1](x - x_0) + a_2(x - x_0)(x - x_1)$$
$$+ \cdots + a_n(x - x_0)(x - x_1) \cdots (x - x_{n-1}).$$

The constants $a_2, a_3, \ldots, a_n$ in P_n can be consecutively obtained in a manner similar to the evaluation of a_0 and a_1, but the algebraic manipulation becomes tedious. The evaluation of a_2 is considered in Exercise 6. For a general procedure of evaluating these constants, the reader should refer to the previously mentioned book by Hildebrand [64].

As might be expected from the evaluation of a_0 and a_1, the required constants are:

$$a_k = f[x_0, x_1, x_2, \ldots, x_k],$$

for each $k = 0, 1, \ldots, n$; so P_n can be rewritten as

$$P_n(x) = f[x_0] + f[x_0, x_1](x - x_0)$$
$$+ f[x_0, x_1, x_2](x - x_0)(x - x_1) + \cdots$$
$$+ f[x_0, x_1, \ldots, x_n](x - x_0)(x - x_1) \cdots (x - x_{n-1}),$$

or as

$$(3.14) \qquad P_n(x) = f[x_0] + \sum_{k=1}^{n} f[x_0, x_1, \ldots, x_k](x - x_0) \cdots (x - x_{k-1}).$$

Equation (3.14) is known as **Newton's interpolatory divided-difference formula**.

An interesting presentation on divided differences is considered in Powell [101], pp. 46–50. This presentation begins with the definition of the nth divided difference $f[x_0, x_1, \ldots, x_n]$ as a_n in Eq. (3.10) and shows that as a consequence Eq. (3.13) is satisfied.

The determination of the divided differences from tabulated data points is outlined in Table 3.6. Two fourth and one fifth difference could also be determined from this data.

TABLE 3.6

x	$f(x)$	First divided differences	Second divided differences	Third divided differences
x_0	$f[x_0]$			
		$f[x_0, x_1] = \dfrac{f[x_1] - f[x_0]}{x_1 - x_0}$		
x_1	$f[x_1]$		$f[x_0, x_1, x_2] = \dfrac{f[x_1, x_2] - f[x_0, x_1]}{x_2 - x_0}$	
		$f[x_1, x_2] = \dfrac{f[x_2] - f[x_1]}{x_2 - x_1}$		$f[x_0, x_1, x_2, x_3] = \dfrac{f[x_1, x_2, x_3] - f[x_0, x_1, x_2]}{x_3 - x_0}$
x_2	$f[x_2]$		$f[x_1, x_2, x_3] = \dfrac{f[x_2, x_3] - f[x_1, x_2]}{x_3 - x_1}$	
		$f[x_2, x_3] = \dfrac{f[x_3] - f[x_2]}{x_3 - x_2}$		$f[x_1, x_2, x_3, x_4] = \dfrac{f[x_2, x_3, x_4] - f[x_1, x_2, x_3]}{x_4 - x_1}$
x_3	$f[x_3]$		$f[x_2, x_3, x_4] = \dfrac{f[x_3, x_4] - f[x_2, x_3]}{x_4 - x_2}$	
		$f[x_3, x_4] = \dfrac{f[x_4] - f[x_3]}{x_4 - x_3}$		$f[x_2, x_3, x_4, x_5] = \dfrac{f[x_3, x_4, x_5] - f[x_2, x_3, x_4]}{x_5 - x_2}$
x_4	$f[x_4]$		$f[x_3, x_4, x_5] = \dfrac{f[x_4, x_5] - f[x_3, x_4]}{x_5 - x_3}$	
		$f[x_4, x_5] = \dfrac{f[x_5] - f[x_4]}{x_5 - x_4}$		
x_5	$f[x_5]$			

The Mean Value Theorem applied to Eq. (3.12) shows that when f' exists, $f[x_0, x_1] = f'(\xi)$ for some number ξ between x_0 and x_1. The following theorem generalizes this result.

THEOREM 3.6   Suppose that $f \in C^n[a, b]$ and $x_0, x_1, \ldots, x_n$ are distinct numbers in $[a, b]$. Then a number ξ in (a, b) exists with

$$f[x_0, x_1, \ldots, x_n] = \frac{f^{(n)}(\xi)}{n!}.$$

PROOF   Let

$$g(x) = f(x) - P_n(x).$$

Since $f(x_i) = P_n(x_i)$ for each $i = 0, 1, \ldots, n$, g has $n + 1$ distinct zeros in $[a, b]$. The Generalized Rolle's Theorem implies that a number ξ in (a, b) exists with $g^{(n)}(\xi) = 0$, so

$$0 = f^{(n)}(\xi) - P_n^{(n)}(\xi).$$

Since $P_n(x)$ is a polynomial of degree n whose leading coefficient is $f[x_0, x_1, \ldots, x_n]$,

$$P_n^{(n)}(x) = f[x_0, x_1, \ldots, x_n] \cdot n!$$

As a consequence,

$$f[x_0, x_1, \ldots, x_n] = \frac{f^{(n)}(\xi)}{n!}. \qquad \square$$

The following algorithm computes the Lagrange polynomial for f at $x_0, x_1, \ldots, x_n$ by using divided differences and Eq. (3.14).

Newton's Interpolatory Divided-Difference Formula Algorithm 3.2

To obtain the divided-difference coefficients of the interpolatory polynomial P on the $(n+1)$-distinct numbers $x_0, x_1, \ldots, x_n$ for the function f:

INPUT numbers $x_0, x_1, \ldots, x_n$; values $f(x_0), f(x_1), \ldots, f(x_n)$ as the first column $Q_{0,0}$, $Q_{1,0}, \ldots, Q_{n,0}$ of Q.

OUTPUT the numbers $Q_{0,0}, Q_{1,1}, \ldots, Q_{n,n}$ where

$$P(x) = \sum_{i=0}^{n} Q_{i,i} \prod_{j=0}^{i-1} (x - x_j).$$

Step 1   For $i = 1, 2, \ldots, n$
　　　　　For $j = 1, 2, \ldots, i$

$$\text{set } Q_{i,j} = \frac{Q_{i,j-1} - Q_{i-1,j-1}}{x_i - x_{i-j}}.$$

Step 2   OUTPUT $(Q_{0,0}, Q_{1,1}, \ldots, Q_{n,n})$;　　　　$(Q_{i,i}$ is $f[x_0, x_1, \ldots, x_i].)$
　　　　　STOP.

The form of the output can be modified to produce all the divided differences. This is done in the following example.

EXAMPLE 1   The Bessel function of the first kind of order zero was considered in Example 3 of Section 3.2. In that example, various interpolating polynomials were used to approximate $f(1.5)$, using the data in the first three columns of Table 3.7. The remaining entries of Table 3.7 contain divided differences computed using Algorithm 3.2.

The coefficients of the Newton forward divided-difference form of the interpolatory polynomial are along the diagonal in the table. The polynomial is

$$P_4(x) = 0.7651977 - 0.4837057(x - 1.0) - 0.1087339(x - 1.0)(x - 1.3)$$
$$+ 0.0658784(x - 1.0)(x - 1.3)(x - 1.6)$$
$$+ 0.0018251(x - 1.0)(x - 1.3)(x - 1.6)(x - 1.9).$$

TABLE 3.7

i	x_i	$f[x_i]$	$f[x_{i-1}, x_i]$	$f[x_{i-2}, x_{i-1}, x_i]$	$f[x_{i-3}, \ldots, x_i]$	$f[x_{i-4}, \ldots, x_i]$
0	1.0	0.7651977				
			-0.4837057			
1	1.3	0.6200860		-0.1087339		
			-0.5489460		0.0658784	
2	1.6	0.4554022		-0.0494433		0.0018251
			-0.5786120		0.0680685	
3	1.9	0.2818186		0.0118183		
			-0.5715210			
4	2.2	0.1103623				

It is easily verified that $P_4(1.5) = 0.5118200$, which agrees with the result in Section 3.2, Example 3. ∎

When $x_0, x_1, \ldots, x_n$ are arranged consecutively with equal spacing, Eq. (3.14) can be expressed in a simplified form. Introducing the notation $h = x_{i+1} - x_i$ for each $i = 0, 1, \ldots, n-1$ and $x = x_0 + sh$, the difference $x - x_i$ can be written as $x - x_i = (s - i)h$; so Eq. (3.14) becomes

$$P_n(x) = P_n(x_0 + sh) = f[x_0] + shf[x_0, x_1] + s(s-1)h^2 f[x_0, x_1, x_2]$$
$$+ \cdots + s(s-1) \cdots (s-n+1)h^n f[x_0, x_1, \ldots, x_n]$$
$$= \sum_{k=0}^{n} s(s-1) \cdots (s-k+1)h^k f[x_0, x_1, \ldots, x_k].$$

Using binomial-coefficient notation

$$\binom{s}{k} = \frac{s(s-1) \cdots (s-k+1)}{k!},$$

we can express $P_n(x)$ compactly as

$$(3.15) \qquad P_n(x) = P_n(x_0 + sh) = \sum_{k=0}^{n} \binom{s}{k} k! h^k f[x_0, x_1, \ldots, x_k].$$

This formula is called the **Newton forward divided-difference formula**. Another form, called the **Newton forward-difference formula**, is constructed by making use of the forward difference notation Δ introduced in Definition 2.11, page 63. With this notation

$$f[x_0, x_1] = \frac{f(x_1) - f(x_0)}{x_1 - x_0} = \frac{1}{h} \Delta f(x_0),$$

$$f[x_0, x_1, x_2] = \frac{1}{2h}[\Delta f(x_1) - \Delta f(x_0)] = \frac{1}{2h^2} \Delta^2 f(x_0),$$

and, in general,

$$f[x_0, x_1, \ldots, x_k] = \frac{1}{k! h^k} \Delta^k f(x_0).$$

Consequently, Eq. (3.15) becomes

$$P_n(x) = \sum_{k=0}^{n} \binom{s}{k} \Delta^k f(x_0).$$

If the interpolating nodes are reordered as $x_n, x_{n-1}, \ldots, x_0$, a formula similar to Eq. (3.14) results:

$$P_n(x) = f[x_n] + f[x_{n-1}, x_n](x - x_n) + f[x_{n-2}, x_{n-1}, x_n](x - x_n)(x - x_{n-1})$$
$$+ \cdots + f[x_0, \ldots, x_n](x - x_n)(x - x_{n-1}) \cdots (x - x_1).$$

Using equal spacing with $x = x_n + sh$ and $x = x_i + (s + n - i)h$ produces

$$P_n(x) = P_n(x_n + sh)$$
$$= f[x_n] + shf[x_{n-1}, x_n] + s(s + 1)h^2 f[x_{n-2}, x_{n-1}, x_n] + \cdots$$
$$+ s(s + 1) \cdots (s + n - 1)h^n f[x_0, x_1, \ldots, x_n].$$

This form is called the **Newton backward divided-difference formula**. It is used to derive a more commonly applied formula known as the **Newton backward-difference formula**. To discuss this formula, we need the following definition.

DEFINITION 3.7   Given the sequence $\{p_n\}_{n=0}^{\infty}$, define the backward difference ∇p_n by

$$\nabla p_n \equiv p_n - p_{n-1} \qquad \text{for } n \geq 1.$$

Higher powers are defined recursively by

$$\nabla^k p_n = \nabla^{k-1}(\nabla p_n) \qquad \text{for } k \geq 2.$$

Definition 3.7 implies that

$$f[x_{n-1}, x_n] = \frac{1}{h}\nabla f(x_n), \qquad f[x_{n-2}, x_{n-1}, x_n] = \frac{1}{2h^2}\nabla^2 f(x_n),$$

and in general

$$f[x_{n-k}, \ldots, x_{n-1}, x_n] = \frac{1}{k!h^k}\nabla^k f(x_n).$$

Consequently,

$$P_n(x) = f[x_n] + s\nabla f(x_n) + \frac{s(s + 1)}{2}\nabla^2 f(x_n) + \cdots$$
$$+ \frac{s(s + 1) \cdots (s + n - 1)}{n!}\nabla^n f(x_n).$$

Extending the binomial coefficient notation to include negative numbers, we let

$$\binom{-s}{k} = \frac{-s(-s - 1) \cdots (-s - k + 1)}{k!}$$

so
$$\binom{-s}{k} = (-1)^k \frac{s(s+1)\cdots(s+k-1)}{k!}$$

and
$$P_n(x) = f(x_n) + (-1)^1 \binom{-s}{1} \nabla f(x_n) + (-1)^2 \binom{-s}{2} \nabla^2 f(x_n) + \cdots$$
$$+ (-1)^n \binom{-s}{n} \nabla^n f(x_n)$$

to obtain

(3.16)
$$P_n(x) = \sum_{k=0}^n (-1)^k \binom{-s}{k} \nabla^k f(x_n).$$

Equation (3.16) is called the **Newton backward-difference formula**.

EXAMPLE 2   Consider the table of data given in Example 1. The divided-difference table corresponding to this data is shown in Table 3.8.

TABLE 3.8

		First divided differences	Second divided differences	Third divided differences	Fourth divided differences
1.0	0.7651977				
		−0.4837057			
1.3	0.6200860		−0.1087339		
		−0.5489460		0.0658784	
1.6	0.4554022		−0.0494433		0.0018251
		−0.5786120		0.0680685	
1.9	0.2818186		0.0118183		
		−0.5715210			
2.2	0.1103623				

If an approximation to $f(1.1)$ is required, the reasonable choice for $x_0, x_1, \ldots, x_n$ would be $x_0 = 1.0$, $x_1 = 1.3$, $x_2 = 1.6$, $x_3 = 1.9$, and $x_4 = 2.2$, since this choice makes the greatest possible use of the data points closest to $x = 1.1$ and also makes use of the fourth divided difference. This implies that $h = 0.3$ and $s = \frac{1}{3}$, so the Newton forward divided-difference formula is used with the divided differences that have a *solid* underline in the table.

$$P_4(1.1) = P_4(1.0 + \tfrac{1}{3}(0.3)) = 0.7651997 + \tfrac{1}{3}(0.3)(-0.4837057)$$
$$+ \tfrac{1}{3}(-\tfrac{2}{3})(0.3)^2(-0.1087339)$$
$$+ \tfrac{1}{3}(-\tfrac{2}{3})(-\tfrac{5}{3})(0.3)^3(0.0658784)$$
$$+ \tfrac{1}{3}(-\tfrac{2}{3})(-\tfrac{5}{3})(-\tfrac{8}{3})(0.3)^4(0.0018251)$$
$$= 0.7196480.$$

To approximate a value when x is close to the end of the tabulated values, say, $x = 2.0$, we again would like to make maximum use of the data points closest to x.

This requires using the Newton backward divided-difference formula with $s = -\frac{2}{3}$ and the divided differences in the table that have a *dashed* underline:

$$
\begin{aligned}
P_4(2.0) = P_4(2.2 - \tfrac{2}{3}(0.3)) = {} & 0.1103623 - \tfrac{2}{3}(0.3)(-0.5715210) \\
& - \tfrac{2}{3}(\tfrac{1}{3})(0.3)^2(0.0118183) \\
& - \tfrac{2}{3}(\tfrac{1}{3})(\tfrac{4}{3})(0.3)^3(0.0680685) \\
& - \tfrac{2}{3}(\tfrac{1}{3})(\tfrac{4}{3})(\tfrac{7}{3})(0.3)^4(0.0018251) \\
= {} & 0.2238754. \qquad \blacksquare
\end{aligned}
$$

The Newton formulas are not appropriate for approximating a value, x, that lies near the center of the table, since employing either the backward or forward method in such a way that the highest-order difference is involved will not allow x_0 to be close to x. A number of divided-difference formulas are available in this instance, each of which has situations when it can be used to maximum advantage. These methods are known as **centered-difference formulas**. Because of the number of such methods, we will present only one, Stirling's method, and again refer the interested reader to Hildebrand [64] for a complete presentation.

For the centered-difference formulas we choose x_0 near the point being approximated, label the points directly below x_0 as $x_1, x_2, \ldots$ and those directly above as $x_{-1}, x_{-2}, \ldots$. Using this convention, Stirling's formula is given by:

$$(3.17) \quad P_n(x) = P_{2m+1}(x) = f[x_0] + \frac{sh}{2}(f[x_{-1}, x_0] + f[x_0, x_1]) + s^2 h^2 f[x_{-1}, x_0, x_1]$$

$$+ \frac{s(s^2-1)h^3}{2}(f[x_{-1}, x_0, x_1, x_2] + f[x_{-2}, x_{-1}, x_0, x_1])$$

$$+ \cdots + s^2(s^2-1)(s^2-4)\cdots(s^2-(m-1)^2)h^{2m}f[x_{-m}, \ldots, x_m]$$

$$+ \frac{s(s^2-1)\cdots(s^2-m^2)h^{2m+1}}{2}(f[x_{-m}, \ldots, x_{m+1}] + f[x_{-m-1}, \ldots, x_m])$$

if $n = 2m + 1$ is odd, and if $n = 2m$ is even, by the same formula with the deletion of the last term. The entries used for this formula are *circled* in Table 3.9.

TABLE 3.9

x	$f(x)$	First divided differences	Second divided differences	Third divided differences	Fourth divided differences
x_{-2}	$f[x_{-2}]$				
		$f[x_{-2}, x_{-1}]$			
x_{-1}	$f[x_{-1}]$		$f[x_{-2}, x_{-1}, x_0]$		
		$f[x_{-1}, x_0]$		$f[x_{-2}, x_{-1}, x_0, x_1]$	
x_0	$f[x_0]$		$f[x_{-1}, x_0, x_1]$		$f[x_{-2}, x_{-1}, x_0, x_1, x_2]$
		$f[x_0, x_1]$		$f[x_{-1}, x_0, x_1, x_2]$	
x_1	$f[x_1]$		$f[x_0, x_1, x_2]$		
		$f[x_1, x_2]$			
x_2	$f[x_2]$				

EXAMPLE 3   Consider the table of data that was given in the previous examples. To use Stirling's formula to approximate $f(1.5)$ with $x_0 = 1.6$, we use the *underlined* entries in the difference Table 3.10.

TABLE 3.10

x	$f(x)$	First divided differences	Second divided differences	Third divided differences	Fourth divided differences
1.0	0.7651977				
		−0.4837057			
1.3	0.6200860		−0.1087339		
		−0.5489460		0.0658784	
1.6	0.4554022		−0.0494433		0.0018251
		−0.5786120		0.0680685	
1.9	0.2818186		0.0118183		
		−0.5715210			
2.2	0.1103623				

The formula with $h = 0.3$, $x_0 = 1.6$, and $s = -\frac{1}{3}$ becomes

$$f(1.5) \approx P_4(1.6 + (-\tfrac{1}{3})(0.3)) = 0.4554022 + (-\tfrac{1}{3})(\tfrac{0.3}{2})(-0.5489460 - 0.5786120)$$
$$+ (-\tfrac{1}{3})^2(0.3)^2(-0.0494433)$$
$$+ \tfrac{1}{2}(-\tfrac{1}{3})((-\tfrac{1}{3})^2 - 1)(0.3)^3(0.0658784 + 0.0680685)$$
$$+ (-\tfrac{1}{3})^2((-\tfrac{1}{3})^2 - 1)(0.3)^4(0.0018251)$$
$$= 0.5118200. \qquad \blacksquare$$

Exercise Set 3.4

1. Use Algorithm 3.2 to construct the interpolating polynomial of degree four for the unequally spaced points given in the following table:

x	$f(x)$
0.0	−6.00000
0.1	−5.89483
0.3	−5.65014
0.6	−5.17788
1.0	−4.28172

2. Suppose the data $f(1.1) = -3.99583$ is added to Exercise 1. Construct the interpolating polynomial of degree five.

3. Approximate $f(0.05)$ using the following data and the Newton forward divided-difference formula:

x	0.0	0.2	0.4	0.6	0.8
$f(x)$	1.00000	1.22140	1.49182	1.82212	2.22554

4. Using the data of Exercise 3 and the Newton backward divided-difference formula, approximate $f(0.65)$.

5. Using the data of Exercise 3 and Stirling's formula, approximate $f(0.43)$.

6. Given

$$P_n(x) = f[x_0] + f[x_0, x_1](x - x_0) + a_2(x - x_0)(x - x_1)$$
$$+ a_3(x - x_0)(x - x_1)(x - x_2) + \cdots$$
$$+ a_n(x - x_0)(x - x_1) \cdots (x - x_{n-1}).$$

Use $P_n(x_2)$ to show that $a_2 = f[x_0, x_1, x_2]$.

7. Show that

$$f[x_0, x_1, \ldots, x_n, x] = \frac{f^{(n+1)}(\xi(x))}{(n+1)!}$$

for some $\xi(x)$. [*Hint*: From Eq. (3.5),

$$f(x) = P_n(x) + \frac{f^{(n+1)}(\xi(x))}{(n+1)!}(x - x_0) \cdots (x - x_n).$$

Considering the interpolation polynomial of degree $n + 1$ on $x_0, x_1, \ldots, x_n, x$, we have

$$f(x) = P_{n+1}(x) = P_n(x) + f[x_0, x_1, \ldots, x_n, x](x - x_0) \cdots (x - x_n).]$$

8. The following population table was given in the introduction to this chapter and studied in Exercise 15 of Section 3.2.

Year	1930	1940	1950	1960	1970	1980
Population (in thousands)	123,203	131,669	150,697	179,323	203,212	226,505

Use an appropriate divided difference method to approximate:

a) The population in the year 1965.   b) The population in the year 2000.

3.5 Hermite Interpolation

The set of **osculating polynomials** is a generalization of both the Taylor polynomials and the Lagrange polynomials. These polynomials have the property that, given $n + 1$ distinct numbers $x_0, x_1, \ldots, x_n$ and nonnegative integers $m_0, m_1, \ldots, m_n$, the osculating polynomial approximating a function $f \in C^m[a, b]$, where $m = \max\{m_0, m_1, \ldots, m_n\}$ and $x_i \in [a, b]$ for each $i = 0, \ldots, n$, is the polynomial of least degree with the property that it agrees with the function f and all of its derivatives of order less than or equal to m_i at x_i for each $i = 0, 1, \ldots, n$. The degree of this osculating polynomial will be at most

$$M = \sum_{i=0}^{n} m_i + n,$$

since the number of conditions to be satisfied is $\sum_{i=0}^{n} m_i + (n+1)$, and a polynomial of degree M has $M+1$ coefficients that can be used to satisfy these conditions. For the sake of clarity, the formal definition of an osculating polynomial follows.

DEFINITION 3.8   Let $x_0, x_1, \ldots, x_n$ be $n+1$ distinct numbers in $[a, b]$ and m_i be a nonnegative integer associated with x_i for $i = 0, 1, \ldots, n$. Let

$$m = \max_{0 \le i \le n} m_i \qquad \text{and} \qquad f \in C^m[a, b].$$

The osculating polynomial approximating f is the polynomial P of least degree such that

$$\frac{d^k P(x_i)}{dx^k} = \frac{d^k f(x_i)}{dx^k} \qquad \text{for each } i = 0, 1, \ldots, n \text{ and } k = 0, 1, \ldots, m_i.$$

Note that when $n = 0$, the osculating polynomial approximating f is simply the Taylor polynomial of degree m_0 for f at x_0. When $m_i = 0$ for $i = 0, 1, \ldots, n$, the osculating polynomial is the polynomial interpolating f on $x_0, x_1, \ldots, x_n$, that is, the Lagrange polynomial.

The situation that occurs when $m_i = 1$ for each $i = 0, 1, \ldots, n$ gives a class of polynomials called **Hermite polynomials**. For a given function f, these polynomials not only agree with f at $x_0, x_1, \ldots, x_n$, but, since their first derivatives agree with those of f, they have the same "shape" as the function at $(x_i, f(x_i))$ in the sense that the *tangent lines* to the polynomial and the function agree. We will restrict our study of osculating polynomials to this situation, and consider first a theorem that describes precisely the form of the polynomials of Hermite type.

THEOREM 3.9   If $f \in C^1[a, b]$ and $x_0, \ldots, x_n \in [a, b]$ are distinct, the unique polynomial of least degree agreeing with f and f' at $x_0, \ldots, x_n$ is a polynomial of degree at most $2n+1$ given by

(3.18) $$H_{2n+1}(x) = \sum_{j=0}^{n} f(x_j) H_{n,j}(x) + \sum_{j=0}^{n} f'(x_j) \hat{H}_{n,j}(x),$$

where

(3.19) $$H_{n,j}(x) = [1 - 2(x - x_j) L'_{n,j}(x_j)] L^2_{n,j}(x)$$

and

(3.20) $$\hat{H}_{n,j}(x) = (x - x_j) L^2_{n,j}(x).$$

In this context, $L_{n,j}$ denotes the jth Lagrange coefficient polynomial of degree n defined in Eq. (3.4).

Moreover, if $f \in C^{(2n+2)}[a, b]$, then

(3.21) $$f(x) - H_{2n+1}(x) = \frac{(x - x_0)^2 \cdots (x - x_n)^2}{(2n+2)!} f^{(2n+2)}(\xi)$$

for some ξ with $a < \xi < b$.

PROOF

To show that

(3.22) $\qquad H_{2n+1}(x_k) = f(x_k) \qquad$ and $\qquad H'_{2n+1}(x_k) = f'(x_k)$

for each $k = 0, 1, \ldots, n$, it suffices to show that $H_{n,j}$ and $\hat{H}_{n,j}$, as defined by Eqs. (3.19) and (3.20), respectively, satisfy conditions (a)–(d):

a) $\quad H_{n,j}(x_k) = \begin{cases} 0, & \text{if } j \neq k, \\ 1, & \text{if } j = k; \end{cases}$

b) $\quad \dfrac{d}{dx} H_{n,j}(x_k) = 0, \quad$ for all k;

c) $\quad \hat{H}_{n,j}(x_k) = 0 \quad$ for all k;

d) $\quad \dfrac{d}{dx} \hat{H}_{n,j}(x_k) = \begin{cases} 0, & \text{if } j \neq k, \\ 1, & \text{if } j = k. \end{cases}$

(*See Figs. 3.6 and 3.7.*)

FIGURE 3.6

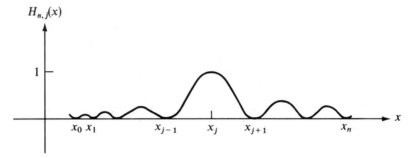

FIGURE 3.7

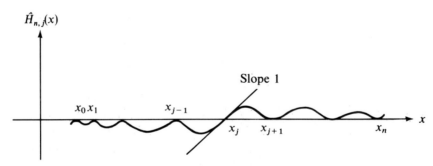

These conditions ensure that when either H_{2n+1} or H'_{2n+1} is evaluated at one of the given numbers the appropriate value is obtained since

$$H_{2n+1}(x_k) = \sum_{j=0}^{n} f(x_j) H_{n,j}(x_k) + \sum_{j=0}^{n} f'(x_j) \hat{H}_{n,j}(x_k)$$

$$= f(x_k) \cdot 1 + \sum_{\substack{j=0 \\ j \neq k}}^{n} f(x_j) \cdot 0 + \sum_{j=0}^{n} f'(x_j) \cdot 0 = f(x_k),$$

and

$$H'_{2n+1}(x_k) = \sum_{j=0}^{n} f(x_j) H'_{n,j}(x_k) + \sum_{j=0}^{n} f'(x_j) \hat{H}'_{n,j}(x_k)$$

$$= \sum_{j=0}^{n} f(x_j) \cdot 0 + f'(x_k) \cdot 1 + \sum_{\substack{j=0 \\ j \neq k}}^{n} f'(x_j) \cdot 0 = f'(x_k),$$

for each $k = 0, 1, \ldots, n$.

Considering first the polynomial $\hat{H}_{n,j}$, requirements (c) and (d) imply that $\hat{H}_{n,j}$ must have a double root at x_k for $k \neq j$ and a single root at x_j. A polynomial of degree at most $(2n+1)$ satisfying these conditions and also having a derivative with a value of one at x_j is

$$\hat{H}_{n,j}(x) = \frac{(x-x_0)^2 \cdots (x-x_{j-1})^2 (x-x_j)(x-x_{j+1})^2 \cdots (x-x_n)^2}{(x_j-x_0)^2 \cdots (x_j-x_{j-1})^2 (1)(x_j-x_{j+1})^2 \cdots (x_j-x_n)^2} = (x-x_j)L_{n,j}^2(x).$$

Conditions (a) and (b) imply that x_k, for each $k \neq j$, must be a double root of $H_{n,j}(x)$ and any polynomial of degree at most $(2n+1)$ satisfying conditions (a) and (b) is given by:

$$H_{n,j}(x) = (x-x_0)^2 \cdots (x-x_{j-1})^2 (x-x_{j+1})^2 \cdots (x-x_n)^2 (\hat{a}x + \hat{b})$$

for some constants $\hat{a}$ and $\hat{b}$. Adjusting the constants $\hat{a}$ and $\hat{b}$ by letting

$$a = \hat{a} \prod_{\substack{i=0 \\ i \neq j}}^{n} (x_i - x_j)^2 \qquad \text{and} \qquad b = \hat{b} \prod_{\substack{i=0 \\ i \neq j}}^{n} (x_i - x_j)^2$$

gives
$$H_{n,j}(x) = L_{n,j}^2(x)(ax+b).$$

Condition (a) implies that

(3.23) $$1 = H_{n,j}(x_j) = L_{n,j}^2(x_j)(ax_j + b) = ax_j + b$$

and using condition (b) and Eq. (3.23) gives

$$0 = \frac{dH_{n,j}(x_j)}{dx} = 2L_{n,j}(x_j)L'_{n,j}(x_j)(ax_j + b) + L_{n,j}^2(x_j)(a)$$

$$= 2L'_{n,j}(x_j)(ax_j + b) + a$$

$$= 2L'_{n,j}(x_j)(1) + a,$$

or $$a = -2L'_{n,j}(x_j).$$

From Eq. (3.23),

$$b = 1 - ax_j = 1 + 2L'_{n,j}(x_j) \cdot x_j;$$

hence, $$ax + b = -2L'_{n,j}(x_j)x + 1 + 2L'_{n,j}(x_j)x_j = 1 - 2(x - x_j)L'_{n,j}(x_j)$$

and $$H_{n,j}(x) = (ax+b)L_{n,j}^2(x) = [1 - 2(x-x_j)L'_{n,j}(x_j)]L_{n,j}^2(x).$$

We have constructed the $H_{n,j}$'s and $\hat{H}_{n,j}$'s as given by Equations (3.19) and (3.20) respectively. Note that, since each $H_{n,j}$ and each $\hat{H}_{n,j}$ is of degree at most $(2n+1)$, H_{2n+1} is of degree at most $2n+1$. Moreover, conditions (a), (b), (c), and (d) imply that Eq. (3.22) is satisfied.

The uniqueness of this polynomial and the calculation of the error term are considered in Exercise 8.   ☐

EXAMPLE 1   Use the polynomial of least degree that agrees with the data listed in Table 3.11 for the Bessel function of the first kind of order zero to find an approximation of $f(1.5)$.

TABLE 3.11

k	x_k	$f(x_k)$	$f'(x_k)$
0	1.3	0.6200860	−0.5220232
1	1.6	0.4554022	−0.5698959
2	1.9	0.2818186	−0.5811571

First compute the Lagrange polynomials and their derivatives:

$$L_{2,0}(x) = \frac{(x-x_1)(x-x_2)}{(x_0-x_1)(x_0-x_2)} = \tfrac{50}{9}x^2 - \tfrac{175}{9}x + \tfrac{152}{9}, \qquad L'_{2,0}(x) = \tfrac{100}{9}x - \tfrac{175}{9};$$

$$L_{2,1}(x) = \frac{(x-x_0)(x-x_2)}{(x_1-x_0)(x_1-x_2)} = \tfrac{-100}{9}x^2 + \tfrac{320}{9}x - \tfrac{247}{9}, \qquad L'_{2,1}(x) = \tfrac{-200}{9}x + \tfrac{320}{9};$$

and

$$L_{2,2}(x) = \frac{(x-x_0)(x-x_1)}{(x_2-x_0)(x_2-x_1)} = \tfrac{50}{9}x^2 - \tfrac{145}{9}x + \tfrac{104}{9}, \qquad L'_{2,2}(x) = \tfrac{100}{9}x - \tfrac{145}{9}.$$

The polynomials $H_{2,j}$ and $\hat{H}_{2,j}$ are then:

$$H_{2,0}(x) = [1 - 2(x-1.3)(-5)](\tfrac{50}{9}x^2 - \tfrac{175}{9}x + \tfrac{152}{9})^2$$
$$= (10x - 12)(\tfrac{50}{9}x^2 - \tfrac{175}{9}x + \tfrac{152}{9})^2,$$

$$H_{2,1}(x) = 1 \cdot (\tfrac{-100}{9}x^2 + \tfrac{320}{9}x - \tfrac{247}{9})^2,$$

and $\qquad H_{2,2}(x) = 10(2-x)(\tfrac{50}{9}x^2 - \tfrac{145}{9}x + \tfrac{104}{9})^2;$

and $\qquad \hat{H}_{2,0}(x) = (x-1.3)(\tfrac{50}{9}x^2 - \tfrac{175}{9}x + \tfrac{152}{9})^2,$

$$\hat{H}_{2,1}(x) = (x-1.6)(\tfrac{-100}{9}x^2 + \tfrac{320}{9}x - \tfrac{247}{9})^2,$$

and $\qquad \hat{H}_{2,2}(x) = (x-1.9)(\tfrac{50}{9}x^2 - \tfrac{145}{9}x + \tfrac{104}{9})^2.$

Finally,

$$H_5(x) = 0.6200860 H_{2,0}(x) + 0.4554022 H_{2,1}(x) + 0.2818186 H_{2,2}(x)$$
$$- 0.5220232 \hat{H}_{2,0}(x) - 0.5698959 \hat{H}_{2,1}(x) - 0.5811571 \hat{H}_{2,2}(x)$$

and $\qquad H_5(1.5) = 0.6200860(\tfrac{64}{27}) + 0.4554022(\tfrac{64}{81}) + 0.2818186(\tfrac{5}{81})$

$$- 0.5220232(\tfrac{4}{405}) - 0.5698959(\tfrac{-32}{405}) - 0.5811571(\tfrac{-2}{405})$$

$$= 0.5118277,$$

a result that is accurate to the places listed.   ∎

Although Theorem 3.9 provides a complete description of the Hermite polynomials, it is clear from Example 1 that the need to determine and evaluate the Lagrange polynomials and their derivatives makes the procedure tedious even for small values of n. An alternative method for generating Hermite approximations has as its basis the Newton interpolatory divided-difference formula (3.14) for the Lagrange polynomial at $x_0, x_1, \ldots, x_n$:

$$P_n(x) = f[x_0] + \sum_{k=1}^{n} f[x_0, x_1, \ldots, x_k](x - x_0) \cdots (x - x_{k-1})$$

and the connection between the nth divided difference and the nth derivative of f, as outlined in Theorem 3.6.

Suppose that n distinct numbers $x_0, x_1, \ldots, x_n$ are given together with their values of f and f'. First define a new sequence $z_0, z_1, \ldots, z_{2n+1}$ by

$$z_{2i} = z_{2i+1} = x_i \qquad \text{for each } i = 0, 1, \ldots, n.$$

Now construct the divided difference table in the form of Table 3.6 that uses $z_0, z_1, \ldots,$ z_{2n+1}.

Since $z_{2i} = z_{2i+1} = x_i$ for each i, $f[z_{2i}, z_{2i+1}]$ cannot be defined by the basic relation (3.12). If, however, we assume, based on Theorem 3.6, that the reasonable substitution in this situation is $f[z_{2i}, z_{2i+1}] = f'(x_i)$, we can use the entries

$$f'(x_0), f'(x_1), \ldots, f'(x_n)$$

in place of the undefined first divided differences

$$f[z_0, z_1], f[z_2, z_3], \ldots, f[z_{2n}, z_{2n+1}].$$

The remaining divided differences are produced as usual and the appropriate divided differences employed in Newton's interpolatory divided-difference formula. Table 3.12 shows the entries that are used for the first three divided-difference columns when determining the Hermite polynomial H_5 for x_0, x_1, and x_2. The remaining entries are generated in the same manner as in Table 3.6.

TABLE 3.12

z	$f(z)$	First divided differences	Second divided differences
$z_0 = x_0$	$f[z_0] = f(x_0)$		
		$f[z_0, z_1] = f'(x_0)$	
$z_1 = x_0$	$f[z_1] = f(x_0)$		$f[z_0, z_1, z_2] = \dfrac{f[z_1, z_2] - f[z_0, z_1]}{z_2 - z_0}$
		$f[z_1, z_2] = \dfrac{f[z_2] - f[z_1]}{z_2 - z_1}$	
$z_2 = x_1$	$f[z_2] = f(x_1)$		$f[z_1, z_2, z_3] = \dfrac{f[z_2, z_3] - f[z_1, z_2]}{z_3 - z_1}$
		$f[z_2, z_3] = f'(x_1)$	
$z_3 = x_1$	$f[z_3] = f(x_1)$		$f[z_2, z_3, z_4] = \dfrac{f[z_3, z_4] - f[z_2, z_3]}{z_4 - z_2}$
		$f[z_3, z_4] = \dfrac{f[z_4] - f[z_3]}{z_4 - z_3}$	
$z_4 = x_2$	$f[z_4] = f(x_2)$		$f[z_3, z_4, z_5] = \dfrac{f[z_4, z_5] - f[z_3, z_4]}{z_5 - z_3}$
		$f[z_4, z_5] = f'(x_2)$	
$z_5 = x_2$	$f[z_5] = f(x_2)$		

EXAMPLE 2   The entries in Table 3.13 use the data given in Example 1. The underlined entries are the given data; the remainder are generated by the standard divided-difference formula (3.13).

$$\begin{aligned}
H_5(1.5) = &\ 0.6200860 + (1.5 - 1.3)(-0.5220232) + (1.5 - 1.3)^2(-0.0897427) \\
&+ (1.5 - 1.3)^2(1.5 - 1.6)(0.0663657) + (1.5 - 1.3)^2(1.5 - 1.6)^2(0.0026663) \\
&+ (1.5 - 1.3)^2(1.5 - 1.6)^2(1.5 - 1.9)(-0.0027738) \\
= &\ 0.5118277.
\end{aligned}$$

TABLE 3.13

1.3	0.6200860					
1.3	0.6200860	−0.5220232	−0.0897427			
1.6	0.4554022	−0.5489460	−0.0698330	0.0663657	0.0026663	
1.6	0.4554022	−0.5698959	−0.0290537	0.0679655	0.0010020	−0.0027738
1.9	0.2818186	−0.5786120	−0.0084837	0.0685667		
1.9	0.2818186	−0.5811571				

Algorithm 3.3 generates the coefficients for the Hermite polynomials using the Newton interpolatory divided-difference formula. The structure of the algorithm is slightly different from the discussion, to take advantage of efficiency of computation.

Hermite Interpolation Algorithm 3.3

To obtain the coefficients of the Hermite interpolating polynomial H on the $(n + 1)$-distinct numbers $x_0, \ldots, x_n$ for the function f:

INPUT numbers $x_0, x_1, \ldots, x_n$; values $f(x_0), f(x_1), \ldots, f(x_n)$ and $f'(x_0), f'(x_1), \ldots, f'(x_n)$.

OUTPUT the numbers $Q_{0,0}, Q_{1,1}, \ldots, Q_{2n+1, 2n+1}$ where

$$H(x) = Q_{0,0} + Q_{1,1}(x - x_0) + Q_{2,2}(x - x_0)^2 + Q_{3,3}(x - x_0)^2(x - x_1)$$
$$+ Q_{4,4}(x - x_0)^2(x - x_1)^2 + \cdots$$
$$+ Q_{2n+1, 2n+1}(x - x_0)^2(x - x_1)^2 \cdots (x - x_{n-1})^2(x - x_n).$$

Step 1   For $i = 0, 1, \ldots, n$ do Steps 2 and 3.

Step 2   Set $z_{2i} = x_i$;
$z_{2i+1} = x_i$;
$Q_{2i,0} = f(x_i)$;
$Q_{2i+1,0} = f(x_i)$;
$Q_{2i+1,1} = f'(x_i)$.

Step 3   If $i \neq 0$ then set

$$Q_{2i,1} = \frac{Q_{2i,0} - Q_{2i-1,0}}{z_{2i} - z_{2i-1}}.$$

Step 4   For $i = 2, 3, \ldots, 2n + 1$
for $j = 2, 3, \ldots, i$ set $Q_{i,j} = \dfrac{Q_{i,j-1} - Q_{i-1,j-1}}{z_i - z_{i-j}}.$

Step 5   OUTPUT $(Q_{0,0}, Q_{1,1}, \ldots, Q_{2n+1, 2n+1})$;
STOP.

The technique used in Algorithm 3.3 can be extended for use in determining other osculating polynomials. A concise discussion of the procedures involved can be found in Powell [101], pp. 53–57.

Exercise Set 3.5

1. Use Hermite interpolation to find an approximation to the following:

a) $f(2.5)$ given

x	$f(x)$	$f'(x)$
2.2	0.5207843	−0.0014878
2.4	0.5104147	−0.1004889
2.6	0.4813306	0.1883635

b) $f(5.3)$ given

x	$f(x)$	$f'(x)$
5.0	2.168861	−1.495067
5.2	1.797350	−1.266911
5.4	1.488591	−1.070309

c) $f(0)$ given

x	$f(x)$	$f'(x)$
−0.3	−0.20431	0.32213
−0.1	−0.08993	0.79731
0.1	0.11007	1.20269
0.3	0.39569	1.67787

d) $f(1.25)$ given

x	$f(x)$	$f'(x)$
1.1	0.48603	2.90986
1.2	0.86160	4.91788
1.3	1.59751	11.06798
1.4	3.76155	41.13928

e) $f(0.5)$ given

x	$f(x)$	$f'(x)$
0.2	0.9798652	0.20271
0.4	0.9177710	0.42279
0.6	0.8080348	0.68414
0.8	0.6386093	1.02964
1.0	0.3843735	1.55741

f) $f(0.2)$ given

x	$f(x)$	$f'(x)$
0.1	1.2314028	2.64281
0.3	1.9121188	4.24424
0.4	2.3855409	5.25108
0.5	2.9682818	6.43656
0.6	3.6801169	7.84023

2. With $f(x) = 3xe^x - e^{2x}$, approximate $f(1.03)$ by the Hermite interpolating polynomial of degree at most three, using $x_0 = 1$ and $x_1 = 1.05$. Compare the actual error to the error bound obtained from Eq. (3.21).

3. Repeat Exercise 2, with the Hermite interpolating polynomial of degree at most five, using $x_0 = 1$, $x_1 = 1.05$, and $x_2 = 1.07$.

4. Let $f(x) = 3xe^x - e^{2x}$. Does the interpolating polynomial approximation of degree two, $P_2(1.03)$, computed in Exercise 4 of Section 3.2 give a better approximation of $f(1.03)$ than

the Hermite polynomial approximation of degree three, $H_3(1.03)$, computed in Exercise 2 of this section? Compare the theoretical error estimates for

$$|P_2(1.03) - f(1.03)| \qquad \text{and} \qquad |H_3(1.03) - f(1.03)|.$$

5. a) Let $f(x) = e^x$. Find $H_5(x)$, the Hermite polynomial approximating f, using the data points $x_0 = 0$, $x_1 = 1$, and $x_2 = 2$. Compare $H_5(0.25)$ to $f(0.25)$ and to $P(0.25)$ as given in Exercise 9 of Section 3.2.

    b) Use Theorem 3.9 to estimate $|H_5(0.25) - f(0.25)|$.

6. a) Use the following values to construct a Hermite interpolatory polynomial to approximate sin 0.34

x	$\sin x$	$D_x(\sin x) = \cos x$
0.30	0.29552	0.95534
0.32	0.31457	0.94924
0.35	0.34290	0.93937

    b) Determine an error bound for the approximation in part (a) and compare to the actual error.

7. Add sin $0.33 = 0.32404$ and cos $0.33 = 0.94604$ to the data in Exercise 6 and redo the calculations in that exercise.

8. a) Show that $H_{2n+1}(x)$ is the unique polynomial of least degree agreeing with f and f' at $x_0, \ldots, x_n$. [*Hint*: Assume that P is another such polynomial and consider $D = H_{2n+1} - P$ and D' at $x_0, x_1, \ldots, x_n$.]

    b) Derive the error term in Theorem 3.9. [*Hint*: Use the same method as in the Lagrange error derivation, Theorem 3.3 defining

$$g(t) = f(t) - H_{2n+1}(t) - \frac{(t - x_0)^2 \cdots (t - x_n)^2}{(x - x_0)^2 \cdots (x - x_n)^2}[f(x) - H_{2n+1}(x)]$$

and using the fact that $g'(t)$ has $(2n + 2)$ *distinct* zeros in $[a, b]$.]

9. The following table lists data for the function described by $f(x) = e^{0.1x^2}$. Approximate $f(1.25)$ by using $H_5(1.25)$ and $H_3(1.25)$ where H_5 uses the nodes $x_0 = 1$, $x_2 = 2$, and $x_3 = 3$ and where H_3 uses the nodes $\bar{x}_0 = 1$ and $\bar{x}_1 = 1.5$. Find error bounds for these approximations.

x	$f(x) = e^{0.1x^2}$	$f'(x) = 0.2xe^{0.1x^2}$
$x_0 = \bar{x}_0 = 1$	1.105170918	0.2210341836
$\bar{x}_1 = 1.5$	1.252322716	0.3756968148
$x_1 = 2$	1.491824698	0.5967298792
$x_2 = 3$	2.459603111	1.475761867

10. A car traveling along a straight road is clocked at a number of points. The data from the observations is given in the following table where the time is in seconds, the distance in feet, and the speed in feet per seconds. Use a Hermite polynomial to predict the position of the car and its speed when $t = 10$ sec.

Time	0	3	5	8	13
Distance	0	225	383	623	993
Speed	75	77	80	74	72

3.6 Cubic Spline Interpolation*

The previous sections of this chapter were concerned with the approximation of arbitrary functions on closed intervals by the use of polynomials. Although this method of approximation is appropriate in many circumstances, the oscillatory nature of high-degree polynomials and the property that a fluctuation over a small portion of the interval can induce large fluctuations over the entire range restrict their use when approximating many of the functions that arise in actual physical situations.

An alternative approach that can be used to obtain interpolatory functions is to divide the interval into a collection of subintervals and construct a (generally) different approximating polynomial on each subinterval. Approximation by functions of this type is called **piecewise polynomial approximation**.

The simplest type of piecewise polynomial approximation is called piecewise linear interpolation and consists of joining a set of data points

$$\{(x_0, f(x_0)), (x_1, f(x_1)), \ldots, (x_n, f(x_n))\}$$

by a series of straight lines. This is the method of interpolation used in elementary courses involving the study of trigonometric or logarithmic functions when intermediate values are required from a collection of tabulated values.

The disadvantage of approaching an approximation problem using functions of this type is that at each of the endpoints of the subintervals, there is no assurance of differentiability, which, in a geometrical context, means that the interpolating function is not "smooth" at these points. Often it is clear from physical conditions that such a smoothness condition is required and in these cases the approximating function must be continuously differentiable.

A procedure that one could follow is to use a piecewise polynomial of Hermite type. For example, if the values of the function f and of f' are known at each of the points $x_0 < x_1 < \cdots < x_n$, a Hermite polynomial of degree three could be used on each of the subintervals $[x_0, x_1]$, $[x_1, x_2]$, $\ldots$, $[x_{n-1}, x_n]$ to obtain a function that is continuously differentiable on the interval $[x_0, x_n]$. To determine the appropriate Hermite cubic polynomial on a given interval is simply a matter of computing the function $H_3(x)$ for that interval. Since the Lagrange interpolating polynomials needed to determine H_3 are of first degree, this can be accomplished without great difficulty.

*The proofs of the theorems in this section rely on results in Chapter 6.

The Hermite cubic polynomials are commonly used in application problems to study the motion of particles moving in space. To consider this application, however, the theory must be extended to multivariate functions. A reference to this material is Schultz [119], pp. 29–39. The difficulty with using Hermite piecewise polynomials for general interpolation problems concerns the need to know the derivative of the function being approximated. Usually data is known about the function, but not about its derivative.

The remainder of this section considers approximation using piecewise polynomials that require no derivative information, except at the endpoints of the interval on which the function is being approximated.

The simplest type of differentiable piecewise polynomial function on an entire interval $[x_0, x_n]$ is the function obtained by fitting a quadratic polynomial between each successive pair of nodes. This is done by constructing a quadratic on $[x_0, x_1]$ agreeing with the function at x_0 and x_1, another quadratic on $[x_1, x_2]$ agreeing with the function at x_1 and x_2, and so on. Since a general quadratic polynomial has three arbitrary constants—the constant term, the coefficient of x, and the coefficient of x^2— and only two conditions are required to fit the data at the endpoints of each subinterval, flexibility exists that allows the quadratic to be chosen so that, in addition, the interpolant has a continuous derivative on $[x_0, x_n]$. The difficulty with this procedure arises when there is a need to specify that the derivative of the interpolant agrees with that of the function at the endpoints x_0 and x_n. In this case, it can be shown that there are not a sufficient number of constants to ensure that the conditions will be satisfied. More elaboration on this difficulty and ways to circumvent it can be found in Exercises 16 and 17.

The most common piecewise polynomial approximation using cubic polynomials between each successive pair of nodes is called **cubic spline interpolation**. A general cubic polynomial involves four constants; so there is sufficient flexibility in the cubic spline procedure to ensure not only that the interpolant is continuously differentiable on the interval, but also that it has a continuous second derivative on the interval. The construction of the cubic spline does not, however, assume that the derivatives of the interpolant agree with those of the function, even at the nodes. (*See Fig. 3.8.*)

DEFINITION 3.10   Given a function f defined on $[a, b]$ and a set of numbers, called nodes, $a = x_0 < x_1 < \cdots < x_n = b$, a cubic spline interpolant, S, for f is a function that satisfies the following conditions:

a) S is a cubic polynomial, denoted S_j, on the subinterval $[x_j, x_{j+1}]$ for each $j = 0, 1, \ldots, n-1$;

b) $S(x_j) = f(x_j)$ for each $j = 0, 1, \ldots, n$;

c) $S_{j+1}(x_{j+1}) = S_j(x_{j+1})$ for each $j = 0, 1, \ldots, n-2$;

d) $S'_{j+1}(x_{j+1}) = S'_j(x_{j+1})$ for each $j = 0, 1, \ldots, n-2$;

e) $S''_{j+1}(x_{j+1}) = S''_j(x_{j+1})$ for each $j = 0, 1, \ldots, n-2$;

f) one of the following set of boundary conditions is satisfied:

(3.24)   (i) $S''(x_0) = S''(x_n) = 0$ (Free boundary)

(3.25)   (ii) $S'(x_0) = f'(x_0)$ and $S'(x_n) = f'(x_n)$ (Clamped boundary)

FIGURE 3.8

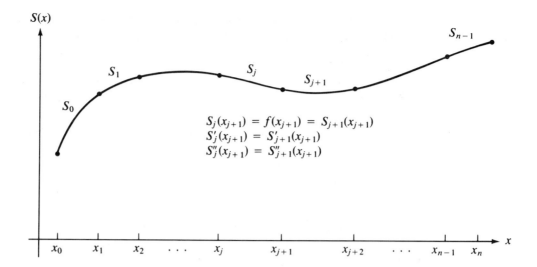

$$S_j(x_{j+1}) = f(x_{j+1}) = S_{j+1}(x_{j+1})$$
$$S_j'(x_{j+1}) = S_{j+1}'(x_{j+1})$$
$$S_j''(x_{j+1}) = S_{j+1}''(x_{j+1})$$

Although cubic splines can be defined with other boundary conditions, the conditions given are sufficient for our purposes. When the free boundary conditions occur, the spline is called a **natural spline** and its graph approximates the shape that a long flexible rod would assume if forced to go through each of the data points $\{(x_0, f(x_0)), (x_1, f(x_1)), \ldots, (x_n, f(x_n))\}$.

In general, the clamped boundary conditions will lead to more accurate approximations since they include more information about the function; however, for this type of boundary condition to hold, it is necessary to have either the values of the derivative at the endpoints or an accurate approximation to those values.

To construct the cubic-spline interpolant for a given function f, the conditions in the definition can be applied to the cubic polynomials.

$$S_j(x) = a_j + b_j(x - x_j) + c_j(x - x_j)^2 + d_j(x - x_j)^3$$

for each $j = 0, 1, \ldots, n - 1$.

Clearly,

$$S_j(x_j) = a_j = f(x_j)$$

and if condition (c) is applied,

$$a_{j+1} = S_{j+1}(x_{j+1}) = S_j(x_{j+1})$$
$$= a_j + b_j(x_{j+1} - x_j) + c_j(x_{j+1} - x_j)^2 + d_j(x_{j+1} - x_j)^3$$

for each $j = 0, 1, \ldots, n - 2$.

Since the term $(x_{j+1} - x_j)$ will be used repeatedly in this development, it is convenient to introduce the simpler notation,

$$h_j = x_{j+1} - x_j,$$

for each $j = 0, 1, \ldots, n - 1$. If we also define $a_n = f(x_n)$, it can be seen that the equation

(3.26) $$a_{j+1} = a_j + b_j h_j + c_j h_j^2 + d_j h_j^3$$

holds for each $j = 0, 1, \ldots, n-1$.

In a similar manner, define $b_n = S'(x_n)$ and observe that

$$S_j'(x) = b_j + 2c_j(x - x_j) + 3d_j(x - x_j)^2$$

implies $S_j'(x_j) = b_j$ for each $j = 0, 1, \ldots, n-1$. Applying condition (d),

(3.27) $$b_{j+1} = b_j + 2c_j h_j + 3d_j h_j^2$$

for each $j = 0, 1, \ldots, n-1$.

Another relation between the coefficients of S_j can be obtained by defining $c_n = S''(x_n)/2$ and applying condition (e). In this case

(3.28) $$c_{j+1} = c_j + 3d_j h_j$$

for each $j = 0, 1, \ldots, n-1$.

Solving for d_j in Eq. (3.28) and substituting this value into Eqs. (3.26) and (3.27) gives the new equations

(3.29) $$a_{j+1} = a_j + b_j h_j + \frac{h_j^2}{3}(2c_j + c_{j+1})$$

and

(3.30) $$b_{j+1} = b_j + h_j(c_j + c_{j+1})$$

for each $j = 0, 1, \ldots, n-1$.

The final relationship involving the coefficients is obtained by solving the appropriate equation in the form of equation (3.29), first for b_j,

(3.31) $$b_j = \frac{1}{h_j}(a_{j+1} - a_j) - \frac{h_j}{3}(2c_j + c_{j+1}),$$

and then, with a reduction of the index, for b_{j-1},

$$b_{j-1} = \frac{1}{h_{j-1}}(a_j - a_{j-1}) - \frac{h_{j-1}}{3}(2c_{j-1} + c_j).$$

Substituting these values into the equation derived from Eq. (3.30), when the index is reduced by one, gives the linear system of equations

(3.32) $$h_{j-1}c_{j-1} + 2(h_{j-1} + h_j)c_j + h_j c_{j+1} = \frac{3}{h_j}(a_{j+1} - a_j) - \frac{3}{h_{j-1}}(a_j - a_{j-1})$$

for each $j = 1, 2, \ldots, n-1$. This system involves, as unknowns, only $\{c_j\}_{j=0}^n$ since the values of $\{h_j\}_{j=0}^{n-1}$ and $\{a_j\}_{j=0}^n$ are given by the spacing of the nodes $\{x_j\}_{j=0}^n$ and the values of f at the nodes.

Note that once the values of $\{c_j\}_{j=0}^n$ are known it is a simple matter to find the remainder of the constants $\{b_j\}_{j=0}^{n-1}$ from Eq. (3.31), and $\{d_j\}_{j=0}^{n-1}$ from Eq. (3.28) and to construct the cubic polynomials $\{S_j\}_{j=0}^{n-1}$.

The major question that arises in connection with this construction is whether the values of $\{c_j\}_{j=0}^n$ can be found using the system of equations given in (3.32) and if so, whether these values are unique. The following theorems indicate that, when either of the boundary conditions given in part (f) of the definition are imposed, the answer to both questions is affirmative. The proofs of these theorems require material from linear algebra. This material is discussed in Chapter 6.

THEOREM 3.11   If f is a function that is defined on $[a, b]$, then f has a unique natural spline interpolant, that is, a unique spline interpolant satisfying the free boundary conditions $S''(a) = S''(b) = 0$.

PROOF   With the usual notation, $a = x_0 < x_1 < \cdots < x_n = b$, the boundary conditions in this case imply that $c_n = S''(x_n)/2 = 0$ and that

$$0 = S''(x_0) = 2c_0 + 6d_0(x_0 - x_0);$$

so $c_0 = 0$.

The two equations $c_0 = 0$ and $c_n = 0$ together with the equations in (3.32) produce a linear system described by the vector equation $A\mathbf{x} = \mathbf{b}$, where A is the $(n + 1)$-by-$(n + 1)$ matrix

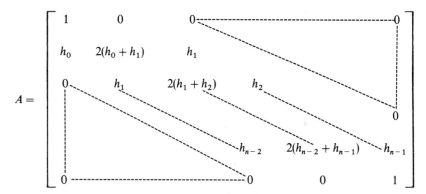

and $\mathbf{b}$ and $\mathbf{x}$ are the vectors

$$\mathbf{b} = \begin{bmatrix} 0 \\ \dfrac{3}{h_1}(a_2 - a_1) - \dfrac{3}{h_0}(a_1 - a_0) \\ \vdots \\ \dfrac{3}{h_{n-1}}(a_n - a_{n-1}) - \dfrac{3}{h_{n-2}}(a_{n-1} - a_{n-2}) \\ 0 \end{bmatrix} \quad \text{and} \quad \mathbf{x} = \begin{bmatrix} c_0 \\ c_1 \\ \vdots \\ c_n \end{bmatrix}.$$

The matrix A is strictly diagonally dominant, so it satisfies the hypotheses of Theorem 6.20, page 335. Therefore, the linear system has a unique solution for $c_0, c_1, \ldots, c_n$. $\square$

The solution to the cubic spline problem with the boundary conditions $S''(x_0) = S''(x_n) = 0$ can be obtained by applying the following algorithm.

Natural Cubic Spline Algorithm 3.4

To construct the cubic spline interpolant S for the function f, defined at the numbers $x_0 < x_1 < \cdots < x_n$, satisfying $S''(x_0) = S''(x_n) = 0$:

INPUT n: $x_0, x_1, \ldots, x_n$; either generate $a_i = f(x_i)$ for $i = 0, 1, \ldots, n$ or input a_i for $i = 0, 1, \ldots, n$.

OUTPUT a_j, b_j, c_j, d_j for $j = 0, 1, \ldots, n - 1$.
(*Note:* $S(x) = S_j(x) = a_j + b_j(x - x_j) + c_j(x - x_j)^2 + d_j(x - x_j)^3$ for $x_j \le x < x_{j+1}$.)

Step 1   For $i = 0, 1, \ldots, n - 1$ set $h_i = x_{i+1} - x_i$.

Step 2   For $i = 1, 2, \ldots, n - 1$ set

$$\alpha_i = \frac{3[a_{i+1}h_{i-1} - a_i(x_{i+1} - x_{i-1}) + a_{i-1}h_i]}{h_{i-1}h_i}.$$

Step 3   Set $l_0 = 1$;   (*Steps 3, 4, 5, and part of Step 6 solve a tridiagonal linear system using Algorithm 6.7.*)
$\mu_0 = 0$;
$z_0 = 0$.

Step 4   For $i = 1, 2, \ldots, n - 1$
set $l_i = 2(x_{i+1} - x_{i-1}) - h_{i-1}\mu_{i-1}$;
$\mu_i = h_i/l_i$;
$z_i = (\alpha_i - h_{i-1}z_{i-1})/l_i$.

Step 5   Set $l_n = 1$;
$z_n = 0$;
$c_n = 0$.

Step 6   For $j = n - 1, n - 2, \ldots, 0$
set $c_j = z_j - \mu_j c_{j+1}$;
$b_j = (a_{j+1} - a_j)/h_j - h_j(c_{j+1} + 2c_j)/3$;
$d_j = (c_{j+1} - c_j)/(3h_j)$.

Step 7   OUTPUT $(a_j, b_j, c_j, d_j$ for $j = 0, 1, \ldots, n - 1)$;
STOP.

THEOREM 3.12   If f is a function that is defined on $[a, b]$, then f has a unique spline interpolant satisfying the clamped boundary conditions $S'(a) = f'(a)$ and $S'(b) = f'(b)$.

PROOF   It can be seen, using the fact that $S'(a) = S'(x_0) = b_0$, that Eq. (3.31) with $j = 0$, implies

$$f'(a) = \frac{a_1 - a_0}{h_0} - \frac{h_0}{3}(2c_0 + c_1).$$

Consequently,

$$2h_0c_0 + h_0c_1 = \frac{3}{h_0}(a_1 - a_0) - 3f'(a).$$

Similarly,

$$f'(b) = b_n = b_{n-1} + h_{n-1}(c_{n-1} + c_n),$$

so Eq. (3.31) with $j = n - 1$, implies that

$$f'(b) = \frac{a_n - a_{n-1}}{h_{n-1}} - \frac{h_{n-1}}{3}(2c_{n-1} + c_n) + h_{n-1}(c_{n-1} + c_n)$$

$$= \frac{a_n - a_{n-1}}{h_{n-1}} + \frac{h_{n-1}}{3}(c_{n-1} + 2c_n)$$

and

$$h_{n-1}c_{n-1} + 2h_{n-1}c_n = 3f'(b) - \frac{3}{h_{n-1}}(a_n - a_{n-1}).$$

Equations (3.32), together with the equations

$$2h_0 c_0 + h_0 c_1 = \frac{3}{h_0}(a_1 - a_0) - 3f'(a)$$

and

$$h_{n-1}c_{n-1} + 2h_{n-1}c_n = 3f'(b) - \frac{3}{h_{n-1}}(a_n - a_{n-1}),$$

determine the linear system $A\mathbf{x} = \mathbf{b}$ where

$$A = \begin{bmatrix} 2h_0 & h_0 & 0 & & & & \cdots & & & 0 \\ h_0 & 2(h_0 + h_1) & h_1 & & & & & & & \\ 0 & h_1 & 2(h_1 + h_2) & h_2 & & & & & & 0 \\ & & & \ddots & \ddots & \ddots & & & & \\ & & & & & h_{n-2} & 2(h_{n-2} + h_{n-1}) & h_{n-1} \\ 0 & & & \cdots & & & 0 & h_{n-1} & 2h_{n-1} \end{bmatrix}$$

$$\mathbf{b} = \begin{bmatrix} \frac{3}{h_0}(a_1 - a_0) - 3f'(a) \\[2mm] \frac{3}{h_1}(a_2 - a_1) - \frac{3}{h_0}(a_1 - a_0) \\[2mm] \vdots \\[2mm] \frac{3}{h_{n-1}}(a_n - a_{n-1}) - \frac{3}{h_{n-2}}(a_{n-1} - a_{n-2}) \\[2mm] 3f'(b) - \frac{3}{h_{n-1}}(a_n - a_{n-1}) \end{bmatrix} \quad \text{and} \quad \mathbf{x} = \begin{bmatrix} c_0 \\ c_1 \\ \vdots \\ c_n \end{bmatrix}.$$

This matrix A is also strictly diagonally dominant, so it satisfies the conditions of Theorem 6.20, page 335. Therefore, the linear system has a unique solution for $c_0, c_1, \ldots, c_n$. $\square$

Clamped Cubic Spline Algorithm 3.5

To construct the cubic spline interpolant S for the function f, defined at the numbers $x_0 < x_1 < \cdots < x_n$, satisfying $S'(x_0) = f'(x_0)$ and $S'(x_n) = f'(x_n)$:

INPUT n; $x_0, x_1, \ldots, x_n$; either generate $a_i = f(x_i)$ for $i = 0, 1, \ldots, n$ or input a_i for $i = 0, 1, \ldots, n$; $FPO = f'(x_0)$; $FPN = f'(x_n)$.

OUTPUT a_j, b_j, c_j, d_j for $j = 0, 1, \ldots, n-1$.
(*Note*: $S(x) = a_j + b_j(x - x_j) + c_j(x - x_j)^2 + d_j(x - x_j)^3$ for $x_j \leq x < x_{j+1}$.)

Step 1   For $i = 0, 1, \ldots, n-1$ set $h_i = x_{i+1} - x_i$.

Step 2   Set $\alpha_0 = 3(a_1 - a_0)/h_0 - 3FPO$;
$\quad\quad\quad \alpha_n = 3FPN - 3(a_n - a_{n-1})/h_{n-1}$.

Step 3   For $i = 1, 2, \ldots, n-1$

$$\text{set } \alpha_i = \frac{3[a_{i+1}h_{i-1} - a_i(x_{i+1} - x_{i-1}) + a_{i-1}h_i]}{h_{i-1}h_i}.$$

Step 4   Set $l_0 = 2h_0$;   (*Steps 4, 5, 6, and part of Step 7 solve a triagonal linear system using Algorithm 6.7.*)

$\quad\quad\quad \mu_0 = 0.5$;
$\quad\quad\quad z_0 = \alpha_0/l_0$.

Step 5   For $i = 1, 2, \ldots, n-1$
$\quad\quad\quad$ set $l_i = 2(x_{i+1} - x_{i-1}) - h_{i-1}\mu_{i-1}$;
$\quad\quad\quad \mu_i = h_i/l_i$;
$\quad\quad\quad z_i = (\alpha_i - h_{i-1}z_{i-1})/l_i$.

Step 6   Set $l_n = h_{n-1}(2 - \mu_{n-1})$;
$\quad\quad\quad z_n = (\alpha_n - h_{n-1}z_{n-1})/l_n$;
$\quad\quad\quad c_n = z_n$.

Step 7   For $j = n-1, n-2, \ldots, 0$
$\quad\quad\quad$ set $c_j = z_j - \mu_j c_{j+1}$;
$\quad\quad\quad b_j = (a_{j+1} - a_j)/h_j - h_j(c_{j+1} + 2c_j)/3$;
$\quad\quad\quad d_j = (c_{j+1} - c_j)/(3h_j)$.

Step 8   OUTPUT $(a_j, b_j, c_j, d_j$ for $j = 0, 1, \ldots, n-1)$;
$\quad\quad\quad$ STOP.

EXAMPLE 1   We will determine a cubic spline interpolant to represent the curve constituting the upper portion of the well-known figure shown in Fig. 3.9.

It is first necessary to set up a grid to describe the points on the curve, as shown in Fig. 3.10. Because the curve does not have a continuous derivative at the points where the paw meets the head and where the head meets the back, separate splines must be used for each of these regions. Using the scale in Fig. 3.10 gives the data in

FIGURE 3.9

FIGURE 3.10

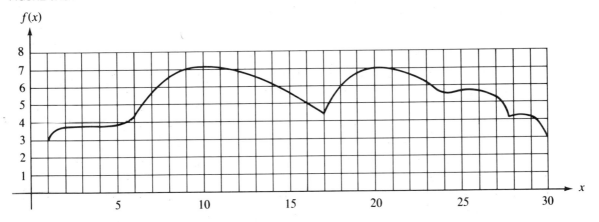

Table 3.14, where the values of x are chosen to be more concentrated where abrupt changes occur. (*See Fig. 3.11.*)

Using Algorithm 3.4 to construct the natural splines corresponding to this data gives the results in Tables 3.15, 3.16, and 3.17.

The graph of this piecewise spline function is shown in Fig. 3.12.

TABLE 3.14

Curve 1			Curve 2			Curve 3		
i	x_i	$f(x_i)$	i	x_i	$f(x_i)$	i	x_i	$f(x_i)$
0	1	3.0	0	17	4.5	0	27.7	4.1
1	2	3.7	1	20	7.0	1	28	4.3
2	5	3.9	2	23	6.1	2	29	4.1
3	6	4.2	3	24	5.6	3	30	3.0
4	7	5.7	4	25	5.8			
5	8	6.6	5	27	5.2			
6	10	7.1	6	27.7	4.1			
7	13	6.7						
8	17	4.5						

FIGURE 3.11

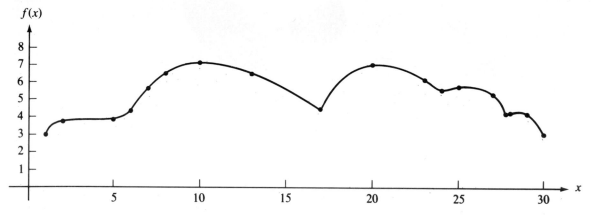

TABLE 3.15

			Spline 1		
j	x_j	$a_j = f(x_j)$	b_j	c_j	d_j
0	1	3.0	0.786	0.0	−0.086
1	2	3.7	0.529	−0.257	0.034
2	5	3.9	−0.086	0.052	0.334
3	6	4.2	1.019	1.053	−0.572
4	7	5.7	1.408	−0.664	0.156
5	8	6.6	0.547	−0.197	0.024
6	10	7.1	0.049	−0.052	−0.003
7	13	6.7	−0.342	−0.078	0.007
8	17	4.5			

TABLE 3.16

			Spline 2		
j	x_j	$a_j = f(x_j)$	b_j	c_j	d_j
0	17	4.5	1.106	0.0	−0.030
1	20	7.0	0.289	−0.272	0.025
2	23	6.1	−0.660	−0.044	0.204
3	24	5.6	−0.137	0.567	−0.230
4	25	5.8	0.306	−0.124	−0.089
5	27	5.2	−1.263	−0.660	0.314
6	27.7	4.1			

TABLE 3.17

			Spline 3		
j	x_j	$a_j = f(x_j)$	b_j	c_j	d_j
0	27.7	4.1	0.749	0.0	−0.910
1	28	4.3	0.503	−0.819	0.116
2	29	4.1	−0.787	−0.470	0.157
3	30	3.0			

FIGURE 3.12

Free spline ———————

Original curve — — — — —

If we approximate the derivative at each of the endpoints of the three smooth pieces of the curve, three clamped boundary-condition splines can be used to approximate the curve. (*See Fig. 3.13.*)

FIGURE 3.13

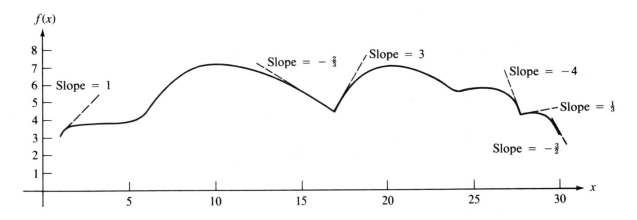

The data for this situation are the same as those involving the natural splines, with the additional assumptions on the derivatives at the endpoints, and are given in Table 3.18.

Using Algorithm 3.5 to construct the clamped boundary-condition splines corresponding to this data gives the results shown in Tables 3.19, 3.20, and 3.21. The graph of this piecewise spline function is shown in Fig. 3.14.

TABLE 3.18

	Curve 1				Curve 2				Curve 3		
i	x_i	$f(x_i)$	$f'(x_i)$	i	x_i	$f(x_i)$	$f'(x_i)$	i	x_i	$f(x_i)$	$f'(x_i)$
0	1	3.0	1.0	0	17	4.5	3.0	0	27.7	4.1	0.33
1	2	3.7		1	20	7.0		1	28	4.3	
2	5	3.9		2	23	6.1		2	29	4.1	
3	6	4.2		3	24	5.6		3	30	3.0	−1.5
4	7	5.7		4	25	5.8					
5	8	6.6		5	27	5.2					
6	10	7.1		6	27.7	4.1	−4.0				
7	13	6.7									
8	17	4.5	−0.67								

TABLE 3.19

				Spline 1		
j	x_j	$a_j = f(x_j)$	b_j	c_j	d_j	$f'(x_j)$
0	1	3.0	1.0	−0.347	0.049	1.0
1	2	3.7	0.447	−0.206	0.027	
2	5	3.9	−0.074	0.033	0.342	
3	6	4.2	1.016	1.058	−0.575	
4	7	5.7	1.409	−0.665	0.156	
5	8	6.6	0.547	−0.196	0.024	
6	10	7.1	0.048	−0.053	−0.003	
7	13	6.7	−0.339	−0.076	0.006	
8	17	4.5				−0.67

TABLE 3.20

				Spline 2		
j	x_j	$a_j = f(x_j)$	b_j	c_j	d_j	$f'(x_j)$
0	17	4.5	3.0	−1.101	0.126	3.0
1	20	7.0	−0.198	0.035	−0.023	
2	23	6.1	−0.609	−0.172	0.280	
3	24	5.6	−0.111	0.669	−0.357	
4	25	5.8	0.154	−0.403	0.088	
5	27	5.2	−0.401	0.126	−2.568	
6	27.7	4.1				−4.0

The reason the natural splines approximate the original curve better than the splines with the clamped boundary conditions is that the derivatives at the endpoints of the sections of the curve have been inaccurately estimated. In this example, the natural spline has assumed a better estimate for these derivatives by using the assumption that the second derivative at each end is zero, than were given for the clamped boundary-condition splines. ∎

TABLE 3.21

			Spline 3			
j	x_j	$a_j = f(x_j)$	b_j	c_j	d_j	$f'(x_j)$
0	27.7	4.1	0.330	2.262	-3.800	0.33
1	28	4.3	0.661	-1.157	0.296	
2	29	4.1	-0.765	-0.269	-0.065	
3	30	3.0				-1.5

FIGURE 3.14

Clamped spline ————

Original curve — — — —

Because using the clamped boundary conditions generally is preferred when approximating functions by cubic splines, the derivative of the function often must be estimated at the endpoints of the interval. In the case where the nodes are equally spaced near both endpoints, approximations can be obtained by using Eq. (4.15) or any of the other appropriate formulas given in Sections 4.1 and 4.2. In the case of unequally spaced nodes, the problem is considerably more difficult.

To conclude this section, we list an error-bound formula for the cubic spline with clamped boundary conditions. The proof of this result can be found in Schultz [119], pp. 57–58. A fourth-order error-bound result also holds in the case of free boundary conditions, but is more difficult to express. (See Birkhoff and de Boor [9], pp. 827–835.)

THEOREM 3.13

Let $f \in C^4[a, b]$ with $\max_{a \le x \le b} |f^{(4)}(x)| \le M$. If S is the unique cubic spline interpolant to f with respect to the nodes $a = x_0 < x_1 < \cdots < x_n = b$, which satisfies $S'(a) = f'(a)$ and $S'(b) = f'(b)$, then

(3.33)
$$\max_{a \le x \le b} |f(x) - S(x)| \le \frac{5M}{384} \max_{0 \le j \le n-1} (x_{j+1} - x_j)^4.$$

Exercise Set 3.6

1. Use free cubic spline interpolation to find an approximation to the following:

   a) $f(2.5)$ given

x	$f(x)$
2.2	0.5207843
2.4	0.5104147
2.6	0.4813306

   b) $f(5.3)$ given

x	$f(x)$
5.0	2.168861
5.2	1.797350
5.4	1.488591

   c) $f(0)$ given

x	$f(x)$
-0.3	-0.20431
-0.1	-0.08993
0.1	0.11007
0.3	0.39569

   d) $f(1.25)$ given

x	$f(x)$
1.1	0.48603
1.2	0.86160
1.3	1.59751
1.4	3.76155

   e) $f(0.5)$ given

x	$f(x)$
0.2	0.9798652
0.4	0.9177710
0.6	0.8080348
0.8	0.6386093
1.0	0.3843735

   f) $f(0.2)$ given

x	$f(x)$
0.1	1.2314028
0.3	1.9121188
0.4	2.3855409
0.5	2.9682818
0.6	3.6801169

2. Repeat Exercise 1 using clamped cubic spline interpolation and the fact that

   a) $f'(2.2) = -0.0014878$, $f'(2.6) = -0.1883635$. d) $f'(1.1) = 2.90986$, $f'(1.4) = 41.13928$.

   b) $f'(5.0) = -1.495067$, $f'(5.4) = -1.070309$.   e) $f'(0.2) = 0.20271$, $f'(1.0) = 1.55741$.

   c) $f'(-0.3) = 0.32213$, $f'(0.3) = 1.67787$.   f) $f'(0.1) = 2.64281$, $f'(0.6) = 7.84023$.

3. Construct a free cubic spline to approximate $f(x) = \cos \pi x$ by using the values given by $f(x)$ at $x = 0, 0.25, 0.5, 0.75$, and 1.0. Integrate the spline over $[0, 1]$, and compare the result to $\int_0^1 \cos \pi x \, dx = 0$. Use the derivatives of the spline to approximate $f'(0.5)$ and $f''(0.5)$. Compare these approximations to the actual values.

4. Construct a free cubic spline to approximate $f(x) = e^{-x}$ by using the values given by $f(x)$ at $x = 0, 0.25, 0.75$, and 1.0. Integrate the spline over $[0, 1]$, and compare the result to $\int_0^1 e^{-x} dx = (1/e)(e - 1)$. Use the derivatives of the spline to approximate $f'(0.5)$ and $f''(0.5)$. Compare the approximations to the actual values.

5. Repeat Exercise 3, constructing instead the clamped cubic spline with $f'(0) = f'(1) = 0$.

6. Repeat Exercise 4, constructing instead the clamped cubic spline with $f'(0) = -1$, $f'(1) = -e^{-1}$.

7. a) Use the following values to construct a free cubic spline to approximate sin 0.34.

   b) Determine the error for the approximation in part (a).

   c) Repeat part (a) using a clamped cubic spline.

   d) Repeat part (b) for the clamped cubic spline constructed in part (c).

x	$\sin x$	$D_x(\sin x) = \cos x$
0.30	0.29552	0.95534
0.32	0.31457	0.94924
0.35	0.34290	0.93937

   e) Use the spline constructed in part (a) to approximate cos 0.34.

   f) Use the spline constructed in part (c) to approximate cos 0.34.

   g) Use the spline constructed in part (a) to approximate

   $$\int_{0.30}^{0.35} \sin x \, dx.$$

   h) Use the spline constructed in part (c) to approximate

   $$\int_{0.30}^{0.35} \sin x \, dx.$$

8. Add $\sin 0.33 = 0.32404$ and $\cos 0.33 = 0.94604$ to the data in Exercise 7 and redo the calculations in that exercise.

9. Use the cubic spline with clamped boundary conditions to approximate the function $f(x) = 3xe^x - e^{2x}$ at $x = 1.03$ using the data

x	1.0	1.02	1.04	1.06
$f(x)$	0.76578939	0.79536678	0.82268817	0.84752226

   Estimate the error, using Eq. (3.33), and compare to the actual error.

10. Given the partition $x_0 = 0$, $x_1 = 0.05$, $x_2 = 0.1$ of $[0, 0.1]$ and $f(x) = e^{2x}$:

    a) Find the cubic spline s with clamped boundary conditions that interpolates f.

    b) Find an approximation for $\int_0^{0.1} e^{2x} \, dx$ by evaluating $\int_0^{0.1} s(x) \, dx$.

    c) Use Theorem 3.13 to estimate $\max_{0 \leq x \leq 0.1} |f(x) - s(x)|$ and

    $$\left| \int_0^{0.1} f(x) \, dx - \int_0^{0.1} s(x) \, dx \right|.$$

d) Determine the cubic spline S with free boundary conditions and compare $S(0.02)$, $s(0.02)$, and $e^{0.04} = 1.04081077$.

11. Given the partition $x_0 = 0$, $x_1 = 0.05$, $x_2 = 0.1$ of $[0, 0.1]$, find the piecewise linear interpolating function F for $f(x) = e^{2x}$. Approximate $\int_0^{0.1} e^{2x} \, dx$ with $\int_0^{0.1} F(x) \, dx$. Compare the results to those of Exercise 10.

12. Suppose the following data is given for a function f: $f(0) = 0$, $f(0.1) = -0.1$, $f(0.2) = -0.3$, $f(0.3) = -0.5$, $f(0.4) = -0.4$, $f(0.5) = -0.2$, $f(0.6) = 0$, $f(0.7) = 0.2$, $f(0.8) = 0.3$, $f(0.9) = 0.7$, $f(1) = 0.9$. Find the spline that interpolates f with free boundary conditions with respect to the partition $x_j = 0.1j$ for $j = 0, 1, \ldots, 10$.

13. Extend Algorithms 3.4 and 3.5 to include as output the first and second derivatives of the spline at the nodes.

14. Extend Algorithms 3.4 and 3.5 to include as output the integral of the spline over the interval $[x_0, x_n]$.

15. Let $f \in C^2[a, b]$, and let the nodes $a = x_0 < x_1 < \cdots < x_n = b$ be given. Derive an error estimate similar to that in Theorem 3.13 for the piecewise linear interpolating function F. Use this estimate to derive error bounds for Exercise 11.

16. Let f be defined on $[a, b]$, and let the nodes $a = x_0 < x_1 < x_2 = b$ be given. A quadratic-spline interpolating function S consists of the quadratic polynomial

$$S_0(x) = a_0 + b_0(x - x_0) + c_0(x - x_0)^2 \qquad \text{on } [x_0, x_1]$$

and the quadratic polynomial

$$S_1(x) = a_1 + b_1(x - x_1) + c_1(x - x_1)^2 \qquad \text{on } [x_1, x_2]$$

such that

i) $S(x_0) = f(x_0)$, $S(x_1) = f(x_1)$, and $S(x_2) = f(x_2)$,

ii) $S \in C^1[x_0, x_2]$.

Show that conditions (i) and (ii) lead to five equations in the six unknowns a_0, b_0, c_0, a_1, b_1, and c_1. The problem is to decide what additional condition to impose to make the solution unique, for example,

$$f'(x_0) = S'(x_0) \qquad \text{or} \qquad f'(x_2) = S'(x_2).$$

Does the condition

$$S \in C^2[x_0, x_2]$$

lead to a meaningful solution?

17. As suggested in a paper by Kammerer, Reddien, and Varga [74], useful quadratic interpolatory splines can be constructed. Given a function f on $[a, b]$ and the points $a = x_0 < x_1 < \cdots < x_n = b$, the quadratic spline satisfies the following:

i) $s(x_0) = f(x_0)$,

$$s\left(\frac{x_i + x_{i+1}}{2}\right) = f\left(\frac{x_i + x_{i+1}}{2}\right) \qquad \text{for each } i = 0, 1, \ldots, n - 1,$$

$$s(x_n) = f(x_n);$$

ii) $s \in C^1[a, b]$;

iii) for each $i = 0, 1, \ldots, n-1$,

$$s(x) = a_i + b_i(x - x_i) + c_i(x - x_i)^2, \quad x_i \leq x \leq x_{i+1}.$$

a) Show that conditions (i), (ii), and (iii) lead to the equations:

$$a_0 = f(x_0);$$

$$a_i + \tfrac{1}{2}h_i b_i + \tfrac{1}{4}h_i^2 c_i = f(x_i + \tfrac{1}{2}h_i) \qquad \text{for each } i = 0, 1, \ldots, n-1;$$

$$a_{n-1} + h_{n-1}b_{n-1} + h_{n-1}^2 c_{n-1} = f(x_n);$$

$$a_i = a_{i-1} + b_{i-1}h_{i-1} + c_{i-1}h_{i-1}^2 \qquad \text{for each } i = 1, 2, \ldots, n-1;$$

$$b_i = b_{i-1} + 2c_{i-1}h_{i-1} \qquad \text{for each } i = 1, 2, \ldots, n-1.$$

b) Show that the results of part (a) lead to the equations:

$$\tfrac{3}{8}h_0 b_0 + \tfrac{1}{8}h_0 b_1 = f\left(x_0 + \frac{h_0}{2}\right) - f(x_0);$$

$$\tfrac{1}{8}h_{i-1}b_{i-1} + \tfrac{3}{8}(h_i + h_{i\ 1})b_i + \tfrac{1}{8}h_i b_{i+1} = f\left(x_i + \frac{h_i}{2}\right) - f\left(x_{i-1} + \frac{h_{i-1}}{2}\right)$$

$$\text{for each } i = 1, 2, \ldots, n-2;$$

$$\tfrac{1}{8}h_{n-2}b_{n-2} + (\tfrac{3}{8}h_{n-2} + \tfrac{1}{3}h_{n-1})b_{n-1}$$

$$= (\tfrac{4}{3})f\left(x_{n-1} + \frac{h_{n-1}}{2}\right) - f\left(x_{n-2} + \frac{h_{n-2}}{2}\right) - \tfrac{1}{3}f(x_n);$$

where $h_i = x_{i+1} - x_i$ for each $i = 0, 1, \ldots, n-1$.

c) Devise an algorithm similar to Algorithms 3.4 and 3.5 to find $b_0, \ldots, b_{n-1}$.

d) Devise formulas for $a_0, \ldots, a_{n-1}, c_0, \ldots, c_{n-1}$, assuming that $b_0, \ldots, b_{n-1}$ are known.

e) Let $f(x) = e^x$, $x_0 = 0$, $x_1 = 0.2$, $x_2 = 0.6$, and $x_3 = 0.9$. Find the quadratic interpolating spline s for f, and compute $s(0.5)$. Is $s(0.5)$ a good approximation to $f(0.5)$?

f) Repeat Exercise 9, using the quadratic interpolating spline, and compare your results to those obtained in Exercise 9. Is it reasonable to believe that

$$|f(x) - s(x)| = O(h^k), \qquad \text{for } k = 2, \text{ or } k = 4?$$

18. Carry out in detail the following derivation of an alternate formula for cubic splines. Given a partition

$$a = x_0 < x_1 < \cdots < x_n = b \qquad \text{of } [a, b],$$

the cubic spline s, which interpolates f, is a cubic polynomial $s_j(x)$ on each $[x_j, x_{j+1}]$. Thus, we may write

$$s''(x) = s_j''(x) = \frac{a_j(x_{j+1} - x)}{h_j} + \frac{a_{j+1}(x - x_j)}{h_j}$$

on $[x_j, x_{j+1}]$ for $j = 0, 1, \ldots, n-1$; that is, $s_j''(x)$ is linear on each subinterval. Show s'' is continuous on $[a, b]$. Using the notation $s_j = s(x_j)$ and $f_j = f(x_j)$ for $j = 0, 1, \ldots, n-1$, integrate $s_j''(x)$ twice and evaluate the constants of integration by using $s_j = f_j$ and $s_{j+1} = f_{j+1}$ to obtain

$$s_j(x) = a_j \frac{(x_{j+1} - x)^3}{6h_j} + \frac{a_{j+1}}{6h_j}(x - x_j)^3$$

$$+ \left(f_{j+1} - \frac{a_{j+1}h_j^2}{6} \right) \frac{(x - x_j)}{h_j} + \left(f_j - \frac{a_j h_j^2}{6} \right) \frac{(x_{j+1} - x)}{h_j}.$$

Now obtain a system of equations for the a_j's from the requirement that s' is continuous on $[a, b]$.

19. Define

$$(x - \xi)_+^3 = \begin{cases} (x - \xi)^3, & x > \xi, \\ 0, & x \le \xi. \end{cases}$$

Let $a = x_0 < x_1 < x_2 < x_3 = b$ and define

$$S(x) = c_1(x - x_1)_+^3 + c_2(x - x_2)_+^3, \qquad \text{where } x \in [a, b].$$

Show that $S \in C^2[a, b]$.

20. Let $P(x) = a_0 + a_1 x + a_2 x^2 + a_3 x^3$ be any cubic polynomial on $[a, b]$ and $a = x_0 < x_1 < \cdots < x_n = b$ be given nodes. Show that

$$S(x) = P(x) + \sum_{j=1}^{n-1} c_j(x - x_j)_+^3$$

is a cubic spline, in that S is a cubic polynomial on $[x_i, x_{i+1}]$ for each $i = 0, 1, \ldots, n-1$ and $S \in C^2[a, b]$.

21. Let P be a polynomial of degree at most three, $\{x_j\}_{j=0}^n$ be a partition of $[a, b]$ and

$$s(x) = P(x) + \sum_{j=1}^{n-1} c_j(x - x_j)_+^3.$$

Show that the third derivative of s exists and is continuous on $[a, b]$ if and only if $c_j = 0$ for $j = 1, 2, \ldots, n$.

22. Exercise 10 of Section 3.5 lists observed data concerning the speed and distance of a car traveling along a straight road. Use a clamped cubic spline to predict the position of the car and its speed at $t = 10$ sec. Use the spline to determine whether the car ever exceeds a 55 mph speed limit on the road, and if so, what is the first time the car exceeds this speed. What is the predicted maximum speed for the car?

23. It is suspected that the high amounts of tannin in mature oak leaves inhibits the growth of the winter moth (Operophtera bromata L., Geometridæ) larvae that extensively damage these trees in certain years. The following table lists the average weight of two samples of larvae at times in the first 28 days after birth. The first sample was reared on young oak leaves, while the second sample was reared on mature leaves from the same tree.

a) Use a free cubic spline to approximate the average weight curve for each sample.

b) Find an approximate maximum average weight for each sample by determining the maximum of the spline.

Day	0	6	10	13	17	20	28
Sample 1 Average Weight (mg.)	6.67	17.33	42.67	37.33	30.10	29.31	28.74
Sample 2 Average Weight (mg.)	6.67	16.11	18.89	15.00	10.56	9.44	8.89

24. The 1979 Kentucky Derby was won by a horse named Spectacular Bid in a time of $2:02\frac{2}{5}$ (2 min, $2\frac{2}{5}$ sec) for the $1\frac{1}{4}$ mile race. Times at the quarter mile, half mile, and mile poles were $25\frac{2}{5}$, $49\frac{2}{5}$, and $1:37\frac{3}{5}$.

   a) Use these values together with the starting time to construct a free cubic spline for Spectacular Bid's race.

   b) Use the spline to predict the time at the three-quarter mile pole, and compare this to the actual time of $1:12\frac{2}{5}$.

   c) Use the spline to approximate Spectacular Bid's starting speed and speed at the finish line.

25. The data in the following table lists the population of the United States for the years 1930 to 1980 and was discussed in the introduction to this chapter, as well as in Exercise 15 of Section 3.2 and Exercise 8 of Section 3.4.

Year	1930	1940	1950	1960	1970	1980
Population (in thousands)	123,203	131,669	150,697	179,323	203,212	226,505

Find a free cubic spline agreeing with this data, and use the spline to predict the population in the years 1920, 1965, and 2000. Compare your approximations with those previously obtained. If you had to make a choice, which interpolation procedure would you choose?

4

Numerical Differentiation and Integration

A sheet of corrugated roofing is to be constructed using a machine that presses a flat sheet of aluminum into one whose cross section has the form of a sine wave.

Suppose a corrugated sheet 4 feet long is needed, the height of each wave is 1 inch from the center line, and each wave has a period of approximately 2π inches. The problem of finding the length of the initial flat sheet is one of determining the arc length of the curve given by $f(x) = \sin x$ from $x = 0$ inches to $x = 48$ inches. From calculus we know that this length can be expressed

$$L = \int_0^{48} \sqrt{1 + \left(\frac{df(x)}{dx}\right)^2}\, dx = \int_0^{48} \sqrt{1 + (\cos x)^2}\, dx,$$

so the problem reduces to evaluating this integral. Although the sine function is one of the most common mathematical functions, the calculation of its arc length gives rise to what is called an elliptic integral of the second kind, which cannot be evaluated by ordinary methods. Approximation methods will be developed in this chapter which will reduce problems of this type to elementary exercises. This particular problem is considered in Exercise 19 of Section 4.4 and in Exercise 5 of Section 4.5.

Because of the many applications that involve the derivatives and integrals of functions, it should be expected that approximation involving these concepts is of interest.

In the introduction to Chapter 3, we mentioned that one reason for using the class of algebraic polynomials to approximate an arbitrary set of data is that, given any continuous function defined on a closed interval, there exists a polynomial which is arbitrarily close to the function at every point in the interval. Another property that this class possesses is that the derivatives and integrals of polynomials are quite easily obtained and evaluated. It should not be surprising, then, that most procedures for approximating integrals and derivatives commence with algebraic polynomials approximating the function.

4.1 Numerical Differentiation

To introduce the subject of numerical differentiation, suppose we are given a function $f \in C^2[a, b]$ and an arbitrary point x_0 in $[a, b]$. A method is needed to approximate $f'(x_0)$. With $x_1 = x_0 + h$ for some $h \neq 0$, small enough to ensure that $x_1 \in [a, b]$, compute $P_{0,1}(x)$. Using the notation in Section 3.3 and the results of Theorem 3.3,

$$f(x) = P_{0,1}(x) + \frac{(x - x_0)(x - x_1)}{2!} f''(\xi(x))$$

$$= \frac{f(x_0)(x - x_0 - h)}{-h} + \frac{f(x_0 + h)(x - x_0)}{h}$$

$$+ \frac{(x - x_0)(x - x_0 - h)}{2} f''(\xi(x))$$

for some $\xi(x)$ in $[a, b]$. Differentiating gives

$$(4.1) \qquad f'(x) = \frac{f(x_0 + h) - f(x_0)}{h} + D_x\left[\frac{(x - x_0)(x - x_0 - h)}{2} f''(\xi(x))\right]$$

$$= \frac{f(x_0 + h) - f(x_0)}{h} + \frac{2(x - x_0) - h}{2} f''(\xi(x))$$

$$+ \frac{(x - x_0)(x - x_0 - h)}{2} D_x(f''(\xi(x))).$$

The difficulty with this formula for approximating $f'(x)$ for arbitrary values of x is that we have no information about $D_x f''(\xi(x)) = f'''(\xi(x)) \cdot \xi'(x)$, so the truncation error cannot be estimated. When x is x_0, however, this term is zero and the formula simplifies to

$$(4.2) \qquad f'(x_0) = \frac{f(x_0 + h) - f(x_0)}{h} - \frac{h}{2} f''(\xi).$$

For small values of h, the difference quotient $[f(x_0 + h) - f(x_0)]/h$ can be used to approximate $f'(x_0)$ with an error bounded by $(h/2)M$, where M is a bound on $f''(x)$

for $x \in [a, b]$. This formula is known as the **forward-difference formula** if $h > 0$ and the **backward-difference formula** if $h < 0$.

EXAMPLE 1   Let $f(x) = \ln x$ and $x_0 = 1.8$. The quotient

$$\frac{f(1.8 + h) - f(1.8)}{h}, \qquad h > 0,$$

will be used to approximate $f'(1.8)$ with error

$$\frac{|hf''(\xi)|}{2} = \frac{|h|}{2\xi^2} \le \frac{|h|}{2(1.8)^2} \qquad \text{where } 1.8 < \xi < 1.8 + h.$$

The results in Table 4.1 are produced when $h = 0.1, 0.01,$ and 0.001.

TABLE 4.1

h	$f(1.8 + h)$	$\dfrac{f(1.8 + h) - f(1.8)}{h}$	$\dfrac{|h|}{2(1.8)^2}$
0.1	0.64185389	0.5406722	0.0154321
0.01	0.59332685	0.5540180	0.0015432
0.001	0.58834207	0.5554013	0.0001543

Since $f'(x) = 1/x$, the exact value of $f'(1.8)$ is $0.55\overline{5}$, and the error bounds given are appropriate. ∎

Suppose that $\{x_0, x_1, \ldots, x_n\}$ are $(n + 1)$ distinct numbers in some interval I and $f \in C^{n+1}(I)$. From Theorem 3.3,

$$f(x) = \sum_{k=0}^{n} f(x_k)L_k(x) + \frac{(x - x_0) \cdots (x - x_n)}{(n + 1)!} f^{(n+1)}(\xi(x))$$

for some $\xi(x)$ in I, where $L_k(x)$ denotes the kth Lagrange coefficient polynomial for f at $x_0, x_1, \ldots, x_n$. Differentiating this expression gives

$$(4.3) \qquad f'(x) = \sum_{k=0}^{n} f(x_k)L_k'(x) + D_x\left[\frac{(x - x_0) \cdots (x - x_n)}{(n + 1)!}\right] f^{(n+1)}(\xi(x))$$

$$+ \frac{(x - x_0) \cdots (x - x_n)}{(n + 1)!} D_x[f^{(n+1)}(\xi(x))].$$

Again we have the problem of estimating the truncation error unless x is one of the numbers x_k. In this case the term involving $D_x[f^{(n+1)}(\xi(x))]$ is zero and the formula becomes

$$(4.4) \qquad f'(x_k) = \sum_{j=0}^{n} f(x_j)L_j'(x_k) + \frac{f^{(n+1)}(\xi(x))}{(n + 1)!} \prod_{\substack{j=0 \\ j \ne k}}^{n} (x_k - x_j).$$

Equation (4.4) is called an $(n + 1)$-point formula to approximate $f'(x_k)$, since a linear combination of the $(n + 1)$ values $f(x_j)$ is used for $j = 0, 1, \ldots, n$.

In general, using more evaluation points in Eq. (4.4) produces greater accuracy, although the number of functional evaluations and growth of rounding error discourages this somewhat. The most common formulas are those involving three and five evaluation points.

We will first derive some useful three-point formulas and consider aspects of their rounding errors. Since

$$L_0(x) = \frac{(x - x_1)(x - x_2)}{(x_0 - x_1)(x_0 - x_2)}, \qquad L_0'(x) = \frac{2x - x_1 - x_2}{(x_0 - x_1)(x_0 - x_2)}.$$

Similarly,

$$L_1'(x) = \frac{2x - x_0 - x_2}{(x_1 - x_0)(x_1 - x_2)} \qquad \text{and} \qquad L_2'(x) = \frac{2x - x_0 - x_1}{(x_2 - x_0)(x_2 - x_1)}.$$

Hence, from Eq. (4.4)

$$(4.5) \qquad f'(x_j) = f(x_0)\left[\frac{2x_j - x_1 - x_2}{(x_0 - x_1)(x_0 - x_2)}\right] + f(x_1)\left[\frac{2x_j - x_0 - x_2}{(x_1 - x_0)(x_1 - x_2)}\right]$$
$$+ f(x_2)\left[\frac{2x_j - x_0 - x_1}{(x_2 - x_0)(x_2 - x_1)}\right] + \tfrac{1}{6}f^{(3)}(\xi_j) \prod_{\substack{i=0 \\ i \neq j}}^{2} (x_j - x_i),$$

for each $j = 0, 1, 2$, where the notation ξ_j indicates that this point depends on x_j.

The three formulas from Eq. (4.5) become especially useful if the nodes are equally spaced, that is, when

$$x_1 = x_0 + h \qquad \text{and} \qquad x_2 = x_0 + 2h \qquad \text{for some } h \neq 0.$$

We will assume equally-spaced nodes throughout the remainder of this section. Using Eq. (4.5) with $x_j = x_0$, $x_1 = x_0 + h$, and $x_2 = x_0 + 2h$ gives

$$f'(x_0) = \frac{1}{h}[-\tfrac{3}{2}f(x_0) + 2f(x_1) - \tfrac{1}{2}f(x_2)] + \frac{h^2}{3} f^{(3)}(\xi_0).$$

Doing the same for $x_j = x_1$ gives

$$f'(x_1) = \frac{1}{h}[-\tfrac{1}{2}f(x_0) + \tfrac{1}{2}f(x_2)] - \frac{h^2}{6} f^{(3)}(\xi_1),$$

and for $x_j = x_2$,

$$f'(x_2) = \frac{1}{h}[\tfrac{1}{2}f(x_0) - 2f(x_1) + \tfrac{3}{2}f(x_2)] + \frac{h^2}{3} f^{(3)}(\xi_2).$$

Since $x_1 = x_0 + h$ and $x_2 = x_0 + 2h$, these formulas can also be expressed as

$$(4.6) \qquad f'(x_0) = \frac{1}{h}[-\tfrac{3}{2}f(x_0) + 2f(x_0 + h) - \tfrac{1}{2}f(x_0 + 2h)] + \frac{h^2}{3} f^{(3)}(\xi_0),$$

$$(4.7) \qquad f'(x_0 + h) = \frac{1}{h}[-\tfrac{1}{2}f(x_0) + \tfrac{1}{2}f(x_0 + 2h)] - \frac{h^2}{6} f^{(3)}(\xi_1), \quad \text{and}$$

$$(4.8) \qquad f'(x_0 + 2h) = \frac{1}{h}\left[\tfrac{1}{2}f(x_0) - 2f(x_0 + h) + \tfrac{3}{2}f(x_0 + 2h)\right] + \frac{h^2}{3}f^{(3)}(\xi_2).$$

As a matter of convenience, the variable substitution x_0 for $x_0 + h$ is used in Eq. (4.7) to change this formula to an approximation for $f'(x_0)$. A similar change, x_0 for $x_0 + 2h$, is used in Eq. (4.8). This gives three formulas for approximating $f'(x_0)$:

$$(4.9) \qquad f'(x_0) = \frac{1}{2h}\left[-3f(x_0) + 4f(x_0 + h) - f(x_0 + 2h)\right] + \frac{h^2}{3}f^{(3)}(\xi_0),$$

$$(4.10) \qquad f'(x_0) = \frac{1}{2h}\left[-f(x_0 - h) + f(x_0 + h)\right] - \frac{h^2}{6}f^{(3)}(\xi_1), \quad \text{and}$$

$$(4.11) \qquad f'(x_0) = \frac{1}{2h}\left[f(x_0 - 2h) - 4f(x_0 - h) + 3f(x_0)\right] + \frac{h^2}{3}f^{(3)}(\xi_2).$$

Finally, note that since Eq. (4.11) can be obtained from Eq. (4.9) by simply replacing h by $-h$, there are actually only two formulas:

$$(4.12) \qquad f'(x_0) = \frac{1}{2h}\left[-3f(x_0) + 4f(x_0 + h) - f(x_0 + 2h)\right] + \frac{h^2}{3}f^{(3)}(\xi_0),$$

where ξ_0 lies between x_0 and $x_0 + 2h$, and

$$(4.13) \qquad f'(x_0) = \frac{1}{2h}\left[f(x_0 + h) - f(x_0 - h)\right] - \frac{h^2}{6}f^{(3)}(\xi_1),$$

where ξ_1 lies between $(x_0 - h)$ and $(x_0 + h)$.

The error in Eq. (4.13) is approximately half the error in Eq. (4.12). This is reasonable since in Eq. (4.13) data is being examined on both sides of x_0 and on only one side in Eq. (4.12). The approximation in Eq. (4.12) is useful near the ends of the interval I since information about f outside the interval is not necessarily available. Note also that in Eq. (4.13), f needs to be evaluated at only two points, whereas in Eq. (4.12) three points are needed. Figure 4.1 gives an illustration of the approximation produced from Eq. (4.13).

FIGURE 4.1

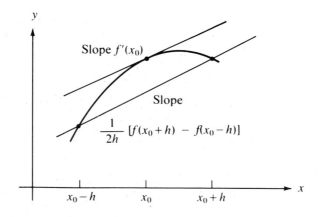

The methods presented in Eqs. (4.12) and (4.13) are known as **three-point formulas**, even though $f(x_0)$ does not appear in Eq. (4.13). Similarly, there are methods, known as **five-point formulas**, which involve evaluating the function at more points but whose error term is of the form $O(h^4)$.

The derivation of Eq. (4.14) will be considered in Section 4.2.

$$(4.14) \qquad f'(x_0) = \frac{1}{12h} [f(x_0 - 2h) - 8f(x_0 - h) + 8f(x_0 + h)$$

$$- f(x_0 + 2h)] + \frac{h^4}{30} f^{(5)}(\xi).$$

Another five-point formula that is useful, particularly with regard to clamped cubic spline interpolation, is

$$(4.15) \qquad f'(x_0) = \frac{1}{12h} [-25f(x_0) + 48f(x_0 + h) - 36f(x_0 + 2h)$$

$$+ 16f(x_0 + 3h) - 3f(x_0 + 4h)] + \frac{h^4}{5} f^{(5)}(\xi),$$

where ξ lies between x_0 and $x_0 + 4h$. Left endpoint approximations can be found using this formula with $h > 0$ and right endpoint approximations with $h < 0$.

EXAMPLE 2   Given in Table 4.2 are values for $f(x) = xe^x$.

TABLE 4.2

x	$f(x)$
1.8	10.889365
1.9	12.703199
2.0	14.778112
2.1	17.148957
2.2	19.855030

Since $f'(x) = (x + 1)e^x$, $f'(2.0) = 22.167168$. Approximating $f'(2.0)$ using the various three- and five-point formulas produces the following results.

Three-Point Formulas

Using (4.12) with $h = 0.1$: $\frac{1}{0.2} [-3f(2.0) + 4f(2.1) - f(2.2)] = 22.032310,$

Using (4.12) with $h = -0.1$: $\frac{1}{-0.2} [-3f(2.0) + 4f(1.9) - f(1.8)] = 22.054525,$

Using (4.13) with $h = 0.1$: $\frac{1}{0.2} [f(2.1) - f(1.9)] = 22.228790,$

Using (4.13) with $h = 0.2$: $\dfrac{1}{0.4}[f(2.2) - f(1.8)] = 22.414163$.

Five-Point Formula

Using (4.14) with $h = 0.1$ (only formula applicable):

$$\frac{1}{1.2}[f(1.8) - 8f(1.9) + 8f(2.1) - f(2.2)] = 22.166999.$$

The errors in the formulas are approximately 1.35×10^{-1}, 1.13×10^{-1}, -6.16×10^{-2}, -2.47×10^{-1}, and 1.69×10^{-4}, respectively. Clearly, the five-point formula gives the superior result. Note also that the error from Eq. (4.13) with $h = 0.1$ is approximately half of the magnitude of the error produced using Eq. (4.12) with either $h = 0.1$ or $h = -0.1$. ∎

Methods can also be derived to find approximations to higher derivatives of a function using only tabulated values of the function at various points. The methods involved become algebraically tedious, so only a representative procedure will be presented here.

Expanding a function f in a third-degree Taylor polynomial about a point x_0 and evaluating at $x_0 + h$ and $x_0 - h$, we obtain

$$(4.16) \qquad f(x_0 + h) = f(x_0) + f'(x_0)h + \tfrac{1}{2}f''(x_0)h^2 + \tfrac{1}{6}f'''(x_0)h^3 + \tfrac{1}{24}f^{(4)}(\xi_1)h^4$$

and

$$(4.17) \qquad f(x_0 - h) = f(x_0) - f'(x_0)h + \tfrac{1}{2}f''(x_0)h^2 - \tfrac{1}{6}f'''(x_0)h^3 + \tfrac{1}{24}f^{(4)}(\xi_{-1})h^4,$$

where $x_0 - h < \xi_{-1} < x_0 < \xi_1 < x_0 + h$.

If we add Eqs. (4.16) and (4.17) we obtain

$$(4.18) \qquad f(x_0 + h) + f(x_0 - h) = 2f(x_0) + f''(x_0)h^2 + \tfrac{1}{24}[f^{(4)}(\xi_1) + f^{(4)}(\xi_{-1})]h^4$$

or

$$(4.19) \qquad f''(x_0) = \frac{1}{h^2}[f(x_0 - h) - 2f(x_0) + f(x_0 + h)] - \frac{h^2}{24}[f^{(4)}(\xi_1) + f^{(4)}(\xi_{-1})].$$

If $f^{(4)}$ is continuous on $[x_0 - h, x_0 + h]$, the Intermediate-Value Theorem permits this to be rewritten as

$$(4.20) \qquad f''(x_0) = \frac{1}{h^2}[f(x_0 - h) - 2f(x_0) + f(x_0 + h)] - \frac{h^2}{12}f^{(4)}(\xi)$$

for some ξ, $x_0 - h < \xi < x_0 + h$.

EXAMPLE 3

Using the data given in Example 2 for $f(x) = xe^x$, we can use Eq. (4.20) to approximate $f''(2.0)$. Since $f''(x) = (x + 2)e^x$, the exact value is $f''(2.0) = 29.556224$. Using (4.20) with $h = 0.1$:

$$f''(2.0) \approx \frac{1}{0.01} [f(1.9) - 2f(2.0) + f(2.1)] = 29.593200.$$

Using (4.20) with $h = 0.2$:

$$f''(2.0) \approx \frac{1}{0.04} [f(1.8) - 2f(2.0) + f(2.2)] = 29.704275.$$

The errors are approximately -3.70×10^{-2} and -1.48×10^{-1}, respectively.   ∎

A particularly important subject in the study of numerical differentiation is the effect of round-off error. Let us examine Eq. (4.13):

$$f'(x_0) = \frac{1}{2h} [f(x_0 + h) - f(x_0 - h)] - \frac{h^2}{6} f^{(3)}(\xi_1)$$

more closely. Suppose that, in evaluating $f(x_0 + h)$ and $f(x_0 - h)$, we have encountered rounding errors $e(x_0 + h)$ and $e(x_0 - h)$; that is, our computed values $\tilde{f}(x_0 + h)$ and $\tilde{f}(x_0 - h)$ are related to the true values $f(x_0 + h)$ and $f(x_0 - h)$ by the formulas

$$f(x_0 + h) = \tilde{f}(x_0 + h) + e(x_0 + h)$$

and

$$f(x_0 - h) = \tilde{f}(x_0 - h) + e(x_0 - h).$$

In this case, the error in approximation,

$$f'(x_0) - \frac{\tilde{f}(x_0 + h) - \tilde{f}(x_0 - h)}{2h} = \frac{e(x_0 + h) - e(x_0 - h)}{2h} - \frac{h^2}{6} f^{(3)}(\xi_1),$$

will have a part due to rounding and a part due to truncating. If we assume that the rounding errors $e(x_0 \pm h)$ are bounded by some number $\varepsilon > 0$, and that the third derivative of f is bounded by a number $M > 0$, then

$$\left| f'(x_0) - \frac{\tilde{f}(x_0 + h) - \tilde{f}(x_0 - h)}{2h} \right| \leq \frac{\varepsilon}{h} + \frac{h^2}{6} M.$$

If h is small, the error due to rounding, ε/h, may be large. In practice, then, it is seldom advantageous to let h be too small since the rounding errors become significant.

EXAMPLE 4

Consider approximating $f'(0.900)$ for $f(x) = \sin x$, using the values in Table 4.3. The true value is $\cos(0.900) = 0.62161$.

Using the formula

$$f'(0.900) \approx \frac{f(0.900 + h) - f(0.900 - h)}{2h}$$

with different values of h gives the approximations in Table 4.4.

TABLE 4.3

x	$\sin x$	x	$\sin x$
0.800	0.71736	0.901	0.78395
0.850	0.75128	0.902	0.78457
0.880	0.77074	0.905	0.78643
0.890	0.77707	0.910	0.78950
0.895	0.78021	0.920	0.79560
0.898	0.78208	0.950	0.81342
0.899	0.78270	1.000	0.84147

TABLE 4.4

h	Approximation to $f'(0.900)$	Error
0.001	0.62500	0.00339
0.002	0.62250	0.00089
0.005	0.62200	0.00039
0.010	0.62150	−0.00011
0.020	0.62150	−0.00011
0.050	0.62140	−0.00021
0.100	0.62055	−0.00106

It appears that an optimal choice for h lies between 0.005 and 0.05. Suppose we perform some analysis on the error term,

$$\frac{\varepsilon}{h} + \frac{h^2}{6} M.$$

Letting

$$e(h) = \frac{\varepsilon}{h} + \frac{h^2}{6} M,$$

we can verify (see Exercise 18) that a minimum for e occurs at $h = \sqrt[3]{3\varepsilon/M}$, where

$$M = \max_{x \in [0.800, 1.00]} |f'''(x)| = \max_{x \in [0.800, 1.00]} |\cos x| \approx 0.69671.$$

Since values of f are given to five decimal places, it is reasonable to assume that $\varepsilon = 0.000005$. Therefore, the optimal choice of h would be approximately

$$h = \sqrt[3]{\frac{3(0.000005)}{0.69671}} \approx 0.028,$$

which is consistent with our results.

In practice, however, we would not be able to compute an actual optimal h to use in approximating the derivative, since we would have no knowledge about the third derivative of the function. ∎

Although we have only considered the rounding-error problems that are presented by the three-point formula Eq. (4.13), similar difficulties occur with all the differentiation formulas. The reason for the problems can be traced to the need to divide by a power of h. As we found in Section 1.2 (see, in particular, Example 4), division by small numbers tends to exaggerate rounding error and should be avoided if possible. In the case of numerical differentiation, it is impossible to avoid the problem entirely, although the higher-order methods will reduce the difficulty.

Remember that, as an approximation method, numerical differentiation is not stable since small values of h can lead to large errors. This is the first method we have studied with this property, and this operation is avoided whenever possible. The formulas derived, however, are necessary and have application in approximating the solutions of ordinary and partial-differential equations.

Exercise Set 4.1

1. Use either Eq. (4.12) or Eq. (4.13) to determine approximations that will complete the following tables:

a)

x	$f(x)$	$f'(x)$
−0.3	−0.20431	
−0.1	−0.08993	
0.1	0.11007	
0.3	0.39569	

b)

x	$f(x)$	$f'(x)$
1.1	0.48603	
1.2	0.86160	
1.3	1.59751	
1.4	3.76155	

2. Consider the following table of data:

x	0.2	0.4	0.6	0.8	1.0
$f(x)$	0.9798652	0.9177710	0.8080348	0.6386093	0.3843735

a) Use Eq. (4.15) to approximate $f'(0.2)$.

b) Use Eq. (4.15) to approximate $f'(1.0)$.

c) Use Eq. (4.14) to approximate $f'(0.6)$.

3. Consider the table of data:

x	0.2	0.4	0.6	0.8	1.0
$f(x)$	0.9798652	0.9177710	0.8080348	0.6386093	0.3843735

a) Use all appropriate formulas to approximate $f'(0.4)$ and $f''(0.4)$.

b) Use all appropriate formulas to approximate $f'(0.6)$ and $f''(0.6)$.

4. Let $f(x) = x^3 e^{x^2} - \sin x$. For $h = 0.1$ and $h = 0.01$, approximate $f'(2.19)$, using Eqs. (4.12) and (4.13).

5. Let $f(x) = \cos \pi x$. Use Eq. (4.14) and the values of $f(x)$ at $x = 0$, 0.25, 0.75, and 1.0 to approximate $f'(0.5)$. Compare this result to the exact value and to the approximation found in Exercise 3 of Section 3.6. Find a bound for the error.

6. Let $f(x) = e^{-x}$. Use Eq. (4.14) and the values of $f(x)$ at $x = 0$, 0.25, 0.75, and 1.0 to approximate $f'(0.5)$. Compare this result to the exact value and to the approximation found in Exercise 4 of Section 3.6. Find a bound for the error.

7. Let $f(x) = 2^x \sin x$. Approximate $f'(1.05)$ using $h = 0.05$ and $h = 0.01$ in Eq. (4.13) with the following data:

x	1.0	1.04	1.06	1.10
$f(x)$	1.6829420	1.7732994	1.8188014	1.9103448

8. Repeat Exercise 7 using four-digit rounding arithmetic, and compare the results to those previously obtained.

9. Let $f(x) = \cos \pi x$. Use Eq. (4.20) and the values of $f(x)$ at $x = 0.25$, 0.5, and 0.75 to approximate $f''(0.5)$. Compare this result to the exact value and to the approximation found in Exercise 3 of Section 3.6. Explain why this method is particularly accurate for this problem. Find a bound for the error.

10. Let $f(x) = 3xe^x - \cos x$. Using the following data and Eq. (4.20), approximate $f''(1.3)$ with $h = 0.1$ and $h = 0.01$.

x	1.20	1.29	1.30	1.31	1.40
$f(x)$	11.59006	13.78176	14.04276	14.30741	16.86187

Compare your results to $f''(1.3)$.

11. Derive an $O(h^4)$ five-point formula to approximate $f'(x_0)$ that uses $f(x_0 - h)$, $f(x_0)$, $f(x_0 + h)$, $f(x_0 + 2h)$, and $f(x_0 + 3h)$.
[Hint: Consider the expression $Af(x_0 - h) + Bf(x_0 + h) + Cf(x_0 + 2h) + Df(x_0 + 3h)$. Expand in fifth-degree Taylor polynomials and choose A, B, C, and D appropriately.]

12. Use the formula derived in Exercise 11 and the data of Exercise 2 to approximate $f'(0.4)$ and $f'(0.8)$.

13. Suppose the following data has been experimentally collected.

x	1.00	1.01	1.02
$f(x)$	1.27	1.32	1.38

a) Approximate $f'(1.005)$ and $f'(1.015)$ using Eq. (4.13).

b) Approximate $f''(1.01)$, using Eq. (4.13) and the results of part (a).

c) Suppose the data is accurate to within ± 0.005. Find the maximum error due to the inaccurate data in parts (a) and (b).

14. With $x_1 = x_0 + h$ and $x_2 = x_0 + 3h$, use Eq. (4.5) to derive an approximation to $f'(x_0)$. Using this formula, repeat Exercise 4.

15. In Exercise 10 of Section 3.5, data was given describing a car traveling on a straight road. That problem asked to predict the position and speed of the car when $t = 10$ seconds. Use the following times (including $t = 10$) and positions at those times, and either Eq. (4.12) or Eq. (4.13) to predict the speed at each time listed.

Time	0	3	5	8	10	13
Distance	0	225	383	623	742	993

16. For the formula

$$f'(x_0) = \frac{f(x_0 + h) - f(x_0)}{h} - \frac{h}{2}f''(\xi_0),$$

analyze the rounding errors as in Example 4. Find an optimal $h > 0$ for the function given in Example 2.

17. In a circuit with impressed voltage $\mathscr{E}(t)$ and inductance L, Kirchhoff's first law gives the relationship

$$\mathscr{E} = L\frac{di}{dt} + Ri$$

where R is the resistance in the circuit and i the current. Suppose we measure the current for several values of t and obtain:

t	1.00	1.01	1.02	1.03	1.04
i	3.10	3.12	3.14	3.18	3.24

where t is measured in seconds, i in amperes, the inductance L is a constant 0.98 henries, and the resistance is 0.142 ohms. Approximate the voltage $\mathscr{E}$ at the values $t = 1.00$, 1.01, 1.02, 1.03, and 1.04, using the appropriate three-point formulas.

18. Consider the function

$$e(h) = \frac{\varepsilon}{h} + \frac{h^2}{6}M,$$

where M is a bound for

$$\left|\frac{d^3(\sin x)}{dx^3}\right| \quad \text{on } [0.800, 1.00].$$

Show that $e(h)$ has a minimum at $\sqrt[3]{3\varepsilon/M}$.

19. To construct a cubic spline with clamped boundary conditions, we need values for f' at both ends of the interval.

a) Repeat Exercise 9 of Section 3.6, using the clamped boundary conditions. Approximate $f'(x_0)$ and $f'(x_3)$ using Eq. (4.12). Compare your results to those of Exercise 9 of Section 3.6.

b) Repeat Exercise 12 of Section 3.6, using the clamped boundary conditions. Approximate $f'(x_0)$ and $f'(x_{10})$ using Eq. (4.15).

20. a) Using the derivative of the cubic spline constructed in Exercise 9 of Section 3.6, approximate $f'(1.03)$.

b) Using the derivative of the cubic constructed in Exercise 19(a), approximate $f'(1.03)$ and compare the answer to the answer in part (a).

21. By expanding the function f in a fourth-degree Taylor polynomial about x_0 and evaluating at $x_0 \pm h$ and $x_0 \pm 2h$, derive a method for approximating $f'''(x_0)$, whose error term is of order h^2.

22. Use the data of Exercise 10 and the formula derived in Exercise 21 to approximate $f'''(1.3)$. Compare this result to the actual value.

4.2 Richardson's Extrapolation

A technique known as Richardson's Extrapolation is frequently employed to generate results of high accuracy by using low-order formulas. Although the name attached to the method refers to a paper written by L. F. Richardson and J. A. Gaunt [109] in 1927, the idea behind the technique is at least as old as Archimedes (ca. 200 B.C.). An interesting article regarding the history and application of extrapolation was written by Joyce [73] in 1971 in SIAM review.

To examine the extrapolation technique, suppose $N(h)$ is a formula that produces approximations of order $O(h^2)$ to an unknown value M. In addition, assume that the error form for the approximation of $N(h)$ to M can be expressed as

$$(4.21) \qquad M = N(h) + K_1 h^2 + O(h^4),$$

where K_1 is a constant. Replacing h by $h/2$ in Eq. (4.21) gives a new, and presumably more accurate, approximation $N(h/2)$ that satisfies

$$(4.22) \qquad M = N\left(\frac{h}{2}\right) + K_1 \frac{h^2}{4} + O\left(\left(\frac{h}{2}\right)^4\right).$$

Multiplying Eq. (4.22) by 4 and subtracting Eq. (4.21) gives

$$3M = 4N\left(\frac{h}{2}\right) - N(h) + O(h^4)$$

or

$$(4.23) \qquad M = \frac{4N\left(\frac{h}{2}\right) - N(h)}{3} + O(h^4).$$

To facilitate the discussion, we define $N_1(h) = N(h)$ and

$$(4.24) \qquad N_2(h) = \frac{4N_1\left(\frac{h}{2}\right) - N_1(h)}{3}.$$

Using this notation, the approximations to M shown in Table 4.5 can be generated. The entry $N_1(h/2^k)$ will be of order $O((h/2^k)^2)$ while $N_2(h/2^k)$ has the higher order $O((h/2^k)^4)$.

TABLE 4.5

$$N_1(h)$$

$$N_1\left(\frac{h}{2}\right) \qquad N_2(h)$$

$$N_1\left(\frac{h}{4}\right) \qquad N_2\left(\frac{h}{2}\right)$$

$$N_1\left(\frac{h}{8}\right) \qquad N_2\left(\frac{h}{4}\right)$$

If, in addition, a number K_2 exists so that Eq. (4.21) can be expressed as

$$M = N(h) + K_1 h^2 + K_2 h^4 + O(h^6),$$

the extrapolation table can be continued to a third column containing the entries $N_3(h)$, $N_3(h/2)$, ..., by defining

$$(4.25) \qquad N_3(h) = \frac{4^2 N_2\left(\frac{h}{2}\right) - N_2(h)}{4^2 - 1} = \frac{16 N_2\left(\frac{h}{2}\right) - N_2(h)}{15}.$$

The entry $N_3(h/2^k)$ will be an $O((h/2^k)^6)$ approximation to M.

This process can be extended to m such columns provided that the error form for the approximation of $N(h)$ to M can be expressed as

$$M = N(h) + \sum_{j=1}^{m-1} K_j h^{2j} + O(h^{2m})$$

for some collection of constants K_j. The $O(h^{2j})$ approximations are generated recursively by the formula:

$$(4.26) \qquad N_j(h) = \frac{4^{j-1} N_{j-1}\left(\frac{h}{2}\right) - N_{j-1}(h)}{4^{j-1} - 1},$$

for each $j = 2, 3, ..., m$.

EXAMPLE 1   The centered difference formula in Eq. (4.13) to approximate $f'(x_0)$ can be expressed with an error formula

$$f'(x_0) = \frac{1}{2h} [f(x_0 + h) - f(x_0 - h)] - \frac{h^2}{6} f'''(x_0) - \frac{h^4}{120} f^{(5)}(x_0) + O(h^6).$$

Using the notation in the preceding discussion, $M = f'(x_0)$,

$$K_1 = -\frac{1}{6}f'''(x_0), \quad K_2 = -\frac{1}{120}f^{(5)}(x_0), \quad \text{and} \quad N(h) = \frac{1}{2h}[f(x_0 + h) - f(x_0 - h)].$$

Suppose that $x_0 = 2.0$, $h = 0.2$, and $f(x) = xe^x$. Then

$$N(0.2) = \frac{1}{0.4}[f(2.2) - f(1.8)] = 22.414160,$$

$$N(0.1) = 22.228786,$$

and $N(0.05) = 22.182564.$

The extrapolation table for this data is shown in Table 4.6. The exact value of $f'(x)$ $= xe^x + e^x$ at $x_0 = 2.0$ is 22.167168.

TABLE 4.6

$N_1(0.2) = 22.414160$		
$N_1(0.1) = 22.228786$	$N_2(0.2) = \dfrac{4N_1(0.1) - N_1(0.2)}{3} = 22.166995$	
$N_1(0.05) = 22.182564$	$N_2(0.1) = \dfrac{4N_1(0.05) - N_1(0.1)}{3} = 22.167157$	$N_3(0.2) = \dfrac{16N_2(0.1) - N_2(0.2)}{15}$ $= 22.167168$

∎

Since each column beyond the first in the extrapolation table is obtained by a simple averaging process, the technique can produce high-order approximations with minimal computational cost and round-off error. Although, as k increases the round-off error in $N_1(h/2^k)$ will generally increase for the reasons mentioned at the end of Section 4.1.

In Section 4.1 we discussed both three- and five-point methods for approximating $f'(x_0)$ given various functional values of f. The three-point methods were derived by differentiating a Lagrange interpolating polynomial for f. The five-point methods can be obtained in a similar manner, but the derivation is tedious. Extrapolation can be used to derive these formulas more easily.

Suppose we expand the function f in a Taylor polynomial of degree four about x_0. Then

$$f(x) = f(x_0) + f'(x_0)(x - x_0) + \frac{1}{2}f''(x_0)(x - x_0)^2 + \frac{1}{6}f'''(x_0)(x - x_0)^3$$

$$+ \frac{1}{24}f^{(4)}(x_0)(x - x_0)^4 + \frac{1}{120}f^{(5)}(\xi)(x - x_0)^5,$$

for some number ξ between x and x_0. Evaluating f at $x_0 + h$ and $x_0 - h$:

(4.27)
$$f(x_0 + h) = f(x_0) + f'(x_0)h + \frac{1}{2}f''(x_0)h^2 + \frac{1}{6}f'''(x_0)h^3$$

$$+ \frac{1}{24}f^{(4)}(x_0)h^4 + \frac{1}{120}f^{(5)}(\xi_1)h^5,$$

and

(4.28)
$$f(x_0 - h) = f(x_0) - f'(x_0)h + \frac{1}{2}f''(x_0)h^2 - \frac{1}{6}f'''(x_0)h^3$$

$$+ \frac{1}{24}f^{(4)}(x_0)h^4 - \frac{1}{120}f^{(5)}(\xi_2)h^5,$$

where $x_0 - h < \xi_2 < x_0 < \xi_1 < x_0 + h$. Subtracting Eq. (4.28) from Eq. (4.27) produces

(4.29) $\quad f(x_0 + h) - f(x_0 - h) = 2hf'(x_0) + \dfrac{h^3}{3}f'''(x_0) + \dfrac{h^5}{120}[f^{(5)}(\xi_1) + f^{(5)}(\xi_2)].$

If $f^{(5)}$ is continuous on $[x_0 - h, x_0 + h]$, the Intermediate Value Theorem implies that a number $\tilde{\xi}$ in $(x_0 - h, x_0 + h)$ exists with

$$f^{(5)}(\tilde{\xi}) = \frac{1}{2}[f^{(5)}(\xi_1) + f^{(5)}(\xi_2)].$$

As a consequence, Eq. (4.29) can be solved for $f'(x_0)$ to give the $O(h^2)$ approximation:

(4.30)
$$f'(x_0) = \frac{1}{2h}[f(x_0 + h) - f(x_0 - h)] - \frac{h^2}{6}f'''(x_0) - \frac{h^4}{120}f^{(5)}(\tilde{\xi}).$$

Although the approximation in Eq. (4.30) is the same as that given in the three-point formula in Eq. (4.13), the unknown evaluation point occurs now in $f^{(5)}$, rather than in f'''. Extrapolation takes advantage of this by first replacing h in Eq. (4.30) by $2h$ to give the new formula

(4.31)
$$f'(x_0) = \frac{1}{4h}[f(x_0 + 2h) - f(x_0 - 2h)] - \frac{4h^2}{6}f'''(x_0) - \frac{16h^4}{120}f^{(5)}(\hat{\xi}),$$

where $\hat{\xi}$ is between $x_0 - 2h$ and $x_0 + 2h$.

Multiplying Eq. (4.30) by 4 and subtracting Eq. (4.31) produces

(4.32)
$$3f'(x_0) = \frac{2}{h}[f(x_0 + h) - f(x_0 - h)] - \frac{1}{4h}[f(x_0 + 2h) - f(x_0 - 2h)]$$

$$- \frac{h^4}{30}f^{(5)}(\tilde{\xi}) + \frac{2h^4}{15}f^{(5)}(\hat{\xi}).$$

If $f^{(5)}$ is continuous on $[x_0 - 2h, x_0 + 2h]$, it can be shown that $f^{(5)}(\tilde{\xi})$ and $f^{(5)}(\hat{\xi})$ can be replaced by a common value $f^{(5)}(\xi)$. Using this result and dividing by 3 produces the five-point formula

$$f'(x_0) = \frac{1}{12h}[f(x_0 - 2h) - 8f(x_0 - h) + 8f(x_0 + h) - f(x_0 + 2h)] + \frac{h^4}{30}f^{(5)}(\xi).$$

This is the five-point formula given as Eq. (4.14).

Other formulas for first and higher derivatives can be derived in a similar manner. Some of these formulas are considered in the exercises.

Extrapolation can be applied whenever the truncation error for a formula has the form

$$\sum_{j=1}^{m-1} K_j h^{\alpha_j} + O(h^{\alpha_m})$$

for a collection of constants K_j and when $\alpha_1 < \alpha_2 < \alpha_3 < \cdots < \alpha_m$. An example using extrapolation when $\alpha_i = i$ is considered in Exercise 8.

The technique of extrapolation will be introduced throughout the text. The most prominent applications occur for approximating integrals in Section 4.6 and for determining approximate solutions to differential equations in Section 5.8.

Exercise Set 4.2

1. Apply the extrapolation process described in Example 1 to determine $N_3(h)$, an approximation to $f'(x_0)$, for the following functions and step-sizes:

    a) $f(x) = \ln x, x_0 = 1.0, h = 0.4$.   b) $f(x) = x + e^x, x_0 = 0.0, h = 0.4$.

    c) $f(x) = 2^x \sin x, x_0 = 1.05, h = 0.4$.   d) $f(x) = x^3 \cos x, x_0 = 2.3, h = 0.4$.

2. Add another line to the extrapolation table in Exercise 1 to obtain the approximation $N_4(h)$.

3. Repeat Exercise 1 using four-digit rounding arithmetic.

4. Repeat Exercise 2 using four-digit rounding arithmetic.

5. The following data gives approximations to the integral

$$M = \int_0^\pi \sin x \, dx.$$

$$N_1(h) = 1.570796 \quad N_1\left(\frac{h}{4}\right) = 1.974232 \quad N_1\left(\frac{h}{2}\right) = 1.896119 \quad N_1\left(\frac{h}{8}\right) = 1.993570.$$

    Assuming $M = N_1(h) + K_1 h^2 + K_2 h^4 + K_3 h^6 + K_4 h^8 + O(h^{10})$, construct an extrapolation table to determine $N_4(h)$.

6. The following data can be used to approximate the integral

$$M = \int_0^{3\pi/2} \cos x \, dx.$$

$$N_1(h) = 2.356194 \quad N_1\left(\frac{h}{4}\right) = -0.8815732 \quad N_1\left(\frac{h}{2}\right) = -0.4879837 \quad N_1\left(\frac{h}{8}\right) = -0.9709157.$$

    Assume a formula exists of the type given in Exercise 5 and determine $N_4(h)$.

7. Show that the five-point formula in Eq. (4.14) applied to $f(x) = xe^x$ at $x_0 = 2.0$ gives $N_2(0.2)$ in Table 4.6 when $h = 0.2$ and $N_2(0.1)$ when $h = 0.1$.

8. The forward-difference formula given in Eq. (4.2) can be expressed as

$$f'(x_0) = \frac{1}{h}[f(x_0 + h) - f(x_0)] - \frac{h}{2}f''(x_0) - \frac{h^2}{6}f'''(x_0) + O(h^3).$$

Use extrapolation to derive a formula that is $O(h^3)$.

9. Suppose an extrapolation table

$N_1(h)$

$N_1\left(\dfrac{h}{2}\right)$ $N_2(h)$

$N_1\left(\dfrac{h}{4}\right)$ $N_2\left(\dfrac{h}{2}\right)$ $N_3(h)$

---

has been constructed to approximate the number M with $M = N_1(h) + K_1 h^2 + K_2 h^4 + K_3 h^6$.

a) Show that the linear interpolating polynomial $P_{0,1}(h)$ through $(h^2, N_1(h))$ and $(h^2/4, N_1(h/2))$ satisfies $P_{0,1}(0) = N_2(h)$. Similarly, $P_{1,2}(0) = N_2(h/2)$.

b) Show that the linear interpolating polynomial $P_{0,2}(h)$ through $(h^4, N_2(h))$ and $(h^4/16, N_2(h/2))$ satisfies $P_{0,2}(0) = N_3(h)$.

4.3 Elements of Numerical Integration

The need often arises for evaluating the definite integral of a function that has no explicit antiderivative or whose antiderivative has values that are not easily obtained. The basic method involved in approximating $\int_a^b f(x)\,dx$ is called **numerical quadrature** and uses a sum of the type

$$\sum_{i=0}^{n} a_i f(x_i)$$

to approximate $\int_a^b f(x)\,dx$.

The methods of quadrature we discuss in this section are based on the interpolation polynomials given in Chapter 3. To proceed, we first select a set of distinct nodes $\{x_0, \dots, x_n\}$ from the interval $[a, b]$. If P_n is the Lagrange interpolating polynomial

$$P_n(x) = \sum_{i=0}^{n} f(x_i)L_i(x),$$

we integrate P_n and its truncation error term over $[a, b]$ to obtain the quadrature formula

$$\int_a^b f(x)\, dx = \int_a^b \sum_{i=0}^n f(x_i) L_i(x)\, dx + \int_a^b \prod_{i=0}^n (x - x_i) \frac{f^{(n+1)}(\xi(x))}{(n+1)!}\, dx$$

$$= \sum_{i=0}^n a_i f(x_i) + \frac{1}{(n+1)!} \int_a^b \prod_{i=0}^n (x - x_i) f^{(n+1)}(\xi(x))\, dx,$$

where $\xi(x)$ is in $[a, b]$ for each x and

$$a_i = \int_a^b L_i(x)\, dx, \qquad \text{for each } i = 0, 1, \ldots, n.$$

Before discussing the general situation of quadrature formulas, we will consider formulas produced by using first- and second-degree Lagrange polynomials with equally spaced nodes. The formulas are the **trapezoidal rule** and **Simpson's rule**, formulas often discussed in calculus courses.

To derive the trapezoidal rule for approximating $\int_a^b f(x)\, dx$, let $x_0 = a$, $x_1 = b$, $h = b - a$ and use the Lagrange polynomial:

$$P_1(x) = \frac{(x - x_1)}{(x_0 - x_1)} f(x_0) + \frac{(x - x_0)}{(x_1 - x_0)} f(x_1).$$

Thus,

(4.33)
$$\int_a^b f(x)\, dx = \int_{x_0}^{x_1} \left[\frac{(x - x_1)}{(x_0 - x_1)} f(x_0) + \frac{(x - x_0)}{(x_1 - x_0)} f(x_1) \right] dx$$

$$+ \frac{1}{2} \int_{x_0}^{x_1} f''(\xi(x))(x - x_0)(x - x_1)\, dx.$$

Since $(x - x_0)(x - x_1)$ does not change sign on $[x_0, x_1]$, the Weighted Mean Value Theorem for Integrals (Theorem 1.10, p. 4) can be applied to the error term and

$$\int_{x_0}^{x_1} f''(\xi(x))(x - x_0)(x - x_1)\, dx = f''(\xi) \int_{x_0}^{x_1} (x - x_0)(x - x_1)\, dx$$

$$= f''(\xi) \left[\frac{x^3}{3} - \frac{(x_1 + x_0)}{2} x^2 + x_0 x_1 x \right]_{x_0}^{x_1}$$

$$= -\frac{h^3}{6} f''(\xi).$$

Consequently, Eq. (4.33) implies that

$$\int_a^b f(x)\, dx = \left[\frac{(x - x_1)^2}{2(x_0 - x_1)} f(x_0) + \frac{(x - x_0)^2}{2(x_1 - x_0)} f(x_1) \right]_{x_0}^{x_1} - \frac{h^3}{12} f''(\xi)$$

$$= \frac{(x_1 - x_0)}{2} \left[f(x_0) + f(x_1) \right] - \frac{h^3}{12} f''(\xi)$$

or:

Trapezoidal Rule

$$\int_a^b f(x)\,dx = \frac{h}{2}[f(x_0)+f(x_1)] - \frac{h^3}{12}f''(\xi).$$

The reason for calling this formula the trapezoidal rule is that when f is a function with positive values, $\int_a^b f(x)\,dx$ is approximated by the area in the trapezoid shown in Figure 4.2.

FIGURE 4.2

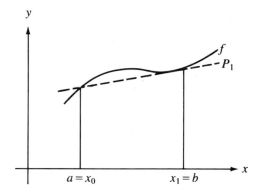

Note that the trapezoidal rule gives the exact result when applied to any function whose second derivative is identically zero, that is, any polynomial of degree one or less.

Simpson's rule results from integrating over $[a, b]$ the second-degree Lagrange polynomial with nodes $x_0 = a$, $x_2 = b$, and $x_1 = a + h$, where $h = (b - a)/2$. (*See Figure 4.3.*)

FIGURE 4.3

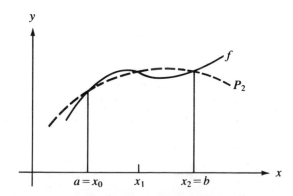

Therefore

$$\int_a^b f(x)\,dx = \int_{x_0}^{x_2} \left[\frac{(x-x_1)(x-x_2)}{(x_0-x_1)(x_0-x_2)} f(x_0) + \frac{(x-x_0)(x-x_2)}{(x_1-x_0)(x_1-x_2)} f(x_1) \right.$$

$$\left. + \frac{(x-x_0)(x-x_1)}{(x_2-x_0)(x_2-x_1)} f(x_2) \right] dx$$

$$+ \int_{x_0}^{x_2} \frac{(x-x_0)(x-x_1)(x-x_2)}{6} f^{(3)}(\xi(x))\,dx.$$

Deriving Simpson's rule in this manner, however, provides only an $O(h^4)$ error term involving $f^{(3)}$. By approaching the problem in another way, a higher-order term involving $f^{(4)}$ can be derived.

Suppose that f is expanded in a third-degree Taylor polynomial about x_1. Then, for each x in $[x_0, x_2]$, a number $\xi(x)$ in (x_0, x_2) exists with

$$f(x) = f(x_1) + f'(x_1)(x-x_1) + \frac{f''(x_1)}{2}(x-x_1)^2$$

$$+ \frac{f'''(x_1)}{6}(x-x_1)^3 + \frac{f^{(4)}(\xi(x))}{24}(x-x_1)^4.$$

and

(4.34)
$$\int_{x_0}^{x_2} f(x)\,dx = f(x_1)(x_2-x_0) + \left[\frac{f'(x_1)}{2}(x-x_1)^2 \right.$$

$$\left. + \frac{f''(x_1)}{6}(x-x_1)^3 + \frac{f'''(x_1)}{24}(x-x_1)^4 \right]_{x_0}^{x_2}$$

$$+ \frac{1}{24} \int_{x_0}^{x_2} f^{(4)}(\xi(x))(x-x_1)^4\,dx.$$

Since $(x-x_1)^4$ is never negative on $[x_0, x_2]$, the Weighted Mean Value Theorem for Integrals implies that

$$\frac{1}{24} \int_{x_0}^{x_2} f^{(4)}(\xi(x))(x-x_1)^4\,dx = \frac{f^{(4)}(\xi_1)}{24} \int_{x_0}^{x_2} (x-x_1)^4\,dx = \frac{f^{(4)}(\xi_1)}{120}(x-x_1)^5 \bigg]_{x_0}^{x_2}$$

for some number ξ_1 in (x_0, x_2).

However, $h = x_2 - x_1 = x_1 - x_0$, so

$$(x_2-x_1)^2 - (x_0-x_1)^2 = (x_2-x_1)^4 - (x_0-x_1)^4 = 0$$

while

$$(x_2-x_1)^3 - (x_0-x_1)^3 = 2h^3 \qquad \text{and} \qquad (x_2-x_1)^5 - (x_0-x_1)^5 = 2h^5.$$

Consequently, Eq. (4.34) can be rewritten as

$$\int_{x_0}^{x_2} f(x)\,dx = 2hf(x_1) + \frac{h^3}{3} f''(x_1) + \frac{f^{(4)}(\xi_1)}{60} h^5.$$

If we now replace $f''(x_1)$ by the approximation given in Eq. (4.20) of Section 4.1 we have

$$\int_{x_0}^{x_2} f(x)\,dx = 2hf(x_1) + \frac{h^3}{3}\left\{ \frac{1}{h^2}[f(x_0) - 2f(x_1) + f(x_2)] - \frac{h^2}{12}f^{(4)}(\xi_2)\right\} + \frac{f^{(4)}(\xi_1)}{60}h^5$$

$$= \frac{h}{3}[f(x_0) + 4f(x_1) + f(x_2)] - \frac{h^5}{12}\left[\frac{1}{3}f^{(4)}(\xi_2) - \frac{1}{5}f^{(4)}(\xi_1)\right].$$

It can be shown that the values ξ_1 and ξ_2 in this expression can be replaced by a common value ξ in (x_0, x_2), hence

Simpson's Rule:

$$\int_{x_0}^{x_2} f(x)\,dx = \frac{h}{3}[f(x_0) + 4f(x_1) + f(x_2)] - \frac{h^5}{90}f^{(4)}(\xi).$$

Since the error term involves the fourth derivative of f, Simpson's rule will give the exact results when applied to any polynomial of degree three or less.

EXAMPLE 1   The trapezoidal rule for a function f on the interval $[0, 2]$ is

$$\int_0^2 f(x)\,dx \approx f(0) + f(2),$$

while Simpson's rule for f on $[0, 2]$ is

$$\int_0^2 f(x)\,dx \approx \tfrac{1}{3}[f(0) + 4f(1) + f(2)].$$

The results to three places for some elementary functions are summarized in Table 4.7.

TABLE 4.7

$f(x)$	x^2	x^4	$1/(x+1)$	$\sqrt{1+x^2}$	$\sin x$	e^x
Exact value	2.667	6.400	1.099	2.958	1.416	6.389
Trapezoidal	4.000	16.000	1.333	3.326	0.909	8.389
Simpson's	2.667	6.667	1.111	2.964	1.425	6.421

∎

The standard derivation of quadrature error formulas is based on determining the class of polynomials for which these formulas produce exact results. The following definition is used to facilitate the discussion of this derivation.

DEFINITION 4.1   The **degree of accuracy**, or **precision**, of a quadrature formula is the positive integer n such that $E(P_k) = 0$ for all polynomials P_k of degree less than or equal to n, but for which $E(P_{n+1}) \neq 0$ for some polynomial of degree $(n + 1)$.

Definition 4.1 implies that the trapezoidal and Simpson's rules have degree of precision one and three, respectively.

Integration and summation are linear operations; that is,

$$\int_a^b (\alpha f(x) + \beta g(x))\,dx = \alpha \int_a^b f(x)\,dx + \beta \int_a^b g(x)\,dx$$

and

$$\sum_{i=0}^n (\alpha f(x_i) + \beta g(x_i)) = \alpha \sum_{i=0}^n f(x_i) + \beta \sum_{i=0}^n g(x_i),$$

for each pair of integrable functions f and g and each pair of real constants α and β. This implies (see Exercise 10) that the degree of precision of a quadrature formula is n if and only if $E(x^k) = 0$, for all $k = 0, 1, \ldots, n$ but $E(x^{n+1}) \neq 0$.

The trapezoidal and Simpson's rules are examples of a class of methods known as Newton–Cotes formulas. There are two types of Newton–Cotes formulas, open formulas and closed formulas.

The $n + 1$ point **closed Newton–Cotes formula** uses nodes $x_i = x_0 + ih$, for $i = 0$, $1, \ldots, n$ where $x_0 = a$, $x_n = b$ and $h = (b - a)/n$. The formula assumes the form

$$\int_a^b f(x)\,dx \approx \sum_{i=0}^n a_i f(x_i),$$

where

$$a_i = \int_{x_0}^{x_n} L_i(x)\,dx = \int_{x_0}^{x_n} \prod_{\substack{j=0 \\ j \neq i}}^n \frac{(x - x_j)}{(x_i - x_j)}\,dx.$$

The following theorem details the error analysis associated with the closed Newton–Cotes formulas. For a proof of this theorem, see Isaacson and Keller [67], page 313.

THEOREM 4.2　Suppose that $\sum_{i=0}^n a_i f(x_i)$ denotes the $n + 1$ point closed Newton–Cotes formula with $x_0 = a$, $x_n = b$, and $h = (b - a)/n$. There exists $\xi \in [a, b]$ for which:

(4.35)
$$\int_a^b f(x)\,dx = \sum_{i=0}^n a_i f(x_i) + \frac{h^{n+3} f^{(n+2)}(\xi)}{(n+2)!} \int_0^n t^2(t-1)\cdots(t-n)\,dt$$

if n is even and $f \in C^{n+2}[a, b]$, and

(4.36)
$$\int_a^b f(x)\,dx = \sum_{i=0}^n a_i f(x_i) + \frac{h^{n+2} f^{(n+1)}(\xi)}{(n+1)!} \int_0^n t(t-1)\cdots(t-n)\,dt$$

if n is odd and $f \in C^{n+1}[a, b]$.

Note that when n is an even integer the degree of precision is $n + 1$, although the interpolation polynomial is of degree at most n. In case n is odd, the second part of the theorem shows that the degree of precision is only n. Consequently, if n is even and more nodes are to be added to increase precision, no accuracy is gained by adding only one node; *nodes should be added in multiples of two.*

Some of the common closed Newton–Cotes formulas with their error terms are as follows:

$n = 1$: Trapezoidal Rule

(4.37) $$\int_{x_0}^{x_1} f(x)\,dx = \frac{h}{2}[f(x_0)+f(x_1)] - \frac{h^3}{12}f''(\xi) \qquad \text{where } x_0 < \xi < x_1;$$

$n = 2$: Simpson's Rule

(4.38) $$\int_{x_0}^{x_2} f(x)\,dx = \frac{h}{3}[f(x_0) + 4f(x_1) + f(x_2)] - \frac{h^5}{90}f^{(4)}(\xi)$$
$$\text{where } x_0 < \xi < x_2;$$

$n = 3$: Simpson's Three-Eighths Rule

(4.39) $$\int_{x_0}^{x_3} f(x)\,dx = \frac{3h}{8}[f(x_0) + 3f(x_1) + 3f(x_2) + f(x_3)] - \frac{3h^5}{80}f^{(4)}(\xi)$$
$$\text{where } x_0 < \xi < x_3;$$

$n = 4$:

(4.40) $$\int_{x_0}^{x_4} f(x)\,dx = \frac{2h}{45}[7f(x_0) + 32f(x_1) + 12f(x_2) + 32f(x_3) + 7f(x_4)]$$
$$- \frac{8h^7}{945}f^{(6)}(\xi) \qquad \text{where } x_0 < \xi < x_4.$$

In the **open Newton–Cotes formulas**, the nodes $x_i = x_0 + ih$ are used for each $i = 0, 1, \ldots, n$, where $h = (b-a)/(n+2)$ and $x_0 = a + h$. This implies that $x_n = b - h$; so if we label the endpoints by setting $x_{-1} = a$ and $x_{n+1} = b$, the formulas become

$$\int_a^b f(x)\,dx = \int_{x_{-1}}^{x_{n+1}} f(x)\,dx \approx \sum_{i=0}^n a_i f(x_i),$$

where again

$$a_i = \int_a^b L_i(x)\,dx.$$

The following theorem is analogous to Theorem 4.2 and its proof is contained in Isaacson and Keller [67], page 314.

THEOREM 4.3 Suppose that $\sum_{i=0}^n a_i f(x_i)$ denotes the $n+1$ point open Newton–Cotes formula with $x_{-1} = a$, $x_{n+1} = b$, and $h = (b-a)/(n+2)$. There exists $\xi \in (a, b)$ for which:

(4.41) $$\int_a^b f(x)\,dx = \sum_{i=0}^n a_i f(x_i) + \frac{h^{n+3}f^{(n+2)}(\xi)}{(n+2)!}\int_{-1}^{n+1} t^2(t-1)\cdots(t-n)\,dt$$

if n is even and $f \in C^{n+2}[a, b]$, and

(4.42) $$\int_a^b f(x)\,dx = \sum_{i=0}^n a_i f(x_i) + \frac{h^{n+2}f^{(n+1)}(\xi)}{(n+1)!}\int_{-1}^{n+1} t(t-1)\cdots(t-n)\,dt$$

if n is odd and $f \in C^{n+1}[a, b]$.

Some of the common open Newton–Cotes formulas with their error terms are as follows:

$n = 0$:   Midpoint Rule

(4.43)
$$\int_{x_{-1}}^{x_1} f(x)\,dx = 2hf(x_0) + \frac{h^3}{3}f''(\xi) \qquad \text{where } x_{-1} < \xi < x_1;$$

$n = 1$:

(4.44)
$$\int_{x_{-1}}^{x_2} f(x)\,dx = \frac{3h}{2}[f(x_0) + f(x_1)] + \frac{3h^3}{4}f''(\xi) \qquad \text{where } x_{-1} < \xi < x_2;$$

$n = 2$:

(4.45)
$$\int_{x_{-1}}^{x_3} f(x)\,dx = \frac{4h}{3}[2f(x_0) - f(x_1) + 2f(x_2)] + \frac{14h^5}{45}f^{(4)}(\xi)$$

$$\text{where } x_{-1} < \xi < x_3;$$

$n = 3$:

(4.46)
$$\int_{x_{-1}}^{x_4} f(x)\,dx = \frac{5h}{24}[11f(x_0) + f(x_1) + f(x_2) + 11f(x_3)] + \frac{95}{144}h^5 f^{(4)}(\xi),$$

$$\text{where } x_{-1} < \xi < x_4.$$

EXAMPLE 2   Using the closed and open Newton–Cotes formulas listed as (4.37)–(4.40) and (4.43)–(4.46) to approximate $\int_0^{\pi/4} \sin x \, dx = 1 - (\sqrt{2}/2)$ gives the results in Table 4.8.

TABLE 4.8

n	0	1	2	3	4
Closed formulas		0.27768018	0.29293264	0.29291070	0.29289318
Error		0.01521303	0.00003942	0.00001748	0.00000004
Open formulas	0.30055887	0.29798754	0.29351798	0.29286923	
Error	0.00766565	0.00509432	0.00062477	0.00002399	

As illustrated in the preceding example, the closed formulas generally produce results superior to the open formulas of the same order. Consequently, the closed formulas are more frequently used in practice. The open formulas are primarily used for the numerical solution of ordinary differential equations.

Exercise Set 4.3

1. Use the trapezoidal and Simpson's rules to approximate the following integrals. Compare the approximations to the actual value and find a bound for the error in each case, if possible.

   a) $\displaystyle\int_1^2 \ln x \, dx$

   b) $\displaystyle\int_0^{0.1} x^{1/3} dx$

   c) $\displaystyle\int_0^{\pi/3} (\sin x)^2 dx$

   d) $\displaystyle\int_{0.2}^{0.4} e^{3x} \cos 2x \, dx$

   e) $\displaystyle\int_0^{\pi/4} \tan x \, dx$

   f) $\displaystyle\int_{\pi/2}^{3\pi/4} \cot x \, dx$

2. Use the following table to find an approximation to $\int_{1.1}^{1.5} e^x dx$, using:

   a) the trapezoidal rule with $x_0 = 1.1$ and $x_1 = 1.5$;

   b) Simpson's rule with $x_0 = 1.1$, $x_1 = 1.3$, and $x_2 = 1.5$.

x	e^x
1.1	3.0042
1.3	3.6693
1.5	4.4817

3. Approximate the following integrals using formulas (4.38), (4.39), (4.40), (4.44), (4.45), and (4.46). Are the accuracies of the approximations consistent with the error formulas? Which of parts (d) and (e) gives the better approximation?

   a) $\displaystyle\int_0^{0.1} \sqrt{1 + x} \, dx$

   b) $\displaystyle\int_0^{\pi/2} (\sin x)^2 dx$

   c) $\displaystyle\int_{1.1}^{1.5} e^x dx$

   d) $\displaystyle\int_1^{10} \frac{1}{x} dx$

   e) $\displaystyle\int_1^{5.5} \frac{1}{x} dx + \int_{5.5}^{10} \frac{1}{x} dx$

   f) $\displaystyle\int_0^1 x^{1/3} dx$

4. Use the Newton–Cotes closed formula for $n = 3$ and the open formula for $n = 2$ to approximate $\int_1^3 e^{-x/2} dx$. Find a bound for the error in each case and compare the approximations obtained with the actual value 0.7668010.

5. Repeat Exercise 4 using the closed formula for $n = 4$ and open formula for $n = 3$.

6. Given the function f at the following values:

x	1.8	2.0	2.2	2.4	2.6
$f(x)$	3.12014	4.42569	6.04241	8.03014	10.46675

approximate $\int_{1.8}^{2.6} f(x)\,dx$ using all the quadrature formulas of this section that can be applied.

7. Suppose the data of Exercise 6 has rounding errors given by the following table:

x	1.8	2.0	2.2	2.4	2.6
Error in $f(x)$	2×10^{-6}	-2×10^{-6}	-0.9×10^{-6}	-0.9×10^{-6}	2×10^{-6}

Calculate the errors due to rounding in Exercise 6.

8. Use each of the methods given by Eq. (4.37)–(4.40) and (4.43)–(4.46) to obtain an approximation for $\int_0^{1.5} (1 + x)^{-1}\,dx$, and compare the results obtained to the exact value 0.9162907.

9. Derive Simpson's rule with error term by using

$$\int_{x_0}^{x_2} f(x)\,dx = a_0 f(x_0) + a_1 f(x_1) + a_2 f(x_2) + kf^{(4)}(\xi).$$

Find a_0, a_1, and a_2 from the fact that Simpson's rule is exact for $f(x) = x^n$ when $n = 1, 2, 3$. Find k by applying the integration formula with $f(x) = x^4$.

10. Prove the statement following Definition 4.1; that is, show that a quadrature formula has degree of precision n precisely when $E(x^k) = 0$ for all $k = 0, 1, \ldots, n$ and $E(x^{n+1}) \neq 0$.

11. Derive Simpson's three-eighths rule, Eq. (4.39), with error term, by the use of Theorem 4.2.

12. Derive Eq. (4.44) with error term by the use of Theorem 4.3.

4.4 Composite Numerical Integration

The Newton–Cotes formulas are generally unsuitable for use over large integration intervals. High-degree formulas would be required for use over such intervals and the values of the coefficients in these formulas are difficult to obtain. More importantly, the Newton–Cotes formulas are based on interpolatory polynomials that use equally spaced nodes, a procedure we found, in Section 3.6, to be inaccurate over large intervals because of the oscillatory nature of high-degree polynomials. In this section we discuss a piecewise approach to numerical integration that uses the low-order Newton–Cotes formulas. These procedures are among the techniques most often applied in practice.

Consider finding an approximation to $\int_0^4 e^x\,dx$. If Simpson's rule is used with $h = 2$,

$$\int_0^4 e^x\,dx \approx \tfrac{2}{3}(e^0 + 4e^2 + e^4) = 56.76958.$$

Since the exact answer in this case is $e^4 - e^0 = 53.59815$, the error of -3.17143 is far larger than would generally be regarded as acceptable.

To apply a piecewise technique to this problem, divide $[0, 4]$ into $[0, 2]$ and $[2, 4]$ and use Simpson's rule, Eq. (4.38), twice with $h = 1$:

$$\int_0^4 e^x \, dx = \int_0^2 e^x \, dx + \int_2^4 e^x \, dx$$

$$\approx \tfrac{1}{3}[e^0 + 4e + e^2] + \tfrac{1}{3}[e^2 + 4e^3 + e^4]$$

$$= \tfrac{1}{3}[e^0 + 4e + 2e^2 + 4e^3 + e^4] = 53.86385.$$

The actual error has been reduced to -0.26570. Encouraged by our results, we subdivide the intervals $[0, 2]$ and $[2, 4]$ and use Simpson's rule with $h = \tfrac{1}{2}$, giving

$$\int_0^4 e^x \, dx = \int_0^1 e^x \, dx + \int_1^2 e^x \, dx + \int_2^3 e^x \, dx + \int_3^4 e^x \, dx$$

$$\approx \tfrac{1}{6}[e^0 + 4e^{1/2} + e] + \tfrac{1}{6}[e + 4e^{3/2} + e^2]$$

$$+ \tfrac{1}{6}[e^2 + 4e^{5/2} + e^3] + \tfrac{1}{6}[e^3 + 4e^{7/2} + e^4]$$

$$= \tfrac{1}{6}[e^0 + 4e^{1/2} + 2e + 4e^{3/2} + 2e^2 + 4e^{5/2} + 2e^3 + 4e^{7/2} + e^4]$$

$$= 53.61622.$$

The error for this approximation is -0.01807.

A generalization of this procedure is as follows: Subdivide the interval $[a, b]$ into n subintervals and use Simpson's rule on each pair of consecutive subintervals. Since each application of Simpson's rule requires two intervals, n must be an even integer, that is, $n = 2m$ for some integer m. (*See Figure 4.4.*)

FIGURE 4.4

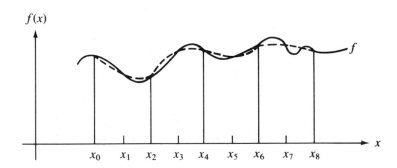

With $h = (b - a)/2m$ and $a = x_0 < x_1 < \cdots < x_{2m} = b$, where $x_j = x_0 + jh$ for each $j = 0, 1, \ldots, 2m$,

$$\int_a^b f(x) \, dx = \sum_{j=1}^m \int_{x_{2j-2}}^{x_{2j}} f(x) \, dx$$

$$= \sum_{j=1}^m \left\{ \frac{h}{3}[f(x_{2j-2}) + 4f(x_{2j-1}) + f(x_{2j})] - \frac{h^5}{90} f^{(4)}(\xi_j) \right\}$$

for some ξ_j with $x_{2j-2} < \xi_j < x_{2j}$. Using the fact that for each $j = 1, 2, \ldots, m - 1, f(x_{2j})$ appears in the term corresponding to the interval $[x_{2j-2}, x_{2j}]$ and also, in the term corresponding to the interval $[x_{2j}, x_{2j+2}]$, this reduces to

$$(4.47) \quad \int_a^b f(x)\,dx = \frac{h}{3}\left[f(x_0) + 2\sum_{j=1}^{m-1} f(x_{2j}) + 4\sum_{j=1}^{m} f(x_{2j-1}) + f(x_{2m}) \right]$$
$$- \frac{h^5}{90}\sum_{j=1}^{m} f^{(4)}(\xi_j).$$

THEOREM 4.4   If $f \in C^4[a, b]$, there exists a $\mu \in (a, b)$ for which **Simpson's composite rule over $n = 2m$ subintervals** of $[a, b]$ can be expressed with error term as

$$(4.48) \quad \int_a^b f(x)\,dx = \frac{h}{3}\left[f(a) + 2\sum_{j=1}^{m-1} f(x_{2j}) + 4\sum_{j=1}^{m} f(x_{2j-1}) + f(b) \right]$$
$$- \frac{(b-a)h^4}{180} f^{(4)}(\mu),$$

where $a = x_0 < x_1 < \cdots < x_{2m} = b$, $h = (b-a)/2m$, and $x_j = x_0 + jh$ for each $j = 0, 1, \ldots, 2m$.

PROOF   From Eq. (4.47) the error associated with this approximation is

$$E(f) = \frac{-h^5}{90}\sum_{j=1}^{m} f^{(4)}(\xi_j),$$

where $x_{2j-2} < \xi_j < x_{2j}$ for each $j = 1, 2, \ldots, m$. Since $f \in C^4[a, b]$, the Extreme Value Theorem implies that $f^{(4)}$ assumes its maximum and minimum in $[a, b]$.

$$\min_{x \in [a, b]} f^{(4)}(x) \le f^{(4)}(\xi_j) \le \max_{x \in [a, b]} f^{(4)}(x);$$

so

$$m \min_{x \in [a, b]} f^{(4)}(x) \le \sum_{j=1}^{m} f^{(4)}(\xi_j) \le m \max_{x \in [a, b]} f^{(4)}(x),$$

and

$$\min_{x \in [a, b]} f^{(4)}(x) \le \frac{1}{m}\sum_{j=1}^{m} f^{(4)}(\xi_j) \le \max_{x \in [a, b]} f^{(4)}(x).$$

By the Intermediate Value Theorem there is a $\mu \in (a, b)$ such that

$$f^{(4)}(\mu) = \frac{1}{m}\sum_{j=1}^{m} f^{(4)}(\xi_j).$$

Thus,

$$E(f) = \frac{-h^5}{90} m f^{(4)}(\mu).$$

Since $h = (b-a)/2m$,

$$E(f) = \frac{-h^4(b-a)}{180} f^{(4)}(\mu). \qquad \square$$

The following algorithm employs Simpson's composite rule on $n = 2m$ subintervals. This is a quadrature algorithm often used in practice.

Simpson's Composite Algorithm 4.1

To approximate the integral $I = \int_a^b f(x)\,dx$:

INPUT endpoints a, b; positive integer m.

OUTPUT approximation XI to I.

Step 1 Set $h = (b - a)/(2m)$.

Step 2 Set $XI0 = f(a) + f(b)$;
  $XI1 = 0$; (*Summation of $f(x_{2i-1})$.*)
  $XI2 = 0$. (*Summation of $f(x_{2i})$.*)

Step 3 For $i = 1, \ldots, 2m - 1$ do Steps 4 and 5.

Step 4 Set $X = a + ih$.

Step 5 If i is even then set $XI2 = XI2 + f(X)$
  else set $XI1 = XI1 + f(X)$.

Step 6 Set $XI = h(XI0 + 2 \cdot XI2 + 4 \cdot XI1)/3$.

Step 7 OUTPUT (XI);
  STOP.

This approach can similarly be applied to any of the lower-order formulas. The extensions of the trapezoidal (*see Fig. 4.5*) and midpoint rules are given without proof in the following two theorems. Since the trapezoidal rule requires only one interval for application, there is no restriction on the integer n in this case.

FIGURE 4.5

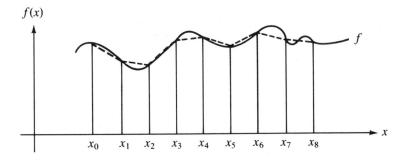

THEOREM 4.5 Let $f \in C^2[a, b]$. With $h = (b - a)/n$ and $x_j = a + jh$ for each $j = 0, 1, \ldots, n$, the **trapezoidal rule for n subintervals** is

$$(4.49) \qquad \int_a^b f(x)\,dx = \frac{h}{2}\left[f(a) + f(b) + 2\sum_{j=1}^{n-1} f(x_j) \right] - \frac{(b-a)h^2}{12} f''(\mu)$$

for some $\mu \in (a, b)$.

THEOREM 4.6

Let $f \in C^2[a, b]$. With $h = (b - a)/(2m + 2)$ and $x_j = a + (j + 1)h$ for each $j = -1, 0, \ldots,$ $2m + 1$, **the midpoint rule for $n = 2m$ subintervals is**

(4.50)
$$\int_a^b f(x)\,dx = 2h \sum_{j=0}^{m} f(x_{2j}) + \frac{b - a}{6} h^2 f''(\mu)$$

for some $\mu \in (a, b)$.

EXAMPLE 1

Consider approximating $\int_0^\pi \sin x\,dx$ with an error of at most 0.00002, using Simpson's composite rule. Applying Eq. (4.47) to the integral $\int_0^\pi \sin x\,dx$:

$$\int_0^\pi \sin x\,dx = \frac{h}{3} \left[2 \sum_{j=1}^{m-1} \sin x_{2j} + 4 \sum_{j=1}^{m} \sin x_{2j-1} \right] - \frac{\pi h^4}{180} \sin \mu.$$

Since the truncation error is required to be less than 0.00002, the inequality

$$\left| \frac{\pi h^4}{180} \sin \mu \right| \leq \frac{\pi h^4}{180} = \frac{\pi^5}{2880 m^4} \leq 0.00002$$

is used to determine m and h. Completing these calculations gives $m \geq 9$. If $m = 10$, then $n = 20$ and $h = \pi/20$, so the formula obtained from Eq. (4.48) becomes

$$\int_0^\pi \sin x\,dx \approx \frac{\pi}{60} \left[2 \sum_{j=1}^{9} \sin\left(\frac{j\pi}{10}\right) + 4 \sum_{j=1}^{10} \sin\left(\frac{(2j-1)\pi}{20}\right) \right].$$

To be assured of this degree of accuracy using the composite trapezoidal rule requires that

$$\left| \frac{\pi h^2}{12} \sin \mu \right| \leq \frac{\pi h^2}{12} = \frac{\pi^3}{12 n^2} < 0.00002$$

or that $n \geq 360$. Since this is many more calculations than are needed for Simpson's rule, it is clear that it would be undesirable to use the trapezoidal rule on this problem. For comparison purposes, the trapezoidal rule with $n = 20$ and $h = \pi/20$ gives

$$\int_0^\pi \sin x\,dx \approx \frac{\pi}{40} \left[2 \sum_{j=1}^{19} \sin\left(\frac{j\pi}{20}\right) + \sin 0 + \sin \pi \right] = \frac{\pi}{40} \left[2 \sum_{j=1}^{19} \sin\left(\frac{j\pi}{20}\right) \right].$$

Simpson's rule gives an answer 2.00000679 and the trapezoidal rule gives an answer 1.9958860. The exact answer is 2; so Simpson's rule with $n = 2m = 20$ gave an answer well within the required error bound, whereas the trapezoidal rule with $n = 20$ clearly did not. ∎

An important property shared by all the Newton–Cotes composite integration techniques is a stability with respect to round-off error. To demonstrate this feature, suppose we apply the composite Simpson's rule with $n = 2m$ subintervals to a function f on $[a, b]$ and determine the maximum bound for the round-off error. Assume that $f(x_i)$ is approximated by $\tilde{f}(x_i)$ and that

$$f(x_i) = \tilde{f}(x_i) + e_i, \qquad \text{for each } i = 0, 1, \ldots, n,$$

where e_i denotes the round-off error associated with using $\tilde{f}(x_i)$ to approximate $f(x_i)$. Then the accumulated error, ERR, in the composite Simpson's rule is

$$\text{ERR} = \left| \frac{h}{3}\left[e_0 + 2 \sum_{j=1}^{m-1} e_{2j} + 4 \sum_{j=1}^{m} e_{2j-1} + e_{2m} \right] \right|$$

$$\leq \frac{h}{3}\left[|e_0| + 2 \sum_{j=1}^{m-1} |e_{2j}| + 4 \sum_{j=1}^{m} |e_{2j-1}| + |e_{2m}| \right].$$

If the round-off errors are uniformly bounded by ε, then

$$\text{ERR} \leq \frac{h}{3}[\varepsilon + 2(m-1)\varepsilon + 4(m)\varepsilon + \varepsilon] = \frac{h}{3}6m\varepsilon = 2mh\varepsilon.$$

But $2mh = b - a$, so

$$\text{ERR} \leq (b-a)\varepsilon,$$

a bound independent of h. This implies that the procedure is stable as h approaches zero. Recall that this was not true in the case of the numerical differentiation procedures considered in Section 4.1.

Exercise Set 4.4

1. Use the extended trapezoidal rule with the indicated values of n to approximate the following definite integrals. Compare the approximations to the exact result.

   a) $\displaystyle\int_1^3 \frac{dx}{x}$; $n=4$.

   b) $\displaystyle\int_0^2 x^3\,dx$; $n=4$.

   c) $\displaystyle\int_0^3 x\sqrt{1+x^2}\,dx$; $n=6$.

   d) $\displaystyle\int_0^1 \sin \pi x\,dx$; $n=6$.

   e) $\displaystyle\int_0^{2\pi} x\sin x\,dx$; $n=8$.

   f) $\displaystyle\int_0^1 x^2 e^x\,dx$; $n=8$.

2. Use the midpoint rule for $n = 2m$ subintervals to approximate the integrals in Exercise 1 with the given values of n.

3. Repeat Exercise 2 using Simpson's composite rule.

4. Exercise 8 of Section 4.3 required that all the methods of that section be used to calculate an approximation to

$$\int_0^{1.5} (1+x)^{-1}\,dx.$$

   With $n = 10$:

   a) Approximate the integral with Simpson's composite rule.

   b) Approximate the integral with the extended trapezoidal rule.

   c) Approximate the integral with the extended midpoint rule.

d) Compare the approximations in parts (a), (b), and (c) to the values obtained in Exercise 8 of Section 4.3.

5. To approximate $\int_0^2 x^2 e^{-x^2} dx$,

a) Use the extended trapezoidal rule with $n = 8$.

b) Use Simpson's composite rule with $n = 2m = 8$.

c) Use the extended midpoint rule with $n = 2m = 8$.

6. Approximate $\int_1^{10} \ln x\, dx$ to within 10^{-4} using Simpson's composite rule.

7. Consider $\int_0^{\pi/4} \tan x\, dx$.

a) Use the extended trapezoidal rule with $n = 4$ and $n = 8$ to approximate the integral.

b) Find a bound for the error in each case in part (a) and compare the approximations to the actual value.

c) Determine the values of n and h needed for the approximation to be within 10^{-8}.

8. Repeat Exercise 7 using Simpson's composite rule.

9. Repeat Exercise 7 using the extended midpoint rule.

10. Repeat Exercise 7 for the integral $\int_{\pi/2}^{3\pi/4} \cot x\, dx$.

11. Repeat Exercise 10 using Simpson's composite rule.

12. Repeat Exercise 10 using the extended midpoint rule.

13. Determine the values of n and h needed to approximate $\int_1^3 e^x \sin x\, dx$ to within 10^{-4} and find the approximation:

a) Use Simpson's composite rule.

b) Use the extended trapezoidal rule.

c) Use the extended midpoint rule.

14. Derive the extended trapezoidal rule, Eq. (4.49), and write an algorithm for the procedure.

15. Derive the extended midpoint rule, Eq. (4.50), and write an algorithm for the procedure.

16. Let f be defined by

$$f(x) = \begin{cases} x^3 + 1, & 0 \le x \le 0.1, \\ 1.001 + 0.03(x - 0.1) + 0.3(x - 0.1)^2 + 2(x - 0.1)^3, & 0.1 \le x \le 0.2, \\ 1.009 + 0.15(x - 0.2) + 0.9(x - 0.2)^2 + 2(x - 0.2)^3, & 0.2 \le x \le 0.3. \end{cases}$$

a) Investigate the continuity of the derivatives of f.

b) Approximate $\int_0^{0.3} f(x)\, dx$, using the extended trapezoidal rule with $n = 6$, and estimate the error using Eq. (4.49).

c) Approximate $\int_0^{0.3} f(x)\, dx$, using Simpson's composite rule with $n = 6$. Are the results more accurate than in part (b)?

17. A particle of mass m moving through a fluid is subjected to a viscous resistance R, which is a function of the velocity v. The relationship between the resistance R, velocity v, and time t is given by the equation

$$t = \int_{v(t_0)}^{v(t)} \frac{m}{R(u)} du.$$

Suppose that $R(v) = -v\sqrt{v}$ for a particular fluid, where R is in newtons and v is in meters/second. If $m = 10$ kg and $v(0) = 10$ m/sec, approximate the time required for the particle to slow to $v = 5$ m/sec.

a) Use Simpson's composite rule with $h = 0.25$.

b) Use the extended trapezoidal rule with $h = 0.25$.

c) Compare these approximations to the actual value.

18. To simulate the thermal characteristics of disk brakes (see following figure), D. A. Secrist and R. W. Hornbeck [121] needed to approximate numerically the "area averaged lining temperature," T, of the brake pad from the equation

$$T = \frac{\int_{r_e}^{r_o} T(r) r \theta_p \, dr}{\int_{r_e}^{r_o} r \theta_p \, dr}$$

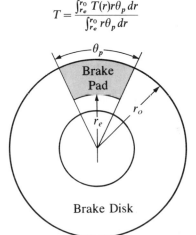

where r_e represents the radius at which the pad–disk contact begins, r_o represents the outside radius of the pad–disk contact, θ_p represents the angle subtended by sector brake pads, and $T(r)$ is the temperature at each point of the pad, obtained numerically from analyzing the heat equation (see Section 11.3). If $r_e = 0.308$ ft, $r_o = 0.478$ ft, $\theta_p = 0.7051$ radians, and the temperatures given in the following table have been calculated at the various points on the disk, find an approximation for T using Simpson's composite rule.

r(feet)	$T(r)(^\circ F)$	r(feet)	$T(r)(^\circ F)$	r(feet)	$T(r)(^\circ F)$
0.308	640	0.376	1034	0.444	1204
0.325	794	0.393	1064	0.461	1222
0.342	885	0.410	1114	0.478	1239
0.359	943	0.427	1152		

19. Use Simpson's composite rule to find an approximation to within 10^{-4} to the value of the integral considered in the application opening this chapter:

$$\int_0^{48} \sqrt{1+(\cos x)^2} \, dx.$$

20. **Finite-Jump Discontinuity** Let f be defined by

$$f(x) = \begin{cases} x^2 - 1, & 0 \le x \le 0.5, \\ e^x \sin x, & 0.5 < x \le 1. \end{cases}$$

a) Is f continuous on $[0, 1]$?

b) Evaluate

$$\int_0^1 f(x) \, dx.$$

c) Approximate the integral in part (b) using Simpson's composite rule with $h = 0.05$.

d) Approximate the integral using Eq. (4.49) with $h = 0.04$.

e) Approximate

$$\int_0^{0.5} f(x) \, dx + \int_{0.5}^1 f(x) \, dx$$

using Eq. (4.48) on $[0, 0.5]$ and on $[0.5, 1]$ with $h = 0.05$, and compare your results to those in part (c).

f) Which part (c), (d), or (e), gives the best approximation? Can an error formula be applied in any of the three parts?

21. **Infinite Discontinuities** The integral

$$\int_a^b \frac{dx}{(x-a)^p}$$

converges if and only if $p < 1$. If the function f can be written in the form

$$f(x) = \frac{g(x)}{(x-a)^p},$$

where $g \in C[a, b]$, the integral

$$\int_a^b f(x) \, dx$$

also exists. If $g \in C^{n+1}[a, b]$ for some $n \ge 0$, the Taylor polynomial P of degree n given by

$$P(x) = \sum_{k=0}^n \frac{g^{(k)}(a)}{k!} (x-a)^k$$

can be constructed. The integral of f can then be approximated by

$$\int_a^b f(x) \, dx = \int_a^b \frac{g(x) - P(x)}{(x-a)^p} \, dx + \int_a^b \frac{P(x)}{(x-a)^p} \, dx,$$

where the function $(g(x) - P(x))/(x - a)^p$ has n derivatives at $x = a$. The integral

$$\int_a^b \frac{g(x) - P(x)}{(x - a)^p}\, dx$$

can be approximated by any of the quadrature formulas considered in Sections 4.3 and 4.4, while the integral

$$\int_a^b \frac{P(x)}{(x - a)^p}\, dx$$

can be explicitly evaluated.

a) Show that

$$\int_a^b \frac{P(x)}{(x - a)^p}\, dx = \sum_{k=0}^n \frac{g^{(k)}(a)}{k!(k + 1 - p)}(b - a)^{k + 1 - p}.$$

b) Approximate the integral

$$\int_0^1 \frac{e^x}{\sqrt{x}}\, dx,$$

using Simpson's composite rule with $h = 0.05$. Can you apply the error formula in Eq. (4.48)?

c) Approximate the integral

$$\int_0^1 \frac{e^{-x}}{\sqrt{1 - x}}\, dx$$

using the extended trapezoidal rule with $h = 0.05$.

22. **Infinite Limits of Integration**   The integral $\int_a^\infty f(x)\, dx$, $a > 0$, if it exists, can often be approximated using a quadrature formula after the change in variable $t = x^{-1}$.

a) Show that

$$\int_a^\infty f(x)\, dx = \int_0^{1/a} t^{-2} f\left(\frac{1}{t}\right) dt.$$

b) Apply part (a) and the extended midpoint rule of Eq. (4.50) with $h = 0.05$, to approximate

$$\int_1^\infty x^{-2} \sin x\, dx.$$

c) Apply part (a) to approximate

$$\int_0^\infty \sqrt{x} e^{-x}\, dx,$$

using the extended midpoint rule of Eq. (4.50) with $h = 0.05$.

d) Apply part (a) to approximate

$$\int_{-\infty}^\infty \frac{dx}{1 + x^2},$$

using the extended midpoint rule of Eq. (4.50) with $h = 0.05$.

23. Suppose a body of mass m is traveling vertically upward starting at the surface $x = R$ of the earth. If all resistance except gravity is neglected, then the escape velocity v is given by

$$v^2 = 2gR \int_1^\infty z^{-2}\, dz, \qquad \text{where } z = \frac{x}{R},$$

and g is the gravitational field strength at the earth's surface. If $g = 0.00609$ miles/sec^2 and $R = 3960$ miles, approximate the escape velocity v using Exercise 2.2 and the extended midpoint rule with $h = 0.1$.

4.5 Adaptive Quadrature Methods

The composite formulas require the use of equally spaced nodes. For many problems this is not an important restriction, but it is inappropriate when integrating a function on an interval that contains both regions with large functional variation and regions with small functional variation. In this situation a smaller step size is needed for the large variation regions than for those with less variation, if the approximation error is to be uniformly distributed. An efficient technique for this type of problem is one that can distinguish the amount of functional variation and adapt the step size to the varying requirements of the problem. Such methods are appropriately named **Adaptive Quadrature Methods**. The method we will discuss is based on Simpson's composite rule, but the technique is easily modified to the other composite procedures.

Suppose that we wish to approximate $\int_a^b f(x)\,dx$ to within a specified tolerance $\varepsilon > 0$. The first step in the procedure is to apply Simpson's rule with step size $h = (b - a)/2$. This results in

$$(4.51) \qquad \int_a^b f(x)\,dx = S(a, b) - \frac{h^5}{90} f^{(4)}(\mu), \qquad \text{for some } \mu \text{ in } (a, b)$$

where

$$S(a, b) = \frac{h}{3}[f(a) + 4f(a + h) + f(b)].$$

The next step is to determine a way to estimate the accuracy of our approximation, in particular, one that does not require determining $f^{(4)}(\mu)$. To accomplish this, we first apply Simpson's composite rule to the problem with $m = 2$ and step size $(b - a)/4 = h/2$. Thus,

$$(4.52) \quad \int_a^b f(x)\,dx = \frac{h}{6}\left[f(a) + 4f\left(a + \frac{h}{2}\right) + 2f(a + h) + 4f\left(a + \frac{3h}{2}\right) + f(b) \right]$$
$$- \left(\frac{h}{2}\right)^4 \frac{(b - a)}{180} f^{(4)}(\tilde{\mu}),$$

for some $\tilde{\mu}$ in (a, b). To simplify notation, let

$$S\left(a, \frac{a + b}{2}\right) = \frac{h}{6}\left[f(a) + 4f\left(a + \frac{h}{2}\right) + f(a + h) \right]$$

and
$$S\left(\frac{a+b}{2}, b\right) = \frac{h}{6}\left[f(a+h) + 4f\left(a + \frac{3h}{2}\right) + f(b)\right].$$

Then Eq. (4.52) can be rewritten as

(4.53)
$$\int_a^b f(x)\,dx = S\left(a, \frac{a+b}{2}\right) + S\left(\frac{a+b}{2}, b\right) - \frac{1}{16}\left(\frac{h^5}{90}\right) f^{(4)}(\tilde{\mu}).$$

The error estimation is derived by assuming that $\mu = \tilde{\mu}$ or, more precisely, that $f^{(4)}(\mu)$ $= f^{(4)}(\tilde{\mu})$. The success of the technique depends on the accuracy of this assumption. If this is an accurate assumption, then Eqs. (4.51) and (4.53) imply that

$$S\left(a, \frac{a+b}{2}\right) + S\left(\frac{a+b}{2}, b\right) - \frac{1}{16}\left(\frac{h^5}{90}\right) f^{(4)}(\mu) \approx S(a, b) - \frac{h^5}{90} f^{(4)}(\mu),$$

so
$$\frac{h^5}{90} f^{(4)}(\mu) \approx \frac{16}{15}\left[S(a, b) - S\left(a, \frac{a+b}{2}\right) - S\left(\frac{a+b}{2}, b\right)\right].$$

Using this estimate in Eq. (4.53) produces the error estimation

(4.54)
$$\left| \int_a^b f(x)\,dx - S\left(a, \frac{a+b}{2}\right) - S\left(\frac{a+b}{2}, b\right)\right|$$
$$\approx \frac{1}{15}\left| S(a, b) - S\left(a, \frac{a+b}{2}\right) - S\left(\frac{a+b}{2}, b\right)\right|.$$

This implies that $S(a, (a+b)/2) + S((a+b)/2, b)$ approximates $\int_a^b f(x)\,dx$ fifteen times better than it agrees with the known value $S(a, b)$. As a consequence, $S(a, (a+b)/2)$ $+ S((a+b)/2, b)$ will approximate $\int_a^b f(x)\,dx$ to within ε provided that the two approximations $S(a, (a+b)/2) + S((a+b)/2, b)$ and $S(a, b)$ differ by less than 15ε, that is, if

(4.55)
$$\left| S(a, b) - S\left(a, \frac{a+b}{2}\right) - S\left(\frac{a+b}{2}, b\right)\right| < 15\varepsilon,$$

then

(4.56)
$$\left| \int_a^b f(x)\,dx - S\left(a, \frac{a+b}{2}\right) - S\left(\frac{a+b}{2}, b\right)\right| < \varepsilon.$$

In this case

$$S\left(a, \frac{a+b}{2}\right) + S\left(\frac{a+b}{2}, b\right)$$

is assumed to be a sufficiently accurate approximation to $\int_a^b f(x)\,dx$. When the inequality (4.55) does not hold, the error estimation procedure is applied individually to the subintervals $[a, (a+b)/2]$ and $[(a+b)/2, b]$ to determine if the approximation to the integral on each subinterval is within a tolerance of $\varepsilon/2$. If so, sum the approximations to produce an approximation to $\int_a^b f(x)\,dx$ within the tolerance ε. If the approximation on one of the subintervals fails to be within the tolerance $\varepsilon/2$, that subinterval is itself subdivided and each of its subintervals analyzed to determine if the approximation on that subinterval is accurate to within $\varepsilon/4$. This halving procedure is continued until each portion is within the required tolerance.

The following algorithm details this adaptive quadrature procedure for Simpson's rule. Some technical difficulties arise which require the implementation of the method to differ slightly from the preceding discussion. Note in Step 1 that the tolerance has been set at 10ε rather than the 15ε figure shown in inequality (4.55). This bound is chosen conservatively to compensate for error in the assumption $f^{(4)}(\mu) = f^{(4)}(\tilde{\mu})$. In problems when $f^{(4)}$ is known to be widely varying, it would be reasonable to lower this bound even further.

The procedure listed in the algorithm always first approximates the integral on the leftmost subinterval in a subdivision. This requires introducing a procedure for efficiently storing and recalling previously computed functional evaluations for the nodes in the right half subintervals. Steps 3, 4, and 5 contain a stacking procedure with an indicator to keep track of the data that will be required for calculating the approximation on the subinterval immediately adjacent and to the right of the subinterval on which the approximation is being generated.

Adaptive Quadrature Algorithm 4.2

To approximate the integral $I = \int_a^b f(x)\,dx$ to within a given tolerance $TOL > 0$:

INPUT endpoints a, b; tolerance TOL; limit N to number of levels.

OUTPUT approximation APP or message that N is exceeded.

Step 1   Set $APP = 0$;
$\quad\quad\quad i = 1$;
$\quad\quad\quad TOL_i = 10\ TOL$;
$\quad\quad\quad a_i = a$;
$\quad\quad\quad h_i = (b - a)/2$;
$\quad\quad\quad FA_i = f(a)$;
$\quad\quad\quad FC_i = f(a + h_i)$;
$\quad\quad\quad FB_i = f(b)$;
$\quad\quad\quad S_i = h_i(FA_i + 4FC_i + FB_i)/3$;   (*Approximation from Simpson's method for entire interval.*)

$\quad\quad\quad L_i = 1$.

Step 2   While $i > 0$ do Steps 3–5.

Step 3   Set $FD = f(a_i + h_i/2)$;
$\quad\quad\quad FE = f(a_i + 3h_i/2)$;
$\quad\quad\quad S1 = h_i(FA_i + 4FD + FC_i)/6$;   (*Approximations from Simpson's method for halves of subintervals.*)

$\quad\quad\quad S2 = h_i(FC_i + 4FE + FB_i)/6$;
$\quad\quad\quad v_1 = a_i$;   (*Save data at this level.*)
$\quad\quad\quad v_2 = FA_i$;
$\quad\quad\quad v_3 = FC_i$;

$$v_4 = FB_i;$$
$$v_5 = h_i;$$
$$v_6 = TOL_i;$$
$$v_7 = S_i;$$
$$v_8 = L_i.$$

Step 4 Set $i = i - 1$.   (*Delete the level.*)

Step 5 If $|S1 + S2 - v_7| < v_6$
  then set $APP = APP + (S1 + S2)$
  else
    if $(v_8 \geq N)$
      then
        OUTPUT ('LEVEL EXCEEDED');   (*Procedure fails.*)
        STOP.
      else   (*Add one level.*)
        set $i = i + 1$;   (*Data for right half subinterval.*)
          $a_i = v_1 + v_5;$
          $FA_i = v_3;$
          $FC_i = FE;$
          $FB_i = v_4;$
          $h_i = v_5/2;$
          $TOL_i = v_6/2;$
          $S_i = S2;$
          $L_i = v_8 + 1;$
        set $i = i + 1$;   (*Data for left half subinterval.*)
          $a_i = v_1;$
          $FA_i = v_2;$
          $FC_i = FD;$
          $FB_i = v_3;$
          $h_i = h_{i-1};$
          $TOL_i = TOL_{i-1};$
          $S_i = S1;$
          $L_i = L_{i-1}.$

Step 6 OUTPUT (APP);   (*APP approximates I to within TOL.*)
  STOP.

EXAMPLE 1   The graph of the function $f(x) = (100/x^2) \sin(10/x)$ for x in $[1, 3]$ is shown in Fig. 4.6. Using the Adaptive Quadrature Algorithm 4.2 with tolerance 10^{-4} to approximate $\int_1^3 f(x)\,dx$ produces -1.426014, a result that is accurate to within 1.4×10^{-6}. The approximation required that Simpson's rule with $m = 2$ be performed on the 23 subintervals whose endpoints are shown on the horizontal axis in Fig. 4.6. The total number of functional evaluations required for this approximation is 93.

FIGURE 4.6

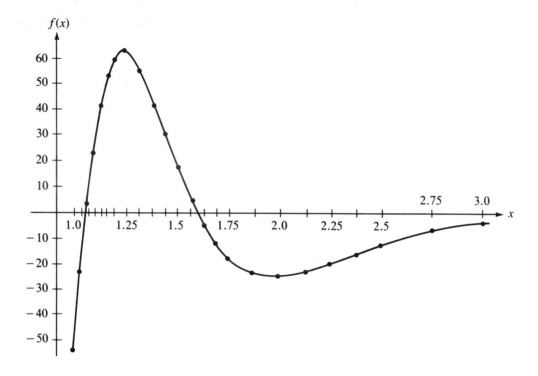

Using Simpson's composite rule to approximate this integral with $h = \frac{1}{64}$ requires 255 functional evaluations and gives the approximation -1.426059, a result that differs from the actual value by 2.4×10^{-6}. ∎

Exercise Set 4.5

1. Use the Adaptive Quadrature procedure to find approximations to within 10^{-3} for the definite integrals in Exercise 1 of Section 4.4. Do not use a computer program to generate these results.

2. Use the Adaptive Quadrature Algorithm to approximate $\int_{1}^{10} \ln x \, dx$ to within 10^{-4}. Compare your approximation to the result in Exercise 6 of Section 4.4. Compare the number of nodes required in each case.

3. Use the Adaptive Quadrature Algorithm to approximate $\int_{0}^{2} x^2 e^{-x^2} dx$ to within 10^{-6}. Compare the results to those generated in Exercise 5 of Section 4.4.

4. Use the Adaptive Quadrature Algorithm to approximate $\int_{1}^{3} e^x \sin x \, dx$ to within 10^{-6}. Compare your approximation to the result in Exercise 13 of Section 4.4. Compare the number of nodes required in each case.

5. Use the Adaptive Quadrature Algorithm to approximate $\int_{0}^{48} \{1 + (\cos x)^2\}^{1/2} dx$ to within 10^{-4}. Does this technique have any advantage in this instance? (Compare the result and the number of functional evaluations with those of Exercise 19 in Section 4.4.)

6. Approximate the following integrals to within 3×10^{-6} using Adaptive Quadrature without a computer program.

a) $\displaystyle\int_0^{\pi/4} \tan x \, dx$ b) $\displaystyle\int_{\pi/2}^{3\pi/4} \cot x \, dx$

7. Use the Adaptive Quadrature Algorithm to approximate $\int_0^1 x^{1/3} \, dx$ to within 10^{-4}. Compare your approximation to the result in Exercise 3(f) of Section 4.3. Compare the number of nodes required in each case.

8. Use the Adaptive Quadrature Algorithm to approximate $\int_{0.1}^2 \sin(1/x) \, dx$ to within 10^{-3}. Sketch the graph of $f(x) = \sin(1/x)$ on $[0.1, 2]$.

9. The differential equation

$$mu''(t) + ku(t) = F_0 \cos \omega t$$

describes a spring-mass system with mass m, spring constant k, and no applied damping. The term $F_0 \cos \omega t$ describes a periodic external force applied to the system. The solution to the equation when the system is initially at rest ($u'(0) = u(0) = 0$) is

$$u(t) = -\frac{2F_0}{m(\omega_0^2 - \omega^2)} \sin \frac{(\omega_0 - \omega)}{2} t \sin \frac{(\omega_0 + \omega)}{2} t, \quad \text{where } \omega_0 = \sqrt{\frac{k}{m}} \neq \omega.$$

Sketch the graph of u when $m = 1$, $k = 9$, $F_0 = 1$, $\omega = 2$, and $t \in [0, 2\pi]$. Use both the Adaptive Quadrative method and Simpson's composite rule to determine $\int_0^{2\pi} u(t) \, dt$ to within 10^{-4}.

10. If the term $cu'(t)$ is added to the left side of the motion equation in Exercise 9, the resulting differential equation describes a spring-mass system that is damped with damping constant c. The solution to this equation, when the solution is initially at rest is

$$u(t) = c_1 e^{r_1 t} + c_2 e^{r_2 t} + \frac{F_0}{\sqrt{m^2(\omega_0^2 - \omega^2) + c^2 \omega^2}} \cos(\omega t - \delta),$$

where

$$\delta = \text{Arctan}\left(\frac{c\omega}{m(\omega_0^2 - \omega^2)}\right), \quad r_1 = \frac{-c + \sqrt{c^2 - 4\omega_0^2 m^2}}{2m},$$

and

$$r_2 = \frac{-c - \sqrt{c^2 - 4\omega_0^2 m^2}}{2m}.$$

Sketch the graph of u when $m = 1$, $k = 9$, $F_0 = 1$, $c = 1$, $\omega = 2$ and $t \in [0, 2\pi]$. Use both the Adaptive Quadrature method and Simpson's composite rule to determine $\int_0^{2\pi} u(t) \, dt$ to within 10^{-4}.

4.6 Romberg Integration

Although the trapezoidal rule is the easiest Newton–Cotes formula to apply, we have shown in the previous sections that it lacks the degree of accuracy generally required. Romberg integration is a method that has wide application because it uses the trapezoidal rule to give preliminary approximations, and then applies the Richardson extrapolation process discussed in Section 4.2 to obtain improvements of the approximations.

To begin the presentation of the Romberg integration scheme, recall (Theorem 4.5) that the extended trapezoidal rule for approximating the integral of a function f on an interval $[a, b]$ using m subintervals is

$$\int_a^b f(x)\,dx = \frac{h}{2}\left[f(a) + f(b) + 2 \sum_{j=1}^{m-1} f(x_j) \right] - \frac{(b-a)}{12} h^2 f''(\mu),$$

where $a < \mu < b$, $h = (b-a)/m$ and $x_j = a + jh$ for each $j = 0, 1, \ldots, m$.

The first step in the Romberg process involves obtaining the trapezoidal rule approximations with $m_1 = 1$, $m_2 = 2$, $m_3 = 4, \ldots, m_n = 2^{n-1}$ where n is some positive integer. The values of the step size h_k corresponding to m_k will be $h_k = (b-a)/m_k = (b-a)/2^{k-1}$, and with this notation the trapezoidal rule becomes

$$(4.57) \quad \int_a^b f(x)\,dx = \frac{h_k}{2}\left[f(a) + f(b) + 2 \left(\sum_{i=1}^{2^{k-1}-1} f(a + ih_k) \right) \right] - \frac{(b-a)}{12} h_k^2 f''(\mu_k),$$

where μ_k is a number in (a, b).

If the notation $R_{k,1}$ is introduced to denote the portion of Eq. (4.57) that is used for the trapezoidal approximation, then

$$R_{1,1} = \frac{h_1}{2}[f(a) + f(b)] = \frac{(b-a)}{2}[f(a) + f(b)];$$

$$R_{2,1} = \frac{h_2}{2}[f(a) + f(b) + 2f(a + h_2)]$$

$$= \frac{(b-a)}{4}\left[f(a) + f(b) + 2f\left(a + \frac{(b-a)}{2} \right) \right]$$

$$= \tfrac{1}{2}[R_{1,1} + h_1 f(a + \tfrac{1}{2}h_1)];$$

$$R_{3,1} = \frac{h_3}{2}\left\{ f(a) + f(b) + 2\left[f\left(a + \frac{(b-a)}{4} \right) \right.\right.$$

$$\left.\left. + f\left(a + \frac{(b-a)}{2} \right) + f\left(a + \frac{3(b-a)}{4} \right) \right] \right\}$$

$$= \frac{(b-a)}{8}\left\{ f(a) + f(b) + 2\left[f\left(a + \frac{(b-a)}{4} \right) \right.\right.$$

$$\left.\left. + f\left(a + \frac{(b-a)}{2} \right) + f\left(a + \frac{3(b-a)}{4} \right) \right] \right\}$$

$$= \frac{1}{2}\left\{ R_{2,1} + h_2\left[f\left(a + \frac{h_2}{2} \right) + f\left(a + \frac{3h_2}{2} \right) \right] \right\};$$

and, in general,

$$(4.58) \qquad R_{k,1} = \frac{1}{2}\left[R_{k-1,1} + h_{k-1} \sum_{i=1}^{2^{k-2}} f\left(a + \left(i - \frac{1}{2}\right) h_{k-1}\right)\right]$$

for each $k = 2, 3, \ldots, n$.

For the derivation of Eq. (4.58), the reader is referred to Exercises 11 and 12.

EXAMPLE 1 Using Eq. (4.58) to perform the first step of the Romberg integration scheme for approximating $\int_0^\pi \sin x \, dx$ with $n = 6$ leads to:

$$R_{1,1} = \frac{\pi}{2}[\sin 0 + \sin \pi] = 0;$$

$$R_{2,1} = \frac{1}{2}\left[R_{1,1} + \pi \sin \frac{\pi}{2}\right] = 1.57079633;$$

$$R_{3,1} = \frac{1}{2}\left[R_{2,1} + \frac{\pi}{2}\left(\sin \frac{\pi}{4} + \sin \frac{3\pi}{4}\right)\right] = 1.89611890;$$

$$R_{4,1} = \frac{1}{2}\left[R_{3,1} + \frac{\pi}{4}\left(\sin \frac{\pi}{8} + \sin \frac{3\pi}{8} + \sin \frac{5\pi}{8} + \sin \frac{7\pi}{8}\right)\right] = 1.97423160;$$

$$R_{5,1} = 1.99357034, \qquad \text{and}$$

$$R_{6,1} = 1.99839336. \qquad \blacksquare$$

Since the correct value for this integral is 2, it is clear that, although the calculations involved are not difficult, the convergence is slow.

To speed the convergence, the Richardson extrapolation procedure will now be performed. It can be shown, although not easily (see Ralston and Rabinowitz [103], pp. 123–126 and accompanying exercises for a presentation), that if $f \in C^4[a, b]$, the extended trapezoidal rule given in Eq. (4.57) can be written with an alternate error term in the form:

$$(4.59) \qquad \int_a^b f(x)\,dx = \frac{h_k}{2}\left[f(a) + f(b) + 2\left(\sum_{i=1}^{2^{k-1}-1} f(a + ih_k)\right)\right]$$

$$- \frac{h_k^2}{12}[f'(b) - f'(a)] + \frac{(b-a)h_k^4}{720} f^{(4)}(\mu_k),$$

for each $k = 1, 2, \ldots, n$ and some $a < \mu_k < b$.

With the extended trapezoidal rule in this form, we can eliminate the term involving h_k^2 by combining the equations

$$(4.60) \qquad \int_a^b f(x)\,dx = R_{k-1,1} - \frac{h_{k-1}^2}{12}[f'(b) - f'(a)]$$

$$+ \frac{(b-a)h_{k-1}^4}{720} f^{(4)}(\mu_{k-1}) \qquad \text{and}$$

$$(4.61) \qquad \int_a^b f(x)\,dx = R_{k,1} - \frac{h_k^2}{12}[f'(b) - f'(a)] + \frac{(b-a)h_k^4}{720} f^{(4)}(\mu_k)$$

$$= R_{k,1} - \frac{h_{k-1}^2}{48}[f'(b) - f'(a)] + \frac{(b-a)h_k^4}{720} f^{(4)}(\mu_k)$$

to obtain

$$\int_a^b f(x)\,dx = \frac{4R_{k,1} - R_{k-1,1}}{3} + \frac{(b-a)}{2160}[4h_k^4 f^{(4)}(\mu_k) - h_{k-1}^4 f^{(4)}(\mu_{k-1})]$$

$$= \frac{4R_{k,1} - R_{k-1,1}}{3} + O(h_k^4).$$

It is an easy matter to show (see Exercise 10) that the approximation obtained by this technique is actually the approximation given by the Simpson's composite rule with $h = h_k$, so the error of order h_k^4 is expected.

To continue the Romberg scheme, define

$$R_{k,2} = \frac{4R_{k,1} - R_{k-1,1}}{3}$$

for each $k = 2, 3, \ldots, n$, and apply the Richardson extrapolation procedure to these values. In general, if $f \in C^{2n+2}[a, b]$, then for each $k = 1, 2, \ldots, n$, the composite trapezoidal rule on k subintervals can be expressed with an error term similar to that in Eq. (4.59):

$$(4.62) \qquad \int_a^b f(x)\,dx = \frac{h_k}{2}\left[f(a) + f(b) + 2\left(\sum_{i=1}^{2^{k-1}-1} f(a + ih_k) \right) \right]$$

$$+ \sum_{i=1}^k K_i h_k^{2i} + O(h_k^{2k+2}),$$

where K_i for each i is a constant independent of h_k and depends only on $f^{(2i-1)}(a)$ and $f^{(2i-1)}(b)$. (See Ralston and Rabinowitz [103], p. 140.) In this case the Romberg scheme can be continued to generate

$$R_{i,j} = \frac{4^{j-1}R_{i,j-1} - R_{i-1,j-1}}{4^{j-1} - 1}$$

for each $i = 2, 3, 4, \ldots, n$, and $j = 2, \ldots, i$.

The values of $R_{i,j}$ with larger j index correspond to successively higher-order Newton–Cotes formulas. The truncation error associated with $R_{i,j}$ is $O(h_i^{2j})$ and involves an evaluation of $f^{(2i+2)}$. The approximations are often presented in a table of the form of Table 4.9.

TABLE 4.9

$R_{1,1}$				
$R_{2,1}$	$R_{2,2}$			
$R_{3,1}$	$R_{3,2}$	$R_{3,3}$		
$R_{4,1}$	$R_{4,2}$	$R_{4,3}$	$R_{4,4}$	
$\vdots$	$\vdots$	$\vdots$	$\ddots$	
$R_{n,1}$	$R_{n,2}$	$R_{n,3}$	$\cdots$	$R_{n,n}$

An elegant summability theorem of Silverman and Toeplitz can be used to show that the terms along the diagonal will converge to the integral provided the values of $R_{n,1}$ converge to this number. A proof of this result, together with necessary conditions for this convergence, can be found in Ralston and Rabinowitz [103], pp. 123–126. It is generally expected that the diagonal sequence $\{R_{n,n}\}_{n=1}^{\infty}$ converges much more rapidly than $\{R_{n,1}\}_{n=1}^{\infty}$.

The Romberg technique has the additional desirable feature that it allows an entire new row in the table to be calculated by simply doing one application of the trapezoidal rule and then using the previously calculated values to obtain the succeeding entries in the row. The method generally used to construct a table of this type incorporates this feature by calculating the entries row by row, that is, in the order $R_{1,1}, R_{2,1}, R_{2,2}, R_{3,1}, R_{3,2}, R_{3,3}$, etc. The following algorithm describes this technique in detail.

Romberg Algorithm 4.3

To approximate the integral $I = \int_a^b f(x)\,dx$, select an integer $n > 0$.

INPUT endpoints a, b; integer n.

OUTPUT an array R.   ($R_{n,n}$ is the approximation to I. Computed by rows; only 2 rows saved in storage.)

Step 1   Set $h = b - a$;
$$R_{1,1} = h(f(a) + f(b))/2.$$

Step 2   OUTPUT ($R_{1,1}$).

Step 3   For $i = 2, \ldots, n$ do Steps 4–8.

Step 4   Set $R_{2,1} = \dfrac{1}{2}\left[R_{1,1} + h \displaystyle\sum_{k=1}^{2^{i-2}} f(a + (k - 0.5)h) \right]$.   (*Approximation from trapezoidal method.*)

Step 5   For $j = 2, \ldots, i$

set $R_{2,j} = \dfrac{4^{j-1}R_{2,j-1} - R_{1,j-1}}{4^{j-1} - 1}$.   (*Extrapolation.*)

Step 6   OUTPUT ($R_{2,j}$ for $j = 1, 2, \ldots, i$).

Step 7   Set $h = h/2$.

Step 8   For $j = 1, 2, \ldots, i$ set $R_{1,j} = R_{2,j}$.   (*Update row 1 of R.*)

Step 9   STOP.

EXAMPLE 2   In Example 1, the values for $R_{1,1}$ through $R_{n,1}$ were obtained for approximating $\int_0^{\pi} \sin x\,dx$ with $n = 6$. With Algorithm 4.3, the Romberg table is shown in Table 4.10.

TABLE 4.10

0					
1.57079633	2.09439511				
1.89611890	2.00455976	1.99857073			
1.97423160	2.00026917	1.99998313	2.00000555		
1.99357034	2.00001659	1.99999975	2.00000001	1.99999999	
1.99839336	2.00000103	2.00000000	2.00000000	2.00000000	2.00000000

∎

Algorithm 4.3 requires a preset integer n to determine the number of rows to be generated. It is often more useful to prescribe an error tolerance for the approximation and generate n, within some upper bound, until consecutive entries in a row $R_{n,n-1}$ and $R_{n,n}$ agree to within the tolerance. To guard against the possibility that two consecutive row elements agree with each other, but not with the value of the integral being approximated, it is common to generate approximations until not only $|R_{n,n-1} - R_{n,n}|$ is within the tolerance, but also $|R_{n-1,n-2} - R_{n-1,n-1}|$. While not a universal safeguard, this will ensure that two differently generated sets of approximations agree within the specified tolerance before $R_{n,n}$ is accepted as sufficiently accurate.

Romberg integration applied to f on $[a, b]$ relies on the assumption that the composite trapezoidal rule has an error term that can be expressed in the form of Eq. (4.62), that is, we must have $f \in C^{2k+2}[a, b]$ for the kth row to be generated. General-purpose algorithms using Romberg integration include a check at each stage to ensure that this assumption is fulfilled. These methods are known as cautious Romberg algorithms and are described in Johnson [72]. This reference also describes methods for using the Romberg technique as an adaptive procedure, similar to the adaptive Simpson's rule that was discussed in Section 4.4. The adaptive cautious Romberg technique $DCADRE$ is the basic subroutine for evaluating integrals in the software package produced by IMSL. (See Rice [108], pp. 196–216.)

Exercise Set 4.6

1. Use Romberg integration to calculate $R_{3,3}$ for the following definite integrals. Compare your results to those obtained in Exercises 1, 2, and 3 of Section 4.4.

a) $\int_1^3 \dfrac{dx}{x}$

b) $\int_0^2 x^3 \, dx$

c) $\int_0^3 x\sqrt{1+x^2} \, dx$

d) $\int_0^1 \sin \pi x \, dx$

e) $\int_0^{2\pi} x \sin x \, dx$

f) $\int_0^1 x^2 e^x \, dx$

2. Use Romberg integration to calculate $R_{4,4}$ for the following integrals and compare the results with those obtained in Exercise 6 of Section 4.5.

    a) $\displaystyle\int_0^{\pi/4} \sin x\, dx$ b) $\displaystyle\int_{\pi/2}^{3\pi/4} \cos x\, dx$

3. Use Romberg integration to approximate $\int_0^2 x^2 e^{-x^2}\, dx$ until $R_{n,n-1}$ and $R_{n,n}$ agree to within 10^{-6}. Compare the result to that obtained in Exercise 3 of Section 4.5.

4. Use Romberg integration to find approximations to $\int_0^{1.5}(1+x)^{-1}\, dx$, completing the table for $n = 6$. Compare this result with the values obtained in Exercise 4 of Section 4.4 and Exercise 8 of Section 4.3.

5. Use Romberg integration to find an approximation to $\int_1^3 e^x \sin x\, dx$. Complete the table until $R_{n,n-1}$ and $R_{n,n}$ agree to within 10^{-6}. Compare your answers to the exact result and to the result obtained in Exercise 13 of Section 4.4.

6. Approximate $\int_1^{10} \ln x\, dx$ using Romberg integration with $n = 10$.

7. Approximate the following integrals using Romberg integration until $R_{n,n-1}$ and $R_{n,n}$ agree to within 3×10^{-6}. Compare results to those obtained in Exercise 6 of Section 4.5.

    a) $\displaystyle\int_0^{\pi/4} \tan x\, dx$ b) $\displaystyle\int_{\pi/2}^{3\pi/4} \cot x\, dx$

8. Apply Romberg integration to the following integrals until $R_{n,n-1}$ and $R_{n,n}$ agree to within 10^{-5}.

    a) $\displaystyle\int_0^1 x^{1/3}\, dx.$ b) $\displaystyle\int_0^{0.3} f(x)\, dx,$

    where

    $$f(x) = \begin{cases} x^3 + 1, & 0 \le x \le 0.1, \\ 1.001 + 0.03(x-0.1) + 0.3(x-0.1)^2 + 2(x-0.1)^3, & 0.1 < x \le 0.2, \\ 1.009 + 0.15(x-0.2) + 0.9(x-0.2)^2 + 2(x-0.2)^3, & 0.2 < x \le 0.3. \end{cases}$$

9. Using Romberg integration, compute an approximation to $\int_0^{48} \sqrt{1 + (\cos x)^2}\, dx$ that is accurate to within 10^{-4}.

10. Show that the approximation obtained from $R_{k,2}$ is the same as that given by the Simpson's composite rule described in Theorem 4.4 with $h = h_k$.

11. Show that, for any k,

    $$\sum_{i=1}^{2^{k-1}-1} f\left(a + \frac{i}{2}h_{k-1}\right) = \sum_{i=1}^{2^{k-2}} f\left(a + \left(i - \frac{1}{2}\right)h_{k-1}\right) + \sum_{i=1}^{2^{k-2}-1} f(a + ih_{k-1}).$$

12. Use the result of Exercise 11 to verify Eq. (4.58); that is, show that, for all k,

    $$R_{k,1} = \frac{1}{2}\left[R_{k-1,1} + h_{k-1} \sum_{i=1}^{2^{k-2}} f\left(a + \left(i - \frac{1}{2}\right)h_{k-1}\right)\right].$$

4.7 Gaussian Quadrature*

The Newton–Cotes formulas presented in Section 4.3 were derived by integrating the Lagrange interpolating polynomials. Since the error term in the Lagrange interpolating polynomial of degree n involves the $(n+1)$st derivative of the function being approximated, it has been remarked previously that the formula is exact when approximating any polynomial of degree less than or equal to n. Consequently, the Newton–Cotes formulas have degree of precision at least n. In fact, the degree of precision of the odd formulas is exactly n, while the even formulas have degree of precision $(n+1)$.

All the Newton–Cotes formulas require that the values of the function whose integral is to be approximated be known at evenly spaced points, which might be the expected situation if tabulated data for the function was being used. If the function is given explicitly, however, the points for evaluating the function could be chosen in another manner, which leads to increased accuracy of approximation. **Gaussian quadrature** is concerned with choosing the points for evaluation in an optimal manner. It presents a procedure for choosing values $x_1, x_2, \ldots, x_n$ in the interval $[a, b]$ and constants $c_1, c_2, \ldots, c_n$, that are expected to minimize the error obtained in performing the approximation

$$(4.63) \qquad \int_a^b f(x)\,dx \approx \sum_{i=1}^n c_i f(x_i)$$

for an arbitrary function f. In order to measure this accuracy, it is generally assumed that the best choice of these values will be the choice that maximizes the degree of precision for the formula.

Since the values of $c_1, c_2, \ldots, c_n$ are arbitrary and those of $x_1, x_2, \ldots, x_n$ are restricted only in the sense that the function whose integral is being approximated must be defined at these points, there are $2n$ parameters involved, n given by the constants $c_1, c_2, \ldots, c_n$ and n given by $x_1, x_2, \ldots, x_n$.

If the coefficients of a polynomial are also considered as parameters, the class of polynomials of degree at most $(2n-1)$ contains $2n$ parameters and is the largest class of polynomials for which it is reasonable to expect Eq. (4.63) to be exact. In fact, for the proper choice of the values and constants, exactness on this set can be obtained. This implies that Eq. (4.63) can be designed to have degree of precision $(2n-1)$.

Before beginning the study of Gaussian quadrature, it is necessary to discuss some of the material of orthogonal sets of functions that is also presented in Section 7.2. The set of functions $\{\phi_0, \phi_1, \ldots, \phi_n\}$ is said to be **orthogonal** on $[a, b]$ with respect to the continuous **weight function** $w(x) \geq 0$ $(w(x) \not\equiv 0)$, provided

$$\int_a^b \phi_k(x)\phi_j(x)w(x)\,dx$$

is zero when $j \neq k$ and positive when $j = k$.

---

*This section requires some material concerning orthogonal functions, which is also discussed in Section 7.2.

It is easily shown (see Exercises 11 and 12 of Section 7.2) that if $\{\phi_0, \phi_1, ..., \phi_n\}$ is an orthogonal set of polynomials defined on $[a, b]$ and ϕ_i is of degree i for each $i = 0$, $1, ..., n$, then for any polynomial Q of degree at most n, there exist unique constants α_0, $\alpha_1, ..., \alpha_n$ with $Q(x) = \sum_{i=0}^{n} \alpha_i \phi_i(x)$. This result will be used in the proof of the next theorem, a result that is of primary importance in determining the optimal choice of the values $x_1, x_2, ..., x_n$.

THEOREM 4.7   If $\{\phi_0, \phi_1, ..., \phi_n\}$ is a set of orthogonal polynomials defined on $[a, b]$ with respect to the continuous weight function w and ϕ_k is a polynomial of degree k for each $k = 0, 1$, $..., n$, then ϕ_k has k distinct roots, when $k \geq 1$, and these roots lie in the interval (a, b).

PROOF   Since ϕ_0 is a polynomial of degree zero, a constant $C \neq 0$ exists with $\phi_0(x) = C$. This implies that for $k \geq 1$

$$0 = \int_a^b \phi_k(x)\phi_0(x)w(x)\,dx = C \int_a^b \phi_k(x)w(x)\,dx.$$

Since w is a weight function, $w(x) \geq 0$, but $w(x) \not\equiv 0$. Thus, ϕ_k must change sign at least once in (a, b).

Suppose ϕ_k changes sign precisely j times in (a, b), at the points $\{r_i\}_{i=1}^{j}$, where $a < r_1 < r_2 < \cdots < r_j < b$, and that $j < k$. Without loss of generality, it may be assumed that $\phi_k(x) > 0$ on (a, r_1) (see Exercise 10). Consequently, $\phi_k(x) < 0$ on (r_1, r_2), $\phi_k(x) > 0$ on (r_2, r_3), and in general ϕ_k is of opposite sign on each of the adjacent intervals (a, r_1), (r_1, r_2), $..., (r_j, b)$.

Define the jth-degree polynomial, P, by

$$P(x) = (-1)^j \prod_{i=1}^{j} (x - r_i).$$

Note that the sign of P agrees with that of ϕ_k on each of the subintervals $(a, r_1), (r_1, r_2), ..., (r_j, b)$ and, consequently, that $P(x)\phi_k(x) > 0$ on each of these intervals. Since $w(x) \geq 0$ on (a, b) but $w(x) \not\equiv 0$, this implies that

(4.64) $$\int_a^b P(x)\phi_k(x)w(x)\,dx > 0.$$

However, P is a polynomial of degree $j < k$, so

$$P(x) = \sum_{i=0}^{j} \alpha_i \phi_i(x)$$

for some collection of constants $\alpha_0, \alpha_1, ..., \alpha_j$, which implies

$$\int_a^b P(x)\phi_k(x)w(x)\,dx = \sum_{i=0}^{j} \alpha_i \int_a^b \phi_i(x)\phi_k(x)w(x)\,dx = 0$$

in contradiction with inequality (4.64).

The only assumption made in this procedure was that ϕ_k changes sign precisely j times in (a, b), where $j < k$; so this statement must be erroneous. This implies that ϕ_k changes sign at least k times in (a, b). The Intermediate Value Theorem implies that a root exists at each sign change; so ϕ_k must have k distinct roots in (a, b).   □

We will show in Section 7.2 (Example 4) that a class of functions called Legendre polynomials are orthogonal on $[-1, 1]$ with respect to $w(x) \equiv 1$. These polynomials are defined recursively by $P_0(x) \equiv 1$, $P_1(x) = x$, $P_k(x) = (x - B_k)P_{k-1}(x) - C_k P_{k-2}(x)$, for $k = 2, 3, \ldots$ where

$$B_k = \frac{\int_{-1}^{1} x[P_{k-1}(x)]^2 \, dx}{\int_{-1}^{1} [P_{k-1}(x)]^2 \, dx} \quad \text{and} \quad C_k = \frac{\int_{-1}^{1} x P_{k-1}(x) P_{k-2}(x) \, dx}{\int_{-1}^{1} [P_{k-2}(x)]^2 \, dx}.$$

We do not need the explicit representation of the Legendre polynomials, only the knowledge that the polynomial P_n, for each n, has n distinct roots $x_1, x_2, \ldots, x_n$ all of which lie in $(-1, 1)$.

Consider now the approximation of a function f on $[-1, 1]$ by the Lagrange interpolating polynomial at $x_1, x_2, \ldots, x_n$:

$$f(x) = \sum_{i=1}^{n} \prod_{\substack{j=1 \\ j \neq i}}^{n} \frac{(x - x_j)}{(x_i - x_j)} f(x_i) + \frac{f^{(n)}(\xi(x))}{n!} \prod_{i=1}^{n} (x - x_i).$$

Since the error formula involves $f^{(n)}$, the quadrature rule

$$(4.65) \qquad \int_{-1}^{1} f(x) \, dx \approx \sum_{i=1}^{n} c_i f(x_i),$$

where

$$(4.66) \qquad c_i = \int_{-1}^{1} \prod_{\substack{j=1 \\ j \neq i}}^{n} \frac{(x - x_j)}{(x_i - x_j)} \, dx$$

must have degree of precision at least $n - 1$. We wish to show that, in fact, the rule is exact for all polynomials of degree $k \leq 2n - 1$.

Suppose P is one such polynomial. Dividing P by the nth degree Legendre polynomial P_n gives

$$P(x) = Q(x)P_n(x) + R(x),$$

where both Q and R are polynomials of degree less than n.

Since Q is of degree less than n, $Q(x)$ can be written

$$Q(x) = \sum_{i=0}^{n-1} d_i P_i(x)$$

for some constants $\{d_i\}$. However, the Legendre polynomials are orthogonal on $[-1, 1]$, so

$$\int_{-1}^{1} Q(x)P_n(x) \, dx = \sum_{i=0}^{n-1} d_i \int_{-1}^{1} P_i(x)P_n(x) \, dx = 0,$$

and $\qquad \displaystyle\int_{-1}^{1} P(x) \, dx = \int_{-1}^{1} Q(x)P_n(x) \, dx + \int_{-1}^{1} R(x) \, dx = \int_{-1}^{1} R(x) \, dx.$

Since the quadrature rule of Eq. (4.65) has degree of precision at least $n - 1$, the maximum degree of R, this implies that

$$\int_{-1}^{1} P(x) \, dx = \int_{-1}^{1} R(x) \, dx = \sum_{i=1}^{n} c_i R(x_i).$$

However, since x_i for each $i = 1, 2, \ldots, n$ is a root of P_n,

$$P(x_i) = Q(x_i)P_n(x_i) + R(x_i) = 0 + R(x_i) = R(x_i).$$

This implies that the quadrature formula is exact for P (and has degree of precision at least $2n - 1$) since

$$\int_{-1}^{1} P(x)\,dx = \sum_{i=1}^{n} c_i P(x_i).$$

The constants c_i needed for the quadrature rule could be generated from Eq. (4.66), but both these constants and the roots of the Legendre polynomials are extensively tabulated. Table 4.11 lists these values for $n = 2, 3, 4,$ and 5. Others can be found in Stroud and Secrest [134].

TABLE 4.11

n	Roots	Coefficients
2	0.5773502692	1.0000000000
	−0.5773502692	1.0000000000
3	0.7745966692	0.5555555556
	0.0000000000	0.8888888889
	−0.7745966692	0.5555555556
4	0.8611363116	0.3478548451
	0.3399810436	0.6521451549
	−0.3399810436	0.6521451549
	−0.8611363116	0.3478548451
5	0.9061798459	0.2369268850
	0.5384693101	0.4786286705
	0.0000000000	0.5688888889
	−0.5384693101	0.4786286705
	−0.9061798459	0.2369268850

Since the simple linear transformation $t = [1/(b - a)](2x - a - b)$ will translate any interval $[a, b]$ into $[-1, 1]$ provided $b > a$, the Legendre polynomials can be used to approximate

$$(4.67) \qquad \int_{a}^{b} f(x)\,dx = \int_{-1}^{1} f\left(\frac{(b - a)t + b + a}{2}\right)\frac{(b - a)}{2}\,dt$$

for any function that can be evaluated at the required points.

EXAMPLE 1   Consider the problem of finding approximations to $\int_{1}^{1.5} e^{-x^2}\,dx$. Table 4.12 lists the values and associated errors for the Newton–Cotes formulas given in Section 4.3. The exact value of the integral to seven decimal places is 0.1093643.

TABLE 4.12

n	0	1	2	3	4
Closed formulas		0.1183197	0.1093104	0.1093404	0.1093643
Open formulas	0.1048057	0.1063473	0.1094116	0.1093971	

The Gaussian quadrature procedure applied to this problem requires that the integral be first translated into a problem whose interval of integration is $[-1, 1]$. Using Eq. (4.67),

$$\int_1^{1.5} e^{-x^2}\,dx = \frac{1}{4}\int_{-1}^{1} e^{(-(t+5)^2/16)}\,dt.$$

Using the values in Table 4.11, the Gaussian quadrature approximations for this problem are:

$n = 2$:

$$\int_1^{1.5} e^{-x^2}\,dx \approx \tfrac{1}{4}[e^{-(5+0.5773502692)^2/16} + e^{-(5-0.5773502692)^2/16}] = 0.1094003,$$

and

$n = 3$:

$$\int_1^{1.5} e^{-x^2}\,dx \approx \tfrac{1}{4}[(0.5555555556)e^{-(5+0.7745966692)^2/16} + (0.8888888889)e^{-(5)^2/16}$$
$$+ (0.5555555556)e^{-(5-0.7745966692)^2/16}]$$
$$= 0.1093642.$$

For comparison the values obtained by using the Romberg procedure with $n = 4$ are listed in Table 4.13.

TABLE 4.13

0.1183197			
0.1115627	0.1093104		
0.1099114	0.1093610	0.1093643	
0.1095009	0.1093641	0.1093643	0.1093643

∎

Other sets of orthogonal polynomials can be used in the Gaussian quadrature technique; examples of some of these are given in the exercises.

Exercise Set 4.7

1. Use Gaussian quadrature with $n = 2$ to approximate the following definite integrals. Compare the approximations to the results of the corresponding exercises in Sections 4.4 and 4.6.

a) $\int_1^3 \frac{dx}{x}$

b) $\int_0^2 x^3 \, dx$

c) $\int_0^3 x\sqrt{1+x^2} \, dx$

d) $\int_0^1 \sin \pi x \, dx$

e) $\int_0^{2\pi} x \sin x \, dx$

f) $\int_0^1 x^2 e^x \, dx$

2. Use Gaussian quadrature with $n = 3$ to approximate the definite integrals in Exercise 1. Compare these results with previously obtained approximations.

3. Use Gaussian quadrature with $n = 2$, 3, and 4 to find approximations to $\int_1^3 e^x \sin x \, dx$. Compare your answers with the results obtained in Exercise 5 in Section 4.6 and in Exercise 13 of Section 4.4, as well as to the exact value.

4. Use Gaussian quadrature with $n = 2$, 3, and 4 to find approximations to $\int_0^{1.5} (1 + x)^{-1} \, dx$. Compare your answers with the results obtained in Exercise 4 of Section 4.6, Exercise 4 of Section 4.4, and Exercise 8 of Section 4.3, as well as to the exact value.

5. Verify the entries for the values of $n = 2$ and 3 in Table 4.11 by finding the roots of the respective Legendre polynomials. Using the equations given in (4.66), find the coefficients associated with the values.

6. The set of Laguerre polynomials $\{L_0, L_1, L_2, L_3\}$ given by $L_0(x) = 1$, $L_1(x) = 1 - x$, $L_2(x) = x^2 - 4x + 2$, $L_3(x) = -x^3 + 9x^2 - 18x + 6$ are shown in Exercise 8 of Section 7.2 to be orthogonal on the interval $(0, \infty)$ with respect to the weight function e^{-x}. These polynomials can be used to give approximations to $\int_0^\infty e^{-x} f(x) \, dx$, provided this improper integral exists. The derivation parallels that given for the Legendre polynomials presented after the proof of Theorem 4.7. Show that this set of polynomials gives a formula with degree of precision at least three to approximate

$$\int_0^\infty e^{-x} f(x) \, dx.$$

7. Find the roots of the Laguerre polynomials L_1, L_2 and L_3 discussed in Exercise 6, and use the corresponding coefficients obtained from the equations in (4.66) to find approximations to:

$$\int_0^\infty e^{-x} \sin x \, dx \qquad \text{when } n = 2 \quad \text{and} \quad n = 3.$$

8. Approximate $\int_0^\infty \sqrt{x} e^{-x} \, dx$ using the Laguerre polynomials as in Exercise 7, and compare the approximation to that obtained in Exercise 22(c) of Section 4.4.

9. Use the Laguerre polynomials as in Exercise 7 to approximate

$$\int_{-\infty}^\infty \frac{dx}{1 + x^2}$$

and compare the approximation to that obtained in Exercise 22(d) of Section 4.4.

10. In the proof of Theorem 4.7, the statement was made that there was no loss of generality in assuming that $\phi_k(x)$ was positive on (a, r_1). Show that this statement is correct.

4.8 Multiple Integrals

The techniques discussed in the previous sections can be modified in a straightforward manner for use in the approximation of multiple integrals. The first type of multiple integral problem we consider involves the approximation of a double integral

$$\iint_R f(x, y)\, dA,$$

where R is a rectangular region in the plane; that is,

$$R = \{(x, y) | a \leq x \leq b, c \leq y \leq d\}$$

for some constants a, b, c, and d. To illustrate the approximation technique, we employ the composite Simpson's rule, although any of the other composite Newton–Cotes formulas could be used in its place.

Suppose that integers n and m are chosen to determine the step sizes $h = (b - a)/2n$ and $k = (d - c)/2m$. Writing the double integral as an iterated integral

$$\iint_R f(x, y)\, dA = \int_a^b \left(\int_c^d f(x, y)\, dy \right) dx,$$

we first use the composite Simpson's rule to evaluate

$$\int_c^d f(x, y)\, dy,$$

treating x as a constant. Letting $y_j = c + jk$ for each $j = 0, 1, \ldots, 2m$, gives

$$\int_c^d f(x, y)\, dy = \frac{k}{3}\left[f(x, y_0) + 2 \sum_{j=1}^{m-1} f(x, y_{2j}) + 4 \sum_{j=1}^{m} f(x, y_{2j-1}) + f(x, y_{2m}) \right]$$

$$- \frac{(d-c)k^4}{180} \frac{\partial^4 f(x, \mu)}{\partial y^4}$$

for some μ in (c, d). Thus,

$$(4.68) \quad \int_a^b \int_c^d f(x, y)\, dy\, dx = \frac{k}{3} \int_a^b f(x, y_0)\, dx + \frac{2k}{3} \sum_{j=1}^{m-1} \int_a^b f(x, y_{2j})\, dx$$

$$+ \frac{4k}{3} \sum_{j=1}^{m} \int_a^b f(x, y_{2j-1})\, dx + \frac{k}{3} \int_a^b f(x, y_{2m})\, dx$$

$$- \frac{(d-c)}{180} k^4 \int_a^b \frac{\partial^4 f(x, \mu)}{\partial y^4}\, dx.$$

Simpson's composite rule is now employed on each integral in Eq. (4.68). Letting $x_i = a + ih$ for each $i = 0, 1, \ldots, 2n$ produces for each $j = 0, 1, \ldots, 2m$:

$$\int_a^b f(x, y_j)\, dx = \frac{h}{3}\left[f(x_0, y_j) + 2 \sum_{i=1}^{n-1} f(x_{2i}, y_j) + 4 \sum_{i=1}^{n} f(x_{2i-1}, y_j) + f(x_{2n}, y_j) \right]$$
$$- \frac{(b-a)}{180} h^4 \frac{\partial^4 f}{\partial x^4}(\xi_j, y_j)$$

for some ξ_j in (a, b). The resulting approximation has the form

$$\int_a^b \int_c^d f(x, y)\, dy\, dx \approx \frac{hk}{9}\left[f(x_0, y_0) + 2 \sum_{i=1}^{n-1} f(x_{2i}, y_0) + 4 \sum_{i=1}^{n} f(x_{2i-1}, y_0) \right.$$
$$+ f(x_{2n}, y_0) + 2 \sum_{j=1}^{m-1} f(x_0, y_{2j}) + 4 \sum_{j=1}^{m-1} \sum_{i=1}^{n-1} f(x_{2i}, y_{2j})$$
$$+ 8 \sum_{j=1}^{m-1} \sum_{i=1}^{n} f(x_{2i-1}, y_{2j}) + 2 \sum_{j=1}^{m-1} f(x_{2n}, y_{2j})$$
$$+ 4 \sum_{j=1}^{m} f(x_0, y_{2j-1}) + 8 \sum_{j=1}^{m} \sum_{i=1}^{n-1} f(x_{2i}, y_{2j-1})$$
$$+ 16 \sum_{j=1}^{m} \sum_{i=1}^{n} f(x_{2i-1}, y_{2j-1}) + 4 \sum_{j=1}^{m} f(x_{2n}, y_{2j-1})$$
$$+ f(x_0, y_{2m}) + 2 \sum_{i=1}^{n-1} f(x_{2i}, y_{2m}) + 4 \sum_{i=1}^{n} f(x_{2i-1}, y_{2m})$$
$$\left. + f(x_{2n}, y_{2m}) \right].$$

The error term E is given by

$$E = \frac{-k(b-a)h^4}{540}\left[\frac{\partial^4 f(\xi_0, y_0)}{\partial x^4} + 2 \sum_{j=1}^{m-1} \frac{\partial^4 f(\xi_{2j}, y_{2j})}{\partial x^4} + 4 \sum_{j=1}^{m} \frac{\partial^4 f(\xi_{2j-1}, y_{2j-1})}{\partial x^4} \right.$$
$$\left. + \frac{\partial^4 f(\xi_{2m}, y_{2m})}{\partial x^4} \right] - \frac{(d-c)k^4}{180} \int_a^b \frac{\partial^4 f(x, \mu)}{\partial y^4}\, dx.$$

If $\partial^4 f / \partial x^4$ and $\partial^4 f / \partial y^4$ are continuous on R, the error term can be reexpressed as

$$E = \frac{-k(b-a)h^4}{540}\left[6m \frac{\partial^4 f}{\partial x^4}(\bar{\eta}, \bar{\mu}) \right] - \frac{(d-c)(b-a)}{180} k^4 \frac{\partial^4 f}{\partial y^4}(\hat{\eta}, \hat{\mu})$$
$$= \frac{-(d-c)(b-a)}{180}\left[h^4 \frac{\partial^4 f}{\partial x^4}(\bar{\eta}, \bar{\mu}) + k^4 \frac{\partial^4 f}{\partial y^4}(\hat{\eta}, \hat{\mu}) \right]$$

for some $(\bar{\eta}, \bar{\mu})$ and $(\hat{\eta}, \hat{\mu})$ in R. The proof of this statement results from applying the Weighted Mean Value Theorem to $\partial^4 f / \partial y^4$ and applying a generalization of the Intermediate Value Theorem to $\partial^4 f / \partial x^4$.

EXAMPLE 1   The composite Simpson's rule applied to approximate

$$\int_{1.4}^{2.0} \int_{1.0}^{1.5} \ln(x + 2y)\, dy\, dx$$

with $n = 2$ and $m = 1$ uses the step sizes $h = 0.15$ and $k = 0.25$. The region of integration R is shown in Fig. 4.7 together with the nodes for (x_i, y_j) $i = 0, 1, 2, 3, 4$;

FIGURE 4.7

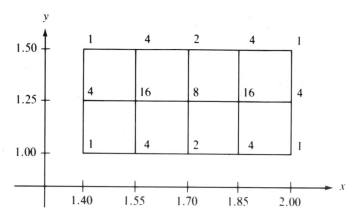

$j = 0,\ 1,\ 2$; and $w_{i,j}$, the coefficients of $f(x_i, y_i) = \ln(x_i + 2y_i)$ in the sum. The approximation is

$$\int_{1.4}^{2.0} \int_{1.0}^{1.5} \ln(x + 2y)\, dy\, dx \approx \frac{(0.15)(0.25)}{9} \sum_{i=0}^{4} \sum_{j=0}^{2} w_{i,j} \ln(x_i + 2y_j) = 0.4295524387.$$

Since

$$\frac{\partial^4 f}{\partial x^4}(x, y) = \frac{-6}{(x + 2y)^4} \quad \text{and} \quad \frac{\partial^4 f}{\partial y^4}(x, y) = \frac{-96}{(x^2 + 2y)^4},$$

the error is bounded by

$$|E| \leq \frac{(0.5)(0.6)}{180} \left[(0.15)^4 \max_{(x, y)\, \text{in } R} \frac{6}{(x + 2y)^4} + (0.25)^4 \max_{(x, y)\, \text{in } R} \frac{96}{(x + 2y)^4} \right] \leq 4.72 \times 10^{-6}.$$

The actual value of the integral to ten decimal places is

$$\int_{1.4}^{2.0} \int_{1.0}^{1.5} \ln(x + 2y)\, dy\, dx = 0.4295545265;$$

so the approximation is accurate to within 2.1×10^{-6}. ∎

The same techniques can be applied for the approximation of triple integrals, as well as higher integrals for functions of more than three variables. The number of functional evaluations required for the approximation is the product of the number of functional evaluations required when the method is applied to each variable. To reduce the number of functional evaluations, more efficient methods such as Gaussian quadrature, Romberg integration, or adaptive quadrature can be incorporated in place of the Newton–Cotes formulas. The following example illustrates the use of Gaussian quadrature for the integral considered in Example 1.

EXAMPLE 2   Consider the double integral given in Example 1. Before employing a Gaussian quadrature technique to approximate this integral, we must translate the region of integration

$$R = \{(x, y) | 1.4 \leq x \leq 2.0,\ 1.0 \leq y \leq 1.5\}$$

into $$\hat{R} = \{(u, v)| -1 \leq u \leq 1, \ -1 \leq v \leq 1\}.$$

A linear transformation that accomplishes this is

$$u = \frac{1}{2.0 - 1.4}(2x - 1.4 - 2.0), \qquad v = \frac{1}{1.5 - 1.0}(2y - 1.0 - 1.5).$$

Employing this change of variables gives an integral on which Gaussian quadrature can be applied

$$\int_{1.4}^{2.0} \int_{1.0}^{1.5} \ln(x + 2y)\, dy\, dx = 0.075 \int_{-1}^{1} \int_{-1}^{1} \ln(0.3u + 0.5v + 4.2)\, dv\, du.$$

The Gaussian quadrature formula for $n = 3$ in both u and v requires that we use the nodes,

$$u_1 = v_1 = 0, \qquad u_0 = v_0 = -0.7745966692, \qquad \text{and} \qquad u_2 = v_2 = 0.7745966692.$$

The associated weights are $w_1 = 0.8888888889$ and $w_0 = w_2 = 0.5555555556$. (See Table 4.12.) Thus,

$$\int_{1.4}^{2.0} \int_{1.0}^{1.5} \ln(x + 2y)\, dy\, dx \approx 0.075 \sum_{i=0}^{2} \sum_{j=0}^{2} w_i w_j \ln(0.3u_i + 0.5v_j + 4.2) = 0.4295545313.$$

Although this result requires only six functional evaluations compared to fifteen for the composite Simpson's rule considered in Example 1, this result is accurate to within 4.8×10^{-9}. ∎

The use of approximation methods for double integrals is not limited to those integrals with rectangular regions of integration. The techniques previously discussed can be modified to approximate integrals of the form

(4.69) $$\int_{a}^{b} \int_{c(x)}^{d(x)} f(x, y)\, dy\, dx$$

or

(4.70) $$\int_{c}^{d} \int_{a(y)}^{b(y)} f(x, y)\, dx\, dy.$$

In fact, integrals on regions not of this type can also be approximated by performing appropriate partitions of the region. (See Exercise 4.)

To describe the technique involved with approximating an integral in the form

$$\int_{a}^{b} \int_{c(x)}^{d(x)} f(x, y)\, dy\, dx,$$

we will use Simpson's rule to integrate with respect to both variables. The step size for the variable x is $h = (b - a)/2$, but the step size for y varies with x. (*See Figure 4.8.*)

$$k(x) = \frac{d(x) - c(x)}{2}.$$

FIGURE 4.8

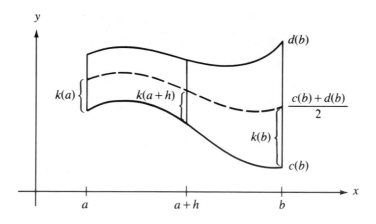

Consequently,

$$\int_a^b \int_{c(x)}^{d(x)} f(x, y)\,dy\,dx \approx \int_a^b \frac{k(x)}{3}[f(x, c(x)) + 4f(x, c(x) + k(x)) + f(x, d(x))]\,dx$$

$$\approx \frac{h}{3}\Bigg\{ \frac{k(a)}{3}[f(a, c(a)) + 4f(a, c(a) + k(a)) + f(a, d(a))]$$

$$+ \frac{4k(a + h)}{3}[f(a + h, c(a + h)) + 4f(a + h, c(a + h)$$

$$+ k(a + h)) + f(a + h, d(a + h))]$$

$$+ \frac{k(b)}{3}[f(b, c(b)) + 4f(b, c(b) + k(b)) + f(b, d(b))] \Bigg\}.$$

The following algorithm applies the composite Simpson's rule to an integral in the form (4.69). Integrals in the form (4.70) can, of course, be handled similarly.

Multiple Integral Algorithm 4.4

To approximate the integral $I = \int_a^b \int_{c(x)}^{d(x)} f(x, y)\,dy\,dx$:

INPUT endpoints a, b; positive integers m, n.

OUTPUT approximation J to I.

Step 1   Set $h = (b - a)/(2n)$.

Step 2   Set $J_1 = 0$;   (*End terms.*)
$\qquad\qquad J_2 = 0$;   (*Even terms.*)
$\qquad\qquad J_3 = 0$.   (*Odd terms.*)

Step 3   For $i = 0, 1, \ldots, 2n$
$\qquad\qquad$ set $x = a + ih$;   (*Composite Simpson's method for fixed x.*)
$\qquad\qquad\qquad HX = (d(x) - c(x))/(2m)$;

$$K_1 = f(x, c(x)) + f(x, d(x)); \quad (End\ terms\ for\ each\ x.)$$
$$K_2 = 0; \quad (Even\ terms\ for\ each\ x.)$$
$$K_3 = 0; \quad (Odd\ terms\ for\ each\ x.)$$
for $j = 1, \dots, 2m - 1$
    set $y = c(x) + jHX$;
        $Z = f(x, y)$;
    if j is even then set $K_2 = K_2 + Z$
              else set $K_3 = K_3 + Z$;
set $L = (K_1 + 2K_2 + 4K_3)HX/3$;
$$\left(L \approx \int_{c(x_i)}^{d(x_i)} f(x_i, y)\, dy\ by\ composite\ Simpson's\ method. \right)$$
if $i = 0$ or $i = 2n$
    then set $J_1 = J_1 + L$
    else if i is even then set $J_2 = J_2 + L$
              else set $J_3 = J_3 + L$.

Step 4   Set $J = (J_1 + 2J_2 + 4J_3)h/3$.

Step 5   OUTPUT (J);
           STOP.

EXAMPLE 3   Algorithm 4.4 applied to the function $f(x, y) = e^{(y/x)}$ on the region $R = \{(x, y) \mid 0.1 \leq x \leq 0.5,\ x^3 \leq y \leq x^2\}$ (see Fig. 4.9) with $n = m = 5$ produces the approximation

$$\int_{0.1}^{0.5} \int_{x^3}^{x^2} e^{(y/x)}\, dy\, dx \approx 0.0333054.$$

The exact value of this integral to seven decimal places is 0.0333056.   ■

FIGURE 4.9

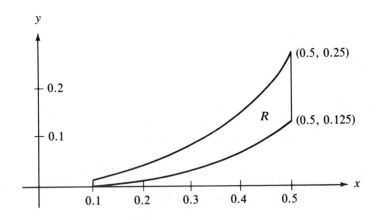

Exercise Set 4.8

1. Use Algorithm 4.4 with $n = m = 3$ to approximate the following double integrals and compare the results to the exact answer.

   a) $\displaystyle\int_{2.1}^{2.2}\int_{1.3}^{1.4} xy^2\,dy\,dx$

   b) $\displaystyle\int_{1.3}^{1.5}\int_{-0.1}^{0.1} \sqrt{x}\,y^2\,dy\,dx$

   c) $\displaystyle\int_{-0.1}^{0.1}\int_{0}^{0.1} xye^{x^2+y^2}\,dx\,dy$

   d) $\displaystyle\int_{0}^{0.1}\int_{0}^{0.1} e^{y-x}\,dx\,dy$

   e) $\displaystyle\int_{2.0}^{2.2}\int_{x}^{2x} (x^2+y^3)\,dy\,dx$

   f) $\displaystyle\int_{1.0}^{1.1}\int_{0}^{x} (x^2+\sqrt{y})\,dy\,dx$

2. Use Algorithm 4.4 with $n = 3$ and $m = 5$ to approximate the following double integrals.

   a) $\displaystyle\int_{0}^{1}\int_{0}^{\pi} (y\sin\sqrt{x})\,dy\,dx$

   b) $\displaystyle\int_{-\pi}^{3\pi/2}\int_{0}^{2\pi} (y\sin x + x\cos y)\,dy\,dx$

3. Use Algorithm 4.4 with $n = m = 7$ to approximate

$$\iint_R e^{-(x+y)}\,dA$$

   for the region R in the plane bounded by the curves $y = x^2$ and $y = \sqrt{x}$.

4. Use Algorithm 4.4 to approximate

$$\iint_R \sqrt{xy+y^2}\,dA,$$

   where R is the region in the plane bounded by the lines $x + y = 6$, $3y - x = 2$, and $3x - y = 2$. First partition R into two regions R_1 and R_2 on which Algorithm 4.4 can be applied. Use $n = m = 3$ on both R_1 and R_2.

5. Use Simpson's rule to approximate the triple integral $\displaystyle\int_{0}^{1}\int_{1}^{2}\int_{0}^{0.5} e^{xyz}\,dx\,dy\,dz$.

6. Modify Algorithm 4.4 to approximate triple integrals. Use this modification and six nodes in each coordinate direction to approximate

$$\iiint_S xy\sin(yz)\,dz\,dy\,dx,$$

   where S is the solid bounded by the coordinate planes and the planes $x = \pi$, $y = \pi/2$, $z = \pi/3$. Compare this approximation to the exact result.

7. Use the algorithm generated in Exercise 6 to approximate

$$\iiint_S \sqrt{xyz}\,dV,$$

when S is the region in the first octant bounded by the cylinder $x^2 + y^2 = 4$, the sphere $x^2 + y^2 + z^2 = 4$, and the plane $x + y + z = 8$. Use eight nodes in each coordinate direction. How many functional evaluations are required for the approximation?

8. Modify Algorithm 4.4 to use the composite trapezoidal rule in place of the composite Simpson's rule. Use this algorithm to approximate the integral

$$\int_0^1 \int_0^1 f(x, y)\, dy\, dx,$$

using the values of f given in the following table.

x \ y	0.00	0.25	0.50	0.75	1.00
0.00	0	0	0	0	0
0.25	0	8.113	8.994	8.113	0
0.50	0	7.005	10.722	8.704	0
0.75	0	4.921	6.779	5.184	0
1.00	0	0	0	0	0

9. Modify Algorithm 4.4 to use the composite trapezoidal rule in the variable x and the composite Simpson's rule in the variable y. Use the algorithm to approximate

$$\int_0^{1.5} \int_0^{1.5} f(x, y)\, dy\, dx,$$

using the values of f given in the following table:

x \ y	0.00	0.25	0.50	0.75	1.00	1.25	1.50
0.00	0	0	0	0	0	0	0
0.50	0	0.2886	0.7582	1.2480	1.4325	1.2754	0.9235
1.00	0	0.2160	0.7163	1.1791	1.3534	1.2049	0.8724
1.50	0	0.0928	0.3078	0.5067	0.5816	0.5178	0.3749

10. Modify Algorithm 4.4 to use Gaussian Quadrature in place of the composite Simpson's rule. Use this algorithm with $n = m = 4$ to approximate

$$\iint_R e^{-(x+y)}\, dA$$

for the region R in the plane bounded by the curves $y = x^2$ and $y = \sqrt{x}$. Compare this approximation to the result obtained in Exercise 3, as well as to the exact result.

11. A plane lamina is defined to be a thin sheet of continuously distributed mass. If ρ is a function describing the density of lamina having the shape of a region R in the xy-plane, then the center of mass of the lamina $(\bar{x}, \bar{y})$ is defined by

$$\bar{x} = \frac{\iint\limits_R x\rho(x, y)\, dA}{\iint\limits_R \rho(x, y)\, dA}, \qquad \bar{y} = \frac{\iint\limits_R y\rho(x, y)\, dA}{\iint\limits_R \rho(x, y)\, dA}.$$

Use Algorithm 4.4 with $n = m = 7$ to find the center of mass of the lamina described by $R = \{(x, y)|0 \le x \le 1, 0 \le y \le \sqrt{1 - x^2}\}$ with density function $\rho(x, y) = e^{-(x^2 + y^2)}$. Compare the approximation to the exact result.

12. The area of a surface described by $z = f(x, y)$ for (x, y) in R is given by

$$\iint\limits_R \sqrt{[f_x(x, y)]^2 + [f_y(x, y)]^2}\, dA.$$

Use Algorithm 4.4 with $n = m = 4$ to find an approximation to the area of the surface on the hemisphere $x^2 + y^2 + z^2 = 9$, $z \ge 0$ that lies above the region in the plane described by $R = \{(x, y)|0 \le x \le 1, 0 \le y \le 1\}$.

5

Initial-Value Problems for Ordinary Differential Equations

The motion of a swinging pendulum under certain simplifying assumptions can be described by the second-order differential equation

$$\frac{d^2\theta}{dt^2} + \frac{g}{L}\sin\theta = 0,$$

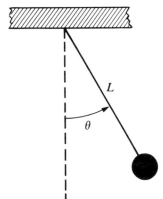

where L is the length of the pendulum, g is the gravitational constant of the earth, and θ is the angle the pendulum makes with the vertical or equilibrium position. If, in addition, we specify the position of the pendulum when the motion begins, $\theta(t_0) = \theta_0$, and its velocity at that point, $\theta'(t_0) = \theta'_0$, we have what is called an initial-value problem.

For small values of θ, the approximation $\theta \approx \sin\theta$ can be used to simplify this problem to the linear equation

$$\frac{d^2\theta}{dt^2} + \frac{g}{L}\theta = 0,$$

which can be solved by a standard differential-equation technique. For larger values of θ, however, approximations of the type discussed in this chapter must be used.

Although any textbook on the subject of ordinary differential equations details a number of methods for explicitly finding solutions to first-order initial-value problems, in practice few of the problems originating from the study of physical phenomena can be solved exactly.

The first part of this chapter is concerned with approximating the solution $y(t)$ to a problem of the form

(5.1)
$$\frac{dy}{dt} = f(t, y) \qquad \text{for } a \le t \le b,$$

subject to an initial condition

(5.2)
$$y(a) = \alpha.$$

Later in the chapter, we deal with the extension of these methods to a system of first-order differential equations in the form

(5.3)
$$\frac{dy_1}{dt} = f_1(t, y_1, y_2, \ldots, y_n),$$

$$\frac{dy_2}{dt} = f_2(t, y_1, y_2, \ldots, y_n),$$

$$\vdots \qquad \qquad \vdots$$

$$\frac{dy_n}{dt} = f_n(t, y_1, y_2, \ldots, y_n),$$

for $a \le t \le b$, subject to the initial conditions

(5.4)
$$y_1(a) = \alpha_1, \quad y_2(a) = \alpha_2, \quad \ldots, \quad y_n(a) = \alpha_n$$

and the relationship of a system of this type to the general nth-order initial-value problem of the form

(5.5)
$$y^{(n)} = f(t, y, y', y'', \ldots, y^{(n-1)})$$

for $a \le t \le b$, subject to the initial conditions

(5.6)
$$y(a) = \alpha_0, \quad y'(a) = \alpha_1, \quad \ldots, \quad y^{(n-1)}(a) = \alpha_{n-1}.$$

5.1 Elementary Theory of Initial-Value Problems

Before attempting to solve an initial-value problem, we would like to know whether a unique solution exists. Also, since problems obtained by observing physical phenomena generally only approximate the actual situation, it is of interest to know whether small changes in the statement of the problem will introduce correspondingly small changes in the solution. This is also important because of the possibility of rounding errors when numerical methods are used.

To discuss these problems we need some definitions and results from the theory of ordinary differential equations. The first definition is an extension of the definition given in Exercise 21 of Section 1.1 to a function involving two variables.

DEFINITION 5.1   A function $f(t, y)$ is said to satisfy a **Lipschitz condition** in the variable y on a set $D \subset R^2$, provided a constant $L > 0$ exists with the property that

$$|f(t, y_1) - f(t, y_2)| \leq L|y_1 - y_2|$$

whenever $(t, y_1), (t, y_2) \in D$.

The constant L is called a **Lipschitz constant** for f.

EXAMPLE 1   If $D = \{(t, y) | 1 \leq t \leq 2, -3 \leq y \leq 4\}$ and $f(t, y) = t|y|$, then for each $(t, y_1), (t, y_2)$ in D

$$|f(t, y_1) - f(t, y_2)| = \big| t|y_1| - t|y_2| \big| = |t| \big| |y_1| - |y_2| \big| \leq 2|y_1 - y_2|.$$

Thus f satisfies a Lipschitz condition on D in the variable y with Lipschitz constant 2. In fact, the smallest value possible for the Lipschitz constant for this problem is $L = 2$, since, for example,

$$|f(2, 1) - f(2, 0)| = |2 - 0| = 2|1 - 0|. \qquad \blacksquare$$

DEFINITION 5.2   A set $D \subset R^2$ is said to be **convex** if whenever (t_1, y_1) and (t_2, y_2) belong to D, $((1 - \lambda)t_1 + \lambda t_2, (1 - \lambda)y_1 + \lambda y_2)$ also belongs to D for each $\lambda, 0 \leq \lambda \leq 1$.

In geometric terms, Definition 5.2 states that a set is convex provided that whenever two points belong to the set, the entire straight-line segment between the points must also belong to the set. The sets we consider in this chapter will generally be of the form $D = \{(t, y) | a \leq t \leq b, -\infty < y < \infty\}$ for some constants a and b. It is easy to verify (see Exercise 5) that such sets are convex.

THEOREM 5.3   Suppose that $f(t, y)$ is defined on a convex set $D \subset R^2$. If a constant $L > 0$ exists with

$$(5.7) \qquad \left| \frac{\partial f}{\partial y}(t, y) \right| \leq L \qquad \text{for all } (t, y) \in D,$$

then f satisfies a Lipschitz condition on D in the variable y with Lipschitz constant L.

The proof of Theorem 5.3 is discussed in Exercise 4, and is similar to the proof of the corresponding result for functions of one variable discussed in Exercise 21 of Section 1.1.

As the next theorem will show, it is often of significant interest to determine whether the function involved in an initial-value problem satisfies a Lipschitz condition in its second variable, and condition (5.7) is generally much easier to apply than the definition. It should be remarked, however, that Theorem 5.3 gives only *sufficient* conditions for a Lipschitz condition to hold; a reexamination of Example 1 will demonstrate that these conditions are definitely *not necessary*.

The following theorem is a version of the fundamental existence and uniqueness theorem for first-order ordinary differential equations. Although the theorem can be proved with the hypothesis reduced somewhat, this form of the theorem is sufficient for our purposes. (The proof of the theorem, in approximately this form, can be found in Birkhoff and Rota [11], pp. 142, 152–155.)

THEOREM 5.4

Suppose that $D = \{(t, y)\,|\,a \le t \le b, \ -\infty < y < \infty\}$, and that $f(t, y)$ is continuous on D. If f satisfies a Lipschitz condition on D in the variable y, then the initial-value problem

$$y' = f(t, y), \qquad a \le t \le b, \quad y(a) = \alpha,$$

has a unique solution $y(t)$ for $a \le t \le b$.

EXAMPLE 2

Consider the initial-value problem

$$y' = 1 + t \sin(ty), \qquad 0 \le t \le 2, \quad y(0) = 0.$$

Holding t constant and applying the Mean Value Theorem to the function

$$f(t, y) = 1 + t \sin(ty),$$

we find that whenever $y_1 < y_2$, a number ξ, $y_1 < \xi < y_2$, exists with

$$t^2 \cos(\xi t) = \frac{\partial}{\partial y} f(t, \xi) = \frac{f(t, y_2) - f(t, y_1)}{y_2 - y_1}.$$

Thus, for all $y_1 < y_2$,

$$|f(t, y_2) - f(t, y_1)| = |y_2 - y_1||t^2 \cos(\xi t)| \le 4|y_2 - y_1|,$$

and f satisfies a Lipschitz condition in the variable y with Lipschitz constant four. Since, additionally, $f(t, y)$ is continuous when $0 \le t \le 2$ and $-\infty < y < \infty$, Theorem 5.4 implies that a unique solution exists to this initial-value problem.

It might be interesting for the reader who has completed a course in differential equations to try to find the exact solution to this problem. ∎

Now that we have, to some extent, taken care of the question of when initial-value problems have unique solutions, we can move to the other question posed earlier in the section. Is there a way to determine whether a particular problem has the property that small changes or perturbations in the statement of the problem introduce correspondingly small changes in the solution? As usual, we need to first give a workable definition to express this concept.

DEFINITION 5.5

The initial-value problem

(5.8)
$$\frac{dy}{dt} = f(t, y), \qquad a \le t \le b, \quad y(a) = \alpha,$$

is said to be a **well-posed problem** if:

i)   a unique solution, $y(t)$, to the problem exists;

ii) there are positive constants ε and k with the property that a unique solution, $z(t)$, to the problem

(5.9) $$\frac{dz}{dt} = f(t, z) + \delta(t), \qquad a \le t \le b, \quad z(a) = \alpha + \varepsilon_0,$$

exists with

$$|z(t) - y(t)| < k\varepsilon \qquad \text{for all } a \le t \le b$$

whenever $|\varepsilon_0| < \varepsilon$ and $|\delta(t)| < \varepsilon$.

The problem specified by Eq. (5.9) is often called a **perturbed problem** associated with the original problem (5.8). The following theorem specifies conditions which ensure that an initial-value problem is well-posed. The proof of this theorem can be found in the previously mentioned reference, Birkhoff and Rota [11], pp. 142–147.

THEOREM 5.6   Suppose $D = \{(t, y) | a \le t \le b \text{ and } -\infty < y < \infty\}$. The initial-value problem

(5.10) $$\frac{dy}{dt} = f(t, y), \qquad a \le t \le b, \quad y(a) = \alpha,$$

is well-posed provided f is continuous and satisfies a Lipschitz condition in the variable y on the set D.

EXAMPLE 3   Let $D = \{(t, y) | 0 \le t \le 1, -\infty < y < \infty\}$ and consider the initial-value problem

(5.11) $$\frac{dy}{dt} = -y + t + 1, \qquad 0 \le t \le 1, \quad y(0) = 1.$$

Since $$\left| \frac{\partial(-y + t + 1)}{\partial y} \right| = 1,$$

Theorem 5.3 implies that $f(t, y) = -y + t + 1$ satisfies a Lipschitz condition on D with Lipschitz constant 1. Since, additionally, f is continuous on D, Theorem 5.6 implies that Eq. (5.11) is a well-posed problem.

Consider the perturbed problem

(5.12) $$\frac{dz}{dt} = -z + t + 1 + \delta, \qquad 0 \le t \le 1, \quad z(0) = 1 + \varepsilon_0,$$

where δ and ε_0 are constants. The solutions to Eqs. (5.11) and (5.12) can be shown to be

$$y(t) = e^{-t} + t \qquad \text{and} \qquad z(t) = (1 + \varepsilon_0 - \delta)e^{-t} + t + \delta,$$

respectively. It is easy to verify that if $|\delta| < \varepsilon$ and $|\varepsilon_0| < \varepsilon$, then

$$|y(t) - z(t)| = |(\delta - \varepsilon_0)e^{-t} - \delta| \le |\varepsilon_0| + |\delta||1 - e^{-t}| \le 2\varepsilon,$$

for all t, which corroborates the result obtained by the use of Theorem 5.6.   ∎

Exercise Set 5.1

1. Use Theorem 5.4, if possible, to show that each of the following initial value problems has a unique solution and find the solution.

    a) $y' = y \cos t$, $0 \le t \le 1$, $y(0) = 1$.

    b) $y' = (2/t)y + t^2 e^t$, $1 \le t \le 2$, $y(1) = 0$.

    c) $y' = -(2/t)y + t^2 e^t$, $1 \le t \le 2$, $y(1) = \sqrt{2}e$.

    d) $y' = \dfrac{4t^3}{y(1 + t^4)}$, $0 \le t \le 1$, $y(0) = 1$.

2. For each choice of $f(t, y)$ given in parts (a)–(d):

    i) Does f satisfy a Lipschitz condition on $D = \{(t, y)|0 \le t \le 1, -\infty < y < \infty\}$?
    ii) Can Theorem 5.6 be used to show that the initial-value problem

    $$y' = f(t, y), \quad 0 \le t \le 1, \quad y(0) = 1,$$

    is well-posed?

    a) $f(t, y) = t^2 y + 1$.   b) $f(t, y) = ty$.

    c) $f(t, y) = 1 - y$.   d) $f(t, y) = -ty + \dfrac{4t}{y}$.

3. For the following initial-value problems, show that the given equation implicitly defines a solution. Approximate $y(2)$, using Newton's method.

    a) $y' = \dfrac{y}{t + 2y \sin y}$, $\quad 1 \le t \le 2$, $\quad y(1) = 1$;

    $$\dfrac{t}{y} - 1 - 2 \int_1^y \dfrac{\sin z}{z}\, dz = 0.$$

    b) $y' = -\dfrac{y \cos t + 2te^y}{\sin t + t^2 e^y + 2}$, $\quad 1 \le t \le 2$, $\quad y(1) = 0$;

    $$y \sin t + t^2 e^y + 2y - 1 = 0.$$

4. Prove Theorem 5.3 by applying the Mean Value Theorem to $f(t, y)$, holding t fixed.

5. Show that, for any constants a and b, the set $D = \{(t, y)|a \le t \le b, -\infty < y < \infty\}$ is convex.

6. Show that the interior of a circle in the plane is convex.

7. Assuming that $\delta(t) = \delta t$ in Eq. (5.9), show directly that the following initial-value problems are well-posed.

a) $y' = 1 - y$, $0 \le t \le 2$, $y(0) = 0$.

b) $y' = t + y$, $0 \le t \le 2$, $y(0) = -1$.

c) $y' = (2/t)y + t^2 e^t$, $1 \le t \le 2$, $y(1) = 0$.

d) $y' = -(2/t)y + t^2 e^t$, $1 \le t \le 2$, $y(1) = \sqrt{2}e$.

5.2 Euler's Method

Although Euler's method is seldom used in practice, the simplicity of its derivation can be used to illustrate the techniques involved in the construction of some of the more advanced techniques, without the cumbersome algebra that accompanies these constructions.

The object of the method is to obtain an approximation to the well-posed initial-value problem

(5.13)
$$\frac{dy}{dt} = f(t, y), \qquad a \le t \le b, \quad y(a) = \alpha.$$

In actuality, a continuous approximation to the solution $y(t)$ will not be obtained; instead approximations to y will be generated at various points, called **mesh points**, in the interval $[a, b]$. Once the approximate solution is obtained at these points, the approximate solution at other points in the interval can be obtained by using one of the interpolation procedures discussed in Chapter 3.

In this section we make the stipulation that the mesh points be equally distributed throughout the interval $[a, b]$. This condition can be ensured by choosing a positive integer N and selecting the mesh points $\{t_0, t_1, t_2, \ldots, t_N\}$ where

$$t_i = a + ih \qquad \text{for each } i = 0, 1, 2, \ldots, N.$$

The common distance between the points, $h = (b - a)/N$, is called the **step size**.

We will use Taylor's Theorem to derive Euler's method. Another method of derivation is presented in Exercise 11.

Suppose that $y(t)$, the unique solution to (5.13), has two continuous derivatives on $[a, b]$, so that for each $i = 0, 1, 2, \ldots, N - 1$, $y(t_{i+1})$ can be written

(5.14)
$$y(t_{i+1}) = y(t_i) + (t_{i+1} - t_i)y'(t_i) + \frac{(t_{i+1} - t_i)^2}{2} y''(\xi_i)$$

for some number ξ_i, where $t_i < \xi_i < t_{i+1}$.

Using the notation $h = t_{i+1} - t_i$,

$$y(t_{i+1}) = y(t_i) + hy'(t_i) + \frac{h^2}{2} y''(\xi_i)$$

and, since $y(t)$ satisfies the differential equation (5.13),

(5.15)
$$y(t_{i+1}) = y(t_i) + hf(t_i, y(t_i)) + \frac{h^2}{2} y''(\xi_i).$$

Euler's method constructs $w_i \approx y(t_i)$ for each $i = 1, 2, \ldots, N$, where

(5.16)
$$w_0 = \alpha,$$
$$w_{i+1} = w_i + hf(t_i, w_i).$$

Equation (5.16) is called the **difference equation** associated with Euler's method. As we will see later in this chapter, the theory and solution of difference equations parallels, in many ways, the theory and solution of differential equations. The algorithmic form of Euler's method is presented as follows.

Euler's Algorithm 5.1

To approximate the solution of the initial-value problem

$$y' = f(t, y), \qquad a \le t \le b, \quad y(a) = \alpha,$$

at $(N + 1)$ equally spaced numbers in the interval $[a, b]$:

INPUT endpoints a, b; integer N; initial condition α.

OUTPUT approximation w to y at the $(N + 1)$ values of t.

Step 1   Set $h = (b - a)/N$;
 $t = a$;
 $w = \alpha$;
 OUTPUT (t, w).

Step 2   For $i = 1, 2, \ldots, N$ do Steps 3, 4.

  Step 3   Set $w = w + hf(t, w)$;   (*Compute w_i.*)
 $t = a + ih$.   (*Compute t_i.*)

  Step 4   OUTPUT (t, w).

Step 5   STOP.

To visualize the geometric interpretation of Euler's method, we first introduce the notation $y_i = y(t_i)$ for each $i = 0, 1, 2, \ldots, N$, and note that, when w_i is a close approximation to y_i, the assumption that the problem is well-posed implies that

$$f(t_i, w_i) \approx y'(t_i) = f(t_i, y(t_i)).$$

One step in the method consequently appears as in Fig. 5.1, and a series of steps appear in Fig. 5.2.

EXAMPLE 1   To find approximations to the initial-value problem

(5.17) $$y' = -y + t + 1, \qquad 0 \le t \le 1, \quad y(0) = 1,$$

suppose $N = 10$ so that $h = 0.1$ and $t_i = 0.1i$.

FIGURE 5.1

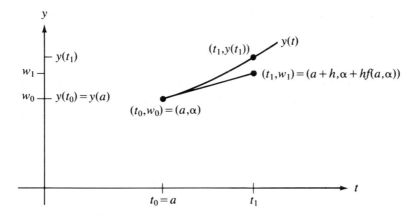

FIGURE 5.2

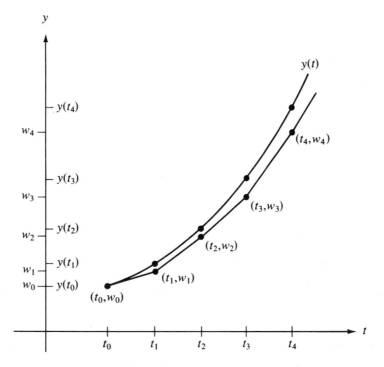

Using the fact that $f(t, y) = -y + t + 1$,

$$w_0 = 1$$

and

$$w_i = w_{i-1} + h(-w_{i-1} + t_{i-1} + 1)$$
$$= w_{i-1} + 0.1(-w_{i-1} + 0.1(i-1) + 1)$$
$$= 0.9w_{i-1} + 0.01(i-1) + 0.1$$

for $i = 1, 2, \ldots, 10$.

The exact solution to (5.17) is $y(t) = t + e^{-t}$. Table 5.1 shows the comparison between the approximate values at t_i and the actual values.   ∎

TABLE 5.1

t_i	w_i	y_i	Error $= |w_i - y_i|$
0.0	1.000000	1.000000	0.0
0.1	1.000000	1.004837	0.004837
0.2	1.010000	1.018731	0.008731
0.3	1.029000	1.040818	0.011818
0.4	1.056100	1.070320	0.014220
0.5	1.090490	1.106531	0.016041
0.6	1.131441	1.148812	0.017371
0.7	1.178297	1.196585	0.018288
0.8	1.230467	1.249329	0.018862
0.9	1.287420	1.306570	0.019150
1.0	1.348678	1.367879	0.019201

Note that the error grows slightly as the value of t_i increases. This controlled error growth is a consequence of the stability of Euler's method, which implies that the errors due to rounding are expected to grow in no worse than a linear manner.

To obtain an error bound for a general case of Euler's method, it is convenient to first present two computational lemmas.

LEMMA 5.7

For all $x \geq -1$ and any positive m,

$$(5.18) \qquad 0 \leq (1 + x)^m \leq e^{mx}.$$

PROOF

Applying Taylor's Theorem with $f(x) = e^x$, $x_0 = 0$, and $n = 1$, gives

$$e^x = 1 + x + \tfrac{1}{2}x^2 e^\xi,$$

where ξ is between x and zero. Thus,

$$0 \leq 1 + x \leq 1 + x + \tfrac{1}{2}x^2 e^\xi = e^x$$

and, since $1 + x \geq 0$, $\qquad 0 \leq (1 + x)^m \leq (e^x)^m = e^{mx}.$ $\qquad\qquad \square$

LEMMA 5.8

If s and t are positive real numbers, $\{a_i\}_{i=0}^k$ is a sequence satisfying $a_0 \geq -t/s$, and

$$(5.19) \qquad a_{i+1} \leq (1 + s)a_i + t \qquad \text{for each } i = 0, 1, 2, \ldots, k,$$

then $\qquad\qquad a_{i+1} \leq e^{(i+1)s}\left(\frac{t}{s} + a_0\right) - \frac{t}{s}.$

PROOF

For a fixed integer i, inequality (5.19) implies that

$$a_{i+1} \leq (1 + s)a_i + t$$
$$\leq (1 + s)[(1 + s)a_{i-1} + t] + t$$
$$\leq (1 + s)\{(1 + s)[(1 + s)a_{i-2} + t] + t\} + t$$
$$\vdots$$
$$\leq (1 + s)^{i+1}a_0 + [1 + (1 + s) + (1 + s)^2 + \cdots + (1 + s)^i]t.$$

But
$$1 + (1 + s) + (1 + s)^2 + \cdots + (1 + s)^i = \sum_{j=0}^{i} (1 + s)^j$$

is a geometric series with ratio $(1 + s)$ and, as such, sums to

$$\frac{1 - (1 + s)^{i+1}}{1 - (1 + s)} = \frac{1}{s}[(1 + s)^{i+1} - 1].$$

Thus,
$$a_{i+1} \leq (1 + s)^{i+1} a_0 + \frac{(1 + s)^{i+1} - 1}{s} t = (1 + s)^{i+1} \left(\frac{t}{s} + a_0\right) - \frac{t}{s},$$

and, by Lemma 5.7,

$$a_{i+1} \leq e^{(i+1)s}\left(\frac{t}{s} + a_0\right) - \frac{t}{s}. \qquad \square$$

THEOREM 5.9 Let $y(t)$ denote the unique solution to the well-posed initial-value problem

$$y' = f(t, y), \qquad a \leq t \leq b, \quad y(a) = \alpha,$$

and $w_0, w_1, \ldots, w_N$ be the approximations generated by Euler's method for some positive integer N.

If f satisfies a Lipschitz condition with constant L on

$$D = \{(t, y) | a \leq t \leq b, -\infty < y < \infty\},$$

and a constant M exists with the property that

$$|y''(t)| \leq M \qquad \text{for all } t \in [a, b],$$

then

(5.20)
$$|y(t_i) - w_i| \leq \frac{hM}{2L}[e^{L(t_i - a)} - 1]$$

for each $i = 0, 1, 2, \ldots, N$.

PROOF When $i = 0$ the result is clearly true, since $y(t_0) = w_0 = \alpha$.
From Eq. (5.15), we have for $i = 0, 1, \ldots, N - 1$,

$$y(t_{i+1}) = y(t_i) + hf(t_i, y(t_i)) + \frac{h^2}{2} y''(\xi_i);$$

and from the equations in (5.16)

$$w_{i+1} = w_i + hf(t_i, w_i).$$

Consequently, using the notation $y_i = y(t_i)$ and $y_{i+1} = y(t_{i+1})$:

$$y_{i+1} - w_{i+1} = y_i - w_i + h[f(t_i, y_i) - f(t_i, w_i)] + \frac{h^2}{2} y''(\xi_i)$$

and
$$|y_{i+1} - w_{i+1}| \leq |y_i - w_i| + h|f(t_i, y_i) - f(t_i, w_i)| + \frac{h^2}{2}|y''(\xi_i)|.$$

The assumptions that f satisfies a Lipschitz condition in the second variable with constant L and that $|y''(t)| \le M$ imply that

$$|y_{i+1} - w_{i+1}| \le |y_i - w_i|(1 + hL) + \frac{h^2 M}{2}.$$

Referring to Lemma 5.8 and letting $a_j = |y_j - w_j|$ for each $j = 0, 1, \ldots, N$, while $s = hL$ and $t = h^2 M/2$, we see that

$$|y_{i+1} - w_{i+1}| \le e^{(i+1)hL}\left(|y_0 - w_0| + \frac{h^2 M}{2hL}\right) - \frac{h^2 M}{2hL}.$$

Since $|y_0 - w_0| = 0$ and $(i+1)h = t_{i+1} - t_0 = t_{i+1} - a$, we have the desired result:

$$|y_{i+1} - w_{i+1}| \le \frac{hM}{2L}(e^{(t_{i+1} - a)L} - 1)$$

for each $i = 0, 1, \ldots, N - 1$.   □

The weakness of Theorem 5.9 lies in the requirement that a bound be known for the second derivative of the solution. Although this condition often prohibits us from obtaining a realistic error bound, it should be noted that if $\partial f/\partial t$ and $\partial f/\partial y$ both exist, then

$$y''(t) = \frac{dy'}{dt}(t) = \frac{df}{dt}(t, y(t))$$

$$= \frac{\partial f}{\partial t}(t, y(t)) + \frac{\partial f}{\partial y}(t, y(t)) \cdot f(t, y(t)),$$

and a bound for $y''(t)$ might be possible to obtain for certain problems without actually knowing $y(t)$.

EXAMPLE 2   Returning to the initial-value problem

$$y' = -y + t + 1, \qquad 0 \le t \le 1, \quad y(0) = 1,$$

considered in Example 1, we see that with $f(t, y) = -y + t + 1$, $\partial f/\partial y = -1$; so Theorem 5.3 implies that f satisfies a Lipschitz condition with $L = 1$.

Also, since in this case we know that the exact solution is $y(t) = t + e^{-t}$, we have $y''(t) = e^{-t}$,

and $|y''(t)| \le e^{-0} = 1$ for all $t \in [0, 1]$.

Using inequality (5.20) with $h = 0.1$ and $M = L = 1$ gives the error bound

$$|y_i - w_i| \le 0.05(e^{t_i} - 1).$$

Table 5.2 lists the actual error found in Example 1, together with this error bound.

Note that, even though the true bound for the second derivative of the solution was used, the error bound is considerably larger than the actual error.

TABLE 5.2

t_i	0.1	0.2	0.3	0.4	0.5	0.6	0.7	0.8	0.9	1.0
Actual error	0.00484	0.00873	0.01182	0.01422	0.01604	0.01737	0.01829	0.01886	0.01915	0.01920
Error bound	0.00526	0.01107	0.01749	0.02459	0.03244	0.04111	0.05069	0.06128	0.07298	0.08591

■

The principal importance of the error-bound formula given in Theorem 5.9 is the fact that the bound depends linearly on the step size h. Consequently, diminishing the step size should give correspondingly greater accuracy to the approximations.

Neglected in the result of Theorem 5.9 is the effect that round-off error plays in the choice of step size. It is clear that as h becomes smaller more calculations are necessary and when finite-digit arithmetic is used, which is certainly the usual case, more round-off error is expected. In actuality then, the difference-equation form

$$w_0 = \alpha,$$

$$w_i = w_{i-1} + hf(t_{i-1}, w_{i-1}) \qquad \text{for each } i = 1, 2, \dots, N,$$

is not used to calculate the approximation to the solution y_i at a mesh point t_i, but instead an equation of the form

(5.21)
$$u_0 = \hat{\alpha},$$

$$u_i = u_{i-1} + hf(t_{i-1}, u_{i-1}) + \delta_i \qquad \text{for each } i = 1, 2, \dots, N,$$

is used, where $\hat{\alpha}$ is the approximation to the initial value α and δ_i denotes the round-off error associated with the computation of $u_{i-1} + hf(t_{i-1}, u_{i-1})$. By using methods similar to those in the proof of Theorem 5.9 (see Exercise 12), we can obtain the following result, which gives an error bound for the finite-digit approximations to y_i obtained using Euler's method.

THEOREM 5.10   Let $y(t)$ denote the unique solution to the well-posed initial-value problem

(5.22)
$$y' = f(t, y), \qquad a \le t \le b, \quad y(a) = \alpha$$

and $u_0, u_1, \dots, u_N$ be the approximations obtained using (5.21). If $|\delta_i| < \delta$ for each $i = 0, 1, \dots, N$, where δ_0 denotes $\hat{\alpha} - \alpha$, and the hypotheses of Theorem 5.9 hold for (5.22), then

(5.23)
$$|y(t_i) - u_i| \le \frac{1}{L}\left(\frac{hM}{2} + \frac{\delta}{h}\right)[e^{L(t_i - a)} - 1] + |\delta_0|e^{L(t_i - a)}$$

for each $i = 0, 1, \dots, N$.

The error bound (5.23) is no longer linear in h and, in fact, since

$$\lim_{h \to 0} \left(\frac{hM}{2} + \frac{\delta}{h} \right) = \infty,$$

the error would be expected to become large for sufficiently small values of h. Ordinary techniques of calculus can be used to determine a lower bound for the step size h. Letting $E(h) = (hM/2) + (\delta/h)$ implies that $E'(h) = (M/2) - (\delta/h^2)$.

If $h > \sqrt{2\delta/M}$, then $E'(h) < 0$ and $E(h)$ is decreasing.

If $h < \sqrt{2\delta/M}$, then $E'(h) > 0$ and $E(h)$ is increasing.

The minimal value of $E(h)$ occurs when

(5.24) $h = \sqrt{2\delta/M};$

decreasing h beyond this value would tend to increase the total error in the approximation.

Normally, however, the value of δ is sufficiently small that this lower bound for h will not affect the operation of Euler's method.

Exercise Set 5.2

1. Use Euler's method to approximate the solution for each of the following initial-value problems.

   a) $y' = \left(\frac{y}{t} \right)^2 + \left(\frac{y}{t} \right), \quad 1 \le t \le 1.2, \quad y(1) = 1$ with $h = 0.1$.

   b) $y' = \sin t + e^{-t}, \quad 0 \le t \le 1, \quad y(0) = 0$ with $h = 0.5$.

   c) $y' = \frac{1}{t}(y^2 + y), \quad 1 \le t \le 3, \quad y(1) = -2$ with $h = 0.5$.

   d) $y' = -ty + \frac{4t}{y}, \quad 0 \le t \le 1, \quad y(0) = 1$ with $h = 0.25$.

2. Use Euler's method with $N = 4$ to approximate the solution for each of the following initial-value problems and compare the approximations to the graph of the exact solution.

   a) $y' = t^2, \quad 0 \le t \le 2, \quad y(0) = 0$.

   b) $y' = ty, \quad 0 \le t \le 2, \quad y(0) = 1$.

   c) $y' = 2t, \quad 0 \le t \le 2, \quad y(0) = 1$.

   d) $y' = -ty, \quad 0 \le t \le 4, \quad y(0) = 4$.

3. Given the initial-value problem

$$y' = \frac{2}{t}y + t^2 e^t, \quad 1 \le t \le 2, \quad y(1) = 0$$

   with exact solution

$$y(t) = t^2(e^t - e):$$

a) Use Euler's method with $h = 0.1$ to approximate the solution and compare it with the actual values of y.

b) Use the answers generated in part (a) and linear interpolation to approximate the following values of y and compare them to the actual values.

i) $y(1.04)$ ii) $y(1.55)$ iii) $y(1.97)$

c) Compute the value of h necessary for $|y(t_i) - w_i| \leq 0.1$, using Eq. (5.20).

4. Given the initial-value problem

$$y' = \frac{1}{t^2} - \frac{y}{t} - y^2, \quad 1 \leq t \leq 2, \quad y(1) = -1$$

with exact solution $y(t) = -1/t$:

a) Use Euler's method with $h = 0.05$ to approximate the solution and compare it with the actual values of y.

b) Use the answers generated in part (a) and linear interpolation to approximate the following values of y and compare them to the actual values.

i) $y(1.052)$ ii) $y(1.555)$ iii) $y(1.978)$

c) Compute the value of h necessary for $|y(t_i) - w_i| \leq 0.05$, using Eq. (5.20).

5. Given the initial-value problem

$$y' = -y + t + 1, \quad 0 \leq t \leq 5, \quad y(0) = 1$$

with exact solution $y(t) = e^{-t} + t$:

a) Approximate $y(5)$ using Euler's method with $h = 0.2$, $h = 0.1$, and $h = 0.05$.

b) Determine the optimal value of h to use in computing $y(5)$, assuming $\delta = 10^{-6}$ and that Eq. (5.24) is valid.

6. Let $E(h) = \dfrac{hM}{2} + \dfrac{\delta}{h}$.

a) For the initial-value problem

$$y' = -y + 1, \qquad 0 \leq t \leq 1, \quad y(0) = 0,$$

compute the value of h to minimize $E(h)$. Assume $\delta = 5 \times 10^{-(n+1)}$ if you will be using n-digit arithmetic in part (c).

b) For the optimal h computed in part (a), use Eq. (5.23) to compute the minimal error obtainable.

c) Compare the actual error obtained using $h = 0.1$ and $h = 0.01$ to the minimal error in part (b). Can you explain the results?

7. Obtain an approximation to the solution of

$$y' = 1 + t \sin(ty), \quad 0 \leq t \leq 2, \quad y(0) = 0$$

using Euler's method with $h = 0.1$. Can an error bound be obtained using Eq. (5.20)?

8. Consider the initial-value problem

$$y' = -10y, \qquad 0 \le t \le 2, \quad y(0) = 1,$$

which has solution $y(t) = e^{-10t}$. What happens when Euler's method is applied to this problem with $h = 0.1$? Does this behavior violate Theorem 5.9?

9. In a book entitled "Looking at History Through Mathematics," Rashevsky [105] (pp. 103–110) considers a model for a problem involving the production of nonconformists in society. Suppose that a society has a population of $x(t)$ individuals at time t, in years, and that all nonconformists who mate with other nonconformists have offspring who are also nonconformists, while a fixed proportion r of all other offspring are also nonconformist. If the birth and death rates for all individuals are assumed to be the constants b and d, respectively, and if conformists and nonconformists mate at random, the problem can be expressed by the differential equations

$$\frac{dx(t)}{dt} = (b-d)x(t) \qquad \text{and} \qquad \frac{dx_n(t)}{dt} = (b-d)x_n(t) + rb(x(t) - x_n(t)),$$

where $x_n(t)$ denotes the number of nonconformists in the population at time t.

a) If the variable $p(t) = x_n(t)/x(t)$ is introduced to represent the proportion of nonconformists in the society at time t, show that these equations can be combined and simplified to the single differential equation

$$\frac{dp(t)}{dt} = rb(1 - p(t)).$$

b) Assuming that $p(0) = 0.01$, $b = 0.02$, $d = 0.015$, and $r = 0.1$, use Algorithm 5.1 to approximate the solution $p(t)$ from $t = 0$ to $t = 50$ when the step size is $h = 1$ year.

c) Solve the differential equation for $p(t)$ exactly, and compare your result in part (b) when $t = 50$ with the exact value at that time.

10. In a circuit with impressed voltage $\mathscr{E}$, and resistance R, inductance L, capacitance C in parallel, the current i satisfies the differential equation

$$\frac{di}{dt} = C\frac{d^2\mathscr{E}}{dt^2} + \frac{1}{R}\frac{d\mathscr{E}}{dt} + \frac{1}{L}\mathscr{E}.$$

Suppose $C = 0.3$ farads, $R = 1.4$ ohms, $L = 1.7$ henries, and the voltage is given by

$$\mathscr{E}(t) = e^{-0.06\pi t} \sin(2t - \pi).$$

If $i(0) = 0$, find the current i for the values $t = 0.1j, j = 0, 1, \ldots, 100$ using Euler's method.

11. For the initial-value problem $y' = f(t, y)$, $a \le t \le b$, with $y(a) = \alpha$.

a) Derive Euler's method by integrating the differential equation from t_i to t_{i+1} and using an appropriate quadrature formula to approximate the integral.

b) Obtain a difference method called the **trapezoidal method** by using the trapezoidal rule as the quadrature formula.

12. Prove Theorem 5.10.

5.3 Higher-Order Taylor Methods

Since the object of numerical techniques is generally to determine sufficiently accurate approximations with minimal effort, it is necessary to have a means of comparing the efficiency of various approximation methods with respect to their use in computation. For difference-equation techniques for solving ordinary differential equations, such as Euler's method, the first measurement device we will consider is called the **local truncation error** of the method. The local truncation error at a specified step measures the amount by which the exact solution to the differential equation fails to satisfy the difference equation being used for the approximation. For Euler's method the local truncation error at the ith step for the problem

$$y' = f(t, y), \qquad a \le t \le b, \quad y(a) = \alpha,$$

is
$$\tau_i = \frac{y_i - y_{i-1}}{h} - f(t_{i-1}, y_{i-1}) \qquad \text{for each } i = 1, 2, \ldots, N,$$

where, as usual, $y_i = y(t_i)$ denotes the exact value of the solution at t_i. This error is called a *local error* because it measures the accuracy of the method at a specific step, assuming that the method was exact at the previous step. As such, it depends on the differential equation, the step size, and the particular step in the approximation.

By considering Eq. (5.15) in the previous section, we see that Euler's method has

$$\tau_i = \frac{h}{2} y''(\xi_i) \qquad \text{for some } \xi_i, \, t_i < \xi_i < t_{i+1}$$

and, when $y''(t)$ is known to be bounded by a constant M on $[a, b]$, this implies

$$|\tau_i| \le \frac{h}{2} M.$$

Recalling Definition 1.18, p. 24, we infer that the truncation error in Euler's method is $O(h)$. This result implies that one way to select difference-equation methods for solving ordinary differential equations is in such a manner that their local truncation errors are $O(h^p)$ for as large a value of p as possible, while keeping the number and complexity of calculations of the methods within a reasonable bound.

Since Euler's method was derived by using Taylor's Theorem with $n = 2$ to approximate the solution of the differential equation, our first attempt to find methods of improving the convergence properties of difference methods will be to extend this technique of derivation to larger values of n.

Assuming that the solution $y(t)$ to the initial-value problem

$$y' = f(t, y), \qquad a \le t \le b, \quad y(a) = \alpha,$$

has $(n + 1)$ continuous derivatives and expanding that solution, $y(t)$, in terms of its nth-degree Taylor polynomial about t_i, we obtain:

(5.25)
$$y(t_{i+1}) = y(t_i) + hy'(t_i) + \frac{h^2}{2}y''(t_i) + \cdots$$

$$+ \frac{h^n}{n!}y^{(n)}(t_i) + \frac{h^{n+1}}{(n+1)!}y^{(n+1)}(\xi_i)$$

for some ξ_i, $t_i < \xi_i < t_{i+1}$.

Successive differentiation of the solution, $y(t)$, gives

$$y'(t) = f(t, y(t)),$$
$$y''(t) = f'(t, y(t)),$$

and, in general,

$$y^{(k)}(t) = f^{(k-1)}(t, y(t)).$$

Substituting these results into Eq. (5.25) gives:

(5.26)
$$y(t_{1+1}) = y(t_i) + hf(t_i, y(t_i)) + \frac{h^2}{2}f'(t_i, y(t_i)) + \cdots$$

$$+ \frac{h^n}{n!}f^{(n-1)}(t_i, y(t_i)) + \frac{h^{n+1}}{(n+1)!}f^{(n)}(\xi_i, y(\xi_i)).$$

The difference method corresponding to Eq. (5.26) is obtained by neglecting the remainder term involving ξ_i. This method is called the **Taylor method of order n**:

(5.27)
$$w_0 = \alpha,$$
$$w_{i+1} = w_i + hT^{(n)}(t_i, w_i) \quad \text{for each } i = 0, 1, \ldots, N-1,$$

where
$$T^{(n)}(t_i, w_i) = f(t_i, w_i) + \frac{h}{2}f'(t_i, w_i) + \cdots + \frac{h^{n-1}}{n!}f^{(n-1)}(t_i, w_i).$$

Note that Euler's method is Taylor's method of order one.

EXAMPLE 1   To apply Taylor's method of orders two and four to the initial-value problem

$$y' = -y + t + 1, \qquad 0 \le t \le 1, \quad y(0) = 1$$

(which was studied in Example 1 of Section 5.2), we must find the first three derivatives of $f(t, y(t)) = -y + t + 1$,

$$f'(t, y(t)) = \frac{d}{dt}(-y + t + 1) = -y' + 1 = y - t - 1 + 1 = y - t,$$

$$f''(t, y(t)) = \frac{d}{dt}(y - t) = y' - 1 = -y + t + 1 - 1 = -y + t,$$

and
$$f'''(t, y(t)) = \frac{d}{dt}(-y + t) = -y' + 1 = y - t - 1 + 1 = y - t;$$

so
$$T^{(2)}(t_i, w_i) = f(t_i, w_i) + \frac{h}{2}f'(t_i, w_i)$$

$$= -w_i + t_i + 1 + \frac{h}{2}(w_i - t_i)$$

$$= \left(1 - \frac{h}{2}\right)(t_i - w_i) + 1$$

and

$$T^{(4)}(t_i, w_i) = f(t_i, w_i) + \frac{h}{2}f'(t_i, w_i) + \frac{h^2}{6}f''(t_i, w_i) + \frac{h^3}{24}f'''(t_i, w_i)$$

$$= -w_i + t_i + 1 + \frac{h}{2}(w_i - t_i) + \frac{h^2}{6}(-w_i + t_i) + \frac{h^3}{24}(w_i - t_i)$$

$$= \left(1 - \frac{h}{2} + \frac{h^2}{6} - \frac{h^3}{24}\right)(t_i - w_i) + 1.$$

The Taylor methods of orders two and four are consequently

$$w_0 = 1,$$

(5.28)
$$w_{i+1} = w_i + h\left[\left(1 - \frac{h}{2}\right)(t_i - w_i) + 1\right]$$

and

$$w_0 = 1,$$

(5.29)
$$w_{i+1} = w_i + h\left[\left(1 - \frac{h}{2} + \frac{h^2}{6} - \frac{h^3}{24}\right)(t_i - w_i) + 1\right]$$

for $i = 0, 1, \ldots, N - 1$, respectively.

Using $h = 0.1$ implies $N = 10$ and $t_i = 0.1i$ for each $i = 1, 2, \ldots, 10$; so the second-order method (5.28) becomes

$$w_0 = 1,$$

(5.30)
$$w_{i+1} = w_i + 0.1\left[\left(1 - \frac{0.1}{2}\right)(0.1i - w_i) + 1\right]$$

$$= 0.905w_i + 0.0095i + 0.1,$$

and the fourth-order method (5.29) becomes

$$w_0 = 1,$$

(5.31)
$$w_{i+1} = w_i + 0.1\left[\left(1 - \frac{0.1}{2} + \frac{0.01}{6} - \frac{0.001}{24}\right)(0.1i - w_i) + 1\right]$$

$$= 0.9048375w_i + 0.00951625i + 0.1$$

for each $i = 0, 1, \ldots, 9$.

Table 5.3 lists the actual values of the solution $y(t) = t + e^{-t}$ together with the approximations obtained from Euler's method in Example 1 of Section 5.2, the results from the Taylor methods of orders two and four, and the actual errors involved with these methods. ■

As might be expected from our study of Euler's method, Taylor's method of order n has local truncation error $O(h^n)$, provided that the solution of the differential

TABLE 5.3

t	Exact value	Euler's method	Error in Euler's method	Taylor's method of order two	Error in Taylor's method of order two	Taylor's method of order four	Error in Taylor's method of order four
0.0	1.0000000000	1.000000	0	1.000000	0	1.0000000000	0
0.1	1.0048374180	1.000000	4.837×10^{-3}	1.005000	1.626×10^{-4}	1.0048375000	8.200×10^{-8}
0.2	1.0187307531	1.010000	8.731×10^{-3}	1.019025	2.942×10^{-4}	1.0187309014	1.483×10^{-7}
0.3	1.0408182207	1.029000	1.182×10^{-2}	1.041218	3.998×10^{-4}	1.0408184220	2.013×10^{-7}
0.4	1.0703200460	1.056100	1.422×10^{-2}	1.070802	4.820×10^{-4}	1.0703202889	2.429×10^{-7}
0.5	1.1065306597	1.090490	1.604×10^{-2}	1.107076	5.453×10^{-4}	1.1065309344	2.747×10^{-7}
0.6	1.1488116361	1.131441	1.737×10^{-2}	1.149404	5.924×10^{-4}	1.1488119344	2.983×10^{-7}
0.7	1.1965853038	1.178297	1.829×10^{-2}	1.197211	6.257×10^{-4}	1.1965856187	3.149×10^{-7}
0.8	1.2493289641	1.230467	1.887×10^{-2}	1.249976	6.470×10^{-4}	1.2493292897	3.256×10^{-7}
0.9	1.3065696597	1.287420	1.915×10^{-2}	1.307228	6.583×10^{-4}	1.3065699912	3.315×10^{-7}
1.0	1.3678794412	1.348678	1.920×10^{-2}	1.368541	6.616×10^{-4}	1.3678797744	3.332×10^{-7}

equation is sufficiently well-behaved. This is seen by noticing that Eq. (5.26) can be rewritten

$$y_{i+1} - y_i - hf(t_i, y_i) - \frac{h^2}{2} f'(t_i, y_i) - \cdots - \frac{h^n}{n!} f^{(n-1)}(t_i, y_i) = \frac{h^{n+1}}{(n+1)!} f^{(n)}(\xi_i, y(\xi_i)),$$

so the local truncation error at the $(i+1)$st step is

$$\tau_{i+1} = \frac{y_{i+1} - y_i}{h} - T^{(n)}(t_i, y_i) = \frac{h^n}{(n+1)!} f^{(n)}(\xi_i, y(\xi_i))$$

for each $i = 0, 1, \ldots, N-1$. If $y \in C^{n+1}[a, b]$, this implies that $y^{(n+1)}(t) = f^{(n)}(t, y(t))$ is bounded on $[a, b]$, and that $\tau_i = O(h^n)$ for each $i = 1, 2, \ldots, N$.

Exercise Set 5.3

1. Use Taylor's method of order two to approximate the solution to each of the following initial-value problems.

   a) $y' = \left(\dfrac{y}{t}\right)^2 + \left(\dfrac{y}{t}\right)$, $1 \le t \le 1.2$, $y(1) = 1$ with $h = 0.1$.

   b) $y' = \sin t + e^{-t}$, $0 \le t \le 1$, $y(0) = 0$ with $h = 0.5$.

   c) $y' = \dfrac{1}{t}(y^2 + y)$, $1 \le t \le 3$, $y(1) = -2$ with $h = 0.5$.

   d) $y' = -ty + \dfrac{4t}{y}$, $0 \le t \le 1$, $y(0) = 1$ with $h = 0.25$.

2. Repeat Exercise 1 using Taylor's method of order four.

3. Use the Taylor methods of orders two and four with $h = 0.1$ to approximate the solutions to the following initial-value problems. Compute the actual error.

a) $y' = t + y, \quad 0 \le t \le 2, \quad y(0) = -1$

b) $y' = 1 - y, \quad 0 \le t \le 2, \quad y(0) = 0$

c) $y' = -y + t + 1, \quad 0 \le t \le 5, \quad y(0) = 1$

d) $y' = 1 + y, \quad 0 \le t \le 2, \quad y(0) = 0$

4. Use the Taylor method of order two with $h = 0.1$ to approximate the solution to

$$y' = 1 + t \sin(ty), \qquad 0 \le t \le 2, \quad y(0) = 0.$$

Compare your answers to those obtained in Exercise 7, Section 5.2.

5. Given the initial-value problem

$$y' = \frac{2}{t}y + t^2 e^t, \quad 1 \le t \le 2, \quad y(1) = 0$$

with exact solution

$$y(t) = t^2(e^t - e):$$

a) Use Taylor's method of order two with $h = 0.1$ to approximate the solution and compare it with the actual values of y.

b) Use the answers generated in part (a) and linear interpolation to approximate y at the following values and compare them to the actual values of y.

    i) $y(1.04)$ *ii)* $y(1.55)$ *iii)* $y(1.97)$

c) Use Taylor's method of order four with $h = 0.1$ to approximate the solution and compare it with the actual values of y.

d) Use the answers generated in part (c) and piecewise cubic Hermite interpolation to approximate y at the following values and compare them to the actual values of y.

    i) $y(1.04)$ *ii)* $y(1.55)$ *iii)* $y(1.97)$

6. Given the initial-value problem

$$y' = \frac{1}{t^2} - \frac{y}{t} - y^2, \quad 1 \le t \le 2, \quad y(1) = -1$$

with exact solution $y(t) = -1/t$:

a) Use Taylor's method of order two with $h = 0.05$ to approximate the solution and compare it with the actual values of y.

b) Use the answers generated in part (a) and linear interpolation to approximate the following values of y and compare them to the actual values.

    i) $y(1.052)$ *ii)* $y(1.555)$ *iii)* $y(1.978)$

c) Use Taylor's method of order four with $h = 0.05$ to approximate the solution and compare it with the actual values of y.

d) Use the answers generated in part (c) and piecewise cubic Hermite interpolation to approximate the following values of y and compare them to the actual values.

   i) $y(1.052)$ ii) $y(1.555)$ iii) $y(1.978)$

7. A projectile of mass $m = 0.11$ kg shot vertically upward with initial velocity $v(0) = 8$ meter/sec is slowed due to the force of gravity $F_g = -mg$ and due to air resistance $F_r = -kv^2$, where $g = 9.8$ meter/sec^2 and $k = 0.002$ kg/meter. The differential equation for the velocity v is given by:

$$mv' = -mg - kv^2.$$

   a) Find the velocity after $0.1, 0.2, \ldots, 1.0$ seconds, using the Taylor methods of orders two and four.

   b) To the nearest tenth of a second, determine when the projectile reaches its maximum height and begins falling. Use the Taylor method of order four.

8. Can the difference method

$$w_{i+1} = w_{i-1} + 2hf(t_i, w_i) \qquad \text{for each } i = 1, 2, \ldots, N-1,$$

with $w_0 = \alpha$ and $w_1 = y(t_1)$ be used with the differential equation $y' = f(t, y)$, $a < t < b$, $y(a) = \alpha$? Find the local truncation error.

5.4 Runge–Kutta Methods

The Taylor methods outlined in the previous section have the desirable property of high-order local truncation error, but the disadvantage of requiring computation and evaluation of the derivatives of $f(t, y)$. This can be a very complicated and time-consuming procedure for many problems and, consequently, the Taylor methods are seldom used in practice. The **Runge–Kutta methods** use the high-order local truncation error of the Taylor methods while eliminating the computation and evaluation of the derivatives of $f(t, y)$. Before presenting the ideas behind their derivation, we need to state Taylor's Theorem in two variables. The proof of this result can be found in any standard book on advanced calculus (see, for example, Fulks [50], page 331).

THEOREM 5.11   Suppose that $f(t, y)$ and all of its partial derivatives of order less than or equal to $n + 1$ are continuous on $D = \{(t, y) | a \leq t \leq b, c \leq y \leq d\}$. Let $(t_0, y_0) \in D$. For every $(t, y) \in D$, there exists ξ between t and t_0 and η between y and y_0 with

$$f(t, y) = P_n(t, y) + R_n(t, y),$$

where

$$P_n(t, y) = f(t_0, y_0) + \left[(t - t_0)\frac{\partial f}{\partial t}(t_0, y_0) + (y - y_0)\frac{\partial f}{\partial y}(t_0, y_0) \right]$$
$$+ \left[\frac{(t - t_0)^2}{2}\frac{\partial^2 f}{\partial t^2}(t_0, y_0) + (t - t_0)(y - y_0)\frac{\partial^2 f}{\partial t\, \partial y}(t_0, y_0) + \frac{(y - y_0)^2}{2}\frac{\partial^2 f}{\partial y^2}(t_0, y_0) \right]$$

$$+ \cdots + \left[\frac{1}{n!} \sum_{j=0}^{n} \binom{n}{j} (t-t_0)^{n-j}(y-y_0)^j \frac{\partial^n f}{\partial t^{n-j} \partial y^j}(t_0, y_0) \right]$$

and

$$R_n(t, y) = \frac{1}{(n+1)!} \sum_{j=0}^{n+1} \binom{n+1}{j}(t-t_0)^{n+1-j}(y-y_0)^j \frac{\partial^{n+1} f}{\partial t^{n+1-j} \partial y^j}(\xi, \eta).$$

P_n is called the **Taylor polynomial of degree n in two variables** for the function f about (t_0, y_0), and $R_n(t, y)$ is the remainder term associated with $P_n(t, y)$.

EXAMPLE 1   The Taylor polynomial of degree three for $f(t, y) = \sin(ty)$ about $(0, \pi)$ is found from

$$P_3(t, y) = f(t_0, y_0) + (t-t_0)\frac{\partial f}{\partial t}(t_0, y_0) + (y-y_0)\frac{\partial f}{\partial y}(t_0, y_0)$$

$$+ \left[\frac{(t-t_0)^2}{2} \frac{\partial^2 f}{\partial t^2}(t_0, y_0) + (t-t_0)(y-y_0)\frac{\partial^2 f}{\partial t \partial y}(t_0, y_0) \right.$$

$$\left. + \frac{(y-y_0)^2}{2} \frac{\partial^2 f}{\partial y^2}(t_0, y_0) \right]$$

$$+ \left[\frac{(t-t_0)^3}{6} \frac{\partial^3 f}{\partial t^3}(t_0, y_0) + \frac{(t-t_0)^2(y-y_0)}{2} \frac{\partial^3 f}{\partial t^2 \partial y}(t_0, y_0) \right.$$

$$\left. + \frac{(t-t_0)(y-y_0)^2}{2} \frac{\partial^2 f}{\partial t \partial y^2}(t_0, y_0) + \frac{(y-y_0)^3}{6} \frac{\partial^3 f}{\partial y^3}(t_0, y_0) \right].$$

Evaluating each of these partial derivatives at $(t_0, y_0) = (0, \pi)$ reduces $P_3(t, y)$ to

$$P_3(t, y) = \pi t + t(y - \pi) - \frac{\pi^3}{6} t^3.$$

This polynomial will give a close approximation to $\sin(ty)$ provided t is close to zero and y is close to π. For example,

$$P_3(0.01, \pi + 0.01) = 0.03151076,$$

while $$\sin(0.01(\pi + 0.01)) = 0.03151071. \qquad \blacksquare$$

The first step in deriving a Runge–Kutta method is to determine values for $a_1, \alpha_1,$ and β_1 with the property that $a_1 f(t + \alpha_1, y + \beta_1)$ approximates

$$T^{(2)}(t, y) = f(t, y) + \frac{h}{2} f'(t, y)$$

with error no greater than $O(h^2)$, the local truncation error for the Taylor method of order two.

Since

$$f'(t, y) = \frac{df}{dt}(t, y) = \frac{\partial f}{\partial t}(t, y) + \frac{\partial f}{\partial y}(t, y) \cdot y'(t) \quad \text{and} \quad y'(t) = f(t, y),$$

this implies

$$(5.32) \qquad T^{(2)}(t, y) = f(t, y) + \frac{h}{2}\frac{\partial f}{\partial t}(t, y) + \frac{h}{2}\frac{\partial f}{\partial y}(t, y) \cdot f(t, y).$$

Expanding $f(t + \alpha_1, y + \beta_1)$ in its Taylor polynomial of degree one about (t, y) implies that:

$$(5.33) \qquad a_1 f(t + \alpha_1, y + \beta_1) = a_1 f(t, y) + a_1\alpha_1 \frac{\partial f}{\partial t}(t, y)$$

$$+ a_1 \beta_1 \frac{\partial f}{\partial y}(t, y) + a_1 \cdot R_1(t + \alpha_1, y + \beta_1),$$

where

$$(5.34) \qquad R_1(t + \alpha_1, y + \beta_1) = \frac{\alpha_1^2}{2}\frac{\partial^2 f}{\partial t^2}(\xi, \eta) + \alpha_1 \beta_1 \frac{\partial^2 f}{\partial t \partial y}(\xi, \eta) + \frac{\beta_1^2}{2}\frac{\partial^2 f}{\partial y^2}(\xi, \eta)$$

for some ξ between t and $t + \alpha_1$, and η between y and $y + \beta_1$.

Matching the coefficients of f and its derivatives in equations (5.32) and (5.33) gives the three equations

$$f(t, y): \quad a_1 = 1; \qquad \frac{\partial f}{\partial t}(t, y): \quad a_1\alpha_1 = \frac{h}{2};$$

and

$$\frac{\partial f}{\partial y}(t, y): \quad a_1\beta_1 = \frac{h}{2}f(t, y).$$

The parameters a_1, α_1, and β_1 are uniquely determined to be

$$a_1 = 1, \qquad \alpha_1 = \frac{h}{2}, \qquad \text{and} \qquad \beta_1 = \frac{h}{2}f(t, y);$$

so

$$T^{(2)}(t, y) = f\left(t + \frac{h}{2}, y + \frac{h}{2}f(t, y)\right) - R_1\left(t + \frac{h}{2}, y + \frac{h}{2}f(t, y)\right)$$

and from Eq. (5.34),

$$R_1\left(t + \frac{h}{2}, y + \frac{h}{2}f(t, y)\right) = \frac{h^2}{8}\frac{\partial^2 f}{\partial t^2}(\xi, \eta) + \frac{h^2}{4}f(t, y)\frac{\partial^2 f}{\partial t \partial y}(\xi, \eta)$$

$$+ \frac{h^2}{8}(f(t, y))^2 \frac{\partial^2 f}{\partial y^2}(\xi, \eta).$$

If all the second-order partial derivatives of f are bounded,

$$R_1\left(t+\frac{h}{2},\, y+\frac{h}{2}f(t,\, y)\right)$$

will be $O(h^2)$, the order of the local truncation error of Taylor's method of order two.

The difference method resulting from replacing $T^{(2)}(t,\, y)$ in Taylor's method of order two by $f(t+(h/2),\, y+(h/2)f(t,\, y))$ is a specific Runge–Kutta method known as the **Midpoint method**.

Midpoint Method

(5.35)
$$\begin{aligned}
w_0 &= \alpha, \\
w_{i+1} &= w_i + hf\left(t_i+\frac{h}{2},\, w_i+\frac{h}{2}f(t_i,\, w_i)\right)
\end{aligned}$$

for each $i = 0,\, 1,\, \ldots,\, N-1$.

Since only three parameters are present in $a_1 f(t+\alpha_1,\, y+\beta_1)$, and all are needed in the match of $T^{(2)}$, we need a more complicated form to satisfy the conditions required for any of the higher-order Taylor methods.

The most appropriate four parameter form for approximating

$$T^{(3)}(t,\, y) = f(t,\, y) + \frac{h}{2}f'(t,\, y) + \frac{h^2}{6}f''(t,\, y),$$

is

(5.36)
$$a_1 f(t,\, y) + a_2 f(t+\alpha_2,\, y+\delta_2 f(t,\, y));$$

and even with this, there is insufficient flexibility to match the term

$$\frac{h^2}{6}\left[\frac{\partial f}{\partial y}(t,\, y)\right]^2 f(t,\, y)$$

resulting from the expansion of $(h^2/6)f''(t,\, y)$. Consequently, the best that can be obtained from using (5.36) are methods with $O(h^2)$ error bound. The fact that (5.36) has four parameters, however, gives a flexibility in their choice so that a number of $O(h^2)$ methods can be derived. The two most important are called the **Modified Euler method**, which corresponds to choosing $a_1 = a_2 = \frac{1}{2}$ and $\alpha_2 = \delta_2 = h$ and has difference-equation form:

Modified Euler Method

(5.37)
$$\begin{aligned}
w_0 &= \alpha, \\
w_{i+1} &= w_i + \frac{h}{2}[f(t_i,\, w_i)+f(t_{i+1},\, w_i+hf(t_i,\, w_i))]
\end{aligned}$$

for each $i = 0,\, 1,\, 2,\, \ldots,\, N-1$,

and the **Heun's method**, which corresponds to $a_1 = \frac{1}{4}$, $a_2 = \frac{3}{4}$, and $\alpha_2 = \delta_2 = \frac{2}{3}h$ and has difference-equation form:

Heun Method

(5.38)
$$w_0 = \alpha,$$
$$w_{i+1} = w_i + \frac{h}{4}\left[f(t_i, w_i) + 3f(t_i + \tfrac{2}{3}h, w_i + \tfrac{2}{3}hf(t_i, w_i))\right]$$

for each $i = 0, 1, 2, ..., N-1$.

Both of these methods are classified as Runge–Kutta methods of order two, the order of their local truncation error.

EXAMPLE 2 Applying any of the Runge–Kutta methods of order two to our usual example,

$$y' = -y + t + 1, \qquad 0 \le t \le 1, \quad y(0) = 1,$$

gives the same difference equation as that of Taylor's method of order two,

$$w_0 = 1,$$
$$w_{i+1} = 0.905w_i + 0.0095i + 0.1,$$

because of the nature of the differential equation.

To compare the various results from these methods we will use, instead, the initial-value problem

$$y' = -y + t^2 + 1, \qquad 0 \le t \le 1, \quad y(0) = 1,$$

which has $y(t) = -2e^{-t} + t^2 - 2t + 3$ as its exact solution. Table 5.4 lists the results of these calculations. ∎

TABLE 5.4

t_i	Exact values	Midpoint method	Error for Midpoint method	Modified Euler method	Error for modified Euler method	Heun's method	Error for Heun's method
0.0	1.0000000	1.0000000		1.0000000		1.0000000	
0.1	1.0003252	1.0002500	7.52×10^{-5}	1.0005000	1.75×10^{-4}	1.0003333	8.10×10^{-6}
0.2	1.0025385	1.0024263	1.12×10^{-4}	1.0029025	3.64×10^{-4}	1.0025850	4.65×10^{-5}
0.3	1.0083636	1.0082458	1.18×10^{-4}	1.0089268	5.63×10^{-4}	1.0084728	1.09×10^{-4}
0.4	1.0193599	1.0192624	9.75×10^{-5}	1.0201288	7.69×10^{-4}	1.0195512	1.91×10^{-4}
0.5	1.0369387	1.0368825	5.62×10^{-5}	1.0379166	9.78×10^{-4}	1.0372272	2.88×10^{-4}
0.6	1.0623767	1.0623787	2.00×10^{-6}	1.0635645	1.19×10^{-3}	1.0627739	3.97×10^{-4}
0.7	1.0968294	1.0969027	7.33×10^{-5}	1.0982259	1.40×10^{-3}	1.0973437	5.14×10^{-4}
0.8	1.1413421	1.1414969	1.55×10^{-4}	1.1429444	1.60×10^{-3}	1.1419794	6.37×10^{-4}
0.9	1.1968607	1.1971047	2.44×10^{-4}	1.1986647	1.80×10^{-3}	1.1976247	7.64×10^{-4}
1.0	1.2642411	1.2645798	3.39×10^{-4}	1.2662416	2.00×10^{-3}	1.2651337	8.93×10^{-4}

Although $T^{(3)}(t, y)$ can be approximated with error $O(h^3)$ by an expression of the form

$$f(t + \alpha_1, y + \delta_1 f(t + \alpha_2, y + \delta_2 f(t, y))),$$

involving four parameters, the algebra involved in the determination of $\alpha_1, \delta_1, \alpha_2$, and δ_2 is quite involved and will not be presented. In fact, the Runge–Kutta method of order three resulting from this expression is not generally used in practice. The most common Runge–Kutta method in use is of order four and, in difference form, is given by:

(5.39)

$$w_0 = \alpha,$$

$$k_1 = hf(t_i, w_i),$$

$$k_2 = hf\left(t_i + \frac{h}{2}, w_i + \frac{1}{2}k_1\right),$$

$$k_3 = hf\left(t_i + \frac{h}{2}, w_i + \frac{1}{2}k_2\right),$$

$$k_4 = hf(t_{i+1}, w_i + k_3),$$

$$w_{i+1} = w_i + \tfrac{1}{6}(k_1 + 2k_2 + 2k_3 + k_4),$$

for each $i = 0, 1, \ldots, N - 1$. This method has local truncation error $O(h^4)$, provided the solution $y(t)$ has five continuous derivatives. The reason for introducing the terminology k_1, k_2, k_3, k_4 in the method is to eliminate the need for successive nesting in the second variable of the function $f(t, y)$ (see Exercise 14).

Runge–Kutta (Order Four) Algorithm 5.2

To approximate the solution of the initial-value problem

$$y' = f(t, y), \qquad a \le t \le b, \quad y(a) = \alpha,$$

at $(N + 1)$ equally spaced numbers in the interval $[a, b]$:

INPUT endpoints a, b; integer N; initial condition α.

OUTPUT approximation w to y at the $(N + 1)$ values of t.

Step 1 Set $h = (b - a)/N$;
     $t = a$;
     $w = \alpha$;
     OUTPUT (t, w).

Step 2 For $i = 1, 2, \ldots, N$ do Steps 3–5.

 Step 3 Set $K_1 = hf(t, w)$;
      $K_2 = hf(t + h/2, w + K_1/2)$;
      $K_3 = hf(t + h/2, w + K_2/2)$;
      $K_4 = hf(t + h, w + K_3)$.

 Step 4 Set $w = w + (K_1 + 2K_2 + 2K_3 + K_4)/6$; (*Compute w_i.*)
      $t = a + ih$. (*Compute t_i.*)

 Step 5 OUTPUT (t, w).

Step 6 STOP.

EXAMPLE 3   Using the Runge–Kutta method of order four to obtain approximations to the solution of the initial-value problem

$$y' = -y + t + 1, \qquad 0 \le t \le 1, \quad y(0) = 1,$$

with $h = 0.1$, $N = 10$, and $t_i = 0.1i$ gives the results and errors listed in Table 5.5. ∎

TABLE 5.5

t_i	Exact values	Runge–Kutta values of order four	Error
0.0	1.0000000000	1.0000000000	
0.1	1.0048374180	1.0048375000	8.200×10^{-8}
0.2	1.0187307531	1.0187309014	1.483×10^{-7}
0.3	1.0408182207	1.0408184220	2.013×10^{-7}
0.4	1.0703200460	1.0703202889	2.429×10^{-7}
0.5	1.1065306597	1.1065309344	2.747×10^{-7}
0.6	1.1488116360	1.1488119344	2.984×10^{-7}
0.7	1.1965853038	1.1965856187	3.149×10^{-7}
0.8	1.2493289641	1.2493292897	3.256×10^{-7}
0.9	1.3065696597	1.3065699912	3.315×10^{-7}
1.0	1.3678794412	1.3678797744	3.332×10^{-7}

The main computational effort in applying the Runge–Kutta methods is the evaluation of f. In the second-order methods the local truncation error is $O(h^2)$, but the cost is two functional evaluations per step. The Runge–Kutta method of order four requires four evaluations per step, but the local truncation error is $O(h^4)$. Butcher [23] has established the relationship between the number of evaluations per step and the order of the local truncation error shown in Table 5.6. Consequently, the methods

TABLE 5.6

Evaluations per step	2	3	4	5	6	7	$n \ge 8$
Best possible local truncation error	$O(h^2)$	$O(h^3)$	$O(h^4)$	$O(h^4)$	$O(h^5)$	$O(h^6)$	$O(h^{n-2})$

of order less than five with smaller step size, are used in preference to the higher-order methods using a larger step size.

One measure to compare the lower-order Runge–Kutta methods is described as follows: Since the Runge–Kutta method of order four requires four evaluations per step, it should give more accurate answers than Euler's method with one-quarter the mesh size (where by mesh size we mean the difference between consecutive mesh points) if it is to be superior. Similarly, if the Runge–Kutta method of order four is to be superior to the second-order Runge–Kutta methods, it should give more accuracy with step size h than a second-order method with step size $\frac{1}{2}h$, because the fourth-order method requires twice as many evaluations per step. An illustration of the superiority of the Runge–Kutta fourth-order method by this measure is shown in the following example.

EXAMPLE 4   For the problem

$$y' = -y + 1, \quad 0 \le t \le 1, \quad y(0) = 0,$$

Euler's method with $h = 0.025$, the Modified Euler's method (5.37) with $h = 0.05$, and the Runge–Kutta fourth-order method (5.39) with $h = 0.1$ will be compared at the mesh points 0.1, 0.2, 0.3, 0.4, and 0.5. The results are given in Table 5.7. For this example the fourth-order method is clearly superior.   ■

TABLE 5.7

t	Euler's method $h = 0.025$	Modified Euler's method $h = 0.05$	Fourth-order Runge–Kutta method $h = 0.1$	Actual value
0.1	0.096312	0.095123	0.09516250	0.095162582
0.2	0.183348	0.181198	0.18126910	0.181269247
0.3	0.262001	0.259085	0.25918158	0.259181779
0.4	0.333079	0.329563	0.32967971	0.329679954
0.5	0.397312	0.393337	0.39346906	0.393469340

Exercise Set 5.4

1. Use the Modified Euler method to approximate the solution to each of the following initial-value problems.

   a) $y' = \left(\dfrac{y}{t}\right)^2 + \left(\dfrac{y}{t}\right), \quad 1 \le t \le 1.2, \quad y(1) = 1$ with $h = 0.1$.

   b) $y' = \sin t + e^{-t}, \quad 0 \le t \le 1, \quad y(0) = 0$ with $h = 0.5$.

   c) $y' = \dfrac{1}{t}(y^2 + y), \quad 1 \le t \le 3, \quad y(1) = -2$ with $h = 0.5$.

   d) $y' = -ty + \dfrac{4t}{y}, \quad 0 \le t \le 1, \quad y(0) = 1$ with $h = 0.25$.

2. Repeat Exercise 1 using the Midpoint method.

3. Repeat Exercise 1 using the Runge–Kutta fourth-order method.

4. Use the Midpoint method, the Modified Euler method, and Heun's method with $h = 0.1$ to approximate the solution to:

$$y' = 1 + t \sin(ty), \qquad 0 \le t \le 2, \quad y(0) = 0.$$

Compare your answers to those obtained in Exercise 4, Section 5.3 and Exercise 7, Section 5.2.

5. Given the initial-value problem

$$y' = \frac{2}{t} y + t^2 e^t, \quad 1 \le t \le 2, \quad y(1) = 0$$

with exact solution

$$y(t) = t^2 (e^t - e):$$

a) Use the Modified Euler method with $h = 0.1$ to approximate the solution and compare it with the actual values of y.

b) Use the answers generated in part (a) and linear interpolation to approximate y at the following values and compare them to the actual values of y.

i) $y(1.04)$ ii) $y(1.55)$ iii) $y(1.97)$

c) Use the Runge–Kutta method of order four with $h = 0.1$ to approximate the solution and compare it with the actual values of y.

d) Use the answers generated in part (c) and piecewise cubic Hermite interpolation to approximate y at the following values and compare them to the actual values of y.

i) $y(1.04)$ ii) $y(1.55)$ iii) $y(1.97)$

6. Given the initial-value problem

$$y' = \frac{1}{t^2} - \frac{y}{t} - y^2, \quad 1 \le t \le 2, \quad y(1) = -1$$

with exact solution $y(t) = -1/t$:

a) Use the Modified Euler method with $h = 0.05$ to approximate the solution and compare it with the actual values of y.

b) Use the answers generated in part (a) and linear interpolation to approximate the following values of y and compare them to the actual values.

i) $y(1.052)$ ii) $y(1.555)$ iii) $y(1.978)$

c) Use the Runge–Kutta method of order four with $h = 0.05$ to approximate the solution and compare with the actual values of y.

d) Use the answers generated in part (c) and piecewise cubic Hermite interpolation to approximate the following values of y and compare to the actual values.

i) $y(1.052)$ ii) $y(1.555)$ iii) $y(1.978)$

7. Use Euler's method with $h = 0.025$, Heun's method with $h = 0.05$, and the Runge–Kutta fourth-order method with $h = 0.1$ for the following problems.

a) $y' = \dfrac{1}{t^2} - \dfrac{y}{t}$, $1 \le t \le 2$, $y(1) = -1$.

b) $y' = t + y$, $0 \le t \le 2$, $y(0) = -1$.

c) $y' = -2y + 2t^2 + 2t$, $0 \le t \le 1$, $y(0) = 1$.

d) $y' = 1 - y$, $0 \le t \le 2$, $y(0) = 0$.

Which method appears to be superior for each problem?

8. Use the Midpoint method for the initial-value problem

$$y' = 50t^2 - 50y + 2t,   0 \le t \le 1,   y(0) = \tfrac{1}{3}$$

with (a) $h = 0.1$, (b) $h = 0.025$, and (c) $h = 0.01$. Find the exact solution. Are the results consistent with the local truncation error?

9. Use the Runge–Kutta method of order four for the initial-value problem

$$y' = 50t^2 - 50y + 2t,   0 \le t \le 1,   y(0) = \tfrac{1}{3}$$

with (a) $h = 0.1$, (b) $h = 0.025$, and (c) $h = 0.01$. Find the exact solution. Are the results consistent with the local truncation error?

10. Show that the Midpoint method, the Modified Euler method, and Heun's method give the same approximations to the initial-value problem

$$y' = -y + t + 1,   0 \le t \le 1,   y(0) = 1$$

for any choice of h. Why?

11. A liquid of low viscosity, such as water, flows from an inverted conical tank with circular orifice at the rate

$$\frac{dx}{dt} = -0.6 \pi r^2 \sqrt{2g} \frac{\sqrt{x}}{A(x)},$$

where r is the radius of the orifice, x is the height of the liquid level from the vertex of the cone, and $A(x)$ is the area of the cross section of the tank x units above the orifice. Suppose $r = 0.1$ feet, $g = 32$ feet/sec^2, and the tank has an initial water level of 8 feet and initial volume of $512\,\pi/3$ cubic feet.

a) Find $A(x)$.

b) Compute the water level after 10 seconds, using the Runge–Kutta fourth-order method with $h = 0.1$.

12. The irreversible chemical reaction in which two molecules of solid potassium dichromate ($K_2Cr_2O_7$), two molecules of water (H_2O), and three atoms of solid sulfur (S) combine to yield three molecules of the gas sulfur dioxide (SO_2), four molecules of solid potassium hydroxide (KOH), and two molecules of solid chromic oxide (Cr_2O_3) can be represented symbolically by the stoichiometric equation:

$$2K_2Cr_2O_7 + 2H_2O + 3S \longrightarrow 4KOH + 2Cr_2O_3 + 3SO_2.$$

If n_1 molecules of $K_2Cr_2O_7$, n_2 molecules of H_2O and n_3 molecules of S are originally available, the following differential equation describes the amount $x(t)$ of KOH after time t:

$$\frac{dx}{dt} = k\left(n_1 - \frac{x}{2}\right)^2 \left(n_2 - \frac{x}{2}\right)^2 \left(n_3 - \frac{3x}{4}\right)^3$$

where k is the velocity constant of the reaction. If $k = 6.22 \times 10^{-19}$, $n_1 = n_2 = 1000$, and $n_3 = 1500$, how many units of potassium hydroxide will have been formed after two seconds? Use the Runge–Kutta fourth-order method with $h = 0.1$.

13. Show that the difference method

$$w_0 = \alpha,$$

$$w_{i+1} = w_i + a_1 f(t_i, w_i) + a_2 f(t_i + \alpha_2, w_i + \delta_2 f(t_i, w_i))$$

for each $i = 0, 1, \ldots, N - 1$, cannot have local truncation error $O(h^3)$ for any choice of constants $a_1, a_2, \alpha_2,$ and δ_2.

14. The Runge–Kutta method of order four can be written in the form

$$w_0 = \alpha,$$

$$w_{i+1} = w_i + \frac{h}{6} f(t_i, w_i) + \frac{h}{3} f(t_i + \alpha_1 h, w_i + \delta_1 f(t_i, w_i))$$

$$+ \frac{h}{3} f(t_i + \alpha_2 h, w_i + \delta_2 hf(t_i + \gamma_2 h, w_i + \gamma_3 hf(t_i, w_i)))$$

$$+ \frac{h}{6} f(t_i + \alpha_3 h, w_i + \delta_3 hf(t_i + \gamma_4 h, w_i + \gamma_5 hf(t_i + \gamma_6 h, w_i + \gamma_7 hf(t_i, w_i)))).$$

Find the values of the constants

$$\alpha_1, \quad \alpha_2, \quad \alpha_3, \quad \delta_1, \quad \delta_2, \quad \delta_3, \quad \gamma_2, \quad \gamma_3, \quad \gamma_4, \quad \gamma_5, \quad \gamma_6, \quad \gamma_7.$$

5.5 Error Control and the Runge–Kutta–Fehlberg Method

An ideal difference-equation method

(5.40) $$w_{i+1} = w_i + h_i \phi(t_i, h_i, w_i), \qquad i = 0, 1, \ldots, N - 1,$$

for approximating the solution, $y(t)$, to the initial-value problem

(5.41) $$y' = f(t, y), \qquad a \le t \le b, \quad y(a) = \alpha,$$

would have the property that, whenever a tolerance $\varepsilon > 0$ was given, the minimal number of mesh points would be used to ensure that the **global error**, $|y(t_i) - w_i|$, does not exceed ε for any $i = 0, 1, 2, \ldots, N$. Having a minimal number of mesh points and also controlling the global error of a difference method is, not unsurprisingly, inconsistent with the points being equally spaced in the interval. In this section we examine a technique that can be used to control the error of a difference method in an efficient manner by the appropriate choice of mesh points.

Since we cannot generally determine the global error of a difference method, we work instead with the local truncation error of the method. In Section 5.10 we present a result, Theorem 5.18, that demonstrates, under suitable hypotheses on the initial-value problem and difference method, a connection between the local truncation error of the method and the global error. This result essentially states that a bound on the local truncation error produces a corresponding bound on the global error.

To illustrate the technique of estimating the local truncation error, suppose that a local-error estimate is desired for Euler's method,

(5.42)
$$w_0 = \alpha,$$
$$w_{i+1} = w_i + hf(t_i, w_i),$$

a method with local truncation error of order $O(h)$ given by

$$\tau_{i+1} = \frac{y(t_{i+1}) - y(t_i)}{h} - f(t_i, y(t_i)).$$

The Modified Euler method,

(5.43)
$$\tilde{w}_0 = \alpha,$$
$$\tilde{w}_{i+1} = \tilde{w}_i + \frac{h}{2}[f(t_i, \tilde{w}_i) + f(t_{i+1}, \tilde{w}_i + hf(t_i, \tilde{w}_i))],$$

has local truncation error $\tilde{\tau}_{i+1}$ of order $O(h^2)$.

If $w_i \approx y(t_i) \approx \tilde{w}_i$, then

$$y(t_{i+1}) - w_{i+1} = y(t_{i+1}) - w_i - hf(t_i, w_i)$$
$$\approx y(t_{i+1}) - y(t_i) - hf(t_i, y(t_i))$$
$$= h\tau_{i+1}.$$

So

$$\tau_{i+1} \approx \frac{1}{h}[y(t_{i+1}) - w_{i+1}]$$

$$= \frac{1}{h}[y(t_{i+1}) - \tilde{w}_{i+1}] + \frac{1}{h}(\tilde{w}_{i+1} - w_{i+1})$$

$$\approx \tilde{\tau}_{i+1} + \frac{1}{h}(\tilde{w}_{i+1} - w_{i+1}).$$

But τ_{i+1} is $O(h)$ while $\tilde{\tau}_{i+1} = O(h^2)$, so the most significant portion of τ_{i+1} must be attributed to $(\tilde{w}_{i+1} - w_{i+1})/h$. Consequently

(5.44)
$$\tau_{i+1} \approx \frac{1}{h}(\tilde{w}_{i+1} - w_{i+1})$$

can be used to approximate the local truncation error of Euler's method.

We will now determine how the estimation of the local truncation error of a difference method can be used to advantage in approximating the optimal step size to control the global error. Suppose two difference methods are available for the approximation of the solution to the initial-value problem (5.41) and that one of these methods,

(5.45)
$$w_0 = \alpha,$$
$$w_{i+1} = w_i + h_i \phi(t_i, h_i, w_i),$$

has local truncation error τ_{i+1} of order $O(h^n)$, while the other method

(5.46)
$$\tilde{w}_0 = \alpha,$$
$$\tilde{w}_{i+1} = \tilde{w}_i + h_i \tilde{\phi}(t_i, h_i, \tilde{w}_i),$$

has local truncation error $\tilde{\tau}_{i+1}$ of order $O(h^{n+1})$.

By using the same analysis as in Euler's method, we have

(5.47)
$$\tau_{i+1} \approx \frac{1}{h}(\tilde{w}_{i+1} - w_{i+1}).$$

However, τ_{i+1} is of order $O(h^n)$; so a constant k exists with

(5.48)
$$\tau_{i+1} \approx kh^n.$$

To estimate τ_{i+1} we use the two relationships (5.47) and (5.48):

(5.49)
$$kh^n \approx \frac{1}{h}(\tilde{w}_{i+1} - w_{i+1}).$$

The design of the procedure is to use this estimate to choose an appropriate step size. To accommodate this, let us consider the truncation error with h replaced by qh where q is positive but bounded above and away from zero. We denote this truncation error by $\tau_{i+1}(qh)$. By (5.48) and (5.49),

$$\tau_{i+1}(qh) \approx k(qh)^n = q^n(kh^n) \approx \frac{q^n}{h}(\tilde{w}_{i+1} - w_{i+1}).$$

To bound $\tau_{i+1}(qh)$ by ε, we choose q so that

$$\frac{q^n}{h}|\tilde{w}_{i+1} - w_{i+1}| \approx |\tau_{i+1}(qh)| \leq \varepsilon;$$

that is, so that

(5.50)
$$q \leq \left(\frac{\varepsilon h}{|\tilde{w}_{i+1} - w_{i+1}|}\right)^{1/n}.$$

One popular technique that uses inequality (5.50) for error control is called the **Runge–Kutta–Fehlberg method** presented by Fehlberg in 1970, [46]. This technique consists of using a Runge–Kutta method with local truncation error of order five,

$$\tilde{w}_{i+1} = w_i + \tfrac{16}{135}k_1 + \tfrac{6656}{12825}k_3 + \tfrac{28561}{56430}k_4 - \tfrac{9}{50}k_5 + \tfrac{2}{55}k_6,$$

to estimate the local error in a Runge–Kutta method of order four,

$$w_{i+1} = w_i + \tfrac{25}{216}k_1 + \tfrac{1408}{2565}k_3 + \tfrac{2197}{4104}k_4 - \tfrac{1}{5}k_5,$$

where $\qquad k_1 = hf(t_i, w_i),$

$$k_2 = hf\left(t_i + \frac{h}{4},\, w_i + \tfrac{1}{4}k_1\right),$$

$$k_3 = hf\left(t_i + \frac{3h}{8},\, w_i + \tfrac{3}{32}k_1 + \tfrac{9}{32}k_2\right),$$

$$k_4 = hf\left(t_i + \frac{12h}{13},\, w_i + \tfrac{1932}{2197}k_1 - \tfrac{7200}{2197}k_2 + \tfrac{7296}{2197}k_3\right),$$

$$k_5 = hf\left(t_i + h,\, w_i + \tfrac{439}{216}k_1 - 8k_2 + \tfrac{3680}{513}k_3 - \tfrac{845}{4104}k_4\right),$$

$$k_6 = hf\left(t_i + \frac{h}{2},\, w_i - \tfrac{8}{27}k_1 + 2k_2 - \tfrac{3544}{2565}k_3 + \tfrac{1859}{4104}k_4 - \tfrac{11}{40}k_5\right).$$

A clear advantage to this method is that only six evaluations of f are required per step, whereas arbitrary Runge–Kutta methods of order four and five used together would require ten evaluations of f per step, four evaluations for the fourth-order method and an additional six evaluations for the fifth-order method. (See Table 5.6, page 226.)

In the error-control theory, an initial value of h at the ith step was used to find the first values of w_{i+1} and $\tilde{w}_{i+1}$, which led to the determination of q for that step, and then the calculations were repeated. This procedure requires twice the number of functional evaluations per step as without the error control. In practice, the value of q to be used is chosen somewhat differently in order to make the increased cost worthwhile. The value of q determined at the ith step is used for two purposes:

1) To reject the initial choice of h at the ith step if necessary and repeat the calculations using qh, and
2) To predict an appropriate initial choice of h for the $(i + 1)$st step.

Because of the penalty (in terms of functional evaluations) that must be paid if many of the steps are repeated, q tends to be chosen rather conservatively; in fact, for the Runge–Kutta–Fehlberg method with $n = 4$, the usual choice is

$$q = \left(\frac{\varepsilon h}{2|\tilde{w}_{i+1} - w_{i+1}|}\right)^{1/4} = 0.84\left(\frac{\varepsilon h}{|\tilde{w}_{i+1} - w_{i+1}|}\right)^{1/4}.$$

The following algorithm uses the Runge–Kutta–Fehlberg method with error control. Step 9 is added to eliminate large modifications in step size. This is done to avoid spending too much time with very small step sizes in regions with irregularities in the derivatives of y, and to avoid large step sizes, which may result in skipping sensitive regions nearby. In some instances the step-size increase procedure is omitted from the algorithm and the step-size decrease procedure is modified so that it is incorporated only when needed to bring the error under control.

Runge–Kutta–Fehlberg Algorithm 5.3

To approximate the solution of the initial-value problem

$$y' = f(t, y), \qquad a \le t \le b, \quad y(a) = \alpha,$$

with local truncation error within a given tolerance:

INPUT endpoints a, b; initial condition α; tolerance TOL; maximum step size $hmax$; minimum step size $hmin$.

OUTPUT t, w, h where w approximates $y(t)$ and step size h was used or a message that minimum step size exceeded.

Step 1   Set $t = a$;

$w = \alpha$;

$h = hmax$;

OUTPUT (t, w).

Step 2   While $(t < b)$ do Steps 3–11.

Step 3   Set $K_1 = hf(t, w)$;

$K_2 = hf(t + \frac{1}{4}h, w + \frac{1}{4}K_1)$;

$K_3 = hf(t + \frac{3}{8}h, w + \frac{3}{32}K_1 + \frac{9}{32}K_2)$;

$K_4 = hf(t + \frac{12}{13}h, w + \frac{1932}{2197}K_1 - \frac{7200}{2197}K_2 + \frac{7296}{2197}K_3)$;

$K_5 = hf(t + h, w + \frac{439}{216}K_1 - 8K_2 + \frac{3680}{513}K_3 - \frac{845}{4104}K_4)$;

$K_6 = hf(t + \frac{1}{2}h, w - \frac{8}{27}K_1 + 2K_2 - \frac{3544}{2565}K_3 + \frac{1859}{4104}K_4 - \frac{11}{40}K_5)$.

Step 4   Set $R = |\frac{1}{360}K_1 - \frac{128}{4275}K_3 - \frac{2197}{75240}K_4 + \frac{1}{50}K_5 + \frac{2}{55}K_6|/h$.
$(R = |\tilde{w}_{i+1} - w_{i+1}|/h.)$

Step 5   Set $\delta = 0.84(TOL/R)^{1/4}$.

Step 6   If $R \le TOL$ then do Steps 7 and 8.

Step 7   Set $t = t + h$;   (*Approximation accepted.*)
$w = w + \frac{25}{216}K_1 + \frac{1408}{2565}K_3 + \frac{2197}{4104}K_4 - \frac{1}{5}K_5$.

Step 8   OUTPUT (t, w, h).

Step 9   If $\delta \le 0.1$ then set $h = 0.1h$
else if $\delta \ge 4$ then set $h = 4h$
else set $h = \delta h$.   (*Calculate new h.*)

Step 10   If $h > hmax$ then set $h = hmax$.

Step 11   If $h < hmin$ then
OUTPUT ('minimum h exceeded');
(*Procedure completed unsuccessfully.*)
STOP.

Step 12   (*The procedure is complete.*)
STOP.

EXAMPLE 1   Algorithm 5.3 will be used to approximate the solution to the initial-value problem

$$y' = -y + t + 1, \qquad 0 \le t \le 1, \quad y(0) = 1.$$

The input included tolerance $TOL = 5 \times 10^{-5}$, a maximum step size of $hmax = 0.1$ and a minimum step size of $hmin = 0.02$. Initially, h was set to $(TOL)^{1/4}$, and if a maximum step size had not been imposed, h would have been increased beyond 0.1. Six digits of precision were used for all computations, and the results are shown in Table 5.8.   ■

TABLE 5.8

i	t_i	h_i	w_i	R_i	$y(t_i)$	$\lvert y(t_i) - w_i\rvert$
1	0.08408963	0.08408963	1.003437	9.674×10^{-8}	1.003437	0
2	0.1840896	0.1	1.015948	2.398×10^{-7}	1.015950	2×10^{-6}
3	0.2840896	0.1	1.036785	1.420×10^{-7}	1.036787	2×10^{-6}
4	0.3840897	0.1	1.065155	1.863×10^{-7}	1.065159	4×10^{-6}
5	0.4840897	0.1	1.100341	1.257×10^{-7}	1.100347	6×10^{-6}
6	0.5840897	0.1	1.141696	1.490×10^{-7}	1.141702	6×10^{-6}
7	0.6840897	0.1	1.188632	2.002×10^{-7}	1.188638	6×10^{-6}
8	0.7840898	0.1	1.240616	1.839×10^{-7}	1.240624	8×10^{-6}
9	0.8840898	0.1	1.297170	1.350×10^{-7}	1.297179	9×10^{-6}
10	0.9840898	0.1	1.357858	8.615×10^{-8}	1.357868	1×10^{-5}

Exercise Set 5.5

1. Use the Runge–Kutta–Fehlberg Algorithm 5.3 with $TOL = 10^{-4}$ to approximate the solution to the following initial-value problems.

   a) $y' = \left(\dfrac{y}{t}\right)^2 + \left(\dfrac{y}{t}\right), \quad 1 \le t \le 1.2, \quad y(1) = 1$ with $hmax = 0.05.$

   b) $y' = \sin t + e^{-t}, \quad 0 \le t \le 1, \quad y(0) = 0$ with $hmax = 0.25.$

   c) $y' = \dfrac{1}{t}(y^2 + y), \quad 1 \le t \le 3, \quad y(1) = -2$ with $hmax = 0.5.$

   d) $y' = t^2, \quad 0 \le t \le 2, \quad y(0) = 0$ with $hmax = 0.5.$

2. Use the Runge–Kutta–Fehlberg Algorithm 5.3 to approximate the solutions to the following initial-value problems. Compare the results to the actual values.

   a) $y' = 1 - y, \quad 0 \le t \le 1, \quad y(0) = 0;$ use $TOL = 10^{-6}$ and $hmax = 0.2.$

   b) $y' = -y + t + 1, \quad 0 \le t \le 5, \quad y(0) = 2;$ use $TOL = 10^{-4}$ and $hmax = 0.2.$

   c) $y' = \dfrac{2}{t}y + t^2 e^t, \quad 1 \le t \le 2, \quad y(1) = 0;$ use $TOL = 10^{-4}$ and $hmax = 0.2.$

   d) $y' = 1 + y^2, \quad 0 \le t \le \pi/4, \quad y(0) = 0;$ use $TOL = 10^{-4}$ and $hmax = 0.2.$

3. Use the Runge–Kutta–Fehlberg Algorithm 5.3 to approximate the solutions to the following initial-value problems.

   a) $y' = 2|t - 2|y$, $0 \le t \le 3$, $y(0) = e^{-4}$; use $TOL = 10^{-4}$ and $hmax = 0.2$.

   b) $y' = 1 + t \sin(ty)$, $0 \le t \le 2$, $y(0) = 0$; use $TOL = 10^{-4}$ and $hmax = 0.2$.

   c) $y' = 50t^2 - 50y + 2t$, $0 \le t \le 1$, $y(0) = \frac{1}{3}$; use $TOL = 10^{-3}$ and $hmax = 0.1$.

   d) $y' = -2y + 2t^2 + 2t$, $0 \le t \le 1$, $y(0) = 1$; use $TOL = 10^{-6}$ and $hmax = 0.2$.

4. Modify the Runge–Kutta–Fehlberg Algorithm 5.3 so that the calculations stop with the final value of t at b.

5. In the theory of the spread of contagious disease (see Bailey [3] or [4]), a relatively elementary differential equation can be used to predict the number of infective individuals in the population at any time, provided appropriate simplification assumptions are made. In particular, let us assume that all individuals in a fixed population have an equally likely chance of being infected and once infected remain in that state. If we let $x(t)$ denote the number of susceptible individuals at time t and $y(t)$ denote the number of infectives, it is reasonable to assume that the rate at which the number of infectives changes is proportional to the product of $x(t)$ and $y(t)$ since the rate depends on both the number of infectives and the number of susceptibles present at that time. If the population is large enough to assume that $x(t)$ and $y(t)$ are continuous variables, the problem can be expressed as

$$\frac{dy}{dt}(t) = kx(t)\, y(t),$$

where k is a constant and $x(t) + y(t) = m$, the total population. This equation can be rewritten involving only $y(t)$ as

$$\frac{dy}{dt}(t) = ky(t)(m - y(t)).$$

   a) Assuming that $m = 100{,}000$, $y(0) = 1000$, $k = 2 \times 10^{-6}$, and that time is measured in days, find an approximation to the number of infective individuals at the end of 30 days by using Algorithm 5.3 as modified in Exercise 4 with $TOL = 1$.

   b) The differential equation in part (a) is called a **Bernoulli equation** and can be transformed into a linear differential equation in $z(t)$ by letting $z(t) = (y(t))^{-1}$. Use this technique to find the exact solution to the equation, under the same assumptions as in part (a), and compare the true value of $y(t)$ to the approximation given there. What is $\lim_{t \to \infty} y(t)$? Does this agree with your intuition?

6. In the previous exercise, all infected individuals remained in the population to spread the disease. A more realistic proposal is to introduce a third variable $z(t)$ to represent the number of individuals who are removed from the affected population at a given time t, by isolation, recovery and consequent immunity, or death. This quite naturally complicates the problem, but it can be shown (see Bailey [4]) that an approximate solution can be given in the form

$$x(t) = x(0)e^{-(k_1/k_2)z(t)} \quad \text{and} \quad y(t) = m - x(t) - z(t),$$

where k_1 is the infective rate, k_2 is the removal rate, and $z(t)$ is determined from the differential equation

$$\frac{dz}{dt}(t) = k_2(m - z(t) - x(0)e^{-(k_1/k_2)z(t)}).$$

The authors are not aware of any technique for solving this problem directly, so a numerical procedure must be applied. Use Algorithm 5.3 to find an approximation to $z(30)$, $y(30)$, and $x(30)$ assuming that $m = 100{,}000$, $x(0) = 99{,}000$, $k_1 = 2 \times 10^{-6}$, and $k_2 = 10^{-4}$, and taking $TOL = 1$.

7. a) Construct an algorithm similar to Algorithm 5.3 using Euler's method and Modified Euler's method.

    b) Repeat Exercise 1 using the algorithm constructed in part (a), with $TOL = 0.1$.

5.6 Multistep Methods

The methods discussed previously in the chapter are called **one-step methods** since the approximation for the mesh point t_{i+1} involves information from only one of the previous mesh points, t_i. Although these methods generally use functional evaluation information at points between t_i and t_{i+1}, they do not retain that information for direct use in future approximations. All the information used by these methods is consequently obtained within the interval over which the solution is being approximated.

Since the approximate solution is available at each of the mesh points $t_0, t_1, \ldots,$ t_i before the approximation at t_{i+1} is obtained, and because the error $|w_j - y(t_j)|$ tends to increase with j, it seems reasonable to develop methods that use this more accurate previous data when approximating the solution at t_{i+1}.

Methods using the approximation at more than one previous mesh point to determine the approximation at the next point are called **multistep** methods. The precise definition of these methods follows, together with the definition of the two types of multistep methods.

DEFINITION 5.12   A **multistep method** for solving the initial-value problem

(5.51) $y' = f(t, y), \qquad a \le t \le b, \quad y(a) = \alpha,$

is one whose difference equation for finding the approximation w_{i+1} at the mesh point t_{i+1} can be represented by the following equation, where m is an integer greater than 1:

(5.52) $w_{i+1} = a_{m-1}w_i + a_{m-2}w_{i-1} + \cdots + a_0 w_{i+1-m}$
$$+ h[b_m f(t_{i+1}, w_{i+1}) + b_{m-1} f(t_i, w_i)$$
$$+ \cdots + b_0 f(t_{i+1-m}, w_{i+1-m})]$$

for $i = m - 1, m, \ldots, N - 1$, where the starting values

$$w_0 = \alpha, \; w_1 = \alpha_1, \; w_2 = \alpha_2, \; \ldots, \; w_{m-1} = \alpha_{m-1}$$

are specified and $h = (b - a)/N$.

When $b_m = 0$, the method is called an **explicit** or **open method** and Eq. (5.52) gives w_{i+1} explicitly in terms of previously determined values. When $b_m \neq 0$, the method is called an **implicit** or **closed method** since w_{i+1} occurs on both sides of Eq. (5.52) and is determined only in an implicit manner.

EXAMPLE 1   The equations

(5.53)
$$w_0 = \alpha, \quad w_1 = \alpha_1, \quad w_2 = \alpha_2, \quad w_3 = \alpha_3,$$
$$w_{i+1} = w_i + \frac{h}{24}[55f(t_i, w_i) - 59f(t_{i-1}, w_{i-1})$$
$$+ 37f(t_{i-2}, w_{i-2}) - 9f(t_{i-3}, w_{i-3})]$$

for each $i = 3, 4, \ldots, N - 1$ define an explicit four-step method known as the **fourth-order Adams–Bashforth technique**. The equations

(5.54)
$$w_0 = \alpha, \quad w_1 = \alpha_1, \quad w_2 = \alpha_2,$$
$$w_{i+1} = w_i + \frac{h}{24}[9f(t_{i+1}, w_{i+1}) + 19f(t_i, w_i)$$
$$- 5f(t_{i-1}, w_{i-1}) + f(t_{i-2}, w_{i-2})]$$

for each $i = 2, 3, \ldots, N - 1$ define an implicit three-step method known as the **fourth-order Adams–Moulton technique**. ∎

The starting values in either (5.53) or (5.54) must be specified, generally by assuming $w_0 = \alpha$, and generating the remaining values by either a Runge–Kutta method or some other one-step technique.

To apply an implicit method such as (5.54) directly, we must solve the implicit equation for w_{i+1}. It is not clear that this can be done in general, or that a unique solution for w_{i+1} will always be obtained, as we will see later in this section.

To begin the derivation of the multistep methods, note that the solution to the initial-value problem (5.51), if integrated over the interval $[t_i, t_{i+1}]$, has the property that

$$y(t_{i+1}) - y(t_i) = \int_{t_i}^{t_{i+1}} y'(t)\,dt = \int_{t_i}^{t_{i+1}} f(t, y(t))\,dt.$$

Consequently,

(5.55)
$$y(t_{i+1}) = y(t_i) + \int_{t_i}^{t_{i+1}} f(t, y(t))\,dt.$$

Since we cannot integrate $f(t, y(t))$ without knowing $y(t)$, the solution to the problem, we instead integrate an interpolating polynomial, P, that is determined by some of the previously obtained data points (t_0, w_0), (t_1, w_1), $\ldots$, (t_i, w_i). When we assume, in addition, that $y(t_i) \approx w_i$, Eq. (5.55) becomes:

$$(5.56) \qquad y(t_{i+1}) \approx w_i + \int_{t_i}^{t_{i+1}} P(t)\, dt.$$

Although any form of the interpolating polynomial can be used for the derivation, it is most convenient to use the Newton backward-difference formula (3.16) given on page 105. This form of the polynomial gives the greatest prominence to the most recently computed approximations. To derive an Adams–Bashforth explicit m-step technique, we form the backward-difference polynomial P_{m-1} through $(t_i, y(t_i))$, $(t_{i-1}, y(t_{i-1}))$, $\ldots$, $(t_{i+1-m}, y(t_{i+1-m}))$.

Since P_{m-1} is an interpolatory polynomial of degree $m-1$, some number ξ_i in (t_{i+1-m}, t_i) exists with

$$f(t, y(t)) = P_{m-1}(t) + \frac{f^{(m)}(\xi_i, y(\xi_i))}{m!}(t - t_i)(t - t_{i-1}) \cdots (t - t_{i+1-m}).$$

Introducing the variable substitution $t = t_i + sh$, $dt = h\,ds$ into $P_{m-1}(t)$ and the error term, implies that

$$\int_{t_i}^{t_{i+1}} f(t, y(t))\, dt = \int_{t_i}^{t_{i+1}} \sum_{k=0}^{m-1} (-1)^k \binom{-s}{k} \nabla^k f(t_i, y(t_i))\, dt$$

$$+ \int_{t_i}^{t_{i+1}} \frac{f^{(m)}(\xi_i, y(\xi_i))}{m!}(t - t_i)(t - t_{i-1}) \cdots (t - t_{i+1-m})\, dt$$

$$= \sum_{k=0}^{m-1} \nabla^k f(t_i, y(t_i)) h(-1)^k \int_0^1 \binom{-s}{k}\, ds$$

$$+ \frac{h^{m+1}}{m!} \int_0^1 s(s+1) \cdots (s+m-1) f^{(m)}(\xi_i, y(\xi_i))\, ds.$$

The integrals $(-1)^k \int_0^1 \binom{-s}{k}\, ds$ for various values of k are easily evaluated and listed in Table 5.9. For example, when $k = 3$,

$$(-1)^3 \int_0^1 \binom{-s}{3}\, ds = -\int_0^1 \frac{(-s)(-s-1)(-s-2)}{1 \cdot 2 \cdot 3}\, ds$$

$$= \frac{1}{6} \int_0^1 (s^3 + 3s^2 + 2s)\, ds$$

$$= \frac{1}{6}\left[\frac{s^4}{4} + s^3 + s^2\right]_0^1 = \frac{1}{6}\left(\frac{9}{4}\right) = \frac{3}{8}.$$

TABLE 5.9

k	0	1	2	3	4	5
$(-1)^k \int_0^1 \binom{-s}{k}\, ds$	1	$\dfrac{1}{2}$	$\dfrac{5}{12}$	$\dfrac{3}{8}$	$\dfrac{251}{720}$	$\dfrac{95}{288}$

As a consequence

$$(5.57) \qquad \int_{t_i}^{t_{i+1}} f(t, y(t))\, dt = h\left[f(t_i, y(t_i)) + \frac{1}{2}\nabla f(t_i, y(t_i)) + \frac{5}{12}\nabla^2 f(t_i, y(t_i)) + \cdots \right]$$
$$+ \frac{h^{m+1}}{m!}\int_0^1 s(s+1)\cdots(s+m-1)f^{(m)}(\xi_i, y(\xi_i))\, ds.$$

Since $s(s+1)\cdots(s+m-1)$ does not change sign on $[0, 1]$, the Weighted Mean Value Theorem for Integrals can be used to deduce that for some number μ_i, $t_i < \mu_i < t_{i+1}$, the error term in Eq. (5.57) becomes

$$\frac{h^{m+1}}{m!}\int_0^1 s(s+1)\cdots(s+m-1)f^{(m)}(\xi_i, y(\xi_i))\, ds = \frac{h^{m+1}f^{(m)}(\mu_i, y(\mu_i))}{m!}\int_0^1 s(s+1)\cdots(s+m+1)\, ds$$

or

$$(5.58) \qquad h^{m+1}f^{(m)}(\mu_i, y(\mu_i))(-1)^m \int_0^1 \binom{-s}{m}\, ds.$$

Since $y(t_{i+1}) - y(t_i) = \displaystyle\int_{t_i}^{t_{i+1}} f(t, y(t))\, dt$, Eq. (5.57) can be written as

$$(5.59) \quad y(t_{i+1}) = y(t_i) + h\left[f(t_i, y(t_i)) + \frac{1}{2}\nabla f(t_i, y(t_i)) + \frac{5}{12}\nabla^2 f(t_i, y(t_i)) + \cdots \right]$$
$$+ h^{m+1}f^{(m)}(\mu_i, y(\mu_i))(-1)^m \int_0^1 \binom{-s}{m}\, ds.$$

EXAMPLE 2   To determine the three-step Adams–Bashforth technique, consider Eq. (5.59) with $m = 3$:

$$y(t_{i+1}) \approx y(t_i) + h\left[f(t_i, y(t_i)) + \frac{1}{2}\nabla f(t_i, y(t_i)) + \frac{5}{12}\nabla^2 f(t_i, y(t_i)) \right]$$
$$= y(t_i) + h\left\{ f(t_i, y(t_i)) + \frac{1}{2}[f(t_i, y(t_i)) - f(t_{i-1}, y(t_{i-1}))] \right.$$
$$\left. + \frac{5}{12}[f(t_i, y(t_i)) - 2f(t_{i-1}, y(t_{i-1})) + f(t_{i-2}, y(t_{i-2}))] \right\}$$
$$= y(t_i) + \frac{h}{12}[23f(t_i, y(t_i)) - 16f(t_{i-1}, y(t_{i-1})) + 5f(t_{i-2}, y(t_{i-2}))].$$

The three-step Adams–Bashforth method is consequently

$$w_0 = \alpha, \quad w_1 = \alpha_1, \quad w_2 = \alpha_2$$

$$w_{i+1} = w_i + \frac{h}{12}[23f(t_i, w_i) - 16f(t_{i-1}, w_{i-1}) + 5f(t_{i-2}, w_{i-2})]$$

for $i = 2, 3, \ldots, n-1$.   ∎

Multistep methods can also be derived by using Taylor series. An example of the procedure involved is considered in Exercise 6. A derivation using a Lagrange interpolating polynomial is discussed in Exercise 7(a).

The local truncation error for multistep methods can be defined analogously to that of one-step methods. This error provides a measure of how the solution to the differential equation fails to solve the difference equation.

DEFINITION 5.13 If $y(t)$ is the solution to the initial-value problem

(5.60) $$y' = f(t, y), \qquad a \leq t \leq b, \quad y(a) = \alpha,$$

and

(5.61) $$w_{i+1} = a_m w_i + a_{m-1} w_{i-1} + \cdots + a_0 w_{i-m}$$
$$+ h[b_{m+1} f(t_{i+1}, w_{i+1}) + b_m f(t_i, w_i) + \cdots + b_0 f(t_{i-m}, w_{i-m})]$$

is the $(i+1)$st step in a multistep method, the **local truncation error** at this step, τ_{i+1}, is

(5.62) $$\tau_{i+1} = \frac{y(t_{i+1}) - a_m y(t_i) - \cdots - a_0 y(t_{i-m})}{h}$$
$$- [b_{m+1} f(t_{i+1}, y(t_{i+1})) + \cdots + b_0 f(t_{i-m}, y(t_{i-m}))]$$

for each $i = m, m+1, \ldots, N-1$.

EXAMPLE 3 To determine the local truncation error for the three-step Adams–Bashforth method derived in Example 2, consider the form of the error given in Eq. (5.58):

$$h^4 f^{(3)}(\mu_i, y(\mu_i))(-1)^3 \int_0^1 \binom{-s}{3} ds = \frac{3h^4}{8} f^{(3)}(\mu_i, y(\mu_i)).$$

Using the fact that $f^{(3)}(\mu_i, y(\mu_i)) = y^{(4)}(\mu_i)$ and the difference equation derived in Example 2,

$$\tau_{i+1} = \frac{y(t_{i+1}) - y(t_i)}{h} - \frac{1}{12}[23f(t_i, y(t_i)) - 16f(t_{i-1}, y(t_{i-1})) + 5f(t_{i-2}, y(t_{i-2}))]$$

$$= \frac{1}{h}\left[\frac{3h^4}{8} f^{(3)}(\mu_i, y(\mu_i))\right] = \frac{3h^3}{8} y^{(4)}(\mu_i). \qquad ■$$

Some of the explicit multistep methods together with their required starting values and local truncation errors are given as follows. The derivation of these techniques is similar to the procedure in Examples 2 and 3.

Adams–Bashforth Two-Step Method

(5.63) $$w_0 = \alpha, \quad w_1 = \alpha_1$$
$$w_{i+1} = w_i + \frac{h}{2}[3f(t_i, w_i) - f(t_{i-1}, w_{i-1})],$$

where $i = 1, 2, \ldots, N - 1$; local truncation error $\tau_{i+1} = \frac{5}{12} y'''(\mu_i) h^2$.

Adams–Bashforth Three-Step Method

(5.64)
$$w_0 = \alpha, \quad w_1 = \alpha_1, \quad w_2 = \alpha_2,$$
$$w_{i+1} = w_i + \frac{h}{12} [23 f(t_i, w_i) - 16 f(t_{i-1}, w_{i-1}) + 5 f(t_{i-2}, w_{i-2})]$$

where $i = 2, 3, \ldots, N - 1$; local truncation error $\tau_{i+1} = \frac{3}{8} y^{(4)}(\mu_i) h^3$.

Adams–Bashforth Four-Step Method

(5.65)
$$w_0 = \alpha, \quad w_1 = \alpha_1, \quad w_2 = \alpha_2, \quad w_3 = \alpha_3,$$
$$w_{i+1} = w_i + \frac{h}{24} [55 f(t_i, w_i) - 59 f(t_{i-1}, w_{i-1}) + 37 f(t_{i-2}, w_{i-2})$$
$$- 9 f(t_{i-3}, w_{i-3})]$$

where $i = 3, 4, \ldots, N - 1$; local truncation error $\tau_{i+1} = \frac{251}{720} y^{(5)}(\mu_i) h^4$.

Adams–Bashforth Five-Step Method

(5.66)
$$w_0 = \alpha, \quad w_1 = \alpha_1, \quad w_2 = \alpha_2, \quad w_3 = \alpha_3, \quad w_4 = \alpha_4,$$
$$w_{i+1} = w_i + \frac{h}{720} [1901 f(t_i, w_i) - 2774 f(t_{i-1}, w_{i-1})$$
$$+ 2616 f(t_{i-2}, w_{i-2}) - 1274 f(t_{i-3}, w_{i-3}) + 251 f(t_{i-4}, w_{i-4})]$$

where $i = 4, 5, \ldots, N - 1$; local truncation error $\tau_{i+1} = \frac{95}{288} y^{(6)}(\mu_i) h^5$.

Implicit methods are derived by using $(t_{i+1}, f(t_{i+1}, y(t_{i+1})))$ as an additional interpolation node in the approximation of the integral

$$\int_{t_i}^{t_{i+1}} f(t, y(t)) \, dt.$$

Some of the more common implicit methods are listed as follows.

Adams–Moulton Two-Step Method

(5.67)
$$w_0 = \alpha, \quad w_1 = \alpha_1,$$
$$w_{i+1} = w_i + \frac{h}{12} [5 f(t_{i+1}, w_{i+1}) + 8 f(t_i, w_i) - f(t_{i-1}, w_{i-1})]$$

where $i = 1, 2, \ldots, N - 1$; local truncation error $\tau_{i+1} = -\frac{1}{24} y^{(4)}(\mu_i) h^3$.

Adams–Moulton Three-Step Method

(5.68)
$$w_0 = \alpha, \quad w_1 = \alpha_1, \quad w_2 = \alpha_2,$$
$$w_{i+1} = w_i + \frac{h}{24} [9 f(t_{i+1}, w_{i+1}) + 19 f(t_i, w_i) - 5 f(t_{i-1}, w_{i-1}) + f(t_{i-2}, w_{i-2})],$$

where $i = 2, 3, \ldots, N - 1$; local truncation error $\tau_{i+1} = -\frac{19}{720} y^{(5)}(\mu_i) h^4$.

Adams–Moulton Four-Step Method

$$(5.69) \quad w_0 = \alpha, \quad w_1 = \alpha_1, \quad w_2 = \alpha_2, \quad w_3 = \alpha_3,$$

$$w_{i+1} = w_i + \frac{h}{720} [251 f(t_{i+1}, w_{i+1}) + 646 f(t_i, w_i)$$

$$- 264 f(t_{i-1}, w_{i-1}) + 106 f(t_{i-2}, w_{i-2}) - 19 f(t_{i-3}, w_{i-3})]$$

where $i = 3, 4, \ldots, N - 1$; local truncation error $\tau_{i+1} = -\frac{3}{160} y^{(6)}(\mu_i) h^5$.

It is interesting to compare an m-step Adams–Bashforth explicit method to an $(m - 1)$-step Adams–Moulton implicit method. Both require m evaluations of f per step, and both have the terms $y^{(m+1)}(\mu_i) h^m$ in their local truncation errors. In general, the coefficients of the terms involving f and in the local truncation error are smaller for the Adams–Moulton methods. This leads to greater stability for the implicit methods and smaller rounding errors. This is illustrated in the next example.

EXAMPLE 4   The initial-value problem

$$y' = -y + t + 1, \quad 0 \le t \le 1, \quad y(0) = 1,$$

will be considered, the approximations being given by the Adams–Bashforth four-step method, (5.65), and the Adams–Moulton three-step method, (5.68), both using $h = 0.1$ and the values from the exact solution $y(t) = e^{-t} + t$ as their starting values.

The Adams–Bashforth method has the difference equation

$$w_{i+1} = w_i + \frac{h}{24} [55 f(t_i, w_i) - 59 f(t_{i-1}, w_{i-1}) + 37 f(t_{i-2}, w_{i-2})$$

$$- 9 f(t_{i-3}, w_{i-3})], \quad \text{for } i = 3, 4, \ldots, 9,$$

which, when simplified using $f(t, y) = -y + t + 1$, $h = 0.1$, and $t_i = 0.1i$, becomes

$$w_{i+1} = \tfrac{1}{24} [18.5 w_i + 5.9 w_{i-1} - 3.7 w_{i-2} + 0.9 w_{i-3} + 0.24i + 2.52]$$

for $i = 3, 4, \ldots, 9$.

The Adams–Moulton method has the difference equation

$$w_{i+1} = w_i + \frac{h}{24} [9 f(t_{i+1}, w_{i+1}) + 19 f(t_i, w_i) - 5 f(t_{i-1}, w_{i-1})$$

$$+ f(t_{i-2}, w_{i-2})], \quad \text{for } i = 2, 3, \ldots, 9,$$

which reduces to

$$w_{i+1} = \tfrac{1}{24} [-0.9 w_{i+1} + 22.1 w_i + 0.5 w_{i-1} - 0.1 w_{i-2} + 0.24i + 2.52],$$

for $i = 2, 3, \ldots, 9$. To use this method we can solve explicitly for w_{i+1}, which gives

$$w_{i+1} = \frac{1}{24.9} [22.1 w_i + 0.5 w_{i-1} - 0.1 w_{i-2} + 0.24i + 2.52] \quad \text{for } i = 2, 3, \ldots, 9.$$

Using the exact values from $y(t) = e^{-t} + t$ for α, α_1, and α_2 in the Adams–Bashforth case and for α and α_1 in the Adams–Moulton case gives the results in Table 5.10. ∎

TABLE 5.10

t_i	Adams–Bashforth w_i	Error	Adams–Moulton w_i	Error
0.3	starting value	—	1.0408180061	2.146×10^{-7}
0.4	1.0703229200	2.874×10^{-6}	1.0703196614	3.846×10^{-7}
0.5	1.1065354755	4.816×10^{-6}	1.1065301384	5.213×10^{-7}
0.6	1.1488184077	6.772×10^{-6}	1.1488110076	6.285×10^{-7}
0.7	1.1965933934	8.090×10^{-6}	1.1965845932	7.106×10^{-7}
0.8	1.2493381564	9.192×10^{-6}	1.2493281927	7.714×10^{-7}
0.9	1.3065796139	9.954×10^{-6}	1.3065688456	8.141×10^{-7}
1.0	1.3678899580	1.052×10^{-5}	1.3678785994	8.418×10^{-7}

In Example 4 the Adams–Moulton method gave considerably better results than the Adams–Bashforth method of the same order. Although this is generally the case, the implicit methods have the inherent weakness of having to first convert the method algebraically to an explicit representation for w_{i+1}. That this procedure can become difficult, if not impossible, can be seen by considering the rather elementary initial-value problem

$$y' = e^y, \quad 0 \le t \le 0.25, \quad y(0) = 1.$$

Since $f(t, y) = e^y$, the three-step Adams–Moulton method has

$$w_{i+1} = w_i + \frac{h}{24}[9e^{w_{i+1}} + 19e^{w_i} - 5e^{w_{i-1}} + e^{w_{i-2}}]$$

as its difference equation and this cannot be solved exactly for w_{i+1}.

In practice, implicit multistep methods are not used as just described; rather, they are used to improve approximations obtained by explicit methods. The combination of an explicit and implicit technique is called a **predictor-corrector method**.

If a fourth-order method for solving an initial-value problem is needed, the first step is to calculate the starting values w_0, w_1, w_2, and w_3 for the four-step Adams–Bashforth method by using the Runge–Kutta method of order four, given in Algorithm 5.2. The next step is to calculate an approximation, $w_4^{(0)}$, to $y(t_4)$ using the Adams–Bashforth method

$$w_4^{(0)} = w_3 + \frac{h}{24}[55f(t_3, w_3) - 59f(t_2, w_2) + 37f(t_1, w_1) - 9f(t_0, w_0)].$$

This approximation is improved by use of the three-step Adams–Moulton method

$$w_4^{(1)} = w_3 + \frac{h}{24}[9f(t_4, w_4^{(0)}) + 19f(t_3, w_3) - 5f(t_2, w_2) + f(t_1, w_1)].$$

The value $w_4^{(1)}$ is then used as the approximation to $y(t_4)$ and the technique of using the Adams–Bashforth method as a predictor and the Adams–Moulton method as a corrector repeated to find $w_5^{(0)}$ and $w_5^{(1)}$, the initial and final approximations to $y(t_5)$, etc.

In theory, improved approximations to $y(t_{i+1})$ can be obtained by iterating the Adams–Moulton formula

$$w_{i+1}^{(k+1)} = w_i + \frac{h}{24}[9f(t_{i+1}, w_{i+1}^{(k)}) + 19f(t_i, w_i) - 5f(t_{i-1}, w_{i-1}) + f(t_{i-2}, w_{i-2})].$$

In practice, since $\{w_{i+1}^{(k+1)}\}$ converges to the actual approximation given by the implicit formula rather than to the solution $y(t_{i+1})$, it is more efficient to use a reduction in the step size if improved accuracy is needed.

The following algorithm is based on the fourth-order Adams–Bashforth method as predictor and one iteration of the Adams–Moulton method as corrector, with the starting values obtained via the fourth-order Runge–Kutta method.

Adams Fourth-Order Predictor-Corrector Algorithm 5.4

To approximate the solution of the initial-value problem

$$y' = f(t, y), \qquad a \le t \le b, \quad y(a) = \alpha,$$

at $(N + 1)$ equally spaced numbers in the interval $[a, b]$:

INPUT endpoints a, b; integer N; initial condition α.

OUTPUT approximation w to y at the $(N + 1)$ values of t.

Step 1 Set $h = (b - a)/N$;
   $t_0 = a$;
   $w_0 = \alpha$;
   OUTPUT (t_0, w_0).

Step 2 For $i = 1, 2, 3$ do Steps 3–5. (*Compute starting values using Runge–Kutta method.*)

Step 3 Set $K_1 = hf(t_{i-1}, w_{i-1})$;
   $K_2 = hf(t_{i-1} + h/2, w_{i-1} + K_1/2)$;
   $K_3 = hf(t_{i-1} + h/2, w_{i-1} + K_2/2)$;
   $K_4 = hf(t_{i-1} + h, w_{i-1} + K_3)$.

Step 4 Set $w_i = w_{i-1} + (K_1 + 2K_2 + 2K_3 + K_4)/6$;
   $t_i = a + ih$.

Step 5 OUTPUT (t_i, w_i).

Step 6 For $i = 4, \ldots, N$ do Steps 7–10.

Step 7 Set $t = a + ih$;
   $w = w_3 + h[55f(t_3, w_3) - 59f(t_2, w_2) + 37f(t_1, w_1)$
   $- 9f(t_0, w_0)]/24$; (*Predict w_i.*)

$$w = w_3 + h[9f(t, w) + 19f(t_3, w_3) - 5f(t_2, w_2)$$
$$+ f(t_1, w_1)]/24. \quad (Correct\ w_i.)$$

Step 8 OUTPUT (t, w).

Step 9 For $j = 0, 1, 2$
   set $t_j = t_{j+1}$; *(Prepare for next iteration.)*
   $w_j = w_{j+1}$.

Step 10 Set $t_3 = t$;
   $w_3 = w$.

Step 11 STOP.

EXAMPLE 5   Table 5.11 lists the results obtained by using Algorithm 5.4 for the initial-value problem

$$y' = -y + t + 1, \qquad 0 \le t \le 1, \quad y(0) = 1,$$

TABLE 5.11

t_i	w_i	$\|y(t_i) - w_i\|$
0.4	1.0703199182	1.278×10^{-7}
0.5	1.1065302684	3.923×10^{-7}
0.6	1.1488110326	6.035×10^{-7}
0.7	1.1965845314	7.724×10^{-7}
0.8	1.2493280604	9.043×10^{-7}
0.9	1.3065686568	1.003×10^{-6}
1.0	1.3678783660	1.075×10^{-6}

with $N = 10$. Although the results here are not as accurate as those in Example 4, which used only the corrector (i.e., the Adams–Moulton method), recall that in order for the method to be applied in that example it was first necessary to change to an explicit representation for w_{i+1}. Moreover, the exact starting values were used in that case.   ■

Other multistep methods can be derived by using integration of interpolating polynomials over intervals of the form $[t_j, t_{i+1}]$, for $j \le i - 1$, to obtain an approximation to $y(t_{i+1})$. One particular method that results when a Newton backward polynomial is integrated over $[t_{i-3}, t_{i+1}]$ is an explicit method called **Milne's method**:

$$(5.70) \qquad w_{i+1} = w_{i-3} + \frac{4h}{3}[2f(t_i, w_i) - f(t_{i-1}, w_{i-1}) + 2f(t_{i-2}, w_{i-2})],$$

which has local truncation error $(14/45)h^4 y^{(5)}(\xi_i)$ where $t_{i-3} < \xi_i < t_{i+1}$.

This method is occasionally used as a predictor for an implicit method called **Simpson's method**,

(5.71) $\qquad w_{i+1} = w_{i-1} + \dfrac{h}{3}[f(t_{i+1}, w_{i+1}) + 4f(t_i, w_i) + f(t_{i-1}, w_{i-1})],$

which has local truncation error $-(h^4/90)y^{(5)}(\xi_i)$ for $t_{i-1} < \xi_i < t_{i+1}$ and is obtained by integrating a Newton backward polynomial over $[t_{i-1}, t_{i+1}]$.

Although the local truncation error involved with a predictor–corrector method of the Milne–Simpson type is generally smaller than that of the Adams–Bashforth–Moulton method, the technique has limited use because of problems in stability, which do not occur with the Adams procedure. More elaboration on this difficulty is contained in Section 5.10.

Exercise Set 5.6

1. Use Algorithm 5.4 to approximate the solutions to the following initial value problems.

   a) $y' = \left(\dfrac{y}{t}\right)^2 + \left(\dfrac{y}{t}\right),\quad 1 \le t \le 1.2,\quad y(1) = 1$ with $h = 0.05$.

   b) $y' = \sin t + e^{-t},\quad 0 \le t \le 1,\quad y(0) = 0$ with $h = 0.25$.

   c) $y' = \dfrac{1}{t}(y^2 + y),\quad 1 \le t \le 3,\quad y(1) = -2$ with $h = 0.5$.

   d) $y' = -ty + \dfrac{4t}{y},\quad 0 \le t \le 1,\quad y(0) = 1$ with $h = 0.25$.

2. Use the Adams–Bashforth methods (5.63), (5.64), (5.65), and (5.66) to approximate the solutions to the following initial-value problems. Use a comparable Runge–Kutta method for starting values.

   a) $y' = \left(\dfrac{y}{t}\right)^2 + \left(\dfrac{y}{t}\right),\quad 1 \le t \le 1.2,\quad y(1) = 1$ with $h = 0.02$.

   b) $y' = \sin t + e^{-t},\quad 0 \le t \le 1,\quad y(0) = 0$ with $h = 0.1$.

   c) $y' = \dfrac{1}{t}(y^2 + y),\quad 1 \le t \le 3,\quad y(1) = -2$ with $h = 0.2$.

   d) $y' = -ty + \dfrac{4t}{y},\quad 0 \le t \le 1,\quad y(0) = 1$ with $h = 0.1$.

3. Use the Adams–Bashforth method (5.53) and the Adams–Moulton method (5.54) to approximate the solutions to the following initial-value problems.

   a) $y' = 1 - y,\quad 0 \le t \le 1,\quad y(0) = 0;$ use $h = 0.1$.

   b) $y' = \dfrac{1}{t^2} - \dfrac{y}{t} - y^2,\quad 1 \le t \le 2,\quad y(1) = -1;$ use $h = 0.05$ and compare the results to

   Exercise 6, Section 5.4.

c) $y' = \dfrac{2}{t} y + t^2 e^t$, $1 \le t \le 2$, $y(1) = 0$;   use $h = 0.1$, and compare the results to Exercise 5, Section 5.4.

d) $y' = 50t^2 - 50y + 2t$, $0 \le t \le 1$, $y(0) = \frac{1}{3}$;   use $h = 0.1, 0.025, 0.01$, and compare the results to Exercise 8, Section 5.4.

4. The three-step Adams–Moulton method for the differential equation

$$y' = e^y, \qquad 0 \le t \le 0.20, \quad y(0) = 1,$$

gives the difference equation:

$$w_{i+1} = w_i + \frac{h}{24}[9e^{w_{i+1}} + 19e^{w_i} - 5e^{w_{i-1}} - e^{w_{i-2}}].$$

a) Show that the difference equation has a unique solution for certain values of $h > 0$.

b) With $h = 0.01$, obtain w_i by functional iteration for $i = 3, \dots, 20$ using exact starting values w_0, w_1, and w_2. At each step use w_i to initially approximate w_{i+1}. The solution is

$$y(t) = \ln \left| \frac{e}{1 - et} \right|.$$

c) Will Newton's method speed the convergence over functional iteration?

5. Use Algorithm 5.4 to approximate the solution to

$$y' = -5y, \quad 0 \le t \le 2, \quad y(0) = 1,$$

with $h = 0.1$. Repeat the procedure with $h = 0.05$. Are the answers consistent with the local truncation error?

6. Derive (5.64) by the following method. Set

$$y(t_{i+1}) = y(t_i) + ahf(t_i, y(t_i)) + bhf(t_{i-1}, y(t_{i-1})) + chf(t_{i-2}, y(t_{i-2})).$$

Expand $y(t_{i+1})$, $f(t_{i-2}, y(t_{i-2}))$, and $f(t_{i-1}, y(t_{i-1}))$ in Taylor series about $(t_i, y(t_i))$, and equate the coefficients of h, h^2 and h^3 to obtain a, b, and c.

7. a) Derive Eq. (5.63) by using the Lagrange form of the interpolating polynomial.

    b) Derive Eq. (5.65) by using Newton's backward difference form of the interpolating polynomial.

8. Derive Eq. (5.67) and its local truncation error by using an appropriate form of an interpolating polynomial.

9. Use the Milne–Simpson predictor–corrector method to approximate the solution to

$$y' = -5y, \quad 0 \le t \le 2, \quad y(0) = 1,$$

with $h = 0.1$. Repeat the procedure with $h = 0.05$. Are the answers consistent with the local truncation error?

10. Derive Simpson's method by applying Simpson's rule to the integral

$$y(t_{i+1}) - y(t_{i-1}) = \int_{t_{i-1}}^{t_{i+1}} f(t, y(t))\, dt.$$

11. Derive the local truncation errors of Milne's and Simpson's methods.

5.7 Variable Step-Size Multistep Methods

As in the case of the one-step methods, when two approximations are available for the same value the possibility exists for approximating the error in one of the methods. This error approximation can then be used to adjust the step size to control the local truncation error of the method. This, as we will see in Theorem 5.18 of Section 5.10, is sufficient to ensure control of the global error of the method.

To demonstrate the error control procedure, we will construct a variable step-size predictor–corrector method using the four-step Adams–Bashforth method as predictor and the three-step Adams–Moulton method as corrector.

The Adams–Bashforth four-step method comes from the relation

$$y(t_{i+1}) = y(t_i) + \frac{h}{24}[55f(t_i, y(t_i)) - 59f(t_{i-1}, y(t_{i-1}))$$

$$+ 37f(t_{i-2}, y(t_{i-2})) - 9f(t_{i-3}, y(t_{i-3}))] + \tfrac{251}{720}y^{(5)}(\hat{\mu}_i)h^5$$

for some $t_{i-3} < \hat{\mu}_i < t_{i+1}$. The assumption that the approximations $w_0, w_1, \ldots, w_i$ are all exact gives

(5.72)
$$\frac{y(t_{i+1}) - w_{i+1}^{(0)}}{h} = \tfrac{251}{720}y^{(5)}(\hat{\mu}_i)h^4.$$

A similar analysis of the Adams–Moulton three-step method leads to

(5.73)
$$\frac{y(t_{i+1}) - w_{i+1}}{h} \approx -\tfrac{19}{720}y^{(5)}(\tilde{\mu}_i)h^4, \qquad \text{for some } t_{i-2} < \tilde{\mu}_i < t_{i+1}.$$

To proceed further we must make the assumption that for small values of h, $y^{(5)}(\hat{\mu}_i) \approx y^{(5)}(\tilde{\mu}_i)$. (The effectiveness of the error control technique will directly depend on this assumption.) If we subtract Eq. (5.73) from Eq. (5.72) we have

$$\frac{w_{i+1} - w_{i+1}^{(0)}}{h} = \frac{h^4}{720}[251\, y^{(5)}(\hat{\mu}_i) + 19\, y^{(5)}(\tilde{\mu}_i)] = \tfrac{3}{8}h^4 y^{(5)}(\tilde{\mu}_i),$$

so
$$y^{(5)}(\tilde{\mu}_i) \approx \frac{8}{3h^5}(w_{i+1} - w_{i+1}^{(0)}).$$

Using this result to eliminate the term involving $h^4 y^{(5)}$ from (5.73) gives the approximation to the error

(5.74)
$$\frac{|y(t_{i+1}) - w_{i+1}|}{h} \approx \frac{19h^4}{720} \cdot \frac{8}{3h^5}|w_{i+1} - w_{i+1}^{(0)}| = \frac{19|w_{i+1} - w_{i+1}^{(0)}|}{270h}.$$

Suppose we now reconsider (5.73) with a new step size qh generating new approximations $\hat{w}^{(0)}_{i+1}$ and $\hat{w}_{i+1}$. The object is to choose q so that the truncation error given in (5.73) is bounded by a prescribed tolerance ε. If we assume the value $y^{(5)}(\mu)$ in (5.73) associated with qh can also be approximated using (5.74), we need to choose q so that

$$\frac{|y(t_i + qh) - \hat{w}_{i+1}|}{qh} = \frac{19}{720}|y^{(5)}(\mu)|q^4h^4 \approx \frac{19}{720}\left[\frac{8}{3h^5}|w_{i+1} - w^{(0)}_{i+1}|\right]q^4h^4$$

and

$$\frac{|y(t_i + qh) - \hat{w}_{i+1}|}{qh} \approx \frac{19}{270}\frac{|w_{i+1} - w^{(0)}_{i+1}|}{h}q^4 < \varepsilon.$$

Consequently, we will choose q so that

$$q < \left(\frac{270}{19}\frac{h\varepsilon}{|w_{i+1} - w^{(0)}_{i+1}|}\right)^{1/4} \approx 2\left(\frac{h\varepsilon}{|w_{i+1} - w^{(0)}_{i+1}|}\right)^{1/4}.$$

A number of approximation assumptions have been made in this development, so in actual practice q is chosen conservatively, usually as

$$q = 1.5\left(\frac{h\varepsilon}{|w_{i+1} - w^{(0)}_{i+1}|}\right)^{1/4}.$$

A change in step size for a multistep method is much more costly in terms of functional evaluations than for a one-step method, since new equally spaced starting values must be computed. As a consequence, it is common practice to ignore the step-size change whenever the local truncation error is between $\varepsilon/10$ and ε, that is, when

$$\frac{\varepsilon}{10} < \frac{|y(t_{i+1}) - w_{i+1}|}{h} \approx \frac{19|w_{i+1} - w^{(0)}_{i+1}|}{270h} < \varepsilon.$$

In addition, q is generally given an upper bound to ensure that a single unusually accurate approximation does not result in too large a step size. Algorithm 5.5 incorporates this safeguard with an upper bound of 4.

It should be emphasized that since the multistep methods require equal step sizes for the starting values, any change in step size necessitates recalculating new starting values at that point. This will be done by calling a Runge–Kutta sub-algorithm (see Algorithm 5.2).

Adams Variable Step-Size Predictor-Corrector Algorithm 5.5

To approximate the solution of the initial-value problem

$$y' = f(t, y), \qquad a \le t \le b, \quad y(a) = \alpha.$$

with local truncation error within a given tolerance:

INPUT endpoints a, b; initial condition α; tolerance TOL; maximum step size $hmax$; minimum step size $hmin$.

OUTPUT i, t_i, w_i, h where at the ith step w_i approximates $y(t_i)$, and step size h was used or a message that the minimum step size was exceeded.

Step 1   Set up a subalgorithm for the Runge–Kutta fourth-order method to be called $RK4(h, v_0, x_0, v_1, x_1, v_2, x_2, v_3, x_3)$, which accepts as input a step size h and starting values $v_0 \approx y(x_0)$ and returns $\{(x_j, v_j)|j = 1, 2, 3\}$ defined by the following:

for $j = 1, 2, 3$

set $K_1 = hf(x_{j-1}, v_{j-1})$;

$K_2 = hf(x_{j-1} + h/2, v_{j-1} + K_1/2)$;

$K_3 = hf(x_{j-1} + h/2, v_{j-1} + K_2/2)$;

$K_4 = hf(x_{j-1} + h, v_{j-1} + K_3)$;

$v_j = v_{j-1} + (K_1 + 2K_2 + 2K_3 + K_4)/6$;

$x_j = x_0 + jh$.

Step 2   Set $t_0 = a$;

$w_0 = \alpha$;

$h = hmax$;

OUTPUT (t_0, w_0).

Step 3   Call $RK4(h, w_0, t_0, w_1, t_1, w_2, t_2, w_3, t_3)$;

Set $NFLAG = 1$;   (*Indicates computation from RK4.*)

$MFLAG = 1$;   (*MFLAG = 0 indicates acceptable computation.*)

$i = 4$;

$t = t_3 + h$.

Step 4   While $(t_{i-1} \leq b$ or $MFLAG = 1)$ do Steps 5–27.   (*Ensures that algorithm stops with an acceptable value.*)

Step 5   Set $WP = w_{i-1} + \dfrac{h}{24}[55f(t_{i-1}, w_{i-1}) - 59f(t_{i-2}, w_{i-2})$

$+ 37f(t_{i-3}, w_{i-3}) - 9f(t_{i-4}, w_{i-4})]$;   (*Predict w_i.*)

$WC = w_{i-1} + \dfrac{h}{24}[9f(t, WP) + 19f(t_{i-1}, w_{i-1})$

$- 5f(t_{i-2}, w_{i-2}) + f(t_{i-3}, w_{i-3})]$;   (*Correct w_i.*)

$\sigma = 19|WC - WP|/(270h)$.

Step 6   If $\sigma \leq TOL$ then do Steps 7–10.

Step 7   Set $w_i = WC$;   (*Result accepted.*)

$t_i = t$;

$MFLAG = 0$.

Step 8   If $NFLAG = 1$ then for $j = i - 3, i - 2, i - 1, i$

OUTPUT (j, t_j, w_j, h);

(*Previous results also accepted.*)

else OUTPUT (i, t_i, w_i, h).

(*Previous results already accepted.*)

Step 9   Set $i = i + 1$;

$NFLAG = 0$.

Step 10   If $\sigma \leq 0.1 \, TOL$ then do Steps 11–16.
                    (Increase h.)

Step 11   Set $HOLD = h$.

Step 12   Set $q = (TOL/(2\sigma))^{1/4}$.

Step 13   If $q > 4$ then set $h = 4h$
                else set $h = qh$.

Step 14   If $h > hmax$ then set $h = hmax$.

Step 15   If $t_{i-1} + 3h \geq b$ then set $h = HOLD$.
            (Avoid terminating with change in step size.)

Step 16   If $h \neq HOLD$ then do Steps 17–19.

     Step 17   Set $NFLAG = 1$.

     Step 18   Call $RK4(h, w_{i-1}, t_{i-1}, w_i, t_i, w_{i+1}, t_{i+1}, w_{i+2}, t_{i+2})$.

     Step 19   Set $i = i + 3$.

else do Steps 20 through 26.
(Result rejected.)

Step 20   Set $MFLAG = 1$.

Step 21   Set $q = (TOL/(2\sigma))^{1/4}$.

Step 22   If $q < 0.1$ then set $h = 0.1h$
                else set $h = qh$.

Step 23   If $h < hmin$ then
            OUTPUT ('*hmin* exceeded');
            (Procedure fails.)
            STOP.

Step 24   If $NFLAG = 1$ then set $i = i - 3$.
            (Previous results also rejected.)

Step 25   Call $RK4(h, w_{i-1}, t_{i-1}, w_i, t_i, w_{i+1}, t_{i+1}, w_{i+2}, t_{i+2})$.

Step 26   Set $i = i + 3$;
               $NFLAG = 1$.

Step 27   Set $t = t_{i-1} + h$.

Step 28   STOP.

EXAMPLE 1   Table 5.12 lists the results obtained using Algorithm 5.5 to find approximations to the solution of the initial-value problem

$$y' = -y + t + 1, \qquad 0 \leq t \leq 1, \quad y(0) = 1.$$

Included in the input was tolerance $TOL = 5 \times 10^{-6}$, maximum step size $hmax = 0.2$ and minimum step size $hmin = 0.02$. The computations were performed using seven digits of precision.   ∎

TABLE 5.12

i	t_i	h_i	w_i	$y(t_i)$	$\|y(t_i) - w_i\|$
0	0	—	1.000000	1.000000	0
1		0.1999999	rejected		
1	0.1072788	0.1072788	1.005553	1.005553	0
2	0.2145577	0.1072788	1.021454	1.021455	1×10^{-6}
3	0.3218366	0.1072788	1.046651	1.046652	1×10^{-6}
4	0.4291155	0.1072788	1.080198	1.080200	2×10^{-6}
5	0.5363944	0.1072788	1.121245	1.121247	2×10^{-6}
6	0.6436733	0.1072788	1.169029	1.169032	3×10^{-6}
7	0.7509521	0.1072788	1.222866	1.222868	2×10^{-6}
8	0.8582310	0.1072788	1.282138	1.282141	3×10^{-6}
9	0.9655099	0.1072788	1.346295	1.346298	3×10^{-6}
10	1.072788	0.1072788	1.414837	1.414841	4×10^{-6}

Standard software packages contain variable step-size methods that also adjust the order of the method. The DGEAR program in the IMSL library, for example, uses variable step-size Adams' methods to obtain specified accuracy while controlling the computational cost. (See, Rice [108], pp. 291–292.)

Exercise Set 5.7

1. Use Algorithm 5.5 with $TOL = 10^{-4}$ to approximate the solutions to the following initial-value problems:

   a) $y' = \left(\dfrac{y}{t}\right)^2 + \left(\dfrac{y}{t}\right)$, $1 \le t \le 1.2$, $y(1) = 1$ with $hmax = 0.05$.

   b) $y' = \sin t + e^{-t}$, $0 \le t \le 1$, $y(0) = 0$ with $hmax = 0.25$.

   c) $y' = \dfrac{1}{t}(y^2 + y)$, $1 \le t \le 3$, $y(1) = -2$ with $hmax = 0.5$.

   d) $y' = -ty + \dfrac{4t}{y}$, $0 \le t \le 1$, $y(0) = 1$ with $hmax = 0.25$.

2. Use Algorithm 5.5 for the following problems, and compare your results to those obtained in Exercise 2 of Section 5.5.

   a) $y' = 1 - y$, $0 \le t \le 1$, $y(0) = 0$;   use $TOL = 10^{-6}$ and $hmax = 0.2$.

   b) $y' = -y + t + 1$, $0 \le t \le 5$, $y(0) = 2$;   use $TOL = 10^{-4}$ and $hmax = 0.2$.

c) $y' = \dfrac{2}{t} y + t^2 e^t$, $1 \le t \le 2$, $y(1) = 0$;   use $TOL = 10^{-4}$ and $hmax = 0.2$.

d) $y' = 1 + y^2$, $0 \le t \le \pi/4$, $y(0) = 0$;   use $TOL = 10^{-4}$ and $hmax = 0.2$.

3. Use Algorithm 5.5 for the following problems, and compare your results to those obtained in Exercise 3 of Section 5.5.

a) $y' = 2|t - 2|y$, $0 \le t \le 3$, $y(0) = e^{-4}$;   use $TOL = 10^{-4}$ and $hmax = 0.2$.

b) $y' = 1 + t \sin(ty)$, $0 \le t \le 2$, $y(0) = 0$;   use $TOL = 10^{-4}$ and $hmax = 0.2$.

c) $y' = 50t^2 - 50y + 2t$, $0 \le t \le 1$, $y(0) = \frac{1}{3}$;   use $TOL = 10^{-3}$ and $hmax = 0.1$.

d) $y' = -2y + 2t^2 + 2t$, $0 \le t \le 1$, $y(0) = 1$;   use $TOL = 10^{-6}$ and $hmax = 0.2$.

4. An electrical circuit consists of a capacitor of constant capacitance $C = 1.1$ farads in series with a resistor of constant resistance $R_0 = 2.1$ ohms. A voltage $\mathscr{E}(t) = 110 \sin t$ is applied at time $t = 0$. When the resistor heats up, the resistance becomes a function of the current i,

$$R(t) = R_0 + ki \qquad \text{where } k = 0.9,$$

and the differential equation for i becomes

$$\left(1 + \frac{2k}{R_0} i\right) \frac{di}{dt} + \frac{1}{R_0 C} i = \frac{1}{R_0} \frac{d\mathscr{E}}{dt}.$$

Find the current i after 2 seconds, assuming $i(0) = 0$ and using Algorithm 5.5 with $TOL = 10^{-3}$.

5. Write an algorithm for the Adams–Bashforth and Adams–Moulton predictor–corrector method incorporating a change of step size by halving or doubling h depending on the error estimate.

6. Repeat Exercise 2 using the algorithm constructed in Exercise 5.

5.8 Extrapolation Methods

The technique of extrapolation was introduced in Sections 4.2 and 4.6 to derive accurate approximations from low-order formulas applied to numerical differenti-ation and integration. Extrapolation can also be used to develop efficient procedures for approximating the solution to initial-value problems associated with ordinary differential equations. To review and describe the extrapolation technique, we first use the procedure with Euler's method. Extrapolation will then be incorporated into a difference scheme to obtain an extrapolation procedure whose idea is credited to Gragg [57].

For the initial-value problem

(5.75) $y' = f(t, y)$, $a \le t \le b$, $y(a) = \alpha$

Euler's method with the step size $h > 0$ is given by

$$w_0 = \alpha,$$

and

(5.76) $\qquad w_{i+1} = w_i + hf(t_i, w_i), \qquad$ for each $i = 0, 1, \ldots, N-1,$

where $N = (b-a)/h$ and $t_i = a + ih$. A careful analysis of the global error $y(t_i) - w_i$ (see Gear [52], page 15) leads to the existence of a function $\delta(t)$ with the property that

(5.77) $\qquad\qquad y(t_i) = w_i + h\delta(t_i) + O(h^2)$

for each $i = 1, 2, \ldots, N$. The important point being that δ is independent of h.

Since the extrapolation process involves approximations using different step sizes, the notation $w(t, h)$ will be used to represent the approximation to $y(t)$ using the step size h. To illustrate the procedure, select two step sizes h_0 and $h_1 < h_0$. Since extrapolation can only be applied to approximate y at a particular value of t, we will consider approximating $y(b)$. Suppose

$$q_0 = \frac{b-a}{h_0} \qquad \text{and} \qquad q_1 = \frac{b-a}{h_1}$$

are both integers and Eq. (5.76) is used first with $h = h_0$ and $N = q_0$ to obtain $w(b, h_0)$ and then with $h = h_1$ and $N = q_1$ to obtain $w(b, h_1)$. This produces

(5.78) $\qquad\qquad y(b) = w(b, h_0) + h_0\delta(b) + O(h_0^2)$

and

(5.79) $\qquad\qquad y(b) = w(b, h_1) + h_1\delta(b) + O(h_1^2).$

Multiplying Eq. (5.79) by h_0 and Eq. (5.78) by h_1, and subtracting the resulting equations eliminates the term involving $\delta(b)$. Solving for $y(b)$ gives the $O(h_0^2)$ approximation for $y(b)$:

(5.80) $\qquad\qquad y(b) = \dfrac{h_0 w(b, h_1) - h_1 w(b, h_0)}{h_0 - h_1} + O(h_0^2).$

The error term in (5.80) is actually

$$\frac{h_0 O(h_1^2) - h_1 O(h_0^2)}{h_0 - h_1}.$$

In Exercise 4 you are asked to show that this term is $O(h_0^2)$.

EXAMPLE 1 We will apply Euler's method with extrapolation to the initial-value problem

$$y' = -y + t + 1, \qquad 0 \le t \le 1, \quad y(0) = 1,$$

using $h_0 = 0.1$ and $h_1 = 0.05$. The difference formula for $h = h_0$ is

$$w_0 = 1$$

$$w_{i+1} = 0.9w_i + 0.1t_i + 0.1, \qquad \text{for each } i = 0, 1, \ldots, 9,$$

and for $h = h_1$ is

$$w_0 = 1,$$

$$w_{i+1} = 0.95w_i + 0.05t_i + 0.05, \quad \text{for each } i = 0, 1, ..., 19.$$

Because of the choice of h_0 and h_1, extrapolation can be used at the common nodes $t_i = 0.1i$ for $i = 0, 1, ..., 10$. The results and the values of the actual solution $y(t) = t + e^{-t}$ are given in Table 5.13.   ∎

TABLE 5.13

$t = t_i$	$w(t, 0.1)$	$w(t, 0.05)$	$2w(t, 0.05) - w(t, 0.1)$	$y(t)$	Error
0.0	1.000000	1.000000	1.000000	1.000000	0
0.1	1.000000	1.002500	1.005000	1.004837	1.63×10^{-4}
0.2	1.010000	1.014506	1.019012	1.018731	2.81×10^{-4}
0.3	1.029000	1.035092	1.041184	1.040818	3.66×10^{-4}
0.4	1.056100	1.063420	1.070740	1.070320	4.20×10^{-4}
0.5	1.090490	1.098737	1.106984	1.106531	4.53×10^{-4}
0.6	1.131441	1.140360	1.149279	1.148812	4.67×10^{-4}
0.7	1.178297	1.187675	1.197053	1.196585	4.68×10^{-4}
0.8	1.230467	1.240126	1.249785	1.249329	4.56×10^{-4}
0.9	1.287420	1.297214	1.307008	1.306570	4.38×10^{-4}
1.0	1.348678	1.358485	1.368292	1.367879	4.13×10^{-4}

To apply extrapolation to a difference method, the method must have a particular type of error expansion. Suppose that a difference method of the form

$$(5.81) \qquad \begin{aligned} w_0 &= \alpha \\ w_{i+1} &= w_i + h\phi(t_i, w_i, h), \quad \text{for each } i = 0, 1, ..., N-1, \end{aligned}$$

has the property that, if $y \in C^{2p+2}[a, b]$, then

$$(5.82) \qquad y(t) = w(t, h) + \sum_{k=1}^{p} \delta_k(t)h^{2k} + O(h^{2p+2})$$

for some $p > 0$, where the functions δ_k are independent of h for each $k = 1, 2, ..., p$. Select a basic step size h and integers $q_0 < q_1 < q_2$, and set $h_j = h/q_j$ for each $j = 0, 1, 2$. For $t = a + h$, compute $w(t, h_0)$, $w(t, h_1)$ and $w(t, h_2)$ from (5.81), and use (5.82) to obtain

$$(5.83) \qquad y(t) = w(t, h_0) + \delta_1(t)h_0^2 + \delta_2(t)h_0^4 + \cdots + \delta_p(t)h_0^{2p} + O(h_0^{2p+2}),$$

$$(5.84) \qquad y(t) = w(t, h_1) + \delta_1(t)h_1^2 + \delta_2(t)h_1^4 + \cdots + \delta_p(t)h_1^{2p} + O(h_1^{2p+2}), \text{ and}$$

$$(5.85) \qquad y(t) = w(t, h_2) + \delta_1(t)h_2^2 + \delta_2(t)h_2^4 + \cdots + \delta_p(t)h_2^{2p} + O(h_2^{2p+2}).$$

Multiplying Eq. (5.84) by h_0^2, Eq. (5.83) by h_1^2, and subtracting the resulting equations eliminates the term involving $\delta_1(t)$. Solving for $y(t)$ gives

(5.86)
$$y(t) = \frac{h_0^2 w(t, h_1) - h_1^2 w(t, h_0)}{h_0^2 - h_1^2} - \delta_2(t) h_0^2 h_1^2$$
$$- \cdots - \delta_p(t) h_0^2 h_1^2 \frac{h_0^{2p-2} - h_1^{2p-2}}{h_0^2 - h_1^2} + O(h_0^{2p+2}).$$

Using equations (5.85) and (5.84) in a similar manner results in

(5.87)
$$y(t) = \frac{h_1^2 w(t, h_2) - h_2^2 w(t, h_1)}{h_1^2 - h_2^2} - \delta_2(t) h_1^2 h_2^2$$
$$- \cdots - \delta_p(t) h_1^2 h_2^2 \frac{h_1^{2p-2} - h_2^{2p-2}}{h_1^2 - h_2^2} + O(h_1^{2p+2}).$$

Since the algebraic expressions become even more complicated in subsequent calculations, we will simplify the notation by letting

$$y_{i,1} = w(t, h_{i-1}) \qquad \text{for each } i = 1, 2, 3$$

and
$$y_{i,2} = \frac{h_{i-2}^2 y_{i,1} - h_{i-1}^2 y_{i-1,1}}{h_{i-2}^2 - h_{i-1}^2} \qquad \text{for each } i = 2, 3.$$

Combining equations (5.86) and (5.87) to eliminate $\delta_2(t)$ yields

$$y(t) = \frac{h_0^2 y_{3,2} - h_2^2 y_{2,2}}{h_0^2 - h_2^2} + \delta_4(t) h_0^2 h_1^2 h_2^2 + \cdots + O(h_0^{2p+2}).$$

If $y_{3,3}$ is defined by

$$y_{3,3} = \frac{h_0^2 y_{3,2} - h_2^2 y_{2,2}}{h_0^2 - h_2^2},$$

the extrapolation Table 5.14 can be formed.

TABLE 5.14

$y_{1,1} = w(t, h_0)$		
$y_{2,1} = w(t, h_1)$	$y_{2,2} = \dfrac{h_0^2 y_{2,1} - h_1^2 y_{1,1}}{h_0^2 - h_1^2}$	
$y_{3,1} = w(t, h_2)$	$y_{3,2} = \dfrac{h_1^2 y_{3,1} - h_2^2 y_{2,1}}{h_1^2 - h_2^2}$	$y_{3,3} = \dfrac{h_0^2 y_{3,2} - h_2^2 y_{2,2}}{h_0^2 - h_2^2}.$

This table is similar to an iterated interpolation table presented in Section 3.3, since the construction consists of forming the linear interpolating polynomials $P_{0,1}(h^2)$ through $(h_0^2, w(t, h_0))$ and $(h_1^2, w(t, h_1))$ and $P_{1,2}(h^2)$ through $(h_1^2, w(t, h_1))$ and $(h_2^2, w(t, h_2))$. Evaluating at $h = 0$ gives $P_{0,1}(0) = y_{2,2}$ and $P_{1,2}(0) = y_{3,2}$. Performing linear iterated interpolation on $y_{2,2}$ and $y_{3,2}$ gives $y_{3,3}$. This process can be continued by selecting h_3 and computing $y_{4,1}, y_{4,2}, y_{4,3}$ and $y_{4,4}$. (See Exercise 9 of Section 4.2.)

The extrapolation process relies on finding a difference method with error expansion of the form (5.82). A modification of the midpoint method

$$w_{i+1} = w_{i-1} + 2hf(t_i, w_i)$$

with an end-correction can be shown to have the required error expansion. A verification of this can be found in Gragg [57].

The following algorithm uses the extrapolation technique with a sequence of integers $q_0 = 2, q_1 = 3, q_2 = 4, q_3 = 6, q_4 = 8, q_5 = 12, q_6 = 16, q_7 = 24$. A basic step size h is selected, and the method progresses by using $h_j = h/q_j$ for each $j = 0, \ldots, 7$ to approximate $y(t + h)$. The error is controlled by requiring that the approximations $y_{1,1}, y_{2,2}, \ldots$ be computed until $|y_{i,i} - y_{i-1,i-1}|$ is less than a given tolerance. If the tolerance is not achieved by $i = 8$, then h is reduced, and the process is reapplied. If $y_{i,i}$ is found to be acceptable, then w_1 is set to $y_{i,i}$, and computations begin again for the approximation of w_2.

Extrapolation Algorithm 5.6

To approximate the solution of the initial value problem

$$y' = f(t, y), \qquad a \le t \le b, \quad y(a) = \alpha,$$

with local error within a given tolerance:

INPUT endpoints a, b; initial condition α; tolerance TOL; level limit $p \le 8$; maximum step size $hmax$; minimum step size $hmin$.

OUTPUT T, W, h where W approximates $y(t)$ and step size h was used or a message that minimum step size exceeded.

Step 1   Initialize the array $NK = (2, 3, 4, 6, 8, 12, 16, 24)$.

Step 2   Set $TO = a$;
$\qquad\qquad WO = \alpha$;
$\qquad\qquad h = hmax$.

Step 3   For $i = 1, 2, \ldots, 7$
$\qquad\qquad$ for $j = 1, \ldots, i$
$\qquad\qquad\qquad$ set $Q_{i,j} = (NK_{i+1}/NK_j)^2$. $(Q_{i,j} = h_j^2/h_{i+1}^2.)$

Step 4   While $(TO < b)$ do Steps 5–21.

Step 5   Set $k = 1$;
$\qquad\qquad FLAG = 0$.   (When desired accuracy is achieved FLAG is set to 1.)

Step 6   While $(k \le p$ and $FLAG = 0)$ do Steps 7–14.

Step 7   Set $HK = h/NK_k$;
$\qquad\qquad T = TO$;
$\qquad\qquad W2 = WO$;
$\qquad\qquad W3 = W2 + HK \cdot f(T, W2)$;   (Euler first step.)
$\qquad\qquad T = TO + HK$.

Step 8   For $j = 1, ..., NK_k - 1$
         set $W1 = W2$;
             $W2 = W3$;
             $W3 = W1 + 2 \cdot HK \cdot f(T, W2)$;   (*Midpoint method.*)
             $T = TO + (j + 1) \cdot HK$.

Step 9   Set $y_k = [W3 + W2 + HK \cdot f(T, W3)]/2$.
          (*Smoothing to compute $y_{k,1}$.*)

Step 10   If $k \geq 2$ then do Steps 11–13.

(*Note:* $y_{k-1} \equiv y_{k-1,1}, \ y_{k-2} \equiv y_{k-2,2}, \ ..., \ y_1 \equiv y_{k-1,k-1}$ *since only
previous row of table is saved.*)

Step 11   Set $j = k$;
           $v = y_1$.   (*Save $y_{k-1,k-1}$.*)

Step 12   While $(j \geq 2)$ do

$$\text{set } y_{j-1} = y_j + \frac{y_j - y_{j-1}}{Q_{k-1,j-1} - 1};$$

          (*Extrapolation to compute $y_{j-1} \equiv y_{k,k-j+2}$*).

$$\left(\text{Note:} \quad y_{j-1} = \frac{h_{j-1}^2 y_j - h_k^2 y_{j-1}}{h_{j-1}^2 - h_k^2}.\right)$$

          $j = j - 1$.

Step 13   If $|y_1 - v| \leq TOL$ then set $FLAG = 1$.
          (*y_1 accepted as new w.*)

Step 14   Set $k = k + 1$.

Step 15   Set $k = k - 1$.

Step 16   If $FLAG = 0$ then do Steps 17 and 18.

Step 17   Set $h = h/2$.   (*New value for w rejected, decrease h.*)

Step 18   If $h < hmin$ then
         OUTPUT ('Minimum h exceeded, method fails');
         STOP.
                else do Steps 19–21.

Step 19   Set $WO = y_1$;   (*New value for w accepted.*)
          $TO = TO + h$.

Step 20   OUTPUT (TO, WO, h).

Step 21   If $(k \leq 3$ and $h < hmax/2)$ then set $h = 2h$.
         (*Increase h if possible.*)

Step 22   STOP.

EXAMPLE 2   Consider the initial-value problem

(5.88) $y' = -y + t + 1,$ $0 \le t \le 1,$ $y(0) = 1,$

which has solution $y(t) = t + e^{-t}$. Extrapolation Algorithm 5.6 will be applied to (5.88) with $h = 0.1$ and $TOL = 10^{-5}$. In the computation of w_1, Table 5.15 is obtained.

TABLE 5.15

$y_{1,1} = 1.004874$		
$y_{2,1} = 1.004854$	$y_{2,2} = 1.004837$	
$y_{3,1} = 1.004845$	$y_{3,2} = 1.004834$	$y_{3,3} = 1.004832$

The computation stopped with $w_1 = y_{3,3}$ because $|y_{3,3} - y_{2,2}| \le 10^{-5}$. All computations were performed using six digits of precision. The complete set of approximations is given in Table 5.16. In each case $y_{3,3}$ was accepted because $|y_{3,3} - y_{2,2}| \le 10^{-5}$. However, each w_i is accurate to within 2×10^{-5} of the actual value.   ∎

TABLE 5.16

i	t_i	w_i
0	0.0	1.000000
1	0.1	1.004832
2	0.2	1.018721
3	0.3	1.040806
4	0.4	1.070306
5	0.5	1.106516
6	0.6	1.148796
7	0.7	1.196569
8	0.8	1.249311
9	0.9	1.306552
10	1.0	1.367859

The proof that the method presented in Algorithm 5.6 converges involves results from summability theory, and can be found in the original paper of Gragg [57]. A number of other extrapolation procedures are available, some of which use variable step-size techniques, and research in this area is quite active. For additional procedures based on the extrapolation process, see Bulirsch and Stoer [19], [20], [21], or the text by Stetter [128]. The methods used by Bulirsch and Stoer involve interpolation with rational functions instead of the polynomial interpolation used in the Gragg procedure. An extrapolation method, DREBS, based on the Bulirsch-Stoer routine, is included in IMSL.

Exercise Set 5.8

1. Use Algorithm 5.6 with $TOL = 10^{-4}$ to approximate the solutions to the following initial value problems:

a) $y' = \left(\dfrac{y}{t}\right)^2 + \left(\dfrac{y}{t}\right)$, $1 \le t \le 1.2$, $y(1) = 1$ with $hmax = 0.05$.

b) $y' = \sin t + e^{-t}$, $0 \le t \le 1$, $y(0) = 0$ with $hmax = 0.25$.

c) $y' = \dfrac{1}{t}(y^2 + y)$, $1 \le t \le 3$, $y(1) = -2$ with $hmax = 0.5$.

d) $y' = -ty + \dfrac{4t}{y}$, $0 \le t \le 1$, $y(0) = 1$ with $hmax = 0.25$.

2. Use the Extrapolation Algorithm 5.6 for the following initial-value problems, and compare the results with those obtained in Exercise 2 of Sections 5.5 and 5.7.

a) $y' = 1 - y$, $0 \le t \le 1$, $y(0) = 0$;   use $TOL = 10^{-6}$ and $hmax = 0.2$.

b) $y' = -y + t + 1$, $0 \le t \le 5$, $y(0) = 2$;   use $TOL = 10^{-4}$ and $hmax = 0.2$.

c) $y' = \dfrac{2}{t}y + t^2 e^t$, $1 \le t \le 2$, $y(1) = 0$;   use $TOL = 10^{-4}$ and $hmax = 0.2$.

d) $y' = 1 + y^2$, $0 \le t \le \pi/4$, $y(0) = 0$;   use $TOL = 10^{-4}$ and $hmax = 0.2$.

3. Use the Extrapolation Algorithm 5.6 for the following initial-value problems, and compare the results with those obtained in Exercise 3 of Sections 5.5 and 5.7.

a) $y' = 2|t - 2|y$, $0 \le t \le 3$, $y(0) = e^{-4}$;   use $TOL = 10^{-4}$ and $hmax = 0.2$.

b) $y' = 1 + t \sin(ty)$, $0 \le t \le 2$, $y(0) = 0$;   use $TOL = 10^{-4}$ and $hmax = 0.2$.

c) $y' = 50t^2 - 50y + 2t$, $0 \le t \le 1$, $y(0) = \frac{1}{3}$;   use $TOL = 10^{-3}$ and $hmax = 0.1$.

d) $y' = -2y + 2t^2 + 2t$, $0 \le t \le 1$, $y(0) = 1$;   use $TOL = 10^{-6}$ and $hmax = 0.2$.

4. Show that the error term in Eq. (5.80) is $O(h_0^2)$.

5. Let $P(t)$ be the number of individuals in a population at time t, measured in years. If the average birth rate b is constant and the average death rate d is proportional to the size of the population (due to overcrowding), then the growth rate of the population is given by the **logistic equation**

$$\frac{dP(t)}{dt} = bP(t) - k[P(t)]^2$$

where $d = kP(t)$. Suppose $P(0) = 50{,}976$, $b = 2.9 \times 10^{-2}$, and $k = 1.4 \times 10^{-7}$. Find the population after five years, using Algorithm 5.6.

5.9 Higher-Order Equations and Systems of Differential Equations

This section contains an introduction to the numerical solution of higher-order differential equations subject to initial conditions. The techniques we discuss are limited to those that transform a higher-order equation into a system of first-order differential equations. Before discussing the transformation procedure, some remarks are needed concerning systems which involve first-order differential equations.

An **mth-order system** of first-order initial-value problems can be expressed in the form

(5.89)
$$\frac{du_1}{dt} = f_1(t, u_1, u_2, \ldots, u_m),$$

$$\frac{du_2}{dt} = f_2(t, u_1, u_2, \ldots, u_m),$$

$$\vdots$$

$$\frac{du_m}{dt} = f_m(t, u_1, u_2, \ldots, u_m)$$

for $a \le t \le b$, with the initial conditions

(5.90)
$$u_1(a) = \alpha_1,$$

$$u_2(a) = \alpha_2,$$

$$\vdots$$

$$u_m(a) = \alpha_m.$$

The object is to find m functions $u_1, u_2, \ldots, u_m$ that satisfy the system of differential equations as well as the initial conditions.

To discuss existence and uniqueness of solutions to systems of equations, it is first necessary to extend the definition of Lipschitz condition to functions of several variables.

DEFINITION 5.14   The function $f(t, y_1, \ldots, y_m)$, defined on the set

$$D = \{(t, u_1, \ldots, u_m) | a \le t \le b, -\infty < u_i < \infty, \text{for each } i = 1, 2, \ldots, m\}$$

is said to satisfy a **Lipschitz condition** on D in the variables $u_1, u_2, \ldots, u_m$ if a constant $L > 0$ exists with the property that

(5.91)
$$|f(t, u_1, \ldots, u_m) - f(t, z_1, \ldots, z_m)| \le L \sum_{j=1}^{m} |u_j - z_j|$$

for all $(t, u_1, u_2, \ldots, u_m)$ and $(t, z_1, \ldots, z_m)$ in D.

By using the Mean Value Theorem, it can be shown that if f and its first partial derivatives are continuous in D and if

$$\left| \frac{\partial f(t, u_1, \ldots, u_m)}{\partial u_i} \right| \le L$$

for each $i = 1, 2, \ldots, m$ and all $(t, u_1, \ldots, u_m)$ in D, then f satisfies a Lipschitz condition on D with Lipschitz constant L (see Birkhoff–Rota [11], p. 141). A basic existence and uniqueness theorem follows. Its proof can be found in Birkhoff–Rota [11], pp. 152–154.

THEOREM 5.15   Suppose $D = \{(t, u_1, u_2, \ldots, u_m) | a \leq t \leq b, -\infty < u_i < \infty, \text{ for each } i = 1, 2, \ldots, m\}$

and let $f_i(t, u_1, \ldots, u_m)$, for each $i = 1, 2, \ldots, m$, be continuous on D and satisfy a Lipschitz condition there. The system of first-order differential equations (5.89), subject to the initial conditions (5.90), has a unique solution $u_1(t), \ldots, u_m(t)$ for $a \leq t \leq b$.

Methods to solve systems of first-order differential equations are simply generalizations of the methods for a single first-order equation presented earlier in this chapter. For example, the classical Runge–Kutta method of order four given by

$$w_0 = \alpha,$$

$$k_1 = hf(t_i, w_i),$$

$$k_2 = hf\left(t_i + \frac{h}{2}, w_i + \tfrac{1}{2}k_1\right),$$

$$k_3 = hf\left(t_i + \frac{h}{2}, w_i + \tfrac{1}{2}k_2\right),$$

$$k_4 = hf(t_{i+1}, w_i + k_3),$$

and $w_{i+1} = w_i + \tfrac{1}{6}[k_1 + 2k_2 + 2k_3 + k_4]$ for each $i = 0, 1, \ldots, N-1,$

used to solve the first-order initial-value problem

$$y' = f(t, y), \qquad a \leq t \leq b, \quad y(a) = \alpha,$$

can be generalized as follows. Let an integer $N > 0$ be chosen and set $h = (b - a)/N$. Partition the interval $[a, b]$ into N subintervals with the mesh points

$$t_j = a + jh \qquad \text{for each } j = 0, 1, \ldots, N.$$

Use the notation w_{ij} to denote an approximation to $u_i(t_j)$ for each $j = 0, 1, \ldots, N$, and $i = 1, 2, \ldots, m$; that is, w_{ij} will approximate the ith solution $u_i(t)$ of (5.89) at the jth mesh point t_j. For the initial conditions, set

(5.92)
$$\begin{aligned} w_{1,0} &= \alpha_1, \\ w_{2,0} &= \alpha_2, \\ &\vdots \\ w_{m,0} &= \alpha_m. \end{aligned}$$

If we assume that the values $w_{1,j}, w_{2,j}, \ldots, w_{m,j}$ have been computed, we obtain $w_{1,j+1}, w_{2,j+1}, \ldots, w_{m,j+1}$ by first calculating

(5.93) $k_{1,i} = hf_i(t_j, w_{1,j}, w_{2,j}, \ldots, w_{m,j})$ for each $i = 1, 2, \ldots, m$;

(5.94) $k_{2,i} = hf_i\left(t_j + \dfrac{h}{2}, w_{1,j} + \tfrac{1}{2}k_{1,1}, w_{2,j} + \tfrac{1}{2}k_{1,2}, \ldots, w_{m,j} + \tfrac{1}{2}k_{1,m}\right)$

for each $i = 1, 2, \ldots, m$;

(5.95) $\qquad k_{3,i} = hf_i\left(t_j + \dfrac{h}{2}, w_{1,j} + \tfrac{1}{2}k_{2,1}, w_{2,j} + \tfrac{1}{2}k_{2,2}, \ldots, w_{m,j} + \tfrac{1}{2}k_{2,m}\right)$

for each $i = 1, 2, \ldots, m$;

(5.96) $\qquad k_{4,i} = hf_i(t_j + h, w_{1,j} + k_{3,1}, w_{2,j} + k_{3,2}, \ldots, w_{m,j} + k_{3,m})$

for each $i = 1, 2, \ldots, m$; and then

(5.97) $\qquad w_{i,j+1} = w_{i,j} + \tfrac{1}{6}[k_{1,i} + 2k_{2,i} + 2k_{3,i} + k_{4,i}]$

for each $i = 1, 2, \ldots, m$. Note that $k_{1,1}, k_{1,2}, \ldots, k_{1,m}$ must all be computed before $k_{2,1}$ can be determined. In general, each $k_{l,1}, k_{l,2}, \ldots, k_{l,m}$ must be computed before any of the expressions $k_{l+1,i}$.

Runge–Kutta for Systems of Differential Equations Algorithm 5.7

To approximate the solution of the mth-order system of first-order initial-value problems

$$u'_j = f_j(t, u_1, u_2, \ldots, u_m), \quad j = 1, 2, \ldots, m$$
$$a \leq t \leq b, \quad u_j(a) = \alpha_j, \quad j = 1, 2, \ldots, m$$

at $(n + 1)$ equally spaced numbers in the interval $[a, b]$:

INPUT endpoints a, b; number of equations m; integer N; initial conditions $\alpha_1, \ldots, \alpha_m$.

OUTPUT approximations w_j to $u_j(t)$ at the $(N + 1)$ values of t.

Step 1   Set $h = (b - a)/N$;
        $t = a$.

Step 2   For $j = 1, 2, \ldots, m$ set $w_j = \alpha_j$.

Step 3   OUTPUT $(t, w_1, w_2, \ldots, w_m)$.

Step 4   For $i = 1, 2, \ldots, N$ do steps 5–11.

    Step 5   For $j = 1, 2, \ldots, m$ set
        $k_{1,j} = hf_j(t, w_1, w_2, \ldots, w_m)$.

    Step 6   For $j = 1, 2, \ldots, m$ set
$$k_{2,j} = hf_j\left(t + \frac{h}{2}, w_1 + \tfrac{1}{2}k_{1,1}, w_2 + \tfrac{1}{2}k_{1,2}, \ldots, w_m + \tfrac{1}{2}k_{1,m}\right).$$

    Step 7   For $j = 1, 2, \ldots, m$ set
$$k_{3,j} = hf_j\left(t + \frac{h}{2}, w_1 + \tfrac{1}{2}k_{2,1}, w_2 + \tfrac{1}{2}k_{2,2}, \ldots, w_m + \tfrac{1}{2}k_{2,m}\right).$$

    Step 8   For $j = 1, 2, \ldots, m$ set
        $k_{4,j} = hf_j(t + h, w_1 + k_{3,1}, w_2 + k_{3,2}, \ldots, w_m + k_{3,m})$.

Step 9   For $j = 1, 2, \ldots, m$ set
$$w_j = w_j + (k_{1,j} + 2k_{2,j} + 2k_{3,j} + k_{4,j})/6.$$

Step 10   Set $t = a + ih$.

Step 11   OUTPUT $(t, w_1, w_2, \ldots, w_m)$.

Step 12   STOP.

EXAMPLE 1   Kirchhoff's Law states that the sum of all instantaneous voltage changes around a closed circuit is zero. This law implies that the current $I(t)$ in a closed circuit containing a resistance of R ohms, a capacitance of C farads, an inductance of L henrys, and a voltage source of $E(t)$ volts must satisfy the equation

$$LI'(t) + RI(t) + \frac{1}{C} \int I(t)\, dt = E(t).$$

The currents $I_1(t)$ and $I_2(t)$ in the left and right loops, respectively, of the circuit shown in Figure 5.3 are the solutions to the system of equations:

(5.98) $$2I_1(t) + 6[I_1(t) - I_2(t)] + 2I_1'(t) = 12,$$

(5.99) $$\frac{1}{0.5} \int I_2(t)\, dt + 4I_2(t) + 6[I_2(t) - I_1(t)] = 0.$$

FIGURE 5.3

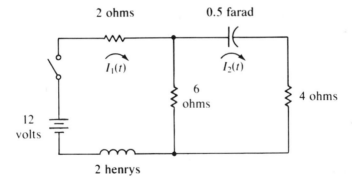

If we assume that the switch in the circuit is closed at time $t = 0$, then differentiating Eq. (5.99) and substituting Eq. (5.98) into the resulting equation gives the system:

$$I_1' = f_1(t, I_1, I_2) = -4I_1 + 3I_2 + 6, \quad I_1(0) = 0$$
$$I_2' = f_2(t, I_1, I_2) = 0.6I_1' - 0.2I_2 = -2.4I_1 + 1.6I_2 + 3.6, \quad I_2(0) = 0.$$

The exact solution to this system can be shown to be

$$I_1(t) = -3.375e^{-2t} + 1.875e^{-0.4t} + 1.5,$$
$$I_2(t) = -2.25e^{-2t} + 2.25e^{-0.4t}.$$

Suppose we apply the Runge–Kutta fourth-order method to this system with $h = 0.1$:

Since $w_{1,0} = I_1(0) = 0$ and $w_{2,0} = I_2(0) = 0$,

$$k_{1,1} = hf_1(t_0, w_{1,0}, w_{2,0}) = 0.1 f_1(0, 0, 0)$$
$$= 0.1[-4(0) + 3(0) + 6] = 0.6,$$

$$k_{2,1} = hf_2(t_0, w_{1,0}, w_{2,0}) = 0.1 f_2(0, 0, 0)$$
$$= 0.1[-2.4(0) + 1.6(0) + 3.6] = 0.36,$$

$$k_{1,2} = hf_1(t_0 + \tfrac{1}{2}h, w_{1,0} + \tfrac{1}{2}k_{1,1}, w_{2,0} + \tfrac{1}{2}k_{2,1}) = 0.1 f_1(0.05, 0.3, 0.18)$$
$$= 0.1[-4(0.3) + 3(0.18) + 6] = 0.534,$$

$$k_{2,2} = hf_2(t_0 + \tfrac{1}{2}h, w_{1,0} + \tfrac{1}{2}k_{1,1}, w_{2,0} + \tfrac{1}{2}k_{2,1}) = 0.1 f_2(0.05, 0.3, 0.18)$$
$$= 0.1[-2.4(0.3) + 1.6(0.18) + 3.6] = 0.3168.$$

Generating the remaining entries in a similar manner produces

$$k_{1,3} = (0.1)f_1(0.05, 0.267, 0.1584) = 0.54072,$$

$$k_{2,3} = (0.1)f_2(0.05, 0.267, 0.1584) = 0.321264,$$

$$k_{1,4} = (0.1)f_1(0.1, 0.54072, 0.321264) = 0.4800912,$$

and $$k_{2,4} = (0.1)f_2(0.1, 0.54072, 0.321264) = 0.28162944.$$

As a consequence,

$$I_1(0.1) \approx w_{1,1} = w_{1,0} + \tfrac{1}{6}[k_{1,1} + 2k_{1,2} + 2k_{1,3} + k_{1,4}]$$
$$= 0 + \tfrac{1}{6}[0.6 + 2(0.534) + 2(0.54072) + 0.4800912]$$
$$= 0.5382550$$

and $$I_2(0.1) \approx w_{2,1} = w_{2,0} + \tfrac{1}{6}[k_{2,1} + 2k_{2,2} + 2k_{2,3} + k_{2,4}] = 0.3196259.$$

The remaining entries in Table 5.17 are generated in a similar manner. ∎

TABLE 5.17

t_j	$w_{1,j}$	$w_{2,j}$	$\lvert I_1(t_j) - w_{1,j}\rvert$	$\lvert I_2(t_j) - w_{2,j}\rvert$
0.0	0	0	0	0
0.1	0.5382550	0.3196259	0.8285×10^{-5}	0.5603×10^{-5}
0.2	0.9684983	0.5687817	0.1514×10^{-4}	0.9596×10^{-5}
0.3	1.310717	0.7607328	0.1907×10^{-4}	0.1216×10^{-4}
0.4	1.581263	0.9063208	0.2098×10^{-4}	0.1311×10^{-4}
0.5	1.793505	1.014402	0.2193×10^{-4}	0.1240×10^{-4}

To convert a general mth-order differential equation of the form

$$y^{(m)}(t) = f(t, y, y', \ldots, y^{(m-1)}), \qquad a \le t \le b,$$

with initial conditions $y(a) = \alpha_1$, $y'(a) = \alpha_2$, ..., $y^{(m-1)}(a) = \alpha_m$ into a system of equations in the form (5.89) and (5.90), let $u_1(t) = y(t)$, $u_2(t) = y'(t)$, ..., and $u_m(t) = y^{(m-1)}(t)$. Using this notation, we obtain the first-order system

$$\frac{du_1}{dt} = \frac{dy}{dt} = u_2,$$

$$\frac{du_2}{dt} = \frac{dy'}{dt} = u_3,$$

$$\vdots$$

$$\frac{du_{m-1}}{dt} = \frac{dy^{(m-2)}}{dt} = u_m,$$

and

$$\frac{du_m}{dt} = \frac{dy^{(m-1)}}{dt} = y^{(m)} = f(t, y, y', \ldots, y^{(m-1)}) = f(t, u_1, u_2, \ldots, u_m),$$

with initial conditions

$$u_1(a) = y(a) = \alpha_1,$$

$$u_2(a) = y'(a) = \alpha_2,$$

$$\vdots$$

$$u_m(a) = y^{(m-1)}(a) = \alpha_m.$$

EXAMPLE 2   Consider the second-order differential equation

(5.100) $$y'' - 2y' + 2y = e^{2t} \sin t, \qquad 0 \le t \le 1,$$

with initial conditions $y(0) = -0.4$, $y'(0) = -0.6$. With $u_1(t) = y(t)$ and $u_2(t) = y'(t)$, (5.100) is transformed into the system

(5.101) $$u_1'(t) = u_2(t),$$
$$u_2'(t) = e^{2t} \sin t - 2u_1(t) + 2u_2(t),$$

with initial conditions

(5.102) $$u_1(0) = -0.4,$$
$$u_2(0) = -0.6.$$

The Runge–Kutta fourth-order method will be used to approximate the solution to this problem using $h = 0.1$. The initial conditions give $w_{1,0} = -0.4$ and $w_{2,0} = -0.6$. Using equations (5.93) through (5.96) with $j = 1$,

$$k_{1,1} = hf_1(t_0, w_{1,0}, w_{2,0}) = hw_{2,0} = -0.06,$$

$$k_{1,2} = hf_2(t_0, w_{1,0}, w_{2,0})$$
$$= h[e^{2t_0} \sin t_0 - 2w_{1,0} + 2w_{2,0}] = -0.04,$$

$$k_{2,1} = hf_1\left(t_0 + \frac{h}{2}, w_{1,0} + \tfrac{1}{2}k_{1,1}, w_{2,0} + \tfrac{1}{2}k_{1,2}\right)$$
$$= h[w_{2,0} + \tfrac{1}{2}k_{1,2}] = -0.062,$$

$$k_{2,2} = hf_2\left(t_0 + \frac{h}{2}, w_{1,0} + \tfrac{1}{2}k_{1,1}, w_{2,0} + \tfrac{1}{2}k_{1,2}\right)$$

$$= h[e^{2(t_0+0.05)}\sin(t_0+0.05) - 2(w_{1,0} + \tfrac{1}{2}k_{1,1}) + 2(w_{2,0} + \tfrac{1}{2}k_{1,2})]$$

$$= -0.03247644757,$$

$$k_{3,1} = h[w_{2,0} + \tfrac{1}{2}k_{2,2}] = -0.06162382238,$$

$$k_{3,2} = h[e^{2(t_0+0.05)}\sin(t_0+0.05) - 2(w_{1,0} + \tfrac{1}{2}k_{2,1}) + 2(w_{2,0} + \tfrac{1}{2}k_{2,2})]$$

$$= -0.03152409237,$$

$$k_{4,1} = h[w_{2,0} + k_{3,2}] = -0.06315240924,$$

$$k_{4,2} = h[e^{2(t_0+0.1)}\sin(t_0+0.1) - 2(w_{1,0} + k_{3,1}) + 2(w_{2,0} + k_{3,2})]$$

$$= -0.02178637298;$$

so $$w_{1,1} = w_{1,0} + \tfrac{1}{6}[k_{1,1} + 2k_{2,1} + 2k_{3,1} + k_{4,1}] = -0.4617333423,$$

and $$w_{2,1} = w_{2,0} + \tfrac{1}{6}[k_{1,2} + 2k_{2,2} + 2k_{3,2} + k_{4,2}] = -0.6316312421.$$

The value $w_{1,1}$ approximates $u_1(0.1) = y(0.1) = 0.2e^{2(0.1)}[\sin 0.1 - 2\cos 0.1]$ and $w_{2,1}$ approximates $u_2(0.1) = y'(0.1) = 0.2e^{2(0.1)}[4\sin 0.1 - 3\cos 0.1]$.

The set of values $w_{1,j}$ and $w_{2,j}$ for $j = 0, 1, \ldots, 10$, are presented in Table 5.18 and the actual value of $u_1(t) = 0.2e^{2t}[\sin t - 2\cos t]$ is also given. Values of $u_2(t)$ are not given but can be easily obtained, since $u_2 = u_1'$.   ∎

TABLE 5.18

t_j	$w_{1,j}$	$w_{2,j}$	$y(t_j) = u_1(t_j)$	$|y(t_j) - w_{1,j}|$
0.0	−0.40000000	−0.60000000	−0.40000000	0
0.1	−0.46173334	−0.63163124	−0.46173297	3.7×10^{-7}
0.2	−0.52555988	−0.64014895	−0.52555905	8.3×10^{-7}
0.3	−0.58860144	−0.61366381	−0.58860005	1.39×10^{-6}
0.4	−0.64661231	−0.53658203	−0.64661028	2.03×10^{-6}
0.5	−0.69356666	−0.38873810	−0.69356395	2.71×10^{-6}
0.6	−0.72115190	−0.14438087	−0.72114849	3.41×10^{-6}
0.7	−0.71815295	0.22899702	−0.71814890	4.05×10^{-6}
0.8	−0.66971133	0.77199180	−0.66970677	4.56×10^{-6}
0.9	−0.55644290	0.15347815	−0.55643814	4.76×10^{-6}
1.0	−0.35339886	0.25787663	−0.35339436	4.50×10^{-6}

The other one-step methods can similarly be extended to systems. If methods such as the Runge–Kutta–Fehlberg method are extended with error control, then each component of the numerical solution $(w_{1j}, w_{2j}, \ldots, w_{mj})$ must be examined for accuracy. If any of the components fail to be sufficiently accurate, the entire numerical solution $(w_{1j}, w_{2j}, \ldots, w_{mj})$ must be recomputed.

The multistep methods and predictor–corrector techniques can also be extended easily to systems. Again, if error control is used, each component must be accurate. The extension of the extrapolation technique to systems can also be done, but the notation becomes quite involved.

Convergence theorems and error estimates for systems are very similar to those which will be considered in Section 5.10 for the single equations, except that the bounds are given in terms of vector norms, a topic considered in Chapter 8. [A good reference for these theorems is Gear [52], pp. 45–72.]

Exercise Set 5.9

1. Use Algorithm 5.7 to approximate the solutions of the following systems of first-order differential equations.

   a) $u'_1 = 3u_1 + 2u_2$, $0 \le t \le 1$, $u_1(0) = 0$;
      $u'_2 = 4u_1 + u_2$, $0 \le t \le 1$, $u_2(0) = 1$.
      Use $h = 0.1$ and compare the results to the actual solution
      $u_1(t) = \frac{1}{3}(e^{5t} - e^{-t})$ and $u_2(t) = \frac{1}{3}(e^{5t} + 2e^{-t})$.

   b) $u'_1 = -4u_1 - 2u_2 + \cos t + 4 \sin t$, $0 \le t \le 2$, $u_1(0) = 0$;
      $u'_2 = 3u_1 + u_2 - 3 \sin t$, $0 \le t \le 2$, $u_2(0) = -1$.
      Use $h = 0.1$ and compare the results to the actual solution
      $u_1(t) = 2e^{-t} - 2e^{-2t} + \sin t$ and $u_2(t) = -3e^{-t} + 2e^{-2t}$.

   c) $u'_1 = u_2$, $0 \le t \le 1$, $u_1(0) = 3$;
      $u'_2 = -u_1 + 2e^{-t} + 1$, $0 \le t \le 1$, $u_2(0) = 0$;
      $u'_3 = -u_1 + e^{-t} + 1$, $0 \le t \le 1$, $u_3(0) = 1$.
      Use $h = 0.1$ and compare the results to the actual solution
      $u_1(t) = \cos t + \sin t + e^{-t} + 1$, $u_2(t) = -\sin t + \cos t - e^{-t}$,   and $u_3(t) = -\sin t + \cos t$.

   d) $u'_1 = u_2 - u_3 + t$, $0 \le t \le 1$, $u_1(0) = 1$;
      $u'_2 = 3t^2$, $0 \le t \le 1$, $u_2(0) = 1$;
      $u'_3 = u_2 + e^{-t}$, $0 \le t \le 1$, $u_3(0) = -1$.
      Use $h = 0.1$ and compare the results to the actual solution
      $u_1(t) = -0.05t^5 + 0.25t^4 + t + 2 - e^{-t}$, $u_2(t) = t^3 + 1$,   and $u_3(t) = 0.25t^4 + t - e^{-t}$.

2. Use Algorithm 5.7 to approximate the solutions of the following higher-order differential equations.

   a) $y'' + 2ty' + t^2 y = e^t$, $0 \le t \le 2$, $y(0) = 1$, $y'(0) = -1$.
      Use $h = 0.1$ and compare the results to the approximation generated by the first five terms of the power series of the actual solution.

   b) $t^2 y'' - 2ty' + 2y = t^3 \ln t$, $1 \le t \le 2$, $y(1) = 1$, $y'(1) = 0$.
      Use $h = 0.05$ and compare the results to the actual solution $y(t) = \frac{7}{4}t + \frac{t^3}{2}\ln t - \frac{3}{4}t^3$.

   c) $y''' + 2y'' - y' - 2y = e^t$, $0 \le t \le 3$, $y(0) = 1$, $y'(0) = 2$, $y''(0) = 0$.
      Use $h = 0.2$ and compare the results to the actual solution.

   d) $y''' = -6y^4$, $1 \le t \le 1.9$, $y(1) = -1$, $y'(1) = -1$, $y''(1) = -2$.
      Use $h = 0.05$ and compare the results to the actual solution $y(t) = (t - 2)^{-1}$.

3. Change the Adams Fourth-Order Predictor–Corrector Algorithm 5.4 to obtain approxi-

mate solutions to systems of first-order equations and repeat Exercise 1 using this algorithm.

4. Change the Runge–Kutta–Fehlberg Algorithm 5.3 to approximate solutions of systems of first-order equations and repeat Exercise 1 using this algorithm with $TOL = 10^{-4}$ and $hmax = 0.1$.

5. The study of mathematical models for predicting the population dynamics of competing species has its origin in independent works published in the early part of this century by A. J. Lotka and V. Volterra. Consider the problem of predicting the population of two species, one of which is a predator, whose population at time t is $x_2(t)$, feeding on the other, which is called a prey, whose population is $x_1(t)$. We will assume that the prey always has an adequate food supply and that its birth rate at any time is proportional to the number of prey alive at that time; that is, birth rate (prey) $= k_1 x_1(t)$. The death rate of the prey depends on both the number of prey and predators alive at that time. For simplicity we will assume the death rate (prey) $= k_2 x_1(t)x_2(t)$. The birth rate of the predator, on the other hand, depends on its food supply, $x_1(t)$, as well as on the number of predators available for reproduction purposes. For this reason we assume that the birth rate (predator) $= k_3 x_1(t)x_2(t)$. The death rate of the predator will be taken as simply proportional to the number of predators alive at the time; that is, death rate (predator) $= k_4 x_2(t)$.

    Since $x_1'(t)$ and $x_2'(t)$ represent the change in the prey and predator populations, respectively, with respect to time, the problem is expressed by the system of nonlinear differential equations

$$x_1'(t) = k_1 x_1(t) - k_2 x_1(t)x_2(t) \quad \text{and} \quad x_2'(t) = k_3 x_1(t)x_2(t) - k_4 x_2(t).$$

    Use the fourth-order Runge-Kutta method with $h = 0.5$ to solve this system, assuming that the initial population of the prey is 1000 and of the predators is 200, and that the constants are $k_1 = 3$, $k_2 = 0.002$, $k_3 = 0.0006$, and $k_4 = 0.5$. Sketch a graph of the solutions to this problem, plotting both populations with time, and describe the physical phenomena represented. Is there a stable solution to this population model? If so, for what values of x_1 and x_2 is the solution stable?

6. In Exercise 5, we considered the problem of predicting the population in a predator–prey model. Another problem of this type is concerned with two species competing for the same food supply. If the number of the species alive at time t are denoted by $x_1(t)$ and $x_2(t)$, it is often assumed that, while the birth rate of each of the species is simply proportional to the number of the species alive at that time, the death rate of each species depends on the population of both species. We will assume that the population of a particular pair of species is described by the equations

$$\frac{dx_1(t)}{dt} = x_1(t)[4 - 0.0003x_1(t) - 0.0004x_2(t)]$$

    and

$$\frac{dx_2(t)}{dt} = x_2(t)[2 - 0.0002x_1(t) - 0.0001x_2(t)].$$

    If it is known that the initial population of each species is 10,000, find the solution to this system. Is there a stable solution to this population model; if so, for what values of x_1 and x_2 is the solution stable?

5.10 Stability

A number of different methods have been presented in this chapter for approximating the solution to an initial-value problem. Although numerous other techniques are available for this purpose, we have chosen the methods described here because generally they satisfied three criteria: first, their development is clear enough so that the first-year student in numerical analysis can understand how and why they work; second, most of the more advanced and complex techniques are based on one or more of the procedures described here; and third, one or more of the methods will give satisfactory results for most of the problems that are encountered by undergraduate students in science and engineering. The topic we discuss in this section is the reason these methods might be expected to give satisfactory results when some similar methods would not.

Before we begin this discussion, we need to present two definitions concerned with the convergence of one-step difference-equation methods to the solution of the differential equation as the step size decreases.

DEFINITION 5.16   A difference-equation method with local truncation error τ_i at the ith step is said to be **consistent** with the differential equation it approximates if

$$\lim_{h \to 0} \max_{1 \leq i \leq N} |\tau_i| = 0.$$

Note that this definition is essentially a "local" definition since, for each of the values τ_i, we are comparing the exact value $f(t_i, y_i)$ to the difference-equation approximation of y'. A more realistic means of analyzing the effects of making h small is to determine the "global" effect of the method, which is the maximum error of the method over the entire range of the approximation, assuming only that the method gives the exact result at the initial value. The following definition describes a method that has convergence in this sense.

DEFINITION 5.17   A difference-equation method is said to be **convergent** with respect to the differential equation it approximates if

$$\lim_{h \to 0} \max_{1 \leq i \leq N} |y_i - w_i| = 0,$$

where $y_i = y(t_i)$ denotes the exact value of the solution of the differential equation and w_i is the approximation obtained from the difference method at the ith step.

Examining inequality (5.20) on page 209 in the error-bound formula for Euler's method, it can be seen that under the hypotheses of Theorem 5.9,

$$\max_{1 \leq i \leq N} |y_i - w_i| \leq \frac{Mh}{2L} |e^{L(b-a)} - 1|,$$

so Euler's method is convergent with respect to a differential equation satisfying the conditions of this theorem and the rate of convergence is $O(h)$.

A one-step method will be consistent precisely when the difference equation for the method approaches the differential equation when the step size goes to zero; that is, the local truncation error approaches zero as the step size approaches zero. The definition of convergence has a similar connotation. A method is convergent precisely when the solution to the difference equation approaches the solution to the differential equation as the step size goes to zero.

The other error-bound type of problem that exists when using difference methods to approximate solutions to differential equations is a consequence of not using exact results. In practice, neither the initial conditions nor the arithmetic that is subsequently performed is represented exactly, because of the round-off error associated with finite-digit arithmetic. In Section 5.2 we saw that this consideration can lead to difficulties even for the convergent Euler's method. In order to at least partially analyze this situation, we will try to determine which methods are stable, in the sense that small changes or perturbations in the initial conditions produce correspondingly small changes in the subsequent approximations; that is, a **stable** method is one that depends *continuously* on the initial data.

Since the concept of stability of a one-step difference equation is somewhat analogous to the condition of a differential equation being well-posed, it is not surprising that the Lipschitz condition appears here as it did in the corresponding theorem for differential equations, Theorem 5.6 (p. 203).

Part (i) of the following theorem concerns the stability of a one-step method. The proof of this result is not difficult and is considered in Exercise 1. Part (ii) of Theorem 5.18 concerns sufficient conditions for a consistent method to be convergent. Part (iii) justifies the remark made in Section 5.5 about controlling the global error of a method by controlling its local error and implies that when the local truncation error has the rate of convergence $O(h^n)$, the global error will have the same rate of convergence. The proofs of parts (ii) and (iii) can be found within the material presented by Gear [52] on pp. 57–58.

THEOREM 5.18   Suppose the initial-value problem

(5.103) $$y' = f(t, y), \qquad a \le t \le b, \quad y(a) = \alpha,$$

is approximated by a one-step difference method in the form

(5.104)
$$w_0 = \alpha,$$
$$w_{i+1} = w_i + h\phi(t_i, w_i, h).$$

If a number $h_0 > 0$ exists and $\phi(t, w, h)$ is continuous and satisfies a Lipschitz condition in the variable w on the set

$$D = \{(t, w, h) | a \le t \le b, \ -\infty < w < \infty, 0 \le h \le h_0\},$$

then

i)   the method is stable;
ii) the difference method is convergent if and only if it is consistent; that is, if and only if

$$\phi(t, y, 0) = f(t, y) \qquad \text{for all } a \le t \le b;$$

iii) if the local truncation error, τ_i, satisfies

$$|\tau_i| \le \tau(h) \qquad \text{for all } i = 1, 2, \ldots, N$$

whenever $0 \le h \le h_0$, then

$$|y(t_i) - w_i| \le \frac{\tau(h)}{L} e^{L(t_i - a)}$$

for each $i = 1, 2, \ldots, N$ where L denotes the Lipschitz constant.

EXAMPLE 1 Consider the modified Euler method given by

$$w_0 = \alpha,$$

$$w_{i+1} = w_i + \frac{h}{2}[f(t_i, w_i) + f(t_{i+1}, w_i + hf(t_i, w_i))] \qquad \text{for } i = 0, 1, \ldots, N - 1.$$

For this method

$$\phi(t, w, h) = \tfrac{1}{2}f(t, w) + \tfrac{1}{2}f(t + h, w + hf(t, w)).$$

Letting $h = 0$,

$$\phi(t, w, 0) = \tfrac{1}{2}f(t, w) + \tfrac{1}{2}f(t + 0, w + 0 \cdot f(t, w)) = f(t, w),$$

so the consistency condition expressed in Theorem 5.18, part (ii), holds.

If f satisfies a Lipschitz condition on $\{(t, w) | a \le t \le b, -\infty < w < \infty\}$ with constant L, then, since

$$\phi(t, w, h) - \phi(t, \bar{w}, h) = \tfrac{1}{2}f(t, w) + \tfrac{1}{2}f(t + h, w + hf(t, w))$$
$$- \tfrac{1}{2}f(t, \bar{w}) - \tfrac{1}{2}f(t + h, \bar{w} + hf(t, \bar{w})),$$

the Lipschitz condition on f leads to

$$|\phi(t, w, h) - \phi(t, \bar{w}, h)| \le \tfrac{1}{2}L|w - \bar{w}| + \tfrac{1}{2}L|w + hf(t, w) - \bar{w} - hf(t, \bar{w})|$$
$$\le L|w - \bar{w}| + \tfrac{1}{2}L|hf(t, w) - hf(t, \bar{w})|$$
$$\le L|w - \bar{w}| + \tfrac{1}{2}hL^2|w - \bar{w}|$$
$$= (L + \tfrac{1}{2}hL^2)|w - \bar{w}|.$$

Therefore ϕ satisfies a Lipschitz condition in w on the set

$$\{(t, w, h) | a \le t \le b, -\infty < w < \infty, 0 \le h \le h_0\}$$

for any $h_0 > 0$ with constant

$$L' = (L + \tfrac{1}{2}h_0 L^2).$$

Finally, if f is continuous on $\{(t, w) | a \le t \le b, -\infty < w < \infty\}$, then ϕ is continuous on

$$\{(t, w, h) | a \le t \le b, -\infty < w < \infty, 0 \le h \le h_0\};$$

so Theorem 5.18 implies that the modified Euler method is convergent and stable. Moreover, we have seen that for this method the local truncation error is $O(h^2)$; so the convergence of the modified Euler method also has rate $O(h^2)$. ∎

The problems involved with consistency, convergence, and stability of multistep methods are compounded because of the number of approximations involved at each step. In the one-step methods, the approximation w_{i+1} depends directly only on the previous approximation w_i, whereas the multistep methods use at least two of the previous approximations, and the methods that are commonly used involve more.

The general multistep method for approximating the solution to the initial-value problem

$$(5.105) \qquad y' = f(t, y), \qquad a \leq t \leq b, \quad y(a) = \alpha,$$

can be written in the form

$$w_0 = \alpha, \quad w_1 = \alpha_1, \quad \ldots, \quad w_{m-1} = \alpha_{m-1},$$

$$(5.106) \qquad w_{i+1} = a_{m-1} w_i + a_{m-2} w_{i-1} + \cdots + a_0 w_{i+1-m}$$
$$+ hF(t_i, h, w_{i+1}, w_i, \ldots, w_{i+1-m})$$

for each $i = m-1, m, \ldots, N-1$, where $a_0, a_1, \ldots, a_{m-1}$ are constants and, as usual, $h = (b-a)/N$ and $t_i = a + ih$.

The local truncation error for a multistep method expressed in the form (5.106) is

$$\tau_{i+1} = \frac{y(t_{i+1}) - a_{m-1} y(t_i) - \cdots - a_0 y(t_{i+1-m})}{h} - F(t_i, h, y(t_{i+1}), y(t_i), \ldots, y(t_{i+1-m}))$$

for each $i = m-1, m, \ldots, N-1$; and, as in the one-step methods, measures how much the solution y to the differential equation fails to satisfy the difference equation.

For the four-step Adams–Bashforth method (5.65), we have seen that

$$\tau_{i+1} = \frac{251}{720} y^{(5)}(\mu_i) h^4 \qquad \text{for } t_{i-3} < \mu_i < t_{i+1},$$

while the local truncation error for the three-step Adams–Moulton method (5.68) is

$$\tau_{i+1} = \frac{-19}{720} y^{(5)}(\mu_i) h^4, \qquad t_{i-2} < \mu_i < t_{i+1},$$

provided, of course that $y \in C^5[a, b]$.

Throughout the analysis two assumptions will be made concerning the function F:

i) if $f \equiv 0$ (that is, the differential equation is homogeneous), then $F \equiv 0$ also;
ii) F satisfies a type of Lipschitz condition with respect to the sequence $\{w_j\}_{j=i+1-m}^{i+1}$, in the sense that a constant C exists with

$$|F(t_i, h, w_{i+1}, \ldots, w_{i+1-m}) - F(t_i, h, v_{i+1}, \ldots, v_{i+1-m})| \leq C \sum_{j=0}^{m} |w_{i+1-j} - v_{i+1-j}|$$

for each $i = m-1, m, \ldots, N$, and sequences $\{w_j\}_{j=0}^N$ and $\{v_j\}_{j=0}^N$.

The Adams–Bashforth and Adams–Moulton methods satisfy both of these conditions, provided f satisfies a Lipschitz condition. (See Exercise 2.)

The concept of convergence for multistep methods is the same as that for one-step methods; that is, a multistep method is **convergent** if the solution to the difference equation approaches the solution to the differential equation as the step size approaches zero. This means that $\lim_{h\to 0} \max_{0\le i\le N}|w_i - y(t_i)| = 0$.

For consistency, however, a slightly different situation occurs. Again, we will want a method to be **consistent** provided that the difference equation approaches the differential equation as the step size approaches zero; that is, the local truncation error must approach zero at each step as the step size approaches zero. The additional condition occurs because of the number of starting values required for multistep methods. Since usually only the first starting value, $w_0 = \alpha$, is exact, it is necessary to require that the errors in all the starting values $\{\alpha_i\}$ approach zero as the step size approaches zero; that is, both

(5.107) $$\lim_{h\to 0} |\tau_i| = 0, \qquad \text{for all } i = m, m+1, \ldots, N, \qquad \text{and}$$

(5.108) $$\lim_{h\to 0} |\alpha_i - y(t_i)| = 0, \qquad \text{for all } i = 1, 2, \ldots, m-1,$$

must be true in order for a multistep method in the form (5.106) to be consistent.

Note that (5.108) implies that a multistep method will not be consistent unless the one-step method for generating the starting values is also consistent.

The following theorem for multistep methods is similar to Theorem 5.18, part (iii), and gives a relationship between the local truncation error and global error of a multistep method. It provides the theoretical justification for attempting to control global error by controlling local truncation error. The proof of a slightly more general form of this theorem can be found in Isaacson and Keller [67].

THEOREM 5.19   Suppose the initial-value problem

$$y' = f(t, y), \qquad a \le t \le b, \quad y(a) = \alpha,$$

is approximated by an Adams predictor–corrector method with an m step Adams–Bashforth predictor equation

$$w_{i+1} = w_i + h[b_{m-1}f(t_i, w_i) + \cdots + b_0 f(t_{i+1-m}, w_{i+1-m})]$$

with local truncation error τ_{i+1} and an $m-1$ step Adams–Moulton corrector equation

$$w_{i+1} = w_i + h[b'_{m-1}f(t_{i+1}, w_{i+1}) + b'_{m-2}f(t_i, w_i)$$
$$+ \cdots + b'_0 f(t_{i+2-m}, w_{i+2-m})]$$

with local truncation error τ'_{i+1}. In addition, suppose that $f(t, y)$ and $f_y(t, y)$ are continuous on $D = \{(t, y) | a \le t \le b \text{ and } -\infty < y < \infty\}$ and that f_y is bounded. Then the local truncation error σ_{i+1} of the predictor-corrector method is

$$\sigma_{i+1} = \tau'_{i+1} + h\tau_{i+1} b'_{m-1} \frac{\partial f}{\partial y}(t_{i+1}, \theta_{i+1}),$$

where θ_{i+1} is a number between zero and $h\tau_{i+1}$. Moreover, there exist constants k_1 and k_2 such that

$$|y(t_i) - w_i| \leq \left[\max_{0 \leq j \leq m-1} |w_j - y(t_j)| + \sigma k_1 \right] e^{k_2(t_i - a)},$$

where $\sigma = \max_{m \leq j \leq N} |\sigma_j|$.

Before discussing some connections between consistency, convergence, and stability for multistep methods, we must consider in more detail the difference equation for a multistep method. In doing this we will discover the reason for choosing the Adams methods as our standard multistep methods.

Associated with the difference equation given in (5.106),

$$w_0 = \alpha, \quad w_1 = \alpha_1, \quad \ldots, \quad w_{m-1} = \alpha_{m-1},$$

and

$$w_{i+1} = a_{m-1} w_i + a_{m-2} w_{i-1} + \cdots + a_0 w_{i+1-m} + h F(t_i, h, w_{i+1}, w_i, \ldots, w_{i+1-m}),$$

is a polynomial, called the **characteristic polynomial** of the method, given by

(5.109) $$p(\lambda) = \lambda^m - a_{m-1} \lambda^{m-1} - a_{m-2} \lambda^{m-2} - \cdots - a_1 \lambda - a_0.$$

The magnitudes of the roots of the characteristic polynomial of a multistep method are associated with the stability of the method with respect to round-off error. To see this, consider applying the standard multistep method (5.106) to the trivial initial-value problem

(5.110) $$y' \equiv 0, \quad y(a) = \alpha.$$

This problem has exact solution $y(t) \equiv \alpha$.

By examining Eqs. (5.55) and (5.56) on pages 238 and 239 it can be seen that any multistep method will, in theory, produce the exact solution $w_n = \alpha$ for all n. The only deviation from the exact solution for any multistep method is due to the inherent round-off error associated with the calculations involved in the method.

The right side of the differential equation in (5.110) has $f(t, y) \equiv 0$, so by assumption (i) on page 274, $F(t_i, h, w_{i+1}, w_{i+2}, \ldots, w_{i+1-m}) = 0$, in the difference equation (5.106). As a consequence, the standard form for the difference equation becomes

(5.111) $$w_{i+1} = a_{m-1} w_i + a_{m-2} w_{i-1} + \cdots + a_0 w_{i+1-m}.$$

Suppose λ is one of the roots of the characteristic equation associated with (5.106). Then $w_n = \lambda^n$ for each n is a solution to (5.111) since

(5.112)

$$\lambda^{i+1} - a_{m-1} \lambda^i - a_{m-2} \lambda^{i-1} - \cdots - a_0 \lambda^{i+1-m} = \lambda^{i+1-m} [\lambda^m - a_{m-1} \lambda^{m-1} - \cdots - a_0] = 0.$$

In fact, if $\lambda_1, \lambda_2, \ldots, \lambda_m$ are distinct roots of the characteristic polynomial for (5.106), it can be shown that *every* solution to (5.111) can be expressed in the form

(5.113) $$w_n = \sum_{i=1}^{m} c_i \lambda_i^n$$

for some unique collection of constants $c_1, c_2, \ldots, c_m$. The form is slightly different when the roots are not distinct, but the effects are similar.

Note that $\lambda = 1$ will always be a solution of the characteristic equation since, when $\alpha \neq 0$, the solution $w_n = \alpha$ to (5.111) implies from (5.112) that

$$0 = \alpha - \alpha a_{m-1} - \alpha a_{m-2} - \cdots - \alpha a_0 = \alpha[1 - a_{m-1} - a_{m-2} - \cdots - a_0].$$

As a consequence we assume that $\lambda_1 = 1$ and $c_1 = \alpha$ in the representation (5.113) for w_n. Thus every solution to (5.111) for solving the initial-value problem can be expressed as

$$(5.114) \qquad\qquad w_n = \alpha + \sum_{i=2}^{m} c_i \lambda_i^n,$$

where $\lambda_2, \lambda_3, \ldots, \lambda_m$ are the remaining distinct roots of the characteristic equation and, in theory, the constants $c_2, c_3, \ldots, c_m$ are all zero.

In practice, the constants $c_2, c_3, \ldots, c_m$ will not be zero due to round-off error. As a consequence, round-off error would be expected to grow exponentially with n unless $|\lambda_i| \leq 1$ for each of the roots $\lambda_2, \lambda_3, \ldots, \lambda_m$. The smaller the magnitude of these roots, the more stable the method will be with respect to the growth of round-off error.

We made the simplifying assumption in deriving (5.114) that all of the roots of the characteristic equation are distinct. The situation is similar when multiple roots occur. For example, if $\lambda_k = \lambda_{k+1} = \cdots = \lambda_{k+p}$ for some k and p, it simply requires replacing the sum

$$c_k \lambda_k^n + c_{k+1} \lambda_{k+1}^n + \cdots + c_{k+p} \lambda_{k+p}^n$$

in (5.114) by

$$c_k \lambda_k^n + c_{k+1} n \lambda_k^{n-1} + c_{k+2} n(n-1) \lambda_k^{n-2} + \cdots + c_{k+p}[n(n-1) \cdots (n-p+1)] \lambda_k^{n-p}.$$

(See Henrici [62], pp. 119–245.) Although the form of the solution is modified, the round-off effect if $|\lambda_k| > 1$ remains the same.

Although it seems that we have considered only the special case of approximating initial-value problems of the form (5.110), the stability characteristics for this equation determine the stability for the more general situation when $f(t, y)$ is not identically zero. This is due to the fact that the solution to the homogeneous equation (5.110) is embedded in the solution to any equation. The following definition is a consequence of this discussion.

DEFINITION 5.20   Let $\lambda_1, \lambda_2, \ldots, \lambda_m$ denote the (not necessarily distinct) roots of the characteristic polynomial equation

$$p(\lambda) = \lambda^m - a_{m-1} \lambda^{m-1} - \cdots - a_1 \lambda - a_0 = 0$$

associated with the multistep difference method

$$w_0 = \alpha, \quad w_1 = \alpha_1, \quad \ldots, \quad w_{m-1} = \alpha_{m-1},$$

and

$$w_{i+1} = a_{m-1}w_i + a_{m-2}w_{i-1} + \cdots + a_0 w_{i+1-m} + hF(t_i, h, w_{i+1}, w_i, \ldots, w_{i+1-m}).$$

If $|\lambda_i| \le 1$ for each $i = 1, 2, \ldots, m$, and all roots with absolute value 1 are simple roots, then the difference method is said to satisfy the **root condition**.

Methods that satisfy the root condition and have $\lambda = 1$ as the only root of the characteristic equation of magnitude one are called **strongly stable**. Methods with more than one root of magnitude one are called **weakly stable**, while those having roots of magnitude greater than one are **unstable**.

Consistency and convergence of a multistep method are closely related to the round-off stability of the method. The next theorem details these connections. For the proof of this result and the theory on which it is based the reader is urged to see Isaacson and Keller [67].

THEOREM 5.21   A multistep method of the form

$$w_0 = \alpha, \quad w_1 = \alpha_1, \quad \ldots, \quad w_{m-1} = \alpha_{m-1}$$

and

$$w_{i+1} = a_{m-1}w_i + a_{m-2}w_{i-1} + \cdots + a_0 w_{i+1-m} + hF(t_i, h, w_{i+1}, w_i, \ldots, w_{i+1-m})$$

is stable if and only if it satisfies the root condition. Moreover, if the difference method is consistent with the differential equation, then the method is stable if and only if it is convergent.

EXAMPLE 2   We have seen that the fourth-order Adams–Bashforth method can be expressed as

$$w_{i+1} = w_i + hF(t_i, h, w_{i+1}, w_i, \ldots, w_{i-3}),$$

where
$$F(t_i, h, w_{i+1}, w_i, \ldots, w_{i-3}) = \tfrac{1}{24}[55f(t_i, w_i) - 59f(t_{i-1}, w_{i-1})$$
$$+ 37f(t_{i-2}, w_{i-2}) - 9f(t_{i-3}, w_{i-3})];$$

so $m = 4$, $a_0 = 0$, $a_1 = 0$, $a_2 = 0$, and $a_3 = 1$.

The characteristic polynomial for the Adams–Bashforth method is consequently,

$$p(\lambda) = \lambda^4 - \lambda^3 = \lambda^3(\lambda - 1),$$

which has roots $\lambda_1 = 1$, $\lambda_2 = 0$, $\lambda_3 = 0$, and $\lambda_4 = 0$. It satisfies the root condition and is strongly stable.

The Adams–Moulton method has a similar characteristic polynomial, $P(\lambda) = \lambda^3 - \lambda^2$, and is also strongly stable.   ∎

EXAMPLE 3   The implicit multistep method given by

$$w_{i+1} = w_{i-1} + \frac{h}{3}[f(t_{i+1}, w_{i+1}) + 4f(t_i, w_i) + f(t_{i-1}, w_{i-1})]$$

was introduced in Section 5.6 as the fourth-order Simpson's method. Since the characteristic polynomial for this method, $p(\lambda) = \lambda^2 - 1$, has zeros $\lambda_1 = 1$ and $\lambda_2 = -1$, the method satisfies the root condition, but is only weakly stable.

Consider the initial-value problem

$$(5.115) \qquad y' = -6y + 6, \qquad 0 \le t \le 1.5, \quad y(0) = 2.$$

With $h = 0.1$ and $t_i = 0.1i$, the method can be explicitly expressed as

$$w_{i+1} = \frac{-0.8w_i + 0.8w_{i-1} + 1.2}{1.2} \qquad \text{for each } i = 1, 2, \ldots, 14.$$

For comparison purposes, the strongly stable Adams–Moulton three-step method

$$v_{i+1} = v_i + \frac{h}{24}[9f(t_{i+1}, v_{i+1}) + 19f(t_i, v_i) - 5f(t_{i-1}, v_{i-1}) + f(t_{i-2}, v_{i-2})]$$

will also be used. For problem (5.115) the Adams–Moulton difference equation is

$$v_{i+1} = \frac{12.6v_i + 3v_{i-1} - 0.6v_{i-2} + 14.4}{29.4} \qquad \text{for each } i = 2, 3, \ldots, 14.$$

Using exact values for the starting values w_0, w_1, v_0, v_1, and v_2, the results w_i and v_i along with the exact solution $y_i = e^{-6t_i} + 1$, and absolute errors $|y_i - w_i|$ and $|y_i - v_i|$ are shown in Table 5.19. It is easy to see that the errors for the fourth-order Simpson's method tend to grow in magnitude, but that the errors for the fourth-order Adams–Moulton method diminish. ■

TABLE 5.19

t_i	w_i	$|w_i - y_i|$	v_i	$|v_i - y_i|$	y_i
0.0	—	—	—	—	2.000000
0.1	—	—	—	—	1.548810
0.2	1.300792	4.015×10^{-4}	—	—	1.301194
0.3	1.165344	4.578×10^{-5}	1.164674	6.237×10^{-4}	1.165298
0.4	1.090298	4.189×10^{-4}	1.090107	6.094×10^{-4}	1.090717
0.5	1.050029	2.432×10^{-4}	1.049273	5.131×10^{-4}	1.049786
0.6	1.026845	4.768×10^{-4}	1.026950	3.719×10^{-4}	1.027322
0.7	1.015455	4.606×10^{-4}	1.014738	2.565×10^{-4}	1.014994
0.8	1.007593	6.361×10^{-4}	1.008060	1.688×10^{-4}	1.008229
0.9	1.005240	7.248×10^{-4}	1.004407	1.078×10^{-4}	1.004515
1.0	1.001567	9.108×10^{-4}	1.002409	6.867×10^{-5}	1.002478
1.1	1.002447	1.087×10^{-3}	1.001317	4.196×10^{-5}	1.001359
1.2	0.999413	1.332×10^{-3}	1.000720	2.575×10^{-5}	1.000745
1.3	1.002021	1.613×10^{-3}	1.000393	1.526×10^{-5}	1.000409
1.4	0.998260	1.964×10^{-3}	1.000214	9.537×10^{-6}	1.000224
1.5	1.002507	2.384×10^{-3}	1.000116	6.676×10^{-6}	1.000123

The reason for choosing the Adams–Bashforth–Moulton as our standard fourth-order predictor–corrector technique in Section 5.6 over the Milne–Simpson method of the same order is that both the Adams–Bashforth and Adams–Moulton methods are strongly stable and, hence, more likely to give accurate approximations to a wider class of problems than is the predictor–corrector based on the Milne and Simpson techniques, both of which are weakly stable.

Exercise Set 5.10

1. To prove Theorem 5.18, part (i), show that the hypotheses imply that there exists a constant $K > 0$ such that

$$|u_i - v_i| \le K|u_0 - v_0| \qquad \text{for each } 1 \le i \le N,$$

whenever $\{u_i\}_{i=1}^N$ and $\{v_i\}_{i=1}^N$ satisfy (5.104).

2. Show that a constant C exists so that:

$$F(t_i, h, w_{i+1}, \ldots, w_{i+1-m}) = 0, \qquad \text{if } f \equiv 0,$$

and

$$|F(t_i, h, w_{i+1}, \ldots, w_{i+1-m}) - F(t_i, h, v_{i+1}, \ldots, v_{i+1-m})| \le C \sum_{j=0}^{m} |w_{i+1-j} - v_{i+1-j}|$$

for the Adams–Bashforth and Adams–Moulton methods of order four.

3. Show that the Runge–Kutta fourth-order method is consistent.

4. Consider the differential equation

$$y' = f(t, y), \qquad a \le t \le b, \quad y(a) = \alpha.$$

a) Show that

$$y'(t_i) = \frac{-3y(t_i) + 4y(t_{i+1}) - y(t_{i+2})}{2h} + \frac{h^2}{3} y'''(\xi_i),$$

for some ξ_i, where $t_i < \xi_i < t_{i+2}$.

b) Part (a) suggests the difference method

$$w_{i+2} = 4w_{i+1} - 3w_i - 2hf(t_i, w_i), \qquad \text{for } i = 0, 1, \ldots, N-2.$$

Use this method to solve

$$y' = 1 - y, \qquad 0 \le t \le 1, \quad y(0) = 0,$$

with $h = 0.1$. Use the starting values $w_0 = 0$ and $w_1 = y(t_1) = 1 - e^{-0.1}$.

c) Repeat part (b) with $h = 0.01$ and $w_1 = 1 - e^{-0.01}$.

d) Analyze this method for consistency, stability, and convergence.

5. Given the multistep method

$$w_{i+1} = -\tfrac{3}{2}w_i + 3w_{i-1} - \tfrac{1}{2}w_{i-2} + 3hf(t_i, w_i), \qquad \text{for } i = 2, \ldots, N-1,$$

with starting values w_0, w_1, w_2:

a) Find the local truncation error.

b) Comment on consistency, stability, and convergence.

6. Obtain an approximate solution to the differential equation

$$y' = -y, \qquad 0 \le t \le 10, \quad y(0) = 1$$

using Milne's method with $h = 0.1$ and then $h = 0.01$, with starting values $w_0 = 1$ and $w_1 = e^{-h}$ in both cases. How does decreasing h from $h = 0.1$ to $h = 0.01$ affect the number of correct digits in the approximate solutions at $t = 1$ and $t = 10$?

7. Investigate stability for the difference method

$$w_{i+1} + 4w_i - 5w_{i-1} = 2h[f(t_i, w_i) + 2hf(t_{i-1}, w_{i-1})],$$

for $i = 1, 2, ..., N-1$, with starting values w_0, w_1.

8. Consider the problem $y' = 0, 0 \le t \le 10, y(0) = 0$, which has solution $y \equiv 0$. If the difference method of Exercise 4 is applied to the problem, then

$$w_{i+1} = 4w_i - 3w_{i-1}, \qquad \text{for } i = 1, 2, ..., N-1,$$

$$w_0 = 0,$$

$$w_1 = \alpha_1.$$

Suppose $w_1 = \alpha_1 = \varepsilon$, where ε is a small rounding error. Compute w_i exactly for $i = 2, 3, ..., 6$ to find how the error ε is propagated.

9. The fourth-order Adams–Moulton method applied to the differential equation

$$y' = \lambda y, \qquad t \ge 0, \quad y(0) = 1$$

is given by

$$w_{i+1} = w_i + \frac{9h\lambda}{24} w_{i+1} + \frac{19h\lambda}{24} w_i - \frac{5h\lambda}{24} w_{i-1} + \frac{h\lambda}{24} w_{i-2}.$$

a) Find the characteristic polynomial for this difference equation.

b) If β_1, β_2, and β_3 are the zeros of the characteristic polynomial, the general solution to the difference equation is

$$w_i = c_1 \beta_1^i + c_2 \beta_2^i + c_3 \beta_3^i.$$

If $|\beta_j| > 1$ for some j, then β_j^i grows exponentially as i increases, so that the Adams–Moulton method becomes unstable. Find the zeros of the characteristic polynomial for various values of $h\lambda < 0$ to experimentally determine a value of h beyond which there is a zero β_j with $|\beta_j| > 1$.

10. Repeat Exercise 9 using Simpson's method.

5.11 Stiff Differential Equations

Significant difficulties can occur when standard numerical techniques are applied to approximate the solution of a differential equation when the exact solution contains terms of the form $e^{\lambda t}$, where λ is a complex number with negative real part. This term decays to zero with increasing t, but its approximation generally will not have this property, unless restrictions are placed on the step size of the method. The problem is particularly acute when the exact solution consists of a steady-state term that does not grow significantly with t, together with a transient term that decays rapidly to zero. In such a problem, the numerical method should approximate the steady state portion of the solution, but unless extreme care is taken, the error associated with the decaying transient portion can dominate the calculations and produce meaningless results.

Problems involving rapidly decaying transient solutions occur naturally in a wide variety of applications, including the study of spring and damping systems, the analysis of control systems, and problems in chemical kinetics. These are all examples of a class of problems called **stiff systems** of differential equations.

EXAMPLE 1   The system of initial-value problems

$$u_1' = 9u_1 + 24u_2 + 5\cos t - \tfrac{1}{3}\sin t, \qquad u_1(0) = \tfrac{4}{3};$$

$$u_2' = -24u_1 - 51u_2 - 9\cos t + \tfrac{1}{3}\sin t, \qquad u_2(0) = \tfrac{2}{3};$$

has the unique solution

$$u_1(t) = 2e^{-3t} - e^{-39t} + \tfrac{1}{3}\cos t,$$

$$u_2(t) = -e^{-3t} + 2e^{-39t} - \tfrac{1}{3}\cos t.$$

The transient term e^{-39t} in the solution causes this system to be stiff. Applying Algorithm 5.7, the Runge–Kutta fourth-order method for systems, gives results listed in Table 5.20. Stability results when $h = 0.05$, but clearly does not when $h = 0.1$.   ∎

TABLE 5.20

t	$w_1(t)$ $h = 0.05$	$w_1(t)$ $h = 0.1$	$u_1(t)$	$w_2(t)$ $h = 0.05$	$w_2(t)$ $h = 0.1$	$u_2(t)$
0.1	1.712219	−2.645169	1.793061	−0.8703152	7.844527	−1.032001
0.2	1.414070	−18.45158	1.423901	−0.8550148	38.87631	−0.8746809
0.3	1.130523	−87.47221	1.131575	−0.7228910	176.4828	−0.7249984
0.4	0.9092763	−934.0722	0.9094086	−0.6079475	789.3540	−0.6082141
0.5	0.7387506	−1760.016	0.7387877	−0.5155810	3520.999	−0.5156575
0.6	0.6056833	−7848.550	0.6057094	−0.4403558	15697.84	−0.4404108
0.7	0.4998361	−34989.63	0.4998603	−0.3773540	69979.87	−0.3774038
0.8	0.4136490	−155979.4	0.4136714	−0.3229078	311959.5	−0.3229535
0.9	0.3415939	−695332.0	0.3416143	−0.2743673	1390664.	−0.2744088
1.0	0.2796568	−3099671.	0.2796748	−0.2298511	6199352.	−0.2298877

Although stiffness is usually associated with systems of differential equations, the approximation characteristics of a particular numerical method applied to a stiff system can be predicted by examining the error produced when the method is applied to a simple *test equation*,

(5.116) $$y' = \lambda y, \qquad y(0) = \alpha$$

when λ is a negative real number. The solution to this equation contains the transient

$e^{\lambda t}$ and the steady-state form is zero, so the approximation characteristics of a method are easy to determine. (A more complete discussion of the round-off error associated with stiff systems requires examining the test equation when λ is a complex number with negative imaginary part; see Gear [52], p. 222.)

Suppose we first consider Euler's method applied to the test equation. Letting $h = (b - a)/N$ and $t_j = jh$, $j = 1, 2, \ldots, N$, equation (5.16) on page 206 implies that

$$w_0 = \alpha,$$

$$w_{j+1} = w_j + h(\lambda w_j)$$
$$= (1 + h\lambda)w_j, \qquad \text{so}$$

(5.117) $\qquad\qquad w_{j+1} = (1 + h\lambda)^{j+1} w_0, \qquad \text{for } j = 0, 1, \ldots, N - 1.$

Since the exact solution is $y(t) = \alpha e^{\lambda t}$, the absolute error is

$$|y(t_j) - w_j| = |e^{\lambda h j} - (1 + h\lambda)^j||\alpha|,$$

and the accuracy is determined by how well the term $1 + h\lambda$ approximates

$$e^{h\lambda} = 1 + h\lambda + (h\lambda)^2/2! + (h\lambda)^3/3! + \cdots.$$

When $\lambda < 0$, the exact solution $e^{\lambda h j}$ decays to zero, but by (5.117), the approximation will have this property only if $|1 + h\lambda| < 1$. This effectively restricts the step size h for Euler's method to satisfy $h < 2/|\lambda|$.

Suppose now that a rounding error δ_0 is introduced in the initial condition for Euler's method,

$$w_0 = \alpha + \delta_0.$$

At the jth step the rounding error is

$$\delta_j = (1 + h\lambda)^j \delta_0.$$

The condition for the control of the growth of rounding error is consequently the same as the condition for controlling the absolute error, $h < 2/|\lambda|$.

The situation is similar for other one-step methods. In general, a function Q exists with the property that the difference method, when applied to the test equation, gives

(5.118) $\qquad\qquad\qquad w_{i+1} = Q(h\lambda)w_i.$

The accuracy of the method depends upon how well $Q(h\lambda)$ approximates $e^{h\lambda}$, and the error grows intolerably unless $|Q(h\lambda)| \leq 1$. Any nth order Taylor method, for example, will have stability with regard to both the growth of rounding error and absolute error provided h is chosen to satisfy

$$\left| 1 + h\lambda + \left(\frac{1}{2}\right)h^2\lambda^2 + \cdots + \left(\frac{1}{n!}\right)h^n\lambda^n \right| < 1.$$

Exercise 4 examines the specific case when the method is the classical fourth-order Runge–Kutta method, a Taylor method of order four.

When a multistep method of the form (5.106) is applied to the test equation, the result is

$$w_{j+1} = a_{m-1}w_j + \cdots + a_0 w_{j+1-m} + h\lambda(b_m w_{j+1} + b_{m-1}w_j + \cdots + b_0 w_{j+1-m})$$

for $j = m - 1, \ldots, N - 1$ or

$$(1 - h\lambda b_m)w_{j+1} - (a_{m-1} + h\lambda b_{m-1})w_j - \cdots - (a_0 + h\lambda b_0)w_{j+1-m} = 0.$$

Associated with this homogeneous difference equation is a characteristic polynomial

$$Q(z, h\lambda) = (1 - h\lambda b_m)z^m - (a_{m-1} + h\lambda b_{m-1})z^{m-1} - \cdots - (a_0 + h\lambda b_0).$$

This polynomial is similar to the characteristic polynomial for the method defined in (5.109), but also incorporates the test equation. The theory here parallels the stability discussion in Section in 5.10.

Suppose $w_0, \ldots, w_{m-1}$ are given, and, for fixed $h\lambda$, let $\beta_1, \ldots, \beta_m$ be the zeros of the polynomial $Q(z, h\lambda)$. If $\beta_1, \ldots, \beta_m$ are distinct, then constants $c_1, \ldots, c_m$ exist with

(5.119)
$$w_j = \sum_{k=1}^{m} c_k(\beta_k)^j \qquad \text{for } j = 0, \ldots, N.$$

(If $Q(z, h\lambda)$ has multiple roots, w_j is similarly defined. See page 277.) If w_j is to accurately approximate $y(t_j) = e^{\lambda h j} = (e^{h\lambda})^j$, then all zeros β_k must satisfy $|\beta_k| < 1$, otherwise certain choices of α will result in $c_k \neq 0$, and the term $c_k(\beta_k)^j$ will not decay to zero.

EXAMPLE 2   The test differential equation

$$y' = -30y, \qquad 0 \le t \le 1.5, \quad y(0) = \tfrac{1}{3}$$

has exact solution $y = \tfrac{1}{3}e^{-30t}$. Using $h = 0.1$ for Euler's Algorithm 5.1, Runge–Kutta Fourth-Order Algorithm 5.2, and the Adams Predictor–Corrector Algorithm 5.4 gives the results at $t = 1.5$ in Table 5.21.   ■

TABLE 5.21

Exact solution	9.54173×10^{-21}
Euler's method	-1.09225×10^4
Runge–Kutta method	3.95730×10^1
Predictor–Corrector method	8.03840×10^5

The inaccuracies in Example 2 are due to the fact that $|Q(h\lambda)| > 1$ for Euler's method and the Runge–Kutta method, and that $Q(z, h\lambda)$ has roots with modulus exceeding one for the predictor–corrector method. To apply these methods to this problem, the step size must be reduced. To describe the amount of step-size reduction that is required, we need the following definition.

DEFINITION 5.22   The **region R of absolute stability** for a one-step method is defined by $R = \{h\lambda \in \mathbb{C} \,|\, |Q(h\lambda)| < 1\}$ and for a multistep method by $R = \{h\lambda \in \mathbb{C} \,|\, |\beta_k| < 1$ for all roots β_k of $Q(z, h\lambda)\}$.

From equations (5.118) and (5.119), it is clear that a method can be applied effectively to a stiff equation only if $h\lambda$ is in the region of absolute stability of the method, which for a given problem places a restriction on the size of h. Even though the exponential term in the exact solution decays quickly to zero, λh must remain within the region of absolute stability throughout the interval of t values for the approximation to decay to zero and the growth of error to be under control. This means that while normally h could be increased because of truncation error considerations, the absolute stability criterion forces h to remain small. Variable step size methods are especially vulnerable to this problem, since an examination of the local truncation error might indicate that the step size could increase, which would inadvertently result in λh being outside the region of absolute stability.

Since the region of absolute stability of a method is generally the critical factor in producing accurate approximations for stiff systems, numerical methods have been sought with as large a region of absolute stability as possible. A numerical method is said to be **A-stable** if its region R of absolute stability contains the left half-plane $\{h\lambda \in \mathbb{C} \,|\, \text{Re}(h\lambda) < 0\}$. The implicit **Trapezoidal method**, given by

(5.120)
$$w_0 = \alpha,$$
$$w_{j+1} = w_j + \frac{h}{2}[f(t_{j+1}, w_{j+1}) + f(t_j, w_j)], \qquad 0 \le j \le N - 1,$$

is an A-stable method (see Exercise 6). It is the only A-stable multistep method.

EXAMPLE 3   The trapezoidal method and the classical Runge–Kutta fourth-order method will be applied to the test equation

$$y' = -30y, \qquad 0 \le t \le 1.5, \quad y(0) = \tfrac{1}{3};$$

with $h = 0.3, 0.1, 0.05, 0.01$. The results in Table 5.22 compare the values obtained with the exact solution at $t = 1.5$. The Runge–Kutta method is more accurate than the trapezoidal method when $h\lambda$ is inside the region of absolute stability since it is a fourth-order method and the trapezoidal method is only second order.   ∎

TABLE 5.22

h	Trapezoidal	Values at $t = 1.5$ Runge–Kutta	Exact
0.3	-3.47861×10^{-2}	7.10213×10^{10}	9.54173×10^{-21}
0.1	-1.09227×10^{-11}	3.95748×10^{1}	
0.05	1.47890×10^{-26}	4.25227×10^{-18}	
0.01	6.77706×10^{-21}	9.57905×10^{-21}	

All the techniques commonly used for stiff systems are implicit multistep methods. Generally, w_{i+1} is obtained by solving a nonlinear equation or nonlinear system iteratively, often by Newton's method.

Consider, for example, the implicit trapezoidal method:

$$w_{j+1} = w_j + \frac{h}{2}[f(t_j, w_j) + f(t_{j+1}, w_{j+1})].$$

Having computed t_j, t_{j+1}, and w_j, we need to determine w_{j+1}, the solution to

(5.121) $$F(y) = y - w_j - \frac{h}{2}[f(t_j, w_j) + f(t_{j+1}, y)] = 0.$$

To approximate this solution, select $w_{j+1}^{(0)}$, usually as w_j, and generate $w_{j+1}^{(k)}$ by applying Newton's method to (5.120):

$$w_{j+1}^{(k)} = w_j^{(k)} - \frac{F(w_j^{(k)})}{F'(w_j^{(k)})} = w_{j+1}^{(k-1)} - \frac{w_{j+1}^{(k-1)} - w_j - \frac{h}{2}[f(t_j, w_j) + f(t_{j+1}, w_{j+1}^{(k-1)})]}{1 - \frac{h}{2}f_y(t_{j+1}, w_{j+1}^{(k-1)})}$$

until $|w_{j+1}^{(k)} - w_{j+1}^{(k-1)}|$ is sufficiently small. Normally only three or four iterations per step are required.

The Secant method could be used as an alternative to Newton's method in Eq. (5.121), but then two distinct initial approximations to w_{j+1} are required. The usual practice is to let $w_{j+1}^{(0)} = w_j$ and obtain $w_{j+1}^{(1)}$ from some explicit multistep method. When a system of stiff equations is involved, a generalization is required for either Newton's or the Secant method. These topics will be considered in Chapter 9.

Trapezoidal with Newton Iteration Algorithm 5.8

To approximate the solution of the initial-value problem

$$y' = f(t, y), \qquad a \le t \le b, \qquad y(a) = \alpha$$

at $(N + 1)$ equally spaced numbers in the interval $[a, b]$:

INPUT endpoints a, b; integer N; initial condition α; tolerance TOL; maximum number of iterations M at any one step.

OUTPUT approximation w to y at the $(N + 1)$ values of t or a message of failure.

Step 1   Set $h = (b - a)/N$;
   $t = a$;
   $w = \alpha$;
   OUTPUT (t, w).

Step 2   For $i = 1, 2, \ldots, N$ do Steps 3–7.

Step 3   Set $k_1 = w + \frac{h}{2}f(t, w)$;
   $w_0 = k_1$;
   $j = 1$;
   $FLAG = 0$.

Step 4   While $FLAG = 0$ do Steps 5–6.

Step 5   Set $w = w_0 - \dfrac{w_0 - \dfrac{h}{2} f(t + h, w_0) - k_1}{1 - \dfrac{h}{2} f_y(t + h, w_0)}$.

Step 6   If $|w - w_0| < TOL$ then set $FLAG = 1$
   else set $j = j + 1$;
     $w_0 = w$;
     if $j > M$ then
        OUTPUT ('MAXIMUM NUMBER OF
           ITERATIONS EXCEEDED');

        STOP.

Step 7   Set $t = a + ih$;
   OUTPUT (t, w).

Step 8   STOP.

We have presented here only a small amount of what the reader frequently encountering stiff equations should know. We recommend for further details that Gear [52], especially the program DIFSUB on pp. 158–166, Lambert [78], or Shampine and Gear [122] be consulted. Some packaged programs for stiff systems are available, for example, the previously mentioned program of Gear and SDBASIC due to Enright [40]. Enright, Hull, and Lindberg [41] compare some of the common packages for stiff systems.

The Choice of a Method for Solving Initial-Value Problems

In closing this chapter, it seems appropriate also to give some recommendations on methods to use for solving nonstiff initial-value problems. Most of these recommendations are based on the results presented in papers by Hull, Enright, Fellen, and Sedgwick [66], published in 1972, by Enright, Hull, and Lindberg [41] in 1975, and by Enright and Hull [42], in 1976. Additional discussion of these topics can also be found in references [52] and [78]. Any updates of these results are likely to appear in the *SIAM Journal on Numerical Analysis*.

When the function being evaluated is relatively simple, that is, does not require many time-consuming manipulations, an extrapolation procedure such as Algorithm 5.6 is most efficient, while the Adams-based predictor–corrector method is favored when the evaluation of the function is complicated. The use of the Runge–Kutta procedures should be restricted to finding starting values for the Adams method, or to problems where the function is easy to evaluate and the accuracy needed is small, about 10^{-4}.

Exercise Set 5.11

1. Solve the following "stiff" initial-value problems using (*i*) Euler's method, (*ii*) Runge–Kutta fourth-order method, and (*iii*) the Trapezoidal method.

   a) $y' = -5y$, $0 \le t \le 1$, $y(0) = e$.
      Use $h = 0.1$ and compare to the actual solution $y(t) = e^{1-5t}$.

   b) $y' = -7(y - t) + 1$, $0 \le t \le 1$, $y(0) = 3$.
      Use $h = 0.1$ and compare to the actual solution $y(t) = t + 3e^{-7t}$.

   c) $y' = -20(y - t^2) + 2t$, $0 \le t \le 1$, $y(0) = \frac{1}{3}$.
      Use $h = 0.05$ for $0 \le t \le 0.2$ and $h = 0.1$ for $0.2 \le t \le 1$ and compare to the actual solution $y(t) = t^2 + \frac{1}{3}e^{-20t}$.

   d) $y' = -20y + 20 \sin t + \cos t$, $0 \le t \le 1$, $y(0) = 1$.
      Use $h = 0.01$ for $0 \le t \le 0.2$ and $h = 0.05$ for $0.2 \le t \le 1$ and compare to the actual solution $y(t) = e^{-20t} + \sin t$.

   e) $y' = (50/y) - 50y$, $0 \le t \le 1$, $y(0) = \sqrt{2}$.
      Use $h = 0.05$ for $0 \le t \le 0.2$ and $h = 0.1$ for $0.2 \le t \le 1$ and compare to the actual solution $y(t) = [1 + e^{-100t}]^{1/2}$.

   f) $u_1' = 32u_1 + 66u_2 + \frac{2}{3}t + \frac{2}{3}$, $0 \le t \le 1$, $u_1(0) = \frac{1}{3}$;
      $u_2' = -66u_1 - 133u_2 - \frac{1}{3}t - \frac{1}{3}$, $0 \le t \le 1$, $u_2(0) = \frac{1}{3}$.
      Use $h = 0.01$ for $0 \le t \le 0.1$ and $h = 0.1$ for $0.1 \le t \le 1$ and compare to the actual solution

   $$u_1(t) = \frac{2}{3}t + \frac{2}{3}e^{-t} - \frac{1}{3}e^{-100t} \qquad \text{and} \qquad u_2(t) = -\frac{1}{3}t - \frac{1}{3}e^{-t} + \frac{2}{3}e^{-100t}.$$

2. Repeat Exercise 1(a) and 1(b) using (*i*) Algorithm 5.4 and (*ii*) Algorithm 5.5 with $TOL = 10^{-3}$ and $hmax = 0.1$.

3. Discuss consistency, stability, and convergence for the Trapezoidal method

   $$w_{i+1} = w_i + \frac{h}{2}[f(t_{i+1}, w_{i+1}) + f(t_i, w_i)] \qquad \text{for } i = 0, 1, \ldots, N - 1$$

   with $w_0 = \alpha$ applied to the differential equation

   $$y' = f(t, y), \qquad a \le t \le b, \quad y(a) = \alpha.$$

4. Show that the fourth-order Runge–Kutta method,

   $$w_{i+1} = w_i + \tfrac{1}{6}(k_1 + 2k_2 + 2k_3 + k_4),$$

   $$k_1 = hf(t_i, w_i),$$

   $$k_2 = hf(t_i + h/2, w_i + k_1/2),$$

   $$k_3 = hf(t_i + h/2, w_i + k_2/2).$$

   $$k_4 = hf(t_i + h, w_i + k_3),$$

   when applied to the differential equation $y' = \lambda y$ can be written in the form

   $$w_{i+1} = (1 + h\lambda + \tfrac{1}{2}(h\lambda)^2 + \tfrac{1}{6}(h\lambda)^3 + \tfrac{1}{24}(h\lambda)^4)w_i.$$

5. The backward Euler one-step method is defined by

$$w_{i+1} = w_i + hf(t_{i+1}, w_{i+1}) \qquad \text{for } i = 0, \ldots, N-1.$$

a) Show that $Q(h\lambda) = 1/(1 - h\lambda)$ for the backward Euler method.

b) Apply the backward Euler method to the differential equations given in Exercise 1. Use Newton's method to solve for w_{i+1}.

6. a) Show that the Trapezoidal method (5.119) is A-stable.

b) Show that the backward Euler method described in Exercise 5 is A-stable.

7. In Exercise 9 of Section 5.2 the differential equation

$$\frac{dp(t)}{dt} = rb(1 - p(t))$$

was obtained as a model for studying the proportion $p(t)$ of nonconformists in a society whose birth rate was b and where r represented the rate at which offspring would become nonconformists when at least one of their parents was a conformist. That exercise required that an approximation for $p(t)$ be found by using Euler's method, Algorithm 5.1, for integral values of t when given $p(0) = 0.01$, $b = 0.02$, and $r = 0.1$, and then the approximation for $p(50)$ be compared with the actual value. Explain the reason for the large error in this approximation. Use the Trapezoidal method to obtain another approximation for $p(50)$, again assuming that $h = 1$ year.

6

Direct Methods for Solving Linear Systems

Kirchhoff's laws of electrical circuits state that the net flow of current through each junction of a circuit is zero, and that the net voltage drop around each closed loop of the circuit is zero. Suppose that a potential of V volts is applied between the points A and G in the circuit and v_b, v_c, v_d, v_e, and v_f are the potentials at the points B, C, D, E, and F, respectively. Using G as a reference point, Kirchhoff's laws imply that these potentials satisfy the following system of linear equations:

$$
\begin{aligned}
31v_b - 10v_c &\qquad\qquad\quad - 6v_f = 15 \text{ V,} \\
2v_b - 8v_c &+ 3v_d + 3v_e \qquad\quad = 0, \\
v_c &- 3v_d + 2v_e \qquad\quad = 0, \\
2v_c &+ 4v_d - 7v_e + \quad v_f = 0, \\
12v_b &\quad\ + 15v_d \qquad\quad - 47v_f = 0.
\end{aligned}
$$

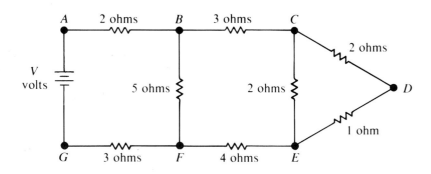

The solution of systems of this type will be considered in this chapter.

Linear systems of equations are associated with many problems in engineering and science, as well as with applications of mathematics to the social sciences and the quantitative study of business and economic problems.

In this chapter, direct techniques are considered to solve the linear system

$$
\begin{aligned}
E_1: &\quad a_{11}x_1 + a_{12}x_2 + \cdots + a_{1n}x_n = b_1, \\
E_2: &\quad a_{21}x_1 + a_{22}x_2 + \cdots + a_{2n}x_n = b_2, \\
&\quad \vdots \qquad \vdots \qquad \vdots \qquad\qquad \vdots \qquad \vdots \\
E_n: &\quad a_{n1}x_1 + a_{n2}x_2 + \cdots + a_{nn}x_n = b_n,
\end{aligned}
$$

(6.1)

for $x_1, \ldots, x_n$, given the a_{ij} for each $i, j = 1, 2, \ldots, n$, and b_i, for each $i = 1, 2, \ldots, n$. Direct techniques are methods that give an answer in a fixed number of steps, subject only to rounding errors. In the presentation it is also necessary to introduce some elementary notions from the subject of linear algebra.

Methods of approximating the solution to linear systems by iterative methods will be discussed in Chapter 8.

6.1 Linear Systems of Equations

To solve a linear system such as (6.1), three operations are permitted on the equations:

1) Equation E_i can be multiplied by any nonzero constant λ and the resulting equation used in place of E_i. This operation will be denoted $(\lambda E_i) \rightarrow (E_i)$.
2) Equation E_j can be multiplied by any constant λ, added to equation E_i, and the resulting equation used in place of E_i. This operation will be denoted $(E_i + \lambda E_j) \rightarrow (E_i)$.
3) Equations E_i and E_j can be transposed in order. This operation will be denoted $(E_i) \leftrightarrow (E_j)$.

By a sequence of the operations given above, a linear system can be transformed to a more easily solved linear system with the same set of solutions. The sequence of operations will be illustrated in the next example.

EXAMPLE 1   The four equations

$$
\begin{aligned}
E_1: &\quad x_1 + x_2 \qquad\quad + 3x_4 = \quad 4, \\
E_2: &\quad 2x_1 + x_2 - x_3 + x_4 = \quad 1, \\
E_3: &\quad 3x_1 - x_2 - x_3 + 2x_4 = -3, \\
E_4: &\quad -x_1 + 2x_2 + 3x_3 - x_4 = \quad 4,
\end{aligned}
$$

(6.2)

will be solved for the unknowns x_1, x_2, x_3, and x_4. The first step is to use equation E_1 to eliminate the unknown x_1 from E_2, E_3, and E_4 by performing $(E_2 - 2E_1) \rightarrow (E_2)$, $(E_3 - 3E_1) \rightarrow (E_3)$, and $(E_4 + E_1) \rightarrow (E_4)$. The resulting system is:

$$
\begin{aligned}
E_1: &\quad x_1 + x_2 \qquad\; + 3x_4 = \quad 4, \\
E_2: &\quad\;\; -x_2 - x_3 - 5x_4 = -\;7, \\
E_3: &\quad\;\; -4x_2 - x_3 - 7x_4 = -15, \\
E_4: &\qquad\;\; 3x_2 + 3x_3 + 2x_4 = \quad 8,
\end{aligned}
$$

(6.3)

where the new equations are, for simplicity, again labeled E_1, E_2, E_3, and E_4.

In the new system, E_2 is used to eliminate x_2 from E_3 and E_4 by the operations $(E_3 - 4E_2) \to (E_3)$ and $(E_4 + 3E_2) \to (E_4)$, resulting in the system

$$
\begin{aligned}
E_1: &\quad x_1 + x_2 \qquad\; + 3x_4 = \quad 4, \\
E_2: &\qquad -x_2 - x_3 - 5x_4 = -\;7, \\
E_3: &\qquad\qquad\;\; 3x_3 + 13x_4 = \quad 13, \\
E_4: &\qquad\qquad\qquad\;\; -13x_4 = -13.
\end{aligned}
$$

(6.4)

The system of equations (6.4) is now in **triangular** or **reduced form** and can easily be solved for the unknowns by a **backward-substitution** process. Noting that E_4 implies $x_4 = 1$, E_3 can be solved for x_3:

$$
x_3 = \tfrac{1}{3}(13 - 13x_4) = \tfrac{1}{3}(13 - 13) = 0.
$$

Continuing, E_2 gives:

$$
x_2 = -(-7 + 5x_4 + x_3) = -(-7 + 5 + 0) = 2;
$$

and E_1 gives

$$
x_1 = 4 - 3x_4 - x_2 = 4 - 3 - 2 = -1.
$$

The solution to (6.4) is therefore $x_1 = -1$, $x_2 = 2$, $x_3 = 0$, and $x_4 = 1$. It can easily be verified that these values also solve the equations in (6.2). ∎

When performing the calculations of Example 1, we did not need to write out the full equations at each step or to carry the variables x_1, x_2, x_3, and x_4 through the calculations since they always remained in the same column. The only variation from system to system occurred in the coefficients of the unknowns and in the values on the right side of the equations. For this reason, a linear system is often replaced by a *matrix*, which contains all the information about the system that is necessary to determine its solution, but in a compact form.

DEFINITION 6.1   An *n* **by** *m* **matrix** is a rectangular array of elements with n rows and m columns in which not only is the value of an element important, but also its position in the array.

The notation for an $n \times m$ (n by m) matrix will be a capital letter such as A for the matrix and lowercase letters with double subscripts, such as a_{ij}, to refer to the entry at the intersection of the ith row and jth column; that is,

$$A = (a_{ij}) = \begin{bmatrix} a_{11} & a_{12} & \cdots & a_{1m} \\ a_{21} & a_{22} & \cdots & a_{2m} \\ \vdots & \vdots & & \vdots \\ a_{n1} & a_{n2} & \cdots & a_{nm} \end{bmatrix}.$$

EXAMPLE 2   The matrix

$$A = \begin{bmatrix} 2 & -1 & 7 \\ 3 & 1 & 0 \end{bmatrix}$$

is a 2×3 matrix with $a_{11} = 2$, $a_{12} = -1$, $a_{13} = 7$, $a_{21} = 3$, $a_{22} = 1$, and $a_{23} = 0$. ∎

The $1 \times n$ matrix

$$A = \begin{bmatrix} a_{11} & a_{12} & \cdots & a_{1n} \end{bmatrix}$$

is called an ***n*-dimensional row vector**, and an $n \times 1$ matrix

$$A = \begin{bmatrix} a_{11} \\ a_{21} \\ \vdots \\ a_{n1} \end{bmatrix}$$

is called an ***n*-dimensional column vector**. Usually the unnecessary subscript is omitted for vectors and a boldface lowercase letter used for notation. Thus,

$$\mathbf{x} = \begin{bmatrix} x_1 \\ x_2 \\ \vdots \\ x_n \end{bmatrix}$$

denotes a column vector and

$$\mathbf{y} = \begin{bmatrix} y_1, & y_2, & \ldots, & y_n \end{bmatrix}$$

a row vector.

An n by $(n + 1)$ matrix can be used to represent the linear system

$$a_{11}x_1 + a_{12}x_2 + \cdots + a_{1n}x_n = b_1,$$
$$a_{21}x_1 + a_{22}x_2 + \cdots + a_{2n}x_n = b_2,$$
$$\vdots \qquad \vdots \qquad\qquad \vdots \qquad \vdots$$
$$a_{n1}x_1 + a_{n2}x_2 + \cdots + a_{nn}x_n = b_n,$$

by first constructing

$$A = \begin{bmatrix} a_{11} & a_{12} & \cdots & a_{1n} \\ a_{21} & a_{22} & \cdots & a_{2n} \\ \vdots & \vdots & & \vdots \\ a_{n1} & a_{n2} & \cdots & a_{nn} \end{bmatrix} \quad \text{and} \quad \mathbf{b} = \begin{bmatrix} b_1 \\ b_2 \\ \vdots \\ b_n \end{bmatrix}$$

and then combining these matrices to form the **augmented matrix**:

$$[A, \mathbf{b}] = \begin{bmatrix} a_{11} & a_{12} & \cdots & a_{1n} & \vdots & b_1 \\ a_{21} & a_{22} & \cdots & a_{2n} & \vdots & b_2 \\ \vdots & \vdots & & \vdots & \vdots & \vdots \\ a_{n1} & a_{n2} & \cdots & a_{nn} & \vdots & b_n \end{bmatrix},$$

where the dotted line is used to separate the coefficients of the unknowns from the values on the right-hand side of the equations.

EXAMPLE 3   Repeating the operations involved in Example 1 with the matrix notation results in considering first the augmented matrix associated with system (6.2):

(6.5)
$$\begin{bmatrix} 1 & 1 & 0 & 3 & \vdots & 4 \\ 2 & 1 & -1 & 1 & \vdots & 1 \\ 3 & -1 & -1 & 2 & \vdots & -3 \\ -1 & 2 & 3 & -1 & \vdots & 4 \end{bmatrix}.$$

Performing the operations associated with $(E_2 - 2E_1) \rightarrow (E_2)$, $(E_3 - 3E_1) \rightarrow (E_3)$, and $(E_4 + E_1) \rightarrow (E_4)$ in system (6.2) is accomplished by manipulating the respective rows of the augmented matrix (6.5), which becomes the matrix corresponding to the system (6.3):

(6.6)
$$\begin{bmatrix} 1 & 1 & 0 & 3 & \vdots & 4 \\ 0 & -1 & -1 & -5 & \vdots & -7 \\ 0 & -4 & -1 & -7 & \vdots & -15 \\ 0 & 3 & 3 & 2 & \vdots & 8 \end{bmatrix}.$$

Performing the final manipulations results in the augmented matrix corresponding to system (6.4):

(6.7)
$$\begin{bmatrix} 1 & 1 & 0 & 3 & \vdots & 4 \\ 0 & -1 & -1 & -5 & \vdots & -7 \\ 0 & 0 & 3 & 13 & \vdots & 13 \\ 0 & 0 & 0 & -13 & \vdots & -13 \end{bmatrix}.$$

■

This matrix can now be transformed into its corresponding linear system (6.4) and solutions for x_1, x_2, x_3, and x_4 obtained. The procedure involved in this process is called **Gaussian elimination with backward substitution**. In subsequent sections, we will consider conditions on the linear system under which the method can be used successfully.

Exercise Set 6.1

1. Solve the following linear systems using the elimination method.

   a) $x_1 - 2x_2 = 3,$
       $2x_1 + x_2 = 4.$

   b) $2x_1 - 3x_2 = 7,$
       $x_1 + 2x_2 = 5.$

   c) $x_1 + x_2 - x_3 = 3,$
       $2x_1 - x_2 + 3x_3 = 0,$
       $-x_1 - 2x_2 + x_3 = -5.$

   d) $3x_1 - x_2 + 2x_3 = -3,$
       $x_1 + x_2 + x_3 = -4,$
       $2x_1 + x_2 - x_3 = -3.$

   e) $x_1 + 3x_2 - x_3 + x_4 = 5,$
       $2x_2 + x_3 + x_4 = 7,$
       $x_3 + 2x_4 = 6,$
       $3x_4 = 9.$

   f) $2x_1 - x_2 + 3x_3 - 4x_4 = 6,$
       $x_2 - x_3 - x_4 = 7,$
       $x_3 = 6,$
       $2x_4 = 5.$

   g) $x_1 + \frac{1}{2}x_2 + \frac{1}{3}x_3 = \frac{11}{6},$
       $\frac{1}{2}x_1 + \frac{1}{3}x_2 + \frac{1}{4}x_3 = \frac{13}{12},$
       $\frac{1}{3}x_1 + \frac{1}{4}x_2 + \frac{1}{5}x_3 = \frac{47}{60}.$

   h) $\frac{1}{5}x_1 + \frac{1}{6}x_2 + \frac{1}{7}x_3 = 1,$
       $\frac{1}{6}x_1 + \frac{1}{7}x_2 + \frac{1}{8}x_3 = 0,$
       $\frac{1}{7}x_1 + \frac{1}{8}x_2 + \frac{1}{9}x_3 = -1.$

2. For each of the following systems, obtain a solution graphically, if possible, and then try to solve the system by the methods in this section.

   a) $x_1 + 2x_2 = 3,$
       $x_1 - x_2 = 0.$

   b) $x_1 + 2x_2 = 3,$
       $-2x_1 - 4x_2 = 6.$

   c) $x_1 + 2x_2 = 3,$
       $2x_1 + 4x_2 = 6.$

   d) $0 \cdot x_1 + x_2 = 3,$
       $2x_1 - x_2 = 7.$

3. Use the procedures of this section to try to solve the following linear systems. Explain why the procedure fails in each case for which no solutions can be obtained.

   a) $x_1 + x_2 - x_3 = 3,$
       $-x_1 + x_2 + x_3 = 2,$
       $x_1 + 3x_2 - x_3 = 8.$

   b) $x_1 + x_2 - x_3 = 3,$
       $-x_1 + x_2 + x_3 = 2,$
       $x_1 + 3x_2 - x_3 = 6.$

   c) $x_1 - 2x_2 + x_3 = 5,$
       $-x_1 + 2x_2 + x_3 = 4,$
       $3x_1 - 2x_2 = 6.$

   d) $3x_1 + x_2 - 5x_3 = 14,$
       $2x_1 + x_2 - 3x_3 = 5,$
       $-x_1 - x_2 - x_3 = 4.$

   e) $x_1 + x_2 + x_4 = 2,$
       $2x_1 + x_2 - x_3 + x_4 = 1,$
       $4x_1 - x_2 - 2x_3 + 2x_4 = 0,$
       $3x_1 - x_2 - x_3 + 2x_4 = -3.$

   f) $x_1 + x_2 + x_4 = 2,$
       $2x_1 + x_2 - x_3 + x_4 = 1,$
       $-x_1 + 2x_2 + 3x_3 - x_4 = 4,$
       $3x_1 - x_2 - x_3 + 2x_4 = -3.$

4. What can be said, from a geometrical standpoint, about the following sets of equations?

   a) $2x_1 + x_2 = -1,$
       $x_1 + x_2 = 2,$
       $x_1 - 3x_2 = 5.$

   b) $2x_1 + x_2 + x_3 = 1,$
       $2x_1 + 4x_2 - x_3 = -1.$

5. Suppose that in a biological system there are n species of animals and m sources of food. Let x_j represent the population of the jth species for each $j = 1, \ldots, n$; b_i represent the available daily supply of the ith food; and a_{ij} represent the amount of the ith food consumed on the average by a member of the jth species. The linear system

$$a_{11}x_1 + a_{12}x_2 + \cdots + a_{1n}x_n = b_1,$$
$$a_{21}x_1 + a_{22}x_2 + \cdots + a_{2n}x_n = b_2,$$
$$\vdots \qquad \vdots \qquad \qquad \vdots \qquad \vdots$$
$$a_{m1}x_1 + a_{m2}x_2 + \cdots + a_{mn}x_n = b_m,$$

represents an equilibrium where there is a daily supply of food to precisely meet the average daily consumption of each species.

a) Let $A = (a_{ij}) = \begin{bmatrix} 1 & 2 & 0 & 3 \\ 1 & 0 & 2 & 2 \\ 0 & 0 & 1 & 1 \end{bmatrix}$,

$\mathbf{x} = (x_j) = [1000, 500, 350, 400]$, and $\mathbf{b} = (b_i) = [3500, 2700, 900]$. Is there sufficient food to satisfy the average daily consumption?

b) What is the maximum number of animals of each species that could be individually added to the system with the supply of food still meeting the consumption?

c) If species 1 became extinct, how much of an individual increase of each of the remaining species could be supported?

d) If species 2 also became extinct, how much of an individual increase of each of the remaining species could be supported?

6. Show that the operations

i) $(\lambda E_i) \rightarrow (E_i)$

ii) $(E_i + \lambda E_j) \rightarrow (E_i)$

iii) $(E_i) \leftrightarrow (E_j)$

do not change the solution set of a linear system.

6.2 Gaussian Elimination and Backward Substitution

The general Gaussian elimination procedure applied to the linear system

$$
\begin{aligned}
E_1: &\quad a_{11}x_1 + a_{12}x_2 + \cdots + a_{1n}x_n = b_1, \\
E_2: &\quad a_{21}x_1 + a_{22}x_2 + \cdots + a_{2n}x_n = b_2, \\
&\quad \vdots \qquad \vdots \qquad \vdots \qquad \qquad \vdots \qquad \vdots \\
E_n: &\quad a_{n1}x_1 + a_{n2}x_2 + \cdots + a_{nn}x_n = b_n,
\end{aligned}
$$

(6.8)

is handled in a manner similar to the procedure followed in Example 3 of Section 6.1. We form the augmented matrix $\tilde{A}$:

$$(6.9) \qquad \tilde{A} = [A, \mathbf{b}] = \begin{bmatrix} a_{11} & a_{12} & \cdots & a_{1n} & \vdots & a_{1,n+1} \\ a_{21} & a_{22} & \cdots & a_{2n} & \vdots & a_{2,n+1} \\ \vdots & \vdots & & \vdots & \vdots & \vdots \\ a_{n1} & a_{n2} & \cdots & a_{nn} & \vdots & a_{n,n+1} \end{bmatrix},$$

where A denotes the matrix formed by the coefficients and the entries in the $(n + 1)$st column are the values of $\mathbf{b}$, that is, $a_{i,n+1} = b_i$ for each $i = 1, 2, \ldots, n$.

Provided $a_{11} \neq 0$, the operations corresponding to $(E_j - (a_{j1}/a_{11})E_1) \rightarrow (E_j)$ are performed for each $j = 2, 3, \ldots, n$ to eliminate the coefficient of x_1 in each of these rows. Although the entries in rows 2, 3, ..., n are expected to change, for ease of notation, we will again denote the entry in the ith row and the jth column by a_{ij}. With this in mind, we follow a sequential procedure for $i = 2, 3, \ldots, n - 1$ and perform the operation $(E_j - (a_{ji}/a_{ii})E_i) \rightarrow (E_j)$ for each $j = i + 1, i + 2, \ldots, n$, provided $a_{ii} \neq 0$. This will eliminate (that is, change the coefficient to zero) x_i in each row below the ith for all values of $i = 1, 2, \ldots, n - 1$. The resulting matrix will have the form:

$$\tilde{\tilde{A}} = \begin{bmatrix} a_{11} & a_{12} & \cdots & a_{1n} & \vdots & a_{1,n+1} \\ 0 & a_{22} & \cdots & a_{2n} & \vdots & a_{2,n+1} \\ \vdots & & \ddots & \vdots & \vdots & \vdots \\ 0 & \cdots\cdots & 0 & a_{nn} & \vdots & a_{n,n+1} \end{bmatrix},$$

where, as was just mentioned, the values of a_{ij} are not expected to agree with those in the original matrix $\tilde{A}$. This matrix represents a linear system with the same solution set as system (6.8). Since the equivalent linear system is triangular:

$$a_{11}x_1 + a_{12}x_2 + \cdots + a_{1n}x_n = a_{1,n+1},$$
$$a_{22}x_2 + \cdots + a_{2n}x_n = a_{2,n+1},$$
$$\vdots \qquad \vdots$$
$$a_{nn}x_n = a_{n,n+1},$$

the backward substitution can be performed. Solving the nth equation for x_n gives:

$$x_n = \frac{a_{n,n+1}}{a_{nn}}.$$

Solving the $(n - 1)$st equation for x_{n-1} and using x_n yields

$$x_{n-1} = \frac{[a_{n-1,n+1} - a_{n-1,n}x_n]}{a_{n-1,n-1}};$$

and continuing this process, we obtain:

$$x_i = \frac{[a_{i,n+1} - a_{in}x_n - a_{i,n-1}x_{n-1} - \cdots - a_{i,i+1}x_{i+1}]}{a_{ii}}$$

$$= \frac{[a_{i,n+1} - \sum_{j=i+1}^{n} a_{ij}x_j]}{a_{ii}} \qquad \text{for each } i = n - 1, n - 2, \ldots, 2, 1.$$

The Gaussian elimination procedure can be presented more precisely, although more intricately, by forming a sequence of augmented matrices $\tilde{A}^{(1)}$, $\tilde{A}^{(2)}$, ..., $\tilde{A}^{(n)}$ where $\tilde{A}^{(1)}$ is the matrix $\tilde{A}$ given in (6.9) and $\tilde{A}^{(k)}$ for each $k = 2, 3, ..., n$ has entries $a_{ij}^{(k)}$ where:

$$
a_{ij}^{(k)} = \begin{cases} a_{ij}^{(k-1)} & \text{when } i = 1, 2, ..., k-1 \text{ and } j = 1, 2, ..., n+1, \\ 0 & \text{when } i = k, k+1, ..., n \text{ and } j = 1, 2, ..., k-1, \\ a_{ij}^{(k-1)} - \dfrac{a_{i,k-1}^{(k-1)}}{a_{k-1,k-1}^{(k-1)}} a_{k-1,j}^{(k-1)} & \text{when } i = k, k+1, ..., n \text{ and } j = k, k+1, ..., n+1. \end{cases}
$$

Thus

$$
\tilde{A}^{(k)} = \begin{bmatrix} a_{11}^{(1)} & a_{12}^{(1)} & a_{13}^{(1)} & \cdots & a_{1,k-1}^{(1)} & a_{1k}^{(1)} & \cdots & a_{1n}^{(1)} & \vdots & a_{1,n+1}^{(1)} \\ 0 & a_{22}^{(2)} & a_{23}^{(2)} & \cdots & a_{2,k-1}^{(2)} & a_{2k}^{(2)} & \cdots & a_{2n}^{(2)} & \vdots & a_{2,n+1}^{(2)} \\ & & & & \vdots & \vdots & & \vdots & \vdots & \vdots \\ & & & & a_{k-1,k-1}^{(k-1)} & a_{k-1,k}^{(k-1)} & \cdots & a_{k-1,n}^{(k-1)} & \vdots & a_{k-1,n+1}^{(k-1)} \\ & & & & 0 & a_{kk}^{(k)} & \cdots & a_{kn}^{(k)} & \vdots & a_{k,n+1}^{(k)} \\ & & & & & \vdots & & \vdots & \vdots & \vdots \\ 0 & \cdots & & \cdots & 0 & a_{nk}^{(k)} & \cdots & a_{nn}^{(k)} & \vdots & a_{n,n+1}^{(k)} \end{bmatrix}
$$

represents the equivalent linear system for which the variable x_{k-1} has just been eliminated from equations $E_k, E_{k+1}, ..., E_n$.

The procedure will not work if any of the elements $a_{11}^{(1)}, a_{22}^{(2)}, a_{33}^{(3)}, ..., a_{n-1,n-1}^{(n-1)}$, $a_{nn}^{(n)}$ are zero for, in this case, the step

$$
\left(E_i - \frac{a_{i,k}^{(k)}}{a_{kk}^{(k)}} E_k \right) \to E_i
$$

either cannot be performed (this occurs if one of $a_{11}^{(1)}, ..., a_{n-1,n-1}^{(n-1)}$ is zero), or the backward substitution cannot be accomplished (in the case $a_{nn}^{(n)} = 0$). This does not necessarily mean that the linear system is not solvable, but that the technique of solution must be altered. An illustration is given in the following example.

EXAMPLE 1　　　Consider the linear system

$$
\begin{aligned}
E_1: & \quad x_1 - x_2 + 2x_3 - x_4 = -8, \\
E_2: & \quad 2x_1 - 2x_2 + 3x_3 - 3x_4 = -20, \\
E_3: & \quad x_1 + x_2 + x_3 = -2, \\
E_4: & \quad x_1 - x_2 + 4x_3 + 3x_4 = 4.
\end{aligned}
$$

The augmented matrix is:

$$
\tilde{A} = \tilde{A}^{(1)} = \begin{bmatrix} 1 & -1 & 2 & -1 & \vdots & -8 \\ 2 & -2 & 3 & -3 & \vdots & -20 \\ 1 & 1 & 1 & 0 & \vdots & -2 \\ 1 & -1 & 4 & 3 & \vdots & 4 \end{bmatrix},
$$

and, performing the operations

$$(E_2 - 2E_1) \to (E_2), \qquad (E_3 - E_1) \to (E_3), \qquad \text{and} \qquad (E_4 - E_1) \to (E_4),$$

we write:

$$\tilde{A}^{(2)} = \left[\begin{array}{rrrr:r} 1 & -1 & 2 & -1 & -8 \\ 0 & 0 & -1 & -1 & -4 \\ 0 & 2 & -1 & 1 & 6 \\ 0 & 0 & 2 & 4 & 12 \end{array} \right].$$

Since the element $a_{22}^{(2)}$, called the **pivot element**, is zero, the procedure cannot continue in its present form. But the operation $(E_i) \leftrightarrow (E_j)$ is permitted, so a search is made of the elements $a_{32}^{(2)}$ and $a_{42}^{(2)}$ for the first nonzero element. Since $a_{32}^{(2)} \neq 0$, the operation $(E_2) \leftrightarrow (E_3)$ is performed to obtain a new matrix

$$\tilde{A}^{(2)'} = \left[\begin{array}{rrrr:r} 1 & -1 & 2 & -1 & -8 \\ 0 & 2 & -1 & 1 & 6 \\ 0 & 0 & -1 & -1 & -4 \\ 0 & 0 & 2 & 4 & 12 \end{array} \right].$$

Since x_2 is already eliminated from E_3 and E_4, $\tilde{A}^{(3)}$ will be $\tilde{A}^{(2)'}$ and the computation can continue with the operation $(E_4 + 2E_3) \to (E_4)$, giving

$$\tilde{A}^{(4)} = \left[\begin{array}{rrrr:r} 1 & -1 & 2 & -1 & -8 \\ 0 & 2 & -1 & 1 & 6 \\ 0 & 0 & -1 & -1 & -4 \\ 0 & 0 & 0 & 2 & 4 \end{array} \right].$$

Finally, the backward substitution can be applied:

$$x_4 = \frac{4}{2} = 2,$$

$$x_3 = \frac{[-4 - (-1)x_4]}{-1} = 2,$$

$$x_2 = \frac{[6 - x_4 - (-1)x_3]}{2} = 3,$$

$$x_1 = \frac{[-8 - (-1)x_4 - 2x_3 - (-1)x_2]}{1} = -7. \qquad \blacksquare$$

Example 1 illustrates what is done if $a_{kk}^{(k)} = 0$ for some $k = 1, 2, \ldots, n - 1$. The kth column of $\tilde{A}^{(k-1)}$ from the kth row to the nth row is searched for the first nonzero entry. If $a_{pk}^{(k)} \neq 0$ for some p, $k + 1 \leq p \leq n$, then the operation $(E_k) \leftrightarrow (E_p)$ is performed to obtain $\tilde{A}^{(k-1)'}$. The procedure can then be continued to form $\tilde{A}^{(k)}$, and so on. If $a_{pk}^{(k)} = 0$ for $p = k, k + 1, \ldots, n$, it can be shown (see Theorem 6.12, p. 322) that the linear system does not have a unique solution and the procedure stops. Finally, if $a_{nn}^{(n)} = 0$, the linear system does not have a unique solution and again the procedure stops. To summarize the entire Gaussian elimination with backward substitution, we present the following algorithm.

Gaussian Elimination with Backward Substitution Algorithm 6.1

To solve the $n \times n$ linear system

$$E_1: \quad a_{11}x_1 + a_{12}x_2 + \cdots + a_{1n}x_n = a_{1,n+1}$$
$$E_2: \quad a_{21}x_1 + a_{22}x_2 + \cdots + a_{2n}x_n = a_{2,n+1}$$
$$\vdots \qquad \vdots \qquad \vdots \qquad\qquad \vdots \qquad\quad \vdots$$
$$E_n: \quad a_{n1}x_1 + a_{n2}x_2 + \cdots + a_{nn}x_n = a_{n,n+1}:$$

INPUT number of unknowns and equations n; augmented matrix $A = (a_{ij})$ where $1 \le i \le n$ and $1 \le j \le n+1$.

OUTPUT solution $x_1, x_2, \ldots, x_n$ or message that linear system has no unique solution.

Step 1   For $i = 1, \ldots, n-1$ do Steps 2–4.   (*Elimination process.*)

Step 2   Let p be the smallest integer with $i \le p \le n$ and $a_{pi} \ne 0$.
If no integer p can be found
then OUTPUT ('no unique solution exists');
STOP.

Step 3   If $p \ne i$ then perform $(E_p) \leftrightarrow (E_i)$.

Step 4   For $j = i+1, \ldots, n$ do Steps 5 and 6.

Step 5   Set $m_{ji} = a_{ji}/a_{ii}$.

Step 6   Perform $(E_j - m_{ji}E_i) \to (E_j)$.

Step 7   If $a_{nn} = 0$ then OUTPUT ('no unique solution exists');
STOP.

Step 8   Set $x_n = a_{n,n+1}/a_{nn}$.   (*Start backward substitution.*)

Step 9   For $i = n-1, \ldots, 1$ set $x_i = \left[a_{i,n+1} - \sum_{j=i+1}^{n} a_{ij}x_j \right] \Big/ a_{ii}$.

Step 10   OUTPUT $(x_1, \ldots, x_n)$;   (*Procedure completed successfully.*)
STOP.

EXAMPLE 2   The purpose of this example is to show what can happen if Algorithm 6.1 fails. The computations will be done simultaneously on two linear systems:

$$
\begin{aligned}
x_1 + x_2 + x_3 + x_4 &= 7, \\
x_1 + x_2 \phantom{{}+x_3} + 2x_4 &= 8, \\
2x_1 + 2x_2 + 3x_3 \phantom{{}+2x_4} &= 10, \\
-x_1 - x_2 - 2x_3 + 2x_4 &= 0,
\end{aligned}
\qquad \text{and} \qquad
\begin{aligned}
x_1 + x_2 + x_3 + x_4 &= 7, \\
x_1 + x_2 \phantom{{}+x_3} + 2x_4 &= 5, \\
2x_1 + 2x_2 + 3x_3 \phantom{{}+2x_4} &= 10, \\
-x_1 - x_2 - 2x_3 + 2x_4 &= 0.
\end{aligned}
$$

These systems produce matrices

$$\tilde{A} = \begin{bmatrix} 1 & 1 & 1 & 1 & \vdots & 7 \\ 1 & 1 & 0 & 2 & \vdots & 8 \\ 2 & 2 & 3 & 0 & \vdots & 10 \\ -1 & -1 & -2 & 2 & \vdots & 0 \end{bmatrix} \quad \text{and} \quad \tilde{A} = \begin{bmatrix} 1 & 1 & 1 & 1 & \vdots & 7 \\ 1 & 1 & 0 & 2 & \vdots & 5 \\ 2 & 2 & 3 & 0 & \vdots & 10 \\ -1 & -1 & -2 & 2 & \vdots & 0 \end{bmatrix}.$$

Since $a_{11} \neq 0$, the steps in eliminating x_1 from E_2, E_3, and E_4 give

$$m_{ji} = m_{j1} = a_{j1}/1 = a_{j1}.$$
$$j = 2, \quad m_{21} = 1,$$
$$j = 3, \quad m_{31} = 2,$$
$$j = 4, \quad m_{41} = -1,$$
$$(E_2 - E_1) \rightarrow (E_2),$$
$$(E_3 - 2E_1) \rightarrow (E_3),$$
$$(E_4 + E_1) \rightarrow (E_4).$$

The matrices become:

$$\tilde{A} = \begin{bmatrix} 1 & 1 & 1 & 1 & \vdots & 7 \\ 0 & 0 & -1 & 1 & \vdots & 1 \\ 0 & 0 & 1 & -2 & \vdots & -4 \\ 0 & 0 & -1 & 3 & \vdots & 7 \end{bmatrix} \quad \text{and} \quad \tilde{A} = \begin{bmatrix} 1 & 1 & 1 & 1 & \vdots & 7 \\ 0 & 0 & -1 & 1 & \vdots & -2 \\ 0 & 0 & 1 & -2 & \vdots & -4 \\ 0 & 0 & -1 & 3 & \vdots & 7 \end{bmatrix}.$$

At this point $a_{22} = a_{23} = a_{24} = 0$. The algorithm requires that the procedure be halted, and no solution to either system is obtained.

To examine more closely the reason for difficulty, perform $(E_4 + E_3) \rightarrow (E_4)$ to obtain:

$$\begin{bmatrix} 1 & 1 & 1 & 1 & \vdots & 7 \\ 0 & 0 & -1 & 1 & \vdots & 1 \\ 0 & 0 & 1 & -2 & \vdots & -4 \\ 0 & 0 & 0 & 1 & \vdots & 3 \end{bmatrix} \quad \text{and} \quad \begin{bmatrix} 1 & 1 & 1 & 1 & \vdots & 7 \\ 0 & 0 & -1 & 1 & \vdots & -2 \\ 0 & 0 & 1 & -2 & \vdots & -4 \\ 0 & 0 & 0 & 1 & \vdots & 3 \end{bmatrix}.$$

Writing the equations for each system gives:

$$
\begin{aligned}
x_1 + x_2 + x_3 + x_4 &= 7, \\
-x_3 + x_4 &= 1, \\
x_3 - 2x_4 &= -4, \\
x_4 &= 3,
\end{aligned}
\quad \text{and} \quad
\begin{aligned}
x_1 + x_2 + x_3 + x_4 &= 7, \\
-x_3 + x_4 &= -2, \\
x_3 - 2x_4 &= -4, \\
x_4 &= 3.
\end{aligned}
$$

Performing backward substitution in each system yields:

$$
\begin{aligned}
x_4 &= 3, \\
x_3 &= -4 + 2x_4 = 2,
\end{aligned}
\quad \text{and} \quad
\begin{aligned}
x_4 &= 3, \\
x_3 &= -4 + 2x_4 = 2.
\end{aligned}
$$

If the backward substitution is continued to the second equation in each case, the difference in the two systems becomes apparent because

$$-x_3 + x_4 = 1, \qquad \text{while} \qquad -x_3 + x_4 = -2,$$
$$\text{implies} \quad 1 = 1, \qquad\qquad \text{implies} \quad 1 = -2.$$

The first linear system has an infinite number of solutions $x_4 = 3$, $x_3 = 2$, x_2 arbitrary, and $x_1 = 2 - x_2$, whereas the other leads to a contradiction, and no solution exists. In both cases, however, there is no unique solution as we conclude from Algorithm 6.1. ∎

Although Algorithm 6.1 can be viewed as the construction of the augmented matrices $\tilde{A}^{(1)}, \ldots, \tilde{A}^{(n)}$, the computations can be performed in a computer using only one n by $(n + 1)$ array for storage by simply replacing at each step the previous value of a_{ij} by the new one. It is also advantageous to store the multipliers m_{ji} in the locations of a_{ji} since a_{ji} has the value zero for each $i = 1, 2, \ldots, n - 1$, and $j = i + 1, i + 2, \ldots, n$. Thus, A can be overwritten by the multipliers below the main diagonal and by the nonzero entries of $\tilde{A}^{(n)}$ on and above the main diagonal. These values can be used to solve other linear systems involving the original matrix A, as we will see in Section 6.7.

When comparing techniques for solving linear systems, we need to consider other concepts in addition to the amount of storage required. One of these is the effect of round-off error and another is the amount of time required to complete the calculations. Both of these topics depend on the number of arithmetic operations that need to be performed in solving a routine problem. In general, the amount of time required to perform a multiplication or division on a computer is about the same, and is considerably greater than that required to perform an addition or subtraction. The actual differences in execution time, however, depend on the particular computing system being used. To demonstrate the procedure involved with counting operations for a given method, we will count the operations required to solve a typical linear system of n equations in n unknowns using Algorithm 6.1. We will keep the count of the additions/subtractions separate from the count of the multiplications/divisions because of the time differential.

No arithmetic operations are performed until Steps 5 and 6 in the algorithm. Step 5 requires that $(n - i)$ divisions be performed. The replacement of the equation E_j by $(E_j - m_{ji}E_i)$ in Step 6 requires that m_{ji} be multiplied by each term in E_i resulting in a total of $(n - i)(n - i + 1)$ multiplications. After this is completed, each term of the resulting equation is subtracted from the corresponding term in E_j. This requires $(n - i)(n - i + 1)$ subtractions. For each $i = 1, 2, \ldots, n - 1$, the operations required in Steps 5 and 6 are

Multiplications/divisions

$$(n - i) + (n - i)(n - i + 1) = (n - i)(n - i + 2);$$

Additions/subtractions

$$(n - i)(n - i + 1).$$

The total number of operations required by these steps is obtained by summing the operation counts for each i. Recalling that

$$\sum_{j=1}^{m} 1 = m, \quad \sum_{j=1}^{m} j = \frac{m(m+1)}{2}, \quad \text{and} \quad \sum_{j=1}^{m} j^2 = \frac{m(m+1)(2m+1)}{6},$$

Multiplications/divisions

$$\sum_{i=1}^{n-1} (n-i)(n-i+2) = (n^2 + 2n) \sum_{i=1}^{n-1} 1 - 2(n+1) \sum_{i=1}^{n-1} i + \sum_{i=1}^{n-1} i^2$$

$$= (n^2 + 2n)(n-1) - 2(n+1)\frac{(n-1)n}{2}$$

$$+ \frac{(n-1)(n)(2n-1)}{6}$$

$$= \frac{2n^3 + 3n^2 - 5n}{6}.$$

Additions/subtractions

$$\sum_{i=1}^{n-1} (n-i)(n-i+1) = (n^2 + n) \sum_{i=1}^{n-1} 1 - (2n+1) \sum_{i=1}^{n-1} i + \sum_{i=1}^{n-1} i^2$$

$$= (n^2 + n)(n-1) - (2n+1)\frac{(n-1)n}{2}$$

$$+ \frac{(n-1)(n)(2n-1)}{6}$$

$$= \frac{n^3 - n}{3}.$$

The only other steps in Algorithm 6.1 that require arithmetic operations are Steps 8 and 9. Step 8 requires one division. Step 9 requires $(n-i)$ multiplications and $(n-i-1)$ additions for each summation term, and then one subtraction and one division. The total number of operations in Steps 8 and 9 is

Multiplications/divisions

$$1 + \sum_{i=1}^{n-1} ((n-i)+1) = \frac{n^2+n}{2}.$$

Additions/subtractions

$$\sum_{i=1}^{n-1} ((n-i-1)+1) = \frac{n^2-n}{2}.$$

The total arithmetic operations in Algorithm 6.1 is therefore

Multiplications/divisions

$$\frac{2n^3 + 3n^2 - 5n}{6} + \frac{n^2 + n}{2} = \frac{n^3 + 3n^2 - n}{3}.$$

Additions/subtractions

$$\frac{n^3 - n}{3} + \frac{n^2 - n}{2} = \frac{2n^3 + 3n^2 - 5n}{6}.$$

Since the total number of multiplications and divisions is approximately $\frac{1}{3}n^3$, and similarly for additions and subtractions, the amount of computation and the time required will increase with n in proportion to n^3. Table 6.1 indicates the growth of computations with increasing n.

TABLE 6.1

n	Multiplications/divisions	Additions/subtractions
3	17	11
10	430	375
50	44,150	42,875
100	343,300	338,250

Exercise Set 6.2

1. Solve the following linear systems using Gaussian elimination with backward substitution and two-digit rounding arithmetic. Do not reorder the equations. (The exact solution to each system is $x_1 = 1$, $x_2 = -1$, $x_3 = 3$.)

   a) $4x_1 + x_2 - x_3 = 6,$
   $2x_1 + 5x_2 + 2x_3 = 3,$
   $x_1 + 2x_2 + 4x_3 = 11.$

   b) $x_1 + 2x_2 + 4x_3 = 11,$
   $4x_1 + x_2 - x_3 = 6,$
   $2x_1 + 5x_2 + 2x_3 = 3.$

   c) $4x_1 + x_2 + 2x_3 = 9,$
   $2x_1 + 4x_2 - x_3 = -5,$
   $x_1 + x_2 - 3x_3 = -9.$

   d) $2x_1 + 4x_2 - x_3 = -5,$
   $x_1 + x_2 - 3x_3 = -9,$
   $4x_1 + x_2 + 2x_3 = 9.$

2. Use Algorithm 6.1 and exact arithmetic to solve the following linear systems, if possible, and determine whether row interchanges are necessary.

   a) $x_1 - x_2 + 3x_3 = 2,$
   $3x_1 - 3x_2 + x_3 = -1,$
   $x_1 + x_2 = 3.$

   b) $x_2 + 4x_3 = 0,$
   $x_1 - x_2 - x_3 = 0.375,$
   $x_1 - x_2 + 2x_3 = 0.$

   c) $2x_1 - 1.5x_2 + 3x_3 = 1,$
   $-x_1 + 2x_3 = 3,$
   $4x_1 - 4.5x_2 + 5x_3 = 1.$

   d) $2x_1 - x_2 + x_3 = -1,$
   $3x_1 + 3x_2 + 9x_3 = 0,$
   $3x_1 + 3x_2 + 5x_3 = 4.$

e) $\quad 2x_1 \qquad\qquad\qquad = 3,$

$\qquad x_1 + 1.5x_2 \qquad\qquad = 4.5,$

$\qquad - 3x_2 + 0.5x_3 \qquad = -6.6,$

$\qquad 2x_1 - 2x_2 + x_3 + x_4 = 0.8.$

f) $\quad x_1 - \frac{1}{2}x_2 + x_3 \qquad = 4,$

$\qquad 2x_1 - x_2 - x_3 + x_4 = 5,$

$\qquad x_1 + x_2 \qquad\qquad = 2,$

$\qquad x_1 - \frac{1}{2}x_2 + x_3 + x_4 = 5.$

3. Use Algorithm 6.1 and single precision arithmetic on a computer to solve the following linear systems, if possible.

a) $\quad \frac{1}{4}x_1 + \frac{1}{5}x_2 + \frac{1}{6}x_3 = 9,$

$\qquad \frac{1}{3}x_1 + \frac{1}{4}x_2 + \frac{1}{5}x_3 = 8,$

$\qquad \frac{1}{2}x_1 + x_2 + 2x_3 = 8.$

b) $\quad 3.333x_1 + 15920x_2 - 10.333x_3 = 15913,$

$\qquad 2.222x_1 + 16.71x_2 + 9.612x_3 = 28.544,$

$\qquad 1.5611x_1 + 5.1791x_2 + 1.6852x_3 = 8.4254.$

c) $\quad 4.01x_1 + 1.23x_2 + 1.43x_3 - 0.73x_4 = 5.94,$

$\qquad 1.23x_1 + 7.41x_2 + 2.41x_3 + 3.02x_4 = 14.07,$

$\qquad 1.43x_1 + 2.41x_2 + 5.79x_3 - 1.11x_4 = 8.52,$

$\qquad -0.73x_1 + 3.02x_2 - 1.11x_3 + 6.41x_4 = 7.59.$

d) $\quad x_1 + \frac{1}{2}x_2 + \frac{1}{3}x_3 + \frac{1}{4}x_4 = \frac{1}{6},$

$\qquad \frac{1}{2}x_1 + \frac{1}{3}x_2 + \frac{1}{4}x_3 + \frac{1}{5}x_4 = \frac{1}{7},$

$\qquad \frac{1}{3}x_1 + \frac{1}{4}x_2 + \frac{1}{5}x_3 + \frac{1}{6}x_4 = \frac{1}{8},$

$\qquad \frac{1}{4}x_1 + \frac{1}{5}x_2 + \frac{1}{6}x_3 + \frac{1}{7}x_4 = \frac{1}{9}.$

4. **Gauss–Jordan Method** This method, used to solve the linear system (6.8), can be described as follows. Use the ith equation to eliminate not only x_i from the equations E_{i+1}, E_{i+2}, ..., E_n, as was done in the Gaussian elimination method, but also from $E_1, E_2, ...,$ E_{i-1}. Upon reducing $[A, \mathbf{b}]$ to:

$$\begin{bmatrix} a_{11}^{(1)} & 0 & \cdots & 0 & \vdots & a_{1,n+1}^{(1)} \\ 0 & a_{22}^{(2)} & \ddots & \vdots & \vdots & a_{2,n+1}^{(2)} \\ \vdots & \ddots & \ddots & 0 & \vdots & \vdots \\ 0 & \cdots & 0 & a_{nn}^{(n)} & \vdots & a_{n,n+1}^{(n)} \end{bmatrix},$$

the solution is obtained by setting

$$x_i = \frac{a_{i,n+1}^{(i)}}{a_{ii}^{(i)}}$$

for each $i = 1, 2, ..., n$. This procedure circumvents the backward substitution in the Gaussian elimination. Construct an algorithm for the Gauss–Jordan procedure patterned after that of Algorithm 6.1.

5. Use the Gauss–Jordan method and two-digit rounding arithmetic to solve the systems in Exercise 1.

6. Repeat Exercise 3 using the Gauss–Jordan method.

7. a) Show that the Gauss–Jordan method requires

$$\frac{n^3}{2} + n^2 - \frac{n}{2} \quad \text{multiplications/divisions} \qquad \text{and} \qquad \frac{n^3}{2} - \frac{n}{2} \quad \text{additions/subtractions.}$$

   b) Make a table comparing the required operations for the Gauss–Jordan and Gaussian elimination methods for $n = 3, 10, 50, 100$. Which method requires less computation?

8. Consider the following Gaussian-elimination–Gauss–Jordan hybrid method for solving the system (6.8). First, apply the Gaussian-elimination technique to reduce the system to triangular form shown on page 297. Then use the nth equation to eliminate the coefficients of x_n in each of the first $n - 1$ rows. After this is completed, use the $n - 1$st equation to eliminate the coefficients of x_{n-1} in the first $n - 2$ rows, etc. The system will eventually appear as the reduced system in Exercise 4.

   a) Determine the number of multiplications/divisions and additions/subtractions required to solve an $n \times n$ system using this method.

   b) Make a table comparing the required operations for the Gaussian elimination, Gauss–Jordan, and hybrid methods for $n = 3, 10, 50, 100$.

9. Use the hybrid method described in Exercise 8 and two-digit arithmetic to solve the systems in Exercise 1.

10. Repeat Exercise 3 using the method described in Exercise 8.

6.3 Linear Algebra and Matrix Inversion

The concept of a matrix was introduced in the previous two sections as a convenient method of expressing and manipulating a linear system. The material in this section is the algebra associated with matrices and the way it can be used to solve problems involving linear systems.

DEFINITION 6.2   Two matrices A and B are said to be **equal** if both are of the same size, say $m \times n$, and if $a_{ij} = b_{ij}$ for each $i = 1, 2, \ldots, m$, and $j = 1, 2, \ldots, n$.

EXAMPLE 1   Let

$$A = \begin{bmatrix} 2 & -1 & 7 \\ 3 & 1 & 0 \end{bmatrix}, \qquad B = \begin{bmatrix} 2 & -1 & 7 \\ 3 & 1 & 0 \end{bmatrix}, \qquad C = \begin{bmatrix} 2 & 3 \\ -1 & 1 \\ 7 & 0 \end{bmatrix}.$$

Then $A = B$, but $A \neq C$. ∎

Two important operations performed on matrices are the sum of two matrices and the multiplication of a matrix by a real number.

DEFINITION 6.3   If A and B are both $m \times n$ matrices, then the **sum** of A and B, denoted $A + B$, is the $m \times n$ matrix whose entries are $a_{ij} + b_{ij}$, for each $i = 1, 2, \ldots, m$, and $j = 1, 2, \ldots, n$.

DEFINITION 6.4   If A is an $m \times n$ matrix and λ a real number, then the **scalar product** of λ and A, denoted λA, is the $m \times n$ matrix whose entries are λa_{ij} for each $i = 1, 2, \ldots, m$ and $j = 1, 2, \ldots, n$.

EXAMPLE 2   Let

$$A = \begin{bmatrix} 2 & -1 & 7 \\ 3 & 1 & 0 \end{bmatrix}, \qquad B = \begin{bmatrix} 4 & 2 & -8 \\ 0 & 1 & 6 \end{bmatrix}, \qquad C = \begin{bmatrix} 1 & 2 & 3 \\ 4 & 5 & 6 \end{bmatrix}$$

and $\lambda = -2$, $\mu = 3$. Then

$$A + B = \begin{bmatrix} 2+4 & -1+2 & 7-8 \\ 3+0 & 1+1 & 0+6 \end{bmatrix} = \begin{bmatrix} 6 & 1 & -1 \\ 3 & 2 & 6 \end{bmatrix};$$

$$B + A = \begin{bmatrix} 6 & 1 & -1 \\ 3 & 2 & 6 \end{bmatrix};$$

$$(A + B) + C = \begin{bmatrix} (2+4)+1 & (-1+2)+2 & (7-8)+3 \\ (3+0)+4 & (1+1)+5 & (0+6)+6 \end{bmatrix} = \begin{bmatrix} 7 & 3 & 2 \\ 7 & 7 & 12 \end{bmatrix};$$

$$A + (B + C) = \begin{bmatrix} 7 & 3 & 2 \\ 7 & 7 & 12 \end{bmatrix};$$

$$\lambda(A + B) = \begin{bmatrix} -2(6) & -2(1) & -2(-1) \\ -2(3) & -2(2) & -2(6) \end{bmatrix} = \begin{bmatrix} -12 & -2 & 2 \\ -6 & -4 & -12 \end{bmatrix};$$

$$\lambda A + \lambda B = \begin{bmatrix} -4 & 2 & -14 \\ -6 & -2 & 0 \end{bmatrix} + \begin{bmatrix} -8 & -4 & 16 \\ 0 & -2 & -12 \end{bmatrix} = \begin{bmatrix} -12 & -2 & 2 \\ -6 & -4 & -12 \end{bmatrix};$$

$$(\lambda + \mu)A = (-2 + 3)A = \begin{bmatrix} 2 & -1 & 7 \\ 3 & 1 & 0 \end{bmatrix};$$

$$\lambda A + \mu A = -2A + 3A = \begin{bmatrix} -4 & 2 & -14 \\ -6 & -2 & 0 \end{bmatrix} + \begin{bmatrix} 6 & -3 & 21 \\ 9 & 3 & 0 \end{bmatrix}$$

$$= \begin{bmatrix} 2 & -1 & 7 \\ 3 & 1 & 0 \end{bmatrix};$$

$$\lambda(\mu A) = (-2)(3A) = \begin{bmatrix} -12 & 6 & -42 \\ -18 & -6 & 0 \end{bmatrix};$$

$$(\lambda \mu)A = [(-2)(3)]A = \begin{bmatrix} -12 & 6 & -42 \\ -18 & -6 & 0 \end{bmatrix};$$

and $$1A = \begin{bmatrix} 2 & -1 & 7 \\ 3 & 1 & 0 \end{bmatrix} = A.$$ ∎

Letting the matrix all of whose entries are zero be denoted simply by O, and $-A$ be the matrix whose entries are $-a_{ij}$, we can list the following general properties of matrix addition and scalar multiplication. These properties are sufficient to classify the set of all $m \times n$ matrices with real entries as a **vector space** over the field of real numbers. (See Noble and Daniel, [92], pp. 107–109.)

THEOREM 6.5   Let A, B, and C be $m \times n$ matrices and λ and μ real numbers. The following properties of addition and scalar multiplication hold:

a) $A + B = B + A$,
b) $(A + B) + C = A + (B + C)$,
c) $A + O = O + A = A$,
d) $A + (-A) = -A + A = O$,
e) $\lambda(A + B) = \lambda A + \lambda B$,
f) $(\lambda + \mu)A = \lambda A + \mu A$,
g) $\lambda(\mu A) = (\lambda \mu)A$,
h) $1A = A$.

Properties (a), (b), (e), (f), (g), and (h) were demonstrated in Example 2 for the special matrices given there. All of the properties can be easily verified. (See Exercise 5.)

DEFINITION 6.6   Let A be an $m \times n$ matrix and B an $n \times p$ matrix. The **matrix product** of A and B, denoted AB, is an $m \times p$ matrix C whose entries c_{ij} are given by

(6.10)
$$c_{ij} = \sum_{k=1}^{n} a_{ik}b_{kj} = a_{i1}b_{1j} + a_{i2}b_{2j} + \cdots + a_{in}b_{nj}$$

for each $i = 1, 2, \ldots, m$, and $j = 1, 2, \ldots, p$.

The computation of c_{ij} can be viewed as the multiplication of the entries of the ith row of A with corresponding entries in the jth column of B, followed by a summation. That is,

$$[a_{i1}, a_{i2}, \ldots, a_{in}] \begin{bmatrix} b_{1j} \\ b_{2j} \\ \vdots \\ b_{nj} \end{bmatrix} = c_{ij},$$

where
$$c_{ij} = a_{i1}b_{1j} + a_{i2}b_{2j} + \cdots + a_{in}b_{nj} = \sum_{k=1}^{n} a_{ik}b_{kj}.$$

This explains why the number of columns of A must equal the number of rows of B in order for the product AB to be defined.

The following example should serve to further clarify the matrix multiplication process.

EXAMPLE 3   Let

$$A = \begin{bmatrix} 2 & 1 & -1 \\ 3 & 1 & 2 \\ 0 & -2 & -3 \end{bmatrix}, \quad B = \begin{bmatrix} 3 & 2 \\ -1 & 1 \\ 6 & 4 \end{bmatrix},$$

$$C = \begin{bmatrix} 2 & 1 & 0 \\ -1 & 3 & 2 \end{bmatrix}, \quad \text{and} \quad D = \begin{bmatrix} 1 & -1 & 1 \\ 2 & -1 & 2 \\ 3 & 0 & 3 \end{bmatrix}.$$

Then,

$$AD = \begin{bmatrix} 2 & 1 & -1 \\ 3 & 1 & 2 \\ 0 & -2 & -3 \end{bmatrix} \begin{bmatrix} 1 & -1 & 1 \\ 2 & -1 & 2 \\ 3 & 0 & 3 \end{bmatrix} = \begin{bmatrix} 1 & -3 & 1 \\ 11 & -4 & 11 \\ -13 & 2 & -13 \end{bmatrix},$$

and $$DA = \begin{bmatrix} 1 & -1 & 1 \\ 2 & -1 & 2 \\ 3 & 0 & 3 \end{bmatrix} \begin{bmatrix} 2 & 1 & -1 \\ 3 & 1 & 2 \\ 0 & -2 & -3 \end{bmatrix} = \begin{bmatrix} -1 & -2 & -6 \\ 1 & -3 & -10 \\ 6 & -3 & -12 \end{bmatrix},$$

which shows that $AD \neq DA$.

Further,

$$BC = \begin{bmatrix} 3 & 2 \\ -1 & 1 \\ 6 & 4 \end{bmatrix} \begin{bmatrix} 2 & 1 & 0 \\ -1 & 3 & 2 \end{bmatrix} = \begin{bmatrix} 4 & 9 & 4 \\ -3 & 2 & 2 \\ 8 & 18 & 8 \end{bmatrix},$$

and $$CB = \begin{bmatrix} 2 & 1 & 0 \\ -1 & 3 & 2 \end{bmatrix} \begin{bmatrix} 3 & 2 \\ -1 & 1 \\ 6 & 4 \end{bmatrix} = \begin{bmatrix} 5 & 5 \\ 6 & 9 \end{bmatrix}.$$

Here BC and CB are not even the same size.

Finally,

$$AB = \begin{bmatrix} 2 & 1 & -1 \\ 3 & 1 & 2 \\ 0 & -2 & -3 \end{bmatrix} \begin{bmatrix} 3 & 2 \\ -1 & 1 \\ 6 & 4 \end{bmatrix} = \begin{bmatrix} -1 & 1 \\ 20 & 15 \\ -16 & -14 \end{bmatrix},$$

while $BA = \begin{bmatrix} 3 & 2 \\ -1 & 1 \\ 6 & 4 \end{bmatrix} \begin{bmatrix} 2 & 1 & -1 \\ 3 & 1 & 2 \\ 0 & -2 & -3 \end{bmatrix}$ cannot be computed.   ∎

DEFINITION 6.7   A **diagonal** matrix of order n is a matrix $D = (d_{ij})$ with the property that $d_{ij} = 0$ whenever $i \neq j$. The **identity matrix of order n**, $I_n = (\delta_{ij})$, is the diagonal matrix with entries

$$\delta_{ij} = \begin{cases} 1 & \text{if } i = j, \\ 0 & \text{if } i \neq j. \end{cases}$$

When the size of I_n is clear, this matrix is sometimes simply written as I.

EXAMPLE 4   The identity matrix of order three is

$$I_3 = \begin{bmatrix} 1 & 0 & 0 \\ 0 & 1 & 0 \\ 0 & 0 & 1 \end{bmatrix}.$$

If A is any other 3×3 matrix, then

$$AI_3 = \begin{bmatrix} a_{11} & a_{12} & a_{13} \\ a_{21} & a_{22} & a_{23} \\ a_{31} & a_{32} & a_{33} \end{bmatrix} \begin{bmatrix} 1 & 0 & 0 \\ 0 & 1 & 0 \\ 0 & 0 & 1 \end{bmatrix} = \begin{bmatrix} a_{11} & a_{12} & a_{13} \\ a_{21} & a_{22} & a_{23} \\ a_{31} & a_{32} & a_{33} \end{bmatrix} = A. \qquad \blacksquare$$

The identity matrix can be seen to commute with A; that is, the order of multiplication does not matter, $I_3 A = A = AI_3$.

In Example 3 it was seen that the property $AB = BA$ is not generally true for matrix multiplication. Some of the properties involving matrix multiplication that do hold are presented in the next theorem.

THEOREM 6.8   Let A be an $n \times m$ matrix, B an $m \times k$ matrix, C a $k \times p$ matrix, D an $m \times k$ matrix, and λ a real number. The following properties hold:

a) $A(BC) = (AB)C$;   b) $I_m B = B$ and $BI_k = B$;
c) $A(B + D) = AB + AD$;   d) $\lambda(AB) = (\lambda A)B = A(\lambda B)$.

PROOF   The verification of the property in part (a) is presented to show the method involved. The other parts can be shown in a similar manner. To show $A(BC) = (AB)C$, compute the i, j-entry of each side of the equation. BC is an $m \times p$ matrix with i, j-entry

$$(BC)_{ij} = \sum_{l=1}^{k} b_{il} c_{lj}.$$

Thus, $A(BC)$ is an $n \times p$ matrix with entries:

$$[A(BC)]_{ij} = \sum_{s=1}^{m} a_{is}(BC)_{sj} = \sum_{s=1}^{m} a_{is}\left(\sum_{l=1}^{k} b_{sl} c_{lj} \right) = \sum_{s=1}^{m} \sum_{l=1}^{k} a_{is} b_{sl} c_{lj}.$$

Similarly, AB is an $n \times k$ matrix with entries:

$$(AB)_{ij} = \sum_{s=1}^{m} a_{is} b_{sj},$$

so $(AB)C$ is an $n \times p$ matrix with entries:

$$[(AB)C]_{ij} = \sum_{l=1}^{k} (AB)_{il}c_{lj} = \sum_{l=1}^{k} \left(\sum_{s=1}^{m} a_{is}b_{sl} \right) c_{lj} = \sum_{l=1}^{k} \sum_{s=1}^{m} a_{is}b_{sl}c_{lj} = \sum_{s=1}^{m} \sum_{l=1}^{k} a_{is}b_{sl}c_{lj}.$$

This implies $[A(BC)]_{ij} = [(AB)C]_{ij}$ for each $i = 1, 2, \ldots, n$, and $j = 1, 2, \ldots, p$, so $A(BC) = (AB)C$.

$\square$

With the definition of matrix multiplication, the relationship of linear systems to linear algebra can be discussed. The linear system

$$a_{11}x_1 + a_{12}x_2 + \cdots + a_{1n}x_n = b_1,$$
$$a_{21}x_1 + a_{22}x_2 + \cdots + a_{2n}x_n = b_2,$$
$$\vdots \qquad \vdots \qquad \qquad \vdots \qquad \vdots$$
$$a_{n1}x_1 + a_{n2}x_2 + \cdots + a_{nn}x_n = b_n,$$

can be viewed as the matrix equation

$$A\mathbf{x} = \mathbf{b},$$

where

$$A = \begin{bmatrix} a_{11} & a_{12} & \cdots & a_{1n} \\ a_{21} & a_{22} & \cdots & a_{2n} \\ \vdots & \vdots & & \vdots \\ a_{n1} & a_{n2} & \cdots & a_{nn} \end{bmatrix}, \qquad \mathbf{x} = \begin{bmatrix} x_1 \\ x_2 \\ \vdots \\ x_n \end{bmatrix}, \qquad \text{and} \qquad \mathbf{b} = \begin{bmatrix} b_1 \\ b_2 \\ \vdots \\ b_n \end{bmatrix}.$$

Also related to linear systems is the concept of the **inverse of a matrix**.

DEFINITION 6.9   An $n \times n$ matrix A is said to be **nonsingular** if an $n \times n$ matrix A^{-1} exists with $AA^{-1} = A^{-1}A = I$. The matrix A^{-1} is called the **inverse** of A. A matrix without an inverse is called **singular**.

EXAMPLE 5   Let

$$A = \begin{bmatrix} 1 & 2 & -1 \\ 2 & 1 & 0 \\ -1 & 1 & 2 \end{bmatrix} \qquad \text{and} \qquad B = \tfrac{1}{9} \begin{bmatrix} -2 & 5 & -1 \\ 4 & -1 & 2 \\ -3 & 3 & 3 \end{bmatrix}.$$

Since

$$AB = \begin{bmatrix} 1 & 2 & -1 \\ 2 & 1 & 0 \\ -1 & 1 & 2 \end{bmatrix} \cdot \tfrac{1}{9} \begin{bmatrix} -2 & 5 & -1 \\ 4 & -1 & 2 \\ -3 & 3 & 3 \end{bmatrix} = \begin{bmatrix} 1 & 0 & 0 \\ 0 & 1 & 0 \\ 0 & 0 & 1 \end{bmatrix}$$

and

$$BA = \tfrac{1}{9} \begin{bmatrix} -2 & 5 & -1 \\ 4 & -1 & 2 \\ -3 & 3 & 3 \end{bmatrix} \begin{bmatrix} 1 & 2 & -1 \\ 2 & 1 & 0 \\ -1 & 1 & 2 \end{bmatrix} = \begin{bmatrix} 1 & 0 & 0 \\ 0 & 1 & 0 \\ 0 & 0 & 1 \end{bmatrix},$$

A and B are nonsingular; $B = A^{-1}$ and $A = B^{-1}$.

The usefulness of the inverse can be shown by considering the linear system

$$x_1 + 2x_2 - x_3 = 2,$$
$$2x_1 + x_2 \quad\quad = 3,$$
$$-x_1 + x_2 + 2x_3 = 4.$$

First, convert the system to the matrix equation

$$\begin{bmatrix} 1 & 2 & -1 \\ 2 & 1 & 0 \\ -1 & 1 & 2 \end{bmatrix} \begin{bmatrix} x_1 \\ x_2 \\ x_3 \end{bmatrix} = \begin{bmatrix} 2 \\ 3 \\ 4 \end{bmatrix},$$

and then multiply both sides by the inverse

$$\frac{1}{9}\begin{bmatrix} -2 & 5 & -1 \\ 4 & -1 & 2 \\ -3 & 3 & 3 \end{bmatrix} \left(\begin{bmatrix} 1 & 2 & -1 \\ 2 & 1 & 0 \\ -1 & 1 & 2 \end{bmatrix} \begin{bmatrix} x_1 \\ x_2 \\ x_3 \end{bmatrix} \right) = \frac{1}{9}\begin{bmatrix} -2 & 5 & -1 \\ 4 & -1 & 2 \\ -3 & 3 & 3 \end{bmatrix} \begin{bmatrix} 2 \\ 3 \\ 4 \end{bmatrix};$$

so

$$\left(\begin{bmatrix} -\frac{2}{9} & \frac{5}{9} & -\frac{1}{9} \\ \frac{4}{9} & -\frac{1}{9} & \frac{2}{9} \\ -\frac{3}{9} & \frac{3}{9} & \frac{3}{9} \end{bmatrix} \begin{bmatrix} 1 & 2 & -1 \\ 2 & 1 & 0 \\ -1 & 1 & 2 \end{bmatrix} \right) \begin{bmatrix} x_1 \\ x_2 \\ x_3 \end{bmatrix} = \frac{1}{9}\begin{bmatrix} 7 \\ 13 \\ 15 \end{bmatrix}$$

and

$$I_3 \begin{bmatrix} x_1 \\ x_2 \\ x_3 \end{bmatrix} = \begin{bmatrix} x_1 \\ x_2 \\ x_3 \end{bmatrix} = \begin{bmatrix} \frac{7}{9} \\ \frac{13}{9} \\ \frac{15}{9} \end{bmatrix},$$

which gives the solution $x_1 = \frac{7}{9}$, $x_2 = \frac{13}{9}$, and $x_3 = \frac{15}{9}$. ■

In general, the technique in Example 5 for solving $Ax = b$ is not recommended because of the number of operations involved in determining A^{-1}. (See Exercise 6.)

To find a method of computing A^{-1}, assuming its existence, let us look again at matrix multiplication. Let B_j be the jth column of the $n \times n$ matrix B,

$$B_j = \begin{bmatrix} b_{ij} \\ b_{2j} \\ \vdots \\ b_{nj} \end{bmatrix}.$$

Form the product

$$AB_j = \begin{bmatrix} a_{11} & a_{12} & \cdots & a_{1n} \\ a_{21} & a_{22} & \cdots & a_{2n} \\ \vdots & \vdots & & \vdots \\ a_{n1} & a_{n2} & \cdots & a_{nn} \end{bmatrix} \begin{bmatrix} b_{1j} \\ b_{2j} \\ \vdots \\ b_{nj} \end{bmatrix} = \begin{bmatrix} \sum_{k=1}^{n} a_{1k}b_{kj} \\ \sum_{k=1}^{n} a_{2k}b_{kj} \\ \vdots \\ \sum_{k=1}^{n} a_{nk}b_{kj} \end{bmatrix}.$$

If $AB = C$, then the jth column of C is given by:

$$C_j = \begin{bmatrix} c_{1j} \\ c_{2j} \\ \vdots \\ c_{nj} \end{bmatrix} = \begin{bmatrix} \sum_{k=1}^{n} a_{1k} b_{kj} \\ \sum_{k=1}^{n} a_{2k} b_{kj} \\ \vdots \\ \sum_{k=1}^{n} a_{nk} b_{kj} \end{bmatrix}.$$

Hence, the jth column of the product AB is the product of A and the jth column of B.

Suppose that A^{-1} exists and that $A^{-1} = B = (b_{ij})$; then $AB = I$ and

$$AB_j = \begin{bmatrix} 0 \\ \vdots \\ 0 \\ 1 \\ 0 \\ \vdots \\ 0 \end{bmatrix}, \qquad \text{where the value 1 appears in the } j\text{th row.}$$

To find B, we must solve n linear systems in which the jth column of the inverse is the solution of the linear system with righthand side the jth column of I. The next example demonstrates the method involved.

EXAMPLE 6   Let

$$A = \begin{bmatrix} 1 & 2 & -1 \\ 2 & 1 & 0 \\ -1 & 1 & 2 \end{bmatrix}.$$

To compute A^{-1}, we must solve the three linear systems

$$\begin{aligned} x_1 + 2x_2 - x_3 &= 1, & x_1 + 2x_2 - x_3 &= 0, \\ 2x_1 + x_2 &= 0, & 2x_1 + x_2 &= 1, \\ -x_1 + x_2 + 2x_3 &= 0; & -x_1 + x_2 + 2x_3 &= 0; \end{aligned}$$

$$\begin{aligned} x_1 + 2x_2 - x_3 &= 0, \\ 2x_1 + x_2 &= 0, \\ -x_1 + x_2 + 2x_3 &= 1. \end{aligned}$$

Using Gaussian elimination, the computations are conveniently performed on the larger augmented matrix, formed by combining the matrices:

$$\left[\begin{array}{ccc:ccc} 1 & 2 & -1 & 1 & 0 & 0 \\ 2 & 1 & 0 & 0 & 1 & 0 \\ -1 & 1 & 2 & 0 & 0 & 1 \end{array}\right].$$

To elaborate, since the actual coefficient matrix does not change, we must perform the same sequence of row operations for each linear system. First, performing $(E_2 - 2E_1)$ $\rightarrow (E_2)$ and $(E_3 + E_1) \rightarrow (E_3)$,

$$\begin{bmatrix} 1 & 2 & -1 & \vdots & 1 & 0 & 0 \\ 0 & -3 & 2 & \vdots & -2 & 1 & 0 \\ 0 & 3 & 1 & \vdots & 1 & 0 & 1 \end{bmatrix}.$$

Next, performing $(E_3 + E_2) \rightarrow (E_3)$,

$$\begin{bmatrix} 1 & 2 & -1 & \vdots & 1 & 0 & 0 \\ 0 & -3 & 2 & \vdots & -2 & 1 & 0 \\ 0 & 0 & 3 & \vdots & -1 & 1 & 1 \end{bmatrix}.$$

Backward substitution could be performed on each of the three augmented matrices,

$$\begin{bmatrix} 1 & 2 & -1 & \vdots & 1 \\ 0 & -3 & 2 & \vdots & -2 \\ 0 & 0 & 3 & \vdots & -1 \end{bmatrix}, \quad \begin{bmatrix} 1 & 2 & -1 & \vdots & 0 \\ 0 & -3 & 2 & \vdots & 1 \\ 0 & 0 & 3 & \vdots & 1 \end{bmatrix}, \quad \begin{bmatrix} 1 & 2 & -1 & \vdots & 0 \\ 0 & -3 & 2 & \vdots & 0 \\ 0 & 0 & 3 & \vdots & 1 \end{bmatrix},$$

to find all the entries of A^{-1}, but it is often more convenient to use further row reduction. In particular, the operation $(\frac{1}{3}E_3) \rightarrow (E_3)$ yields:

$$\begin{bmatrix} 1 & 2 & -1 & \vdots & 1 & 0 & 0 \\ 0 & -3 & 2 & \vdots & -2 & 1 & 0 \\ 0 & 0 & 1 & \vdots & -\frac{1}{3} & \frac{1}{3} & \frac{1}{3} \end{bmatrix}$$

and $(E_2 - 2E_3) \rightarrow (E_2)$ and $(E_1 + E_3) \rightarrow (E_1)$ produce:

$$\begin{bmatrix} 1 & 2 & 0 & \vdots & \frac{2}{3} & \frac{1}{3} & \frac{1}{3} \\ 0 & -3 & 0 & \vdots & -\frac{4}{3} & \frac{1}{3} & -\frac{2}{3} \\ 0 & 0 & 1 & \vdots & -\frac{1}{3} & \frac{1}{3} & \frac{1}{3} \end{bmatrix}.$$

Performing $(-\frac{1}{3}E_2) \rightarrow (E_2)$,

$$\begin{bmatrix} 1 & 2 & 0 & \vdots & \frac{2}{3} & \frac{1}{3} & \frac{1}{3} \\ 0 & 1 & 0 & \vdots & \frac{4}{9} & -\frac{1}{9} & \frac{2}{9} \\ 0 & 0 & 1 & \vdots & -\frac{1}{3} & \frac{1}{3} & \frac{1}{3} \end{bmatrix};$$

and finally, $(E_1 - 2E_2) \rightarrow (E_1)$ gives

$$\begin{bmatrix} 1 & 0 & 0 & \vdots & -\frac{2}{9} & \frac{5}{9} & -\frac{1}{9} \\ 0 & 1 & 0 & \vdots & \frac{4}{9} & -\frac{1}{9} & \frac{2}{9} \\ 0 & 0 & 1 & \vdots & -\frac{3}{9} & \frac{3}{9} & \frac{3}{9} \end{bmatrix}.$$

The final augmented matrix represents the solutions to the three linear systems

$$\begin{array}{lll} x_1 = -\frac{2}{9}, & x_1 = \frac{5}{9}, & x_1 = -\frac{1}{9}, \\ x_2 = \frac{4}{9}, & x_2 = -\frac{1}{9}, & x_2 = \frac{2}{9}, \\ x_3 = -\frac{3}{9}, & x_3 = \frac{3}{9}, & x_3 = \frac{3}{9}, \end{array}$$

so $\qquad A^{-1} = \begin{bmatrix} -\frac{2}{9} & \frac{5}{9} & -\frac{1}{9} \\ \frac{4}{9} & -\frac{1}{9} & \frac{2}{9} \\ -\frac{3}{9} & \frac{3}{9} & \frac{3}{9} \end{bmatrix} = \frac{1}{9} \begin{bmatrix} -2 & 5 & -1 \\ 4 & -1 & 2 \\ -3 & 3 & 3 \end{bmatrix}.$ ∎

In the last example, we illustrated the computation of A^{-1}. As we saw in that example, it is convenient to set up the larger augmented matrix

$$[A \ \vdots \ I].$$

Upon performing the elimination in accordance with Algorithm 6.1, we obtain an augmented matrix of the form

$$[U \ \vdots \ Y],$$

where U is an $n \times n$ matrix with $u_{ij} = 0$ whenever $i > j$ and Y represents the $n \times n$ matrix obtained by performing the same operations on the identity I that were performed to take A into U. At this point there is a choice between n applications of the backward-substitution algorithm or further reduction to

$$[I \ \vdots \ A^{-1}].$$

If straightforward backward substitution is used, $\frac{4}{3}n^3 - \frac{1}{3}n$ multiplications and divisions and $\frac{4}{3}n^3 - \frac{3}{2}n^2 + \frac{1}{6}n$ additions and subtractions are required (see Exercise 6(c)). It is also shown in that exercise that the further reduction, which is a special case of the Gauss–Jordan method (see Exercise 4, Section 6.2), requires $\frac{3}{2}n^3 - \frac{1}{2}n$ multiplications and divisions and $\frac{3}{2}n^3 - 2n^2 + \frac{1}{2}n$ additions and subtractions. Table 6.2 indicates the relative amount of computations for varying sizes of n.

TABLE 6.2

	Gauss Elimination		Gauss–Jordan	
n	Mult./div.	Add./sub.	Mult./div.	Add./sub.
	$(4n^3 - n)/3$	$(8n^3 - 9n^2 + n)/6$	$(3n^3 - n)/2$	$(3n^3 - 4n^2 + n)/2$
3	35	23	39	24
10	1330	1185	1495	1305
50	166,650	162,925	187,475	182,525
100	1,333,300	1,318,350	1,499,950	1,480,050

In the actual implementation of either method, special care can be given to note the operations that need *not* be performed, as, for example, a multiplication when one of the multipliers is known to be unity, or a subtraction when the subtrahend is known to be zero. For both methods, the number of multiplications and divisions required can then be reduced to n^3 and the number of additions and subtractions reduced to $n^3 - 2n^2 + n$ (see Exercises 6(d) and 6(e)).

	Both methods, eliminating unnecessary computations	
n	Multi./div.	Add./sub.
3	27	12
10	1000	810
50	125,000	120,050
100	1,000,000	980,100

Exercise Set 6.3

1. Determine which of the following matrices are nonsingular and compute their inverses, if possible:

a) $\begin{bmatrix} 4 & 2 & 6 \\ 3 & 0 & 7 \\ -2 & -1 & -3 \end{bmatrix}$

b) $\begin{bmatrix} 1 & 2 & 0 \\ 2 & 1 & -1 \\ 3 & 1 & 1 \end{bmatrix}$

c) $\begin{bmatrix} 1 & 1 & -1 & 1 \\ 1 & 2 & -4 & -2 \\ 2 & 1 & 1 & 5 \\ -1 & 0 & -2 & -4 \end{bmatrix}$

d) $\begin{bmatrix} 2 & 3 & 1 & 2 \\ -2 & 4 & -1 & 5 \\ 3 & 7 & \frac{3}{2} & 1 \\ 6 & 9 & 3 & 7 \end{bmatrix}$

e) $\begin{bmatrix} 4 & 0 & 0 & 0 \\ 6 & 7 & 0 & 0 \\ 9 & 11 & 1 & 0 \\ 5 & 4 & 1 & 1 \end{bmatrix}$

f) $\begin{bmatrix} 2 & 0 & 1 & 2 \\ 1 & 1 & 0 & 2 \\ 2 & -1 & 3 & 1 \\ 3 & -1 & 4 & 3 \end{bmatrix}$

2. Given the four 3×3 linear systems with the same coefficient matrix:

$$2x_1 - 3x_2 + x_3 = 2, \qquad 2x_1 - 3x_2 + x_3 = 6,$$
$$x_1 + x_2 - x_3 = -1, \qquad x_1 + x_2 - x_3 = 4,$$
$$-x_1 + x_2 - 3x_3 = 0. \qquad -x_1 + x_2 - 3x_3 = 5.$$

$$2x_1 - 3x_2 + x_3 = 0, \qquad 2x_1 - 3x_2 + x_3 = -1,$$
$$x_1 + x_2 - x_3 = 1, \qquad x_1 + x_2 - x_3 = 0,$$
$$-x_1 + x_2 - 3x_3 = -3. \qquad -x_1 + x_2 - 3x_3 = 0.$$

a) Solve the linear systems by applying Gaussian elimination to the augmented matrix

$$\begin{bmatrix} 2 & -3 & 1 & \vdots & 2 & 6 & 0 & -1 \\ 1 & 1 & -1 & \vdots & -1 & 4 & 1 & 0 \\ -1 & 1 & -3 & \vdots & 0 & 5 & -3 & 0 \end{bmatrix}$$

b) Solve the linear systems by applying the Gauss–Jordan method to the augmented matrix in part (a).

c)   Solve the linear system by finding the inverse of

$$A = \begin{bmatrix} 2 & -3 & 1 \\ 1 & 1 & -1 \\ -1 & 1 & -3 \end{bmatrix} \quad \text{and multiplying.}$$

d)   Which method seems easier? Which method requires more operations?

3.   Repeat Exercise 2 using the linear systems:

$$
\begin{array}{rl}
x_1 - x_2 + 2x_3 - x_4 = & 6, \\
x_1 \quad\quad - x_3 + x_4 = & 4, \\
2x_1 + x_2 + 3x_3 - 4x_4 = & -2, \\
-x_2 + x_3 - x_4 = & 5,
\end{array}
\qquad
\begin{array}{rl}
x_1 - x_2 + 2x_3 - x_4 = & 1, \\
x_1 \quad\quad - x_3 + x_4 = & 1, \\
2x_1 + x_2 + 3x_3 - 4x_4 = & 2, \\
-x_2 + x_3 - x_4 = & -1.
\end{array}
$$

4.   In a paper entitled "Population Waves," Bernadelli [8] (see also Searle [120]) hypothesizes a type of simplified beetle, which has a natural life span of three years. The female of this species has a survival rate of $\frac{1}{2}$ in the first year of life, a survival rate of $\frac{1}{3}$ from the second to third years, and gives birth to an average of six new females before expiring at the end of the third year. A matrix can be used to show the contribution an individual female beetle makes, in a probabilistic sense, to the female population of the species, by letting a_{ij} in the matrix $A = (a_{ij})$, denote the contribution that a single female beetle of age j will make to the next years' female population of age i; that is,

$$A = \begin{bmatrix} 0 & 0 & 6 \\ \frac{1}{2} & 0 & 0 \\ 0 & \frac{1}{3} & 0 \end{bmatrix}.$$

a)   The contribution that a female beetle will make to the population two years hence could be determined from the entries of A^2, of three years hence from A^3, and so on. Construct A^2 and A^3, and try to make a general statement about the contribution of a female beetle to the population in n years' time for any positive integral value of n.

b)   Use your conclusions from part (a) to describe what will occur in future years to a population of these beetles that initially consists of 6000 female beetles in each of the three age groups.

c)   Construct A^{-1} and describe its significance regarding the population of this species.

5.   Prove Theorem 6.5.

6.   Suppose m linear systems

$$A\mathbf{x}^{(p)} = \mathbf{b}^{(p)}, \qquad p = 1, 2, \ldots, m,$$

are to be solved, each with the coefficient matrix A.

a)   Show that Gaussian elimination applied to the augmented matrix

$$[A: \quad \mathbf{b}^{(1)}\mathbf{b}^{(2)} \cdots \mathbf{b}^{(m)}]$$

requires

$$\frac{n^3}{3} + mn^2 - \frac{n}{3} \quad \text{multiplications/divisions}$$

4(7) + 2 (-14+9) + 6 (-3)
28 - 28 + 18 - 18 = 0

and $\dfrac{n^3}{3} + mn^2 - \dfrac{n^2}{2} - mn + \dfrac{n}{6}$ additions/subtractions.

b) Show that the Gauss–Jordan method (See Exercise 4, Section 6.2) applied to the augmented matrix

$$[A: \ \mathbf{b}^{(1)}\mathbf{b}^{(2)} \cdots \mathbf{b}^{(m)}]$$

requires

$$\tfrac{1}{2}n^3 + mn^2 - \tfrac{1}{2}n \quad \text{multiplications/divisions}$$

and $\tfrac{1}{2}n^3 + (m-1)n^2 + (\tfrac{1}{2} - m)n$ additions/subtractions.

c) For the special case $\mathbf{b}^{(p)} = \begin{bmatrix} 0 \\ \vdots \\ 0 \\ 1 \\ \vdots \\ 0 \end{bmatrix} \leftarrow p\text{th row}$

for each $p = 1, \ldots, m$, with $m = n$, the solution $\mathbf{x}^{(p)}$ is the pth column of A^{-1}. Show that Gaussian elimination requires

$$\tfrac{4}{3}n^3 - \dfrac{n}{3} \quad \text{multiplications/divisions}$$

and $\tfrac{4}{3}n^3 - \tfrac{3}{2}n^2 + \tfrac{1}{6}n$ additions/subtractions

for this application, and that the Gauss–Jordan method requires

$$\tfrac{3}{2}n^3 - \tfrac{1}{2}n \quad \text{multiplications/divisions}$$

and $\tfrac{3}{2}n^3 - 2n^2 + \tfrac{1}{2}n$ additions/subtractions.

d) Construct an algorithm using Gaussian elimination to find A^{-1}, but do not perform multiplications when one of the multipliers is known to be unity, and do not perform additions/subtractions when one of the elements involved is known to be zero. Show that the required computations are reduced to n^3 multiplications/divisions and $n^3 - 2n^2 + n$ additions/subtractions.

e) Repeat part (d), using the Gauss–Jordan method.

f) Show that solving the linear system $A\mathbf{x} = \mathbf{b}$, when A^{-1} is known, still requires n^2 multiplications/divisions and $n^2 - n$ additions/subtractions.

g) Show that solving m linear systems $A\mathbf{x}^{(p)} = \mathbf{b}^{(p)}$ for $p = 1, 2, \ldots, m$, by the method $\mathbf{x}^{(p)} = A^{-1}\mathbf{b}^{(p)}$ requires mn^2 multiplications and $m(n^2 - n)$ additions, if A^{-1} is known.

h) Let A be an $n \times n$ matrix. Compare the number of operations required to solve m linear systems involving A by Gaussian elimination and by first inverting A and then multiplying $A\mathbf{x} = \mathbf{b}$ by A^{-1}, for $n = 3, 10, 50, 100$. Is it ever advantageous to compute A^{-1} for the purpose of solving linear systems?

7. Use the algorithm developed in Exercise 6(d) to find the inverses of the nonsingular matrices in Exercise 1.

8. It is often useful to partition matrices into a collection of submatrices. For example, the matrices

$$A = \begin{bmatrix} 1 & 2 & -1 \\ 3 & -4 & -3 \\ 6 & 5 & 0 \end{bmatrix} \quad \text{and} \quad B = \begin{bmatrix} 2 & -1 & 7 & 0 \\ 3 & 0 & 4 & 5 \\ -2 & 1 & -3 & 1 \end{bmatrix}$$

can be partitioned into

$$\left[\begin{array}{cc:c} 1 & 2 & -1 \\ 3 & -4 & -3 \\ \hdashline 6 & 5 & 0 \end{array} \right] = \begin{bmatrix} A_{11} & \vdots & A_{12} \\ \hdashline A_{21} & \vdots & A_{22} \end{bmatrix}$$

and

$$\left[\begin{array}{ccc:c} 2 & -1 & 7 & 0 \\ 3 & 0 & 4 & 5 \\ \hdashline -2 & 1 & -3 & 1 \end{array} \right] = \begin{bmatrix} B_{11} & \vdots & B_{12} \\ \hdashline B_{21} & \vdots & B_{22} \end{bmatrix}$$

a) Show that the product of A and B in this case is

$$AB = \begin{bmatrix} A_{11}B_{11} + A_{12}B_{21} & \vdots & A_{11}B_{12} + A_{12}B_{22} \\ \hdashline A_{21}B_{11} + A_{22}B_{21} & \vdots & A_{21}B_{12} + A_{22}B_{22} \end{bmatrix}.$$

b) If B were instead partitioned into

$$B = \left[\begin{array}{ccc:c} 2 & -1 & 7 & 0 \\ \hdashline 3 & 0 & 4 & 5 \\ -2 & 1 & -3 & 1 \end{array} \right] = \begin{bmatrix} B_{11} & \vdots & B_{12} \\ \hdashline B_{21} & \vdots & B_{22} \end{bmatrix}$$

would the result in part (a) hold?

c) Make a conjecture concerning the conditions necessary for the result in part (a) to hold in the general case.

9. The study of food chains is an important topic in the determination of the spread and accumulation of environmental pollutants in living matter. Suppose that a food chain has three links. The first link consists of vegetation of types $v_1, v_2, \ldots, v_n$, which provide all the food requirements for herbivores of species $h_1, h_2, \ldots, h_m$ in the second link. The third link consists of carnivorous animals $c_1, c_2, \ldots, c_k$, which depend entirely on the herbivores in the second link for their food supply.

The coordinate a_{ij} of the matrix

$$A = \begin{bmatrix} a_{11} & a_{12} & \cdots & a_{1m} \\ a_{21} & a_{22} & \cdots & a_{2m} \\ \vdots & \vdots & & \vdots \\ a_{n1} & a_{n2} & \cdots & a_{nm} \end{bmatrix}$$

represents the total number of plants of type v_i eaten by the herbivores in the species h_j, while b_{ij} in

$$B = \begin{bmatrix} b_{11} & b_{12} & \cdots & b_{1k} \\ b_{21} & b_{22} & \cdots & b_{2k} \\ \vdots & \vdots & & \vdots \\ b_{m1} & b_{m2} & \cdots & b_{mk} \end{bmatrix}$$

describes the number of herbivores in species h_i which are devoured by the animals of type c_j.

a) Show that the number of plants of type v_i which eventually end up in the animals of species c_j would be given by the entry in the ith row and jth column of the matrix AB.

b) What physical significance is associated with the matrices A^{-1}, B^{-1}, and $(AB)^{-1} = B^{-1}A^{-1}$?

6.4 The Determinant of a Matrix

A fundamental concept of linear algebra that is very useful in determining the existence and uniqueness of solutions to linear systems is the **determinant** of an $n \times n$ matrix. The only approach to defining and computing the determinant given here will be the recursive definition. The determinant of a matrix A will be denoted by det A, but it is also common to use the notation $|A|$. A **submatrix** of a matrix A is a matrix "extracted" from A by deleting certain rows and/or columns of A.

DEFINITION 6.10
a) If $A = [a]$ is a 1×1 matrix, then det $A = a$.
b) The **minor**, M_{ij}, is the determinant of the $(n-1) \times (n-1)$ submatrix of an $n \times n$ matrix A obtained by deleting the ith row and jth column.
c) The **cofactor**, A_{ij}, associated with M_{ij}, is defined to be $A_{ij} = (-1)^{i+j}M_{ij}$.
d) The **determinant** of an $n \times n$ matrix A, where $n > 1$, is given by either

$$(6.11) \qquad \det A = \sum_{j=1}^{n} a_{ij}A_{ij} \qquad \text{for any } i = 1, 2, \ldots, n,$$

or

$$(6.12) \qquad \det A = \sum_{i=1}^{n} a_{ij}A_{ij} \qquad \text{for any } j = 1, 2, \ldots, n.$$

By mathematical induction, it can be shown (see Exercise 4) that, if $n > 1$, using Definition 6.10 to calculate the determinant of a general $n \times n$ matrix requires $n!$ multiplications/divisions and $n! - 1$ additions/subtractions. Even for relatively small values of n, the number of calculations can become unwieldy.

It appears that there are $2n$ different definitions of det A depending on which row or column is expanded. However, it can be shown that all definitions give the same numerical result. (See Noble and Daniel, [92], page 200.) For the case of 2×2 matrices, the following example shows the equivalence of these definitions.

EXAMPLE 1
a) Let $A = \begin{bmatrix} a_{11} & a_{12} \\ a_{21} & a_{22} \end{bmatrix}$. Using Eq. (6.11) with $i = 1$,

$$\det A = a_{11}A_{11} + a_{12}A_{12}$$
$$= a_{11}[(-1)^{1+1}\det(a_{22})] + a_{12}[(-1)^{1+2}\det(a_{21})]$$
$$= a_{22}a_{11} - a_{21}a_{12}.$$

Using Eq. (6.11) with $i = 2$,

$$\det A = a_{21}A_{21} + a_{22}A_{22} = -a_{21}a_{12} + a_{22}a_{11}.$$

Using Eq. (6.12) with $j = 1$,

$$\det A = a_{11}A_{11} + a_{21}A_{21} = a_{11}a_{22} - a_{21}a_{12}.$$

Using Eq. (6.12) with $j = 2$,

$$\det A = a_{12}A_{12} + a_{22}A_{22} = -a_{12}a_{21} + a_{22}a_{11}.$$

In all cases,

$$\det A = \det \begin{bmatrix} a_{11} & a_{12} \\ a_{21} & a_{22} \end{bmatrix} = a_{11}a_{22} - a_{12}a_{21}.$$

b) Let

$$A = \begin{bmatrix} 2 & 0 & 0 \\ 3 & -1 & 1 \\ 4 & 6 & -2 \end{bmatrix}.$$

Using Eq. (6.11) with $i = 3$,

$$\begin{aligned}
\det A &= a_{31}A_{31} + a_{32}A_{32} + a_{33}A_{33} \\
&= 4A_{31} + 6A_{32} - 2A_{33} \\
&= 4(-1)^{3+1} \det \begin{bmatrix} 0 & 0 \\ -1 & 1 \end{bmatrix} + 6(-1)^{3+2} \det \begin{bmatrix} 2 & 0 \\ 3 & 1 \end{bmatrix} \\
&\quad - 2(-1)^{3+3} \det \begin{bmatrix} 2 & 0 \\ 3 & -1 \end{bmatrix} \\
&= -8.
\end{aligned}$$

Clearly, it is most convenient to compute $\det A$ across the row with the most zeros or down the column with the most zeros.

c) Let

$$A = \begin{bmatrix} 2 & -1 & 3 & 0 \\ 4 & -2 & 7 & 0 \\ -3 & -4 & 1 & 5 \\ 6 & -6 & 8 & 0 \end{bmatrix}.$$

To compute $\det A$, it is easiest to use the fourth column, Eq. (6.12) with $j = 4$;

$$\begin{aligned}
\det A &= a_{14}A_{14} + a_{24}A_{24} + a_{34}A_{34} + a_{44}A_{44} \\
&= 5A_{34} \\
&= -5 \det \begin{bmatrix} 2 & -1 & 3 \\ 4 & -2 & 7 \\ 6 & -6 & 8 \end{bmatrix} \\
&= -5 \left\{ 2 \det \begin{bmatrix} -2 & 7 \\ -6 & 8 \end{bmatrix} - (-1) \det \begin{bmatrix} 4 & 7 \\ 6 & 8 \end{bmatrix} + 3 \det \begin{bmatrix} 4 & -2 \\ 6 & -6 \end{bmatrix} \right\} \\
&= -30.
\end{aligned}$$

■

The following properties of determinants are useful in relating linear systems and Gaussian elimination to determinants.

THEOREM 6.11   Suppose A is an $n \times n$ matrix:

a)   If any row or column of A has only zero entries, then det $A = 0$.

b)   If $\tilde{A}$ is obtained from A by the operation $(E_i) \leftrightarrow (E_j)$, with $i \neq j$, then det $\tilde{A} = -\det A$.

c)   If A has two rows the same, then det $A = 0$.

d)   If $\tilde{A}$ is obtained from A by the operation $(\lambda E_i) \to (E_i)$, then det $\tilde{A} = \lambda \det A$.

e)   If $\tilde{A}$ is obtained from A by the operation $(E_i + \lambda E_j) \to (E_i)$, with $i \neq j$, then det $\tilde{A} = \det A$.

f)   If B is also an $n \times n$ matrix, then det $AB = \det A \det B$.

PROOF   a)   Suppose the ith row of A is zero. Expanding along the ith row, we write:

$$\det A = \sum_{j=1}^{n} a_{ij} A_{ij} = \sum_{j=1}^{n} 0 \cdot A_{ij} = 0.$$

b)   The proof can be found in any standard linear algebra text. (See, for example, Noble and Daniel [92], page 201.) The case when n is three is considered in Exercise 3.

c)   Suppose the ith and jth rows are identical. Form $\tilde{A}$ by interchanging the ith and jth rows. By (b), det $\tilde{A} = -\det A$. Since the operation $(E_i) \leftrightarrow (E_j)$ did not change A, in this case det $\tilde{A} = \det A$, which implies det $A = -\det A$ and det $A = 0$.

d)   Expanding det $\tilde{A}$ along the ith row gives

$$\det \tilde{A} = \sum_{k=1}^{n} \lambda a_{ik} A_{ik} = \lambda \sum_{k=1}^{n} a_{ik} A_{ik} = \lambda \det A.$$

e)   Expanding det $\tilde{A}$ along the ith row gives

$$\det A = \sum_{k=1}^{n} (a_{ik} + \lambda a_{jk}) A_{ik} = \sum_{k=1}^{n} a_{ik} A_{ik} + \lambda \sum_{k=1}^{n} a_{jk} A_{ik} = \det A + \lambda \det B,$$

where the matrix B is exactly the same as A except that the ith row of B is the jth row of A. Thus, det $B = 0$ by (c) and det $\tilde{A} = \det A$.

f)   The proof of this result can be found in Noble and Daniel [92], page 204.   □

We now present the key result relating nonsingularity, Gaussian elimination, linear systems, and determinants. The proof of this theorem is not difficult but is laborious, and will not be presented here. (See Noble and Daniel [92].)

THEOREM 6.12   The following statements are equivalent for any $n \times n$ matrix A:

a)   The equation $A\mathbf{x} = \mathbf{0}$ has the unique solution $\mathbf{x} = \mathbf{0}$.

b)   The linear system $A\mathbf{x} = \mathbf{b}$ has a unique solution for any n-dimensional column vector $\mathbf{b}$.

c)   The matrix A is nonsingular; that is, A^{-1} exists.

d)   det $A \neq 0$.

e)   Algorithm 6.1 (Gaussian elimination with row interchanges) can be performed on the linear system $A\mathbf{x} = \mathbf{b}$ for any n-dimensional column vector $\mathbf{b}$.

Exercise Set 6.4

1. Compute the determinants of the following matrices:

   a) $\begin{bmatrix} 1 & 2 & 0 \\ 2 & 1 & -1 \\ 3 & 1 & 1 \end{bmatrix}$

   b) $\begin{bmatrix} 4 & 0 & 1 \\ 2 & 1 & 0 \\ 2 & 2 & 3 \end{bmatrix}$

   c) $\begin{bmatrix} 1 & 1 & -1 & 1 \\ 1 & 2 & -4 & -2 \\ 2 & 1 & 1 & 5 \\ -1 & 0 & -2 & -4 \end{bmatrix}$

   d) $\begin{bmatrix} 2 & 0 & 1 & 2 \\ 1 & 1 & 0 & 2 \\ 2 & -1 & 3 & 1 \\ 3 & -1 & 4 & 3 \end{bmatrix}$

2. Compute det A, det B, det AB, and det BA for

$$A = \begin{bmatrix} 4 & 6 & 1 & -1 \\ 2 & 1 & 0 & \frac{1}{2} \\ 3 & 0 & 0 & 1 \\ 1 & -1 & 1 & 1 \end{bmatrix} \quad \text{and} \quad B = \begin{bmatrix} 1 & 2 & 3 & 4 \\ 0 & 2 & -1 & 1 \\ 0 & 0 & 3 & 2 \\ 0 & 0 & 0 & -1 \end{bmatrix}.$$

3. Let A be a 3×3 matrix. Show that if $\tilde{A}$ is the matrix obtained from A using any of the operations

$$(E_1) \leftrightarrow (E_2), \qquad (E_1) \leftrightarrow (E_3), \qquad \text{or} \qquad (E_2) \leftrightarrow (E_3),$$

   then det $\tilde{A} = -\det A$.

4. Use mathematical induction to show that, when $n > 1$, the evaluation of the determinant of an $n \times n$ matrix requires $n!$ multiplications/divisions and $n! - 1$ additions/subtractions.

5. Prove that AB is nonsingular if and only if one of A and B is nonsingular.

6. The solution by **Cramer's rule** to the linear system

$$a_{11}x_1 + a_{12}x_2 + a_{13}x_3 = b_1,$$
$$a_{21}x_1 + a_{22}x_2 + a_{23}x_3 = b_2,$$
$$a_{31}x_1 + a_{32}x_2 + a_{33}x_3 = b_3,$$

   has

$$x_1 = \frac{1}{D} \det \begin{bmatrix} b_1 & a_{12} & a_{13} \\ b_2 & a_{22} & a_{23} \\ b_3 & a_{32} & a_{33} \end{bmatrix} \equiv \frac{D_1}{D},$$

$$x_2 = \frac{1}{D} \det \begin{bmatrix} a_{11} & b_1 & a_{13} \\ a_{21} & b_2 & a_{23} \\ a_{31} & b_3 & a_{33} \end{bmatrix} \equiv \frac{D_2}{D},$$

   and

$$x_3 = \frac{1}{D} \det \begin{bmatrix} a_{11} & a_{12} & b_1 \\ a_{21} & a_{22} & b_2 \\ a_{31} & a_{32} & b_3 \end{bmatrix} \equiv \frac{D_3}{D},$$

where
$$D = \det \begin{bmatrix} a_{11} & a_{12} & a_{13} \\ a_{21} & a_{22} & a_{23} \\ a_{31} & a_{32} & a_{33} \end{bmatrix}.$$

a)   Find the solution to the linear system

$$\begin{aligned} 2x_1 + 3x_2 - x_3 &= 4, \\ x_1 - 2x_2 + x_3 &= 6, \\ x_1 - 12x_2 + 5x_3 &= 10, \end{aligned}$$

by Cramer's Rule.

b)   Show that the linear system

$$\begin{aligned} 2x_1 + 3x_2 - x_3 &= 4, \\ x_1 - 2x_2 + x_3 &= 6, \\ -x_1 - 12x_2 + 5x_3 &= 9 \end{aligned}$$

does not have a solution. Compute D_1, D_2, and D_3.

c)   Show that the linear system

$$\begin{aligned} 2x_1 + 3x_2 - x_3 &= 4, \\ x_1 - 2x_2 + x_3 &= 6, \\ -x_1 - 12x_2 + 5x_3 &= 10, \end{aligned}$$

has an infinite number of solutions. Compute D_1, D_2, and D_3.

d)   Prove that if a 3×3 linear system with $D = 0$ is to have solutions, then $D_1 = D_2 = D_3 = 0$.

e)   Determine the number of multiplications/divisions and additions/subtractions required for Cramer's rule on a 3×3 system.

7.   a)   Generalize Cramer's rule to an $n \times n$ linear system.

b)   Use the result in Exercise 4 to determine the number of multiplications/divisions and additions/subtractions required for Cramer's rule on an $n \times n$ system.

6.5 Pivoting Strategies

During the derivation of Algorithm 6.1, it was found that obtaining a zero for a pivot element $a_{k,k}^{(k)}$ necessitated a row interchange of the form $(E_k) \leftrightarrow (E_p)$ where $k + 1 \leq p \leq n$ was the smallest integer with $a_{p,k}^{(k)} \neq 0$. In practice it is often desirable to perform row interchanges involving the pivot elements even when they are not zero. When the calculations are performed using finite-digit arithmetic, as would be the case for calculator or computer-generated solutions, a pivot element that is small compared to the entries below it in the same column can lead to substantial round-off error. An illustration of this difficulty is given in the following example.

EXAMPLE 1   The linear system

$$E_1:\quad 0.003000x_1 + 59.14x_2 = 59.17,$$
$$E_2:\qquad 5.291x_1 - 6.130x_2 = 46.78,$$

has the exact solution $x_1 = 10.00$ and $x_2 = 1.000$.

To illustrate the difficulties of round-off error, Gaussian elimination will be performed on this system using four-digit arithmetic with rounding.

The first pivot element is $a_{11}^{(1)} = 0.003000$, and its associated multiplier is

$$m_{21} = \frac{5.291}{0.003000} = 1763.6\overline{6},$$

which rounds to 1764. Performing the operation $(E_2 - m_{21}E_1) \to (E_2)$ and the appropriate rounding,

$$0.003000x_1 + 59.14x_2 = 59.17$$
$$-104300x_2 = -104400.$$

Backward substitution implies

$$x_2 = 1.001 \quad \text{and} \quad x_1 = \frac{59.17 - (59.14)(1.001)}{0.003000} = -10.00.$$

The large error in the numerical solution for x_1 resulted from the small error of 0.001 in solving for x_2. This error was magnified by a factor of 20,000 in the solution of x_1 because of the order in which the calculations were performed. ∎

Example 1 illustrates that difficulties can arise in some cases when the pivot element $a_{k,k}^{(k)}$ is small relative to the entries $a_{i,j}^{(k)}$ for $k \le i \le n$ and $k \le j \le n$. Pivoting strategies in general are accomplished by selecting a new element for the pivot $a_{p,q}^{(k)}$ and interchanging the kth and pth rows, followed by the interchange of the kth and qth columns, if necessary. The simplest strategy is to select the element in the same column that is below the diagonal and has the largest absolute value; that is, determine p such that

$$|a_{p,k}^{(k)}| = \max_{k \le i \le n} |a_{i,k}^{(k)}|,$$

and perform $(E_k) \leftrightarrow (E_p)$. In this case no interchange of columns is considered.

EXAMPLE 2   Reconsidering the system

$$E_1:\quad 0.003000x_1 + 59.14x_2 = 59.17,$$
$$E_2:\qquad 5.291x_1 - 6.130x_2 = 46.78.$$

Using the pivoting procedure just described results in first finding

$$\max\{|a_{11}^{(1)}|, |a_{21}^{(1)}|\} = \max\{|0.003000|, |5.291|\} = |5.291| = |a_{21}^{(1)}|.$$

The operation $(E_2) \leftrightarrow (E_1)$ is performed to give the system

$$E_1: \qquad 5.291x_1 - 6.130x_2 = 46.78,$$
$$E_2: \quad 0.003000x_1 + 59.14x_2 = 59.17.$$

The multiplier for this system is

$$m_{21} = \frac{a_{21}^{(1)}}{a_{11}^{(1)}} = 0.0005670,$$

and the operation $(E_2 - m_{21}E_1) \rightarrow (E_2)$ reduces the system to

$$5.291x_1 - 6.130x_2 = 46.78,$$
$$59.14x_2 = 59.14.$$

The four-digit answers resulting from the backward substitution are the correct values $x_1 = 10.00$ and $x_2 = 1.000$. ∎

This technique is known as **maximal column pivoting** or **partial pivoting**, and is described in detail in the following algorithm. The actual row interchanging is simulated in the algorithm by interchanging the content of $NROW$ in Step 5.

Gaussian Elimination with Maximal Column Pivoting Algorithm 6.2

To solve the $n \times n$ linear system

$$E_1: \quad a_{11}x_1 + a_{12}x_2 + \cdots + a_{1n}x_n = a_{1,n+1}$$
$$E_2: \quad a_{21}x_1 + a_{22}x_2 + \cdots + a_{2n}x_n = a_{2,n+1}$$
$$\vdots \qquad \vdots \qquad \vdots \qquad\qquad \vdots \qquad \vdots$$
$$E_n: \quad a_{n1}x_1 + a_{n2}x_2 + \cdots + a_{nn}x_n = a_{n,n+1}:$$

INPUT number of unknowns and equations n; augmented matrix $A = (a_{ij})$ where $1 \le i \le n$ and $1 \le j \le n+1$.

OUTPUT solution $x_1, \ldots, x_n$ or message that linear system has no unique solution.

Step 1   For $i = 1, \ldots, n$ set $NROW(i) = i$.   (*Initialize row pointer.*)

Step 2   For $i = 1, \ldots, n-1$ do Steps 3–6.   (*Elimination process.*)

   Step 3   Let p be the smallest integer with $i \le p \le n$ and
   $$|a(NROW(p), i)| = \max_{i \le j \le n} |a(NROW(j), i)|.$$
   (*Notation:* $a(NROW(i), j) \equiv a_{NROW_i, j}$.)

   Step 4   If $a(NROW(p), i) = 0$ then OUTPUT (no unique solution exists');
   STOP.

Step 5   If $NROW(i) \neq NROW(p)$ then set $NCOPY = NROW(i)$;
$$NROW(i) = NROW(p);$$
$$NROW(p) = NCOPY.$$
(*Simulated row interchange.*)

Step 6   For $j = i + 1, \ldots, n$ do Steps 7 and 8.

    Step 7   Set $m(NROW(j), i) = a(NROW(j), i)/a(NROW(i), i)$.

    Step 8   Perform $(E_{NROW(j)} - m(NROW(j), i)E_{NROW(i)}) \to (E_{NROW(j)})$.

Step 9   If $a(NROW(n), n) = 0$ then OUTPUT ('no unique solution exists');
STOP.

Step 10   Set $x_n = a(NROW(n), n + 1)/a(NROW(n), n)$.   (*Start backward substitution.*)

Step 11   For $i = n - 1, \ldots, 1$

$$\text{set } x_i = \frac{a(NROW(i), n + 1) - \sum_{j=i+1}^{n} a(NROW(i), j)x_j}{a(NROW(i), i)}.$$

Step 12   OUTPUT $(x_1, \ldots, x_n)$;
(*Procedure completed successfully.*)
STOP.

The procedures detailed in this algorithm are sufficient to guarantee that each multiplier m_{ji} has magnitude not exceeding one. Although the maximal column pivoting strategy is sufficient for most linear systems, situations do arise for which this strategy is inadequate.

EXAMPLE 3   The linear system

$$E_1: \quad 30.00x_1 + 591{,}400x_2 = 591{,}700,$$
$$E_2: \quad 5.291x_1 - \quad 6.130x_2 = 46.78,$$

is the same system as that presented in the previous example except that all entries in the first equation are multiplied by 10^4. The procedure described in Algorithm 6.2 with four-digit arithmetic would lead to the same inaccurate results as obtained in Example 1. The maximal value in the first column is 30.00 and the multiplier

$$m_{21} = \frac{5.291}{30.00} = 0.1764$$

leads to the system

$$30.00x_1 + 591{,}400x_2 = 591{,}700,$$
$$- 104{,}300x_2 = - 104{,}400,$$

which has solutions $x_2 = 1.001$ and $x_1 = -10.00$. ∎

A technique known as **scaled-column pivoting** is appropriate for the system in Example 3. The first step in this procedure is to define a scale factor s_i for each row

$$s_i = \max_{j=1,2,\ldots,n} |a_{ij}|.$$

If $s_i = 0$ for any i, Theorems 6.11 and 6.12 imply that no unique solution exists and the procedure halts. The appropriate row interchange to obtain zeros in the first column is determined by choosing the first integer k with

$$\frac{|a_{k1}|}{s_k} = \max_{j=1,2,\ldots,n} \frac{|a_{j1}|}{s_j},$$

and performing $(E_1) \leftrightarrow (E_k)$. The effect of scaling is to ensure that the largest element in each row has a relative magnitude one before the comparison for row interchange is performed. The scaling is done only for comparison purposes, so the division by the scaling factors produces no round-off error into the system.

Applying scaled-column pivoting to Example 3 gives

$$s_1 = \max\{|30.00|, |591400|\} = 591400,$$

and

$$s_2 = \max\{|5.291|, |-6.130|\} = 6.130.$$

Consequently,

$$\frac{|a_{11}|}{s_1} = \frac{30.00}{591400} = 0.5073 \times 10^{-4} \quad \text{and} \quad \frac{|a_{21}|}{s_2} = \frac{5.291}{6.130} = 0.8631$$

and the interchange $(E_1) \leftrightarrow (E_2)$ is made.

Applying Gaussian elimination to the new system

$$5.291x_1 - 6.130x_2 = 46.78$$
$$30.00x_1 + 591400x_2 = 591700$$

will produce the correct results: $x_1 = 10.00$ and $x_2 = 1.000$.

Gaussian Elimination with Scaled Column Pivoting Algorithm 6.3

The only steps in this algorithm that differ from those of Algorithm 6.2 are:

Step 1 For $i = 1, \ldots, n$ set $s_i = \max_{1 \le j \le n} |a_{ij}|$;

if $s_i = 0$ then OUTPUT ('no unique solution exists');
STOP.
set $NROW(i) = i$.

Step 2 For $i = 1, \ldots, n-1$ do Steps 3–6.   (*Elimination process.*)

Step 3 Let p be the smallest integer with $i \le p \le n$ and

$$\frac{|a(NROW(p), i)|}{s(NROW(p))} = \max_{i \le j \le n} \frac{|a(NROW(j), i)|}{s(NROW(j))}.$$

The additional computations required for scaled column pivoting result first from the determination of the scale factors, $(n-1)$ comparisons for each of the n rows, for a total of

$$n(n-1) \quad \text{comparisons.}$$

To determine the correct first interchange, n divisions are performed and $n-1$ comparisons are made. The first interchange determination, then, adds a total of

$$n(n-1)+(n-1) \quad \text{comparisons and} \quad n \quad \text{divisions.}$$

Since the scaling factors are computed only once, the second step requires only

$$(n-2) \quad \text{comparisons and} \quad (n-1) \quad \text{divisions.}$$

Proceeding in a similar manner, the scaled column pivoting procedure adds a total of

(6.13) $$n(n-1)+\sum_{k=2}^{n}(k-1)=\frac{3}{2}n(n-1) \quad \text{comparisons}$$

and

(6.14) $$\sum_{k=2}^{n} k = \frac{n(n+1)}{2}-1 \quad \text{divisions}$$

to the Gaussian elimination procedure. The time required to perform a comparison is comparable to, though slightly more than, that of an addition/subtraction. Since the total time to perform the basic Gaussian elimination procedure is $O(n^3/3)$ multiplications/divisions and $O(n^3/3)$ additions/subtractions, scaled-column pivoting does not add significantly to the computational time required to solve a system for large values of n.

To emphasize the importance of choosing the scale factors only once, consider the amount of additional computational time that would be required if new scale factors were determined each time a row interchange decision was to be made. Then the term $n(n-1)$ in Eq. (6.13) would be replaced by $\sum_{k=2}^{n} k(k-1)$, which can be shown to be $[n(n^2-1)/3]$. As a consequence, this modified scaled-column pivoting technique would add $O(n^3/3)$ comparisons, in addition to the $[n(n+1)/2]-1$ divisions.

If a system warrants the type of pivoting the modified scaled-column pivoting provides, **maximal** (or **total**) **pivoting** should instead be used. Maximal pivoting at the kth step searches all the entries.

$$a_{ij}, \quad \text{for } i=k, \quad k+1, \ldots, n, \quad \text{and } j=k, \quad k+1, \ldots, n$$

to find the entry with the largest magnitude. Both row and column interchanges are performed to bring this entry to the pivot position. The first step of total pivoting requires that n^2-1 comparisons be performed, the second step requires $(n-1)^2-1$ comparisons, and so on. The total additional time required to incorporate maximal pivoting into Gaussian elimination is consequently,

$$\sum_{k=2}^{n}(k^2-1)=\frac{n(n-1)(2n+5)}{6} \quad \text{comparisons.}$$

This figure is comparable to the number required for the modified scaled-column pivoting technique, but no divisions are required. Maximal pivoting is consequently the strategy recommended for the most stubborn systems where the extensive amount of execution time needed for this method can be justified. For problems involving maximal pivoting, see Exercises 6 and 7.

Exercise Set 6.5

1. Solve the following linear systems using:

   i) Gaussian elimination with two-digit rounding arithmetic,
   ii) Gaussian elimination with two-digit rounding arithmetic and maximal column pivoting,
   iii) Gaussian elimination with two-digit rounding arithmetic and scaled-column pivoting,
   iv) exact arithmetic and determine which part, (i), (ii), or (iii), is the most accurate.

   a) $\begin{aligned} x_1 + 2x_2 + 3x_3 &= 1, \\ 2x_1 + 3x_2 + 4x_3 &= -1, \\ 3x_1 + 4x_2 + 6x_3 &= 2. \end{aligned}$

   b) $\begin{aligned} 2x_1 + 4x_2 - x_3 &= -5, \\ x_1 + x_2 - 3x_3 &= -9, \\ 4x_1 + x_2 + 2x_3 &= 9. \end{aligned}$

   c) $\begin{aligned} 0.1x_1 + 0.2x_2 + 0.4x_3 &= 1.1, \\ 4x_1 + x_2 - x_3 &= 6, \\ 2x_1 + 5x_2 + 2x_3 &= 3. \end{aligned}$

   d) $\begin{aligned} 0.04x_1 + 0.01x_2 - 0.01x_3 &= 0.06, \\ 0.2x_1 + 0.5x_2 - 0.2x_3 &= 0.3, \\ x_1 + 2x_2 + 4x_3 &= 11. \end{aligned}$

2. Use Algorithms 6.1, 6.2, and 6.3 to solve the following linear systems using three-digit rounding arithmetic. Which algorithm yields the most accurate results?

   a) $\begin{aligned} 0.03x_1 + 58.9x_2 &= 59.2, \\ 5.31x_1 - 6.10x_2 &= 47.0. \end{aligned}$

   b) $\begin{aligned} 58.9x_1 + 0.03x_2 &= 59.2, \\ -6.10x_1 + 5.31x_2 &= 47.0. \end{aligned}$

   c) $\begin{aligned} x_1 + \tfrac{1}{2}x_2 + \tfrac{1}{3}x_3 &= \tfrac{11}{16}, \\ 5x_1 + \tfrac{10}{3}x_2 + \tfrac{5}{2}x_3 &= \tfrac{65}{6}, \\ \tfrac{100}{3}x_1 + 25x_2 + 20x_3 &= \tfrac{235}{3}. \end{aligned}$

   d) $\begin{aligned} 0.832x_1 + 0.448x_2 + 0.193x_3 &= 1.00, \\ 0.784x_1 + 0.421x_2 - 0.207x_3 &= 0.00, \\ 0.784x_1 - 0.421x_2 + 0.279x_3 &= 0.00. \end{aligned}$

3. Repeat Exercise 2 using three-digit chopping arithmetic.

4. Apply Algorithms 6.1, 6.2, and 6.3 to the following linear systems using four-digit rounding arithmetic. Which algorithm yields the most accurate results?

   a) $\begin{aligned} 58.09x_1 + 1.003x_2 &= 68.12, \\ 321.8x_1 + 5.550x_2 &= 377.3. \end{aligned}$

   b) $\begin{aligned} 1.003x_1 + 58.09x_2 &= 68.12, \\ 5.550x_1 + 321.8x_2 &= 377.3. \end{aligned}$

   c) $\begin{aligned} 1.003x_1 + 58.09x_2 &= 68.12, \\ 321.8x_1 + 5.550x_2 &= 377.3. \end{aligned}$

   d) $\begin{aligned} 321.8x_1 + 5.550x_2 &= 377.3, \\ 100.3x_1 + 5809x_2 &= 6812. \end{aligned}$

5. Apply Algorithms 6.1, 6.2, and 6.3 to the following linear systems using single precision computer arithmetic.

a) $3.3330x_1 + 15920x_2 - 10.333x_3 = 15913,$ b) $x_1 + \frac{1}{2}x_2 + \frac{1}{3}x_3 = \quad 2,$

$2.2220x_1 + 16.710x_2 + 9.6120x_3 = 28.544,$ $\frac{1}{2}x_1 + \frac{1}{3}x_2 + \frac{1}{4}x_3 = -1,$

$1.5611x_1 + 5.1791x_2 + 1.6852x_3 = 8.4254.$ $\frac{1}{3}x_1 + \frac{1}{4}x_2 + \frac{1}{5}x_3 = \quad 0.$

c) $x_1 + \frac{1}{2}x_2 + \frac{1}{3}x_3 + \frac{1}{4}x_4 + \frac{1}{5}x_5 = 1,$ d) $\pi x_1 - \quad ex_2 + \sqrt{2}x_3 - \sqrt{3}x_4 = \quad 1,$

$\frac{1}{2}x_1 + \frac{1}{3}x_2 + \frac{1}{4}x_3 + \frac{1}{5}x_4 + \frac{1}{6}x_5 = 1,$ $\pi^2 x_1 + \quad ex_2 - \quad e^2 x_3 + \quad \frac{3}{7}x_4 = -1,$

$\frac{1}{3}x_1 + \frac{1}{4}x_2 + \frac{1}{5}x_3 + \frac{1}{6}x_4 + \frac{1}{7}x_5 = 1,$ $\sqrt{5}x_1 - \sqrt{6}x_2 + \quad x_3 - 1.1x_4 = \quad 0,$

$\frac{1}{4}x_1 + \frac{1}{5}x_2 + \frac{1}{6}x_3 + \frac{1}{7}x_4 + \frac{1}{8}x_5 = 1,$ $\pi^3 x_1 + \quad e^2 x_2 - \sqrt{7}x_3 + \quad x_4 = \sqrt{2}.$

$\frac{1}{5}x_1 + \frac{1}{6}x_2 + \frac{1}{7}x_3 + \frac{1}{8}x_4 + \frac{1}{9}x_5 = 1.$

6. Construct an algorithm for the maximal pivoting procedure discussed on page 329.

7. Use the maximal pivoting algorithm developed in Exercise 6 to obtain solutions to

a)   Exercise 1,   b)   Exercise 2,   c)   Exercise 3,

d)   Exercise 4,   e)   Exercise 5.

6.6 Special Types of Matrices

In this section some additional material on matrices is presented. The first type of matrix we consider is produced naturally when Gaussian elimination is performed on a linear system.

DEFINITION 6.13   An **upper-triangular** $n \times n$ matrix U has for each j entries

$$u_{ij} = 0 \qquad \text{for each } i = j+1, j+2, \ldots, n;$$

and a **lower-triangular** matrix L has for each j entries

$$l_{ij} = 0 \qquad \text{for each } i = 1, 2, \ldots, j-1.$$

(A **diagonal** matrix is both upper and lower triangular.)

To evaluate the determinant of an arbitrary matrix can require considerable manipulation. A matrix in triangular form, however, has an easily calculated determinant.

THEOREM 6.14   If $A = (a_{ij})$ is an $n \times n$ matrix that is either upper triangular or lower triangular (or diagonal), then $\det A = \prod_{i=1}^{n} a_{ii}$.

The proof of this result consists of expanding the matrix and each submatrix about either the first row or first column.

EXAMPLE 1   Reconsider Examples 1 and 3 of Section 6.1, in which the linear system

$$
\begin{aligned}
x_1 + x_2 \quad\;\; + 3x_4 &= 4, \\
2x_1 + x_2 - x_3 + x_4 &= 1, \\
3x_1 - x_2 - x_3 + 2x_4 &= -3, \\
-x_1 + 2x_2 + 3x_3 - x_4 &= 4
\end{aligned}
$$

was reduced to the equivalent system

$$
\begin{bmatrix}
1 & 1 & 0 & 3 & \vdots & 4 \\
0 & -1 & -1 & -5 & \vdots & -7 \\
0 & 0 & 3 & 13 & \vdots & 13 \\
0 & 0 & 0 & -13 & \vdots & -13
\end{bmatrix}.
$$

Let U be the 4×4 upper-triangular matrix

$$
U = \begin{bmatrix}
1 & 1 & 0 & 3 \\
0 & -1 & -1 & -5 \\
0 & 0 & 3 & 13 \\
0 & 0 & 0 & -13
\end{bmatrix},
$$

which is the result of Gaussian elimination performed on A. For $i = 1, 2, 3$, define m_{ji} for each $j = i + 1, i + 2, \ldots, 4$ to be the number used in the elimination step $(E_j - m_{ji}E_i) \rightarrow E_j$; that is, $m_{21} = 2$, $m_{31} = 3$, $m_{41} = -1$, $m_{32} = 4$, $m_{42} = -3$, and $m_{43} = 0$. If L is defined to be the 4×4 lower-triangular matrix with entries l_{ij} given by

$$
l_{ij} = \begin{cases}
0, & \text{when } i = 1, 2, \ldots, j - 1, \\
1, & \text{when } i = j, \\
m_{ij}, & \text{when } i = j + 1, j + 2, \ldots, n,
\end{cases}
$$

then

$$
L = \begin{bmatrix}
1 & 0 & 0 & 0 \\
2 & 1 & 0 & 0 \\
3 & 4 & 1 & 0 \\
-1 & -3 & 0 & 1
\end{bmatrix};
$$

and it is easy to verify that

$$
LU = \begin{bmatrix}
1 & 0 & 0 & 0 \\
2 & 1 & 0 & 0 \\
3 & 4 & 1 & 0 \\
-1 & -3 & 0 & 1
\end{bmatrix}
\begin{bmatrix}
1 & 1 & 0 & 3 \\
0 & -1 & -1 & -5 \\
0 & 0 & 3 & 13 \\
0 & 0 & 0 & -13
\end{bmatrix}
$$

$$
= \begin{bmatrix}
1 & 1 & 0 & 3 \\
2 & 1 & -1 & 1 \\
3 & -1 & -1 & 2 \\
-1 & 2 & 3 & -1
\end{bmatrix} = A.
$$

■

The results obtained in Example 1 are true in general and are given in the following theorem. The proof of this result is considered in Exercise 12.

THEOREM 6.15   If the Gaussian elimination procedure (Algorithm 6.1) can be performed on the system $A\mathbf{x} = \mathbf{b}$ without row interchanges, then the matrix A can be factored into the product of a lower-triangular matrix L with an upper-triangular matrix U:

$$A = LU,$$

where $U = (u_{ij})$ and $L = (l_{ij})$ are defined for each j by:

$$u_{ij} = \begin{cases} a_{ij}^{(i)}, & \text{when } i = 1, 2, \ldots, j, \\ 0, & \text{when } i = j+1, j+2, \ldots, n, \end{cases}$$

and

$$l_{ij} = \begin{cases} 0, & \text{when } i = 1, 2, \ldots, j-1, \\ 1, & \text{when } i = j, \\ m_{ij}, & \text{when } i = j+1, j+2, \ldots, n, \end{cases}$$

where $a_{ij}^{(i)}$ is the i, j-element of the final matrix obtained in the Gaussian elimination method, and m_{ij} is the multiplier.

If row interchanges must be performed for the procedure to work, then A can be factored into LU, where U is the same as in Theorem 6.15, but in general, L is not lower triangular. This topic is further elaborated upon in the exercises (in particular, see Exercises 13 and 14).

The problem of computing the determinant of a matrix can be simplified by first reducing the matrix to triangular form and then using Theorem 6.14 to find the determinant of the triangular matrix.

EXAMPLE 2   Let

$$A = \begin{bmatrix} 1 & 1 & 0 & 3 \\ 2 & 1 & -1 & 1 \\ -1 & 2 & 3 & -1 \\ 3 & -1 & -1 & 2 \end{bmatrix}.$$

By the three permitted types of operation, the matrix A will be reduced to an upper-triangular matrix. First perform $(E_2 - 2E_1) \rightarrow (E_2)$, $(E_3 + E_1) \rightarrow (E_3)$, and $(E_4 - 3E_1) \rightarrow (E_4)$ to obtain:

$$\tilde{A}_1 = \begin{bmatrix} 1 & 1 & 0 & 3 \\ 0 & -1 & -1 & -5 \\ 0 & 3 & 3 & 2 \\ 0 & -4 & -1 & -7 \end{bmatrix}.$$

By Theorem 6.11(e), $\det A = \det \tilde{A}_1$. Forming $\tilde{A}_2$ from $\tilde{A}_1$ by the operations $(E_3 + 3E_2) \rightarrow (E_3)$ and $(E_4 - 4E_2) \rightarrow (E_4)$,

$$\tilde{A}_2 = \begin{bmatrix} 1 & 1 & 0 & 3 \\ 0 & -1 & -1 & -5 \\ 0 & 0 & 0 & -13 \\ 0 & 0 & 3 & 13 \end{bmatrix},$$

and, again, $\det \tilde{A}_2 = \det \tilde{A}_1 = \det A$. Let $\tilde{A}_3$ be formed from $\tilde{A}_2$ by $(E_3) \leftrightarrow (E_4)$, thus:

$$\tilde{A}_3 = \begin{bmatrix} 1 & 1 & 0 & 3 \\ 0 & -1 & -1 & -5 \\ 0 & 0 & 3 & 13 \\ 0 & 0 & 0 & -13 \end{bmatrix}.$$

By Theorem 6.14, $\det \tilde{A}_3 = (1)(-1)(3)(-13) = 39$, and since $\tilde{A}_3$ was formed from $\tilde{A}_2$ by a row interchange,

$$\det A = \det \tilde{A}_2 = -\det \tilde{A}_3 = -39. \qquad \blacksquare$$

DEFINITION 6.16   The **transpose** of an $m \times n$ matrix A, denoted A^t, is an $n \times m$ matrix whose entries are $(A^t)_{ij} = (A)_{ji}$. A matrix whose transpose is itself is said to be **symmetric**.

EXAMPLE 3   The matrices

$$A = \begin{bmatrix} 7 & 2 & 0 \\ 3 & 5 & -1 \\ 0 & 5 & -6 \end{bmatrix}, \quad B = \begin{bmatrix} 2 & 4 & 7 \\ 3 & -5 & -1 \end{bmatrix}, \quad C = \begin{bmatrix} 6 & 4 & 3 \\ 4 & -2 & 0 \\ 3 & 0 & 1 \end{bmatrix}$$

have transposes

$$A^t = \begin{bmatrix} 7 & 3 & 0 \\ 2 & 5 & 5 \\ 0 & -1 & -6 \end{bmatrix}, \quad B^t = \begin{bmatrix} 2 & 3 \\ 4 & -5 \\ 7 & -1 \end{bmatrix}, \quad C^t = \begin{bmatrix} 6 & 4 & 3 \\ 4 & -2 & 0 \\ 3 & 0 & 1 \end{bmatrix}.$$

Since C^t and C have the same entries, C is symmetric. $\qquad \blacksquare$

THEOREM 6.17   The following operations involving the transpose of a matrix hold whenever the operation is possible:

1. $(A^t)^t = A$,
2. $(A + B)^t = A^t + B^t$,
3. $(AB)^t = B^t A^t$,
4. If A^{-1} exists, $(A^{-1})^t = (A^t)^{-1}$,
5. $\det A^t = \det A$.

The proof of these statements is elementary and is left to Exercise 4.

DEFINITION 6.18   An $n \times n$ matrix is called a **band matrix** if integers p and q, $1 < p, q < n$, exist with the property that $a_{ij} = 0$ whenever $i + p \leq j$ or $j + q \leq i$. The **band width** for a matrix of this type is defined to be $w = p + q - 1$.

The matrix A defined in Example 3 is a band matrix with $p = q = 2$ and band width 3.

The definition of band matrix forces those matrices to concentrate all their nonzero entries about the diagonal. Two special cases of band matrices that occur often in practice have $p = q = 2$ and $p = q = 4$. The matrix of band width 3, occurring when $p = q = 2$, has already been encountered in connection with the study of cubic spline approximations in Section 3.6. These matrices are often called **tridiagonal** since they have the form

$$A = \begin{bmatrix} a_{11} & a_{12} & 0 & \cdots & \cdots & 0 \\ a_{21} & a_{22} & a_{23} & & & \vdots \\ 0 & a_{32} & a_{33} & a_{34} & & 0 \\ \vdots & & & & & a_{n-1,n} \\ 0 & \cdots & \cdots & 0 & a_{n,n-1} & a_{nn} \end{bmatrix}.$$

Tridiagonal matrices will be discussed again in Chapter 10 in connection with the study of piecewise linear approximations to boundary-value problems. The case of $p = q = 4$ will also be used for the solution of boundary-value problems, when the approximating functions assume the form of cubic splines.

Another useful class of matrices is described in the following definition.

DEFINITION 6.19   The $n \times n$ matrix A is said to be **strictly diagonally dominant** in case

(6.15)
$$|a_{ii}| > \sum_{\substack{j=1, \\ j \neq i}}^{n} |a_{ij}|$$

holds for each $i = 1, 2, \ldots, n$.

The matrix A described in Example 3 is strictly diagonally dominant since $|7| > |2| + |0|$, $|5| > |3| + |-1|$, and $|-6| > |0| + |5|$. It is interesting to note that A^t is not strictly diagonally dominant, however.

THEOREM 6.20   If A is a strictly diagonally dominant $n \times n$ matrix, then A is nonsingular. Moreover, Gaussian elimination can be performed on any linear system of the form $A\mathbf{x} = \mathbf{b}$ to obtain its unique solution without row or column interchanges, and the computations are stable with respect to the growth of rounding errors.

PROOF   To show that A is nonsingular, consider the linear system described by $A\mathbf{x} = \mathbf{0}$, and suppose that a nonzero solution $\mathbf{x} = (x_i)$ to this system exists. In this case, for some k,

$0 < |x_k| = \max_{1 \le j \le n} |x_j|$. Since $\sum_{j=1}^{n} a_{ij}x_j = 0$ for each $i = 1, 2, \ldots, n$, when $i = k$

$$a_{kk}x_k = -\sum_{\substack{j=1, \\ j \ne k}}^{n} a_{kj}x_j.$$

This implies

$$|a_{kk}||x_k| \le \sum_{\substack{j=1, \\ j \ne k}}^{n} |a_{kj}||x_j| \quad \text{or} \quad |a_{kk}| \le \sum_{\substack{j=1, \\ j \ne k}}^{n} |a_{kj}|\frac{|x_j|}{|x_k|} \le \sum_{\substack{j=1, \\ j \ne k}}^{n} |a_{kj}|,$$

in contradiction to the strict diagonal dominance of A. Consequently, the only solution to $A\mathbf{x} = \mathbf{0}$ is $\mathbf{x} = \mathbf{0}$, a condition shown in Theorem 6.12 to be equivalent to the nonsingularity of A.

The proof that Gaussian elimination can be performed without interchanges is considered in Exercise 16. The demonstration of stability for this procedure can be found in Wendroff [140]. □

Note that Theorem 6.20 can be used to give elementary proofs of Theorem 3.11 and of Theorem 3.12 of Section 3.6 when applied to the matrices given in those theorems.

The final special class of matrices to be discussed in this section is called **positive definite**.

DEFINITION 6.21   A symmetric $n \times n$ matrix A is called **positive definite** if $\mathbf{x}^t A\mathbf{x} > 0$ for every n-dimensional column vector $\mathbf{x} \ne \mathbf{0}$.

To be precise, the definition should specify that the 1×1 matrix generated by the operation $\mathbf{x}^t A\mathbf{x}$ has a positive value for its only entry, since the operation is performed as follows:

$$\mathbf{x}^t A\mathbf{x} = [x_1, x_2, \ldots, x_n] \begin{bmatrix} a_{11} & a_{12} & \cdots & a_{1n} \\ a_{21} & a_{22} & \cdots & a_{2n} \\ \vdots & \vdots & & \vdots \\ a_{n1} & a_{n2} & \cdots & a_{nn} \end{bmatrix} \begin{bmatrix} x_1 \\ x_2 \\ \vdots \\ x_n \end{bmatrix}$$

$$= [x_1, x_2, \ldots, x_n] \begin{bmatrix} \sum_{j=1}^{n} a_{1j}x_j \\ \sum_{j=1}^{n} a_{2j}x_j \\ \vdots \\ \sum_{j=1}^{n} a_{nj}x_j \end{bmatrix} = \left[\sum_{i=1}^{n} \sum_{j=1}^{n} a_{ij}x_i x_j \right].$$

EXAMPLE 4   The matrix

$$A = \begin{bmatrix} 2 & -1 & 0 \\ -1 & 2 & -1 \\ 0 & -1 & 2 \end{bmatrix}$$

is positive definite, for suppose $\mathbf{x}$ is any three-dimensional column vector, then

$$\mathbf{x}^t A \mathbf{x} = [x_1, x_2, x_3] \begin{bmatrix} 2 & -1 & 0 \\ -1 & 2 & -1 \\ 0 & -1 & 2 \end{bmatrix} \begin{bmatrix} x_1 \\ x_2 \\ x_3 \end{bmatrix}$$

$$= [x_1, x_2, x_3] \begin{bmatrix} 2x_1 - x_2 \\ -x_1 + 2x_2 - x_3 \\ -x_2 + 2x_3 \end{bmatrix}$$

$$= [2x_1^2 - 2x_1 x_2 + 2x_2^2 - 2x_2 x_3 + 2x_3^2]$$
$$= [x_1^2 + (x_1 - x_2)^2 + (x_2 - x_3)^2 + x_3^2]$$

and
$$x_1^2 + (x_1 - x_2)^2 + (x_2 - x_3)^2 + x_3^2 > 0,$$

unless $x_1 = x_2 = x_3 = 0$. ∎

It should be clear from Example 4 that using the definition to determine whether a matrix is positive definite can be extremely tedious. Fortunately, there are more easily verified criteria, which will be presented in Chapter 8, for identifying members of this important class. Conditions that can be used to eliminate certain matrices from consideration are discussed in Exercise 10.

The next theorem parallels Theorem 6.20, which dealt with strictly diagonally dominant matrices.

THEOREM 6.22 If A is a positive definite $n \times n$ matrix, then A is nonsingular. Moreover, Gaussian elimination can be performed on any linear system of the form $A\mathbf{x} = \mathbf{b}$ to obtain its unique solution without row or column interchanges, and the computations are stable with respect to the growth of rounding errors.

PROOF If $\mathbf{x} \neq \mathbf{0}$ is a vector that satisfies $A\mathbf{x} = \mathbf{0}$, then $\mathbf{x}^t A \mathbf{x} = 0$. This contradicts the assumption that A is positive definite. Consequently, $A\mathbf{x} = \mathbf{0}$ has only the zero solution and Theorem 6.12 implies that A is nonsingular.

To show the validity of the initial statement concerning Gaussian elimination we need to use concepts discussed in Exercise 15. Let $\hat{A}$ denote a leading principal submatrix of A for an arbitrary k, $k = 1, 2, \ldots, n$ (the definition of leading principal submatrix is given in Exercise 15). If $\hat{A}$ is not positive definite, then a k-dimensional column vector $\hat{\mathbf{x}} \neq \mathbf{0}$ exists with $\hat{\mathbf{x}}^t \hat{A} \hat{\mathbf{x}} \leq \mathbf{0}$. Construct an n-dimensional column vector $\mathbf{x} \neq \mathbf{0}$ from $\hat{\mathbf{x}}$ by placing zeros in the last $(n - k)$ coordinates. Since

$$\mathbf{x}^t A \mathbf{x} = \hat{\mathbf{x}}^t \hat{A} \hat{\mathbf{x}} \leq 0,$$

A is not positive definite and we have a contradiction. Consequently every principal submatrix of A is positive definite and hence nonsingular, and the result follows from Exercise 15.

The proof of stability of the Gaussian elimination procedure can be found in Wendroff [140], beginning on page 120. □

Exercise Set 6.6

1. Determine which of the following matrices are

   i) symmetric
   ii) singular
   iii) strictly diagonally dominant
   iv) positive definite

   a) $\begin{bmatrix} 2 & 1 \\ 1 & 3 \end{bmatrix}$

   b) $\begin{bmatrix} -2 & 1 \\ 1 & -3 \end{bmatrix}$

   c) $\begin{bmatrix} 2 & 1 & 0 \\ 0 & 3 & 0 \\ 1 & 0 & 4 \end{bmatrix}$

   d) $\begin{bmatrix} 2 & 1 & 0 \\ 0 & 3 & 2 \\ 1 & 2 & 4 \end{bmatrix}$

   e) $\begin{bmatrix} 4 & 2 & 6 \\ 3 & 0 & 7 \\ -2 & -1 & -3 \end{bmatrix}$

   f) $\begin{bmatrix} 2 & -1 & 0 \\ -1 & 4 & 2 \\ 0 & 2 & 2 \end{bmatrix}$

   g) $\begin{bmatrix} 4 & 0 & 0 & 0 \\ 6 & 7 & 0 & 0 \\ 9 & 11 & 1 & 0 \\ 5 & 4 & 1 & 1 \end{bmatrix}$

   h) $\begin{bmatrix} 2 & 3 & 1 & 2 \\ -2 & 4 & -1 & 5 \\ 3 & 7 & 1.5 & 1 \\ 6 & 9 & 3 & 7 \end{bmatrix}$

2. Verify the statements in Theorem 6.17 for the matrices

$$A = \begin{bmatrix} 4 & 6 & 1 & -1 \\ 2 & 1 & 0 & 0.5 \\ 3 & 0 & 0 & 1 \\ 1 & -1 & 1 & 1 \end{bmatrix}, \quad B = \begin{bmatrix} 1 & 2 & 3 & 4 \\ 0 & 2 & -1 & 1 \\ 0 & 0 & 3 & 2 \\ 0 & 0 & 0 & -1 \end{bmatrix}.$$

3. Prove the following statements or provide counterexamples to show they are not true:

   a) The product of two symmetric matrices is symmetric.

   b) The inverse of a nonsingular symmetric matrix is a nonsingular symmetric matrix.

   c) If A and B are $n \times n$ matrices, then $(AB)^t = A^t B^t$.

4. Prove the statements presented in Theorem 6.17.

5. Suppose that A and B are strictly diagonally dominant $n \times n$ matrices.

   a) Is $-A$ strictly diagonally dominant?

   b) Is A^t strictly diagonally dominant?

   c) Is $A + B$ strictly diagonally dominant?

   d) Is A^2 strictly diagonally dominant?

   e) Is $A - B$ strictly diagonally dominant?

6. Suppose that A and B are positive definite $n \times n$ matrices.

   a) Is $-A$ positive definite?

   b) Is A^t positive definite?

   c) Is $A + B$ positive definite?

   d) Is A^2 positive definite?

   e) Is $A - B$ positive definite?

7. Suppose A and B commute, that is, $AB = BA$. Must A^t and B^t also commute?

8. Construct a matrix A that is nonsymmetric but for which $x^t A x > 0$ for all $x \neq 0$.

9. An $n \times n$ matrix A is called **diagonally dominant** if

$$|a_{ii}| \geq \sum_{\substack{j=1, \\ j \neq i}}^{n} |a_{ij}| \qquad \text{for each } i = 1, 2, \ldots, n.$$

   a) Find a singular diagonally dominant matrix with no zero entries.

   b) Does a singular diagonally dominant matrix exist that has no zero entries and for which strict inequality holds in inequality (6.15) for all but one row?

10. Suppose A is an $n \times n$ positive definite matrix. Show that:

    a) $a_{ii} > 0$ for each $i = 1, 2, \ldots, n$.

    b) $\max_{1 \leq i \leq n} |a_{ii}| \geq \max_{1 \leq i, j \leq n} |a_{ij}|$.

    c) $(a_{ij})^2 < a_{ii} a_{jj}$ for each $i, j = 1, 2, \ldots, n$, with $i \neq j$.

11. Show that if the matrix A can be factored in the form given in Theorem 6.15 where, in addition, $U = L^t$, then A must be positive definite.

12. Let $m_{ji} = a_{ji}^{(i)}/a_{ii}^{(i)}$ for each $j = i + 1, i + 2, \ldots, n$, and define

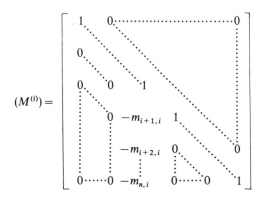

for each $i = 1, 2, \ldots, n - 1$.

    a) Show that $\tilde{A}^{(2)} = M^{(1)} \tilde{A}^{(1)}$; that is, the operations $(E_j - m_{j1} E_1) \rightarrow (E_j)$ for each $j = 2, 3, \ldots, n$ are equivalent to the matrix multiplications $M^{(1)} A$ and $M^{(1)} b$.

b) Assuming no row interchanges are necessary, show that $\tilde{A}^{(k)} = M^{(k-1)}M^{(k-2)} \cdots M^{(1)}\tilde{A}^{(1)}$ for each $k = 2, \dots, n$.

c) Show that $M^{(i)}$ is nonsingular for each i, and that:

$$(M^{(i)})^{-1} = \begin{bmatrix} 1 & 0 & \cdots & & & & & & 0 \\ 0 & & & & & & & & \\ 0 & 0 & 1 & & & & & & \\ \vdots & & 0 & m_{i+1,i} & 1 & & & & \\ & & & m_{i+2,i} & 0 & & & 0 \\ 0 & \cdots & 0 & m_{n,i} & 0 & \cdots & 0 & 1 \end{bmatrix}$$

d) Let $M = M^{(n-1)}M^{(n-2)} \cdots M^{(1)}$. Show that M is lower triangular and nonsingular and that $U = MA$, where U is the upper-triangular matrix formed in Gauss elimination.

e) Show that

$$L = M^{-1} = \begin{bmatrix} 1 & & 0 & \cdots & 0 \\ m_{21} & 1 & & & \\ \vdots & & & & 0 \\ m_{n1} & \cdots & & m_{n,n-1} & 1 \end{bmatrix}$$

and that A can be factored into the product of a lower-triangular matrix and an upper-triangular matrix.

13. **A permutation matrix** P is a matrix that has precisely one entry whose value is one in each column and each row, and all of whose other entries are zero. For example,

$$P = \begin{bmatrix} 1 & 0 & 0 \\ 0 & 0 & 1 \\ 0 & 1 & 0 \end{bmatrix}$$

is a 3×3 permutation matrix.

a) Show that the operation $(E_i) \leftrightarrow (E_j)$ performed on an $n \times n$ matrix A is equivalent to the multiplication PA, where P is the $n \times n$ permutation matrix formed by applying $(E_i) \leftrightarrow (E_j)$ on I_n.

b) What effect does the multiplication AP have when P is the permutation matrix obtained in (a)?

c) Show that $P^{-1} = P^t$.

d) If the row interchanges required in Algorithm 6.1 to reduce A are also applied to I, resulting in the matrix P, show that $LU = PA$. This implies $A = (P^{-1}L)U$, but $P^{-1}L$ is not necessarily lower triangular.

14. Let

$$A = \begin{bmatrix} 1 & 1 & 0 & 3 \\ 2 & 1 & -1 & 1 \\ -1 & 2 & 3 & -1 \\ 3 & -1 & -1 & 2 \end{bmatrix}.$$

Use the discussion in Exercise 13 to obtain a factorization of A in the form $A = (P^{-1}L)U$.

15. Given an $n \times n$ matrix A, a **leading principal submatrix** of A is defined to be a submatrix of the form

$$\begin{bmatrix} a_{11} & a_{12} & \cdots & a_{1k} \\ a_{21} & a_{22} & \cdots & a_{2k} \\ \vdots & \vdots & & \vdots \\ a_{k1} & a_{k2} & \cdots & a_{kk} \end{bmatrix}$$

where $1 \le k \le n$. Show that Gaussian elimination can be performed on A without row interchanges if and only if all leading principal submatrices of A are nonsingular. [*Hint:* Fix $1 \le k \le n$ and recall, from Exercise 12 that $A^{(k)} = M^{(k-1)} \cdots M^{(1)}A$. Partition each of these matrices vertically between the kth and $(k+1)$st columns and horizontally between the kth and $(k+1)$st rows, to obtain:

$$\begin{bmatrix} A_{11}^{(k)} & A_{12}^{(k)} \\ A_{21}^{(k)} & A_{22}^{(k)} \end{bmatrix} = \begin{bmatrix} M_{11}^{(k-1)} & 0 \\ M_{21}^{(k-1)} & M_{22}^{(k-1)} \end{bmatrix}\begin{bmatrix} M_{11}^{(k-2)} & 0 \\ M_{21}^{(k-2)} & M_{22}^{(k-2)} \end{bmatrix} \cdots \begin{bmatrix} M_{11}^{(1)} & 0 \\ M_{21}^{(1)} & M_{22}^{(1)} \end{bmatrix}\begin{bmatrix} A_{11} & A_{12} \\ A_{21} & A_{22} \end{bmatrix}.$$

Show that A_{11} nonsingular implies $A_{11}^{(k)}$ nonsingular. If Gaussian elimination can be performed without row interchanges, then $A = LU$. Use this, with the consideration of the partitioned product, to obtain the result.]

16. Use Exercise 15 and mathematical induction to prove that the Gaussian elimination procedure can be performed to find the unique solution to $A\mathbf{x} = \mathbf{b}$ without row or column interchanges whenever A is a strictly diagonally dominant matrix.

17. In a paper by Dorn and Burdick [39], it is reported that the average wing length which resulted from mating three mutant varieties of fruit flies (*Drosophila melanogaster*) can be expressed in the symmetric matrix form

$$A = \begin{bmatrix} 1.59 & 1.69 & 2.13 \\ 1.69 & 1.31 & 1.72 \\ 2.13 & 1.72 & 1.85 \end{bmatrix},$$

where a_{ij} denotes the average wing length of an offspring resulting from the mating of a male of type i with a female of type j.

a) What physical significance is associated with the symmetry of this matrix?

b) Is this matrix positive definite? If so, prove it; if not, find a nonzero vector $\mathbf{x}$ for which $\mathbf{x}^t A\mathbf{x} \le 0$.

6.7 Direct Factorization of Matrices

The discussion centering around Theorem 6.15 on page 333 concerned factoring a matrix A in terms of a lower-triangular matrix L and an upper-triangular matrix U. It was shown that this factorization exists whenever the linear system $A\mathbf{x} = \mathbf{b}$ can be solved uniquely by Gaussian elimination without row or column interchanges. The system $LU\mathbf{x} = A\mathbf{x} = \mathbf{b}$ can then be transformed into the system $U\mathbf{x} = L^{-1}\mathbf{b}$ and, since U is upper triangular, backward substitution can be applied. Although the specific forms of L and U can be obtained from the Gaussian elimination process, it is desirable to find a more direct method for their determination so that if many systems are to be solved using A, only a forward and backward substitution need be performed (see Steps 11–14 of Algorithm 6.5). To illustrate a procedure for calculating the entries of these matrices, let us consider an example.

EXAMPLE 1   Consider the strictly diagonally dominant 4×4 matrix

$$A = \begin{bmatrix} 6 & 2 & 1 & -1 \\ 2 & 4 & 1 & 0 \\ 1 & 1 & 4 & -1 \\ -1 & 0 & -1 & 3 \end{bmatrix}.$$

Theorems 6.20 (p. 335) and 6.15 (p. 333) guarantee that A can be factored in the form $A = LU$, where:

$$L = \begin{bmatrix} l_{11} & 0 & 0 & 0 \\ l_{21} & l_{22} & 0 & 0 \\ l_{31} & l_{32} & l_{33} & 0 \\ l_{41} & l_{42} & l_{43} & l_{44} \end{bmatrix} \quad \text{and} \quad U = \begin{bmatrix} u_{11} & u_{12} & u_{13} & u_{14} \\ 0 & u_{22} & u_{23} & u_{24} \\ 0 & 0 & u_{33} & u_{34} \\ 0 & 0 & 0 & u_{44} \end{bmatrix}.$$

The 16 known entries in A can be used to partially determine the ten unknown entries in L and the like number in U. If a procedure leading to a unique solution is desired, however, four additional conditions on the entries of L and U are needed. The method to be used in this example arbitrarily requires that $l_{11} = l_{22} = l_{33} = l_{44} = 1$; this is known as **Doolittle's method**. Later in this section, methods requiring that all the diagonal elements of U be one—**Crout's Method**—and requiring that $l_{ii} = u_{ii}$ for each value of i—**Choleski's Method**—will be considered.

The portion of the multiplication of L by U,

$$L \cdot U = \begin{bmatrix} 1 & 0 & 0 & 0 \\ l_{21} & 1 & 0 & 0 \\ l_{31} & l_{32} & 1 & 0 \\ l_{41} & l_{42} & l_{43} & 1 \end{bmatrix} \begin{bmatrix} u_{11} & u_{12} & u_{13} & u_{14} \\ 0 & u_{22} & u_{23} & u_{24} \\ 0 & 0 & u_{33} & u_{34} \\ 0 & 0 & 0 & u_{44} \end{bmatrix},$$

that determines the first row of A, results in the four equations

$$u_{11} = 6, \qquad u_{12} = 2, \qquad u_{13} = 1, \qquad u_{14} = -1.$$

The portion of the multiplication that determines the remaining entries in the first column of A provides equations

$$l_{21}u_{11} = 2, \qquad\qquad l_{21} = \tfrac{1}{3},$$
$$l_{31}u_{11} = 1, \quad \text{and} \quad l_{31} = \tfrac{1}{6},$$
$$l_{41}u_{11} = -1, \qquad\qquad l_{41} = -\tfrac{1}{6}.$$

At this stage the matrices L and U assume the form

$$L = \begin{bmatrix} 1 & 0 & 0 & 0 \\ \tfrac{1}{3} & 1 & 0 & 0 \\ \tfrac{1}{6} & l_{32} & 1 & 0 \\ -\tfrac{1}{6} & l_{42} & l_{43} & 1 \end{bmatrix} \quad \text{and} \quad U = \begin{bmatrix} 6 & 2 & 1 & -1 \\ 0 & u_{22} & u_{23} & u_{24} \\ 0 & 0 & u_{33} & u_{34} \\ 0 & 0 & 0 & u_{44} \end{bmatrix}.$$

The portion of the multiplication that determines the remaining entries in the second row of A leads to the equations

$$\tfrac{2}{3} + u_{22} = 4, \qquad\qquad u_{22} = \tfrac{10}{3},$$
$$\tfrac{1}{3} + u_{23} = 1, \quad \text{so} \quad u_{23} = \tfrac{2}{3},$$
$$-\tfrac{1}{3} + u_{24} = 0, \qquad\qquad u_{24} = \tfrac{1}{3};$$

and that which determines the remaining entries in the second column of A gives

$$\tfrac{2}{6} + \tfrac{10}{3}l_{32} = 1, \qquad\qquad l_{32} = \tfrac{1}{5},$$
$$-\tfrac{2}{6} + \tfrac{10}{3}l_{42} = 0, \quad \text{so} \quad l_{42} = \tfrac{1}{10}.$$

This process is continued, alternating columns and rows between L and U, to finally obtain:

$$L = \begin{bmatrix} 1 & 0 & 0 & 0 \\ \tfrac{1}{3} & 1 & 0 & 0 \\ \tfrac{1}{6} & \tfrac{1}{5} & 1 & 0 \\ -\tfrac{1}{6} & \tfrac{1}{10} & -\tfrac{9}{37} & 1 \end{bmatrix} \quad \text{and} \quad U = \begin{bmatrix} 6 & 2 & 1 & -1 \\ 0 & \tfrac{10}{3} & \tfrac{2}{3} & \tfrac{1}{3} \\ 0 & 0 & \tfrac{37}{10} & -\tfrac{9}{10} \\ 0 & 0 & 0 & \tfrac{191}{74} \end{bmatrix}. \qquad \blacksquare$$

A general procedure for factoring matrices into a product of triangular matrices is contained in the following algorithm. Although new matrices L and U are constructed, the actual values generated can replace the corresponding entries of A that are no longer needed. Thus, the new matrix has entries $a_{ij} = l_{ij}$ for each $i = 2, 3, \ldots, n$ and $j = 1, 2, \ldots, i-1$; and $a_{ij} = u_{ij}$ for each $i = 1, 2, \ldots, n$ and $j = i, i+1, \ldots, n$.

When this algorithm is being used to factor the coefficient matrix of a linear system of equations, difficulties can arise because no pivoting is used to reduce the effect of round-off error. Previous calculations showed that round-off error can be quite significant when finite-digit arithmetic is used, and any efficient algorithm must take this effect into consideration.

Although column interchange is difficult to incorporate into the factorization algorithm, the algorithm can be easily altered to include a row-interchange technique

Direct Factorization Algorithm 6.4

To factor the $n \times n$ matrix $A = (a_{ij})$ into the product of the lower-triangular matrix $L = (l_{ij})$ and the upper-triangular matrix $U = (u_{ij})$, that is, $A = LU$ where the main diagonal of either L or U is given:

INPUT dimension n; the entries a_{ij}, $1 \leq i, j \leq n$, of A; the diagonal $l_{11}, \ldots, l_{nn}$ of L or the diagonal $u_{11}, \ldots, u_{nn}$ of U.

OUTPUT the entries l_{ij}, $1 \leq j \leq i$, $1 \leq i \leq n$ of L and the entries u_{ij}, $i \leq j \leq n$, $1 \leq i \leq n$ of U.

Step 1   Select l_{11} and u_{11} satisfying $l_{11}u_{11} = a_{11}$.
   If $l_{11}u_{11} = 0$ then OUTPUT ('Factorization impossible');
          STOP.

Step 2   For $j = 2, \ldots, n$ set $u_{1j} = a_{1j}/l_{11}$;   (*First row of U.*)
                     $l_{j1} = a_{j1}/u_{11}$.   (*First column of L.*)

Step 3   For $i = 2, \ldots, n-1$ do Steps 4 and 5.

   Step 4   Select l_{ii} and u_{ii} satisfying $l_{ii}u_{ii} = a_{ii} - \sum_{k=1}^{i-1} l_{ik}u_{ki}$.

      If $l_{ii}u_{ii} = 0$ then OUTPUT ('Factorization impossible');
                STOP.

   Step 5   For $j = i+1, \ldots, n$

      set $u_{ij} = \dfrac{1}{l_{ii}}\left[a_{ij} - \sum_{k=1}^{i-1} l_{ik}u_{kj} \right]$;   (*ith row of U.*)

      $l_{ji} = \dfrac{1}{u_{ii}}\left[a_{ji} - \sum_{k=1}^{i-1} l_{jk}u_{ki} \right]$.   (*ith column of L.*)

Step 6   Select l_{nn} and u_{nn} satisfying $l_{nn}u_{nn} = a_{nn} - \sum_{k=1}^{n-1} l_{nk}u_{kn}$.

   (*Note: If $l_{nn}u_{nn} = 0$, then $A = LU$ but A is singular.*)

Step 7   OUTPUT (l_{ij} for $j = 1, \ldots, i$ and $i = 1, \ldots, n$);
   OUTPUT (u_{ij} for $j = i, \ldots, n$ and $i = 1, \ldots, n$);
   STOP.

equivalent to the maximal column pivoting procedure described in Algorithm 6.2. This interchange is sufficient in most cases.

The following algorithm incorporates the factorization procedure of Algorithm 6.4 together with maximal column pivoting and forward and backward substitution to obtain a solution to a linear system of equations. The process involves writing the linear system $A\mathbf{x} = \mathbf{b}$ as $LU\mathbf{x} = \mathbf{b}$. The forward substitution solves the system $L\mathbf{z} = \mathbf{b}$

and the backward substitution solves the system $U\mathbf{x} = \mathbf{z} = L^{-1}\mathbf{b}$. Note that the nonzero entries of L and U can be stored in the corresponding entries of A except for the diagonal of L or U, which must be input.

Direct Factorization with Maximal Column Pivoting Algorithm 6.5

To solve the $n \times n$ linear system $A\mathbf{x} = \mathbf{b}$ in the form

$$
\begin{aligned}
E_1: \quad & a_{11}x_1 + a_{12}x_2 + \cdots + a_{1n}x_n = a_{1,n+1} \\
E_2: \quad & a_{21}x_1 + a_{22}x_2 + \cdots + a_{2n}x_n = a_{2,n+1} \\
& \vdots \qquad \vdots \qquad \vdots \qquad \qquad \vdots \qquad \vdots \\
E_n: \quad & a_{n1}x_1 + a_{n2}x_2 + \cdots + a_{nn}x_n = a_{n,n+1}
\end{aligned}
$$

by factoring A into LU, solving $L\mathbf{z} = \mathbf{b}$ and $U\mathbf{x} = \mathbf{z}$ where the main diagonal of either L or U is given:

INPUT the dimension n; the entries a_{ij}, $1 \le j \le n+1$, $1 \le i \le n$ of the augmented form of A; the diagonal $l_{11}, \ldots, l_{nn}$ of L or the diagonal $u_{11}, \ldots, u_{nn}$ of U.

OUTPUT the solution $x_1, \ldots, x_n$ or a message that the linear system has no unique solution.

Step 1   Let p be the smallest integer such that $1 \le p \le n$ and $|a_{p1}| = \max\limits_{1 \le j \le n} |a_{j1}|$;

  (*Find first pivot element.*)
  If $|a_{p1}| = 0$, then OUTPUT ('No unique solution exists');
                    STOP.

Step 2   If $p \ne 1$ then interchange rows p and 1 in A.

Step 3   Select l_{11} and u_{11} satisfying $l_{11}u_{11} = a_{11}$.

Step 4   For $j = 2, \ldots, n$ set $u_{1j} = a_{1j}/l_{11}$;   (*First row of U.*)
                         $l_{j1} = a_{j1}/u_{11}$.   (*First column of L.*)

Step 5   For $i = 2, \ldots, n-1$ do Steps 6–9.

  Step 6   Let p be the smallest integer such that $i \le p \le n$ and

$$
\left| a_{pi} - \sum_{k=1}^{i-1} l_{pk}u_{ki} \right| = \max_{i \le j \le n} \left| a_{ji} - \sum_{k=1}^{i-1} l_{jk}u_{ki} \right|.
$$

  (*Find ith pivot element.*)
  If the maximum is zero then OUTPUT ('No unique solution exists').

Step 7   If $p \ne i$ then interchange rows p and i in both matrices A and L.

Step 8   Select l_{ii} and u_{ii} satisfying $l_{ii}u_{ii} = a_{ii} - \sum_{k=1}^{i-1} l_{ik}u_{ki}$.

Step 9   For $j = i + 1, \ldots, n$

$$\text{set } u_{ij} = \frac{1}{l_{ii}} \left[a_{ij} - \sum_{k=1}^{i-1} l_{ik} u_{kj} \right]; \quad \text{(ith row of U.)}$$

$$l_{ji} = \frac{1}{u_{ii}} \left[a_{ji} - \sum_{k=1}^{i-1} l_{jk} u_{ki} \right]. \quad \text{(ith column of L.)}$$

Step 10   Set $HOLD = a_{nn} - \sum_{k=1}^{n-1} l_{nk} u_{kn}$;

If $HOLD = 0$ then OUTPUT ('No unique solution exists');
STOP.

Select l_{nn} and u_{nn} satisfying $l_{nn} u_{nn} = a_{nn} - \sum_{k=1}^{n-1} l_{nk} u_{kn}$.

(*Steps* 11 *and* 12 *solve lower-triangular system* $L\mathbf{z} = \mathbf{b}$.)

Step 11   Set $z_1 = a_{1,n+1}/l_{11}$.

Step 12   For $i = 2, \ldots, n$ set $z_i = \frac{1}{l_{ii}} \left[a_{i,n+1} - \sum_{j=1}^{i-1} l_{ij} z_j \right]$.

(*Steps* 13 *and* 14 *solve upper-triangular system* $U\mathbf{x} = \mathbf{z}$.)

Step 13   Set $x_n = z_n/u_{nn}$.

Step 14   For $i = n - 1, \ldots, 1$ set $x_i = \frac{1}{u_{ii}} \left[z_i - \sum_{j=i+1}^{n} u_{ij} x_j \right]$.

Step 15   OUTPUT $(x_1, \ldots, x_n)$;
STOP.

EXAMPLE 2   To illustrate the procedure involved in Algorithm 6.5, consider the linear system

$$\begin{aligned}
1.00x_1 + 0.333x_2 + \ 1.50x_3 - 0.333x_4 &= 3.00 \\
-2.01x_1 + \ 1.45x_2 + 0.500x_3 + \ 2.95x_4 &= 5.40 \\
4.32x_1 - \ 1.95x_2 \qquad\qquad + \ 2.08x_4 &= 0.130 \\
5.11x_1 - \ 4.00x_2 + \ 3.33x_3 - \ 1.11x_4 &= 3.77.
\end{aligned}$$

We will trace the steps of Algorithm 6.5 with $l_{11} = l_{22} = l_{33} = l_{44} = 1.00$, using three-digit rounding arithmetic.

Step 1 $p = 4$, so the first and fourth rows are interchanged. The augmented matrix becomes

$$\begin{bmatrix}
5.11 & -4.00 & 3.33 & -1.11 & \vdots & 3.77 \\
-2.01 & 1.45 & 0.500 & 2.95 & \vdots & 5.40 \\
4.32 & -1.95 & 0.000 & 2.08 & \vdots & 0.130 \\
1.00 & 0.333 & 1.50 & -0.333 & \vdots & 3.00
\end{bmatrix}$$

Step 3   Since $l_{11} = 1.00$, $u_{11} = 5.11$.

Step 4 $u_{12} = -4.00$, $u_{13} = 3.33$, $u_{14} = -1.11$,
$l_{21} = a_{21}/u_{11} = -0.393$, $l_{31} = a_{31}/u_{11} = 0.845$,
$l_{41} = a_{41}/l_{11} = 0.196$.

Step 5 $i = 2$.

Step 6 $|a_{22} - l_{21}u_{12}| = |1.45 - (-0.393)(-4.00)| = |-0.120| = 0.120$,
$|a_{32} - l_{31}u_{12}| = |-1.95 - (0.845)(-4.00)| = |1.43| = 1.43$,
$|a_{42} - l_{41}u_{12}| = |0.333 - (0.196)(-4.00)| = |1.12| = 1.12$.

Thus, $p = 3$.

Step 7   Interchange rows 3 and 2 in A and L.
Thus, $l_{21} = 0.845$, $l_{31} = -0.393$ and A becomes

$$\begin{bmatrix} 5.11 & -4.00 & 3.33 & -1.11 & \vdots & 3.77 \\ 4.32 & -1.95 & 0.000 & 2.08 & \vdots & 0.130 \\ -2.01 & 1.45 & 0.500 & 2.95 & \vdots & 5.40 \\ 1.00 & 0.333 & 1.50 & -0.333 & \vdots & 3.00 \end{bmatrix}$$

Step 8 $l_{22} = 1.00$, $u_{22} = a_{22} - l_{21}u_{12} = -1.95 - (0.845)(-4.00) = 1.43$.

Step 9 $u_{23} = \dfrac{1}{l_{22}}[a_{23} - (l_{21}u_{13})] = [0.000 - (0.845)(3.33)] = -2.81$,

$u_{24} = \dfrac{1}{l_{22}}[a_{24} - (l_{21}u_{14})] = [2.08 - (0.845)(-1.11)] = 3.02$,

$l_{32} = \dfrac{1}{u_{22}}[a_{32} - (l_{31}u_{12})] = \dfrac{1}{1.43}[1.45 - (-0.393)(-4.00)]$
$= -0.120/1.43 = -0.0839$,

$l_{42} = \dfrac{1}{u_{22}}[a_{42} - (l_{41}u_{12})] = \dfrac{1}{1.43}[0.333 - (0.196)(-4.00)]$
$= 1.12/1.43 = 0.783$.

Step 5 $i = 3$.

Step 6 $|a_{33} - (l_{31}u_{13} + l_{32}u_{23})| = |0.500 - ((-0.393)(3.33)$
$+ (-0.0839)(-2.81))|$
$= 1.57$,
$|a_{43} - (l_{41}u_{13} + l_{42}u_{23})| = |1.50 - ((0.196)(3.33) + (0.783)(-2.81))|$
$= 3.05$.
Thus, $p = 4$.

Step 7   Interchange rows 4 and 3 in A and L.
Thus, $l_{31} = 0.196$, $l_{41} = -0.393$,
$l_{32} = 0.783$, $l_{42} = -0.0839$,
and A becomes

$$
\begin{bmatrix}
5.11 & -4.00 & 3.33 & -1.11 & \vdots & 3.77 \\
4.32 & -1.95 & 0.000 & 2.08 & \vdots & 0.130 \\
1.00 & 0.333 & 1.50 & -0.333 & \vdots & 3.00 \\
-2.01 & 1.45 & 0.500 & 2.95 & \vdots & 5.40
\end{bmatrix}
$$

Step 8 $l_{33} = 1$,

$$
\begin{aligned}
u_{33} &= a_{33} - (l_{31}u_{13} + l_{32}u_{23}) \\
&= 1.50 - (0.196)(3.33) + (-0.0839)(-2.81) \\
&= 3.05.
\end{aligned}
$$

Step 9 $u_{34} = \dfrac{1}{l_{33}}[a_{34} - (l_{31}u_{14} + l_{32}u_{24})]$

$$
= [-0.333 - (0.196(-1.11) + 0.783(3.02))] = -2.47,
$$

$$
\begin{aligned}
l_{43} &= \frac{1}{u_{33}}[a_{43} - (l_{41}u_{13} + l_{42}u_{23})] \\
&= \frac{1}{3.05}[0.500 - ((-0.393)(3.33) + (-0.0839)(-2.81))] \\
&= 1.57/3.05 \\
&= 0.515.
\end{aligned}
$$

Step 10 $l_{44} = 1.00$,

$$
\begin{aligned}
u_{44} &= a_{44} - (l_{41}u_{14} + l_{42}u_{24} + l_{43}u_{34}) \\
&= 2.95 - ((-0.393)(-1.11) + (-0.0839)(3.02) + (0.515)(-2.47)) \\
&= 4.04.
\end{aligned}
$$

The factorization is complete:

$$
\begin{bmatrix}
5.11 & -4.00 & 3.33 & -1.11 \\
4.32 & -1.95 & 0.000 & 2.08 \\
1.00 & 0.333 & 1.50 & -0.333 \\
-2.01 & 1.45 & 0.500 & 2.95
\end{bmatrix}
=
$$

$$
\begin{bmatrix}
1.00 & 0 & 0 & 0 \\
0.845 & 1.00 & 0 & 0 \\
0.196 & 0.783 & 1.00 & 0 \\
-0.393 & -0.0839 & 0.515 & 1.00
\end{bmatrix}
\begin{bmatrix}
5.11 & -4.00 & 3.33 & -1.11 \\
0 & 1.43 & -2.81 & 3.02 \\
0 & 0 & 3.05 & -2.47 \\
0 & 0 & 0 & 4.04
\end{bmatrix}
$$

Continuing

Step 11 $z_1 = a_{15}/l_{11} = 3.77$.

Step 12 $z_2 = \dfrac{1}{l_{22}}[a_{25} - l_{21}z_1] = 0.130 - (0.845)(3.77) = -3.06$,

$$
\begin{aligned}
z_3 &= \frac{1}{l_{33}}[a_{35} - (l_{31}z_1 + l_{32}z_2)] = 3.00 - (0.196(3.77) + 0.783(-3.06)) \\
&= 4.66,
\end{aligned}
$$

$$z_4 = \frac{1}{l_{44}}[a_{45} - (l_{41}z_1 + l_{42}z_2 + l_{43}z_3)]$$
$$= 5.40 - ((-0.393)(3.77) + (-0.0839)(-3.06) + (0.515)(4.66))$$
$$= 4.22.$$

Step 13 $x_4 = z_4/u_{44} = 4.22/4.04 = 1.04,$

$$x_3 = \frac{1}{u_{33}}[z_3 - u_{34}x_4] = \frac{1}{3.05}[4.66 - (-2.47)(1.04)] = 2.37,$$

$$x_2 = \frac{1}{u_{22}}[z_2 - (u_{23}x_3 + u_{24}x_4)]$$

$$= \frac{1}{1.43}[-3.06 - ((-2.81)(2.37) + (3.02)(1.04))]$$

$$= 0.460/1.43 = 0.322,$$

$$x_1 = \frac{1}{u_{11}}[z_1 - (u_{12}x_2 + u_{13}x_3 + u_{14}x_4)]$$

$$= \frac{1}{5.11}[3.77 - ((-4.00)(0.322) + (3.33)(2.37) + (-1.11)(1.04))]$$

$$= -1.68/5.11$$
$$= -0.329.$$

The solution is $x_1 = -0.329$, $x_2 = 0.322$, $x_3 = 2.37$, $x_4 = 1.04$.
An application of Algorithm 6.4 would yield the factorization

$$\begin{bmatrix} 1.00 & 0.333 & 1.50 & -0.333 \\ -2.01 & 1.45 & 0.500 & 2.95 \\ 4.32 & -1.95 & 0.000 & 2.08 \\ 5.11 & -4.00 & 3.33 & -1.11 \end{bmatrix} =$$

$$\begin{bmatrix} 1.00 & 0 & 0 & 0 \\ -2.01 & 1.00 & 0 & 0 \\ 4.32 & -1.60 & 1.00 & 0 \\ 5.11 & -2.69 & -6.04 & 1.00 \end{bmatrix} \begin{bmatrix} 1.00 & 0.333 & 1.50 & -0.333 \\ 0 & 2.12 & 3.52 & 2.28 \\ 0 & 0 & -0.850 & 7.17 \\ 0 & 0 & 0 & 50.0 \end{bmatrix}$$

Applying steps 11 through 15 of Algorithm 6.5 then gives the solution $x_1 = -0.370$, $x_2 = 0.236$, $x_3 = 2.42$, $x_4 = 1.03$.

Table 6.3 compares the results of Algorithm 6.5, Algorithm 6.4, and the actual answer to three digits. Note the improved accuracy when row interchanges are included.   ∎

When the matrix is known to be positive definite, a significant improvement in the matrix factorization technique can be made with regard to the number of arithmetic operations required.

TABLE 6.3

	x_1	x_2	x_3	x_4
ALG. 6.5	−0.329	0.322	2.37	1.04
ALG. 6.4	−0.370	0.236	2.42	1.03
Actual	−0.324	0.331	2.37	1.04

THEOREM 6.23

If A is a positive definite $n \times n$ matrix, then A has a factorization of the form $A = LL^t$, where L is a lower-triangular matrix. The factorization can be achieved by applying Algorithm 6.4 with $l_{ii} = u_{ii}$ for each $i = 1, 2, \ldots, n$.

PROOF

Since A is positive definite, Theorems 6.22 and 6.15 imply that A can be factored in the form $A = LU$, where L is lower triangular and U is upper triangular. Positive definiteness also implies (see Exercise 10 of Section 6.6) that $a_{ii} > 0$ for each $i = 1, 2, \ldots, n$. With the notation of Algorithm 6.4, the procedure is started by choosing $l_{11} = u_{11} = \sqrt{a_{11}}$.

Since A is symmetric,

$$l_{j1} = \frac{a_{j1}}{u_{11}} = \frac{a_{1j}}{l_{11}} = u_{1j}$$

for each $j = 2, 3, \ldots, n$ so the entries in the first row of U agree with the corresponding entries in the first column of L.

The proof proceeds by mathematical induction. Assume that $k < n$ and that the entries in the first k rows of U agree with the corresponding entries in the first k columns of L. The proof of the theorem will be complete if it can be established that the entries in the $(k + 1)$st row of U agree with the entries in the $(k + 1)$st column of L.

Since L is lower triangular and U is upper triangular, it is clear that

$$l_{j, k+1} = 0 = u_{k+1, j} \qquad \text{whenever } j = 1, 2, \ldots, k.$$

Choose $l_{k+1, k+1} = u_{k+1, k+1} = (a_{k+1, k+1} - \sum_{j=1}^{k} l_{k+1, j}^2)^{1/2}$, which can be shown to be a real number (see Exercise 19).

Since A is symmetric and $l_{ji} = u_{ij}$ for each $i = 1, 2, \ldots, k$, and $j = 1, 2, \ldots, n$,

$$l_{j, k+1} = \frac{1}{u_{k+1, k+1}} \left[a_{j, k+1} - \sum_{i=1}^{k} l_{ji} u_{i, k+1} \right] = \frac{1}{l_{k+1, k+1}} \left[a_{k+1, j} - \sum_{i=1}^{k} u_{ij} l_{k+1, i} \right] = u_{k+1, j}$$

for each $j = k + 2, k + 3, \ldots, n$. $\qquad \square$

For a positive definite matrix, this theorem can be used to simplify the Factorization Algorithm 6.4. If a linear system is to be solved involving a positive definite matrix, Steps 1–6 of the following algorithm can be substituted for Steps 1–10 of Algorithm 6.5 to take advantage of the simplification that results provided u_{ij} is replaced by l_{ji} in Steps 13 and 14. The factorization procedure is described in the following algorithm.

Choleski's Algorithm 6.6

To factor the positive definite $n \times n$ matrix A into LL^t where L is lower triangular:

INPUT the dimension n; entries a_{ij}, $1 \le i, j \le n$ of A.

OUTPUT the entries l_{ij}, $1 \le j \le i$, $1 \le i \le n$ of L. (*The entries of* $U = L^t$ *are* $u_{ij} = l_{ji}$, $i \le j \le n$, $1 \le i \le n$.)

Step 1   Set $l_{11} = \sqrt{a_{11}}$.

Step 2   For $j = 2, \ldots, n$ set $l_{j1} = a_{j1}/l_{11}$.

Step 3   For $i = 2, \ldots, n-1$ do Steps 4 and 5.

Step 4   Set $l_{ii} = \left[a_{ii} - \sum_{k=1}^{i-1} l_{ik}^2 \right]^{1/2}$.

Step 5   For $j = i+1, \ldots, n$

set $l_{ji} = \dfrac{1}{l_{ii}} \left[a_{ji} - \sum_{k=1}^{i-1} l_{jk} l_{ik} \right]$.

Step 6   Set $l_{nn} = \left[a_{nn} - \sum_{k=1}^{n-1} l_{nk}^2 \right]^{1/2}$.

Step 7   OUTPUT (l_{ij} *for* $j = 1, \ldots, i$ *and* $i = 1, \ldots, n.$);
STOP.

It is not difficult to verify (see Exercise 17) that the solution of a typical linear system involving a positive definite matrix by using Choleski's Algorithm requires

n square roots,

$$\frac{n^3 + 9n^2 + 2n}{6} \quad \text{multiplications/divisions,}$$

and
$$\frac{n^3 + 6n^2 - 7n}{6} \quad \text{additions/subtractions.}$$

This is about half the arithmetic operations that are required in the Gaussian Elimination Algorithm 6.1, a method that is typical with regard to the number of arithmetic operations required. The computational advantage of the Choleski method depends on the number of operations that are required for determining the values of the n square roots which, since it is a linear factor of n, will decrease in significance as n increases.

EXAMPLE 3   To illustrate the steps involved in applying Choleski's method, consider the positive definite matrix

$$A = \begin{bmatrix} 4 & -1 & 1 \\ -1 & 4.25 & 2.75 \\ 1 & 2.75 & 3.5 \end{bmatrix}.$$

Tracing the steps of Algorithm 6.6:

Step 1 $l_{11} = \sqrt{a_{11}} = \sqrt{4} = 2.$

Step 2 $l_{21} = \dfrac{a_{21}}{l_{11}} = \dfrac{-1}{2} = -0.5, \quad l_{31} = \dfrac{a_{31}}{l_{11}} = \dfrac{1}{2} = 0.5.$

Step 3 $i = 2.$

Step 4 $l_{22} = (a_{22} - l_{21}^2)^{1/2} = (4.25 - (-0.5)^2)^{1/2} = 2.$

Step 5 $l_{32} = \dfrac{1}{l_{22}}(a_{32} - l_{31}l_{21}) = \dfrac{1}{2}(2.75 - 0.5(-0.5)) = 1.5.$

Step 6 $l_{33} = (a_{33} - l_{31}^2 - l_{32}^2)^{1/2} = (3.5 - (0.5)^2 - (1.5)^2)^{1/2} = 1.$

Since L is lower triangular and $U = L^t$,

$$l_{12} = l_{13} = l_{23} = 0,$$
$$u_{11} = l_{11} = 2, \quad u_{12} = l_{21} = -0.5, \quad u_{13} = l_{31} = 0.5,$$
$$u_{21} = l_{12} = 0, \quad u_{22} = l_{22} = \quad 2, \quad u_{23} = l_{32} = 1.5,$$
$$u_{31} = l_{13} = 0, \quad u_{32} = l_{23} = \quad 0, \quad u_{33} = l_{33} = 1.$$

So $L = \begin{bmatrix} 2 & 0 & 0 \\ -0.5 & 2 & 0 \\ 0.5 & 1.5 & 1 \end{bmatrix}$ and $U = \begin{bmatrix} 2 & -0.5 & 0.5 \\ 0 & 2 & 1.5 \\ 0 & 0 & 1 \end{bmatrix}.$ ■

The factorization algorithms can be simplified considerably in the case of band matrices because of the large number of zeros appearing in these matrices in regular patterns. It is particularly interesting to observe the form the Crout or Doolittle method assumes in this case. To illustrate the situation, suppose a tridiagonal matrix

$$A = \begin{bmatrix} a_{11} & a_{12} & 0 & \cdots & 0 \\ a_{21} & a_{22} & a_{23} & & 0 \\ 0 & & & & a_{n-1,n} \\ \vdots & & & & \\ 0 & \cdots & 0 & a_{n,n-1} & a_{nn} \end{bmatrix}$$

can be factored into the triangular matrices L and U.

Since A has only $(3n - 2)$ nonzero entries, there are only $(3n - 2)$ conditions to be applied to determine the entries of L and U, provided, of course, that the zero entries of A are also obtained. Suppose that the matrices can actually be found in the form

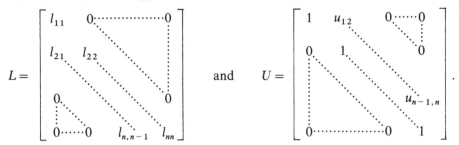

In this form, there are $(2n - 1)$ undetermined entries of L and $(n - 1)$ undetermined entries of U, which totals the number of conditions, $(3n - 2)$, and the zero entries of A are obtained automatically.

The multiplication involved with $A = LU$ gives, in addition to the zero entries, the equations

(6.16) $$a_{11} = l_{11};$$

(6.17) $$a_{i, i-1} = l_{i, i-1} \qquad \text{for each } i = 2, 3, \ldots, n;$$

(6.18) $$a_{ii} = l_{i, i-1} u_{i-1, i} + l_{ii} \qquad \text{for each } i = 2, 3, \ldots, n;$$

and

(6.19) $$a_{i, i+1} = l_{ii} u_{i, i+1} \qquad \text{for each } i = 1, 2, \ldots, n-1.$$

A solution to this system of equations can be found by first obtaining all the nonzero off-diagonal terms in L, using Eq. (6.17), and then using Eqs. (6.19) and (6.18) to alternately obtain the remainder of the entries in U and L, which can be stored in the corresponding entries of A.

A complete algorithm for solving an $n \times n$ system of linear equations whose coefficient matrix is tridiagonal follows.

Crout Reduction for Tridiagonal Linear Systems Algorithm 6.7

To solve the $n \times n$ linear system

$$
\begin{aligned}
E_1: &\quad a_{11}x_1 + a_{12}x_2 &&= a_{1, n+1} \\
E_2: &\quad a_{21}x_1 + a_{22}x_2 + a_{23}x_3 &&= a_{2, n+1} \\
&\quad \vdots \qquad\qquad \vdots &&\quad \vdots \\
E_{n-1}: &\quad a_{n-1, n-2}x_{n-2} + a_{n-1, n-1}x_{n-1} + a_{n-1, n}x_n &&= a_{n-1, n+1} \\
E_n: &\quad a_{n, n-1}x_{n-1} + a_{nn}x_n &&= a_{n, n+1},
\end{aligned}
$$

which is assumed to have a unique solution:

INPUT the dimension n; the entries of A.

OUTPUT the solution $x_1, \ldots, x_n$.

Step 1   Set $l_{11} = a_{11}$;
             $u_{12} = a_{12}/l_{11}$.

Step 2   For $i = 2, \ldots, n-1$ set $l_{i,i-1} = a_{i,i-1}$;   *(ith row of L.)*
             $l_{ii} = a_{ii} - l_{i,i-1}u_{i-1,i}$;
             $u_{i,i+1} = a_{i,i+1}/l_{ii}$.   *((i + 1)st column of U.)*

Step 3   Set $l_{n,n-1} = a_{n,n-1}$;   *(nth row of L.)*
             $l_{nn} = a_{nn} - l_{n,n-1}u_{n-1,n}$.

*(Steps 4, 5 solve Lz = **b**.)*

Step 4   Set $z_1 = a_{1,n+1}/l_{11}$.

Step 5   For $i = 2, \ldots, n$ set $z_i = \dfrac{1}{l_{ii}}[a_{i,n+1} - l_{i,i-1}z_{i-1}]$.

(Steps 6, 7 solve Ux = z.)

Step 6   Set $x_n = z_n$.

Step 7   For $i = n-1, \ldots, 1$ set $x_i = z_i - u_{i,i+1}x_{i+1}$.

Step 8   OUTPUT $(x_1, \ldots, x_n)$;
             STOP.

This algorithm requires only $(5n - 4)$ multiplication/divisions and $(3n - 3)$ addition/subtractions, and consequently has considerable computational advantage over the methods that do not consider the tridiagonality of the matrix, especially for large values of n.

EXAMPLE 4   To illustrate the procedure involved in Algorithm 6.7, consider the tridiagonal system of equations

$$\begin{aligned}
2x_1 - x_2 &= 1, \\
-x_1 + 2x_2 - x_3 &= 0, \\
- x_2 + 2x_3 - x_4 &= 0, \\
- x_3 + 2x_4 &= 1,
\end{aligned}$$

whose augmented matrix is

$$\begin{bmatrix}
2 & -1 & 0 & 0 & \vdots & 1 \\
-1 & 2 & -1 & 0 & \vdots & 0 \\
0 & -1 & 2 & -1 & \vdots & 0 \\
0 & 0 & -1 & 2 & \vdots & 1
\end{bmatrix}.$$

Tracing the steps of Algorithm 6.7:

Step 1 $l_{11} = 2$, $u_{12} = \dfrac{a_{12}}{l_{11}} = -\dfrac{1}{2}$.

Step 2 $i = 2$.

$$l_{21} = a_{21} = -1, \quad l_{22} = a_{22} - l_{21}u_{12} = 2 - (-1)(-\tfrac{1}{2}) = \tfrac{3}{2},$$

$$u_{23} = \frac{a_{23}}{l_{22}} = -\frac{1}{(\tfrac{3}{2})} = -\frac{2}{3},$$

$i = 3$.

$$l_{32} = a_{32} = -1, \quad l_{33} = a_{33} - l_{32}u_{23} = 2 - (-1)\left(-\frac{2}{3}\right) = \frac{4}{3},$$

$$u_{34} = \frac{a_{34}}{l_{33}} = -\frac{3}{4}.$$

Step 3 $l_{43} = a_{43} = -1, \quad l_{44} = a_{44} - l_{43}u_{34} = 2 - (-1)(-\tfrac{3}{4}) = \tfrac{5}{4}.$

Step 4 $z_1 = \dfrac{a_{15}}{l_{11}} = \dfrac{1}{2}.$

Step 5 $z_2 = \dfrac{1}{l_{22}}[a_{25} - l_{21}z_1] = \dfrac{0 - (-1)(\tfrac{1}{2})}{(\tfrac{3}{2})} = \dfrac{1}{3},$

$$z_3 = \frac{1}{l_{33}}[a_{35} - l_{32}z_2] = \frac{0 - (-1)(\tfrac{1}{3})}{(\tfrac{4}{3})} = \frac{1}{4},$$

$$z_4 = \frac{1}{l_{44}}[a_{45} - l_{43}z_3] = \frac{1 - (-1)(\tfrac{1}{4})}{(\tfrac{5}{4})} = 1.$$

Step 6 $x_4 = 1.$

Step 7 $x_3 = z_3 - u_{34}x_4 = \tfrac{1}{4} - (-\tfrac{3}{4}) = 1,$
$x_2 = z_2 - u_{23}x_3 = \tfrac{1}{3} - (-\tfrac{2}{3}) = 1,$
$x_1 = z_1 - u_{12}x_2 = \tfrac{1}{2} - (-\tfrac{1}{2}) = 1.$

The algorithm factored the matrix into

$$
\begin{bmatrix}
2 & -1 & 0 & 0 \\
-1 & 2 & -1 & 0 \\
0 & -1 & 2 & -1 \\
0 & 0 & -1 & 2
\end{bmatrix}
=
\begin{bmatrix}
2 & 0 & 0 & 0 \\
-1 & \tfrac{3}{2} & 0 & 0 \\
0 & -1 & \tfrac{4}{3} & 0 \\
0 & 0 & -1 & \tfrac{5}{4}
\end{bmatrix}
\begin{bmatrix}
1 & -\tfrac{1}{2} & 0 & 0 \\
0 & 1 & -\tfrac{2}{3} & 0 \\
0 & 0 & 1 & -\tfrac{3}{4} \\
0 & 0 & 0 & 1
\end{bmatrix},
$$

and gave the correct solution, $x_1 = x_2 = x_3 = x_4 = 1.$ ∎

Algorithm 6.7 can be applied whenever $l_{ii} \neq 0$ for each $i = 1, 2, \ldots, n$. Two conditions, either of which will ensure that this is true, are that the coefficient matrix of the system is positive definite or that it is strictly diagonally dominant. An additional condition that ensures that this algorithm can be applied is given in the next theorem, whose proof is discussed in Exercise 12.

THEOREM 6.24   Suppose that $A = (a_{ij})$ is tridiagonal with $a_{i,i-1}a_{i,i+1} \neq 0$ for each $i = 2, 3, \ldots, n-1$. If $|a_{11}| > |a_{12}|$, $|a_{ii}| \geq |a_{i,i-1}| + |a_{i,i+1}|$ for each $i = 2, 3, \ldots, n-1$, and $|a_{nn}| > |a_{n,n-1}|$, then A is nonsingular and the values of l_{ii} described in Algorithm 6.7 are nonzero for each $i = 1, 2, \ldots, n$.

The Choice of a Method for Solving a Linear System

When the linear system is small enough to be efficiently accommodated in the main memory of a computer, it is generally most efficient to use a direct technique that minimizes the effect of rounding error. Specifically, Gaussian Elimination with Scaled-Column Pivoting Algorithm 6.3 is appropriate.

Large linear systems with primarily zero entries occurring in regular patterns can generally be efficiently solved by using an iterative procedure such as those discussed in Chapter 8. Systems of this type arise naturally, for example, when finite difference techniques are used to solve boundary value problems, a common application in the numerical solution of partial-differential equations.

It can be very difficult to solve a large linear system that has primarily nonzero entries, or one where the zero entries are not in a predictable pattern. The matrix associated with such a system can be placed in secondary storage in partitioned form (see Exercise 8 of Section 6.3) and portions read into main memory only as needed for calculation. Methods that require secondary storage can be either iterative or direct, but they generally require techniques from the fields of data structures and graph theory. A study of the problems involved in theory and implementation is well beyond the scope of this text. The reader is referred to Bunch and Rose [21], and Rose and Willoughby [110] for a discussion of the current techniques.

Exercise Set 6.7

1. Factor the following matrices into the LU decomposition using Algorithm 6.4 with $l_{ii} = 1$ for all i.

a) $\begin{bmatrix} 2 & -1 & 1 \\ 3 & 3 & 9 \\ 3 & 3 & 5 \end{bmatrix}$;

b) $\begin{bmatrix} 2 & -1.5 & 3 \\ -1 & 0 & 2 \\ 4 & -4.5 & 5 \end{bmatrix}$;

c) $\begin{bmatrix} 1.012 & -2.132 & 3.104 \\ -2.132 & 4.096 & -7.013 \\ 3.104 & -7.013 & 0.014 \end{bmatrix}$;

d) $\begin{bmatrix} 3.107 & 2.101 & 0 \\ 0 & -1.213 & 2.101 \\ 0 & 0 & 2.179 \end{bmatrix}$;

e) $\begin{bmatrix} 2 & 0 & 0 & 0 \\ 1 & 1.5 & 0 & 0 \\ 0 & -3 & 0.5 & 0 \\ 2 & -2 & 1 & 1 \end{bmatrix}$;

f) $\begin{bmatrix} 2.1756 & 4.0231 & -2.1732 & 5.1967 \\ -4.0231 & 6.0000 & 0 & 1.1973 \\ -1.0000 & -5.2107 & 1.1111 & 0 \\ 6.0235 & 7.0000 & 0 & -4.1561 \end{bmatrix}$.

2. Solve the following linear systems using Algorithm 6.5 with $l_{ii} = 1$.

a) $2x_1 - x_2 + x_3 = -1,$
   $3x_1 + 3x_2 + 9x_3 = 0,$
   $3x_1 + 3x_2 + 5x_3 = 4.$

b) $2x_1 = 3,$
   $x_1 + 1.5x_2 = 4.5,$
   $- 3x_2 + 0.5x_3 = -6.6,$
   $2x_1 - 2x_2 + x_3 + x_4 = 0.8.$

c) $1.012x_1 - 2.132x_2 + 3.104x_3 = 1.984,$
   $-2.132x_1 + 4.096x_2 - 7.013x_3 = -5.049,$
   $3.104x_1 - 7.013x_2 + 0.014x_3 = -3.895.$

d) $3.107x_1 + 2.101x_2 = 1.001,$
   $-1.213x_2 + 2.101x_3 = 0.000,$
   $2.179x_3 = 7.013.$

e) $2x_1 - 1.5x_2 + 3x_3 = 1,$
   $-x_1 + 2x_3 = 3,$
   $4x_1 - 4.5x_2 + 5x_3 = -1.$

f) $2.1756x_1 + 4.0231x_2 - 2.1732x_3 + 5.1967x_4 = 17.102,$
   $-4.0231x_1 + 6.0000x_2 + 1.1973x_4 = -6.1593,$
   $-1.0000x_1 - 5.2107x_2 + 1.1111x_3 = 3.0004,$
   $6.0235x_1 + 7.0000x_2 - 4.1561x_4 = 0.0000.$

3. Use Algorithm 6.6 to find a factorization of the form $A = LL^t$ for the following matrices:

a) $A = \begin{bmatrix} 2 & -1 & 0 \\ -1 & 2 & -1 \\ 0 & -1 & 2 \end{bmatrix};$

b) $A = \begin{bmatrix} 4 & 1 & 1 & 1 \\ 1 & 3 & -1 & 1 \\ 1 & -1 & 2 & 0 \\ 1 & 1 & 0 & 2 \end{bmatrix};$

c) $A = \begin{bmatrix} 4 & 1 & -1 & 0 \\ 1 & 3' & -1 & 0 \\ -1 & -1 & 5 & 2 \\ 0 & 0 & 2 & 4 \end{bmatrix};$

d) $A = \begin{bmatrix} 6 & 2 & 1 & -1 \\ 2 & 4 & 1 & 0 \\ 1 & 1 & 4 & -1 \\ -1 & 0 & -1 & 3 \end{bmatrix}.$

4. Modify Algorithm 6.4 so that it can be used to solve linear systems. Repeat Exercise 2 using the modified algorithm, and compare your answers to those obtained in Exercise 2.

5. Solve the following linear systems using Algorithm 6.7:

a) $x_1 - x_2 = 0,$
   $-2x_1 + 4x_2 - 2x_3 = -1,$
   $- x_2 + 2x_3 = 1.5.$

b) $3x_4 + x_2 = -1,$
   $2x_1 + 4x_2 + x_3 = 7,$
   $2x_2 + 5x_3 = 9.$

c) $2x_1 - x_2 = 3,$
   $-x_1 + 2x_2 - x_3 = -3,$
   $- x_2 + 2x_3 = 1.$

d) $0.5x_1 + 0.25x_2 = 0.35,$
   $0.35x_1 + 0.8x_2 + 0.4x_3 = 0.77,$
   $0.25x_2 + x_3 + 0.5x_4 = -0.5,$
   $x_3 - 2x_4 = -2.25.$

6. Modify Algorithm 6.6 so that it can be used to solve linear systems. Use the modified algorithm to solve the following linear systems:

a)
$$2x_1 - x_2 \qquad = 3,$$
$$-x_1 + 2x_2 - x_3 = -3,$$
$$- x_2 + 2x_3 = 1.$$

b)
$$4x_1 + x_2 + x_3 + x_4 = 0.65,$$
$$x_1 + 3x_2 - x_3 + x_4 = 0.05,$$
$$x_1 - x_2 + 2x_3 \qquad = 0,$$
$$x_1 + x_2 \qquad + 2x_4 = 0.5,$$

c)
$$4x_1 + x_2 - x_3 \qquad = 7,$$
$$x_1 + 3x_2 - x_3 \qquad = 8,$$
$$-x_1 - x_2 + 5x_3 + 2x_4 = -4,$$
$$2x_3 + 4x_4 = 6.$$

d)
$$6x_1 + 2x_2 + x_3 - x_4 = 0,$$
$$2x_1 + 4x_2 + x_3 \qquad = 7,$$
$$x_1 + x_2 + 4x_3 - x_4 = -1,$$
$$-x_1 \qquad - x_3 + 3x_4 = -2.$$

7. Tridiagonal matrices are usually labelled by using the notation

$$A = \begin{bmatrix} a_1 & c_1 & 0 & \cdots\cdots & 0 \\ b_2 & a_2 & c_2 & & \vdots \\ 0 & b_3 & & & 0 \\ \vdots & & & & c_{n-1} \\ 0 & \cdots\cdots & 0 & b_n & a_n \end{bmatrix}$$

to emphasize that it is not necessary to consider the entire matrix. Rewrite Algorithm 6.7, using this notation, and changing the notation of the l_{ij} and u_{ij} in a similar manner.

8. Derive an algorithm for the factorization of tridiagonal matrices directly from Algorithm 6.4, using $u_{ii} = 1$ for each $i = 1, 2, \ldots, n$, and compare this procedure with Algorithm 6.7.

9. Derive an algorithm for the factorization of tridiagonal matrices using Algorithm 6.4 with $l_{ii} = 1$ for each $i = 1, 2, \ldots, n$.

10. Let A be the 10×10 tridiagonal matrix given by $a_{ii} = 2$, $a_{i,i+1} = a_{i,i-1} = -1$ for each $i = 2, \ldots, 9$ with $a_{11} = a_{10,10} = 2$, $a_{12} = a_{10,9} = -1$. Let $\mathbf{b}$ be the ten-dimensional column vector given by $b_1 = b_{10} = 1$ and $b_i = 0$ for each $i = 2, 3, \ldots, 9$. Solve $A\mathbf{x} = \mathbf{b}$, using Algorithm 6.7.

11. Repeat Exercise 10 using a method based on the algorithm obtained in Exercise 9.

12. Prove Theorem 6.24. [*Hint*: Show that $|u_{i,i+1}| < 1$ for each $i = 1, 2, \ldots, n-1$, and that $|l_{ii}| > 0$ for each $i = 1, 2, \ldots, n$. Deduce that $\det A = \det L \cdot \det U \neq 0$.]

13. A block (or partitioned) tridiagonal matrix is a matrix of the form

$$A = \begin{bmatrix} A_1 & C_1 & 0 & \cdots\cdots & 0 \\ B_2 & A_2 & C_2 & & \vdots \\ 0 & B_3 & & & 0 \\ \vdots & & & & C_{N-1} \\ 0 & \cdots\cdots & 0 & B_N & A_N \end{bmatrix},$$

where each A_i is an $n_i \times n_i$ matrix, each B_i is an $n_i \times n_{i-1}$ matrix and each C_i is an $n_i \times n_{i+1}$ matrix for some collection of positive integers $n_1, n_2, \ldots, n_N$.

a) Factor A into LU, where

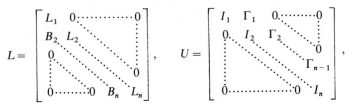

$$L = \begin{bmatrix} L_1 & 0 & \cdots & \cdots & 0 \\ B_2 & L_2 & & & \vdots \\ 0 & & \ddots & & \\ \vdots & & & \ddots & 0 \\ 0 & \cdots & 0 & B_n & L_n \end{bmatrix}, \qquad U = \begin{bmatrix} I_1 & \Gamma_1 & 0 & \cdots & 0 \\ 0 & I_2 & \Gamma_2 & & \vdots \\ \vdots & & \ddots & \ddots & 0 \\ & & & & \Gamma_{n-1} \\ 0 & \cdots & \cdots & 0 & I_n \end{bmatrix},$$

and each L_i is an $n_i \times n_i$ matrix, each Γ_i is an $n_i \times n_{i+1}$ matrix, and I_i denotes the $n_i \times n_i$ identity matrix.

b) Derive a block tridiagonal matrix algorithm, similar to Algorithm 6.7, to solve a linear system of the form $Ax = y$ where y is row partitioned the same as A.

14. Consider the block tridiagonal matrix defined in Exercise 13. Show that if the leading principal submatrices of A of the form

$$A^{(k)} = \begin{bmatrix} A_1 & C_1 & 0 & \cdots & \cdots & 0 \\ B_2 & A_2 & C_2 & & & \vdots \\ 0 & & \ddots & \ddots & & 0 \\ \vdots & & & \ddots & \ddots & C_{k-1} \\ 0 & \cdots & \cdots & 0 & B_k & A_k \end{bmatrix}$$

are all nonsingular for each $k = 1, 2, \ldots, n$, then the matrices $L_1, L_2, \ldots, L_n$ are also nonsingular, and the algorithm obtained in Exercise 13(b) can be performed.

15. Let A be the block tridiagonal matrix defined by

$$A = \begin{bmatrix} A_1 & C_1 & 0 \\ B_2 & A_2 & C_2 \\ 0 & B_3 & A_3 \end{bmatrix}, \qquad \text{where} \qquad A_i = \begin{bmatrix} 4 & -1 & 0 \\ -1 & 4 & -1 \\ 0 & -1 & 4 \end{bmatrix}$$

for each $i = 1, 2, 3$, and

$$B_{i+1} = C_i = \begin{bmatrix} -1 & 0 & 0 \\ 0 & -1 & 0 \\ 0 & 0 & -1 \end{bmatrix}$$

for $i = 1, 2$. Given the nine-dimensional column vector $\mathbf{b}$ with $b_1 = b_9 = 2$ and $b_i = 1$ for each $i = 2, 3, \ldots, 8$, solve $Ax = \mathbf{b}$ by the algorithm obtained in Exercise 13.

16. Show that Algorithm 6.5 requires as many multiplications/divisions and additions/subtractions as Gaussian elimination.

17. Construct the operation count for an $n \times n$ linear system, using Algorithm 6.6.

18. Construct the operation count for an $n \times n$ linear system, using Algorithm 6.7.

19. Show that for any positive definite matrix A, Algorithm 6.5 can be performed with $l_{ii} > 0$ and $u_{ii} > 0$ for each $i = 1, 2, \ldots, n$.

7

Approximation Theory

Hooke's law states that when a force is applied to a spring constructed of uniform material, the length of the spring will be a linear function of the force applied. We can write the linear function as $F(l) = k \cdot l + E$, where $F(l)$ represents the force required to stretch the spring l units, the constant E represents the length of the spring with no force applied, and the constant k is called the spring constant.

Suppose we want to determine the spring constant for a spring that has an initial length of 5.3 inches. We consecutively apply forces of 2, 4, and 6 pounds to the spring and find that its length increases to 7.0, 9.4, and 12.3 inches, respectively. A quick examination shows that the points (0, 5.3), (2, 7.0), (4, 9.4), and (6, 12.3) do not quite lie in a straight line. Although we could simply use one random pair of these data points to approximate the spring constant, it would seem more reasonable to find the line that *best* approximates all the data points, to determine the constant. This type of approximation will be considered in this chapter.

The study of approximation theory involves two general types of problems. One problem arises when a function is given explicitly but we wish to find a "simpler" type of function, such as a polynomial, that can be used to determine approximate values of the given function. The other problem in approximation theory is concerned with fitting functions to given data and finding the "best" function in a certain class that can be used to represent the data.

Both problems have been touched upon in Chapter 3. The Taylor polynomial of degree n about the number x_0 was discussed as an excellent approximation to an $(n+1)$-times differentiable function f in a small neighborhood of x_0. The Lagrange interpolating polynomials, or more generally osculatory polynomials, were discussed both as approximating polynomials and as polynomials to fit certain data. Cubic splines were also discussed in that chapter. In this chapter, limitations to these techniques will be considered and other avenues of approach discussed.

7.1 Discrete Least-Squares Approximation

Consider the problem of estimating the values of a function at nontabulated points, given the experimental data in Table 7.1.

TABLE 7.1

i	x_i	y_i
1	2	2
2	4	11
3	6	28
4	8	40

The methods described in Chapter 3 require that either a Lagrange polynomial of degree three or a cubic spline be constructed that assumes the value y_i at x_i for each $i = 1, 2, 3, 4$. Sketching the graph of the values given in Table 7.1, however, indicates that it would be reasonable to assume that the actual relationship is linear and that no line fits the data exactly because of error in the data collection procedure. (*See Figure 7.1*.)

If this is indeed the case, it would be unreasonable to require that the approximating function agree exactly with the given data; in fact, such an approximating function would introduce oscillations that were not present originally. A better approach for a problem of this type would be to find the "best" (in some sense) line that could be used as an approximating function, even though it might not agree precisely with the data at any point.

The **least-squares** approach to this problem involves determining the best approximating line when the error involved is the sum of the squares of the differences between the values on the approximating line and the given values. Letting $ax_i + b$

FIGURE 7.1

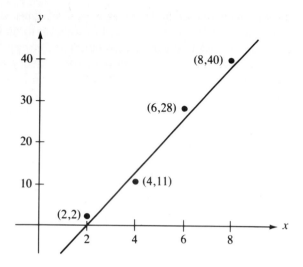

denote the ith value on the approximating line and y_i the ith given value, requires that constants a and b be found that minimize the least-squares error:

$$\sum_{i=1}^{n} [y_i - (ax_i + b)]^2.$$

This particular problem reduces to finding constants a and b that minimize

$$\sum_{i=1}^{4} [y_i - (ax_i + b)]^2 = [2 - (2a + b)]^2 + [11 - (4a + b)]^2$$
$$+ [28 - (6a + b)]^2 + [40 - (8a + b)]^2.$$

If $\sum_{i=1}^{4} [y_i - (ax_i + b)]^2$ is considered to be a function of two variables a and b, an elementary result from multivariate calculus implies that, for a minimum to occur at (a, b), it is necessary for

$$0 = \frac{\partial}{\partial a} \sum_{i=1}^{4} [y_i - (ax_i + b)]^2 \quad \text{and} \quad 0 = \frac{\partial}{\partial b} \sum_{i=1}^{4} [y_i - (ax_i + b)]^2.$$

Consequently,

$$0 = 2(2 - 2a - b)(-2) + 2(11 - 4a - b)(-4)$$
$$+ 2(28 - 6a - b)(-6) + 2(40 - 8a - b)(-8),$$

and
$$0 = 2(2 - 2a - b)(-1) + 2(11 - 4a - b)(-1)$$
$$+ 2(28 - 6a - b)(-1) + 2(40 - 8a - b)(-1),$$

which simplifies to

$$30a + 5b = 134 \quad \text{and} \quad 20a + 4b = 81.$$

The solution to this system of equations is $a = 6.55$ and $b = -12.5$, so the best linear equation in the least-squares sense is

$$y = 6.55x - 12.5.$$

Table 7.2 lists the observed values together with the values obtained using this approximation.

TABLE 7.2

i	x_i	y_i	$6.55x - 12.5$
1	2	2	0.6
2	4	11	13.7
3	6	28	26.8
4	8	40	39.9

The problem of finding the equation of the best linear approximation in the absolute sense requires that values of a and b be found to minimize

$$\max_{i=1,2,3,4} \{|y_i - (ax_i + b)|\}.$$

This is commonly called a **minimax** problem and cannot be handled by elementary techniques.

Another approach to determining the best linear approximation involves finding values of a and b to minimize

$$\sum_{i=1}^{4} |y_i - (ax_i + b)| = |2 - (2a + b)| + |11 - (4a - b)|$$
$$+ |28 - (6a + b)| + |40 - (8a + b)|.$$

This quantity is called the **absolute deviation**. To minimize the absolute deviation, it is necessary that:

$$0 = \frac{\partial}{\partial a} \sum_{i=1}^{4} |y_i - (ax_i + b)| \quad \text{and} \quad 0 = \frac{\partial}{\partial b} \sum_{i=1}^{4} |y_i - (ax_i + b)|.$$

The difficulty with this procedure is that the absolute-value function is not differentiable at zero, and solutions to this pair of equations cannot necessarily be obtained.

The preceding remarks indicate that the least-squares method is the most convenient procedure for determining best linear approximations; fortunately, there are important theoretical considerations that also favor this method. The minimax approach will generally assign too much weight to a bit of data that is badly in error, while the method using absolute deviation simply averages the error at the various points and does not give sufficient weight to a point that is considerably out of line with the approximation. The least-squares approach puts substantially more weight on a point that is out of line with the rest of the data but will not allow that point to completely dominate the approximation.

An additional reason for considering the least-squares approach involves the study of the statistical distribution of error. If the data is known or assumed to have its mean distributed in a linear manner, the values obtained from a linear least-squares procedure are unbiased estimates for the equation that describes the mean. Moreover, the values obtained can be used to calculate an unbiased estimator for the variance associated with the distribution. (An easily readable presentation of the theory involved can be found in Larson [80], pp. 463–481.)

The general problem of fitting the best least-squares line to a collection of data involves minimizing $\sum_{i=1}^{n} [y_i - (ax_i + b)]^2$ with respect to the parameters a and b. For a minimum to occur, it is necessary that

$$0 = \frac{\partial}{\partial a} \sum_{i=1}^{n} [y_i - (ax_i + b)]^2 = 2 \sum_{i=1}^{n} (y_i - ax_i - b)(-x_i)$$

and

$$0 = \frac{\partial}{\partial b} \sum_{i=1}^{n} (y_i - ax_i - b)^2 = 2 \sum_{i=1}^{n} (y_i - ax_i - b)(-1).$$

These equations simplify to what is known as the **normal equations**:

$$a \sum_{i=1}^{n} x_i^2 + b \sum_{i=1}^{n} x_i = \sum_{i=1}^{n} x_i y_i \quad \text{and} \quad a \sum_{i=1}^{n} x_i + b \cdot n = \sum_{i=1}^{n} y_i.$$

The solution to this system of equations is

(7.1)
$$a = \frac{n(\sum_{i=1}^{n} x_i y_i) - (\sum_{i=1}^{n} x_i)(\sum_{i=1}^{n} y_i)}{n(\sum_{i=1}^{n} x_i^2) - (\sum_{i=1}^{n} x_i)^2}$$

and

(7.2)
$$b = \frac{(\sum_{i=1}^{n} x_i^2)(\sum_{i=1}^{n} y_i) - (\sum_{i=1}^{n} x_i y_i)(\sum_{i=1}^{n} x_i)}{n(\sum_{i=1}^{n} x_i^2) - (\sum_{i=1}^{n} x_i)^2}.$$

EXAMPLE 1　　Consider the data presented in the first two columns of Table 7.3. To find the least-squares line approximating this data, extend the table and sum the columns, as shown in the last two columns of Table 7.3.

TABLE 7.3

x_i	y_i	x_i^2	$x_i y_i$
1	1.3	1	1.3
2	3.5	4	7.0
3	4.2	9	12.6
4	5.0	16	20.0
5	7.0	25	35.0
6	8.8	36	52.8
7	10.1	49	70.7
8	12.5	64	100.0
9	13..0	81	117.0
10	15.6	100	156.0
11	16.1	121	177.1
66	97.1	506	749.5

TABLE 7.4

x_i	y_i	$1.517x_i - 0.276$
1	1.3	1.24
2	3.5	2.76
3	4.2	4.28
4	5.0	5.79
5	7.0	7.31
6	8.8	8.83
7	10.1	10.34
8	12.5	11.86
9	13.0	13.38
10	15.6	14.89
11	16.1	16.41

Equations (7.1) and (7.2) imply that

$$a = \frac{11(749.5) - 66(97.1)}{11(506) - (66)^2} = 1.517$$

and

$$b = \frac{506(97.1) - 66(749.5)}{11(506) - (66)^2} = -0.276.$$

The graph of this line together with the data points is shown in Fig. 7.2.

FIGURE 7.2

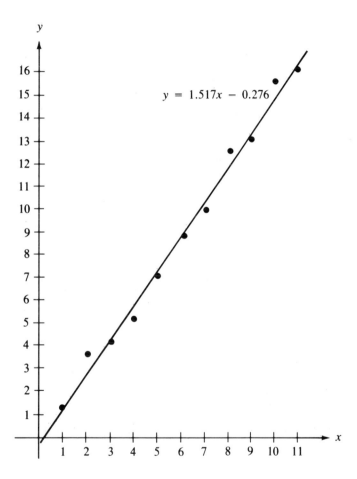

$$y = 1.517x - 0.276$$

The approximate values given by the least-squares technique and the data points are given in Table 7.4. ■

The general problem of approximating a set of data, $\{(x_i, y_i)|i = 0, 1, ..., M\}$, with a polynomial $P_n(x) = \sum_{k=0}^{n} a_k x^k$ of degree $n < M$ using the least-squares procedure is handled in a similar manner and requires choosing the constants $a_0, a_1, \ldots,$ a_n to minimize the least-squares error

$$
\begin{aligned}
E &= \sum_{i=0}^{M} (y_i - P(x_i))^2 \\
&= \sum_{i=0}^{M} y_i^2 - 2 \sum_{i=0}^{M} P(x_i)y_i + \sum_{i=0}^{M} (P(x_i))^2 \\
&= \sum_{i=0}^{M} y_i^2 - 2 \sum_{i=0}^{M} \left(\sum_{j=0}^{n} a_j x_i^j \right) y_i + \sum_{i=0}^{M} \left(\sum_{j=0}^{n} a_j x_i^j \right)^2 \\
&= \sum_{i=0}^{M} y_i^2 - 2 \sum_{j=0}^{n} a_j \left(\sum_{i=0}^{M} y_i x_i^j \right) + \sum_{j=0}^{n} \sum_{k=0}^{n} a_j a_k \left(\sum_{i=0}^{M} x_i^{j+k} \right).
\end{aligned}
$$

As in the linear case, for E to be minimized, it is necessary that $\partial E/\partial a_j = 0$ for each $j = 0, 1, \ldots, n$. Thus, for each j,

$$
0 = \frac{\partial E}{\partial a_j} = -2 \sum_{i=0}^{M} y_i x_i^j + 2 \sum_{k=0}^{n} a_k \sum_{i=0}^{M} x_i^{j+k}.
$$

This gives $n + 1$ equations in the $n + 1$ unknowns, a_j, called the **normal equations**,

$$
\sum_{k=0}^{n} a_k \sum_{i=0}^{M} x_i^{j+k} = \sum_{i=0}^{M} y_i x_i^j, \qquad j = 0, 1, \ldots, n.
$$

It is helpful to write the equations as follows:

$$
a_0 \sum_{i=0}^{M} x_i^0 + a_1 \sum_{i=0}^{M} x_i^1 + a_2 \sum_{i=0}^{M} x_i^2 + \cdots + a_n \sum_{i=0}^{M} x_i^n = \sum_{i=0}^{M} y_i x_i^0,
$$

$$
a_0 \sum_{i=0}^{M} x_i^1 + a_1 \sum_{i=0}^{M} x_i^2 + a_2 \sum_{i=0}^{M} x_i^3 + \cdots + a_n \sum_{i=0}^{M} x_i^{n+1} = \sum_{i=0}^{M} y_i x_i^1
$$

$$
\vdots
$$

$$
a_0 \sum_{i=0}^{M} x_i^n + a_1 \sum_{i=0}^{M} x_i^{n+1} + a_2 \sum_{i=0}^{M} x_i^{n+2} + \cdots + a_n \sum_{i=0}^{M} x_i^{2n} = \sum_{i=0}^{M} y_i x_i^n.
$$

It can be shown, (see Exercise 13), that the normal equations always have a unique solution provided that the x_i, for $i = 0, 1, \ldots, M$, are distinct.

EXAMPLE 2   Fit the data in Table 7.5 with the discrete least-squares polynomial of degree two. For this problem, $n = 2$, $M = 4$, and the three normal equations are:

$$
\begin{aligned}
5a_0 + \quad 2.5a_1 + \quad 1.875a_2 &= 8.7680, \\
2.5a_0 + \quad 1.875a_1 + 1.5625a_2 &= 5.4514, \\
1.875a_0 + 1.5625a_1 + 1.3828a_2 &= 4.4015.
\end{aligned}
$$

TABLE 7.5

i	0	1	2	3	4
x_i	0	0.25	0.5	0.75	1.00
y_i	1.0000	1.2840	1.6487	2.1170	2.7183

The solution to this system is

$$a_0 = 1.0052, \qquad a_1 = 0.8641, \qquad a_2 = 0.8437.$$

Thus, the least-squares polynomial of degree two fitting the preceding data is $P_2(x) = 1.0052 + 0.8641x + 0.8437x^2$, whose graph is shown in Fig. 7.3. At the given values of x_i, we have the approximations shown in Table 7.6.

FIGURE 7.3

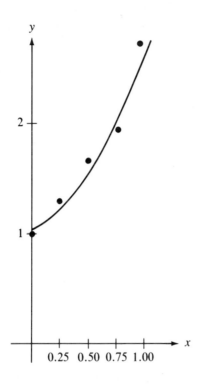

TABLE 7.6

i	0	1	2	3	4
x_i	0	0.25	0.50	0.75	1.00
y_i	1.0000	1.2840	1.6487	2.1170	2.7183
$P(x_i)$	1.0052	1.2740	1.6482	2.1279	2.7130
$y_i - P(x_i)$	−0.0052	0.0100	0.0005	−0.0109	0.0053

The error

$$\sum_{i=0}^{4} |y_i - P(x_i)|^2 = 2.76 \times 10^{-4}$$

is the least that can be obtained by using a quadratic polynomial.   ■

Although least squares using polynomials is the most extensively used procedure, occasionally it is appropriate to assume that the data is exponentially related. This requires the approximating function to be of the form

(7.3) $$y = be^{ax}$$

or

(7.4) $$y = bx^a$$

for some constants a and b. The difficulty with applying the least-squares procedure in a situation of this type comes from attempting to minimize

$$E = \sum_{i=0}^{M} (y_i - be^{ax_i})^2 \quad \text{in the case of Eq. (7.3)}$$

or

$$E = \sum_{i=0}^{M} (y_i - bx_i^a)^2 \quad \text{in the case of Eq. (7.4).}$$

The normal equations associated with these procedures are obtained from either

$$0 = \frac{\partial E}{\partial b} = 2 \sum_{i=0}^{M} (y_i - be^{ax_i})(-e^{ax_i})$$

and

$$0 = \frac{\partial E}{\partial a} = 2 \sum_{i=0}^{M} (y_i - be^{ax_i})(-bx_i e^{ax_i}), \quad \text{in the case of Eq. (7.3),}$$

or

$$0 = \frac{\partial E}{\partial b} = 2 \sum_{i=0}^{M} (y_i - bx_i^a)(-x_i^a)$$

and

$$0 = \frac{\partial E}{\partial a} = 2 \sum_{i=0}^{M} (y_i - bx_i^a)(-b(\ln x_i)x_i^a), \quad \text{in the case of Eq. (7.4).}$$

No exact solution to either of these systems can generally be found.

The method that is usually followed when the data is suspected to be exponentially related is to consider the logarithm of the approximating equation:

(7.5) $$\ln y = \ln b + ax \quad \text{in the case of Eq. (7.3),}$$

and

(7.6) $$\ln y = \ln b + a \ln x \quad \text{in the case of Eq. (7.4).}$$

In either case, a linear problem now appears and solutions for $\ln b$ and a can be obtained by appropriately modifying Eqs. (7.1) and (7.2). It should be remembered, however, that the approximation obtained in this manner is not the least-squares approximation for the original problem and that this approximation can in some cases differ significantly from the least-squares approximation to the original problem. The application in Exercise 12 describes such a problem. This application will be reconsidered as an exercise in Section 9.2, where the exact solution to the exponential least-squares problem is approximated by using methods suitable for solving nonlinear systems of equations.

EXAMPLE 3   Consider the collection of data in the first three columns of Table 7.7.

If x_i is graphed with $\ln y_i$, the data appears to have a linear relation, so it is reasonable to assume an approximation of the form

$$y = be^{ax} \quad \text{or} \quad \ln y = \ln b + ax.$$

Extending the table and summing the columns gives the remaining data in Table 7.7.

TABLE 7.7

i	x_i	y_i	$\ln y_i$	x_i^2	$x_i \ln y_i$
1	1.00	5.10	1.629	1.0000	1.629
2	1.25	5.79	1.756	1.5625	2.195
3	1.50	6.53	1.876	2.2500	2.814
4	1.75	7.45	2.008	3.0625	3.514
5	2.00	8.46	2.135	4.0000	4.270
	7.50		9.404	11.875	14.422

Using Eqs. (7.1) and (7.2),

$$a = \frac{(5)(14.422) - (7.5)(9.404)}{(5)(11.875) - (7.5)^2} = 0.5056,$$

and

$$\ln b = \frac{(11.875)(9.404) - (14.422)(7.5)}{(5)(11.875) - (7.5)^2} = 1.122.$$

Since $b = e^{1.122} = 3.071$, the approximation assumes the form

$$y = 3.071e^{0.5056x},$$

which, at the data points, gives the values in Table 7.8.

TABLE 7.8

i	x_i	y_i	$3.071e^{0.5056x_i}$
1	1.00	5.10	5.09
2	1.25	5.79	5.78
3	1.50	6.53	6.56
4	1.75	7.45	7.44
5	2.00	8.46	8.44

■

Exercise Set 7.1

1. Compute the least-squares polynomial of degree one for the data of Example 2.

2. The data for Example 2 is actually given by the function $f(x) = e^x$. Considering the graph of $f(x) = e^x$ on $[0, 1]$, do you believe that choosing a higher-degree least-squares polynomial would improve upon the results of Example 2? Support your claim by considering least-squares polynomials of degree three and four.

3. Compute the Lagrange interpolating polynomial of degree two for $f(x) = e^x$ on the nodes $x_0 = 0$, $x_1 = 0.5$, and $x_2 = 1.0$. Compare the second-degree polynomial and that of Example 2 to determine which better approximated $f(x) = e^x$ on $[0, 1]$.

4. Find the least-squares polynomials of degree 1, 2, 3, and 4 for the data in the following table.

i	x_i	y_i
0	0	1.0
1	0.15	1.004
2	0.31	1.031
3	0.5	1.117
4	0.6	1.223
5	0.75	1.422

Which degree gives the best least-squares approximation; that is, which gives the smaller error?

5. Given the data:

x_i	y_i
4.0	102.56
4.2	113.18
4.5	130.11
4.7	142.05
5.1	167.53
5.5	195.14
5.9	224.87
6.3	256.73
6.8	299.50
7.1	326.72

a) Construct the least-squares approximation of degree one and compute the error.

b) Construct the least-squares approximation of degree two and compute the error.

c) Construct the least-squares approximation of degree three and compute the error.

d) Construct the least-squares approximation of the form be^{ax} and compute the error.

e) Construct the least-squares approximation of the form bx^a and compute the error.

6. Repeat Exercise 5 for the data:

x_i	y_i
0.2	0.050446
0.3	0.098426
0.6	0.33277
0.9	0.72660
1.1	1.0972
1.3	1.5697
1.4	1.8487
1.6	2.5015

7. The following lists contain homework grades and final examination grade for 30 numerical analysis students. Find the equation of the least-squares line for this data, and use this line to determine the homework grade required to predict minimal A(90%) and D(60%) grades on the final.

Homework	Final	Homework	Final	Homework	Final
302	45	343	83	234	51
325	72	290	74	337	53
285	54	326	76	351	100
339	54	233	57	339	67
334	79	254	45	343	83
322	65	323	83	314	42
331	99	337	99	344	79
279	63	337	70	185	59
316	65	304	62	340	75
347	99	319	66	316	45

8. The following table lists the college grade-point averages of 20 mathematics and computer science majors, together with the scores that these students received on the mathematics portion of the ACT (American College Testing Program) test while in high school. Plot this data, and find the equation of the least-squares line for this data.

ACT score	Grade-point average	ACT score	Grade-point average
28	3.84	29	3.75
25	3.21	28	3.65
28	3.23	27	3.87
27	3.63	29	3.75
28	3.75	21	1.66
33	3.20	28	3.12
28	3.41	28	2.96
29	3.38	26	2.92
23	3.53	30	3.10
27	2.03	24	2.81

9. To determine a relationship between the number of fish and the number of species of fish in samples taken for a portion of the Great Barrier Reef, P. Sale and R. Dybdahl [114] fit a linear least-squares polynomial to the following collection of data, which was collected in samples over a two-year period. Let x be the number of fish in the sample, and y be the number of species in the sample.

x	y	x	y	x	y
13	11	29	12	60	14
15	10	30	14	62	21
16	11	31	16	64	21
21	12	36	17	70	24
22	12	40	13	72	17
23	13	42	14	100	23
25	13	55	22	130	34

Determine the linear least-squares polynomial for this data.

10. The following set of data, presented in March, 1970, to the Senate Antitrust Subcommittee, shows the comparative crash-survivability characteristics of cars in various classes. Find the best least-squares line that approximates this data. (The table shows the percent of accident-involved vehicles in which the most severe injury was fatal or serious.)

Type	Average weight	Percent occurrence
1. Domestic "luxury" regular	4800 lb	3.1
2. Domestic "intermediate" regular	3700 lb	4.0
3. Domestic "economy" regular	3400 lb	5.2
4. Domestic compact	2800 lb	6.4
5. Foreign compact	1900 lb	9.6

11. To determine a functional relationship between the attenuation coefficient and the thickness of a sample of taconite, V. P. Singh [125] fit a collection of data by using a linear least-squares polynomial. The following collection of data is taken from a graph in that paper. Find the best linear least-squares polynomial fitting this data.

Thickness (cm)	Attenuation coefficient db/cm
0.040	26.5
0.041	28.1
0.055	25.2
0.056	26.0
0.062	24.0
0.071	25.0
0.071	26.4
0.078	27.2
0.082	25.6
0.090	25.0
0.092	26.8
0.100	24.8
0.105	27.0
0.120	25.0
0.123	27.3
0.130	26.9
0.140	26.2

12. In a paper dealing with the efficiency of energy utilization of the larvae of the Modest Sphinx moth (*Pachysphinx modesta*), L. Schroeder [117] used the following data to determine a relation between W, the live weight of the larvae in grams, and R, the oxygen consumption of the larvae in ml/hr. For biological reasons, it is assumed that a relationship in the form of Eq. (7.4) exists between W and R.

a)   Find the logarithmic linear least-squares polynomial by using

$$\ln R = \ln b + a \ln W.$$

b)   Compute the error associated with the approximation in part (a):

$$E = \sum_{i=1}^{37} (R_i - bW_i^a)^2.$$

c)   Modify the logarithmic least-squares equation in part (a) by adding the quadratic term $c(\ln W_i)^2$ and determine the logarithmic quadratic least-squares polynomial.

d)   Determine the formula for and compute the error associated with the approximation in part (c).

W	R	W	R	W	R	W	R	W	R
0.017	0.154	0.025	0.23	0.020	0.181	0.020	0.180	0.025	0.234
0.087	0.296	0.111	0.357	0.085	0.260	0.119	0.299	0.233	0.537
0.174	0.363	0.211	0.366	0.171	0.334	0.210	0.428	0.783	1.47
1.11	0.531	0.999	0.771	1.29	0.87	1.32	1.15	1.35	2.48
1.74	2.23	3.02	2.01	3.04	3.59	3.34	2.83	1.69	1.44
4.09	3.58	4.28	3.28	4.29	3.40	5.48	4.15	2.75	1.84
5.45	3.52	4.58	2.96	5.30	3.88			4.83	4.66
5.96	2.40	4.68	5.10					5.53	6.94

13. Show that the normal equations discussed on page 366 resulting from discrete least-squares approximation yield a symmetric and nonsingular matrix and hence have a unique solution. [*Hint*: Let $A = (a_{ij})$ where

$$a_{ij} = \sum_{k=1}^{M} x_k^{i+j-2},$$

and $x_1, x_2, \ldots, x_n$ are distinct points with $n < M$. Suppose A is singular and that $\mathbf{c} \neq \mathbf{0}$ is such the $\mathbf{c}^t A \mathbf{c} = 0$. Show that the nth-degree polynomial whose coefficients are the coordinates of $\mathbf{c}$ has more than n roots, and use this to establish a contradiction.]

7.2 Orthogonal Polynomials and Least-Squares Approximation

The previous section considered the problem of least-squares approximation to fit a collection of data. The other approximation problem mentioned in the introduction concerns the approximation of functions.

Suppose $f \in C[a, b]$ and that a polynomial of degree at most n, P_n, is required that will minimize the error

(7.7)
$$\int_a^b (f(x) - P_n(x))^2 \, dx.$$

To determine a least-squares approximating polynomial, that is, a polynomial to minimize expression (7.7), let

$$P_n(x) = a_n x^n + a_{n-1} x^{n-1} + \cdots + a_1 x + a_0 = \sum_{k=0}^{n} a_k x^k,$$

and define

$$E(a_0, a_1, \ldots, a_n) = \int_a^b \left(f(x) - \sum_{k=0}^{n} a_k x^k \right)^2 \, dx.$$

The problem is to find real coefficients $a_0, \ldots, a_n$ that will minimize E. From the calculus of functions of several variables, a necessary condition for the numbers $a_0, \ldots, a_n$ to minimize E is that

$$\frac{\partial E}{\partial a_j} = 0 \qquad \text{for each } j = 0, 1, \ldots, n.$$

Since $\qquad E = \int_a^b [f(x)]^2 \, dx - 2 \sum_{k=0}^n a_k \int_a^b x^k f(x) \, dx + \int_a^b \left(\sum_{k=0}^n a_k x^k \right)^2 dx,$

$$\frac{\partial E}{\partial a_j} = -2 \int_a^b x^j f(x) \, dx + 2 \sum_{k=0}^n a_k \int_a^b x^{j+k} \, dx.$$

Hence, to find P_n, the $(n+1)$ linear equations

$$(7.8) \qquad \sum_{k=0}^n a_k \int_a^b x^{j+k} \, dx = \int_a^b x^j f(x) \, dx, \qquad j = 0, 1, \ldots, n,$$

must be solved for the $(n+1)$ unknowns a_j, $j = 0, 1, \ldots, n$. These equations are called the **normal equations**. It can be shown that the normal equations always have a unique solution provided $f \in C[a, b]$ and $a \neq b$. (See Exercise 10.)

EXAMPLE 1 Find the least-squares approximating polynomial of degree two for the function $f(x) = \sin \pi x$ on the interval $[0, 1]$. The normal equations for $P_2(x) = a_2 x^2 + a_1 x + a_0$ are given by:

$$a_0 \int_0^1 1 \, dx + a_1 \int_0^1 x \, dx + a_2 \int_0^1 x^2 \, dx = \int_0^1 \sin \pi x \, dx,$$

$$a_0 \int_0^1 x \, dx + a_1 \int_0^1 x^2 \, dx + a_2 \int_0^1 x^3 \, dx = \int_0^1 x \sin \pi x \, dx,$$

$$a_0 \int_0^1 x^2 \, dx + a_1 \int_0^1 x^3 \, dx + a_2 \int_0^1 x^4 \, dx = \int_0^1 x^2 \sin \pi x \, dx.$$

Performing the integration yields

$$a_0 + \tfrac{1}{2} a_1 + \tfrac{1}{3} a_2 = \frac{2}{\pi},$$

$$\tfrac{1}{2} a_0 + \tfrac{1}{3} a_1 + \tfrac{1}{4} a_2 = \frac{1}{\pi},$$

$$\tfrac{1}{3} a_0 + \tfrac{1}{4} a_1 + \tfrac{1}{5} a_2 = \frac{\pi^2 - 4}{\pi^3}.$$

The three equations in three unknowns can be solved to obtain

$$a_0 = \frac{12\pi^2 - 120}{\pi^3} \approx -0.050465, \qquad \text{and} \qquad a_1 = -a_2 = \frac{720 - 60\pi^2}{\pi^3} \approx 4.12251.$$

Consequently the least-squares polynomial approximation of degree two for $f(x) = \sin \pi x$ on $[0, 1]$ is $P_2(x) = -4.12251 x^2 + 4.12251 x - 0.050465$. (*See Fig. 7.4.*)

∎

FIGURE 7.4

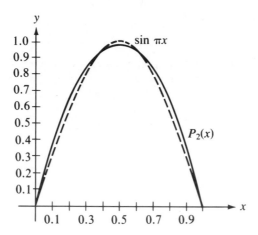

Example 1 serves to illustrate the degree of difficulty in obtaining a least-squares polynomial approximation. One must solve an $(n+1) \times (n+1)$ linear system for the coefficients $a_0, \ldots, a_n$ of P_n. Also, since the coefficients in the linear system are of the form

$$\int_a^b x^{j+k} \, dx = \frac{b^{j+k+1} - a^{j+k+1}}{j+k+1},$$

the linear system does not have a convenient numerical solution because of the presence of rounding errors. The matrix in the linear system is known as the **Hilbert matrix**. This matrix is a classic example for demonstrating round-off error difficulties; no pivoting technique can be used satisfactorily.

Another disadvantage to the technique used in Example 1 is similar to the situation that occurred when the Lagrange polynomials were first introduced in Section 3.2; that is, the calculations that were performed in obtaining the best nth-degree polynomial, P_n, do not lessen the amount of work required to obtain P_{n+1}, the next higher degree polynomial.

A different technique to obtain least-squares approximations will now be considered. This technique turns out to be efficient in computation and also, once P_n is known, it is easy to determine P_{n+1}. To facilitate further discussion, some new concepts will be introduced.

DEFINITION 7.1   The set of functions $\{\phi_0, \ldots, \phi_n\}$ is said to be **linearly independent** on $[a, b]$ where $b > a$ if, whenever

$$c_0\phi_0(x) + c_1\phi_1(x) + \cdots + c_n\phi_n(x) = 0 \qquad \text{for all } x \in [a, b],$$

then $c_0 = c_1 = \cdots = c_n = 0$. Otherwise the set of functions is said to be **linearly dependent**.

THEOREM 7.2   If ϕ_j is a polynomial of degree j, for each $j = 0, 1, \ldots, n$, then $\{\phi_0, \ldots, \phi_n\}$ is linearly independent on any interval $[a, b]$ where $a < b$.

PROOF   Suppose $c_0, \ldots, c_n$ are real numbers for which

$$c_0\phi_0(x) + c_1\phi_1(x) + \cdots + c_n\phi_n(x) = 0 \qquad \text{for all } x \in [a, b].$$

Since $P(x) = \sum_{k=0}^{n} c_k\phi_k(x)$ vanishes on $[a, b]$, P has an infinite number of roots. P is a polynomial of degree at most n, so Corollary 2.16, page 67, implies $P \equiv 0$ and that the coefficient of each power of x vanishes. Since the only polynomial involving the term x^n is ϕ_n, the constant c_n must be zero. Now only ϕ_{n-1} involves the term x^{n-1} so c_{n-1} must also be zero. In a like manner $c_j = 0$ for $j = 0, 1, \ldots, n-2$. This implies that $\{\phi_0, \ldots, \phi_n\}$ is linearly independent.   □

EXAMPLE 2   Let $\phi_0(x) = 2$, $\phi_1(x) = x - 3$, and $\phi_2(x) = x^2 + 2x + 7$. By Theorem 7.2, $\{\phi_0, \phi_1, \phi_2\}$ is linearly independent on any interval $[a, b]$. Suppose $Q(x) = a_0 + a_1 x + a_2 x^2$. To show that there exist constants c_0, c_1, and c_2 such that $Q(x) = c_0\phi_0(x) + c_1\phi_1(x) + c_2\phi_2(x)$, note that

$$1 = \tfrac{1}{2}\phi_0(x),$$

$$x = \phi_1(x) + 3 = \phi_1(x) + \tfrac{3}{2}\phi_0(x),$$

$$x^2 = \phi_2(x) - 2x - 7 = \phi_2(x) - 2[\phi_1(x) + \tfrac{3}{2}\phi_0(x)] - 7[\tfrac{1}{2}\phi_0(x)]$$
$$= \phi_2(x) - 2\phi_1(x) - \tfrac{13}{2}\phi_0(x).$$

Hence,

$$Q(x) = a_0[\tfrac{1}{2}\phi_0(x)] + a_1[\phi_1(x) + \tfrac{3}{2}\phi_0(x)] + a_2[\phi_2(x) - 2\phi_1(x) - \tfrac{13}{2}\phi_0(x)]$$
$$= [\tfrac{1}{2}a_0 + \tfrac{3}{2}a_1 - \tfrac{13}{2}a_2]\phi_0(x) + [a_1 - 2a_2]\phi_1(x) + a_2\phi_2(x). \qquad ∎$$

The situation illustrated in Example 2 holds in a much more general setting. Let Π_n **be the set of all polynomials of degree at most n.** If $\{\phi_0, \phi_1, \ldots, \phi_n\}$ is any collection of linearly independent polynomials in Π_n, then any polynomial in Π_n can be written uniquely as a linear combination of $\phi_0, \phi_1, \ldots, \phi_n$. The proof of this statement is considered in Exercise 11.

To discuss general function approximation requires the introduction of the notions of weight functions and orthogonality.

DEFINITION 7.3   An integrable function w is called a **weight function** on $[a, b]$ if $w(x) \geq 0$ for $x \in [a, b]$, but $w(x) \not\equiv 0$ on any subinterval of $[a, b]$.

The purpose of a weight function is to assign varying degrees of importance to approximations on certain portions of the interval. For example, the weight function

$$w(x) = \frac{1}{\sqrt{1 - x^2}}$$

places less emphasis near the center of the interval $(-1, 1)$ and more emphasis when $|x|$ is near one. This weight function will be used in Example 5.

Suppose $\{\phi_0, \phi_1, ..., \phi_n\}$ is a set of linearly independent functions on $[a, b]$, w is a weight function for $[a, b]$, and for $f \in C[a, b]$ a linear combination

$$P(x) = \sum_{k=0}^{n} a_k \phi_k(x)$$

is sought to minimize the error

$$(7.9) \qquad E(a_0, ..., a_n) = \int_a^b w(x) \left[f(x) - \sum_{k=0}^{n} a_k \phi_k(x) \right]^2 dx.$$

This problem reduces to the situation considered at the beginning of this section in the special case when $w(x) \equiv 1$ and $\phi_k(x) = x^k$ for each $k = 0, 1, ..., n$.

The normal equations associated with Eq. (7.9) are derived from the fact that for each $j = 0, 1, ..., n$,

$$0 = \frac{\partial E}{\partial a_j} = 2 \int_a^b w(x) \left[f(x) - \sum_{k=0}^{n} a_k \phi_k(x) \right] \phi_j(x) \, dx.$$

The system of normal equations can be written:

$$\int_a^b w(x) f(x) \phi_j(x) \, dx = \sum_{k=0}^{n} a_k \int_a^b w(x) \phi_k(x) \phi_j(x) \, dx \qquad \text{for } j = 0, 1, ..., n.$$

Suppose that the functions $\phi_0, \phi_1, ..., \phi_n$ can be chosen so that

$$(7.10) \qquad \int_a^b w(x) \phi_k(x) \phi_j(x) \, dx = \begin{cases} 0, & \text{when } j \neq k, \\ \alpha_k > 0, & \text{when } j \neq k. \end{cases}$$

Then, for each $j = 0, 1, ..., n$.

$$\int_a^b w(x) f(x) \phi_j(x) \, dx = a_j \int_a^b w(x) [\phi_j(x)]^2 \, dx = a_j \alpha_j$$

and

$$a_j = \frac{1}{\alpha_j} \int_a^b w(x) f(x) \phi_j(x) \, dx.$$

Hence the least-squares approximation problem is greatly simplified when the functions $\phi_0, \phi_1, ..., \phi_n$ are chosen to satisfy Eq. (7.10). The remainder of this section will be devoted to studying collections of this type.

DEFINITION 7.4 $\{\phi_0, \phi_1, ..., \phi_n\}$ is said to be an **orthogonal set of functions** for the interval $[a, b]$ with respect to the weight function w if

$$\int_a^b w(x) \phi_j(x) \phi_k(x) \, dx = \begin{cases} 0, & \text{whenever } j \neq k, \\ \alpha_k > 0, & \text{whenever } j = k. \end{cases}$$

If, in addition, $\alpha_k = 1$ for each $k = 0, 1, ..., n$, the set is said to be **orthonormal**.

This definition, together with the remarks preceding it, produces the following theorem.

THEOREM 7.5   If $\{\phi_0, \ldots, \phi_n\}$ is an orthogonal set of functions on an interval $[a, b]$ with respect to the weight function w, then the least-squares approximation to f on $[a, b]$ with respect to w is

$$P(x) = \sum_{k=0}^{n} a_k \phi_k(x),$$

where

$$a_k = \frac{\int_a^b w(x)\phi_k(x)f(x)\,dx}{\int_a^b w(x)[\phi_k(x)]^2\,dx} = \frac{1}{\alpha_k}\int_a^b w(x)\phi_k(x)\,f(x)\,dx.$$

EXAMPLE 3   For each positive integer n, the set of functions $\{\phi_0, \phi_1, \ldots, \phi_{2n-1}\}$ where

$$\phi_0(x) \equiv \frac{1}{\sqrt{2\pi}},$$

$$\phi_k(x) = \frac{1}{\sqrt{\pi}}\cos kx \qquad \text{for each } k = 1, 2, \ldots, n.$$

and

$$\phi_{n+k}(x) = \frac{1}{\sqrt{\pi}}\sin kx \qquad \text{for each } k = 1, 2, \ldots, n-1,$$

is an orthonormal set on $[-\pi, \pi]$ with respect to $w(x) \equiv 1$. (See Exercise 13.) (Some sources include an additional function in the set, $\phi_{2n}(x) = (1/\sqrt{\pi})\sin(nx)$. We will not follow this convention.)

Let $\mathcal{T}_n$ denote the set of all linear combinations of the functions $\phi_0, \phi_1, \ldots,$ ϕ_{2n-1} and call $\mathcal{T}_n$ the set of trigonometric polynomials of degree less than or equal to n. Given $f \in C[-\pi, \pi]$, the least-squares approximation (called a **trigonometric polynomial**) by functions in $\mathcal{T}_n$ is defined by

$$S_n(x) = \sum_{k=0}^{2n-1} a_k \phi_k(x),$$

where

$$a_k = \int_{-\pi}^{\pi} f(x)\phi_k(x)\,dx \qquad \text{for each } k = 0, 1, \ldots, 2n-1.$$

The limit of S_n when $n \to \infty$ is called the **Fourier Series** of f. Fourier series are an extremely useful tool for describing the solution of various ordinary and partial-differential equations that occur in physical situations.

To determine the trigonometric polynomial from $\mathcal{T}_n$ that approximates

$$f(x) = |x| \qquad \text{for } -\pi < x < \pi$$

requires finding

$$a_0 = \int_{-\pi}^{\pi} |x| \frac{1}{\sqrt{2\pi}}\,dx$$

$$= -\frac{1}{\sqrt{2\pi}}\int_{-\pi}^{0} x\,dx + \frac{1}{\sqrt{2\pi}}\int_{0}^{\pi} x\,dx$$

$$= \frac{2}{\sqrt{2\pi}}\int_{0}^{\pi} x\,dx = \frac{\sqrt{2\pi^2}}{2\sqrt{\pi}},$$

$$a_k = \frac{1}{\sqrt{\pi}} \int_{-\pi}^{\pi} |x| \cos kx \, dx = \frac{2}{\sqrt{\pi}} \int_0^{\pi} x \cos kx \, dx$$

$$= \frac{2}{\sqrt{\pi} k^2} [(-1)^k - 1] \qquad \text{for each } k = 1, 2, \ldots, n,$$

and

$$a_{n+k} = \frac{1}{\sqrt{\pi}} \int_{-\pi}^{\pi} |x| \sin kx \, dx$$

$$= 0 \qquad \text{for each } k = 1, 2, \ldots, n - 1.$$

The trigonometric polynomial from $\mathcal{T}_n$ approximating f is therefore

$$S_n(x) = \frac{\pi}{2} + \frac{2}{\pi} \sum_{k=1}^{n} \frac{(-1)^k - 1}{k^2} \cos kx.$$

The Fourier series for f is

$$S(x) = \lim_{n \to \infty} S_n(x) = \frac{\pi}{2} + \frac{2}{\pi} \sum_{k=1}^{\infty} \frac{(-1)^k - 1}{k^2} \cos kx.$$

Since $|\cos kx| \leq 1$ for every k and x, $S(x)$ converges for all real numbers x. ∎

For the remainder of this section, only orthogonal sets of polynomials will be considered. The next theorem, which is based on what is called the **Gram–Schmidt process**, describes how to construct orthogonal polynomials on $[a, b]$ with respect to a weight function w.

THEOREM 7.6

The set of polynomials $\{\phi_0, \phi_1, \ldots, \phi_n\}$ defined in the following way is orthogonal on $[a, b]$ with respect to the weight function w.

$$\phi_0(x) \equiv 1, \qquad \phi_1(x) = x - B_1, \qquad \text{for each } a \leq x \leq b,$$

where

$$B_1 = \frac{\int_a^b xw(x)[\phi_0(x)]^2 \, dx}{\int_a^b w(x)[\phi_0(x)]^2 \, dx};$$

and when $k \geq 2$,

$$\phi_k(x) = (x - B_k)\phi_{k-1}(x) - C_k\phi_{k-2}(x) \qquad \text{for each } a \leq a \leq b,$$

where

$$B_k = \frac{\int_a^b xw(x)[\phi_{k-1}(x)]^2 \, dx}{\int_a^b w(x)[\phi_{k-1}(x)]^2 \, dx}$$

and

$$C_k = \frac{\int_a^b xw(x)\phi_{k-1}(x)\phi_{k-2}(x) \, dx}{\int_a^b w(x)[\phi_{k-2}(x)]^2 \, dx}.$$

Theorem 7.6 provides a recursive procedure for constructing a set of orthogonal polynomials. The proof of this theorem follows by applying the technique of mathematical induction to the degree of the polynomial ϕ_n.

COROLLARY 7.7   For any $n > 0$, the set of polynomials $\{\phi_0, \dots, \phi_n\}$ given in Theorem 7.6 is linearly independent on $[a, b]$ and

$$\int_a^b w(x)\phi_n(x)Q_k(x) \, dx = 0,$$

for any polynomial Q_k of degree $k < n$.

PROOF   Since ϕ_n is a polynomial of degree n, Theorem 7.2 implies that $\{\phi_0, \dots, \phi_n\}$ is a linearly independent set.

Let $Q_k(x)$ be a polynomial of degree k. In Exercise 11, it is shown that there exist numbers $c_0, \dots, c_k$ such that

$$Q_k(x) = \sum_{j=0}^k c_j \phi_j(x).$$

Thus, $\displaystyle \int_a^b w(x)Q_k(x) \, \phi_n(x) \, dx = \sum_{j=0}^k c_j \int_a^b w(x)\phi_j(x)\phi_n(x) \, dx = 0,$

since ϕ_n is orthogonal to ϕ_j for each $j = 0, 1, \dots, k$.   □

EXAMPLE 4   One of the most common sets of orthogonal polynomials is the set of **Legendre polynomials** $\{P_n\}$, which are orthogonal on $[-1, 1]$ with respect to the weight function $w(x) \equiv 1$. The classical definition of the Legendre polynomials requires that $P_n(1) = 1$ for each n, and a recursive relation can be used to generate the polynomials when $n \geq 2$. This normalization will not be needed in our discussion, and the least-squares approximating polynomials generated in either case will be essentially the same. Using the procedure of Theorem 7.6, $P_0(x) \equiv 1$,

so $\displaystyle B_1 = \frac{\int_{-1}^1 x \, dx}{\int_{-1}^1 dx} = 0$ and $P_1(x) = (x - B_1)P_0(x) = x.$

Also, $\displaystyle B_2 = \frac{\int_{-1}^1 x^3 \, dx}{\int_{-1}^1 x^2 \, dx} = 0$ and $\displaystyle C_2 = \frac{\int_{-1}^1 x^2 \, dx}{\int_{-1}^1 1 \, dx} = \frac{1}{3},$

so $P_2(x) = (x - B_2)P_1(x) - C_2 P_0(x) = (x - 0)x - \frac{1}{3} \cdot 1 = x^2 - \frac{1}{3}.$

Higher-degree Legendre polynomials are derived in the same manner. The next three are $P_3(x) = x^3 - (3/5)x$, $P_4(x) = x^4 - (6/7)x^2 + 3/35$, and $P_5(x) = x^5 - (10/9)x^3 + (5/21)x$.   ∎

EXAMPLE 5   Another set of orthogonal polynomials, a set that will be used later in this chapter, is called the set of **Chebyshev polynomials**, $\{T_n\}$. They can be derived by means of Theorem 7.6 for the interval $(-1, 1)$, using the weight function $w(x) = (1 - x^2)^{-1/2}$. We will derive the Chebyshev polynomials differently here, and then show that they satisfy the required orthogonality properties.

For $x \in [-1, 1]$, define

$$T_n(x) = \cos[n \arccos x] \qquad \text{for each } n \geq 0.$$

Introducing the substitution $\theta = \arccos x$ changes this equation to

$$T_n(\theta(x)) \equiv T_n(\theta) = \cos(n\theta) \qquad \text{where } \theta \in [0, \pi].$$

A recurrence relation can be derived by noting that

$$T_{n+1}(\theta) = \cos((n+1)\theta) = \cos(n\theta)\cos\theta - \sin(n\theta)\sin\theta$$

and

$$T_{n-1}(\theta) = \cos((n-1)\theta) = \cos(n\theta)\cos\theta + \sin(n\theta)\sin\theta,$$

so

$$T_{n+1}(\theta) = 2\cos(n\theta)\cos\theta - T_{n-1}(\theta).$$

Returning to the variable x,

(7.11) $$T_{n+1}(x) = 2xT_n(x) - T_{n-1}(x) \qquad \text{for each } 1 \le n.$$

Since

$$T_0(x) = \cos(0 \cdot \arccos x) = 1 \qquad \text{and} \qquad T_1(x) = \cos(1 \arccos x) = x,$$

the Chebyshev polynomials are now easily obtained in a sequential manner by using Eq. (7.11):

$$T_2(x) = 2xT_1(x) - T_0(x) = 2x^2 - 1,$$

$$T_3(x) = 2xT_2(x) - T_1(x) = 4x^3 - 3x,$$

$$T_4(x) = 2xT_3(x) - T_2(x) = 8x^4 - 8x^2 + 1, \qquad \text{and so on.}$$

To show the orthogonality of the Chebyshev polynomials, consider

$$\int_{-1}^1 \frac{T_n(x)T_m(x)}{\sqrt{1-x^2}}\,dx = \int_{-1}^1 \frac{\cos(n \arccos x)\cos(m \arccos x)}{\sqrt{1-x^2}}\,dx.$$

Reintroducing the substitution $\theta = \arccos x$,

$$\int_{-1}^1 \frac{T_n(x)T_m(x)}{\sqrt{1-x^2}}\,dx = \int_{\pi}^0 \frac{\cos(n\theta)\cos(m\theta)}{\sin\theta}(-\sin\theta\,d\theta)$$

$$= -\int_{\pi}^0 \cos(n\theta)\cos(m\theta)\,d\theta$$

$$= \int_0^{\pi} \cos(n\theta)\cos(m\theta)\,d\theta.$$

Suppose $n \ne m$, since $\cos n\theta \cos m\theta = \frac{1}{2}[\cos(n+m)\theta + \cos(n-m)\theta]$,

$$\int_{-1}^1 \frac{T_n(x)T_m(x)\,dx}{\sqrt{1-x^2}} = \frac{1}{2}\int_0^{\pi}\cos((n+m)\theta)\,d\theta + \frac{1}{2}\int_0^{\pi}\cos((n-m)\theta)\,d\theta$$

$$= \left[\frac{1}{2(n+m)}\sin((n+m)\theta) + \frac{1}{2(n-m)}\sin((n-m)\theta)\right]_0^{\pi}$$

$$= 0.$$

By a similar technique (see Exercise 14) it can also be shown that

(7.12)
$$\int_{-1}^{1} \frac{[T_n(x)]^2}{\sqrt{1-x^2}}\, dx = \frac{\pi}{2} \quad \text{for each } n \geq 1. \qquad \blacksquare$$

Exercise Set 7.2

1. Find the least-squares polynomial approximation of degree one to $f(x)$ on the indicated interval if

   a) $f(x) = x^2 - 2x + 3$, $[0, 1]$.   b) $f(x) = x^3 - 1$, $[0, 2]$.

   c) $f(x) = 1/x$, $[1, 3]$.   d) $f(x) = e^{-x}$, $[0, 1]$.

   e) $f(x) = \cos \pi x$, $[0, 1]$.   f) $f(x) = \ln x$, $[1, 2]$.

2. Find the least-squares polynomial approximation of degree two to the functions and intervals in Exercise 1.

3. Find the least-squares polynomial of degree one on the interval $[-1, 1]$ for the functions listed in Exercise 1, parts (a), (b), (d), and (e), using Theorem 7.5 and the Legendre polynomials. Also, compute the error.

4. Find the least-squares polynomial of degree two on the interval $[-1, 1]$ for the functions listed in Exercise 1, parts (a), (b), (d), and (e), using Theorem 7.5 and the Legendre polynomials. Also, compute the error.

5. Find the trigonometric least-squares approximating polynomial for $f(x) = x$ on $[-\pi, \pi]$, using Example 3 with $n = 2$.

6. Find the general least-squares trigonometric polynomial, $S_n(x)$, for

$$f(x) = \begin{cases} -1, & \text{if } -\pi < x < 0, \\ 1, & \text{if } 0 < x < \pi. \end{cases}$$

7. Obtain the least-squares approximating polynomial of degree three for each of the following functions, using Theorems 7.5 and 7.6. [*Note:* The integration required in this exercise is quite time consuming and might best be done numerically.]

   a) $f(x) = \cos x$, $0 \leq x \leq 1$; $w(x) = 1$.

   b) $f(x) = \ln x$, $1 \leq x \leq 2$; $w(x) = 1$.

   c) $f(x) = x^4$, $0 \leq x \leq 1$; $w(x) = e^x$.

   d) $f(x) = e^{-0.1x}$, $0 \leq x \leq 1$; $w(x) = 1$.

8. Use the technique presented in Theorem 7.6 to calculate L_1, L_2, and L_3 where $\{L_0, L_1, L_2, L_3\}$ is an orthogonal set of polynomials defined on $(0, \infty)$ with respect to the weight function $w(x) = e^{-x}$ and $L_0(x) \equiv 1$. The polynomials obtained from this procedure are called the **Laguerre polynomials**.

9. Use the Laguerre polynomials and Theorem 7.5 to compute the least-squares polynomial of degree three on the interval $(0, \infty)$ with respect to the weight function $w(x) = e^{-x}$ for the following functions.

a) $f(x) = x^2$.   b) $f(x) = e^{-x}$.

c) $f(x) = x^3$.   d) $f(x) = e^{-2x}$.

10. Show that the normal equations (7.8) have a unique solution. [*Hint:* Show that the only solution for the function $f(x) \equiv 0$ is $a_j = 0$, $j = 0, 1, \ldots, n$. Multiply Eq. (7.8) by a_j and sum over all j. Interchange the integral sign and the summation sign to obtain $\int_a^b [P(x)]^2 \, dx = 0$. Thus, $P(x) = 0$ or $a_j = 0$ for $j = 0, \ldots, n$. Hence, the coefficient matrix is nonsingular, and there is a unique solution to Eq. (7.8)].

11. Suppose $\{\phi_0, \phi_1, \ldots, \phi_n\}$ is any linearly independent set in Π_n. Show that, for any element $Q \in \Pi_n$, there exist unique constants $c_0, c_1, \ldots, c_n$ such that

$$Q(x) = \sum_{k=0}^n c_k \phi_k(x).$$

12. Show that, if $\{\phi_0, \ldots, \phi_n\}$ is an orthogonal set of functions on $[a, b]$ with respect to the weight function w, then $\{\phi_0, \ldots, \phi_n\}$ is a linearly independent set.

13. Show that the functions $\phi_0(x) = 1/\sqrt{2\pi}$, $\phi_1(x) = (1/\sqrt{\pi})\cos x$, $\ldots$, $\phi_n(x) = (1/\sqrt{\pi}) \cos nx$, $\phi_{n+1}(x) = (1/\sqrt{\pi}) \sin x$, $\ldots$, $\phi_{2n-1}(x) = (1/\sqrt{\pi}) \sin(n-1)x$ are orthogonal on $[-\pi, \pi]$ with respect to $w(x) \equiv 1$. [*Hint:* Use the trigonometric identities for $\cos(mx \pm nx)$ and $\sin(mx \pm nx)$ to simplify the integrals involved.]

14. Show that

$$\int_{-1}^1 \frac{[T_n(x)]^2}{\sqrt{1 - x^2}} \, dx = \frac{\pi}{2} \qquad \text{for each Chebyshev polynomial } T_n(x).$$

7.3 Chebyshev Polynomials and Economization of Power Series

In this section we will continue the study of the Chebyshev polynomials initiated in Example 5 of Section 7.2. This study leads to the following results:

i) an optimal placing of interpolating points to minimize the error in Lagrange interpolation;

ii) a means of reducing the degree of an approximating polynomial with minimal loss of accuracy.

Recall that the definition of the Chebyshev polynomial is

(7.13) $T_n(x) = \cos[n \arccos x]$ for $x \in [-1, 1]$ and $n = 0, 1, 2, \ldots$.

The substitution $\theta = \arccos x$ implies that

(7.14) $T_n(x) = \cos(n\theta)$ for $\theta \in [0, \pi]$ and $n = 0, 1, 2, \ldots$.

With the aid of Eq. (7.14) and various trigonometric identities, it was shown in Example 5 of Section 7.2 that:

(7.15) $T_{n+1}(x) = 2x T_n(x) - T_{n-1}(x)$ for each $n = 1, 2, \ldots$ and $x \in [-1, 1]$.

Using the fact that $T_0(x) = 1$ and $T_1(x) = x$, this recurrence relation implies that, for each $n \geq 1$, T_n is a polynomial of degree n with leading coefficient 2^{n-1}.

The first result in this section concerns the roots of the polynomial T_n.

THEOREM 7.8

The Chebyshev polynomial T_n of degree $n \geq 1$ has n simple zeros in $[-1, 1]$ at

$$(7.16) \qquad \bar{x}_k = \cos\left(\frac{2k-1}{2n}\pi\right) \qquad \text{for each } k = 1, 2, \dots, n.$$

Moreover, T_n assumes its absolute extrema at

$$(7.17) \qquad \bar{x}'_k = \cos\left(\frac{k\pi}{n}\right) \qquad \text{for each } k = 0, \dots, n,$$

with

$$(7.18) \qquad T_n(\bar{x}'_k) = (-1)^k \qquad \text{for each } k = 0, 1, \dots, n.$$

PROOF

If

$$\bar{x}_k = \cos\left(\frac{2k-1}{2n}\pi\right) \qquad \text{for } k = 1, 2, \dots, n,$$

then

$$T_n(\bar{x}_k) = \cos(n \arccos \bar{x}_k) = \cos\left(n \arccos\left(\cos\left(\frac{2k-1}{2n}\pi\right)\right)\right) = \cos\left(\frac{2k-1}{2}\pi\right) = 0;$$

so $\bar{x}_k$ is a zero of T_n for each $k = 1, 2, \dots, n$. Since T_n is a polynomial of degree n, all zeros of T_n must be of this form.

To show the second part of this theorem, note first that

$$T'_n(x) = \frac{d}{dx}[\cos(n \arccos x)] = \frac{n \sin(n \arccos x)}{\sqrt{1 - x^2}},$$

and that, when $1 \leq k \leq n-1$,

$$T'_n(\bar{x}'_k) = \frac{n \sin\left(n \arccos\left(\cos\left(\frac{k\pi}{n}\right)\right)\right)}{\sqrt{1 - \left[\cos\left(\frac{k\pi}{n}\right)\right]^2}} = \frac{n \sin(k\pi)}{\sin\left(\frac{k\pi}{n}\right)} = 0.$$

Since T_n is a polynomial of degree n, T'_n is a polynomial of degree $(n-1)$ and all zeros of T'_n occur at these points. The only other possibilities for extrema of the function T_n occur at the endpoints of the interval $[-1, 1]$; that is, at $\bar{x}'_0 = 1$ and $\bar{x}'_n = -1$. Since for any $k = 0, 1, \dots, n$,

$$T_n(\bar{x}'_k) = \cos\left(n \arccos\left(\cos\left(\frac{k\pi}{n}\right)\right)\right) = \cos(k\pi) = (-1)^k,$$

a maximum occurs at each even value of k and a minimum at each odd value. $\qquad \square$

The monic Chebyshev polynomials (polynomials with leading coefficient 1) $\tilde{T}_n$ can be derived from the Chebyshev polynomials T_n simply by dividing by their leading coefficient 2^{n-1}:

(7.19) $$\tilde{T}_0(x) = 1, \qquad \tilde{T}_n(x) = 2^{1-n} T_n(x) \qquad \text{for each } n \geq 1.$$

It is not difficult to see that these polynomials satisfy the recurrence relation

(7.20) $$\tilde{T}_2(x) = x\tilde{T}_1(x) - \tfrac{1}{2}\tilde{T}_0(x); \qquad \tilde{T}_{n+1}(x) = x\tilde{T}_n(x) - \tfrac{1}{4}\tilde{T}_{n-1}(x),$$

for each $n \geq 2$.

Because of the linear relationship between $\tilde{T}_n$ and T_n, Theorem 7.8 implies that the zeros of $\tilde{T}_n$ also occur at

$$\bar{x}_k = \cos\left(\frac{2k-1}{2n}\pi\right) \qquad \text{for each } k = 1, 2, \ldots, n,$$

and the extreme values of $\tilde{T}_n$ occur at

$$\bar{x}'_k = \cos\left(\frac{k\pi}{n}\right) \qquad \text{for each } k = 0, 1, 2, \ldots, n.$$

At these extreme values, for $n \geq 1$,

(7.21) $$\tilde{T}_n(\bar{x}'_k) = \frac{(-1)^k}{2^{n-1}} \qquad \text{for each } k = 0, 1, 2, \ldots, n.$$

Suppose $\tilde{\Pi}_n$ denotes **the set of all monic polynomials of degree n**. The relation expressed in Eq. (7.21) leads to an important minimization property that distinguishes the polynomials $\tilde{T}_n$ from the other members of $\tilde{\Pi}_n$.

THEOREM 7.9   The polynomials of the form $\tilde{T}_n$, when $n \geq 1$, have the property that

(7.22) $$\frac{1}{2^{n-1}} = \max_{x \in [-1, 1]} |\tilde{T}_n(x)| \leq \max_{x \in [-1, 1]} |P_n(x)| \qquad \text{for all } P_n \in \tilde{\Pi}_n.$$

Equality in (7.22) can occur only if $P_n = \tilde{T}_n$.

PROOF   Suppose $P_n \in \tilde{\Pi}_n$ and

$$\max_{x \in [-1, 1]} |P_n(x)| \leq \frac{1}{2^{n-1}} = \max_{x \in [-1, 1]} |\tilde{T}_n(x)|.$$

Let $Q = \tilde{T}_n - P_n$. Since $\tilde{T}_n$ and P_n are both monic polynomials of degree n, Q is a polynomial of degree at most $(n-1)$. Moreover, at the extreme points of $\tilde{T}_n$,

$$Q(\bar{x}'_k) = \tilde{T}_n(\bar{x}'_k) - P_n(\bar{x}'_k) = \frac{(-1)^k}{2^{n-1}} - P_n(\bar{x}'_k).$$

The fact that

$$|P_n(\bar{x}'_k)| \leq \frac{1}{2^{n-1}}$$

implies that, for $k = 0, 1, \ldots, n$,

$$Q(\bar{x}'_k) \le 0, \qquad \text{when } k \text{ is odd,}$$

and $\qquad\qquad\qquad Q(\bar{x}'_k) \ge 0, \qquad \text{when } k \text{ is even.}$

Since Q is continuous, the Intermediate Value Theorem implies that the polynomial Q must have at least n zeros in the interval $[-1, 1]$. This is clearly impossible unless $Q \equiv 0$. This implies $P_n = \tilde{T}_n$, which establishes the result. $\qquad\qquad\square$

Theorem 7.9 can be used to answer the question of where to place interpolating nodes to minimize the error in Lagrange interpolation. Theorem 3.3 (p. 87), applied to the interval $[-1, 1]$, states that, if $x_0, \ldots, x_n$ are distinct numbers in the interval $[-1, 1]$ and $f \in C^{n+1}[-1, 1]$, then, for each $x \in [-1, 1]$, $\xi(x)$ in $(-1, 1)$ exists with

$$(7.23) \qquad f(x) = P(x) + \frac{f^{(n+1)}(\xi(x))}{(n+1)!}(x - x_0)(x - x_1) \cdots (x - x_n),$$

where P is the Lagrange interpolating polynomial. To minimize the error by shrewd placement of the nodes $x_0, \ldots, x_n$ would, in general, be equivalent to finding the $x_0, \ldots, x_n$ that would minimize the quantity

$$|(x - x_0)(x - x_1) \cdots (x - x_n)|$$

throughout the interval $[-1, 1]$. Since $(x - x_0)(x - x_1) \cdots (x - x_n)$ is a monic polynomial of degree $(n+1)$, Theorem 7.9 implies that this minimum is obtained if and only if

$$(x - x_0)(x - x_1) \cdots (x - x_n) = \tilde{T}_{n+1}(x).$$

When x_k is chosen to be the $(k+1)$st zero of $\tilde{T}_{n+1}$ for each $k = 0, 1, \ldots, n$, that is, when x_k is chosen to be

$$(7.24) \qquad\qquad \bar{x}_{k+1} = \cos \frac{2k+1}{2(n+1)} \pi,$$

the maximum value of $|(x - x_0)(x - x_1) \cdots (x - x_n)|$ will be minimized. Since $\max_{x \in [-1, 1]} |\tilde{T}_{n+1}(x)| = 2^{-n}$, this implies that

$$\frac{1}{2^n} = \max_{x \in [-1, 1]} |(x - \bar{x}_1)(x - \bar{x}_2) \cdots (x - \bar{x}_{n+1})|$$

$$\le \max_{x \in [-1, 1]} |(x - x_0)(x - x_1) \cdots (x - x_n)|,$$

for any choice of $x_0, x_1, \ldots, x_n$ in the interval $[-1, 1]$.

This technique for choosing points that will minimize the interpolating error can be easily extended to a general closed interval $[a, b]$ by using the change of variable $\tilde{x} = \frac{1}{2}[(b - a)x + a + b]$ to transform the numbers $\bar{x}_k$ in the interval $[-1, 1]$ into the corresponding numbers $\tilde{x}_k$ in the interval $[a, b]$.

EXAMPLE 1   Let $f(x) = xe^x$ on $[0, 1.5]$. Two interpolation polynomials of degree at most three will be constructed. First, the equally spaced nodes $x_0 = 0$, $x_1 = 0.5$, $x_2 = 1$, and $x_3 = 1.5$ are used to give

$$L_0(x) = -1.3333x^3 + 4.0000x^2 - 3.6667x + 1,$$

$$L_1(x) = 4.0000x^3 - 10.000x^2 + 6.0000x,$$

$$L_2(x) = -4.0000x^3 + 8.0000x^2 - 3.0000x,$$

$$L_3(x) = 1.3333x^3 - 2.0000x^2 + 0.66667x.$$

With the values listed in Table 7.9, the interpolating polynomial is given by

$$P_3(x) = 1.3875x^3 + 0.057570x^2 + 1.2730x.$$

TABLE 7.9

x	$f(x) = xe^x$
0.0	0.00000
0.5	0.824361
1.0	2.71828
1.5	6.72253

TABLE 7.10

x	$f(x) = xe^x$
$\tilde{x}_0 = 1.44291$	6.10783
$\tilde{x}_1 = 1.03701$	2.92517
$\tilde{x}_2 = 0.46299$	0.73560
$\tilde{x}_3 = 0.05709$	0.060444

For the second interpolating polynomial, shift the zeros $\bar{x}_k = \cos((2k+1)/8)\pi$ $k = 0, 1, 2, 3$ of $\tilde{T}_4$ from $[-1, 1]$ to $[0, 1.5]$, using the linear transformation

$$\tilde{x}_k = 0.75 + 0.75\bar{x}_k$$

to obtain

$$\tilde{x}_0 = 1.44291, \quad \tilde{x}_1 = 1.03701, \quad \tilde{x}_2 = 0.46299, \quad \text{and} \quad \tilde{x}_3 = 0.05709.$$

The Lagrange coefficient polynomials for this set of nodes are then computed as:

$$\tilde{L}_0(x) = 1.8142x^3 - 2.8249x^2 + 1.0264x - 0.049728,$$

$$\tilde{L}_1(x) = -4.3799x^3 + 8.5977x^2 - 3.4026x + 0.16705,$$

$$\tilde{L}_2(x) = 4.3799x^3 - 11.112x^2 + 7.1738x - 0.37415,$$

$$\tilde{L}_3(x) = -1.8142x^3 + 5.3390x^2 - 4.7976x + 1.2568.$$

The functional values required for these polynomials are given in Table 7.10, and the interpolation polynomial of degree at most three is given by

$$\tilde{P}_3(x) = 1.3811x^3 + 0.044445x^2 + 1.3030x - 0.014357.$$

For comparison, Table 7.11 lists various values of x, together with the values of $f(x)$, $P_3(x)$, and $\tilde{P}_3(x)$. It can be seen from this table that, although the error using P_3 is less than using $\tilde{P}_3$ near the middle of the table, the maximum error involved with using $\tilde{P}_3$ is considerably less.   ∎

TABLE 7.11

x	$f(x) = xe^x$	$P_3(x)$	$\|xe^x - P_3(x)\|$	$\tilde{P}_3(x)$	$\|xe^x - \tilde{P}_3(x)\|$
0.15	0.1743	0.1969	0.0226	0.1868	0.0125
0.25	0.3210	0.3435	0.0225	0.3358	0.0148
0.35	0.4967	0.5121	0.0154	0.5064	0.0097
0.65	1.245	1.233	0.0120	1.231	0.0140
0.75	1.588	1.572	0.0160	1.571	0.0170
0.85	1.989	1.976	0.0130	0.973	0.0160
1.15	3.632	3.650	0.0180	3.643	0.0110
1.25	4.363	4.391	0.0280	4.381	0.0180
1.35	5.208	5.237	0.0290	5.224	0.0160

The following corollary follows immediately from Theorem 7.9 and Eq. (7.23).

COROLLARY 7.10   If P is the interpolating polynomial of degree at most n with nodes at the roots of $T_{n+1}(x)$, then

$$\max_{x \in [-1, 1]} |f(x) - P(x)| \leq \frac{1}{2^n(n+1)!} \max_{x \in [-1, 1]} |f^{(n+1)}(x)| \quad \text{for each } f \in C^{n+1}[-1, 1].$$

Chebyshev polynomials can also be used to reduce the degree of an approximating polynomial with a minimal loss of accuracy. This technique is particularly useful when the approximating polynomial being used is a Taylor polynomial. Although Taylor polynomials are very accurate near the number about which they are expanded, their accuracy deteriorates rapidly when they are employed in situations farther from this point. For this reason, a high-degree Taylor polynomial may be needed to achieve a prescribed error tolerance. Because the Chebyshev polynomials have minimum maximum-absolute value that is spread uniformly on an interval, they can be used to reduce the degree of the Taylor polynomial without exceeding the error tolerance. The following example illustrates the technique involved.

EXAMPLE 2   The function $f(x) = e^x$ can be approximated on the interval $[-1, 1]$ by the fourth-degree Taylor polynomial expanded about zero.

$$P_4(x) = 1 + x + \frac{x^2}{2} + \frac{x^3}{6} + \frac{x^4}{24},$$

with an error

$$|R_4(x)| = \frac{|f^{(5)}(\xi(x))||x^5|}{120} \leq \frac{e}{120} \approx 0.023 \quad \text{for } -1 \leq x \leq 1.$$

Suppose that an actual error of 0.05 is tolerable and consider the situation that occurs if we replaced the term in the Taylor polynomial involving x^4 by the equivalent Chebyshev polynomials of degree less than or equal to four. Before continuing with this example, we list the explicit representation of x^k in terms of T_0, T_1, ..., T_k for some of the smaller positive integers k. The verification of Table 7.12 is considered in Exercise 7.

k	T_k	x^k
0	1	T_0
1	x	T_1
2	$2x^2 - 1$	$\frac{1}{2}T_0 + \frac{1}{2}T_2$
3	$4x^3 - 3x$	$\frac{3}{4}T_1 + \frac{1}{4}T_3$
4	$8x^4 - 8x^2 + 1$	$\frac{3}{8}T_0 + \frac{1}{2}T_2 + \frac{1}{8}T_4$
5	$16x^5 - 20x^3 + 5x$	$\frac{5}{8}T_1 + \frac{5}{16}T_3 + \frac{1}{16}T_5$
6	$32x^6 - 48x^4 + 18x^2 - 1$	$\frac{5}{16}T_0 + \frac{15}{32}T_2 + \frac{3}{16}T_4 + \frac{1}{32}T_6$
7	$64x^7 - 112x^5 + 56x^3 - 7x$	$\frac{35}{64}T_1 + \frac{21}{64}T_3 + \frac{7}{64}T_5 + \frac{1}{64}T_7$

Thus,

$$
\begin{aligned}
P_4(x) &= 1 + x + \tfrac{1}{2}x^2 + \tfrac{1}{6}x^3 + \tfrac{1}{24}[\tfrac{3}{8}T_0(x) + \tfrac{1}{2}T_2(x) + \tfrac{1}{8}T_4(x)] \\
&= 1 + x + \tfrac{1}{2}x^2 + \tfrac{1}{6}x^3 + \tfrac{1}{64}T_0(x) + \tfrac{1}{48}T_2(x) + \tfrac{1}{192}T_4(x) \\
&= 1 + x + \tfrac{1}{2}x^2 + \tfrac{1}{6}x^3 + \tfrac{1}{64} + \tfrac{1}{48}(2x^2 - 1) + \tfrac{1}{192}T_4(x) \\
&= (1 + \tfrac{1}{64} - \tfrac{1}{48}) + x + (\tfrac{1}{2} + \tfrac{1}{24})x^2 + \tfrac{1}{6}x^3 + \tfrac{1}{192}T_4(x) \\
&= \tfrac{191}{192} + x + \tfrac{13}{24}x^2 + \tfrac{1}{6}x^3 + \tfrac{1}{192}T_4(x).
\end{aligned}
$$

But
$$
\max_{x \in [-1, 1]} |T_4(x)| = 1, \qquad |\tfrac{1}{192}T_4(x)| \le \tfrac{1}{192} = 0.0053,
$$

and
$$
|R_4(x)| + |\tfrac{1}{192}T_4(x)| \le 0.023 + 0.0053 = 0.0283
$$

is still less than the tolerance 0.05. Consequently, the fourth-degree term $(1/192)T_4(x)$ can be omitted from the approximating polynomial and the desired accuracy will still be obtained. The third-degree polynomial,

$$
P_3(x) = \tfrac{191}{192} + x + \tfrac{13}{24}x^2 + \tfrac{1}{6}x^3,
$$

will give the desired accuracy on the interval $[-1, 1]$.

To attempt to eliminate the third-degree term requires replacing x^3 with $\frac{3}{4}T_1(x) + \frac{1}{4}T_3(x)$ and results in

$$
\begin{aligned}
P_3(x) &= \tfrac{191}{192} + x + \tfrac{13}{24}x^2 + \tfrac{1}{6}[\tfrac{3}{4}T_1(x) + \tfrac{1}{4}T_3(x)] \\
&= \tfrac{191}{192} + \tfrac{9}{8}x + \tfrac{13}{24}x^2 + \tfrac{1}{24}T_3(x).
\end{aligned}
$$

However, $\max_{x \in [-1, 1]} |\tfrac{1}{24}T_3(x)| = 0.0417$, which, when combined with the possible error of 0.0283 obtained previously, gives an error bound of 0.07, and exceeds the tolerance of 0.05. $P_3(x)$ is therefore the lowest-degree polynomial suitable for this approximation. Table 7.13 lists the function and the approximating polynomials at some representative points where $P_2(x)$ represents the polynomial

$$
P_2(x) = \tfrac{191}{192} + \tfrac{9}{8}x + \tfrac{13}{24}x^2.
$$

TABLE 7.13

x	e^x	$P_4(x)$	$P_3(x)$	$P_2(x)$
-0.75	0.47237	0.47412	0.47917	0.45573
-0.25	0.77880	0.77881	0.77604	0.74740
0.00	1.00000	1.00000	0.99479	0.99479
0.25	1.28403	1.28402	1.28125	1.30990
0.75	2.11700	2.11475	2.11979	2.14323

Even though the error bound for the polynomial P_2 was greater than the tolerance of 0.05, note that the tabulated entries are well within that bound.   ∎

Exercise Set 7.3

1. Use the zeros of $\tilde{T}_3$ to construct an interpolating polynomial of degree two for the functions and intervals given.

   a) $f(x) = e^x$, $[-1, 1]$.   b) $f(x) = \sin x$, $[0, \pi]$.

   c) $f(x) = \ln x$, $[1, 2]$.   d) $f(x) = 1 + x$, $[0, 1]$.

2. Use the zeros of $\tilde{T}_4$ to construct an interpolating polynomial of degree at most three for the functions and intervals given in Exercise 1.

3. Find the third-degree interpolating polynomial P_3 to

$$f(x) = \sqrt{1 + x}, \qquad -1 \le x \le 1,$$

   using the roots of T_4 as interpolating nodes. Compute $P_3(0.1)$, and compare to Example 1 of Section 3.1.

4. Find the third-degree interpolating polynomial P_3 to $f(x) = \sqrt{1 + x}$, $0 \le x \le 1$, using the roots of T_4 and the transformation of the interval $[0, 1]$ onto $[-1, 1]$. Compute $P_3(0.1)$, and compare to Exercise 3 and to Example 1 of Section 3.1.

5. Find the sixth-degree Taylor polynomial for xe^x, and use Chebyshev economization to obtain a lesser-degree polynomial approximation while keeping the error less than 0.01 on $[-1, 1]$.

6. Find the sixth-degree Taylor polynomial for $\sin x$, and use Chebyshev economization to obtain a lesser-degree polynomial approximation while keeping the error less than 0.01 on the interval $[-1, 1]$.

7. Derive the entries in Table 7.12.

8. Show that for any positive integers i and j

$$T_i(x)T_j(x) = \tfrac{1}{2}[T_{i+j}(x) + T_{|i-j|}(x)].$$

7.4 Rational Function Approximation

The class of algebraic polynomials has some distinct advantages for use in approximation. There is a sufficient number of polynomials to approximate any continuous function on a closed interval to within an arbitrary tolerance, polynomials are easily evaluated at arbitrary values, and the derivatives and integrals of polynomials exist and are easily determined. The disadvantages of using polynomials for approximation is their tendency for oscillation. This tendency often causes error bounds in polynomial approximation to significantly exceed the average approximation error, since error bounds are determined by the maximum approximation error. To find techniques that decrease approximation error bounds, we will consider methods that more evenly spread the approximation error over the approximation interval. These techniques require the introduction of a new class of approximating functions, the class of **rational functions.**

A rational function r of degree N is a function of the form

$$r(x) = \frac{p(x)}{q(x)}$$

where p and q are polynomials whose degrees sum to N.

Since every polynomial is also a rational function (simply let $q(x) \equiv 1$), approximation by rational functions will give results with no greater error bounds than approximation by polynomials. In fact, rational functions whose numerator and denominator have the same or nearly the same degree generally produce approximation results superior to polynomial methods for the same amount of computation effort. (This statement is based on the assumption that the amount of computation effort required for division is approximately the same as for multiplication. This assumption is valid in many, but not all, computing systems.) Rational functions have the added advantage of permitting efficient approximation of functions that have infinite discontinuities near, but outside, the interval of approximation. Polynomial approximation is generally unacceptable in this situation.

Suppose that r is a rational function of degree $N = n + m$ of the form

$$r(x) = \frac{p(x)}{q(x)} = \frac{p_0 + p_1 x + \cdots + p_n x^n}{q_0 + q_1 x + \cdots + q_m x^m}$$

that will be used to approximate a function f on a closed interval I containing zero. For r to be defined at zero requires that $q_0 \neq 0$. In fact, we can assume that $q_0 = 1$, for if this is not the case we can simply replace $p(x)$ by $p(x)/q_0$ and $q(x)$ by $q(x)/q_0$. Consequently, there are $N + 1$ parameters $q_1, q_2, \ldots, q_m, p_0, p_1, \ldots, p_n$ available for the approximation of f by r.

The **Padé approximation technique** chooses the $N + 1$ parameters so that $f^{(k)}(0) = r^{(k)}(0)$ for each $k = 0, 1, \ldots, N$. Padé approximation is the extension of Taylor polynomial approximation to rational functions. When $n = N$ and $m = 0$, the Padé approximation is the Taylor polynomial of degree N expanded about zero, that is, the Maclaurin polynomial of degree N.

Consider the function

$$f(x) - r(x) = f(x) - \frac{p(x)}{q(x)} = \frac{f(x)q(x) - p(x)}{q(x)} = \frac{f(x)\sum_{i=0}^{m} q_i x^i - \sum_{i=0}^{n} p_i x^i}{q(x)}$$

and suppose f has the Maclaurin series expansion $f(x) = \sum_{i=0}^{\infty} a_i x^i$. Then

(7.25)
$$f(x) - r(x) = \frac{\sum_{i=0}^{\infty} a_i x^i \sum_{i=0}^{m} q_i x^i - \sum_{i=0}^{n} p_i x^i}{q(x)}.$$

The object is to choose the constants $q_1, q_2, \ldots, q_m$ and $p_0, p_1, \ldots, p_n$ so that

$$f^{(k)}(0) - r^{(k)}(0) = 0 \qquad \text{for each } k = 0, 1, \ldots, N.$$

In Section 2.4 (see, in particular, Exercise 14) we found that this is equivalent to $f - r$ having a root of multiplicity $N + 1$ at zero. As a consequence, we want to choose $q_1, q_2, \ldots, q_m$ and $p_0, p_1, \ldots, p_n$ so that the numerator on the right side of Eq. (7.25):

(7.26) $$(a_0 + a_1 x + \cdots)(1 + q_1 x + \cdots + q_m x^m) - (p_0 + p_1 x + \cdots + p_n x^n)$$

has no terms of degree less than or equal to N. To simplify notation, let us define $p_{n+1} = p_{n+2} = \cdots = p_N = 0$ and $q_{m+1} = q_{m+2} = \cdots = q_N = 0$. We can then express the coefficient of x^k in expression (7.26) as

$$\sum_{i=0}^{k} a_i q_{k-i} - p_k;$$

so the rational function for Padé approximation results from the solution of the $N + 1$ linear equations

$$\sum_{i=0}^{k} a_i q_{k-i} - p_k = 0, \qquad k = 0, 1, \ldots, N$$

in the $N + 1$ unknowns $q_1, q_2, \ldots, q_m, p_0, p_1, \ldots, p_n$.

EXAMPLE 1   The Maclaurin series expansion for e^{-x} is

$$\sum_{i=0}^{\infty} \frac{(-1)^i}{i!} x^i.$$

To find the Padé approximation to e^{-x} of degree five with $n = 3$ and $m = 2$ requires choosing $p_0, p_1, p_2, p_3, q_1,$ and q_2 so that the coefficients of x^k for $k = 0, 1, \ldots, 5$ are zero in the expression

$$\left(1 - x + \frac{x^2}{2} - \frac{x^3}{6} + \cdots\right)(1 + q_1 x + q_2 x^2) - (p_0 + p_1 x + p_2 x^2 + p_3 x^3).$$

Expanding and collecting terms produces

$$x^5: \quad -\tfrac{1}{120} + \tfrac{1}{24}q_1 - \tfrac{1}{6}q_2 = 0; \qquad x^2: \quad \tfrac{1}{2} - q_1 + q_2 = p_2;$$

$$x^4: \quad \tfrac{1}{24} - \tfrac{1}{6}q_1 + \tfrac{1}{2}q_2 = 0; \qquad x^1: \quad -1 + q_1 = p_1:$$

$$x^3: \quad -\tfrac{1}{6} + \tfrac{1}{2}q_1 - q_2 = p_3; \qquad x^0: \quad 1 = p_0.$$

The solution to this system is:

$$p_0 = 1, \quad p_1 = -\tfrac{3}{5}, \quad p_2 = \tfrac{3}{20}, \quad p_3 = -\tfrac{1}{60}, \quad q_1 = \tfrac{2}{5}, \quad \text{and} \quad q_2 = \tfrac{1}{20};$$

so the Padé approximation is

$$r(x) = \frac{1 - \tfrac{3}{5}x + \tfrac{3}{20}x^2 - \tfrac{1}{60}x^3}{1 + \tfrac{2}{5}x + \tfrac{1}{20}x^2}.$$

Table 7.14 lists values of $r(x)$ and $P_5(x)$, the fifth-degree Taylor polynomial about $x = 0$. The Padé approximation is clearly superior in this example. ∎

TABLE 7.14

x	e^{-x}	$P_5(x)$	$|e^{-x} - P_5(x)|$	$r(x)$	$|e^{-x} - r(x)|$
0.2	0.81873075	0.81873067	8.64×10^{-8}	0.81873075	7.55×10^{-9}
0.4	0.67032005	0.67031467	5.38×10^{-6}	0.67031963	4.11×10^{-7}
0.6	0.54881164	0.54875200	5.96×10^{-5}	0.54880763	4.00×10^{-6}
0.8	0.44932896	0.44900267	3.26×10^{-4}	0.44930966	1.93×10^{-5}
1.0	0.36787944	0.36666667	1.21×10^{-3}	0.36781609	6.33×10^{-5}

It is also interesting to compare the number of arithmetic operations required for calculations of $P_5(x)$ and $r(x)$ in Example 1. Using nested multiplication, $P_5(x)$ can be expressed as

$$P_5(x) = 1 - x(1 - x(\tfrac{1}{2} - x(\tfrac{1}{6} - x(\tfrac{1}{24} - \tfrac{1}{120}x)))).$$

Assuming that the coefficients of 1, x, x^2, x^3, x^4, and x^5 are represented as decimals, a single calculation of $P_5(x)$ in nested form requires five multiplications and five additions/subtractions. Using nested multiplication $r(x)$ can be expressed as

$$r(x) = \frac{1 - x(\tfrac{3}{5} - x(\tfrac{3}{20} - \tfrac{1}{60}x))}{1 + x(\tfrac{2}{5} + \tfrac{1}{20}x)};$$

so a single calculation of $r(x)$ requires five multiplications, five additions/subtractions, and one division. Hence, computational effort appears to favor the polynomial approximation. However, by re-expressing $r(x)$ by continued division, we can write

$$r(x) = \frac{1 - \tfrac{3}{5}x + \tfrac{3}{20}x^2 - \tfrac{1}{60}x^3}{1 + \tfrac{2}{5}x + \tfrac{1}{20}x^2}$$

$$= \frac{-\tfrac{1}{3}x^3 + 3x^2 - 12x - 20}{x^2 + 8x + 20}$$

$$= -\tfrac{1}{3}x + \tfrac{17}{3} + \frac{(-\tfrac{152}{3}x - \tfrac{280}{3})}{(x^2 + 8x + 20)}$$

$$= -\tfrac{1}{3}x + \tfrac{17}{3} + \frac{-\tfrac{152}{3}}{\left(\dfrac{x^2 + 8x + 20}{x + \tfrac{35}{19}}\right)}$$

or

$$(7.27) \qquad r(x) = -\tfrac{1}{3}x + \tfrac{17}{3} + \cfrac{-\tfrac{152}{3}}{\left(x + \tfrac{117}{19} + \cfrac{\tfrac{3125}{361}}{(x + \tfrac{35}{19})}\right)}.$$

Written in this form, a single calculation of $r(x)$ requires one multiplication, five addition/subtractions, and two divisions. If the amount of computation required for division is approximately the same as for multiplication, the computational effort required for an evaluation of $P_5(x)$ should significantly exceed that required for an evaluation of $r(x)$.

Expressing a rational function approximation in a form such as Eq. (7.27) is called **continued fraction** approximation. This is a classical approximation technique that is of current interest because of the computational efficiency of this representation. It is, however, a specialized technique, and one we will not discuss further. A rather extensive treatment of this subject, and of rational approximation in general, can be found in Ralston and Rabinowitz [103], pp. 285–322.

Although the rational function approximation in Example 1 gave results superior to the polynomial approximation of the same degree, the approximation has a wide variation in accuracy; the approximation at 0.2 is accurate to within 8×10^{-9} while, at 1.0, the approximation and the function agree only to within 7×10^{-5}. This accuracy variation is not unexpected because the Padé approximation is based on a Taylor polynomial representation of e^{-x}, and this representation has a wide variation of accuracy in [0.2, 1.0].

To obtain more uniformly accurate rational function approximations, we will use a class of polynomials that exhibit more uniform behavior on the interval $[-1, 1]$, the set of Chebyshev polynomials. The general Chebyshev rational function approximation method proceeds in the same manner as Padé approximation except that each x^k term in the Padé approximation is replaced by the kth degree Chebyshev polynomial T_k.

Suppose we want to approximate the function f by an Nth degree rational function r written in the form

$$r(x) = \frac{\sum_{k=0}^n p_k T_k(x)}{\sum_{k=0}^m q_k T_k(x)} \qquad \text{where } N = n + m \text{ and } q_0 = 1.$$

Writing $f(x)$ in a series involving Chebyshev polynomials gives

$$f(x) - r(x) = \sum_{k=0}^{\infty} a_k T_k(x) - \frac{\sum_{k=0}^n p_k T_k(x)}{\sum_{k=0}^m q_k T_k(x)}$$

or

$$(7.28) \qquad f(x) - r(x) = \frac{\left[\sum_{k=0}^{\infty} a_k T_k(x)\right]\left[\sum_{k=0}^m q_k T_k(x)\right] - \sum_{k=0}^n p_k T_k(x)}{\sum_{k=0}^m q_k T_k(x)}.$$

The coefficients $q_1, q_2, \ldots, q_m$ and $p_0, p_1, \ldots, p_n$ are chosen so that the numerator on the right-hand side of Eq. (7.28) has zero coefficients for $T_k(x)$ when $k = 0, 1, \ldots, N$, that is, so that

$$(a_0 T_0(x) + a_1 T_1(x) + \cdots)(T_0(x) + q_1 T_1(x) + \cdots + q_m T_m(x))$$
$$- (p_0 T_0(x) + p_1 T_1(x) + \cdots + p_n T_n(x))$$

has no terms of degree less than or equal to N.

Two problems arise with the Chebyshev procedure that make it more difficult to implement than the Padé method. One problem occurs because the product of the polynomial $q(x)$ and the series for $f(x)$ involves products of Chebyshev polynomials. This problem is easily resolved by making use of the relationship

$$(7.29) \qquad T_i(x) T_j(x) = \tfrac{1}{2}[T_{i+j}(x) + T_{|i-j|}(x)].$$

(See Exercise 8 of Section 7.3.) The other problem is more difficult to resolve and involves the computation of the Chebyshev series for $f(x)$. In theory, this is not a difficult problem for if

$$f(x) = \sum_{k=0}^{\infty} a_k T_k(x),$$

then the orthogonality of the Chebyshev polynomials (see Example 5 of Section 7.2 for a verification) implies that

$$a_0 = \frac{1}{\pi} \int_{-1}^{1} \frac{f(x)}{\sqrt{1-x^2}} dx \quad \text{and} \quad a_k = \frac{2}{\pi} \int_{-1}^{1} \frac{f(x) T_k(x)}{\sqrt{1-x^2}} dx \quad \text{when } k \geq 1.$$

Practically, however, these integrals can seldom be evaluated in closed form and a numerical integration technique is required for each evaluation.

EXAMPLE 2   The first five terms of the Chebyshev expansion for e^{-x} can be shown to be

$$\tilde{P}_5(x) = 1.266066 T_0(x) - 1.130318 T_1(x) + 0.271495 T_2(x) - 0.044337 T_3(x)$$
$$+ 0.005474 T_4(x) - 0.000543 T_5(x).$$

To determine the Chebyshev rational approximation of degree five with $n = 3$ and $m = 2$ requires choosing p_0, p_1, p_2, p_3, q_1, and q_2 so that for $k = 0, 1, 2, 3, 4, 5$ the coefficients of $T_k(x)$ are zero in the expansion

$$\tilde{P}_5(x)[T_0(x) + q_1 T_1(x) + q_2 T_2(x)] - [p_0 T_0(x) + p_1 T_1(x) + p_2 T_2(x)].$$

Using the relation (7.29) and collecting terms gives the equations

$$T_0: \qquad 1.266066 - 0.565159 q_1 + 0.1357485 q_2 = p_0$$
$$T_1: \qquad -1.130318 + 1.401814 q_1 - 0.583275 q_2 = p_1$$
$$T_2: \qquad 0.271495 - 0.587328 q_1 + 1.268803 q_2 = p_2$$
$$T_3: \qquad -0.044337 + 0.138485 q_1 - 0.565431 q_2 = p_3$$
$$T_4: \qquad 0.005474 - 0.022440 q_1 + 0.135748 q_2 = 0$$
$$T_5: \qquad -0.000543 + 0.002737 q_1 - 0.022169 q_2 = 0.$$

The solution to this system produces the rational function

$$r_T(x) = \frac{1.055265T_0(x) - 0.613016T_1(x) + 0.077478T_2(x) - 0.004506T_3(x)}{T_0(x) + 0.378331T_1(x) + 0.022216T_2(x)}.$$

Using Table 7.12 to convert to an expression involving powers of x gives

$$r_T(x) = \frac{0.977787 - 0.599499x + 0.154956x^2 - 0.018022x^3}{0.977784 + 0.378331x + 0.044432x^2}.$$

Table 7.15 lists values of $r_T(x)$ and, for comparison purposes, the values of $r(x)$ obtained in Example 1. Note that the approximation given by $r(x)$ is superior to that of $r_T(x)$ for $x = 0.2$ and 0.4, but that the maximum error for $r(x)$ is 6.33×10^{-5} as compared to 9.13×10^{-6} for $r_T(x)$. ∎

TABLE 7.15

x	e^{-x}	$r(x)$	$|e^{-x} - r(x)|$	$r_T(x)$	$|e^{-x} - r_T(x)|$
0.2	0.81873075	0.81873075	7.55×10^{-9}	0.81872510	5.66×10^{-6}
0.4	0.67032005	0.67031963	4.11×10^{-7}	0.67031310	6.95×10^{-6}
0.6	0.54881164	0.54880763	4.00×10^{-6}	0.54881292	1.28×10^{-6}
0.8	0.44932896	0.44930966	1.93×10^{-5}	0.44933809	9.13×10^{-6}
1.0	0.36787944	0.36781609	6.33×10^{-5}	0.36787155	7.89×10^{-6}

The Chebyshev method does not produce the best rational function approximation in the sense of the approximation whose maximum approximation error is minimal. The method can, however, be used as a starting point for an iterative method known as the second Remes' algorithm that converges to the best approximation. A discussion of the techniques involved with this procedure and an improvement on this algorithm can be found in the previously mentioned book by Ralston and Rabinowitz [103] pp. 292–305.

Exercise Set 7.4

1. Determine the Padé approximation of degree five with $n = 2$ and $m = 3$ for $f(x) = e^x$. Compare the results at $x_i = 0.2i$, $i = 1, 2, 3, 4, 5$ from this approximation with those from the fifth-degree Maclaurin polynomial.

2. Repeat Exercise 1 using instead the Padé approximation of degree five with $n = 3$ and $m = 2$. Compare the results at each x_i with those computed in Exercise 1. Compare the rational function obtained with the rational function determined in Example 1.

3. Determine the Padé approximation of degree six with $n = m = 3$ for $f(x) = \sin x$. Compare the results at $x_i = 0.1i$, $i = 0, 1, \ldots, 5$ with the exact results and with the results of the sixth-degree Maclaurin polynomial.

4. Determine the Padé approximations of degree six with (a) $n = 2$, $m = 4$ and (b) $n = 4$, $m = 2$ for $f(x) = \sin x$. Compare the results at each x_i to those obtained in Exercise 3.

5. Table 7.14 lists results of the Padé approximation of degree five with $n = 2$ and $m = 3$, the Maclaurin polynomial of degree five, and the exact values of $f(x) = e^{-x}$ when $x_i = 0.2i$,

$i = 1, 2, 3, 4, 5$. Compare these results with those produced from the other Padé approximations of degree five:

a) $n = 0, m = 5,$ b) $n = 1, m = 4,$

c) $n = 3, m = 2,$ d) $n = 4, m = 1.$

6. Express the following rational functions in continued-fraction form.

a) $\dfrac{x^2 + 3x + 2}{x^2 - x + 1}$ b) $\dfrac{4x^4 + 3x^2 - 7x + 5}{x^5 - 2x^4 + 3x^3 + 6}$

c) $\dfrac{3x^5 - 4x^2 + 8x - 1}{x^4 + 5x^2 + x - 2}$ d) $\dfrac{2x^3 + x^2 - x + 3}{3x^3 + 2x^2 - x + 1}$

7. Use three-digit rounding arithmetic to compute the values of the rational fractions in Exercise 6 for $x = 1.23$

a)   precisely as listed in the exercise,

b)   using nested multiplication in the numerator and denominator,

c)   in continued-fraction form.

Compare to the exact result.

8. Determine the first four terms of the Chebyshev expansion for

a) $\sin x$ b) e^x.

9. Find the Chebyshev rational approximation of degree four with $n = m = 2$ for $f(x) = \sin x$. Compare the results at $x_i = 0.1$, $i = 0, 1, \ldots, 5$ from this approximation with those obtained in Exercise 3 using a sixth-degree Padé approximation.

10. Find the Chebyshev rational approximation of degree four with $n = m = 2$ for $f(x) = e^x$. Compare the results at $x_i = 0.2i$, $i = 1, 2, 3, 4, 5$ from this approximation with those obtained in Exercises 1 and 2 where fifth-degree Padé approximations were used.

7.5 Trigonometric Polynomial Approximation

In Section 7.2 Example 3 we discussed briefly some facts concerning trigonometric polynomials and Fourier series. In particular, for each positive integer n, we defined the set $\mathscr{T}_n$ of trigonometric polynomials of degree less than or equal to n as the set of all linear combinations at $\phi_0, \phi_1, \ldots, \phi_{2n-1}$

where $\phi_0(x) = \dfrac{1}{\sqrt{2\pi}},$

$\phi_k(x) = \dfrac{1}{\sqrt{\pi}} \cos kx$ for each $k = 1, 2, \ldots, n,$

and $\qquad \phi_{n+k}(x) = \dfrac{1}{\sqrt{\pi}} \sin kx \qquad$ for each $k = 1, 2, \ldots, n-1.$

(Some sources include an additional function in the set, $\phi_{2n}(x) = (1/\sqrt{\pi}) \sin nx$. We will not follow this convention.)

The set $\mathscr{T}_n$ is orthogonal on $[-\pi, \pi]$ with respect to the weight function $w(x) \equiv 1$. We also found that the least-squares approximation to a function $f \in C[-\pi, \pi]$ by functions in $\mathscr{T}_n$ is

$$S_n(x) = \sum_{k=0}^{2n-1} a_k \phi_k(x),$$

where $\qquad a_k = \displaystyle\int_{-\pi}^{\pi} f(x) \phi_k(x)\, dx, \qquad$ for each $k = 0, 1, \ldots, 2n-1.$

There is a discrete analog to this situation that is useful for the least-squares approximation and interpolation of large amounts of data when the data is given at equally spaced points.

Suppose that a collection of $2m$ paired data points $\{(x_j, y_j)\}_{j=0}^{2m-1}$ is given with the first elements in the pairs equally spaced within an interval. For convenience we will assume that the interval is $[-\pi, \pi]$ and that

(7.30) $\qquad x_j = -\pi + \left(\dfrac{j}{m}\right)\pi \qquad$ for each $j = 0, 1, \ldots, 2m-1.$

If this were not the case, a simple linear transformation could be used to translate the data into this form. (Example 1 contains a demonstration of the technique).

For a fixed $n < m$, consider the set of functions $\hat{\mathscr{T}}_n = \{\hat{\phi}_0, \hat{\phi}_1, \ldots, \hat{\phi}_{2n-1}\}$ where

(7.31) $$\hat{\phi}_0(x) = \dfrac{1}{\sqrt{2m}},$$

(7.32) $$\hat{\phi}_k(x) = \dfrac{1}{\sqrt{m}} \cos kx \qquad \text{for each } k = 1, 2, \ldots, n,$$

and

(7.33) $$\hat{\phi}_{n+k}(x) = \dfrac{1}{\sqrt{m}} \sin kx \qquad \text{for each } k = 1, 2, \ldots, n-1.$$

The object is to determine the trigonometric polynomial S_n composed of functions in $\hat{\mathscr{T}}_n$ with the property that

$$E(S_n) = \sum_{j=0}^{2m-1} \{y_j - S_n(x_j)\}^2$$
$$= \sum_{j=0}^{2m-1} \left\{ y_j - \left[\dfrac{\alpha_0}{\sqrt{2m}} + \dfrac{\alpha_n}{\sqrt{m}} \cos nx_j + \dfrac{1}{\sqrt{m}} \sum_{k=1}^{n-1} (\alpha_k \cos kx_j + \alpha_{n+k} \sin kx_j) \right] \right\}^2$$

is a minimum. As usual in least-squares approximations, the constants $\alpha_0, \alpha_1, \ldots, \alpha_{2n-1}$ are determined by the condition that this minimum can occur only if $\partial E(S_n)/\partial \alpha_k = 0$ for each $k = 0, 1, \ldots, 2n-1.$

The determination of the constants is simplified by the fact that the set $\mathcal{T}_n$ is orthonormal with respect to summation over the equally spaced points $\{x_j\}_{j=0}^{2m-1}$ in $[-\pi, \pi]$. By this we mean that for each $k \neq l$,

$$\sum_{j=0}^{2m-1} \hat{\phi}_k(x_j) \hat{\phi}_l(x_j) = 0,$$

and for each k,

$$\sum_{j=0}^{2m-1} [\hat{\phi}_k(x_j)]^2 = 1.$$

The demonstration of this orthonormality is not particularly difficult, but it is a tedious process that relies on the trigonometric identities

$$\sin a \cos b = \tfrac{1}{2}[\sin(a+b) + \sin(a-b)],$$

$$\sin a \sin b = \tfrac{1}{2}[\cos(a-b) - \cos(a+b)],$$

$$\cos a \cos b = \tfrac{1}{2}[\cos(a+b) + \cos(a-b)]$$

together with the relations

$$\sum_{j=0}^{2m-1} \cos\frac{kj\pi}{m} = \sum_{j=0}^{2m-1} \sin\frac{kj\pi}{m} = 0,$$

relations that are true for each pair of positive integers k and m. (See Exercise 7 for details and hints for the verification of these results.) The use of these relations produces the least-squares approximation S_n. It is most easily expressed in the form

$$(7.34) \qquad S_n(x) = \frac{a_0}{2} + a_n \cos nx + \sum_{k=1}^{n-1} (a_k \cos kx + a_{n+k} \sin kx),$$

where

$$(7.35) \qquad a_k = \frac{1}{m} \sum_{j=0}^{2m-1} y_j \cos kx_j \qquad \text{for each } k = 0, 1, \ldots, n$$

and

$$(7.36) \qquad a_{n+k} = \frac{1}{m} \sum_{j=0}^{2m-1} y_j \sin kx_j \qquad \text{for each } k = 1, 2, \ldots, n-1.$$

EXAMPLE 1   To find the least-squares approximation from $\mathcal{T}_3$ for the data $\{(x_j, y_j)\}_{j=0}^{11}$, where $x_j = j/6$ and $y_j = e^{-x_j}$, requires that the data points $\{x_j\}_{j=0}^{11}$ first be translated from $[0, 2]$ to $[-\pi, \pi]$. It is easily verified that the required linear transformation is

$$z_j = \pi(x_j - 1),$$

and that the translated data is of the form

$$\{(z_j, e^{-1-z_j/\pi})\}_{j=0}^{11}.$$

The least-squares trigonometric polynomial is consequently

$$S_3(z) = \left[\frac{a_0}{2} + a_3 \cos 3z + \sum_{k=1}^{2} (a_k \cos kz + a_{3+k} \sin kz) \right],$$

where
$$a_k = \tfrac{1}{6} \sum_{j=0}^{11} e^{-1-z_j/\pi} \cos kz_j \qquad \text{for } k = 0, 1, 2, 3$$

and
$$a_{3+k} = \tfrac{1}{6} \sum_{j=0}^{11} e^{-1-z_j/\pi} \sin kz_j \qquad \text{for } k = 1, 2.$$

Evaluating these sums produces the approximation
$$S_3(z) = 0.46936 - 0.15363 \cos z + 0.09553 \cos 2z - 0.08396 \cos 3z$$
$$- 0.24360 \sin z + 0.12143 \sin 2z. \qquad\blacksquare$$

The least-squares approximation on the $2m$ data points $\{(x_j, y_j)\}_{j=0}^{2m-1}$ by a function in $\mathcal{T}_m$ is simply the interpolatory trigonometric polynomial for the data points. It is this form of the trigonometric polynomial that is most useful in application. The only modification that is imposed is the change of S_m to

$$(7.37) \qquad S_m(x) = \frac{a_0 + a_m\cos mx}{2} + \sum_{k=1}^{m-1} [a_k \cos kx + a_{m+k} \sin kx].$$

The interpolation of large amounts of equally spaced data by trigonometric polynomials can produce very accurate results. It is the appropriate approximation technique used in areas such as those involving digital filters, antenna field patterns, quantum mechanics, optics, and many areas involving simulation problems. Until quite recently, however, the method had limited application due to the number of arithmetic calculations required for the determination of the constants in the approximation. The interpolation of $2m$ data points requires approximately $(2m)^2$ multiplications and $(2m)^2$ additions by the direct calculation technique. The approximation of thousands of data points is not unusual in areas requiring trigonometric interpolation; so the direct methods for evaluating the constants require multiplication and addition operations numbering in the millions. The round-off error associated with this number of calculations generally dominates the approximation and makes the results meaningless.

In 1965 a paper written by J. W. Cooley and J. W. Tukey was published in the journal *Mathematics of Computation* [29] which described a different method of calculating the constants in the interpolating trigonometric polynomial. This method requires only $O(m \log_2 m)$ multiplications and $O(m \log_2 m)$ additions, provided m is chosen in an appropriate manner. For a problem with thousands of data points, this reduces the number of calculations to thousands compared to millions for the direct technique. The method had actually been discovered a number of years before the Cooley–Tukey paper appeared, but had gone unnoticed by most researchers until that time. (Brigham [14], pp. 8–9, contains a short, but interesting, historical summary of the method.)

The method described by Cooley and Tukey has come to be known either as the **Cooley–Tukey Algorithm** or the **Fast Fourier Transform (FFT) Algorithm** and has led to a revolution in the use of interpolatory trigonometric polynomials in many scientific areas. The method consists of organizing the problem so that the number of data points being used can be easily factored, particularly into powers of two.

The relationship between the number of data points $2m$ and the degree m of the trigonometric polynomial used in the fast Fourier transform procedure allows some notational simplification. Equation (7.30) implies that the nodes are given by

$$(7.38) \qquad x_j = -\pi + \left(\frac{j}{m}\right)\pi$$

for each $j = 0, 1, \ldots, 2m - 1$. By defining $b_k = a_{m+k}$, when $k = 1, 2, \ldots, m - 1$, the coefficients in Eq. (7.37) can be expressed as

$$(7.39) \qquad a_k = \frac{1}{m} \sum_{j=0}^{2m-1} y_j \cos kx_j,$$

and

$$(7.40) \qquad b_k = a_{m+k} = \frac{1}{m} \sum_{j=0}^{2m-1} y_j \sin kx_j, \qquad \text{for each } k = 0, 1, \ldots, m.$$

Note that b_0 and b_m have been added to the collection in Eq. (7.40), but that as defined both will be zero and not contribute to the resulting sum.

Instead of directly evaluating the constants a_k and b_k, the fast Fourier transform procedure given in Algorithm 7.1 computes the complex coefficients c_k in the formula

$$(7.41) \qquad F(x) = \frac{1}{m} \sum_{k=0}^{2m-1} c_k e^{ikx},$$

where

$$(7.42) \qquad c_k = \sum_{j=0}^{2m-1} y_j e^{\pi i j k/m}, \qquad \text{for each } k = 0, 1, \ldots, 2m - 1.$$

Once the constants c_k have been determined, a_k and b_k can be recovered by using Euler's formula:

$$e^{i\theta} = \cos \theta + i \sin \theta$$

and noting that, for each $k = 0, 1, \ldots, m$,

$$\frac{1}{m} c_k e^{-i\pi k} = \frac{1}{m} \sum_{j=0}^{2m-1} y_j e^{\pi i j k/m} e^{-i\pi k}$$

$$= \frac{1}{m} \sum_{j=0}^{2m-1} y_j e^{ik(-\pi + (\pi j/m))}$$

$$= \frac{1}{m} \sum_{j=0}^{2m-1} y_j (\cos kx_j + i \sin kx_j)$$

$$= a_k + ib_k.$$

The operation reduction feature of the fast Fourier transform results from calculating the coefficients c_k in clusters and uses as a basic relation the fact that for any integer n,

$$e^{n\pi i} = \cos n\pi + i \sin n\pi = (-1)^n.$$

Suppose $m = 2^p$, for some positive integer p. For each $k = 0, 1, \ldots, m-1$,

$$c_k + c_{m+k} = \sum_{j=0}^{2m-1} y_j e^{\pi ijk/m} + \sum_{j=0}^{2m-1} y_j e^{\pi ij(m+k)/m}$$

$$= \sum_{j=0}^{2m-1} y_j e^{\pi ijk/m}(1 + e^{\pi ij}).$$

But

$$1 + e^{\pi ij} = \begin{cases} 2, & \text{if } j \text{ is even,} \\ 0, & \text{if } j \text{ is odd,} \end{cases}$$

so if j is replaced by $2j$ in the index of the sum:

$$c_k + c_{m+k} = 2\sum_{j=0}^{m-1} y_{2j} e^{\pi i(2j)k/m},$$

or

(7.43) $$c_k + c_{m+k} = 2\sum_{j=0}^{m-1} y_{2j} e^{\pi ijk/(m/2)}.$$

In a similar manner

(7.44) $$c_k - c_{m+k} = 2e^{\pi ik/m}\sum_{j=0}^{m-1} y_{2j+1} e^{\pi ijk/(m/2)}.$$

Both c_k and c_{m+k} can be recovered from Eqs. (7.43) and (7.44), so these relations determine all the coefficients c_k.

Note that the sum in Eqs. (7.43) and (7.44) are of the same form as the sum in Eq. (7.42), except that the index m has been replaced by $m/2$.

Let $\mathcal{M}(M)$ denote the number of complex multiplications required to compute $2M$ sums of the form

$$\sum_{j=0}^{2M-1} y_j e^{\pi ijk/M}.$$

If the basic relation in Eq. (7.42) is used to compute the coefficients, the number of operations is

$$\mathcal{M}(m) = (2m)^2 = 4m^2.$$

However, using Eqs. (7.43) and (7.44), the number is reduced to

(7.45) $$2\mathcal{M}(m/2) + m = 2\left[4\left(\frac{m}{2}\right)^2\right] + m = 2m^2 + m.$$

Since the sums in Eqs. (7.43) and (7.44) are of the same form as the original sum and m is a power of 2, the reduction technique can be reapplied to write each of the sums in Eqs. (7.43) and (7.44) as two sums from $j = 0$ to $j = (m/2) - 1$.

Calculating the coefficients using the sums from $j = 0$ to $j = (m/2) - 1$ reduces the number of calculations by replacing $\mathcal{M}(m/2)$ in Eq. (7.45) with $2\mathcal{M}(m/4) + m/2$. The result is that

(7.45) $$2\left[2\mathcal{M}\left(\frac{m}{4}\right) + \frac{m}{2}\right] + m = 4\left[4\left(\frac{m}{4}\right)^2\right] + 2m = m^2 + 2m$$

complex multiplications are now needed for the determination of the coefficients c_k.

If the process is repeated r times, the number of required multiplications is reduced to

$$2^r \mathcal{M}\left(\frac{m}{2^r}\right) + rm = \frac{m^2}{2^{r-2}} + mr.$$

Since $m = 2^p$, $p + 2$ reductions of this type will reduce the number of complex multiplications to

$$\frac{(2^p)^2}{2^p} + m(p + 2) = m + pm + 2m = 3m + m \log_2 m = O(m \log_2 m).$$

Because of the way the calculations are arranged, the number of required complex additions is comparable.

The following algorithm performs the fast Fourier transform when $m = 2^p$ for some positive integer p. Modifications of the technique can be made when m takes other forms. The *IMSL* package provides a number of these.

Fast Fourier Transform Algorithm 7.1

To compute the discrete approximation

$$F(x) = \frac{1}{m} \sum_{k=0}^{2m-1} c_k e^{ikx} = \frac{1}{m} \sum_{k=0}^{2m-1} c_k(\cos kx + i \sin kx) \qquad \text{where } i = \sqrt{-1},$$

for the data $\{(x_j, y_j)\}_{j=0}^{2m-1}$ where $m = 2^p$ and $x_j = -\pi + j\pi/m$ for $j = 0, 1, \ldots, 2m-1$:

INPUT m, p; $y_0, y_1, \ldots, y_{2m-1}$.

OUTPUT   complex numbers $c_0, \ldots, c_{2m-1}$; real numbers $a_0, \ldots, a_m$; $b_1, \ldots, b_{m-1}$.

Step 1   Set $M = m$;
$\quad\quad\quad q = p$;
$\quad\quad\quad \zeta = e^{\pi i/m}$.

Step 2   For $j = 0, 1, \ldots, 2m-1$ set $c_j = y_j$.

Step 3   For $j = 1, 2, \ldots, M$ set $\xi_j = \zeta^j$;
$\quad\quad\quad\quad\quad\quad\quad\quad\quad\quad\quad\quad\quad \xi_{j+M} = -\xi_j$.

Step 4   Set $K = 0$;
$\quad\quad\quad \xi_0 = 1$.

Step 5   For $L = 1, 2, \ldots, p+1$ do Steps 6–12.

Step 6   While $K < 2m-1$ do Steps 7–11.

Step 7   For $j = 1, 2, \ldots, M$ do Steps 8–10.

Step 8   Let $K = k_p \cdot 2^p + k_{p-1} \cdot 2^{p-1} + \cdots + k_1 2 + k_0$; (*Decompose k.*)
$\quad\quad\quad\quad$ set $K_1 = K/2^q = k_p \cdot 2^{p-q} + \cdots + k_{q+1} \cdot 2 + k_q$;
$\quad\quad\quad\quad\quad\quad K_2 = k_q \cdot 2^p + k_{q+1} \cdot 2^{p-1} + \cdots + k_p \cdot 2^q$.

Step 9   Set $\eta = c_{K+M}\zeta_{K_2}$;
$$c_{K+M} = c_K - \eta;$$
$$c_K = c_K + \eta.$$

Step 10   Set $K = K + 1$.

Step 11   Set $K = K + M$.

Step 12   Set $K = 0$;
$$M = M/2;$$
$$q = q - 1.$$

Step 13   While $K < 2m - 1$ do Steps 14–16.

Step 14   Let $K = k_p \cdot 2^p + k_{p-1} \cdot 2^{p-1} + \cdots + k_1 \cdot 2 + k_0$; (*Decompose k.*)
set $j = k_0 \cdot 2^p + k_1 \cdot 2^{p-1} + \cdots + k_{p-1} \cdot 2 + k_p$.

Step 15   If $j > K$ then interchange c_j and c_k.

Step 16   Set $K = K + 1$.

Step 17   Set $a_0 = c_0/m$;
$$a_m = \text{Re}(e^{-m\pi i}c_m/m).$$

Step 18   For $j = 1, \ldots, m - 1$ set $a_j = \text{Re}(e^{-j\pi i}c_j/m)$;
$$b_j = \text{Im}(e^{-j\pi i}c_j/m).$$

Step 19   OUTPUT $(c_0, \ldots, c_{2m-1}; a_0, \ldots, a_m; b_1, \ldots, b_{m-1})$;
STOP.

EXAMPLE 2   To determine the trigonometric interpolating polynomial of degree eight for the data $\{(x_j, y_j)\}_{j=0}^{15}$ where $x_j = j/8$ and $y_j = e^{-x_j}$ requires a transformation of the interval $[0, 2]$ to $[-\pi, \pi]$. The translation is given by

$$z_j = \pi(x_j - 1)$$

so that the input data to Algorithm 7.1 is

$$\{z_j, e^{-1-z_j/\pi}\}_{j=0}^{15}.$$

Table 7.16 lists $\hat{c}_k = e^{-k\pi i}c_k/8$ for $k = 0, 1, \ldots, 15$, generated by Algorithm 7.1. The interpolating polynomial is

$$S_8(x) = 0.4599157 - 0.1347242 \cos x + 0.0765635 \cos 2x + \cdots$$
$$+ 0.02870726 \cos 8x - 0.2463663 \sin x + \cdots - 0.01070678 \sin 7x. \quad \blacksquare$$

More details on the verification of the validity of the fast Fourier transform procedure can be found in Hamming [59], who presents the method from a mathematical approach, or in Bracewell [12] where the presentation is based on methods more likely to be familiar to engineers. Aho, Hopcroft, and Ullman [1],

TABLE 7.16

k	$\mathrm{Re}(\hat{c}_k)$	$\mathrm{Im}(\hat{c}_k)$
0	0.91983150	0.00000000
1	-0.13472420	-0.24636630
2	$0.76563530 \times 10^{-1}$	0.12707350
3	$-0.64875540 \times 10^{-1}$	$-0.79866820 \times 10^{-1}$
4	$0.60761670 \times 10^{-1}$	$0.53622240 \times 10^{-1}$
5	$-0.58912200 \times 10^{-1}$	$-0.35906370 \times 10^{-1}$
6	$0.57990620 \times 10^{-1}$	$0.22283140 \times 10^{-1}$
7	$-0.57547600 \times 10^{-1}$	$-0{\cdot}10706750 \times 10^{-1}$
8	$0.57414520 \times 10^{-1}$	$0.28837380 \times 10^{-6}$
9	$-0.57547800 \times 10^{-1}$	$0.10705270 \times 10^{-1}$
10	$0.57991160 \times 10^{-1}$	$-0.22281700 \times 10^{-1}$
11	$-0.58912600 \times 10^{-1}$	$0.35905730 \times 10^{-1}$
12	$0.60762590 \times 10^{-1}$	$-0.53621170 \times 10^{-1}$
13	$-0.64877270 \times 10^{-1}$	$0.79865270 \times 10^{-1}$
14	$0.76565200 \times 10^{-1}$	-0.12707240
15	-0.13472890	0.24636360

pp. 252–269, is a good reference for a discussion of the computational aspects of the method.

For a presentation of the techniques and related material from the point of view of applied abstract algebra we recommend Laufer [81], pp. 438–465. Modification of the procedure for the case when m is not a power of two can be found in Winograd [144].

Exercise Set 7.5

1. Compute the trigonometric least-squares polynomial of degree three using $m = 4$ for $f(x) = e^x \cos 2x$ on the interval $[-\pi, \pi]$.

2. Repeat Exercise 1 using $m = 8$. Which approximation is better?

3. Compute the trigonometric interpolating polynomial of degree four for $f(x) = x(\pi - x)$ on the interval $[-\pi, \pi]$ using

   a) Eqs. (7.35) and (7.36),   b) Algorithm 7.1.

   Compare the accuracy by evaluating each polynomial at the nodes.

4. a) Compute the trigonometric interpolating polynomial S_4 of degree four for $f(x) = x^2 \sin x$ on the interval $[0, 1]$.

   b) Compute $\int_0^1 S_4(x)\, dx$.

   c) Compare the integral in part (b) to $\int_0^1 x^2 \sin x\, dx$.

5. a) Compute the least-squares trigonometric polynomial S_4 of degree four using $m = 16$, for $f(x) = x^2 \sin x$ on the interval $[0, 1]$.

   b) Compute $\int_0^1 S_4(x)\, dx$.

   c) Compare the integral in part (b) to $\int_0^1 x^2 \sin x\, dx$.

6. Using Algorithm 7.1, compute the trigonometric interpolating polynomial of degree sixteen for $f(x) = x^2 \cos x$ on $[-\pi, \pi]$.

7. a) Show that for any positive integers m and k

$$\sum_{j=0}^{2m-1} \cos \frac{kj\pi}{m} = \sum_{j=0}^{2m-1} \sin \frac{kj\pi}{m} = 0.$$

   [*Hint:*   Let $w = e^{k\pi i/m}$. Then

$$0 = \sum_{j=0}^{2m-1} w^j = \sum_{j=0}^{2m-1} e^{k\pi ij/m} = \sum_{j=0}^{2m-1} \left(\cos \frac{kj\pi}{m} + \sin \frac{kj\pi}{m} \right).]$$

   b) Show that for $l \neq k$,

$$\sum_{j=0}^{2m-1} \cos kx_j \sin lx_j = \sum_{j=0}^{2m-1} \cos kx_j \cos lx_j = \sum_{j=0}^{2m-1} \sin kx_j \sin lx_j = 0$$

   and

$$\sum_{j=0}^{2m-1} (\cos kx_j)^2 = \sum_{j=0}^{2m-1} (\sin kx_j)^2 = m \qquad \text{for } k > 0.$$

8. Show that $c_0, \ldots, c_{2m-1}$ in Algorithm 7.1 are given by

$$\begin{bmatrix} c_0 \\ c_1 \\ c_2 \\ \vdots \\ c_{2m-1} \end{bmatrix} = \begin{bmatrix} 1 & 1 & 1 & \cdots & 1 \\ 1 & \zeta & \zeta^2 & \cdots & \zeta^{2m-1} \\ 1 & \zeta^2 & \zeta^4 & \cdots & \zeta^{4m-2} \\ \vdots & \vdots & \vdots & & \vdots \\ 1 & \zeta^{2m-1} & \zeta^{4m-2} & \cdots & \zeta^{(2m-1)^2} \end{bmatrix} \begin{bmatrix} y_0 \\ y_1 \\ y_2 \\ \vdots \\ y_{2m-1} \end{bmatrix}$$

   where $\zeta = e^{\pi i/m}$.

9. Consider Exercise 8 in the case $m = 2$; i.e.,

$$2\begin{bmatrix} c_0 \\ c_1 \\ c_2 \\ c_3 \end{bmatrix} = \begin{bmatrix} 1 & 1 & 1 & 1 \\ 1 & \zeta & \zeta^2 & \zeta^3 \\ 1 & \zeta^2 & \zeta^4 & \zeta^6 \\ 1 & \zeta^3 & \zeta^6 & \zeta^9 \end{bmatrix} \begin{bmatrix} y_0 \\ y_1 \\ y_2 \\ y_3 \end{bmatrix}.$$

   Show that

$$2\begin{bmatrix} c_0 \\ c_2 \\ c_1 \\ c_3 \end{bmatrix} = \begin{bmatrix} 1 & 1 & 0 & 0 \\ 1 & \zeta^2 & 0 & 0 \\ 0 & 0 & 1 & \zeta \\ 0 & 0 & 1 & \zeta^3 \end{bmatrix} \begin{bmatrix} 1 & 0 & 1 & 0 \\ 0 & 1 & 0 & 1 \\ 1 & 0 & \zeta^2 & 0 \\ 0 & 1 & 0 & \zeta^2 \end{bmatrix} \begin{bmatrix} y_0 \\ y_1 \\ y_2 \\ y_3 \end{bmatrix}.$$

   Compare the preceding equation to a trace of Algorithm 7.1 in the case $m = 2$.

8

Iterative Techniques in Matrix Algebra

At the beginning of Chapter 6 we discussed an electrical circuit problem involving seven resistors and an impressed voltage. This problem led to a 5×5 linear system whose solution gave the potential at each junction in the circuit. In a circuit involving many more resistances the situation would be similar but, of course, the associated linear system would be larger. Even in large circuits, however, we would expect each junction to have only a relatively small number of connections. If we consider the matrix associated with the 5×5 linear system discussed in Chapter 6,

$$
\begin{array}{c c}
 & \begin{array}{c c c c c} v_b & v_c & v_d & v_e & v_f \end{array} \\
\begin{array}{c} v_b \\ v_c \\ v_d \\ v_e \\ v_f \end{array} &
\left[\begin{array}{r r r r r}
31 & -10 & 0 & 0 & -6 \\
2 & -8 & 3 & 3 & 0 \\
0 & 1 & -3 & 2 & 0 \\
0 & 2 & 4 & -7 & 1 \\
12 & 0 & 15 & 0 & -47
\end{array} \right]
\end{array}
$$

we see that a zero occurs in the matrix whenever two junctions are not directly connected. Matrices associated with large electrical circuits generally have a high percentage of zero entries. Matrices of this type are called **sparse**; the appropriate methods for solving large linear systems involving sparse matrices often involve techniques which are iterative rather than direct.

The methods presented in Chapter 6 used direct techniques to solve a system of $n \times n$ linear equations of the form $A\mathbf{x} = \mathbf{b}$. In this chapter, we present some methods that can be used to iteratively solve a system of this type.

8.1 Norms of Vectors and Matrices

Before considering iterative methods for solving linear systems, a method must be determined for quantitatively measuring the distance between vectors in R^n, the set of all column vectors with real components, to determine whether the sequence of vectors that results from using an iterative technique converges to a solution of the system. In actuality, this measure is also needed when the solution is being obtained by the direct methods presented in Chapter 6, since these methods require the performance of a large number of arithmetic operations, and using finite-digit arithmetic leads only to an approximation to an actual solution of the system.

To define a distance in R^n, we use the notion of the **norm** of a vector.

DEFINITION 8.1   A **vector norm** on R^n, the collection of all n-dimensional column vectors with real components, is a function, $\| \cdot \|$, from R^n into R with the following properties:

    i) $\|\mathbf{x}\| \geq 0$ for all $\mathbf{x} \in R^n$,

    ii) $\|\mathbf{x}\| = 0$ if and only if $\mathbf{x} = (0, 0, \ldots, 0)^t \equiv \mathbf{0}$,

    iii) $\|\alpha\mathbf{x}\| = |\alpha|\,\|\mathbf{x}\|$ for all $\alpha \in R$ and $\mathbf{x} \in R^n$,

    iv) $\|\mathbf{x} + \mathbf{y}\| \leq \|\mathbf{x}\| + \|\mathbf{y}\|$ for all $\mathbf{x}, \mathbf{y} \in R^n$.

For our purposes we need only two specific norms on R^n, although a third norm on R^n is presented in Exercise 2.

Since vectors in R^n are column vectors, it is convenient to use the transpose notation presented in Section 6.6 when a vector is represented in terms of its components. For example, the vector

$$\mathbf{x} = \begin{bmatrix} x_1 \\ x_2 \\ \vdots \\ x_n \end{bmatrix}$$

will generally be written $\mathbf{x} = (x_1, x_2, \ldots, x_n)^t$.

DEFINITION 8.2   The l_2 and l_∞ norms for the vector $\mathbf{x} = (x_1, x_2, \ldots, x_n)^t$ are defined by

(8.1)
$$\|\mathbf{x}\|_2 = \left\{ \sum_{i=1}^{n} x_i^2 \right\}^{1/2}$$

and

(8.2)
$$\|\mathbf{x}\|_\infty = \max_{1 \le i \le n} |x_i|.$$

(*See Figs. 8.1 and 8.2.*)

The l_2 norm is often called the **Euclidean norm** of the vector $\mathbf{x}$ since it represents the usual notion of distance from the origin in case $\mathbf{x}$ is in $R^1 \equiv R$, R^2, or R^3.

EXAMPLE 1   The vector $\mathbf{x} = (-1, 1, -2)^t$ in R^3 has norms

$$\|\mathbf{x}\|_2 = \sqrt{(-1)^2 + (1)^2 + (-2)^2} = \sqrt{6}$$

and
$$\|\mathbf{x}\|_\infty = \max\{|-1|, |1|, |-2|\} = 2. \qquad \blacksquare$$

The justification for calling the concept defined by $\|\mathbf{x}\|_\infty = \max_{1 \le i \le n}|x_i|$ a norm on R^n is presented in the following theorem.

FIGURE 8.1

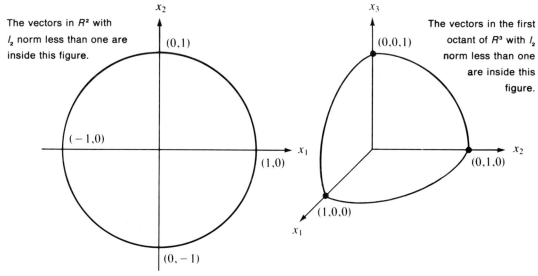

The vectors in R^2 with l_2 norm less than one are inside this figure.

The vectors in the first octant of R^3 with l_2 norm less than one are inside this figure.

THEOREM 8.3   For each $\mathbf{x}$, $\mathbf{y} \in R^n$ and $\alpha \in R$,

i) $\|\mathbf{x}\|_\infty \ge 0$,
ii) $\|\mathbf{x}\|_\infty = 0$ if and only if $\mathbf{x} = \mathbf{0}$,
iii) $\|\alpha\mathbf{x}\|_\infty = |\alpha|\, \|\mathbf{x}\|_\infty$
iv) $\|\mathbf{x} + \mathbf{y}\|_\infty \le \|\mathbf{x}\|_\infty + \|\mathbf{y}\|_\infty$,

so $\|\cdot\|_\infty$ is a norm on R^n.

PROOF   The properties follow easily from similar statements concerning the absolute value function. As an example, consider statement (iv). If $\mathbf{x} = (x_1, x_2, \ldots, x_n)^t$ and $\mathbf{y} = (y_1, y_2, \ldots, y_n)^t$, then

FIGURE 8.2

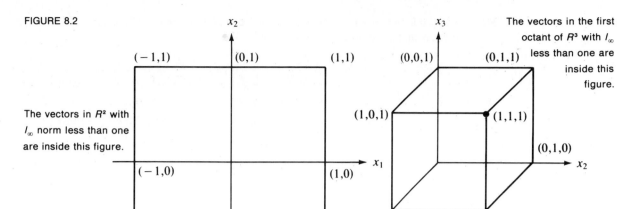

The vectors in R^2 with l_∞ norm less than one are inside this figure.

The vectors in the first octant of R^3 with l_∞ less than one are inside this figure.

$$\| \mathbf{x} + \mathbf{y} \|_\infty = \max_{1 \le i \le n} |x_i + y_i|$$

$$\le \max_{1 \le i \le n} \{|x_i| + |y_i|\}$$

$$\le \max_{1 \le i \le n} |x_i| + \max_{1 \le i \le n} |y_i| = \| \mathbf{x} \|_\infty + \| \mathbf{y} \|_\infty.$$

The other properties are shown in a similar manner.   □

To show that

$$\| \mathbf{x} \|_2 = \left\{ \sum_{i=1}^{n} x_i^2 \right\}^{1/2}$$

describes a norm in R^n is more difficult. The problem occurs in proving that

$$\| \mathbf{x} + \mathbf{y} \|_2 \le \| \mathbf{x} \|_2 + \| \mathbf{y} \|_2 \qquad \text{for each } \mathbf{x}, \mathbf{y} \in R^n.$$

THEOREM 8.4   For each $\mathbf{x}, \mathbf{y} \in R^n$ and $\alpha \in R$.

  i) $\| \mathbf{x} \|_2 \ge 0$,
  ii) $\| \mathbf{x} \|_2 = 0$ if and only if $\mathbf{x} = \mathbf{0}$,
  iii) $\| \alpha \mathbf{x} \|_2 = |\alpha| \| \mathbf{x} \|_2$,
  iv) $\| \mathbf{x} + \mathbf{y} \|_2 \le \| \mathbf{x} \|_2 + \| \mathbf{y} \|_2$.

PROOF   For each $\mathbf{x} = (x_1, x_2, \ldots, x_n)^t$ in R^n,

$$\| \mathbf{x} \|_2 = \left\{ \sum_{i=1}^{n} x_i^2 \right\}^{1/2} \ge 0$$

and $\| \mathbf{x} \|_2 = 0$ if and only if $x_i^2 = 0$ for each $i = 1, 2, \ldots, n$.

Thus, $\|\mathbf{x}\|_2 = 0$ if and only if $\mathbf{x} = \mathbf{0}$, and properties (i) and (ii) are established. To show that (iii) holds, simply note that:

$$\|\alpha\mathbf{x}\|_2 = \left\{\sum_{i=1}^{n} (\alpha x_i)^2\right\}^{1/2} = \left\{\sum_{i=1}^{n} \alpha^2 x_i^2\right\}^{1/2} = (\alpha^2)^{1/2}\left\{\sum_{i=1}^{n} x_i^2\right\}^{1/2} = |\alpha|\,\|\mathbf{x}\|_2.$$

To show property (iv), it is first necessary to prove the following lemma.

LEMMA 8.5   (Cauchy–Buniakowsky–Schwarz Inequality for Sums)   For each $\mathbf{x} = (x_1, x_2, \ldots, x_n)^t$ and $\mathbf{y} = (y_1, y_2, \ldots, y_n)^t$ in R^n,

$$(8.3) \qquad \sum_{i=1}^{n} |x_i y_i| \leq \left\{\sum_{i=1}^{n} x_i^2\right\}^{1/2} \left\{\sum_{i=1}^{n} y_i^2\right\}^{1/2}.$$

PROOF OF LEMMA 8.5   If $\mathbf{y} = \mathbf{0}$ or $\mathbf{x} = \mathbf{0}$, the result is immediate, since both sides of Eq. (8.3) are zero. Suppose $\mathbf{y} \neq \mathbf{0}$ and $\mathbf{x} \neq \mathbf{0}$. For each $\lambda \in R$,

$$0 \leq \|\mathbf{x} - \lambda\mathbf{y}\|_2^2 = \sum_{i=1}^{n} (x_i - \lambda y_i)^2 = \sum_{i=1}^{n} x_i^2 - 2\lambda \sum_{i=1}^{n} x_i y_i + \lambda^2 \sum_{i=1}^{n} y_i^2,$$

and

$$2\lambda \sum_{i=1}^{n} x_i y_i \leq \sum_{i=1}^{n} x_i^2 + \lambda^2 \sum_{i=1}^{n} y_i^2 = \|\mathbf{x}\|_2^2 + \lambda^2 \|\mathbf{y}\|_2^2.$$

Since Theorem 8.4 implies that $\|\mathbf{x}\|_2 > 0$ and $\|\mathbf{y}\|_2 > 0$, choosing $\lambda = \|\mathbf{x}\|_2/\|\mathbf{y}\|_2$ gives

$$\left(2\frac{\|\mathbf{x}\|_2}{\|\mathbf{y}\|_2}\right)\left(\sum_{i=1}^{n} x_i y_i\right) \leq \|\mathbf{x}\|_2^2 + \frac{\|\mathbf{x}\|_2^2}{\|\mathbf{y}\|_2^2}\|\mathbf{y}\|_2^2 = 2\|\mathbf{x}\|_2^2$$

so

$$2 \sum_{i=1}^{n} x_i y_i \leq 2 \|\mathbf{x}\|_2^2 \frac{\|\mathbf{y}\|_2}{\|\mathbf{x}\|_2} = 2\|\mathbf{x}\|_2\|\mathbf{y}\|_2.$$

Thus,

$$\sum_{i=1}^{n} x_i y_i \leq \|\mathbf{x}\|_2\|\mathbf{y}\|_2.$$

Replacing x_i by $-x_i$ whenever $x_i y_i < 0$ and calling the vector $\tilde{\mathbf{x}}$:

$$\sum_{i=1}^{n} |x_i y_i| \leq \|\tilde{\mathbf{x}}\|_2\|\mathbf{y}\|_2 = \|\mathbf{x}\|_2\|\mathbf{y}\|_2 = \left\{\sum_{i=1}^{n} x_i^2\right\}^{1/2}\left\{\sum_{i=1}^{n} y_i^2\right\}^{1/2} \qquad \square$$

PROOF OF THEOREM 8.4 (continued)   For each $\mathbf{x}, \mathbf{y} \in R^n$,

$$\|\mathbf{x} + \mathbf{y}\|_2^2 = \sum_{i=1}^{n} (x_i + y_i)^2 = \sum_{i=1}^{n} x_i^2 + 2 \sum_{i=1}^{n} x_i y_i + \sum_{i=1}^{n} y_i^2$$

(by Lemma 8.5),

$$\leq \|\mathbf{x}\|_2^2 + 2\|\mathbf{x}\|_2\|\mathbf{y}\|_2 + \|\mathbf{y}\|_2^2 = (\|\mathbf{x}\|_2 + \|\mathbf{y}\|_2)^2,$$

so

$$\|\mathbf{x} + \mathbf{y}\|_2 \leq \|\mathbf{x}\|_2 + \|\mathbf{y}\|_2. \qquad \square$$

Since the norm of a vector gives a measure for the distance between the vector and the origin, the **distance between two vectors** can be defined as the norm of the difference of the vectors.

DEFINITION 8.6   If $\mathbf{x} = (x_1, x_2, \ldots, x_n)^t$ and $\mathbf{y} = (y_1, y_2, \ldots, y_n)^t$ are vectors in R^n, the l_2 and l_∞ distances between $\mathbf{x}$ and $\mathbf{y}$ are defined by:

(8.4)
$$\|\mathbf{x} - \mathbf{y}\|_2 = \left\{ \sum_{i=1}^{n} |x_i - y_i|^2 \right\}^{1/2}$$

and

(8.5)
$$\|\mathbf{x} - \mathbf{y}\|_\infty = \max_{1 \le i \le n} |x_i - y_i|.$$

EXAMPLE 2   The linear system

$$3.3330x_1 + 15920x_2 - 10.333x_3 = 15913,$$
$$2.2220x_1 + 16.710x_2 + 9.6120x_3 = 28.544,$$
$$1.5611x_1 + 5.1791x_2 + 1.6852x_3 = 8.4254,$$

has solution $(x_1, x_2, x_3)^t = (1.0000, 1.0000, 1.0000)^t$. If Gaussian elimination is performed in five-digit arithmetic using maximal column pivoting (Algorithm 6.2. p. 326), the solution obtained is

$$\tilde{\mathbf{x}} = (\tilde{x}_1, \tilde{x}_2, \tilde{x}_3)^t = (1.2001, 0.99991, 0.92538)^t.$$

Measurements of $\mathbf{x} - \tilde{\mathbf{x}}$ are given by

$$\|\mathbf{x} - \tilde{\mathbf{x}}\|_\infty = \max\{|1.0000 - 1.2001|, |1.0000 - 0.99991|, |1.0000 - 0.92538|\}$$
$$= \max\{0.2001, 0.00009, 0.07462\}$$
$$= 0.2001,$$

and

$$\|\mathbf{x} - \tilde{\mathbf{x}}\|_2 = (|1.0000 - 1.2001|^2 + |1.0000 - 0.99991|^2 + |1.0000 - 0.92538|^2)^{1/2}$$
$$= [(0.2001)^2 + (0.00009)^2 + (0.07462)^2]^{1/2}$$
$$= 0.21356.$$

Although the components $\tilde{x}_2$ and $\tilde{x}_3$ are good approximations to x_2 and x_3, the component $\tilde{x}_1$ is a poor approximation to x_1 and $|x_1 - \tilde{x}_1|$ dominates the norms. ∎

The concept of distance in R^n can also be used to define a limit of a sequence of vectors in this space.

DEFINITION 8.7   A sequence $\{\mathbf{x}^{(k)}\}_{k=1}^{\infty}$ of vectors in R^n is said to **converge** to $\mathbf{x}$ with respect to the norm $\|\cdot\|$ if, given any $\varepsilon > 0$, there exists an integer $N(\varepsilon)$ such that

$$\|\mathbf{x}^{(k)} - \mathbf{x}\| < \varepsilon \qquad \text{for all } k \ge N(\varepsilon).$$

THEOREM 8.8   The sequence of vectors $\{\mathbf{x}^{(k)}\}$ converges to $\mathbf{x}$ in R^n with respect to $\|\cdot\|_\infty$ if and only if $\lim_{k \to \infty} x_i^{(k)} = x_i$ for each $i = 1, 2, \ldots, n$.

PROOF   Suppose $\{\mathbf{x}^{(k)}\}$ converges to $\mathbf{x}$ with respect to $\|\cdot\|_\infty$. Given any $\varepsilon > 0$ there exists an integer $N(\varepsilon)$ such that for all $k \geqslant N(\varepsilon)$,

$$\max_{i=1,2,\ldots,n} |x_i^{(k)} - x_i| = \|\mathbf{x}^{(k)} - \mathbf{x}\|_\infty < \varepsilon.$$

This implies that $|x_i^{(k)} - x_i| < \varepsilon$ for each $i = 1, 2, \ldots, n$, so $\lim_{k\to\infty} x_i^{(k)} = x_i$ for each i.

Conversely, suppose that $\lim_{k\to\infty} x_i^{(k)} = x_i$ for every $i = 1, 2, \ldots, n$. For a given $\varepsilon > 0$, let $N_i(\varepsilon)$ for each i represent an integer with the property that

$$|x_i^{(k)} - x_i| < \varepsilon$$

whenever $k \geqslant N_i(\varepsilon)$.

Define $N(\varepsilon) = \max_{i=1,2,\ldots,n} N_i(\varepsilon)$. If $k \geqslant N(\varepsilon)$, then

$$\max_{i=1,2,\ldots,n} |x_i^{(k)} - x_i| = \|\mathbf{x}^{(k)} - \mathbf{x}\|_\infty < \varepsilon.$$

This implies that $\{\mathbf{x}^{(k)}\}$ converges to $\mathbf{x}$.   □

EXAMPLE 3   Let $\mathbf{x}^{(k)} \in R^4$ be defined by

$$\mathbf{x}^{(k)} = (x_1^{(k)}, x_2^{(k)}, x_3^{(k)}, x_4^{(k)})^t = \left(1, \, 2 + \frac{1}{k}, \, \frac{3}{k^2}, \, e^{-k} \sin k\right)^t.$$

Since $\lim_{k\to\infty} 1 = 1$, $\lim_{k\to\infty} (2 + 1/k) = 2$, $\lim_{k\to\infty} 3/k^2 = 0$, and $\lim_{k\to\infty} e^{-k} \sin k = 0$, Theorem 8.8 implies that the sequence $\{\mathbf{x}^{(k)}\}$ converges to $(1, 2, 0, 0)^t$ with respect to $\|\cdot\|_\infty$.   ■

To show directly that the sequence in Example 3 converges to $(1, 2, 0, 0)^t$ with respect to the l_2 norm is quite complicated. It is much easier to prove the next theorem and apply it to this special case.

THEOREM 8.9   For each $\mathbf{x} \in R^n$,

$$\|\mathbf{x}\|_\infty \leq \|\mathbf{x}\|_2 \leq \sqrt{n}\|\mathbf{x}\|_\infty.$$

PROOF   Let x_j be a coordinate of $\mathbf{x}$ such that $\|\mathbf{x}\|_\infty = \max_{1 \leq i \leq n}|x_i| = |x_j|$. Then

$$\|\mathbf{x}\|_\infty^2 = |x_j|^2 = x_j^2 \leq \sum_{i=1}^{n} x_i^2 \leq \sum_{i=1}^{n} x_j^2 = nx_j^2 = n\|\mathbf{x}\|_\infty^2.$$

Thus, $$\|\mathbf{x}\|_\infty \leq \left\{\sum_{i=1}^{n} x_i^2\right\}^{1/2} = \|\mathbf{x}\|_2 \leq \sqrt{n}\|\mathbf{x}\|_\infty.$$ □

EXAMPLE 4   In Example 3 we found that the sequence $\{\mathbf{x}^{(k)}\}$ defined by

$$\mathbf{x}^{(k)} = \left(1, \, 2 + \frac{1}{k}, \, \frac{3}{k^2}, \, e^{-k} \sin k\right)^t,$$

converges to $\mathbf{x} = (1, 2, 0, 0)^t$ with respect to $\|\cdot\|_\infty$. Given any $\varepsilon > 0$, there exists an integer $N(\varepsilon/2)$ with the property

$$\|\mathbf{x}^{(k)} - \mathbf{x}\|_\infty < \varepsilon/2$$

whenever $k \geqslant N(\varepsilon/2)$. By Theorem 8.9 this implies that

$$\|\mathbf{x}^{(k)} - \mathbf{x}\|_2 < \sqrt{4}\,\|\mathbf{x}^{(k)} - \mathbf{x}\|_\infty < 2(\varepsilon/2) = \varepsilon$$

when $k \geqslant N(\varepsilon/2)$. So $\{\mathbf{x}^{(k)}\}$ converges to $\mathbf{x}$ with respect to $\|\cdot\|_2$. ∎

It can be shown that all norms on R^n are equivalent with respect to convergence; that is, if $\|\cdot\|$ and $\|\cdot\|'$ are any two norms on R^n and $\{\mathbf{x}^{(k)}\}_{k=1}^\infty$ has the limit $\mathbf{x}$ with respect to $\|\cdot\|$, then $\{\mathbf{x}^{(k)}\}_{k=1}^\infty$ has the limit $\mathbf{x}$ with respect to $\|\cdot\|'$. The proof of this fact for the general case can be found in Ortega [94], page 8. The case for the norms $\|\cdot\|_2$ and $\|\cdot\|_\infty$ is considered in Exercise 4.

It is also necessary to have a method for measuring the distance between two $n \times n$ matrices, which again requires the use of the norm concept.

DEFINITION 8.10   A **matrix norm** on the set of all real $n \times n$ matrices is a real-valued function, $\|\cdot\|$, defined on this set, satisfying for all $n \times n$ matrices A and B and all real numbers α:

i) $\|A\| \geq 0$,
ii) $\|A\| = 0$ if and only if A is O, the matrix with all zero entries,
iii) $\|\alpha A\| = |\alpha|\,\|A\|$,
iv) $\|A + B\| \leq \|A\| + \|B\|$,
v) $\|AB\| \leq \|A\|\,\|B\|$.

A **distance between $n \times n$ matrices** A and B can be defined in the usual manner, as $\|A - B\|$.

Although matrix norms can be obtained in various ways, the only norms we will consider are those that are natural consequences of the vector norms l_2 and l_∞.

The following theorem is not difficult to show, and its proof is left to Exercise 11.

THEOREM 8.11   If $\|\cdot\|$ is any vector norm on R^n, then

$$\|A\| = \max_{\|\mathbf{x}\| = 1} \|A\mathbf{x}\|$$

defines a matrix norm on the set of real $n \times n$ matrices, which is called a **natural norm**.

In this text all matrix norms will be assumed to be natural matrix norms, unless specified otherwise.

The matrix norms we will consider consequently have the forms

$$\|A\|_\infty = \max_{\|\mathbf{x}\|_\infty = 1} \|A\mathbf{x}\|_\infty, \quad \text{the } l_\infty \text{ norm}$$

and

$$\|A\|_2 = \max_{\|\mathbf{x}\|_2 = 1} \|A\mathbf{x}\|_2, \quad \text{the } l_2 \text{ norm}.$$

The l_∞ norm of a matrix has a particularly interesting representation with respect to the entries of the matrix.

THEOREM 8.12   If $A = (a_{ij})$ is an $n \times n$ matrix, then

$$\|A\|_\infty = \max_{1 \le i \le n} \sum_{j=1}^{n} |a_{ij}|.$$

PROOF   Let $\mathbf{x}$ be an n-dimensional column vector with $1 = \|\mathbf{x}\|_\infty = \max_{1 \le i \le n} |x_i|$. Since $A\mathbf{x}$ is also an n-dimensional column vector,

$$\|A\mathbf{x}\|_\infty = \max_{1 \le i \le n} |(A\mathbf{x})_i| = \max_{1 \le i \le n} \left| \sum_{j=1}^{n} a_{ij} x_j \right|$$

$$\le \max_{1 \le i \le n} \sum_{j=1}^{n} |a_{ij}| \max_{1 \le j \le n} |x_j| = \max_{1 \le i \le n} \sum_{j=1}^{n} |a_{ij}| \|\mathbf{x}\|_\infty$$

$$= \max_{1 \le i \le n} \sum_{j=1}^{n} |a_{ij}|,$$

so

$$\|A\mathbf{x}\|_\infty \le \max_{1 \le i \le n} \sum_{j=1}^{n} |a_{ij}|$$

for all $\mathbf{x}$ with $\|\mathbf{x}\|_\infty = 1$. Consequently,

(8.6)
$$\|A\|_\infty = \max_{\|\mathbf{x}\|_\infty = 1} \|A\mathbf{x}\|_\infty \le \max_{1 \le i \le n} \sum_{j=1}^{n} |a_{ij}|.$$

However, if p is the integer, $1 \le p \le n$, with

$$\sum_{j=1}^{n} |a_{pj}| = \max_{1 \le i \le n} \sum_{j=1}^{n} |a_{ij}|,$$

and $\mathbf{x}$ is chosen with

$$x_j = \begin{cases} 1, & \text{if } a_{pj} \ge 0, \\ -1, & \text{if } a_{pj} < 0, \end{cases}$$

then $\|\mathbf{x}\|_\infty = 1$ and $a_{pj} x_j = |a_{pj}|$ for all $j = 1, 2, \ldots, n$. Moreover,

$$\|A\mathbf{x}\|_\infty = \max_{1 \le i \le n} \left| \sum_{j=1}^{n} a_{ij} x_j \right| \ge \left| \sum_{j=1}^{n} a_{pj} x_j \right| = \sum_{j=1}^{n} |a_{pj}| = \max_{1 \le i \le n} \sum_{j=1}^{n} |a_{ij}|.$$

This implies that

$$\|A\|_\infty = \max_{\|\mathbf{x}\|_\infty = 1} \|A\mathbf{x}\|_\infty \ge \max_{1 \le i \le n} \sum_{j=1}^{n} |a_{ij}|,$$

which, together with inequality (8.6), gives

$$\|A\|_\infty = \max_{1 \le i \le n} \sum_{j=1}^{n} |a_{ij}|. \qquad \square$$

EXAMPLE 5   If

$$A = \begin{bmatrix} 1 & 2 & -1 \\ 0 & 3 & -1 \\ 5 & -1 & 1 \end{bmatrix},$$

then

$$\sum_{j=1}^{3} |a_{1j}| = |1| + |2| + |-1| = 4,$$

$$\sum_{j=1}^{3} |a_{2j}| = |0| + |3| + |-1| = 4,$$

and

$$\sum_{j=1}^{3} |a_{3j}| = |5| + |-1| + |1| = 7;$$

so

$$\|A\|_\infty = \max\{4, 4, 7\} = 7. \qquad \blacksquare$$

To investigate the l_2-norm, it is necessary to discuss some further concepts of linear algebra.

DEFINITION 8.13   If A is a real $n \times n$ matrix, the polynomial defined by

$$p(\lambda) = \det(A - \lambda I)$$

is called the **characteristic polynomial** of A.

It is easy to show (see Exercise 13) that p is an nth-degree polynomial with real coefficients and, consequently, has at most n distinct zeros, some of which may be complex. If λ is a zero of p, then, since $\det(A - \lambda I) = 0$, Theorem 6.12 (p. 322) implies that the linear system defined by $(A - \lambda I)\mathbf{x} = \mathbf{0}$ has a solution other than the identically zero solution. We wish to study the zeros of p and the nontrivial solutions corresponding to these systems.

DEFINITION 8.14   If p is the characteristic polynomial of the matrix A, the zeros of p are called **eigenvalues** or characteristic values of the matrix A. If λ is an eigenvalue of A and $\mathbf{x} \neq \mathbf{0}$ has the property that $(A - \lambda I)\mathbf{x} = \mathbf{0}$, then $\mathbf{x}$ is called an **eigenvector** or characteristic vector of A corresponding to the eigenvalue λ.

EXAMPLE 6   Let

$$A = \begin{bmatrix} 1 & 0 & 1 \\ 2 & 2 & 1 \\ -1 & 0 & 0 \end{bmatrix}.$$

To compute the eigenvalues of A consider

$$p(\lambda) = \det(A - \lambda I) = \det \begin{bmatrix} 1-\lambda & 0 & 1 \\ 2 & 2-\lambda & 1 \\ -1 & 0 & -\lambda \end{bmatrix}$$

$$= (1-\lambda) \det \begin{bmatrix} 2-\lambda & 1 \\ 0 & -\lambda \end{bmatrix} + 1 \det \begin{bmatrix} 2 & 2-\lambda \\ -1 & 0 \end{bmatrix}$$

$$= (1-\lambda)(2-\lambda)(-\lambda) + (2-\lambda)$$

$$= (2-\lambda)(\lambda^2 - \lambda + 1).$$

The eigenvalues of A are the solutions of $p(\lambda) = 0$; $\lambda_1 = 2$, $\lambda_2 = \frac{1}{2} + (\sqrt{3}/2)i$, and $\lambda_3 = \frac{1}{2} - (\sqrt{3}/2)i$.

An eigenvector $\mathbf{x}$ of A associated with λ_1 is a solution of the system $(A - \lambda_1 I)\mathbf{x} = \mathbf{0}$:

$$\begin{bmatrix} -1 & 0 & 1 \\ 2 & 0 & 1 \\ -1 & 0 & -2 \end{bmatrix} \begin{bmatrix} x_1 \\ x_2 \\ x_3 \end{bmatrix} = \begin{bmatrix} 0 \\ 0 \\ 0 \end{bmatrix}.$$

Thus,
$$\begin{aligned} -x_1 + x_3 &= 0, \\ 2x_1 + x_3 &= 0, \\ -x_1 - 2x_3 &= 0, \end{aligned}$$

which has a solution with $x_1 = x_3 = 0$ and x_2 arbitrary. In particular, $\mathbf{x} = (0, 1, 0)^t$ is an eigenvector of A corresponding to the eigenvalue $\lambda_1 = 2$.

To find an eigenvector for $\lambda_2 = \frac{1}{2} + (\sqrt{3}/2)i$ requires solving the system

$$\begin{bmatrix} 1 - \left(\frac{1}{2} + \frac{\sqrt{3}}{2}i\right) & 0 & 1 \\ 2 & 2 - \left(\frac{1}{2} + \frac{\sqrt{3}}{2}i\right) & 1 \\ -1 & 0 & -\left(\frac{1}{2} + \frac{\sqrt{3}}{2}i\right) \end{bmatrix} \begin{bmatrix} x_1 \\ x_2 \\ x_3 \end{bmatrix} = \begin{bmatrix} 0 \\ 0 \\ 0 \end{bmatrix}.$$

Using complex arithmetic, it can be shown that one solution is

$$x_1 = -\frac{1}{2} - \frac{\sqrt{3}}{2}i, \qquad x_2 = -\frac{1}{2} + \frac{\sqrt{3}}{2}i, \qquad x_3 = 1;$$

so an eigenvector corresponding to the eigenvalue $\lambda_2 = \frac{1}{2} + (\sqrt{3}/2)i$ is

$$\mathbf{x} = \left(-\frac{1}{2} - \frac{\sqrt{3}}{2}i, \; -\frac{1}{2} + \frac{\sqrt{3}}{2}i, \; 1 \right)^t.$$

In a similar manner,

$$\mathbf{x} = \left(-\frac{1}{2} + \frac{\sqrt{3}}{2}i, \ -\frac{1}{2} - \frac{\sqrt{3}}{2}i, \ 1 \right)^t$$

can be shown to be an eigenvector corresponding to the eigenvalue

$$\lambda_3 = \frac{1}{2} - \frac{\sqrt{3}}{2}i. \qquad\qquad\qquad ■$$

The computation of eigenvalues and eigenvectors of matrices is an important part of numerical linear algebra and will be considered further in Sections 8.4 and 8.5.

DEFINITION 8.15   The **spectral radius** $\rho(A)$ of a matrix A is defined by

$$\rho(A) = \max|\lambda| \qquad \text{where } \lambda \text{ is an eigenvalue of } A.$$

(Recall that, for complex $\lambda = \alpha + \beta i, |\lambda| = \{\alpha^2 + \beta^2\}^{1/2}$.)

For the matrix considered in Example 5

$$\rho(A) = \max\left\{ 2, \left| \frac{1}{2} + \frac{\sqrt{3}}{2}i \right|, \left| \frac{1}{2} - \frac{\sqrt{3}}{2}i \right| \right\} = \max\{2, 1, 1\} = 2.$$

The spectral radius is closely related to the norm of a matrix, as shown in the following theorem.

THEOREM 8.16   If A is a real $n \times n$ matrix, then

i) $[\rho(A^t A)]^{1/2} = \|A\|_2,$
ii) $\rho(A) \le \|A\|$ for any natural norm $\|\cdot\|$.

PROOF   The proof of part (i) requires more information concerning eigenvalues than we presently have available. For the details involved in the proof, see Ortega [94], page 21.

To prove part (ii), suppose λ is an eigenvalue of A with eigenvector $\mathbf{x}$ where $\|\mathbf{x}\| = 1$. (Exercise 12(b) ensures that such an eigenvector exists.) Since $(A - \lambda I)\mathbf{x} = \mathbf{0}$, $A\mathbf{x} = \lambda\mathbf{x}$, so for any natural norm

$$|\lambda| = \|\lambda\mathbf{x}\| = \|A\mathbf{x}\| \le \|A\| \, \|\mathbf{x}\| = \|A\|.$$

Thus, $$\rho(A) = \max|\lambda| \le \|A\|. \qquad\qquad □$$

An interesting and useful result, which is similar to part (ii) of Theorem 8.16, is that for any matrix A and any $\varepsilon > 0$, there exists a norm $\|\cdot\|$ with the property that $\|A\| < \rho(A) + \varepsilon$. Consequently, $\rho(A)$ is the greatest lower bound for the norms on A. The proof of this result can be found in Ortega [94], page 23.

EXAMPLE 7   If

$$A = \begin{bmatrix} 2 & 1 & 0 \\ 1 & 1 & 1 \\ 0 & 1 & 2 \end{bmatrix},$$

then, since A is symmetric, $A^t = A$ and

$$A^t A = A^2 = \begin{bmatrix} 2 & 1 & 0 \\ 1 & 1 & 1 \\ 0 & 1 & 2 \end{bmatrix} \begin{bmatrix} 2 & 1 & 0 \\ 1 & 1 & 1 \\ 0 & 1 & 2 \end{bmatrix} = \begin{bmatrix} 5 & 3 & 1 \\ 3 & 3 & 3 \\ 1 & 3 & 5 \end{bmatrix}.$$

To calculate $\rho(A^t A)$, we need the eigenvalues of $A^t A$. If

$$0 = \det(A^t A - \lambda I) = \det \begin{bmatrix} 5-\lambda & 3 & 1 \\ 3 & 3-\lambda & 3 \\ 1 & 3 & 5-\lambda \end{bmatrix}$$

$$= (5-\lambda)^2(3-\lambda) + 9 + 9 - (3-\lambda) - 9(5-\lambda) - 9(5-\lambda)$$

$$= -\lambda^3 + 13\lambda^2 - 36\lambda$$

$$= -\lambda(\lambda - 4)(\lambda - 9),$$

then λ is 0, 4, or 9. Thus,

$$\|A\|_2 = \sqrt{\rho(A^t A)} = \sqrt{\max\{0, 4, 9\}} = 3. \qquad \blacksquare$$

In studying iterative matrix techniques, it is of particular importance to know when powers of a matrix become small; that is, when all of the entries approach zero. Matrices of this type are called **convergent**.

DEFINITION 8.17   We call an $n \times n$ matrix A **convergent** if

$$\lim_{k \to \infty} (A^k)_{ij} = 0 \qquad \text{for each } i = 1, 2, \dots, n \text{ and } j = 1, 2, \dots, n.$$

EXAMPLE 8   Let

$$A = \begin{bmatrix} \frac{1}{2} & 0 \\ \frac{1}{4} & \frac{1}{2} \end{bmatrix}.$$

Computing powers of A, we obtain:

$$A^2 = \begin{bmatrix} \frac{1}{4} & 0 \\ \frac{1}{4} & \frac{1}{4} \end{bmatrix}, \qquad A^3 = \begin{bmatrix} \frac{1}{8} & 0 \\ \frac{3}{16} & \frac{1}{8} \end{bmatrix}, \qquad A^4 = \begin{bmatrix} \frac{1}{16} & 0 \\ \frac{1}{8} & \frac{1}{16} \end{bmatrix},$$

and, in general,

$$A^k = \begin{bmatrix} (\frac{1}{2})^k & 0 \\ k/2^{k+1} & (\frac{1}{2})^k \end{bmatrix}.$$

Since

$$\lim_{k \to \infty} \left(\frac{1}{2}\right)^k = 0 \quad \text{and} \quad \lim_{k \to \infty} \frac{k}{2^{k+1}} = 0,$$

A is a convergent matrix. Note that $\rho(A) = \frac{1}{2}$, since $\frac{1}{2}$ is the only eigenvalue of A.   ∎

An important connection exists between the spectral radius of a matrix and the convergence of the matrix.

THEOREM 8.18   The following statements are equivalent:

i) A is a convergent matrix;
ii) $\lim_{n \to \infty} \|A^n\| = 0$,   for some natural norm $\| \cdot \|$;
iii) $\rho(A) < 1$;
iv) $\lim_{n \to \infty} A^n \mathbf{x} = \mathbf{0}$, for every $\mathbf{x}$.

The proof of this theorem can be found in Isaacson and Keller [67], page 14,

Exercise Set 8.1

1.   Find $\|\mathbf{x}\|_\infty$ and $\|\mathbf{x}\|_2$ for the following vectors:

a) $\mathbf{x} = (3, -4, 0, \frac{3}{2})^t$.

b) $\mathbf{x} = (2, 1, -3, 4)^t$.

c) $\mathbf{x} = (\sin k, \cos k, 2^k)^t$ for a fixed positive integer k.

d) $\mathbf{x} = (4/(k+1), 2/k^2, k^2 e^{-k})^t$ for a fixed positive integer k.

2.   a)   Verify that the function $\| \cdot \|_1$ defined on R^n by

$$\|\mathbf{x}\|_1 = \sum_{i=1}^n |x_i|,$$

is a norm on R^n.

b)   Find $\|\mathbf{x}\|_1$ for the vectors given in Exercise 1.

3.   Prove that the following sequences are convergent, and find their limits.

a) $\mathbf{x}^{(k)} = (1/k, e^{1-k}, -2/k^2)^t$.

b) $\mathbf{x}^{(k)} = \left(e^{-k} \cos k, k \sin \frac{1}{k}, 3 + k^{-2}\right)^t$.

c) $\mathbf{x}^{(k)} = (ke^{-k^2}, (\cos k)/k, \sqrt{k^2 + k} - k)^t$.

d) $\mathbf{x}^{(k)} = (e^{1/k}, (k^2 + 1)/(1 - k^2), (1/k^2)(1 + 3 + 5 + \cdots + (2k - 1)))^t$.

4.   Show that a sequence $\{\mathbf{x}^{(k)}\}_{k=1}^\infty$ converges to $\mathbf{x} \in R^n$ relative to the l_2 norm if and only if it converges to $\mathbf{x}$ relative to the l_∞ norm.

5. Find $\| \cdot \|_\infty$ for the following matrices, using Theorem 8.12.

a) $\begin{bmatrix} 1 & -1 \\ 2 & 1 \end{bmatrix}$

b) $\begin{bmatrix} 1 & 1 \\ 1 & 1 \end{bmatrix}$

c) $\begin{bmatrix} 10 & 15 \\ 0 & 1 \end{bmatrix}$

d) $\begin{bmatrix} 3 & 1 \\ 2 & 5 \end{bmatrix}$

6. Find $\| \cdot \|_\infty$ for the matrices in Exercise 5 using Theorem 8.11.

7. Define the matrix norm $\| \cdot \|_1$ by

$$\|A\|_1 = \max_{\|\mathbf{x}\|_1 = 1} \|A\mathbf{x}\|_1$$

and show that

$$\|A\|_1 = \max_{1 \le j \le n} \sum_{i=1}^{n} |a_{ij}|.$$

8. Find $\| \cdot \|_1$ for the matrices in Exercise 5 using both equations given in Exercise 7.

9. a) Show that $\| \cdot \|_②$, and $\| \cdot \|_①$ are matrix norms, where

$$\|A\|_② = \left(\sum_{i=1}^{n} \sum_{j=1}^{n} |a_{ij}|^2 \right)^{1/2} \quad \text{and} \quad \|A\|_① = \sum_{i=1}^{n} \sum_{j=1}^{n} |a_{ij}|$$

for any $n \times n$ matrix A.

b) Find $\| \cdot \|_①$ and $\| \cdot \|_②$ for the matrices in Exercise 5.

10. Show that $\| \cdot \|_⊚$, defined by $\|A\|_⊚ = \max_{1 \le i, j \le n} |a_{ij}|$, does not define a matrix norm.

11. Prove Theorem 8.11.

12. For any vector $\mathbf{x} \ne \mathbf{0}$, $\mathbf{x}/\|\mathbf{x}\|$ is a vector whose norm is one. Use this to show that

a) $\|A\mathbf{x}\| \le \|A\| \, \|\mathbf{x}\|$,

b) for any eigenvalue λ an eigenvector $\mathbf{x}$ exists with $\|\mathbf{x}\| = 1$.

13. Show that the characteristic polynomial $p(\lambda) = \det(A - \lambda I)$, for the $n \times n$ matrix A, is an nth-degree polynomial. [*Hint:* Expand $\det(A - \lambda I)$ along the first row and use induction on n.]

14. a) Show that if A is an $n \times n$ matrix, then

$$\det A = \prod_{i=1}^{n} \lambda_i,$$

where $\lambda_1, \ldots, \lambda_n$ are the eigenvalues of A. [*Hint:* Consider $p(0)$.]

b) Show that A is singular if and only if $\lambda = 0$ is an eigenvalue of A.

15. Compute the eigenvalues and associated eigenvectors of the following matrices:

a) $\begin{bmatrix} 2 & -1 \\ -1 & 2 \end{bmatrix}$

b) $\begin{bmatrix} 1 & 1 \\ 0 & 1 \end{bmatrix}$

c) $\begin{bmatrix} 0 & \frac{1}{2} \\ \frac{1}{2} & 0 \end{bmatrix}$

d) $\begin{bmatrix} 2 & 1 & 1 \\ 2 & 3 & 2 \\ 1 & 1 & 2 \end{bmatrix}$

e) $\begin{bmatrix} 2 & 1 & 0 \\ 1 & 2 & 0 \\ 0 & 0 & 3 \end{bmatrix}$

f) $\begin{bmatrix} -1 & 2 & 0 \\ 0 & 3 & 4 \\ 0 & 0 & 7 \end{bmatrix}$

16. Find the spectral radius for each matrix in Exercise 15.

17. Find $\| \cdot \|_2$ for each matrix in Exercise 15

a) using $\sqrt{\rho(A^t A)}$

b) Theorem 8.11.

18. Show that $\| A \|_2 = \rho(A)$ if A is symmetric.

19. Let λ be an eigenvalue of the $n \times n$ matrix A and let $\mathbf{x} \neq \mathbf{0}$ be an associated eigenvector.

a) Show that λ is also an eigenvalue of A^t.

b) Show for any integer $k \geq 1$ that λ^k is an eigenvalue of A^k with eigenvector $\mathbf{x}$.

c) Show that if A^{-1} exists, then $1/\lambda$ is an eigenvalue of A^{-1} with eigenvector $\mathbf{x}$.

d) Generalize parts (b) and (c) to $(A^{-1})^k$ for integers $k \geq 2$.

e) Given the polynomial $q(x) = q_0 + q_1 x + \cdots + q_k x^k$ define $q(A)$ to be the matrix $q(A) = q_0 I + q_1 A + \cdots + q_k A^k$. Show that $q(\lambda)$ is an eigenvalue of $q(A)$ with eigenvector $\mathbf{x}$.

f) Let $\alpha \neq \lambda$ be given. Show that if $A - \alpha I$ is nonsingular, then $1/(\lambda - \alpha)$ is an eigenvalue of $(A - \alpha I)^{-1}$ with eigenvector $\mathbf{x}$.

20. Find matrices A and B for which $\rho(A + B) > \rho(A) + \rho(B)$. (This shows that $\rho(A)$ cannot be a matrix norm.)

21. Show that

$$A_1 = \begin{bmatrix} 1 & 0 \\ \frac{1}{4} & \frac{1}{2} \end{bmatrix}$$

is not convergent, but

$$A_2 = \begin{bmatrix} \frac{1}{2} & 0 \\ 16 & \frac{1}{2} \end{bmatrix}$$

is convergent.

22. Which of the matrices in Exercise 15 are convergent?

23. Show that $(1/\|A^{-1}\|) \leq |\lambda| \leq \|A\|$ for any eigenvalue λ of the nonsingular matrix A where $\| \cdot \|$ is any natural norm.

24. Let S be a positive definite matrix.

a) Show that the eigenvalues of S are positive real numbers.

b) For any $\mathbf{x} \in R^n$ define $\| \mathbf{x} \| = \sqrt{\mathbf{x}^t S \mathbf{x}}$. Show that this defines a norm on R^n.

25. Let S be a real and nonsingular matrix, and let $\| \cdot \|$ be any norm on R^n. Define $\| \cdot \|'$ by $\|x\|' = \|Sx\|$. Show that $\| \cdot \|'$ is also a norm on R^n.

26. In Exercise 17 of Section 6.6, a symmetric matrix

$$A = \begin{bmatrix} 1.59 & 1.69 & 2.13 \\ 1.69 & 1.31 & 1.72 \\ 2.13 & 1.72 & 1.85 \end{bmatrix}$$

was used to describe the average wing lengths of fruit flies which were offspring resulting from the mating of three mutants of the flies. The entry a_{ij} represents the average wing length of a fly that is the offspring of a male fly of type i and a female fly of type j.

a) Find the eigenvalues and associated eigenvectors of this matrix.

b) Use the result in Exercise 24 to answer the question posed in part (b) of Exercise 17 in Section 6.6; that is, is this matrix positive definite?

27. In Exercise 4 of Section 6.3, we assumed that the contribution a female beetle of a certain type made to the future years' female beetle population could be expressed in terms of the matrix

$$A = \begin{bmatrix} 0 & 0 & 6 \\ \frac{1}{2} & 0 & 0 \\ 0 & \frac{1}{3} & 0 \end{bmatrix}$$

where the entry in the ith row and jth column represents the probabilistic contribution of a beetle of age j onto the next year's female population of age i.

a) Does the matrix A have any real eigenvalues? If so, determine them and any associated eigenvectors.

b) If a sample of this species were needed for laboratory test purposes which would have a constant proportion in each age group from year to year, what criteria could be imposed on the initial population to ensure that this would be satisfied?

8.2 Iterative Techniques for Solving Linear Systems

An iterative technique to solve the $n \times n$ linear system $Ax = b$ starts with an initial approximation $x^{(0)}$ to the solution x, and generates a sequence of vectors $\{x^{(k)}\}_{k=0}^{\infty}$ that converges to x. Most of these iterative techniques involve a process that converts the system $Ax = b$ into an equivalent system of the form $x = Tx + c$ for some $n \times n$ matrix T and vector c. After the initial vector $x^{(0)}$ is selected, the sequence of approximate solution vectors is generated by computing

(8.7) $$x^{(k)} = Tx^{(k-1)} + c$$

for each $k = 1, 2, 3, \ldots$. This type of procedure should be reminiscent of the fixed-point iteration studied in Chapter 2.

Iterative techniques are seldom used for solving linear systems of small dimension since the time required for sufficient accuracy exceeds that required for direct techniques such as the Gaussian elimination method. For large systems with a high percentage of zero entries, however, these techniques are efficient in terms of computer storage and time requirements. Systems of this type arise frequently in the numerical solution of boundary-value problems and partial-differential equations.

EXAMPLE 1   The linear system $Ax = b$ given by

$$
\begin{aligned}
E_1: \quad & 10x_1 - x_2 + 2x_3 & = & 6, \\
E_2: \quad & -x_1 + 11x_2 - x_3 + 3x_4 & = & 25, \\
E_3: \quad & 2x_1 - x_2 + 10x_3 - x_4 & = & -11, \\
E_4: \quad & 3x_2 - x_3 + 8x_4 & = & 15
\end{aligned}
$$

has solution $\mathbf{x} = (1, 2, -1, 1)^t$. To convert $Ax = b$ to the form $\mathbf{x} = T\mathbf{x} + \mathbf{c}$, solve equation E_i for x_i, for each $i = 1, 2, 3, 4$, to obtain

$$
\begin{aligned}
x_1 &= \tfrac{1}{10}x_2 - \tfrac{1}{5}x_3 & + \tfrac{3}{5}, \\
x_2 &= \tfrac{1}{11}x_1 & + \tfrac{1}{11}x_3 - \tfrac{3}{11}x_4 + \tfrac{25}{11}, \\
x_3 &= -\tfrac{1}{5}x_1 + \tfrac{1}{10}x_2 & + \tfrac{1}{10}x_4 - \tfrac{11}{10}, \\
x_4 &= -\tfrac{3}{8}x_2 + \tfrac{1}{8}x_3 & + \tfrac{15}{8}.
\end{aligned}
$$

In this example,

$$
T = \begin{bmatrix}
0 & \tfrac{1}{10} & -\tfrac{1}{5} & 0 \\
\tfrac{1}{11} & 0 & \tfrac{1}{11} & -\tfrac{3}{11} \\
-\tfrac{1}{5} & \tfrac{1}{10} & 0 & \tfrac{1}{10} \\
0 & -\tfrac{3}{8} & \tfrac{1}{8} & 0
\end{bmatrix}
\quad \text{and} \quad
\mathbf{c} = \begin{bmatrix}
\tfrac{3}{5} \\
\tfrac{25}{11} \\
-\tfrac{11}{10} \\
\tfrac{15}{8}
\end{bmatrix}.
$$

For an initial approximation let $\mathbf{x}^{(0)} = (0, 0, 0, 0)^t$ and generate $\mathbf{x}^{(1)}$ by:

$$
\begin{aligned}
x_1^{(1)} &= \tfrac{1}{10}x_2^{(0)} - \tfrac{1}{5}x_3^{(0)} & + \tfrac{3}{5} & = \quad 0.6000, \\
x_2^{(1)} &= \tfrac{1}{11}x_1^{(0)} & + \tfrac{1}{11}x_3^{(0)} - \tfrac{3}{11}x_4^{(0)} + \tfrac{25}{11} & = \quad 2.2727, \\
x_3^{(1)} &= -\tfrac{1}{5}x_1^{(0)} + \tfrac{1}{10}x_2^{(0)} & + \tfrac{1}{10}x_4^{(0)} - \tfrac{11}{10} & = -1.1000, \\
x_4^{(1)} &= -\tfrac{3}{8}x_2^{(0)} + \tfrac{1}{8}x_3^{(0)} & + \tfrac{15}{8} & = \quad 1.8750.
\end{aligned}
$$

Additional iterates, $\mathbf{x}^{(k)} = (x_1^{(k)}, x_2^{(k)}, x_3^{(k)}, x_4^{(k)})^t$, are generated in a similar manner and are presented in Table 8.1.

TABLE 8.1

k	0	1	2	3	4	5	6	7	8	9	10
$x_1^{(k)}$	0.0000	0.6000	1.0473	0.9326	1.0152	0.9890	1.0032	0.9981	1.0006	0.9997	1.0001
$x_2^{(k)}$	0.0000	2.2727	1.7159	2.0533	1.9537	2.0114	1.9922	2.0023	1.9987	2.0004	1.9998
$x_3^{(k)}$	0.0000	-1.1000	-0.8052	-1.0493	-0.9681	-1.0103	-0.9945	-1.0020	-0.9990	-1.0004	-0.9998
$x_4^{(k)}$	0.0000	1.8750	0.8852	1.1309	0.9739	1.0214	0.9944	1.0036	0.9989	1.0006	0.9998

The decision to stop after ten iterations is based on

$$\frac{\|\mathbf{x}^{(10)} - \mathbf{x}^{(9)}\|_\infty}{\|\mathbf{x}^{(10)}\|_\infty} = \frac{8.0 \times 10^{-4}}{1.9998} < 10^{-3}.$$

In fact, $\|\mathbf{x}^{(10)} - \mathbf{x}\|_\infty = 0.0002,$ ■

The method of Example 1 is called the **Jacobi iterative method**. It consists of solving the ith equation in $A\mathbf{x} = \mathbf{b}$ for x_i, to obtain, provided $a_{ii} \neq 0$,

$$(8.8) \qquad x_i = \sum_{\substack{j=1 \\ j \neq i}}^{n} \left(-\frac{a_{ij} x_j}{a_{ii}} \right) + \frac{b_i}{a_{ii}} \qquad \text{for } i = 1, 2, \dots, n$$

and generating each $x_i^{(k)}$ from components of $\mathbf{x}^{(k-1)}$ for $k \geq 1$ by

$$(8.9) \qquad x_i^{(k)} = \frac{\sum_{\substack{j=1 \\ j \neq i}}^{n} (-a_{ij} x_j^{(k-1)}) + b_i}{a_{ii}} \qquad \text{for } i = 1, 2, \dots, n.$$

The method can be written in the form $\mathbf{x}^{(k)} = T\mathbf{x}^{(k-1)} + \mathbf{c}$ by splitting A into its diagonal and off-diagonal parts. To see this, let D be the diagonal matrix whose diagonal is the same as A, $-L$ be the strictly lower-triangular part of A, and $-U$ be the strictly upper-triangular part of A. With this notation, A is split into

$$A = \begin{bmatrix} a_{11} & a_{12} & \cdots & a_{1n} \\ a_{21} & a_{22} & \cdots & a_{2n} \\ \vdots & \vdots & & \vdots \\ a_{n1} & a_{n2} & \cdots & a_{nn} \end{bmatrix}$$

$$= \begin{bmatrix} a_{11} & 0 & \cdots & 0 \\ 0 & a_{22} & & \\ & & \ddots & 0 \\ 0 & \cdots & 0 & a_{nn} \end{bmatrix} - \begin{bmatrix} 0 & \cdots & & 0 \\ -a_{21} & & & \\ \vdots & \ddots & & \\ -a_{n1} & \cdots & -a_{n,n-1} & 0 \end{bmatrix} - \begin{bmatrix} 0 & -a_{12} & \cdots & -a_{1n} \\ & \ddots & & \vdots \\ & & \ddots & -a_{n-1,n} \\ 0 & \cdots & & 0 \end{bmatrix}$$

$$= \quad D \qquad\qquad - \quad L \qquad\qquad\qquad - \quad U.$$

The equation $A\mathbf{x} = \mathbf{b}$ or $(D - L - U)\mathbf{x} = \mathbf{b}$ is then transformed into

$$D\mathbf{x} = (L + U)\mathbf{x} + \mathbf{b},$$

and finally

$$(8.10) \qquad \mathbf{x} = D^{-1}(L + U)\mathbf{x} + D^{-1}\mathbf{b}.$$

This results in the matrix form of the Jacobi iterative technique:

$$(8.11) \qquad \mathbf{x}^{(k)} = D^{-1}(L + U)\mathbf{x}^{(k-1)} + D^{-1}\mathbf{b}, \qquad k = 1, 2, \dots.$$

In practice, Eq. (8.9) is used in computation, with Eq. (8.11) being reserved for theoretical purposes.

Jacobi Iterative Algorithm 8.1

To solve $A\mathbf{x} = \mathbf{b}$ given an initial approximation $\mathbf{x}^{(0)}$:

INPUT the number of equations and unknowns n; the entries a_{ij}, $1 \le i, j \le n$ of the matrix A; the entries b_i, $1 \le i \le n$ of the inhomogeneous term $\mathbf{b}$; the entries XO_i, $1 \le i \le n$ of $\mathbf{XO} = \mathbf{x}^{(0)}$; tolerance TOL; maximum number of iterations N.

OUTPUT the approximate solution $x_1, \ldots, x_n$ or a message that the number of iterations was exceeded.

Step 1   Set $k = 1$.

Step 2   While $(k \le N)$ do Steps 3–6.

   Step 3   For $i = 1, \ldots, n$

$$set \; x_i = \frac{-\sum_{\substack{j=1 \\ j \ne i}}^{n} (a_{ij} XO_j) + b_i}{a_{ii}}.$$

   Step 4   If $\| \mathbf{x} - \mathbf{XO} \| < TOL$ then OUTPUT $(x_1, \ldots, x_n)$;
                                 (*Procedure completed successfully.*)
                                 STOP.

   Step 5   Set $k = k + 1$.

   Step 6   For $i = 1, \ldots, n$ set $XO_i = x_i$.

Step 7   OUTPUT ('Maximum number of iterations exceeded');
             (*Procedure completed unsuccessfully.*)
             STOP.

Step 3 of the algorithm requires that $a_{ii} \ne 0$ for each $i = 1, 2, \ldots, n$. If this is not the case, a reordering of the equations can be performed so that no $a_{ii} = 0$, unless the system is singular. To speed convergence, the equations should be arranged so that a_{ii} is as large as possible. (See Exercise 1(e) and 1(f).) This subject will be discussed in more detail later.

Another possible stopping criterion in Step 4 is to iterate until

$$\frac{\| \mathbf{x}^{(k)} - \mathbf{x}^{(k-1)} \|}{\| \mathbf{x}^{(k)} \|}$$

is smaller than some prescribed tolerance $\varepsilon > 0$. For this purpose, any convenient norm can be used, the most usual being the l_∞ norm.

A possible improvement in Algorithm 8.1 is suggested by an analysis of Eq. (8.9). To compute $x_i^{(k)}$, the components of $\mathbf{x}^{(k-1)}$ are used. Since, for $i > 1$, $x_1^{(k)}, \ldots, x_{i-1}^{(k)}$ have already been computed and are supposedly better approximations to the actual solutions $x_1, \ldots, x_{i-1}$ than $x_1^{(k-1)}, \ldots, x_{i-1}^{(k-1)}$, it seems reasonable to compute $x_i^{(k)}$ using these most recently calculated values; that is,

$$(8.12) \qquad x_i^{(k)} = \frac{-\sum_{j=1}^{i-1}(a_{ij}x_j^{(k)}) - \sum_{j=i+1}^{n}(a_{ij}x_j^{(k-1)}) + b_i}{a_{ii}},$$

for each $i = 1, 2, \ldots, n$, instead of Eq. (8.9). An example illustrating this procedure follows.

EXAMPLE 2   The linear system given by

$$
\begin{aligned}
10x_1 - x_2 + 2x_3 &= 6, \\
-x_1 + 11x_2 - x_3 + 3x_4 &= 25, \\
2x_1 - x_2 + 10x_3 - x_4 &= -11, \\
3x_2 - x_3 + 8x_4 &= 15
\end{aligned}
$$

was solved in Example 1 by the Jacobi iterative method. Incorporating Eq. (8.12) into Algorithm 8.1 gives the equations to be used for each $k = 1, 2, \ldots,$

$$
\begin{aligned}
x_1^{(k)} &= \tfrac{1}{10}x_2^{(k-1)} - \tfrac{1}{5}x_3^{(k-1)} &&+ \tfrac{3}{5}, \\
x_2^{(k)} &= \tfrac{1}{11}x_1^{(k)} &&+ \tfrac{1}{11}x_3^{(k-1)} - \tfrac{3}{11}x_4^{(k-1)} + \tfrac{25}{11}, \\
x_3^{(k)} &= -\tfrac{1}{5}x_1^{(k)} + \tfrac{1}{10}x_2^{(k)} &&+ \tfrac{1}{10}x_4^{(k-1)} - \tfrac{11}{10}, \\
x_4^{(k)} &= -\tfrac{3}{8}x_2^{(k)} + \tfrac{1}{8}x_3^{(k)} &&+ \tfrac{15}{8}.
\end{aligned}
$$

Letting $\mathbf{x}^{(0)} = (0, 0, 0, 0)^t$, we generate the vector iterates in Table 8.2.

TABLE 8.2

k	0	1	2	3	4	5
$x_1^{(k)}$	0.0000	0.6000	1.030	1.0065	1.0009	1.0001
$x_2^{(k)}$	0.0000	2.3272	2.037	2.0036	2.0003	2.0000
$x_3^{(k)}$	0.0000	-0.9873	-1.014	-1.0025	-1.0003	-1.0000
$x_4^{(k)}$	0.0000	0.8789	0.9844	0.9983	0.9999	1.0000

Since

$$\frac{\|\mathbf{x}^{(5)} - \mathbf{x}^{(4)}\|_\infty}{\|\mathbf{x}^{(5)}\|_\infty} = \frac{0.0008}{2.000} = 4 \times 10^{-4},$$

$\mathbf{x}^{(5)}$ is accepted as a reasonable approximation to the solution. It is interesting to note that Jacobi's method in Example 1 required twice as many iterations for the same accuracy. ∎

The technique presented in Example 2 is called the **Gauss–Seidel iterative method**. To write this method in matrix form, multiply both sides of Eq. (8.12) by a_{ii} and collect all kth iterate terms to give

$$a_{i1}x_1^{(k)} + a_{i2}x_2^{(k)} + \cdots + a_{ii}x_i^{(k)} = -a_{i,i+1}x_{i+1}^{(k-1)} - \cdots - a_{in}x_n^{(k-1)} + b_i,$$

for each $i = 1, 2, \ldots, n$. Writing all n equations gives:

$$
\begin{aligned}
a_{11}x_1^{(k)} &= -a_{12}x_2^{(k-1)} - a_{13}x_3^{(k-1)} - \cdots - a_{1n}x_n^{(k-1)} + b_1, \\
a_{21}x_1^{(k)} + a_{22}x_2^{(k)} &= \qquad\qquad -a_{23}x_3^{(k-1)} - \cdots - a_{2n}x_n^{(k-1)} + b_2, \\
\vdots \qquad\quad \vdots & \\
a_{n1}x_1^{(k)} + a_{n2}x_2^{(k)} + \cdots + a_{nn}x_n^{(k)} &= \qquad\qquad\qquad\qquad\qquad\qquad\qquad\quad b_n;
\end{aligned}
$$

and it follows that, in matrix form, the Gauss–Seidel method can be represented by:

$$
(8.13) \qquad\qquad (D - L)\mathbf{x}^{(k)} = U\mathbf{x}^{(k-1)} + \mathbf{b}
$$

or

$$
(8.14) \qquad \mathbf{x}^{(k)} = (D - L)^{-1}U\mathbf{x}^{(k-1)} + (D - L)^{-1}\mathbf{b} \qquad \text{for each } k = 1, 2, \ldots.
$$

For the lower-triangular matrix $D - L$ to be nonsingular, it is necessary and sufficient that $a_{ii} \neq 0$ for each $i = 1, 2, \ldots, n$.

Gauss–Seidel Iterative Algorithm 8.2

To solve $A\mathbf{x} = \mathbf{b}$ given an initial approximation $\mathbf{x}^{(0)}$:

INPUT the number of equations and unknowns n; the entries a_{ij}, $1 \leq i, j \leq n$ of the matrix A; the entries b_i, $1 \leq i \leq n$ of $\mathbf{b}$; the entries XO_i, $1 \leq i \leq n$ of $\mathbf{XO} = \mathbf{x}^{(0)}$; tolerance TOL; maximum number of iterations N.

OUTPUT the approximate solution $x_1, \ldots, x_n$ or a message that the number of iterations was exceeded.

Step 1   Set $k = 1$.

Step 2   While ($k \leq N$) do Steps 3–6.

Step 3   For $i = 1, \ldots, n$

$$
\text{set } x_i = \frac{-\sum_{j=1}^{i-1} a_{ij}x_j - \sum_{j=i+1}^{n} a_{ij}XO_j + b_i}{a_{ii}}.
$$

Step 4   If $\| \mathbf{x} - \mathbf{XO} \| < TOL$ then OUTPUT $(x_1, \ldots, x_n)$;
(*Procedure completed successfully.*)
STOP.

Step 5   Set $k = k + 1$.

Step 6   For $i = 1, \ldots, n$ set $XO_i = x_i$.

Step 7   OUTPUT ('Maximum number of iterations exceeded');
(*Procedure completed unsuccessfully.*)
STOP.

The comments following Algorithm 8.1 regarding stopping criteria also apply to the Gauss–Seidel Algorithm 8.2.

The results of Examples 1 and 2 seem to imply that the Gauss–Seidel method is superior to the Jacobi method. This is generally the case but is not always true. In fact, there are linear systems for which the Jacobi method converges and the Gauss–Seidel method does not, and others for which the Gauss–Seidel method converges and the Jacobi method does not. (See Varga [135], page 74.)

To study the convergence of general iteration techniques, we consider the formula

$$\mathbf{x}^{(k)} = T\mathbf{x}^{(k-1)} + \mathbf{c} \qquad \text{for each } k = 1, 2, \dots,$$

where $\mathbf{x}^{(0)}$ is arbitrary. The study requires the following lemma.

LEMMA 8.19

If the spectral radius $\rho(T)$ satisfies $\rho(T) < 1$, then $(I - T)^{-1}$ exists, and

$$(8.15) \qquad (I - T)^{-1} = I + T + T^2 + \cdots.$$

PROOF

For any eigenvalue λ of T, $1 - \lambda$ is an eigenvalue of $I - T$. Since $|\lambda| \le \rho(T) < 1$, it follows that no eigenvalue of $I - T$ could be zero and, consequently, $I - T$ is nonsingular.

Let $S_m = I + T + T^2 + \cdots + T^m$. Then

$$(I - T)S_m = I - T^{m+1}$$

and, since T is convergent, Theorem 8.18 (p. 421), implies that

$$\lim_{m \to \infty} (I - T)S_m = \lim_{m \to \infty} (I - T^{m+1}) = I.$$

Thus, $\lim_{m \to \infty} S_m = (I - T)^{-1}$.   □

THEOREM 8.20

For any $\mathbf{x}^{(0)} \in R^n$, the sequence $\{\mathbf{x}^{(k)}\}_{k=0}^{\infty}$ defined by

$$(8.16) \qquad \mathbf{x}^{(k)} = T\mathbf{x}^{(k-1)} + \mathbf{c} \qquad \text{for each } k \ge 1 \text{ and } \mathbf{c} \ne 0,$$

converges to the unique solution of $\mathbf{x} = T\mathbf{x} + \mathbf{c}$ if and only if $\rho(T) < 1$.

PROOF

From Eq. (8.16),

$$(8.17) \qquad
\begin{aligned}
\mathbf{x}^{(k)} &= T\mathbf{x}^{(k-1)} + \mathbf{c} \\
&= T(T\mathbf{x}^{(k-2)} + \mathbf{c}) + \mathbf{c} \\
&= T^2\mathbf{x}^{(k-2)} + (T + I)\mathbf{c} \\
&\ \ \vdots \\
&= T^k\mathbf{x}^{(0)} + (T^{k-1} + \cdots + T + I)\mathbf{c}.
\end{aligned}$$

Assuming $\rho(T) < 1$, we can use Theorem 8.18 on page 421 and Lemma 8.19 to give

$$(8.18) \qquad
\begin{aligned}
\lim_{k \to \infty} \mathbf{x}^{(k)} &= \lim_{k \to \infty} T^k\mathbf{x}^{(0)} + \lim_{k \to \infty} \left(\sum_{j=0}^{k-1} T^j \right)\mathbf{c} \\
&= 0 \cdot \mathbf{x}^{(0)} + (I - T)^{-1}\mathbf{c} = (I - T)^{-1}\mathbf{c}.
\end{aligned}$$

By Eq. (8.16), $\mathbf{x} = \lim_{k \to \infty} \mathbf{x}^{(k)} = (I - T)^{-1}\mathbf{c}$ will be the unique solution to $\mathbf{x} = T\mathbf{x} + \mathbf{c}$.

To prove the converse, let $\{\mathbf{x}^{(k)}\}$ converge to $\mathbf{x}$ for any $\mathbf{x}^{(0)}$. From Eq. (8.16) it follows that $\mathbf{x} = T\mathbf{x} + \mathbf{c}$, so for each k,

$$\mathbf{x} - \mathbf{x}^{(k)} = T(\mathbf{x} - \mathbf{x}^{(k-1)}) = \cdots = T^k(\mathbf{x} - \mathbf{x}^{(0)}).$$

Hence, for any vector $\mathbf{x}^{(0)}$,

$$\lim_{k \to \infty} T^k(\mathbf{x} - \mathbf{x}^{(0)}) = \lim_{k \to \infty} (\mathbf{x} - \mathbf{x}^{(k)}) = \mathbf{0}.$$

Consequently, if $\mathbf{z}$ is an arbitrary vector and $\mathbf{x}^{(0)} = \mathbf{x} - \mathbf{z}$, then

$$\lim_{k \to \infty} T^k\mathbf{z} = \lim_{k \to \infty} T^k(\mathbf{x} - (\mathbf{x} - \mathbf{z})) = \mathbf{0},$$

which, by Theorem 8.18, implies that $\rho(T) < 1$.   □

The proof of the following corollary is considered in Exercise 7.

COROLLARY 8.21   If $\|T\| < 1$ for any natural matrix norm, then the sequence $\{\mathbf{x}^{(k)}\}_{k=0}^{\infty}$ in Eq. (8.16) converges, for any $\mathbf{x}^{(0)} \in R^n$, to a vector $\mathbf{x} \in R^n$, and the following error bounds hold:

$$(8.19) \qquad \|\mathbf{x} - \mathbf{x}^{(k)}\| \leq \|T\|^k \|\mathbf{x}^{(0)} - \mathbf{x}\|;$$

$$(8.20) \qquad \|\mathbf{x} - \mathbf{x}^{(k)}\| \leq \frac{\|T\|^k}{1 - \|T\|} \|\mathbf{x}^{(1)} - \mathbf{x}^{(0)}\|.$$

To apply the preceding results to the Jacobi or Gauss–Seidel iterative techniques, we need to write the iteration matrices for the Jacobi method, T_j, given in Eq. (8.11) and the Gauss–Seidel method, T_g, given in Eq. (8.14) as

$$T_j = D^{-1}(L + U), \qquad T_g = (D - L)^{-1}U.$$

Should $\rho(T_j)$ or $\rho(T_g)$ be less than one, it is clear that the sequence $\{\mathbf{x}^{(k)}\}_{k=0}^{\infty}$ will converge to the solution $\mathbf{x}$ of $A\mathbf{x} = \mathbf{b}$. For example, the Jacobi scheme has

$$\mathbf{x}^{(k)} = D^{-1}(L + U)\mathbf{x}^{(k-1)} + D^{-1}\mathbf{b},$$

and, if $\{\mathbf{x}^{(k)}\}_{k=0}^{\infty}$ converges to $\mathbf{x}$, then

$$\mathbf{x} = D^{-1}(L + U)\mathbf{x} + D^{-1}\mathbf{b}.$$

This implies that

$$D\mathbf{x} = (L + U)\mathbf{x} + \mathbf{b} \qquad \text{and} \qquad (D - L - U)\mathbf{x} = \mathbf{b}.$$

Since $D - L - U = A$, $\mathbf{x}$ satisfies $A\mathbf{x} = \mathbf{b}$. We can now give easily verified sufficiency conditions for convergence of the Jacobi and Gauss–Seidel methods. (To prove convergence for the Jacobi scheme, see Exercise 8, and for the Gauss–Seidel scheme, see Ortega [94], page 120.)

THEOREM 8.22   If A is strictly diagonally dominant, then for any choice of $\mathbf{x}^{(0)}$ both the Jacobi and Gauss–Seidel methods give sequences $\{\mathbf{x}^{(k)}\}_{k=0}^{\infty}$ that converge to the solution of $A\mathbf{x} = \mathbf{b}$.

The relationship of the rapidity of convergence to the spectral radius of the iteration matrix T can be seen from inequality (8.19). Since (8.19) holds for any natural matrix norm, it follows, from the statement following Theorem 8.16 that

$$(8.21) \qquad \| \mathbf{x}^{(k)} - \mathbf{x} \| \approx \rho(T)^k \| \mathbf{x}^{(0)} - \mathbf{x} \|.$$

Suppose that $\rho(T) < 1$ and that $\mathbf{x}^{(0)} = \mathbf{0}$ is to be used in an iterative technique to approximate $\mathbf{x}$ with relative error at most 10^{-t}. By the estimate (8.21), the relative error after k iterations is approximately $\rho(T)^k$ so accuracy of 10^{-t} is expected if

$$\rho(T)^k \leq 10^{-t},$$

that is, if
$$k \geq \frac{t}{-\log_{10} \rho(T)}.$$

Thus, it is desirable to select the iterative technique with minimal $\rho(T) < 1$ for a particular system $A\mathbf{x} = \mathbf{b}$.

In general, it is not known which of the two techniques, Jacobi or Gauss–Seidel, should be used. However, in a special case the answer is known. The proof of the theorem will not be given here, but can be found in Young [145], pp. 120–127.

THEOREM 8.23   (Stein–Rosenberg)   If $a_{ij} \leq 0$ for each $i \neq j$ and $a_{ii} > 0$ for each $i = 1, 2, \ldots, n$, then one and only one of the following statements holds:

a) $0 < \rho(T_g) < \rho(T_j) < 1$,
b) $1 < \rho(T_j) < \rho(T_g)$,
c) $\rho(T_j) = \rho(T_g) = 0$,
d) $\rho(T_j) = \rho(T_g) = 1$.

For the special case described in Theorem 8.23, we see that when one method gives convergence, then both give convergence, with the Gauss–Seidel method being faster than the Jacobi method.

Since the rate of convergence of a procedure depends on the spectral radius of the matrix associated with the method, one way to select a procedure that will lead to accelerated convergence is to choose a method whose associated matrix has minimal spectral radius. Before describing a procedure for selecting such a method we need to introduce a new means of measuring the amount by which an approximation to the solution to a linear system differs from the true solution to the system. The method makes use of the vector described in the following definition.

DEFINITION 8.24   If $\tilde{\mathbf{x}} \in R^n$ is an approximation to the solution of the linear system defined by $A\mathbf{x} = \mathbf{b}$, the **residual vector** for $\tilde{\mathbf{x}}$ with respect to this system is defined by $\mathbf{r} = \mathbf{b} - A\tilde{\mathbf{x}}$.

In procedures such as the Jacobi or Gauss–Seidel methods, a residual vector is associated with each calculation of an approximation component to the solution vector. The object of the method is to generate a sequence of approximations that will cause the associated residual vectors to converge to zero. Suppose we let

$$\mathbf{r}_i^{(k)} = (r_{1i}^{(k)}, r_{2i}^{(k)}, \dots, r_{ni}^{(k)})^t$$

denote the residual vector for the Gauss–Seidel method corresponding to the approximate solution vector

$$(x_1^{(k)}, x_2^{(k)}, \dots, x_{i-1}^{(k)}, x_i^{(k-1)}, \dots, x_n^{(k-1)})^t.$$

The mth component of $\mathbf{r}_i^{(k)}$ is

(8.22)
$$r_{mi}^{(k)} = b_m - \sum_{j=1}^{i-1} a_{mj} x_j^{(k)} - \sum_{j=i}^{n} a_{mj} x_j^{(k-1)}$$

or

$$r_{mi}^{(k)} = b_m - \sum_{j=1}^{i-1} a_{mj} x_j^{(k)} - \sum_{j=i+1}^{n} a_{mj} x_j^{(k-1)} - a_{mi} x_i^{(k-1)}$$

for each $m = 1, 2, \dots, n$.

In particular, the ith component of $\mathbf{r}_i^{(k)}$ is

$$r_{ii}^{(k)} = b_i - \sum_{j=1}^{i-1} a_{ij} x_j^{(k)} - \sum_{j=i+1}^{n} a_{ij} x_j^{(k-1)} - a_{ii} x_i^{(k-1)};$$

so

(8.23)
$$a_{ii} x_i^{(k-1)} + r_{ii}^{(k)} = b_i - \sum_{j=1}^{i-1} a_{ij} x_j^{(k)} - \sum_{j=i+1}^{n} a_{ij} x_j^{(k-1)}.$$

Recall, however, that in the Gauss–Seidel method, $x_i^{(k)}$ is chosen to be

(8.24)
$$x_i^{(k)} = \frac{1}{a_{ii}} \left[b_i - \sum_{j=1}^{i-1} a_{ij} x_j^{(k)} - \sum_{j=i+1}^{n} a_{ij} x_j^{(k-1)} \right]$$

so Eq. (8.23) can be rewritten as

$$a_{ii} x_i^{(k-1)} + r_{ii}^{(k)} = a_{ii} x_i^{(k)}$$

or

(8.25)
$$x_i^{(k)} = x_i^{(k-1)} + \frac{r_{ii}^{(k)}}{a_{ii}}.$$

We can derive another connection between residual vectors and the Gauss–Seidel technique. By Eq. (8.22), the ith component of $\mathbf{r}_{i+1}^{(k)}$ is

$$r_{i,i+1}^{(k)} = b_i - \sum_{j=1}^{i} a_{ij} x_j^{(k)} - \sum_{j=i+1}^{n} a_{ij} x_j^{(k-1)}$$

$$= b_i - \sum_{j=1}^{i-1} a_{ij} x_j^{(k)} - \sum_{j=i+1}^{n} a_{ij} x_j^{(k-1)} - a_{ii} x_i^{(k)}.$$

Equation (8.24) implies that $r_{i,i+1}^{(k)} = 0$. In a sense, then, the Gauss–Seidel technique is devised to require that the ith component of $\mathbf{r}_{i+1}^{(k)}$ be zero.

Reducing one coordinate of the residual vector to zero, however, is not necessarily the most efficient way to reduce the norm of the vector $\mathbf{r}_{i+1}^{(k)}$. In fact, modifying the Gauss–Seidel procedure as given by Eq. (8.25) to:

$$(8.26) \qquad\qquad x_i^{(k)} = x_i^{(k-1)} + \omega \frac{r_{ii}^{(k)}}{a_{ii}}$$

for certain choices of positive ω will lead to significantly faster convergence.

Methods involving Eq. (8.26) are called **relaxation methods**. For choices of $0 < \omega < 1$, the procedures are called **under-relaxation methods** and can be used to obtain convergence of some systems that are not convergent by the Gauss–Seidel method. For choices $1 < \omega$ the procedures are called **over-relaxation methods**, which can be used to accelerate the convergence for systems that are convergent by the Gauss–Seidel technique. These methods are often abbreviated **SOR** for **Successive Over-Relaxation**, and are particularly useful for solving the linear systems that occur in the numerical solution of certain partial-differential equations.

Before illustrating the advantages of the SOR method, we note that, by using Eq. (8.23), Eq. (8.26) can be reformulated for calculation purposes to read:

$$(8.27) \qquad x_i^{(k)} = (1 - \omega)x_i^{(k-1)} + \frac{\omega}{a_{ii}}\left[b_i - \sum_{j=1}^{i-1} a_{ij}x_j^{(k)} - \sum_{j=i+1}^{n} a_{ij}x_j^{(k-1)}\right].$$

To determine the matrix form of the SOR method we rewrite (8.27) as

$$a_{ii}x_i^{(k)} + \omega \sum_{j=1}^{i-1} a_{ij}x_j^{(k)} = (1 - \omega)a_{ii}x_i^{(k-1)} - \omega \sum_{j=i+1}^{n} a_{ij}x_j^{(k-1)} + \omega b_i$$

so

$$(D - \omega L)\mathbf{x}^{(k)} = [(1 - \omega)D + \omega U]\mathbf{x}^{(k-1)} + \omega \mathbf{b}$$

or

$$(8.28) \qquad \mathbf{x}^{(k)} = (D - \omega L)^{-1}[(1 - \omega)D + \omega U]\mathbf{x}^{(k-1)} + \omega(D - \omega L)^{-1}\mathbf{b}.$$

EXAMPLE 3   The linear system $A\mathbf{x} = \mathbf{b}$ given by

$$\begin{aligned} 4x_1 + 3x_2 \qquad\;\; &= \;\;\; 24, \\ 3x_1 + 4x_2 - \;\; x_3 &= \;\;\; 30, \\ - \;\; x_2 + 4x_3 &= -24, \end{aligned}$$

has the solution $(3, 4, -5)^t$. Gauss–Seidel and the SOR method with $\omega = 1.25$ will be used to solve this system, using $\mathbf{x}^{(0)} = (1, 1, 1)^t$ for both methods. The equations for the Gauss–Seidel method are

$$\begin{aligned} x_1^{(k)} &= -0.75x_2^{(k-1)} + 6, \\ x_2^{(k)} &= -0.75x_1^{(k)} + 0.25x_3^{(k-1)} + 7.5, \\ x_3^{(k)} &= 0.25x_2^{(k)} - 6 \end{aligned}$$

for each $k = 1, 2, \ldots$, and the equations for the SOR method with $\omega = 1.25$ are

$$x_1^{(k)} = -0.25x_1^{(k-1)} - 0.9375x_2^{(k-1)} + 7.5,$$

$$x_2^{(k)} = -0.9375x_1^{(k)} - 0.25x_2^{(k-1)} + 0.3125x_3^{(k-1)} + 9.375,$$

$$x_3^{(k)} = 0.3125x_2^{(k)} - 0.25x_3^{(k-1)} - 7.5.$$

The first seven iterates for each method are listed in Tables 8.3 and 8.4.

TABLE 8.3   Gauss-Seidel

k	0	1	2	3	4	5	6	7
$x_1^{(k)}$	1	5.250000	3.1406250	3.0878906	3.0549316	3.0343323	3.0214577	3.0134110
$x_2^{(k)}$	1	3.812500	3.8828125	3.9267578	3.9542236	3.9713898	3.9821186	3.9888241
$x_3^{(k)}$	1	−5.046875	−5.0292969	−5.0183105	−5.0114441	−5.0071526	−5.0044703	−5.0027940

TABLE 8.4   SOR with $\omega = 1.25$

k	0	1	2	3	4	5	6	7
$x_1^{(k)}$	1	6.312500	2.6223145	3.1333027	2.9570512	3.0037211	2.9963276	3.0000498
$x_2^{(k)}$	1	3.5195313	3.9585266	4.0102646	4.0074838	4.0029250	4.0009262	4.0002586
$x_3^{(k)}$	1	−6.6501465	−4.6004238	−5.0966863	−4.9734897	−5.0057135	−4.9982822	−5.0003486

For the iterates to be accurate to seven decimal places the Gauss–Seidel method required 34 iterations, as opposed to 14 iterations for the over-relaxation method with $\omega = 1.25$.   ∎

The obvious question to ask is how the appropriate value of ω is chosen. Although no complete answer to this question is known for the general $n \times n$ linear system, the following results can be used in certain situations.

THEOREM 8.25   (Kahan)   If $a_{ii} \neq 0$ for each $i = 1, 2, \ldots, n$, then $\rho(T_\omega) \geq |\omega - 1|$. This implies that $\rho(T_\omega) < 1$ only if $0 < \omega < 2$, where

$$T_\omega = (D - \omega L)^{-1}[(1 - \omega)D + \omega U]$$

is the iteration matrix for the SOR method.

The proof of this theorem is considered in Exercise 9. The proof of the next two results can be found in Ortega [94], pp. 123–133. These results will be needed in Chapter 11.

THEOREM 8.26   (Ostrowski–Reich)   If A is a positive definite matrix and $0 < \omega < 2$, then the SOR method converges for any choice of initial approximate solution vector $\mathbf{x}^{(0)}$.

THEOREM 8.27   If A is positive definite and tridiagonal, then $\rho(T_g) = [\rho(T_j)]^2 < 1$, the optimal choice of ω for the SOR method is

$$\omega = \frac{2}{1 + \sqrt{1 - [\rho(T_j)]^2}},$$

and with this choice of ω, $\rho(T_\omega) = \omega - 1$.

EXAMPLE 4   In Example 3 the matrix A was given by

$$A = \begin{bmatrix} 4 & 3 & 0 \\ 3 & 4 & -1 \\ 0 & -1 & 4 \end{bmatrix}.$$

This matrix is positive definite and tridiagonal, so Theorem 8.27 applies. Since

$$T_j = D^{-1}(L + U)$$

$$= \begin{bmatrix} \tfrac{1}{4} & 0 & 0 \\ 0 & \tfrac{1}{4} & 0 \\ 0 & 0 & \tfrac{1}{4} \end{bmatrix} \begin{bmatrix} 0 & -3 & 0 \\ -3 & 0 & 1 \\ 0 & 1 & 0 \end{bmatrix} = \begin{bmatrix} 0 & -0.75 & 0 \\ -0.75 & 0 & 0.25 \\ 0 & 0.25 & 0 \end{bmatrix},$$

we have

$$T_j - \lambda I = \begin{bmatrix} -\lambda & -0.75 & 0 \\ -0.75 & -\lambda & 0.25 \\ 0 & 0.25 & -\lambda \end{bmatrix}$$

so $\det(T_j - \lambda I) = -\lambda(\lambda^2 - 0.625).$

Thus, $\rho(T_j) = \sqrt{0.625}$

and $\omega = \dfrac{2}{1 + \sqrt{1 - \rho(T_g)}} = \dfrac{2}{1 + \sqrt{1 - [\rho(T_j)]^2}} = \dfrac{2}{1 + \sqrt{1 - 0.625}} \approx 1.24.$

This explains the comparatively rapid convergence obtained by using $\omega = 1.25$ in Example 3.   ∎

We close this section with an algorithm for the SOR method.

SOR Algorithm 8.3

To solve $Ax = b$ given the parameter ω and an initial approximation $x^{(0)}$:

INPUT the number of equations and unknowns n; the entries a_{ij}, $1 \le i, j \le n$ of the matrix A; the entries b_i, $1 \le i \le n$ of the inhomogeneous term b; the entries XO_i, $1 \le i \le n$ of $XO = x^{(0)}$; the parameter ω; tolerance TOL; maximum number of iterations N.

OUTPUT the approximate solution $x_1, \ldots, x_n$ or a message that the number of iterations was exceeded.

Step 1   Set $k = 1$.

Step 2   While $(k \le N)$ do Steps 3–6.

    Step 3   For $i = 1, \ldots, n$

$$\text{set } x_i = (1 - \omega)XO_i + \frac{\omega(-\sum_{j=1}^{i-1} a_{ij}x_j - \sum_{j=i+1}^{n} a_{ij}XO_j + b_i)}{a_{ii}}.$$

    Step 4   If $\|\mathbf{x} - \mathbf{XO}\| < TOL$ then OUTPUT $(x_1, \ldots, x_n)$;
                        (*Procedure completed successfully.*)
                        STOP.

    Step 5   Set $k = k + 1$.

    Step 6   For $i = 1, \ldots, n$ set $XO_i = x_i$.

Step 7   OUTPUT ('Maximum number of iterations exceeded');
       (*Procedure completed unsuccessfully.*)
       STOP.

Exercise Set 8.2

1. Find the first two iterations of the Jacobi method for the following linear systems, using $\mathbf{x}^{(0)} = \mathbf{0}$.

a)
$$2x_1 - x_2 + x_3 = -1,$$
$$3x_1 + 3x_2 + 9x_3 = 0,$$
$$3x_1 + 3x_2 + 5x_3 = 4.$$

b)
$$2x_2 + 4x_3 = 0,$$
$$x_1 - x_2 - x_3 = 0.375,$$
$$x_1 - x_2 + 2x_3 = 0.$$

c)
$$10x_1 - x_2 = 9,$$
$$-x_1 + 10x_2 - 2x_3 = 7,$$
$$- 2x_2 + 10x_3 = 6.$$

d)
$$2x_1 = 3,$$
$$x_1 + 1.5x_2 = 4.5,$$
$$- 3x_2 + 0.5x_3 = -6.6,$$
$$2x_1 - 2x_2 + x_3 + x_4 = 0.8.$$

e)
$$2x_1 - x_2 + 10x_3 = -11,$$
$$3x_2 - x_3 + 8x_4 = -11,$$
$$10x_1 - x_2 + 2x_3 = 6,$$
$$-x_1 + 11x_2 - x_3 + 3x_4 = 25.$$

f)
$$10x_1 - x_2 + 2x_3 = 6,$$
$$-x_1 + 11x_2 - x_3 + 3x_4 = 25,$$
$$2x_1 - x_2 + 10x_3 = -11,$$
$$3x_2 - x_3 + 8x_4 = -11.$$

g)
$$4x_1 - 2x_2 = 0,$$
$$-2x_1 + 5x_2 - x_3 = 2,$$
$$- x_2 + 4x_3 + 2x_4 = 3,$$
$$2x_3 + 3x_4 = -2.$$

h)
$$4x_1 - x_2 - x_4 = 0,$$
$$-x_1 + 4x_2 - x_3 - x_5 = 5,$$
$$- x_2 + 4x_3 - x_6 = 0,$$
$$-x_1 + 4x_4 - x_5 = 6,$$
$$- x_2 - x_4 + 4x_5 - x_6 = -2,$$
$$- x_3 - x_5 + 4x_6 = 6.$$

2. Apply Algorithm 8.1 to solve the linear systems in Exercise 1, if possible. Use $TOL = 10^{-2}$ and the maximum number of iterations $N = 25$.

3. Repeat Exercise 1 using the Gauss–Seidel method.

4. Repeat Exercise 2 using Algorithm 8.2.

5. Repeat Exercise 1 using the SOR method with $\omega = 1.2$.

6. Solve the following linear systems using Algorithm 8.3 with the specified values of ω and the same values of TOL and N as in Exercise 2.

   a) Exercise 1(c) with $\omega = 0.5$ and $\omega = 1.1$.

   b) Exercise 1(c) with ω computed using Theorem 8.27.

   c) Exercise 1(g) with ω computed using Theorem 8.27.

   d) Exercise 1(h) with $\omega = 1.334$, $\omega = 1.95$, and $\omega = 0.95$.

7. Prove that

$$\| \mathbf{x}^{(k)} - \mathbf{x} \| \le \| T \|^k \| \mathbf{x}^{(0)} - \mathbf{x} \| \qquad \text{and} \qquad \| \mathbf{x}^{(k)} - \mathbf{x} \| \le \frac{\| T \|^k}{1 - \| T \|} \| \mathbf{x}^{(1)} - \mathbf{x}^{(0)} \|,$$

where T is an $n \times n$ matrix with $\| T \| < 1$ and

$$\mathbf{x}^{(k)} = T\mathbf{x}^{(k-1)} + \mathbf{c}, \; k = 1, 2, \ldots \qquad \text{with } \mathbf{x}^{(0)} \text{ arbitrary}, \; \mathbf{c} \in R^n, \text{ and } \mathbf{x} = T\mathbf{x} + \mathbf{c}.$$

   Apply the bounds to Exercise 1.

8. Show that if A is strictly diagonally dominant, then $\| T_j \|_\infty < 1$.

9. Prove Theorem 8.25. [*Hint:*   If $\lambda_1, \ldots, \lambda_n$ are the eigenvalues of T_ω, then det $T_\omega = \Pi_{i=1}^n \lambda_i$. Since det $D^{-1} = \det(D - \omega L)^{-1}$ and the determinant of a product of matrices is the product of the determinants of the factors, the result follows from Eq. (8.28).]

10. Use the Gauss–Seidel method to solve the linear system given in the introduction to Chapter 6 involving the potential at the junctions of the electrical circuit.

8.3 Error Estimates and Iterative Refinement

It seems intuitively reasonable that if $\tilde{\mathbf{x}}$ is an approximation to the solution $\mathbf{x}$ of $A\mathbf{x} = \mathbf{b}$ and the residual vector $\mathbf{r} = A\tilde{\mathbf{x}} - \mathbf{b}$ has the property that $\| \mathbf{r} \|$ is small, then $\| \mathbf{x} - \tilde{\mathbf{x}} \|$ would be small as well. Although this is often the case, certain special systems, which occur quite often in practice, fail to have this property.

EXAMPLE 1   The linear system $A\mathbf{x} = \mathbf{b}$ given by

$$\begin{bmatrix} 1 & 2 \\ 1.0001 & 2 \end{bmatrix} \begin{bmatrix} x_1 \\ x_2 \end{bmatrix} = \begin{bmatrix} 3 \\ 3.0001 \end{bmatrix}$$

has the unique solution $\mathbf{x} = (1, 1)^t$. The approximation to this solution $\tilde{\mathbf{x}} = (3, 0)^t$ has the residual vector

$$\mathbf{r} = \mathbf{b} - A\tilde{\mathbf{x}} = \begin{bmatrix} 3 \\ 3.0001 \end{bmatrix} - \begin{bmatrix} 1 & 2 \\ 1.0001 & 2 \end{bmatrix} \begin{bmatrix} 3 \\ 0 \end{bmatrix} = \begin{bmatrix} 0 \\ 0.0002 \end{bmatrix},$$

so $\|\mathbf{r}\|_\infty = 0.0002$. Although the norm of the residual vector is small, the approximation $\tilde{\mathbf{x}} = (3, 0)^t$ is obviously quite poor; in fact, $\|\mathbf{x} - \tilde{\mathbf{x}}\|_\infty = 2$. ∎

The difficulty that arose in Example 1 can be explained quite simply by noting that the solution to the system represents the intersection of the lines

$$l_1: \quad x_1 + 2x_2 = 3 \qquad \text{and} \qquad l_2: \quad 1.0001x_1 + 2x_2 = 3.0001.$$

The point $(3, 0)$ lies on l_1 and the lines are nearly parallel. This implies that $(3, 0)$ also lies close to l_2, even though it differs significantly from the intersection point $(1, 1)$. (*See Fig. 8.3.*)

FIGURE 8.3

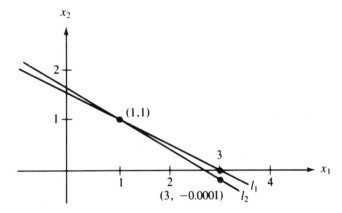

Example 1 was clearly constructed to show the difficulties that might, and in fact do, arise. Had the lines not been nearly parallel, it would be expected that a small residual vector implies an accurate approximation.

In the general situation, we cannot rely on the geometry of the system to give an indication of when problems might occur. We can, however, obtain this information by considering the norms of the matrix A and its inverse.

THEOREM 8.28   If $\tilde{\mathbf{x}}$ is an approximation to the solution of $A\mathbf{x} = \mathbf{b}$ and A is a nonsingular matrix, then for any natural norm,

(8.29) $$\|\mathbf{x} - \tilde{\mathbf{x}}\| \leq \|\mathbf{r}\| \, \|A^{-1}\|$$

and

(8.30) $$\frac{\|\mathbf{x} - \tilde{\mathbf{x}}\|}{\|\mathbf{x}\|} \leq \|A\| \, \|A^{-1}\| \frac{\|\mathbf{r}\|}{\|\mathbf{b}\|}, \qquad \text{provided } \mathbf{x} \neq \mathbf{0} \text{ and } \mathbf{b} \neq \mathbf{0},$$

where $\mathbf{r}$ is the residual vector for $\tilde{\mathbf{x}}$ with respect to the system $A\mathbf{x} = \mathbf{b}$.

PROOF   Since $\mathbf{r} = \mathbf{b} - A\tilde{\mathbf{x}} = A\mathbf{x} - A\tilde{\mathbf{x}}$ and A is nonsingular, $\mathbf{x} - \tilde{\mathbf{x}} = A^{-1}\mathbf{r}$ and Exercise 12 of Section 8.1 implies that

$$\|\mathbf{x} - \tilde{\mathbf{x}}\| = \|A^{-1}\mathbf{r}\| \le \|A^{-1}\| \, \|\mathbf{r}\|.$$

Moreover, since $\mathbf{b} = A\mathbf{x}$, $\|\mathbf{b}\| \le \|A\| \, \|\mathbf{x}\|$; so

$$\frac{\|\mathbf{x} - \tilde{\mathbf{x}}\|}{\|\mathbf{x}\|} \le \frac{\|A\| \, \|A^{-1}\|}{\|\mathbf{b}\|} \|\mathbf{r}\|. \qquad \square$$

Inequalities (8.29) and (8.30) imply that the quantities $\|A^{-1}\|$ and $\|A\| \, \|A^{-1}\|$ can be used to give an indication of the connection between the residual vector and the accuracy of the approximation. In general, the relative error $\|\mathbf{x} - \tilde{\mathbf{x}}\| / \|\mathbf{x}\|$ is of most interest and, by inequality (8.30), this error is bounded by the product of $\|A\| \, \|A^{-1}\|$ with the relative residual for this approximation, $\|\mathbf{r}\|/\|\mathbf{b}\|$. Any convenient norm can be used for this approximation, the only requirement being that it be used consistently throughout.

DEFINITION 8.29   The **condition number** $K(A)$ of the nonsingular matrix A relative to a natural norm $\|\cdot\|$ is defined to be

$$K(A) = \|A\| \, \|A^{-1}\|.$$

With this notation, the inequalities in Theorem 8.27 become

(8.31) $$\|\mathbf{x} - \tilde{\mathbf{x}}\| \le K(A) \frac{\|\mathbf{r}\|}{\|A\|}$$

and

(8.32) $$\frac{\|\mathbf{x} - \tilde{\mathbf{x}}\|}{\|\mathbf{x}\|} \le K(A) \frac{\|\mathbf{r}\|}{\|\mathbf{b}\|}.$$

Since for any nonsingular matrix A

$$1 = \|I\| = \|A \cdot A^{-1}\| \le \|A\| \, \|A^{-1}\| = K(A),$$

the expectation is that the matrix A will be well-behaved (formally called a **well-conditioned matrix**) if $K(A)$ is close to one and not well-behaved (called **ill-conditioned**), when $K(A)$ is significantly greater than one. Behavior in this instance refers to the relative security that a small residual vector implies a correspondingly accurate approximate solution.

EXAMPLE 2   The matrix for the system considered in Example 1 was

$$A = \begin{bmatrix} 1 & 2 \\ 1.0001 & 2 \end{bmatrix},$$

which has $\|A\|_\infty = 3.0001$. This norm would not be considered to be large. However,

$$A^{-1} = \begin{bmatrix} -10000 & 10000 \\ 5000.5 & -5000 \end{bmatrix};$$

$$\|A^{-1}\|_\infty = 20{,}000,$$

and, for the infinity norm, $K(A) = (20{,}000)(3.0001) = 60{,}002$. The size of the condition number for this example would certainly keep us from making hasty accuracy decisions based on the residual of an approximation.   ∎

Although in theory the condition number of a matrix depends totally on the norms of the matrix and its inverse, in practice the calculation of the inverse is subject to round-off error and is dependent on the accuracy with which the calculations are performed. If the operations involve arithmetic with t significant digits of accuracy, the approximate condition number for the matrix A is the norm of the matrix times the norm of the approximation to the inverse of A, which is obtained using t-digit arithmetic. In fact, this condition number even depends on the method used to calculate the inverse of A.

If we assume that the approximate solution to the linear system $A\mathbf{x} = \mathbf{b}$ is being determined using t-digit arithmetic and Gaussian elimination, it can be shown (see Forsythe and Moler [49], pp. 49–51) that the residual vector $\mathbf{r}$ for the approximation $\tilde{\mathbf{x}}$ has the property

$$(8.33) \qquad \|\mathbf{r}\| \approx 10^{-t} \|A\| \, \|\tilde{\mathbf{x}}\|.$$

From this approximate equation, an estimate for the effective condition number in t-digit arithmetic can be obtained without having to invert the matrix A. In actuality, the approximation in Eq. (8.33) assumes that all the arithmetic operations in the Gaussian elimination technique are performed using t-digit arithmetic, but that the operations that are needed to determine the residual are done in double precision (that is, $2t$-digit) arithmetic. This technique does not add significantly to the computational effort, and it eliminates much of the loss of accuracy (see Section 1.2) involved with the subtraction of the nearly equal numbers that occur in the calculation of the residual.

The approximation for the t-digit condition number $K(A)$ comes from consideration of the linear system

$$A\mathbf{y} = \mathbf{r}.$$

The solution to this system can be readily approximated since the multipliers for the Gaussian elimination method have already been calculated and presumably retained. In fact $\tilde{\mathbf{y}}$, the approximate solution of $A\mathbf{y} = \mathbf{r}$, satisfies

$$(8.34) \qquad \tilde{\mathbf{y}} \approx A^{-1}\mathbf{r} = A^{-1}(\mathbf{b} - A\tilde{\mathbf{x}}) = A^{-1}\mathbf{b} - A^{-1}A\tilde{\mathbf{x}} = \mathbf{x} - \tilde{\mathbf{x}};$$

so $\tilde{\mathbf{y}}$ is an estimate of the error in approximating the solution to the original system. Equation (8.33) can consequently be used to deduce that

$$\|\tilde{\mathbf{y}}\| \approx \|\mathbf{x} - \tilde{\mathbf{x}}\| = \|A^{-1}\mathbf{r}\| \leq \|A^{-1}\| \, \|\mathbf{r}\| \approx \|A^{-1}\|(10^{-t}\|A\| \, \|\tilde{\mathbf{x}}\|)$$
$$= 10^{-t}\|\tilde{\mathbf{x}}\| K(A).$$

This gives an approximation for the condition number involved with solving the system $A\mathbf{x} = \mathbf{b}$ using Gaussian elimination and the t-digit type of arithmetic just described:

(8.35) $$K(A) \approx \frac{\|\tilde{\mathbf{y}}\|}{\|\tilde{\mathbf{x}}\|} 10^t.$$

EXAMPLE 3   The linear system given by

$$\begin{bmatrix} 3.3330 & 15920 & -10.333 \\ 2.2220 & 16.710 & 9.6120 \\ 1.5611 & 5.1791 & 1.6852 \end{bmatrix} \begin{bmatrix} x_1 \\ x_2 \\ x_3 \end{bmatrix} = \begin{bmatrix} 15913 \\ 28.544 \\ 8.4254 \end{bmatrix}$$

has the exact solution $\mathbf{x} = (1, 1, 1)^t$.

Using Gaussian elimination and five-digit rounding arithmetic leads successively to the augmented matrices

$$\begin{bmatrix} 3.3330 & 15920 & -10.333 & \vdots & 15913 \\ 0 & -10596 & 16.501 & \vdots & 10580 \\ 0 & -7451.4 & 6.5250 & \vdots & -7444.9 \end{bmatrix}$$

and

$$\begin{bmatrix} 3.3330 & 15920 & -10.333 & \vdots & 15913 \\ 0 & -10596 & 16.501 & \vdots & -10580 \\ 0 & 0 & -5.0790 & \vdots & -4.7000 \end{bmatrix}$$

The approximate solution to this system is

$$\tilde{\mathbf{x}} = (1.2001, 0.99991, 0.92538)^t.$$

The residual vector corresponding to $\tilde{\mathbf{x}}$ is computed in double precision to be

$$\mathbf{r} = \mathbf{b} - A\tilde{\mathbf{x}}$$
$$= \begin{bmatrix} 15913 \\ 28.544 \\ 8.4254 \end{bmatrix} - \begin{bmatrix} 3.3330 & 15920 & -10.333 \\ 2.2220 & 16.710 & 9.6120 \\ 1.5611 & 5.1791 & 1.6852 \end{bmatrix} \begin{bmatrix} 1.2001 \\ 0.99991 \\ 0.92538 \end{bmatrix}$$
$$= \begin{bmatrix} -0.00518 \\ 0.27413 \\ -0.18616 \end{bmatrix};$$

so $$\|\mathbf{r}\|_\infty = 0.27413.$$

The estimate for the condition number given in the preceding discussion is obtained by first solving the system $A\mathbf{y} = \mathbf{r}$:

$$\begin{bmatrix} 3.3330 & 15920 & -10.333 \\ 2.2220 & 16.710 & 9.6120 \\ 1.5611 & 5.1791 & 1.6852 \end{bmatrix} \begin{bmatrix} y_1 \\ y_2 \\ y_3 \end{bmatrix} = \begin{bmatrix} -0.00518 \\ 0.27413 \\ -0.18616 \end{bmatrix},$$

which implies that $\tilde{\mathbf{y}} = (-0.20008, 8.9987 \times 10^{-5}, 0.074607)^t$. Using the estimate given in Eq. (8.35):

(8.36) $$K(A) \approx 10^5 \frac{\|\tilde{\mathbf{y}}\|_\infty}{\|\tilde{\mathbf{x}}\|_\infty} = \frac{10^5(0.20008)}{1.2001} = 16672.$$

To determine the *exact* condition number of A, we first must construct A^{-1}.

Using five-digit rounding arithmetic for the calculations gives the approximation:

$$A^{-1} = \begin{bmatrix} -1.1701 \times 10^{-4} & -1.4983 \times 10^{-1} & 8.5416 \times 10^{-1} \\ 6.2782 \times 10^{-5} & 1.2124 \times 10^{-4} & -3.0662 \times 10^{-4} \\ -8.6631 \times 10^{-5} & 1.3846 \times 10^{-1} & -1.9689 \times 10^{-1} \end{bmatrix}.$$

Theorem 8.12 can be used to show that $\|A^{-1}\|_\infty = 1.0041$ and $\|A\|_\infty = 15934$.

As a consequence, the ill-condition matrix A has

(8.37) $$K(A) = (1.0041)(15934) = 15999.$$

The estimate in (8.36) is quite close to $K(A)$ and requires considerably less computational effort.

Since the actual solution $\mathbf{x} = (1, 1, 1)^t$ is known for this system, we can calculate both

$$\|\mathbf{x} - \tilde{\mathbf{x}}\|_\infty = 0.2001 \qquad \text{and} \qquad \frac{\|\mathbf{x} - \tilde{\mathbf{x}}\|_\infty}{\|\mathbf{x}\|_\infty} = \frac{0.2001}{1} = 0.2001.$$

The error bounds given in Theorem 8.28 for these values are

$$\|\mathbf{x} - \tilde{\mathbf{x}}\|_\infty \le K(A)\frac{\|\mathbf{r}\|_\infty}{\|A\|_\infty} = \frac{(15999)(0.27413)}{15934} = 0.27525$$

and $$\frac{\|\mathbf{x} - \tilde{\mathbf{x}}\|_\infty}{\|\mathbf{x}\|_\infty} \le K(A)\frac{\|\mathbf{r}\|_\infty}{\|\mathbf{b}\|_\infty} = \frac{(15999)(0.27413)}{15913} = 0.27561. \qquad \blacksquare$$

In Eq. (8.34) we used the estimate $\tilde{\mathbf{y}} \approx \mathbf{x} - \tilde{\mathbf{x}}$ where $\tilde{\mathbf{y}}$ is the approximate solution to the system $A\mathbf{y} = \mathbf{r}$. It would be reasonable to suspect from this result that $\tilde{\mathbf{x}} + \tilde{\mathbf{y}}$ is a more accurate approximation to the solution of the linear system $A\mathbf{x} = \mathbf{b}$ than the original approximation $\tilde{\mathbf{x}}$. The method using this assumption is called **iterative refinement**, or iterative improvement, and consists of performing iterations on the system whose right-hand side is the residual vector for successive approximations until satisfactory accuracy results. The procedure is generally used only on systems when it is suspected that the matrix involved is ill-conditioned, since this technique will not significantly improve the approximation for a well-conditioned system.

Iterative Refinement Algorithm 8.4

To approximate the solution to the linear system $A\mathbf{x} = \mathbf{b}$ when A is suspected to be ill-conditioned:

INPUT the number of equations and unknowns n; the entries a_{ij}, $1 \leq i, j \leq n$ of the matrix A; the entries b_i, $1 \leq i \leq n$ of $\mathbf{b}$; the maximum number of iterations N; tolerance TOL.

OUTPUT the approximation $\mathbf{xx} = (xx_1, \ldots, xx_n)^t$ or a message that the number of iterations was exceeded.

Step 0   Solve the system $A\mathbf{x} = \mathbf{b}$ for $x_1, \ldots, x_n$ by Gaussian elimination saving the multipliers m_{ji}, $j = i + 1$, $i + 2, \ldots, n$, $i = 1$, $2, \ldots, n - 1$ and noting row interchanges.

Step 1   Set $k = 1$.

Step 2   While $(k \leq N)$ do Steps 3–8.

    Step 3   For $i = 1, 2, \ldots, n$ (*Calculate* $\mathbf{r}$.)

$$\text{set } r_i = b_i - \sum_{j=1}^{n} a_{ij} x_j.$$

        (*Perform the computation in double-precision arithmetic.*)

    Step 4   Solve the linear system $A\mathbf{y} = \mathbf{r}$ by using Gaussian elimination in the same order as in Step 0.

    Step 5   For $i = 1, \ldots, n$ set $xx_i = x_i + y_i$.

    Step 6   If $\|\mathbf{x} - \mathbf{xx}\|_\infty < TOL$ then OUTPUT ($\mathbf{xx}$);
                               (*Procedure completed successfully.*)
                               STOP.

    Step 7   Set $k = k + 1$.

    Step 8   For $i = 1, \ldots, n$ set $x_i = xx_i$.

Step 9   OUTPUT ('Maximum number of iterations exceeded');
        (*Procedure completed unsuccessfully.*)
        STOP.

A recommended stopping procedure in Step 6 is to iterate until $|y_i^{(k)}| \leq 10^{-t}$ for each $i = 1, 2, \ldots, n$, if t-digit arithmetic is being used. It should be emphasized that the iterative refinement technique will not give satisfactory results for all systems involving ill-conditioned matrices. In particular, if $K(A) \geq 10^t$, the procedure will probably fail, and the only alternative is to use increased precision for the calculations.

EXAMPLE 4   In Example 3 we found the approximation to the problem we have been considering, using five-digit arithmetic and Gaussian elimination, to be

$$\tilde{\mathbf{x}}^{(1)} = (1.2001, 0.99991, 0.92538)^t$$

and the solution to $A\mathbf{y} = \mathbf{r}^{(1)}$ to be

$$\tilde{\mathbf{y}}^{(1)} = (-0.20008, 8.9987 \times 10^{-5}, 0.074607)^t.$$

By Step 5 in the algorithm, this implies that

$$\tilde{\mathbf{x}}^{(2)} = \tilde{\mathbf{x}}^{(1)} + \tilde{\mathbf{y}}^{(1)} = (1.0000, 1.0000, 0.99999)^t,$$

and the actual error in this approximation is

$$\|\mathbf{x} - \tilde{\mathbf{x}}^{(2)}\|_\infty = 1 \times 10^{-5}.$$

Using the suggested stopping technique for the algorithm, we compute $\mathbf{r}^{(2)} = \mathbf{b} - A\tilde{\mathbf{x}}^{(2)}$, and solve the system $A\mathbf{y}^{(2)} = \mathbf{r}^{(2)}$, which gives

$$\tilde{\mathbf{y}}^{(2)} = (1.5002 \times 10^{-9}, 2.0951 \times 10^{-10}, 1.0000 \times 10^{-5})^t.$$

Since $\|\tilde{\mathbf{y}}^{(2)}\|_\infty \le 10^{-5}$, we conclude that

$$\tilde{\mathbf{x}}^{(3)} = \tilde{\mathbf{x}}^{(2)} + \tilde{\mathbf{y}}^{(2)} = (1.0000, 1.0000, 1.0000)^t$$

is sufficiently accurate. This is certainly correct.   ■

Throughout this section it has been assumed that, in the linear system $A\mathbf{x} = \mathbf{b}$, A and $\mathbf{b}$ could be represented exactly. Realistically, the entries a_{ij} and b_j might be altered or perturbed by an amount δa_{ij} and δb_j, causing the linear system

$$(A + \delta A)\mathbf{x} = \mathbf{b} + \delta \mathbf{b}$$

to be solved in place of $A\mathbf{x} = \mathbf{b}$. Normally, if $\|\delta A\|$ and $\|\delta \mathbf{b}\|$ are small, on the order of 10^{-t}, the t-digit arithmetic should yield a solution $\tilde{\mathbf{x}}$ for which $\|\mathbf{x} - \tilde{\mathbf{x}}\|$ is correspondingly small. However, in the case of ill-conditioned systems, we have seen that even if A and $\mathbf{b}$ are represented exactly, rounding errors can cause $\|\mathbf{x} - \tilde{\mathbf{x}}\|$ to be large. The following theorem relates the perturbations of linear systems to the condition number of a matrix. (The proof of this result can be found in Ortega [94], page 33.)

THEOREM 8.30   Suppose A is nonsingular and

$$\|\delta A\| < \frac{1}{\|A^{-1}\|}.$$

The solution $\tilde{\mathbf{x}}$ to $(A + \delta A)\tilde{\mathbf{x}} = \mathbf{b} + \delta \mathbf{b}$ approximates the solution $\mathbf{x}$ of $A\mathbf{x} = \mathbf{b}$ with error estimate

$$(8.38) \qquad \frac{\|\mathbf{x} - \tilde{\mathbf{x}}\|}{\|\mathbf{x}\|} \le \frac{K(A)}{1 - K(A)(\|\delta A\|/\|A\|)} \left(\frac{\|\delta \mathbf{b}\|}{\|\mathbf{b}\|} + \frac{\|\delta A\|}{\|A\|} \right).$$

The estimate in inequality (8.38) states that if the matrix A is well-conditioned, that is, $K(A)$ is not too large, then small changes in A and b produce correspondingly small changes in the solution x. On the other hand, if A is ill-conditioned, then small changes in A and b may produce large changes in x.

This theorem is independent of the particular numerical procedure used to solve $Ax = b$. It can be shown, by means of Wilkinson's backward error analysis (see Wilkinson [141] or [142]), that if Gaussian elimination with pivoting is used to solve $Ax = b$ in t-digit arithmetic, the numerical solution $\tilde{x}$ is the actual solution of a linear system

$$(8.39) \qquad\qquad\qquad (A + \delta A)\tilde{x} = b$$

where

$$(8.40) \qquad\qquad\qquad \|\delta A\|_\infty \le f(n)10^{1-t} \max_{i,j,k} |a_{ij}^{(k)}|.$$

Wilkinson has found in practice that $f(n) \approx n$ and, at worst, $f(n) \le 1.01(n^3 + 3n^2)$.

Exercise Set 8.3

1. Solve the following linear systems using Gaussian elimination and iterative refinement. Estimate $K(A)$.

   a) $3.9x_1 + 1.6x_2 = 5.5,$
      $6.8x_1 + 2.9x_2 = 9.7;$

      use two-digit rounding arithmetic.

   b) $4.56x_1 + 2.18x_2 = 6.74,$
      $2.79x_1 + 1.38x_2 = 4.13;$

      use three-digit rounding arithmetic.

   c) $x_1 + \frac{1}{2}x_2 + \frac{1}{3}x_3 = \frac{11}{6},$

      $5x_1 + \frac{10}{3}x_2 + \frac{5}{2}x_3 = \frac{65}{6},$

      $\frac{100}{3}x_1 + 25x_2 + 20x_3 = \frac{235}{3};$

      use three-digit rounding arithmetic.

   d) $\frac{1}{4}x_1 + \frac{1}{5}x_2 + \frac{1}{6}x_3 = 9,$

      $\frac{1}{3}x_1 + \frac{1}{4}x_2 + \frac{1}{5}x_3 = 8,$

      $\frac{1}{2}x_1 + x_2 + 2x_3 = 8;$

      use single-precision computer arithmetic.

2. Compute the condition numbers of the following matrices relative to $\|\cdot\|_\infty$.

   a) $\begin{bmatrix} 1 & 2 \\ 1.0001 & 2 \end{bmatrix}$

   b) $\begin{bmatrix} 3.9 & 1.6 \\ 6.8 & 2.9 \end{bmatrix}$

   c) $\begin{bmatrix} 4.56 & 2.18 \\ 2.79 & 1.38 \end{bmatrix}$

   d) $\begin{bmatrix} 1.003 & 58.09 \\ 5.550 & 321.8 \end{bmatrix}$

3. Show that if B is singular, then

$$\frac{1}{K(A)} \le \frac{\|A - B\|}{\|A\|}.$$

[*Hint*: There exists a vector $x \ne 0$, with $\|x\| = 1$, such that $Bx = 0$. Derive the estimate using $\|Ax\| \ge \|x\|/\|A^{-1}\|$.]

4. Using Exercise 3, estimate the condition numbers for the following matrices:

a) $\begin{bmatrix} 1 & 2 \\ 1.0001 & 2 \end{bmatrix}$

b) $\begin{bmatrix} 3.9 & 1.6 \\ 6.8 & 2.9 \end{bmatrix}$

5. The linear system

$$\begin{bmatrix} 1 & 2 \\ 1.0001 & 2 \end{bmatrix} \begin{bmatrix} x_1 \\ x_2 \end{bmatrix} = \begin{bmatrix} 3 \\ 3.0001 \end{bmatrix}$$

has solution $(1, 1)^t$. Change A slightly to

$$\begin{bmatrix} 1 & 2 \\ 0.9999 & 2 \end{bmatrix}$$

and consider the linear system

$$\begin{bmatrix} 1 & 2 \\ 0.9999 & 2 \end{bmatrix} \begin{bmatrix} x_1 \\ x_2 \end{bmatrix} = \begin{bmatrix} 3 \\ 3.0001 \end{bmatrix}.$$

Compute the new solution $\tilde{x}$ using five-digit arithmetic, and compare the actual error to the estimate (8.38). Is A ill-conditioned?

6. The linear system $A\mathbf{x} = \mathbf{b}$ given by

$$\begin{bmatrix} 1 & 2 \\ 1.00001 & 2 \end{bmatrix} \begin{bmatrix} x_1 \\ x_2 \end{bmatrix} = \begin{bmatrix} 3 \\ 3.00001 \end{bmatrix}$$

has solution $(1, 1)^t$. Find the solution of the perturbed system, using seven-digit arithmetic

$$\begin{bmatrix} 1 & 2 \\ 1.000011 & 2 \end{bmatrix} \begin{bmatrix} x_1 \\ x_2 \end{bmatrix} = \begin{bmatrix} 3.00001 \\ 3.00003 \end{bmatrix},$$

and compare the actual error to the estimate (8.38). Is A ill-conditioned?

7. Using a computer, analyze the linear system in Exercise 1(d) with respect to the estimate (8.40). Is $f(n)$ closer to n or $1.01(n^3 + 3n^2)$?

8. The $n \times n$ **Hilbert** matrix $H^{(n)}$ defined by

$$H_{ij}^{(n)} = \frac{1}{i+j-1}, \qquad 1 \le i, j \le n$$

is an ill-conditioned matrix that arises in solving the normal equations for the coefficients of the least-squares polynomial (see Example 1 of Section 7.2).

a) Show that

$$[H^{(4)}]^{-1} = \begin{bmatrix} 16 & -120 & 240 & -140 \\ -120 & 1200 & -2700 & 1680 \\ 240 & -2700 & 6480 & -4200 \\ -140 & 1680 & -4200 & 2800 \end{bmatrix},$$

and compute $K(H^{(4)})$ relative to $\| \cdot \|_{\infty}$.

b) Show that

$$[H^{(5)}]^{-1} = \begin{bmatrix} 52 & -300 & 1050 & -1400 & 630 \\ -300 & 4800 & -18900 & 26880 & -12600 \\ 1050 & -18900 & 79380 & -117600 & 56700 \\ -1400 & 26880 & -117600 & 179200 & -88200 \\ 630 & -12600 & 56700 & -88200 & 44100 \end{bmatrix},$$

and compute $K(H^{(5)})$ relative to $\|\cdot\|_\infty$.

c) Solve the linear system

$$H^{(4)} \begin{bmatrix} x_1 \\ x_2 \\ x_3 \\ x_4 \end{bmatrix} = \begin{bmatrix} 1 \\ 0 \\ 0 \\ 1 \end{bmatrix}$$

using three-digit arithmetic, and compare the actual error to that estimated in (8.38).

8.4 Eigenvalues and Eigenvectors

The solution of many physical problems requires the calculation, or at least estimation, of the eigenvalues and corresponding eigenvectors of a matrix associated with a linear system of equations. We have seen (see Definitions 8.13 and 8.14 of Section 8.1) that an $n \times n$ matrix A has precisely n, not necessarily distinct, eigenvalues that are the roots of the polynomial $p(\lambda) = \det(A - \lambda I)$. Theoretically the eigenvalues of A can be obtained by finding the n roots of $p(\lambda)$ and then the associated linear systems can be solved to determine the corresponding eigenvectors. In practice, $p(\lambda)$ is difficult to obtain, and in any case, we have seen in Section 2.6 that it can be difficult to determine the roots of an nth-degree polynomial, except for small values of n. As a consequence it is necessary to construct approximation techniques for finding eigenvalues.

Before considering further results concerning eigenvalues and eigenvectors, we need some definitions and results from linear algebra. All of the general results that will be needed in the remainder of this chapter are listed here for ease of reference. The proofs of these results will not be given, but are available in most standard texts on linear algebra (see, for example, Noble and Daniel [92]).

DEFINITION 8.31   Let $\{v^{(1)}, v^{(2)}, v^{(3)}, \ldots, v^{(k)}\}$ be a set of vectors. The set is said to be **linearly dependent** if numbers $\alpha_1, \alpha_2, \ldots, \alpha_k$ exist, not all zero, with

$$0 = \alpha_1 v^{(1)} + \alpha_2 v^{(2)} + \alpha_3 v^{(3)} + \cdots + \alpha_k v^{(k)}.$$

A set of vectors that is not linearly dependent is called **linearly independent**.

THEOREM 8.32 If $\{\mathbf{v}^{(1)}, \mathbf{v}^{(2)}, \mathbf{v}^{(3)}, \ldots, \mathbf{v}^{(n)}\}$ is a set of n linearly independent vectors in R^n, then any vector $\mathbf{x} \in R^n$ can be written uniquely as

$$\mathbf{x} = \beta_1 \mathbf{v}^{(1)} + \beta_2 \mathbf{v}^{(2)} + \beta_3 \mathbf{v}^{(3)} + \cdots + \beta_n \mathbf{v}^{(n)}$$

for some collection of constants $\beta_1, \beta_2, \ldots, \beta_n$.

Any collection of n linearly independent vectors in R^n is called a **basis** for R^n.

EXAMPLE 1 Let $\mathbf{v}^{(1)} = (1, 0, 0)^t$, $\mathbf{v}^{(2)} = (-1, 1, 1)^t$, and $\mathbf{v}^{(3)} = (0, 4, 2)^t$. If α_1, α_2, and α_3 are numbers with

$$\mathbf{0} = \alpha_1 \mathbf{v}^{(1)} + \alpha_2 \mathbf{v}^{(2)} + \alpha_3 \mathbf{v}^{(3)},$$

then

$$(0, 0, 0)^t = \alpha_1 (1, 0, 0)^t + \alpha_2 (-1, 1, 1)^t + \alpha_3 (0, 4, 2)^t$$
$$= (\alpha_1 - \alpha_2, \alpha_2 + 4\alpha_3, \alpha_2 + 2\alpha_3)^t,$$

so $\alpha_1 - \alpha_2 = 0, \qquad \alpha_2 + 4\alpha_3 = 0, \qquad \text{and} \qquad \alpha_2 + 2\alpha_3 = 0.$

Since the only solution to this system is $\alpha_1 = \alpha_2 = \alpha_3 = 0$, the set $\{\mathbf{v}^{(1)}, \mathbf{v}^{(2)}, \mathbf{v}^{(3)}\}$ is linearly independent in R^3 and is a basis for R^3.

Any vector $\mathbf{x} = (x_1, x_2, x_3)^t$ in R^3 can be written as

$$\mathbf{x} = \beta_1 \mathbf{v}^{(1)} + \beta_2 \mathbf{v}^{(2)} + \beta_3 \mathbf{v}^{(3)}$$

by choosing

$$\beta_1 = x_1 - x_2 + 2x_3, \qquad \beta_2 = 2x_3 - x_2, \qquad \text{and} \qquad \beta_3 = \tfrac{1}{2}(x_2 - x_3). \qquad \blacksquare$$

DEFINITION 8.33 Two $n \times n$ matrices A and B are said to be **similar** if a nonsingular matrix S exists with $A = S^{-1}BS$.

THEOREM 8.34 Suppose A and B are similar $n \times n$ matrices and λ is an eigenvalue of A, with associated eigenvector $\mathbf{x}$. Then λ is also an eigenvalue of B and if $A = S^{-1}BS$, then $S\mathbf{x}$ is an eigenvector associated with λ and the matrix B.

DEFINITION 8.35 A set of vectors $\{\mathbf{v}^{(1)}, \mathbf{v}^{(2)}, \ldots, \mathbf{v}^{(n)}\}$ is called **orthogonal** if $(\mathbf{v}^{(i)})^t \mathbf{v}^{(j)} = 0$ for all $i \neq j$. If, in addition, $(\mathbf{v}^{(i)})^t \mathbf{v}^{(i)} = 1$ for all $i = 1, 2, \ldots, n$, then the set is called **orthonormal**.

THEOREM 8.36 An orthogonal set of vectors that does not contain the zero vector is linearly independent.

THEOREM 8.37 If A is a matrix and $\lambda_1, \ldots, \lambda_k$ are distinct eigenvalues of A with associated eigenvectors $\mathbf{x}^{(1)}, \mathbf{x}^{(2)}, \ldots, \mathbf{x}^{(k)}$, then $\{\mathbf{x}^{(1)}, \mathbf{x}^{(2)}, \ldots, \mathbf{x}^{(k)}\}$ is linearly independent.

DEFINITION 8.38   An $n \times n$ matrix P is said to be an **orthogonal matrix** if $P^{-1} = P^t$.

The terminology in Definition 8.38 follows from the fact that the columns of an orthogonal matrix will form an orthogonal, in fact orthonormal, set of vectors. (See Exercise 14.)

THEOREM 8.39   If A is an $n \times n$ symmetric matrix, then there exists an orthogonal matrix P such that $D = P^{-1}AP = P^t AP$, where D is a diagonal matrix whose diagonal entries are the eigenvalues of A.

Theorem 8.39 implies that every symmetric matrix is similar to a diagonal matrix. This property forms the basis for the results in Section 8.5. The generalization of this theorem to arbitrary nonsymmetric matrices will not be considered in this text, but is an important area in linear algebra known as the Jordan canonical form of a matrix. A special case connected with this result is given in the following.

THEOREM 8.40   (Schur)   Let A be an arbitrary $n \times n$ matrix. A nonsingular matrix U exists with the property that

$$T = U^{-1}AU,$$

where T is an upper-triangular matrix whose diagonal entries consist of the eigenvalues of A.

The matrix U, whose existence is ensured in Theorem 8.40, can actually be chosen to satisfy a very important condition: $\| U\mathbf{x} \|_2 = \| \mathbf{x} \|_2$ for any vector $\mathbf{x}$. Matrices with this property are called **unitary matrices**. Although we will not make use of this norm-preserving property, it does significantly increase the application of the theorem.

THEOREM 8.41   If A is a symmetric $n \times n$ matrix, then there exist n eigenvectors of A that form an orthonormal set.

THEOREM 8.42   The eigenvalues of a symmetric matrix are all real numbers.

This completes the background material that will be needed from linear algebra. The first topic we will consider concerning the approximation of eigenvalues is an important result that gives bounds for these approximations.

THEOREM 8.43   (Gerschgorin Circle Theorem)   Let A be an $n \times n$ matrix and R_i denote the circle in the complex plane with center a_{ii} and radius $\sum_{\substack{j=1, \\ j \neq i}}^{n} |a_{ij}|$; that is,

$$R_i = \left\{ z \in \cancel{C} \,\middle|\, |z - a_{ii}| \leq \sum_{\substack{j=1 \\ j \neq i}}^{n} |a_{ij}| \right\},$$

where $\cancel{C}$ is used to denote the complex plane. The eigenvalues of A are contained within

$R = \bigcup_{i=1}^{n} R_i$ and the union of any k of these circles that do not intersect the remaining $(n - k)$ must contain precisely k (counting multiplicities) of the eigenvalues.

PROOF

The result in Exercise 12 of Section 8.1 implies that an eigenvector $\mathbf{x}$ associated with λ can be found with $\|\mathbf{x}\|_\infty = 1$.

Since $A\mathbf{x} - \lambda\mathbf{x} = \mathbf{0}$, the equivalent component representation is

$$(8.41) \qquad \sum_{j=1}^{n} a_{ij}x_j = \lambda x_i \qquad \text{for each } i = 1, 2, \ldots, n.$$

If k is an integer with $|x_k| = \|\mathbf{x}\|_\infty = 1$, Eq. (8.41), with $i = k$, implies that

$$\sum_{j=1}^{n} a_{kj}x_j = \lambda x_k.$$

Thus

$$\sum_{\substack{j=1, \\ j \neq k}}^{n} a_{kj}x_j = \lambda x_k - a_{kk}x_k,$$

and

$$|(a_{kk} - \lambda)x_k| = \left| \sum_{\substack{j=1, \\ j \neq k}}^{n} - a_{kj}x_j \right| \leq \sum_{\substack{j=1, \\ j \neq k}}^{n} |a_{kj}| |x_j|.$$

Since $|x_k| \geq |x_j|$ for all $j = 1, 2, \ldots, n$,

$$|a_{kk} - \lambda| \leq \sum_{\substack{j=1, \\ j \neq k}}^{n} |a_{kj}| \left| \frac{x_j}{x_k} \right| \leq \sum_{\substack{j=1, \\ j \neq k}}^{n} |a_{kj}|.$$

Thus $\lambda \in R_k$, which proves the first assertion in the theorem. The second part of this theorem requires a clever continuity argument. A quite readable proof is contained in Ortega [92], page 48. $\square$

EXAMPLE 2

For the matrix

$$A = \begin{bmatrix} 4 & 1 & 1 \\ 0 & 2 & 1 \\ -2 & 0 & 9 \end{bmatrix},$$

the circles in the Gerschgorin theorem are (*see Fig. 8.4*)

$$R_1 = \{z \in \mathbb{C} \,|\, |z - 4| \leq 2\},$$
$$R_2 = \{z \in \mathbb{C} \,|\, |z - 2| \leq 1\},$$
$$R_3 = \{z \in \mathbb{C} \,|\, |z - 9| \leq 2\}.$$

Since R_1 and R_2 are disjoint from R_3, there must be precisely two eigenvalues within $R_1 \cup R_2$ and one within R_3. Moreover, since $\rho(A) = \max_{1 \leq i \leq 3} |\lambda_i|$, $7 \leq \rho(A) \leq 11$. ∎

FIGURE 8.4

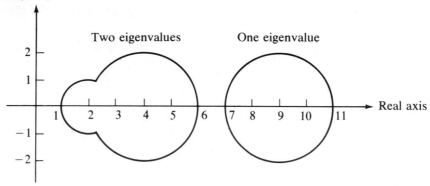

We will now develop a technique for approximating eigenvalues called the **power method**, that is iterative in nature. To apply the power method, we must assume that the $n \times n$ matrix A has n eigenvalues $\lambda_1, \lambda_2, \ldots, \lambda_n$ with an associated collection of eigenvectors $\{\mathbf{v}^{(1)}, \mathbf{v}^{(2)}, \mathbf{v}^{(3)}, \ldots, \mathbf{v}^{(n)}\}$ that is linearly independent. Moreover, we assume A has precisely one eigenvalue that is largest in magnitude and, for this reason, assume λ_1, $\lambda_2, \ldots, \lambda_n$ denote the eigenvalues of A with $|\lambda_1| > |\lambda_2| \geq |\lambda_3| \geq \cdots \geq |\lambda_n| \geq 0$.

If $\mathbf{x}$ is any vector in R^n, the fact that $\{\mathbf{v}^{(1)}, \mathbf{v}^{(2)}, \ldots, \mathbf{v}^{(n)}\}$ is linearly independent implies that constants $\alpha_1, \alpha_2, \ldots, \alpha_n$ exist with

(8.42)
$$\mathbf{x} = \sum_{j=1}^{n} \alpha_j \mathbf{v}^{(j)}.$$

Multiplying both sides of Eq. (8.42) by $A, A^2, \ldots, A^k$, we obtain:

$$A\mathbf{x} = \sum_{j=1}^{n} \alpha_j A\mathbf{v}^{(j)} = \sum_{j=1}^{n} \alpha_j \lambda_j \mathbf{v}^{(j)},$$

(8.43)
$$A^2\mathbf{x} = \sum_{j=1}^{n} \alpha_j \lambda_j A\mathbf{v}^{(j)} = \sum_{j=1}^{n} \alpha_j \lambda_j^2 \mathbf{v}^{(j)},$$

$$\vdots \qquad\qquad \vdots$$

$$A^k\mathbf{x} = \sum_{j=1}^{n} \alpha_j \lambda_j^k \mathbf{v}^{(j)}.$$

If λ_1^k is factored from each term on the right side of the equations in (8.43),

$$A^k\mathbf{x} = \lambda_1^k \sum_{j=1}^{n} \alpha_j \left(\frac{\lambda_j}{\lambda_1}\right)^k \mathbf{v}^{(j)}.$$

The fact that $|\lambda_1| > |\lambda_j|$ for all $j = 2, 3, \ldots, n$ implies that $\lim_{k \to \infty} (\lambda_j/\lambda_1)^k = 0$, and

(8.44)
$$\lim_{k \to \infty} A^k\mathbf{x} = \lim_{k \to \infty} \lambda_1^k \alpha_1 \mathbf{v}^{(1)}.$$

This sequence will converge to zero if $|\lambda_1| < 1$ and diverge if $|\lambda_1| \geq 1$, provided, of course, that $\alpha_1 \neq 0$.

Advantage can be made of the relationship expressed in Eq. (8.44) by scaling the powers of $A^k \mathbf{x}$ in an appropriate manner to ensure that the limit in Eq. (8.44) is finite and nonzero. The scaling begins by choosing $\mathbf{x}$ to be a unit vector $\mathbf{x}^{(0)}$ relative to $\| \cdot \|_\infty$ and a component $x_{p_0}^{(0)}$ of $\mathbf{x}^{(0)}$ with

$$x_{p_0}^{(0)} = 1 = \| \mathbf{x}^{(0)} \|_\infty.$$

Let $\mathbf{y}^{(1)} = A\mathbf{x}^{(0)}$, and define $\mu^{(1)} = y_{p_0}^{(1)}$.

With this notation,

$$\mu^{(1)} = y_{p_0}^{(1)} = \frac{y_{p_0}^{(1)}}{x_{p_0}^{(0)}} = \frac{\alpha_1 \lambda_1 v_{p_0}^{(1)} + \sum_{j=2}^n \alpha_j \lambda_j v_{p_0}^{(j)}}{\alpha_1 v_{p_0}^{(1)} + \sum_{j=2}^n \alpha_j v_{p_0}^{(j)}} = \lambda_1 \left[\frac{\alpha_1 v_{p_0}^{(1)} + \sum_{j=2}^n \alpha_j (\lambda_j/\lambda_1) v_{p_0}^{(j)}}{\alpha_1 v_{p_0}^{(1)} + \sum_{j=2}^n \alpha_j v_{p_0}^{(j)}} \right].$$

Let p_1 be the smallest integer, $1 \le p_1 \le n$, such that

$$|y_{p_1}^{(1)}| = \| \mathbf{y}^{(1)} \|_\infty$$

and define $\mathbf{x}^{(1)}$ by

$$\mathbf{x}^{(1)} = \frac{1}{y_{p_1}^{(1)}} \mathbf{y}^{(1)} = \frac{1}{y_{p_1}^{(1)}} A\mathbf{x}^{(0)}.$$

Then

$$x_{p_1}^{(1)} = 1 = \| \mathbf{x}^{(1)} \|_\infty.$$

Now define

$$\mathbf{y}^{(2)} = A\mathbf{x}^{(1)} = \frac{1}{y_{p_1}^{(1)}} A^2 \mathbf{x}^{(0)}$$

and

$$\mu^{(2)} = y_{p_1}^{(2)} = \frac{y_{p_1}^{(2)}}{x_{p_1}^{(1)}} = \frac{[\alpha_1 \lambda_1^2 v_{p_1}^{(1)} + \sum_{j=2}^n \alpha_j \lambda_j^2 v_{p_1}^{(j)}]/y_{p_1}^{(1)}}{[\alpha_1 \lambda_1 v_{p_1}^{(1)} + \sum_{j=2}^n \alpha_j \lambda_j v_{p_1}^{(j)}]/y_{p_1}^{(1)}}$$

$$= \lambda_1 \left[\frac{\alpha_1 v_{p_1}^{(1)} + \sum_{j=2}^n \alpha_j (\lambda_j/\lambda_1)^2 v_{p_1}^{(j)}}{\alpha_1 v_{p_1}^{(1)} + \sum_{j=2}^n \alpha_j (\lambda_j/\lambda_1) v_{p_1}^{(j)}} \right].$$

Let p_2 be the smallest integer with

$$|y_{p_2}^{(2)}| = \| \mathbf{y}^{(2)} \|_\infty$$

and define

$$\mathbf{x}^{(2)} = \frac{1}{y_{p_2}^{(2)}} \mathbf{y}^{(2)} = \frac{1}{y_{p_2}^{(2)}} A\mathbf{x}^{(1)} = \frac{1}{y_{p_2}^{(2)} y_{p_1}^{(1)}} A^2 \mathbf{x}^{(0)}.$$

In a similar manner, define sequences of vectors $\{\mathbf{x}^{(m)}\}_{m=0}^\infty$ and $\{\mathbf{y}^{(m)}\}_{m=1}^\infty$ and a sequence of scalars $\{\mu^{(m)}\}_{m=1}^\infty$ inductively by

(8.45) $$\mathbf{y}^{(m)} = A\mathbf{x}^{(m-1)}$$

(8.46) $$\mu^{(m)} = y_{p_{m-1}}^{(m)} = \lambda_1 \left[\frac{\alpha_1 v_{p_{m-1}}^{(1)} + \sum_{j=2}^n (\lambda_j/\lambda_1)^m \alpha_j v_{p_{m-1}}^{(j)}}{\alpha_1 v_{p_{m-1}}^{(1)} + \sum_{j=2}^n (\lambda_j/\lambda_1)^{m-1} \alpha_j v_{p_{m-1}}^{(j)}} \right],$$

and

(8.47) $$\mathbf{x}^{(m)} = \frac{\mathbf{y}^{(m)}}{y_{p_m}^{(m)}} = \frac{A^m \mathbf{x}^{(0)}}{\prod_{k=1}^m y_{p_k}^{(k)}},$$

where at each step p_m is used to represent the smallest integer for which

$$|y_{p_m}^{(m)}| = \|\mathbf{y}^{(m)}\|_\infty.$$

By examining Eq. (8.46), we see that since $|\lambda_j/\lambda_1| < 1$ for each $j = 2, 3, \ldots, n$, $\lim_{m \to \infty} \mu^{(m)} = \lambda_1$, provided that $\mathbf{x}^{(0)}$ is chosen so that $\alpha_1 \neq 0$. Moreover, it can be seen that the sequence of vectors $\{\mathbf{x}^{(m)}\}_{m=0}^\infty$ will converge to an eigenvector of norm one associated with λ_1.

Note, however, that the power method in general has the disadvantage that it is unknown at the outset whether or not the matrix has a single dominant eigenvalue. Nor is it known how $\mathbf{x}^{(0)}$ should be chosen so as to ensure that its representation in terms of the eigenvectors of the matrix will contain a nonzero contribution from the eigenvector associated with the dominant eigenvalue, should it exist.

Power Method Algorithm 8.5

To approximate the dominant eigenvalue and an associated eigenvector of the $n \times n$ matrix A given a nonzero vector $\mathbf{x}$:

INPUT dimension n; matrix A; vector $\mathbf{x}$; tolerance TOL; maximum number of iterations N.

OUTPUT approximate eigenvalue μ; approximate eigenvector $\mathbf{x}$ (with $\|\mathbf{x}\|_\infty = 1$) or a message that the maximum number of iterations was exceeded.

Step 1   Set $k = 1$.

Step 2   Find an integer p with $1 \leq p \leq n$ and $|x_p| = \|\mathbf{x}\|_\infty$.

Step 3   Set $\mathbf{x} = \dfrac{1}{x_p}\mathbf{x}$.

Step 4   While $(k \leq N)$ do Steps 5–11.

Step 5   Set $\mathbf{y} = A\mathbf{x}$.

Step 6   Set $\mu = y_p$.

Step 7   Find an integer p with $1 \leq p \leq n$ and $|y_p| = \|\mathbf{y}\|_\infty$.

Step 8   If $y_p = 0$ then OUTPUT ('eigenvector', $\mathbf{x}$);
OUTPUT ('A has eigenvalue 0, select new vector $\mathbf{x}$ and restart');
STOP.

Step 9   Set $ERR = \left\| \mathbf{x} - \dfrac{1}{y_p}\mathbf{y} \right\|_\infty$

$$\mathbf{x} = \frac{1}{y_p}\mathbf{y}.$$

> **Step 10** If $ERR < TOL$ then OUTPUT (μ, **x**);
> *(Procedure completed successfully.)*
> STOP.
>
> **Step 11** Set $k = k + 1$.
>
> **Step 12** OUTPUT ('Maximum number of iterations exceeded');
> STOP.

Choosing, in Step 7, the smallest integer, p_m, for which $|y_{p_m}^{(m)}| = \|\mathbf{y}^{(m)}\|_\infty$ will generally ensure that this index eventually becomes invariant. The rate at which $\{\mu^{(m)}\}_{m=1}^\infty$ converges to λ_1, is determined by the ratios $|\lambda_j/\lambda_1|^m$ for $j = 2, 3, \ldots, n$, and in particular by $|\lambda_2/\lambda_1|^m$; that is, the convergence is of order $O((\lambda_2/\lambda_1)^m)$. Thus, there is a constant k such that for large m

$$|\mu^{(m)} - \lambda_1| \approx k \left| \frac{\lambda_2}{\lambda_1} \right|^m,$$

which implies that

$$\lim_{m \to \infty} \frac{|\mu^{(m+1)} - \lambda_1|}{|\mu^{(m)} - \lambda_1|} \approx \left| \frac{\lambda_2}{\lambda_1} \right|.$$

By Definition 2.7 (p. 53) the sequence $\{\mu^{(m)}\}$ converges linearly to λ_1. The Aitken's Δ^2 procedure discussed in Section 2.5 can consequently be used to speed the convergence. Implementing the Δ^2 procedure in Algorithm 8.5 can be accomplished by modifying the algorithm as follows:

Step 1 Set $k = 1$;
$\mu_0 = 0$;
$\mu_1 = 0$.

Step 6 Set $\mu = y_p$;

$$\tilde{\mu} = \mu_0 - \frac{(\mu_1 - \mu_0)^2}{\mu - 2\mu_1 + \mu_0}.$$

Step 10 If $ERR < TOL$ and $k \geq 4$ then OUTPUT ($\tilde{\mu}$, **x**);
STOP.

Step 11 Set $k = k + 1$;
$\mu_0 = \mu_1$;
$\mu_1 = \mu$.

In actuality, it is not necessary for the matrix to have n distinct eigenvalues for the power method to converge. In fact, if the unique dominant eigenvalue, λ_1, has multiplicity r greater than one and $\mathbf{v}^{(1)}, \mathbf{v}^{(2)}, \ldots, \mathbf{v}^{(r)}$ are linearly independent eigenvectors associated with λ_1, the procedure will still converge to λ_1. The sequence of vectors $\{\mathbf{x}^{(m)}\}_{m=0}^{\infty}$ will in this case converge to an eigenvector of λ_1 of norm one that is a linear combination of $\mathbf{v}^{(1)}, \mathbf{v}^{(2)}, \ldots, \mathbf{v}^{(r)}$ and depends on the choice of the initial vector $\mathbf{x}^{(0)}$.

EXAMPLE 3   The matrix

$$A = \begin{bmatrix} -4 & 14 & 0 \\ -5 & 13 & 0 \\ -1 & 0 & 2 \end{bmatrix}$$

has eigenvalues $\lambda_1 = 6$, $\lambda_2 = 3$, and $\lambda_3 = 2$. The power method described in Algorithm 8.5 will consequently converge.

Let $\mathbf{x}^{(0)} = (1, 1, 1)^t$, then

$$\mathbf{y}^{(1)} = A\mathbf{x}^{(0)} = (10, 8, 1)^t;$$

so

$$\|\mathbf{y}^{(1)}\|_{\infty} = 10, \qquad \mu^{(1)} = y_1^{(1)} = 10, \qquad \text{and} \qquad \mathbf{x}^{(1)} = \frac{\mathbf{y}^{(1)}}{\|\mathbf{y}^{(1)}\|_{\infty}} = (1, 0.8, 0.1)^t.$$

Continuing in this manner leads to the values in Table 8.5 where $\tilde{\mu}^{(m)}$ represents the sequence generated by the Aitken's Δ^2 procedure.

TABLE 8.5

m	$(\mathbf{x}^{(m)})^t$	$\mu^{(m)}$	$\tilde{\mu}^{(m)}$
0	$(1, 1, 1)$	—	
1	$(1, 0.8, 0.1)$	10	6.266667
2	$(1, 0.75, -0.111)$	7.2	6.062473
3	$(1, 0.730769, -0.188034)$	6.5	6.015054
4	$(1, 0.722200, -0.220850)$	6.230769	6.004202
5	$(1, 0.718182, -0.235915)$	6.111000	6.000855
6	$(1, 0.716216, -0.243095)$	6.054546	6.000240
7	$(1, 0.715247, -0.246588)$	6.027027	6.000058
8	$(1, 0.714765, -0.248306)$	6.013453	6.000017
9	$(1, 0.714525, -0.249157)$	6.006711	6.000003
10	$(1, 0.714405, -0.249579)$	6.003352	6.000000
11	$(1, 0.714346, -0.249790)$	6.001675	6.000000
12	$(1, 0.714316, -0.249895)$	6.000837	

An approximation to the dominant eigenvalue, 6, at this stage is either $\tilde{\mu}^{(10)} = 6.000000$ or $\mu^{(12)} = 6.000837$ with approximate unit eigenvector $(1, 0.714316, -0.249895)^t$. ■

When A is a symmetric matrix, a variation in the choice of the vectors $\mathbf{x}^{(m)}$, $\mathbf{y}^{(m)}$, and scalars $\mu^{(m)}$ can be made to significantly improve the rate of convergence of the sequence $\{\mu^{(m)}\}_{m=1}^{\infty}$ to the dominant eigenvalue λ_1. In fact, while the rate of convergence of the general power method is $O((\lambda_2/\lambda_1)^m)$, the rate of convergence of the modified procedure given below for symmetric matrices is $O((\lambda_2/\lambda_1)^{2m})$. (See Isaacson and Keller [67], page 140 ff.)

Power Method for Symmetric Matrices Algorithm 8.6

To approximate the dominant eigenvalue and an associated eigenvector of the $n \times n$ symmetric matrix A given a nonzero vector $\mathbf{x}$:

INPUT dimension n; matrix A; vector $\mathbf{x}$; tolerance TOL; maximum number of iterations N.

OUTPUT approximate eigenvalue μ; approximate eigenvector $\mathbf{x}$ (with $\|\mathbf{x}\|_2 = 1$) or a message that the maximum number of iterations was exceeded.

Step 1   Set $k = 1$;
$$\mathbf{x} = \mathbf{x}/\|\mathbf{x}\|_2.$$

Step 2   While $(k \leq N)$ do Steps 3–8.

   Step 3   Set $\mathbf{y} = A\mathbf{x}$.

   Step 4   Set $\mu = \mathbf{x}^t \mathbf{y}$.

   Step 5   If $\|\mathbf{y}\|_2 = 0$, then OUTPUT ('eigenvector', $\mathbf{x}$);
                        OUTPUT ('A has eigenvalue 0, select new vector $\mathbf{x}$
                                and restart');
                   STOP.

   Step 6   Set $ERR = \left\| \mathbf{x} - \dfrac{y}{\|\mathbf{y}\|_2} \right\|_2$

               $\mathbf{x} = \mathbf{y}/\|\mathbf{y}\|_2.$

   Step 7   If $ERR < TOL$ then OUTPUT $(\mu, \mathbf{x})$;
                        (*Procedure completed successfully.*)
                        STOP.

   Step 8   Set $k = k + 1$.

Step 9   OUTPUT ('Maximum number of iterations exceeded');
         STOP.

EXAMPLE 4   The matrix

$$A = \begin{bmatrix} 4 & -1 & 1 \\ -1 & 3 & -2 \\ 1 & -2 & 3 \end{bmatrix}$$

is symmetric with eigenvalues $\lambda_1 = 6$, $\lambda_2 = 3$, and $\lambda_3 = 1$. Listing the results of the first ten iterations of the power methods presented in Algorithms 8.5 and 8.6 with $\mathbf{y}^{(0)} = \mathbf{x}^{(0)} = (1, 0, 0)^t$ demonstrates the significant improvement obtained by using the latter algorithm. Using Algorithm 8.5, we have the results listed in Table 8.6. The results listed in Table 8.7 are obtained using Algorithm 8.6. ■

TABLE 8.6

m	$(\mathbf{y}^{(m)})^t$	$\mu^{(m)}$	$(\mathbf{x}^{(m)})^t$
0	—	—	(1, 0, 0)
1	(4, −1, 1)	4	(1, −0.25, 0.25)
2	(4.5, −2.25, 2.25)	4.5	(1, −0.5, 0.5)
3	(5, −3.5, 3.5)	5	(1, −0.7, 0.7)
4	(5.4, −4.5, 4.5)	5.4	(1, −0.8333, 0.8333)
5	(5.666, −5.1666, 5.1666)	5.666	(1, −0.911765, 0.911765)
6	(5.823529, −5.558824, 5.558824)	5.823529	(1, −0.954545, 0.954545)
7	(5.909091, −5.772727, 5.772727)	5.909091	(1, −0.976923, 0.976923)
8	(5.953846, −5.884615, 5.884615)	5.953846	(1, −0.988372, 0.988372)
9	(5.976744, −5.941861, 5.941861)	5.976744	(1, −0.994163, 0.994163)
10	(5.988327, −5.970817, 5.970817)	5.988327	(1, −0.997076, 0.997076)

TABLE 8.7

m	$(\mathbf{y}^{(m)})^t$	$\mu^{(m)}$	$(\mathbf{x}^{(m)})^t$
0	(1, 0, 0)	—	(1, 0, 0)
1	(4, −1, 1)	4	(0.942809, −0.235702, 0.235702)
2	(4.242641, −2.121320, 2.121320)	5	(0.816497, −0.408248, 0.408248)
3	(4.082483, −2.857738, 2.857738)	5.666667	(0.710669, −0.497468, 0.497468)
4	(3.837613, −3.198011, 3.198011)	5.909091	(0.646997, −0.539164, 0.539164)
5	(3.666314, −3.342816, 3.342816)	5.976744	(0.612836, −0.558763, 0.558763)
6	(3.568871, −3.406650, 3.406650)	5.994152	(0.595247, −0.568190, 0.568190)
7	(3.517370, −3.436200, 3.436200)	5.998536	(0.586336, −0.572805, 0.572805)
8	(3.490952, −3.450359, 3.450359)	5.999634	(0.581852, −0.575086, 0.575086)
9	(3.477580, −3.457283, 3.457283)	5.999908	(0.579603, −0.576220, 0.576220)
10	(3.470854, −3.460706, 3.460706)	5.999977	(0.578477, −0.576786, 0.576786)

The following error bound for approximating the eigenvalues of a symmetric matrix can be used as a stopping criteria in Step 7 of Algorithm 8.6.

THEOREM 8.44 If A is an $n \times n$ symmetric matrix with eigenvalues $\lambda_1, \lambda_2, \ldots, \lambda_n$ and $\|A\mathbf{x} - \lambda\mathbf{x}\|_2 \le \varepsilon$ for some vector $\mathbf{x}$ with $\|\mathbf{x}\|_2 = 1$ and real number λ, then

$$(8.48) \qquad \min_{1 \le i \le n} |\lambda_i - \lambda| \le \varepsilon.$$

PROOF Let $\mathbf{v}^{(1)}, \mathbf{v}^{(2)}, ..., \mathbf{v}^{(n)}$ denote an orthonormal set of eigenvectors of A associated, respectively, with the eigenvalues $\lambda_1, \lambda_2, ..., \lambda_n$. By Theorem 8.41 $\mathbf{x}$ can be expressed, for some unique set of constants $\alpha_1, \alpha_2, ..., \alpha_n$, as

$$\mathbf{x} = \sum_{i=1}^{n} \alpha_i \mathbf{v}^{(i)}.$$

Thus

$$\|A\mathbf{x} - \lambda\mathbf{x}\|_2^2 = \left\| \sum_{i=1}^{n} \alpha_i(\lambda_i - \lambda)\mathbf{v}^{(i)} \right\|_2^2$$

$$= \sum_{i=1}^{n} |\alpha_i|^2 |\lambda_i - \lambda|^2$$

$$\geq \min_{1 \leq i \leq n} |\lambda_i - \lambda|^2 \sum_{i=1}^{n} |\alpha_i|^2$$

But

$$\sum_{i=1}^{n} |\alpha_i|^2 = \|\mathbf{x}\|_2^2 = 1,$$

so

$$\varepsilon \geq \|A\mathbf{x} - \lambda\mathbf{x}\|_2 \geq \min_{1 \leq i \leq n} |\lambda_i - \lambda|. \qquad \square$$

The **inverse power method** uses the power method as a basis but gives faster convergence. It is used to determine the eigenvalue of A that is closest to a specified number q.

Assume that the matrix A has eigenvalues $\lambda_1, ..., \lambda_n$ with linearly independent eigenvectors $\mathbf{v}^{(1)}, ..., \mathbf{v}^{(n)}$. Consider the matrix $(A - qI)^{-1}$ where $q \neq \lambda_i$ for $i = 1, 2, ..., n$. The eigenvalues of $(A - qI)^{-1}$ are

$$\frac{1}{\lambda_1 - q}, \frac{1}{\lambda_2 - q}, ..., \frac{1}{\lambda_n - q}$$

with eigenvectors $\mathbf{v}^{(1)}, \mathbf{v}^{(2)}, ..., \mathbf{v}^{(n)}$. (See Exercise 19 of Section 8.1.) Applying the power method to $(A - qI)^{-1}$ gives

(8.49) $$\mathbf{y}^{(m)} = (A - qI)^{-1}\mathbf{x}^{(m-1)},$$

(8.50) $$\mu^{(m)} = y_{p_{m-1}}^{(m)} = \frac{y_{p_{m-1}}^{(m)}}{x_{p_{m-1}}^{(m-1)}} = \frac{\displaystyle\sum_{j=1}^{n} \alpha_j \frac{1}{(\lambda_j - q)^m} v_{p_{m-1}}^{(j)}}{\displaystyle\sum_{j=1}^{n} \alpha_j \frac{1}{(\lambda_j - q)^{m-1}} v_{p_{m-1}}^{(j)}},$$

and

(8.51) $$\mathbf{x}^{(m)} = \frac{\mathbf{y}^{(m)}}{y_{p_m}^{(m)}}$$

where, at each step, p_m is used to represents an integer for which $|y_{p_m}^{(m)}| = \|\mathbf{y}^{(m)}\|_\infty$. The sequence $\{\mu^{(m)}\}$ in Eq. (8.50) will converge to

$$\frac{1}{\lambda_k - q} = \max_{1 \leq i \leq n} \frac{1}{|\lambda_i - q|},$$

where λ_k is the eigenvalue of A closest to q.

With k known, Eq. (8.50) can be written

$$(8.52) \qquad \mu^{(m)} = \frac{1}{\lambda_k - q} \left[\frac{\alpha_k v^{(k)}_{p_{m-1}} + \sum\limits_{\substack{j=1, \\ j \neq k}}^{n} \alpha_j \left[\dfrac{\lambda_k - q}{\lambda_j - q} \right]^m v^{(j)}_{p_{m-1}}}{\alpha_k v^{(k)}_{p_{m-1}} + \sum\limits_{\substack{j=1 \\ j \neq k}}^{n} \alpha_j \left[\dfrac{\lambda_k - q}{\lambda_j - q} \right]^{m-1} v^{(j)}_{p_{m-1}}} \right].$$

Thus the choice of q determines the convergence, provided that $1/(\lambda_k - q)$ is a unique dominant eigenvalue of $(A - qI)^{-1}$ (although it may be a multiple eigenvalue). The closer q is to an eigenvalue λ_k of A, the faster the convergence since the convergence is of order

$$O\left(\left| \frac{(\lambda_k - q)^{-1}}{(\lambda - q)^{-1}} \right|^m \right) = O\left(\left| \frac{\lambda_k - q}{\lambda - q} \right|^m \right),$$

where λ represents the eigenvalue of A that is second closest to q.

The determination of $\mathbf{y}^{(m)}$ is obtained from the equation

$$(8.53) \qquad (A - qI)\mathbf{y}^{(m)} = \mathbf{x}^{(m-1)}.$$

To solve this equation, it is recommended that Gaussian elimination with pivoting be used. Although the method requires the solution of an $n \times n$ linear system at each step, the multipliers can be saved to reduce the computation. The selection of q can be based on the Gerschgorin Theorem, Theorem 8.43 (p. 450), or on any other means of localizing an eigenvalue.

The following algorithm computes q from an initial approximation to the eigenvector $\mathbf{x}^{(0)}$ by

$$q = \frac{\mathbf{x}^{(0)t} A \mathbf{x}^{(0)}}{\mathbf{x}^{(0)t} \mathbf{x}^{(0)}}.$$

This choice of q results from the observation that if $\mathbf{x}$ is an eigenvector of A with respect to the eigenvalue λ, then $A\mathbf{x} = \lambda\mathbf{x}$. So $\mathbf{x}^t A\mathbf{x} = \lambda \mathbf{x}^t \mathbf{x}$ and

$$\lambda = \frac{\mathbf{x}^t A \mathbf{x}}{\mathbf{x}^t \mathbf{x}}.$$

If q is close to an eigenvalue, the convergence will be quite rapid, but a pivoting technique should be used in Step 6 to avoid contamination by rounding error.

Algorithm 8.7 with pivoting is often used to approximate an eigenvector when an approximate eigenvalue is known. In this case, q is the approximate eigenvalue.

Inverse Power Method Algorithm 8.7

To approximate an eigenvalue and an associated eigenvector of the $n \times n$ matrix A given a nonzero vector $\mathbf{x}$:

INPUT dimension n; matrix A; vector $\mathbf{x}$; tolerance TOL; maximum number of iterations N.

OUTPUT approximate eigenvalue μ; approximate eigenvector $\mathbf{x}$ (with $\|\mathbf{x}\|_\infty = 1$) or a message that the maximum number of iterations was exceeded.

Step 1   Set $q = \mathbf{x}^t A \mathbf{x} / \mathbf{x}^t \mathbf{x}$.

Step 2   Set $k = 1$.

Step 3   Find the integer p with $1 \le p \le n$ and $|x_p| = \|\mathbf{x}\|_\infty$.

Step 4   Set $\mathbf{x} = \dfrac{1}{x_p}\mathbf{x}$.

Step 5   While $(k \le N)$ do Steps 6–11.

  Step 6   Solve the linear system $(A - qI)\mathbf{y} = \mathbf{x}$.
     If the system does not have a unique solution, then
        OUTPUT ('q is an eigenvalue', q);
        STOP.

  Step 7   Set $\mu = y_p$.

  Step 8   Find an integer p with $1 \le p \le n$ and $|y_p| = \|\mathbf{y}\|_\infty$.

  Step 9   Set $ERR = \left\| \mathbf{x} - \dfrac{1}{y_p}\mathbf{y} \right\|_\infty$

$$\mathbf{x} = \frac{1}{y_p}\mathbf{y}.$$

  Step 10   If $ERR < TOL$ then set $\mu = (1/\mu) + q$;
           OUTPUT $(\mu, \mathbf{x})$;
           (*Procedure completed successfully.*)
           STOP.

Step 11   Set $k = k + 1$.

Step 12   OUTPUT ('Maximum number of iterations exceeded');
        (*Procedure completed unsuccessfully.*)
        STOP.

Since the convergence of the inverse power method is linear, Aitken's Δ^2 procedure can be used to speed convergence. The following example illustrates the fast convergence of the inverse power method if q is close to an eigenvalue. To obtain the other eigenvalues for this example, see Exercise 11.

EXAMPLE 5   The matrix

$$A = \begin{bmatrix} -4 & 14 & 0 \\ -5 & 13 & 0 \\ -1 & 0 & 2 \end{bmatrix}$$

was considered in Example 3. Algorithm 8.5 gave the approximation $\mu^{(12)} = 6.000837$ using $\mathbf{x}^{(0)} = (1, 1, 1)^t$. With $\mathbf{x}^{(0)} = (1, 1, 1)^t$, we have

$$q = \frac{\mathbf{x}^{(0)t} A \mathbf{x}^{(0)}}{\mathbf{x}^{(0)t} \mathbf{x}^{(0)}} = \frac{19}{3} = 6.333333.$$

The results of applying Algorithm 8.7 are listed in Table 8.8.   ■

TABLE 8.8

m	$\mathbf{x}^{(m)t}$	$\mu^{(m)}$
0	$(1, 1, 1)$	
1	$(1, 0.720727, -0.194042)$	6.183183
2	$(1, 0.715518, -0.245052)$	6.017244
3	$(1, 0.714409, -0.249522)$	6.001719
4	$(1, 0.714298, -0.249953)$	6.000175
5	$(1, 0.714287, -0.250000)$	6.000021
6	$(1, 0.714286, -0.249999)$	6.000005

If A is symmetric, then $(A - qI)^{-1}$ will also be symmetric for any real number q, so the power method for symmetric matrices, Algorithm 8.6, can be applied to $(A - qI)^{-1}$ to speed the convergence to

$$O\left(\left|\frac{\lambda_k - q}{\lambda - q}\right|^{2m}\right).$$

Numerous techniques are available for obtaining approximations to the other eigenvalues of a matrix once an approximation to the dominant eigenvalue has been computed. We will restrict our presentation to **deflation techniques**. Other procedures are considered in Exercises 17 and 18.

Deflation techniques involve forming a new matrix B from the original matrix A whose eigenvalues are the same as those of A, except that the dominant eigenvalue of A is replaced by the eigenvalue 0 in B.

THEOREM 8.45   Suppose that $\lambda_1, \lambda_2, \ldots, \lambda_n$ are eigenvalues of A with associated eigenvectors $\mathbf{v}^{(1)}$, $\mathbf{v}^{(2)}, \ldots, \mathbf{v}^{(n)}$, and that λ_1 has multiplicity one. If $\mathbf{x}$ is any vector with the property that $\mathbf{x}^t \mathbf{v}^{(1)} = 1$, then the matrix

(8.54) $$B = A - \lambda_1 \mathbf{v}^{(1)} \mathbf{x}^t$$

has eigenvalues $0, \lambda_2, \lambda_3, \ldots, \lambda_n$ with associated eigenvectors $\mathbf{v}^{(1)}, \mathbf{w}^{(2)}, \mathbf{w}^{(3)}, \ldots, \mathbf{w}^{(n)}$, where $\mathbf{v}^{(i)}$ and $\mathbf{w}^{(i)}$ are related by the equation

$$(8.55) \qquad \mathbf{v}^{(i)} = (\lambda_i - \lambda_1)\mathbf{w}^{(i)} + \lambda_1(\mathbf{x}^t\mathbf{w}^{(i)})\mathbf{v}^{(1)}$$

for each $i = 2, 3, \ldots, n$.

The proof of this result can be found in Wilkinson [142], page 596.

A particularly useful deflation technique, called the **Wielandt's deflation** procedure, results by defining

$$(8.56) \qquad \mathbf{x} = \frac{1}{\lambda_1 v_i^{(1)}} \begin{bmatrix} a_{i1} \\ a_{i2} \\ \vdots \\ a_{in} \end{bmatrix}$$

where $v_i^{(1)}$ is a coordinate of $\mathbf{v}^{(1)}$ which is nonzero, and the values $a_{i1}, a_{i2}, \ldots, a_{in}$ are the entries in the ith row of A.

With this definition,

$$\mathbf{x}^t\mathbf{v}^{(1)} = \frac{1}{\lambda_1 v_i^{(1)}} [a_{i1}, a_{i2}, \ldots, a_{in}] (v_1^{(1)}, v_2^{(1)}, \ldots, v_n^{(1)})^t = \frac{1}{\lambda_1 v_i^{(1)}} \sum_{j=1}^{n} a_{ij}v_j^{(1)},$$

where the sum is the ith coordinate of the product $A\mathbf{v}^{(1)}$. Since $A\mathbf{v}^{(1)} = \lambda_1\mathbf{v}^{(1)}$, this implies that

$$\mathbf{x}^t\mathbf{v}^{(1)} = \frac{1}{\lambda_1 v_i^{(1)}} (\lambda_1 v_i^{(1)}) = 1;$$

so $\mathbf{x}$ satisfies the hypotheses of Theorem 8.45. Moreover, (see Exercise 20) the ith row of $B = A - \lambda_1\mathbf{v}^{(1)}\mathbf{x}^t$ consists entirely of zero entries.

If $\lambda \neq 0$ is any eigenvalue with associated eigenvector $\mathbf{w}$, the relation $B\mathbf{w} = \lambda\mathbf{w}$ implies that the ith coordinate of $\mathbf{w}$ must also be zero. Consequently the ith column of the matrix B makes no contribution to the product $B\mathbf{w} = \lambda\mathbf{w}$. Thus, the matrix B can be replaced by an $(n-1) \times (n-1)$ matrix B' that is obtained by deleting the ith row and column from B. B' will have eigenvalues $\lambda_2, \lambda_3, \ldots, \lambda_n$. If $|\lambda_2| > |\lambda_3|$, the power method can be reapplied to the matrix B' to determine this new dominant eigenvalue and an eigenvector, $\mathbf{w}^{(2)\prime}$, associated with λ_2, with respect to the matrix B'. To find the associated eigenvector for the original matrix A simply requires constructing $\mathbf{w}^{(2)}$ by inserting a zero coordinate between the coordinates $w_{i-1}^{(2)\prime}$ and $w_i^{(2)\prime}$ of the $(n-1)$-dimensional vector $\mathbf{w}^{(2)\prime}$ and then calculating $\mathbf{v}^{(2)}$ by the use of Eq. (8.55).

EXAMPLE 6   From Example 4 we know that the matrix

$$A = \begin{bmatrix} 4 & -1 & 1 \\ -1 & 3 & -2 \\ 1 & -2 & 3 \end{bmatrix}$$

has eigenvalues $\lambda_1 = 6$, $\lambda_2 = 3$, and $\lambda_3 = 1$. Assuming that the dominant eigenvalue $\lambda_1 = 6$ and associated unit eigenvector $\mathbf{v}^{(1)} = (1, -1, 1)^t$ have been calculated, the procedure just outlined for obtaining λ_2 proceeds as follows:

$$\mathbf{x} = \frac{1}{6}\begin{bmatrix} 4 \\ -1 \\ 1 \end{bmatrix} = (\tfrac{2}{3}, -\tfrac{1}{6}, \tfrac{1}{6})^t,$$

$$\mathbf{v}^{(1)}\mathbf{x}^t = \begin{bmatrix} 1 \\ -1 \\ 1 \end{bmatrix} [\tfrac{2}{3}, -\tfrac{1}{6}, \tfrac{1}{6}] = \begin{bmatrix} \tfrac{2}{3} & -\tfrac{1}{6} & \tfrac{1}{6} \\ -\tfrac{2}{3} & \tfrac{1}{6} & -\tfrac{1}{6} \\ \tfrac{2}{3} & -\tfrac{1}{6} & \tfrac{1}{6} \end{bmatrix},$$

and

$$B = A - \lambda_1 \mathbf{v}^{(1)}\mathbf{x}^t = \begin{bmatrix} 4 & -1 & 1 \\ -1 & 3 & -2 \\ 1 & -2 & 3 \end{bmatrix} - 6\begin{bmatrix} \tfrac{2}{3} & -\tfrac{1}{6} & \tfrac{1}{6} \\ -\tfrac{2}{3} & \tfrac{1}{6} & -\tfrac{1}{6} \\ \tfrac{2}{3} & -\tfrac{1}{6} & \tfrac{1}{6} \end{bmatrix}$$

$$= \begin{bmatrix} 0 & 0 & 0 \\ 3 & 2 & -1 \\ -3 & -1 & 2 \end{bmatrix}.$$

Deleting the first row and column gives

$$B' = \begin{bmatrix} 2 & -1 \\ -1 & 2 \end{bmatrix},$$

which has eigenvalues $\lambda_2 = 3$ and $\lambda_3 = 1$. For $\lambda_2 = 3$ the eigenvector, $\mathbf{w}^{(2)'}$, can be obtained by solving the second-order linear system

$$(B' - 3I)\mathbf{w}^{(2)'} = \mathbf{0},$$

resulting in

$$\mathbf{w}^{(2)'} = (1, -1)^t.$$

Thus, $\mathbf{w}^{(2)} = (0, 1, -1)^t$ and, from Eq. (8.55), we have:

$$\mathbf{v}^{(2)} = (3 - 6)(0, 1, -1)^t + 6[(\tfrac{2}{3}, -\tfrac{1}{6}, \tfrac{1}{6})(0, 1, -1)^t](1, -1, 1)^t$$
$$= (-2, -1, 1)^t. \qquad\blacksquare$$

Although this deflation process can in some cases be used to find approximations to all of the eigenvalues and eigenvectors of a matrix, the process is susceptible to rounding error. Techniques based on similarity transformations will be presented in the next section; these methods are generally preferable when approximations to all eigenvalues and eigenvectors are needed.

We close this section with an algorithm for the calculation of the second most dominant eigenvalue and associated eigenvector for a matrix, once the dominant eigenvalue and associated eigenvector have been determined.

Weilandt's Deflation Algorithm 8.8

To approximate the second most dominant eigenvalue and an associated eigenvector of the $n \times n$ matrix A given an approximation λ to the dominant eigenvalue, an approximation $\mathbf{v}$ to a corresponding eigenvector and a vector $\mathbf{x} \in R^{n-1}$:

INPUT dimension n; matrix A; approximate eigenvector $\mathbf{v} \in R^n$; vector $\mathbf{x} \in R^{n-1}$.

OUTPUT approximate eigenvalue μ; approximate eigenvector $\mathbf{u}$ or a message that the method fails.

Step 1   Let i be the smallest integer with $1 \le i \le n$ and $|v_i| = \max\limits_{1 \le j \le n} |v_j|$.

Step 2   If $i \ne 1$ then
   for $k = 1, \ldots, i - 1$
      for $j = 1, \ldots, i - 1$

$$\text{set } b_{kj} = a_{kj} - \frac{v_k}{v_i} a_{ij}.$$

Step 3   If $i \ne 1$ and $i \ne n$ then
   for $k = i, \ldots, n - 1$
      for $j = 1, \ldots, i - 1$

$$\text{set } b_{kj} = a_{k+1,j} - \frac{v_{k+1}}{v_i} a_{ij};$$

$$b_{jk} = a_{j,k+1} - \frac{v_j}{v_i} a_{i,k+1}.$$

Step 4   If $i \ne n$ then
   for $k = i, \ldots, n - 1$
      for $j = i, \ldots, n - 1$

$$\text{set } b_{kj} = a_{k+1,j+1} - \frac{v_{k+1}}{v_i} a_{i,j+1}.$$

Step 5   Perform the power method on the $(n - 1) \times (n - 1)$ matrix $B' = (b_{kj})$ with $\mathbf{x}$ as initial approximation.
   If the method fails, then OUTPUT ('Method fails');
                    STOP.
               else let μ be the approximate eigenvalue and
         $\mathbf{w}' = (w'_1, \ldots, w'_{n-1})^t$ the approximate eigenvector.

Step 6   If $i \ne 1$ then for $k = 1, \ldots, i - 1$ set $w_k = w'_k$.

Step 7   Set $w_i = 0$.

Step 8   If $i \ne n$ then for $k = i + 1, \ldots, n$ set $w_k = w'_{k-1}$.

Step 9   For $k = 1, \ldots, n$

$$\text{set } u_k = (\mu - \lambda)w_k + \left(\sum_{j=1}^{n} a_{ij}w_j\right)\frac{v_k}{v_i}.$$

(*Compute eigenvector using Eq. 8.55.*)

Step 10   OUTPUT $(\mu, \mathbf{u})$;
(*Procedure completed successfully.*)
STOP.

Exercise Set 8.4

1. Find the first three iterations obtained by the power method applied to the following matrices:

a)
$$\begin{bmatrix} 1 & -1 & 0 \\ -2 & 4 & -2 \\ 0 & -1 & 1 \end{bmatrix},$$
use $\mathbf{x}^{(0)} = (1, 0, 0)^t$.

b)
$$\begin{bmatrix} 1 & -1 & 0 \\ -2 & 4 & -2 \\ 0 & -1 & 2 \end{bmatrix},$$
use $\mathbf{x}^{(0)} = (1, 0, 0)^t$.

c)
$$\begin{bmatrix} 4 & 1 & 1 & 1 \\ 1 & 3 & -1 & 1 \\ 1 & -1 & 2 & 0 \\ 1 & 1 & 0 & 2 \end{bmatrix},$$
use $\mathbf{x}^{(0)} = (1, 1, 1, 1)^t$.

d)
$$\begin{bmatrix} 5 & -2 & -0.5 & 1.5 \\ -2 & 5 & 1.5 & -0.5 \\ -0.5 & 1.5 & 5 & -2 \\ 1.5 & -0.5 & -2 & 5 \end{bmatrix},$$
use $\mathbf{x}^{(0)} = (1, 1, 1, 1)^t$.

2. Repeat Exercise 1 using the inverse power method.

3. Find the first three iterations obtained by the symmetric power method applied to the following matrices:

a)
$$\begin{bmatrix} 2 & -1 & 0 \\ -1 & 2 & -1 \\ 0 & -1 & 2 \end{bmatrix},$$
use $\mathbf{x}^{(0)} = (1, 0, 0)^t$.

b)
$$\begin{bmatrix} 4.75 & 2.25 & -0.25 \\ 2.25 & 4.75 & 1.25 \\ -0.25 & 1.25 & 4.75 \end{bmatrix},$$
use $\mathbf{x}^{(0)} = (0, 1, 0)^t$.

c)
$$\begin{bmatrix} 4 & 1 & -1 & 0 \\ 1 & 3 & -1 & 0 \\ -1 & -1 & 5 & 2 \\ 0 & 0 & 2 & 4 \end{bmatrix},$$
use $\mathbf{x}^{(0)} = (0, 1, 0, 0)^t$.

d)
$$\begin{bmatrix} 4 & 1 & 1 & 1 \\ 1 & 3 & -1 & 1 \\ 1 & -1 & 2 & 0 \\ 1 & 1 & 0 & 2 \end{bmatrix},$$
use $\mathbf{x}^{(0)} = (1, 0, 0, 0)^t$.

4. Show that the following $n \times n$ matrices have fewer than n linearly independent eigenvectors.

a) $\begin{bmatrix} 2 & 1 \\ 0 & 2 \end{bmatrix}$

b) $\begin{bmatrix} 1 & 0 & 0 \\ -1 & 0 & 1 \\ -1 & -1 & 2 \end{bmatrix}$

5. Use Theorem 8.43 to determine bounds for the eigenvalues of the following matrices.

a) $\begin{bmatrix} 1 & 0 & 0 \\ -1 & 0 & 1 \\ -1 & -1 & 2 \end{bmatrix}$

b) $\begin{bmatrix} 4 & -1 & 0 \\ -1 & 4 & -1 \\ -1 & -1 & 4 \end{bmatrix}$

c) $\begin{bmatrix} 3 & 2 & 1 \\ 2 & 3 & 0 \\ 1 & 0 & 3 \end{bmatrix}$

d) $\begin{bmatrix} 4.75 & 2.25 & -0.25 \\ 2.25 & 4.75 & 1.25 \\ -0.25 & 1.25 & 4.75 \end{bmatrix}$

6. a) Develop an algorithm to incorporate the inverse power method into the symmetric power method.

   b) Repeat Exercise 3 using the algorithm developed in part (a).

7. Use the power method and Wielandt's deflation to approximate the two most dominant eigenvalues for the matrices in Exercise 1. Iterate until a tolerance of 10^{-4} is achieved or until the number of iterations exceeds 25.

8. Repeat Exercise 7 using Aitken's Δ^2 technique with the power method.

9. Use the symmetric power method to compute the largest eigenvalue (in absolute value) of the matrices given in Exercise 3. Iterate until a tolerance of 10^{-4} is achieved or until the number of iterations exceeds 25.

10. Repeat Exercise 9 using the inverse power method. Find the second dominant eigenvalue using Wielandt's deflation.

11. Apply the inverse power method to the matrix A in Example 5 using $q = 7$, 4, and 0. Explain the results.

12. Show that any four vectors in R^3 are linearly dependent.

13. Let $v_1, \ldots, v_k$ be k orthogonal vectors. Show that $\{v_1, \ldots, v_k\}$ is a linearly independent set.

14. Let P be an orthogonal matrix. Show that the columns of P form an orthonormal set of vectors. Also, show that $\|P\|_2 = 1$ and $\|P^t\|_2 = 1$.

15. Let $v_1, \ldots, v_n$ be n orthonormal vectors in R^n. Let $x \in R^n$. Show that $x = \sum_{k=1}^{n} c_k v_k$ where $c_k = v_k^t x$.

16. Show that if A is an $n \times n$ matrix with n distinct eigenvalues, then A has n linearly independent eigenvectors.

17. Let the eigenvalues of A be $\lambda_1 \geq \cdots \geq \lambda_{n-1} > \lambda_n$ or $\lambda_1 \leq \cdots \leq \lambda_{n-1} < \lambda_n$ with λ_1 dominant.

    a) Show that the eigenvalue λ_n can be approximated by applying the power method to $cI - A$ where c is any number such that $|c| > |\lambda_1|$ and c and λ_1 have the same sign.

    b) Apply this technique to the matrices in Exercise 1.

18. **Annihilation Technique**   Suppose the $n \times n$ matrix A has eigenvalues $\lambda_1, \ldots, \lambda_n$ ordered by

$$|\lambda_1| > |\lambda_2| > |\lambda_3| \geq \cdots \geq |\lambda_n|$$

    with linearly independent eigenvectors $\mathbf{v}^{(1)}, \mathbf{v}^{(2)}, \ldots, \mathbf{v}^{(n)}$.

    a) Show that if the power method is applied with an initial vector $\mathbf{x}^{(0)}$ given by

$$\mathbf{x}^{(0)} = \alpha_2 \mathbf{v}^{(2)} + \alpha_3 \mathbf{v}^{(3)} + \cdots + \alpha_n \mathbf{v}^{(n)},$$

    then the sequence $\{\mu^{(m)}\}$ described in Algorithm 8.5, will converge to λ_2.

    b) Show that, for any vector $\mathbf{x} = \sum_{i=1}^n \alpha_i \mathbf{v}^{(i)}$, the vector $\mathbf{x}^{(0)} = (A - \lambda_1 I)\mathbf{x}$ satisfies the property given in part (a).

    c) Obtain an approximation to λ_2 for the matrices in Exercise 1.

    d) Show that this method can be continued to find λ_3 using $\mathbf{x}^{(0)} = (A - \lambda_2 I)(A - \lambda_1 I)\mathbf{x}$.

19. **Hotelling's Deflation**   Assume that the largest eigenvalue λ_1 in magnitude and an associated eigenvector $\mathbf{v}^{(1)}$ have been obtained for the $n \times n$ symmetric matrix A. Show that the matrix B,

$$B = A - \frac{\lambda_1}{(\mathbf{v}^{(1)})^t \mathbf{v}^{(1)}} \mathbf{v}^{(1)}(\mathbf{v}^{(1)})^t,$$

    has the same eigenvalues $\lambda_2, \ldots, \lambda_n$ as A, except that B has eigenvalue 0 with eigenvector $\mathbf{v}^{(1)}$ instead of eigenvalue λ_1. Use this deflation method to find λ_2 for each matrix in Exercise 3. Theoretically, this method can be continued to find more eigenvalues, but rounding error soon makes the effort worthless.

20. Show that the ith row of $B = A - \lambda_1 \mathbf{v}^{(1)} \mathbf{x}^t$ is zero where λ_1 is the largest eigenvalue of A in absolute value, $\mathbf{v}^{(1)}$ is the associated eigenvector of A for λ_1, and $\mathbf{x}$ is the vector defined in Eq. (8.56).

21. Following along the line of Exercise 4 in Section 6.3 and Exercise 27 in Section 8.1, suppose that a species of beetle has a life span of four years, that a female in the first year has a survival rate of $\frac{1}{2}$, in the second year a survival rate of $\frac{1}{4}$, and in the third year a survival rate of $\frac{1}{8}$. Suppose additionally that a female gives birth, on the average, to two new females in the third year and to four new females in the fourth year. The matrix describing a single female's contribution in one year to the female population in the succeeding year is

$$A = \begin{bmatrix} 0 & 0 & 2 & 4 \\ \frac{1}{2} & 0 & 0 & 0 \\ 0 & \frac{1}{4} & 0 & 0 \\ 0 & 0 & \frac{1}{8} & 0 \end{bmatrix},$$

where again the entry in the ith row and jth column denotes the probabilistic contribution that a female of age j makes on the next year's female population of age i.

a) Use the Gerschgorin Circle Theorem 8.43 to determine a region in the complex plane containing all the eigenvalues of A.

b) Use the power method to determine the dominant eigenvalue of the matrix and its associated eigenvector.

c) Use Algorithm 8.8 to determine any remaining eigenvalues and eigenvectors of A.

d) Find the eigenvalues of A by using the characteristic polynomial of A and the Newton–Raphson method.

e) What is your long-range prediction for the population of these beetles?

22. A **persymmetric matrix** is a matrix that is symmetric about both diagonals; that is, an $N \times N$ matrix $A = (a_{ij})$ is persymmetric if $a_{ij} = a_{ji} = a_{N+1-i, N+1-j}$ for all $i = 1, 2, ..., N$ and $j = 1, 2, ..., N$. A number of problems in communication theory have solutions that involve the eigenvalues and eigenvectors of matrices that are in persymmetric form. For example, the eigenvector corresponding to the minimal eigenvalue of the 4×4 persymmetric matrix

$$A = \begin{bmatrix} 2 & -1 & 0 & 0 \\ -1 & 2 & -1 & 0 \\ 0 & -1 & 2 & -1 \\ 0 & 0 & -1 & 2 \end{bmatrix}$$

gives the unit energy-channel impulse response for a given error sequence of length 2, and subsequently the minimum weight of any possible error sequence.

a) Use the Gerschgorin Circle Theorem (Theorem 8.43) to show that if A is the matrix given above and λ is its minimal eigenvalue, then $|\lambda - 4| = \rho(A - 4I)$ where ρ denotes the spectral radius.

b) Find the minimal eigenvalue of the matrix A by finding all the eigenvalues of $A - 4I$ and computing its spectral radius. Then find the corresponding eigenvector.

c) Find the minimal eigenvalue of the matrix A by applying the power method to $A - 4I$.

d) Use the Gerschgorin Circle Theorem to show that if λ is the minimal eigenvalue of the matrix

$$B = \begin{bmatrix} 3 & -1 & -1 & 1 \\ -1 & 3 & -1 & -1 \\ -1 & -1 & 3 & -1 \\ 1 & -1 & -1 & 3 \end{bmatrix},$$

then $|\lambda - 6| = \rho(B - 6I)$.

e) Try to repeat parts (b) and (c) of this exercise using the matrix B and the result in part (d). What difficulties arise?

23. A linear dynamical system can be represented by the equations

$$\frac{d\mathbf{x}}{dt} = A(t)\mathbf{x}(t) + B(t)\mathbf{u}(t), \qquad \mathbf{y}(t) = C(t)\mathbf{x}(t) + D(t)\mathbf{u}(t)$$

where A is an $n \times n$ variable matrix, B an $n \times r$ variable matrix, C an $m \times n$ variable matrix, D an $m \times r$ variable matrix, $\mathbf{x}$ an n-dimensional vector variable, $\mathbf{y}$ an m-dimensional vector variable, and $\mathbf{u}$ an r-dimensional vector variable. For the system to be stable, the matrix A must have all of its eigenvalues with nonpositive real part for all t.

a) If

$$A(t) = \begin{bmatrix} -1 & 2 & 0 \\ -2.5 & -7 & 4 \\ 0 & 0 & -5 \end{bmatrix},$$

is the system stable?

b) If

$$A(t) = \begin{bmatrix} -1 & 1 & 0 & 0 \\ 0 & -2 & 1 & 0 \\ 0 & 0 & -5 & 1 \\ -1 & -1 & -2 & -3 \end{bmatrix},$$

is the system stable?

8.5 Householder's Method and the QL Algorithm

The deflation methods discussed in the previous section are not generally suitable for calculating all the eigenvalues of a matrix because of rounding error growth. In this section we consider a matrix reduction technique that is designed to determine all the eigenvalues of a symmetric matrix. The method is called the **QL Algorithm** and, because of its stability characteristics, is the type of technique most often used in practice. The method can be modified for use in computing the eigenvalues of nonsymmetric matrices; these modifications are discussed briefly at the end of this section. Further remarks on this topic, as well as a complete general discussion, can be found in Wilkinson [142] and Wilkinson and Reinsch [143].

To apply the QL Algorithm, the symmetric matrix must be in tridiagonal form; that is, the only nonzero entries in the matrix lie either on the diagonal or on the subdiagonals directly above or below the diagonal. Since this is not the usual form of a symmetric matrix, the first problem is to find a tridiagonal symmetric matrix whose eigenvalues agree with those of the original matrix. For this purpose we use a technique known as **Householder's method**.

Suppose that A is an arbitrary $n \times n$ symmetric matrix. Theorem 8.39 implies that an orthogonal matrix Q exists with the property that $T = Q^{-1}AQ = Q^t AQ$ where T is a diagonal matrix. If the matrix Q, and hence T, was easy to determine, the eigenvalue problem would be solved since the eigenvalues of T lie along its diagonal and agree with the eigenvalues of A. The matrix Q, however, is not in general easy to determine.

Instead of attempting to determine the diagonal matrix T, Householder's method computes a tridiagonal matrix with the same eigenvalues as A. The QL Algorithm can be applied to the tridiagonal matrix to resolve the problem. (For an example of a method that determines the diagonal matrix T directly, see Exercise 9.)

Householder's method begins by determining a matrix $P^{(1)}$ with $P^{(1)} = [P^{(1)}]^{-1}$. The matrix $P^{(1)}$ is defined by

$$P^{(1)} = I - 2\mathbf{w}\mathbf{w}^t$$

for a specially defined vector $\mathbf{w}$. This vector

$$\mathbf{w} = (w_1, w_2, \ldots, w_{n-1}, 0)^t$$

must satisfy $\mathbf{w}^t\mathbf{w} = 1$ and, in addition, the components $w_1, w_2, \ldots, w_{n-1}$ are selected so that the resulting matrix $A^{(2)} = P^{(1)}AP^{(1)}$ will have $a_{n,j}^{(2)} = 0$ for each $j = 1, 2, \ldots, n-2$. (By symmetry this also implies that $a_{j,n}^{(2)} = 0$ for each $j = 1, 2, \ldots, n-2$). This imposes $n-1$ conditions on the $n-1$ unknowns $w_1, w_2, \ldots, w_{n-1}$. The last component of $\mathbf{w}$ is chosen as zero to leave invariant the $a_{n,n}$ element of A. In Exercise 6 you are asked to show that $P^{(1)}$ defined in this way must be an orthogonal matrix.

A tridiagonal matrix $A^{(n-1)}$ is produced by repeating this procedure $n-3$ times, where the kth step is described by

$$A^{(k+1)} = P^{(k)}A^{(k)}P^{(k)}$$

for a matrix $P^{(k)}$ chosen to have the properties

(8.57) $$[P^{(k)}]^{-1} = P^{(k)}$$

and

(8.58) $$a_{n-k+1,j}^{(k+1)} = a_{j,n-k+1}^{(k+1)} = 0 \quad \text{for each } j = 1, 2, \ldots, n-k-1.$$

The matrix $A^{(k)}$ will be tridiagonal in its last $k-1$ rows and columns, and it has the form shown in (8.59) where, for convenience of notation, we have let $i = n-k+2$.

(8.59)
$$A^{(k)} = \begin{bmatrix} a_{11} & \cdots & a_{1,i-2} & a_{1,i-1} & 0 & \cdots & \cdots & 0 \\ \vdots & & \vdots & \vdots & & & & \vdots \\ & & & & & 0 & & \\ a_{i-1,1} & \cdots & a_{i-1,i-2} & a_{i-1,i-1} & a_{i-1,i} & & & \vdots \\ 0 & \cdots & 0 & a_{i,i-1} & a_{i,i} & a_{i,i+1} & & 0 \\ \vdots & & & & & & & a_{n-1,n} \\ 0 & \cdots & \cdots & \cdots & 0 & a_{n,n-1} & & a_{n,n} \end{bmatrix}$$

To construct $P^{(k)}$, suppose that $A^{(k)}$ has been formed as shown in Eq. (8.59). Set

$$P^{(k)} = I - 2\mathbf{w}\mathbf{w}^t,$$

where $\mathbf{w}$ is a vector with the property that $\mathbf{w}^t\mathbf{w} = 1$. This ensures that $P^{(k)}$ satisfies Eq.

(8.57). To guarantee that Eq. (8.58) also holds, we choose $w_j = 0$ for each $j = i-1, ..., n$ and w_j for $j = 1, 2, ..., i-2$ so that

$$[A^{(k+1)}]_{n-k+1,j} = [A^{(k+1)}]_{j,n-k+1} = 0 \qquad \text{for each } j = 1, 2, ..., i-2.$$

The precise procedure for the choice of $w_1, w_2, ..., w_{i-2}$ is similar to the choice in $A^{(2)}$ and is detailed in the following algorithm. It requires solving a system of nonlinear equations in a way that round-off error is minimized.

Computational accuracy requires (see Wilkinson and Reinsch [143]) that the larger diagonal elements be initially placed toward the upper left corner of the matrix. This is accomplished by simply interchanging appropriate rows and columns of the matrix.

Householder Algorithm 8.9

To obtain a symmetric tridiagonal matrix $A^{(n-1)}$ similar to the symmetric matrix $A = A^{(1)}$ construct the following matrices $A^{(2)}, A^{(3)}, ..., A^{(n-1)}$ where $A^{(k)} = a_{i,j}^{(k)}$ for each $k = 1, 2, ..., n-1$:

INPUT dimension n; matrix A.

OUTPUT $A^{(n-1)}$.   (*Could over-write A.*)

Step 1   For $k = 1, 2, ..., n-2$ do Steps 2–10.

Step 2   Set $i = n - k + 2$.

Step 3   Set $q = \left(\sum\limits_{j=1}^{i-2} [a_{i-1,j}^{(k)}]^2 \right)^{1/2}$.

Step 4   If $a_{i-1,i-2}^{(k)} = 0$ then set $s = q$
                              else set $s = q a_{i-1,i-2}^{(k)} / |a_{i-1,i-2}^{(k)}|$.

Step 5   Set $RSQ = (q^2 + s a_{i-1,i-2}^{(k)})$.   (*Note: $RSQ = 2r^2$*)

Step 6   Set $v_{i-1} = 0$.   (*Note: $v_i = \cdots = v_n = 0$, but are not needed.*)
             Set $v_{i-2} = a_{i-1,i-2}^{(k)} + s$.
             If $i > 3$ then for $j = 1, ..., i-3$ set $v_j = a_{i-1,j}^{(k)}$.

$$\left(Note: \mathbf{w} = \frac{1}{\sqrt{2RSQ}} \mathbf{v} = \frac{1}{2r} \mathbf{v}. \right)$$

Step 7   For $j = 1, 2, ..., i-1$ set $u_j = \dfrac{1}{RSQ} \sum\limits_{l=1}^{i-2} a_{jl}^{(k)} v_l$.

$$\left(Note: \mathbf{u} = \frac{1}{RSQ} A^{(k)} \mathbf{v} = \frac{1}{2r^2} A^{(k)} \mathbf{v}. \right)$$

Step 8   Set $PROD = \sum_{j=1}^{i-2} v_j u_j$.

$$\left(Note: PROD = \mathbf{v}^t \mathbf{u} = \frac{1}{2r^2} \mathbf{v}^t A^{(k)} \mathbf{v}. \right)$$

Step 9   For $j = 1, 2, ..., i-1$ set $z_j = u_j - \dfrac{PROD}{2RSQ} v_j$.

$$\left(Note: \mathbf{z} = \mathbf{u} - \frac{1}{2RSQ} \mathbf{v}^t \mathbf{u} \mathbf{v} = \mathbf{u} - \frac{1}{4r^2} \mathbf{v}^t \mathbf{u} \mathbf{v} = \mathbf{u} - \mathbf{w}\mathbf{w}^t \mathbf{u} \right.$$

$$\left. = \frac{1}{r} A^{(k)} \mathbf{w} - \mathbf{w}\mathbf{w}^t \frac{1}{r} A^{(k)} \mathbf{w}. \right)$$

Step 10   For $l = 2, ..., i-2$ (*Note: Compute* $A^{(k+1)} = A^{(k)} - \mathbf{v}\mathbf{z}^t - \mathbf{z}\mathbf{v}^t$
$= (I - 2\mathbf{w}\mathbf{w}^t) A^{(k)} (I - 2\mathbf{w}\mathbf{w}^t).$)

for $j = 1, ..., l-1$
   set $a_{lj}^{(k+1)} = a_{lj}^{(k)} - v_l z_j - v_j z_l$;
   $a_{jl}^{(k+1)} = a_{lj}^{(k+1)}$;
set $a_{ll}^{(k+1)} = a_{ll}^{(k)} - 2v_l z_l$;
set $a_{11}^{(k+1)} = a_{11}^{(k)} - 2v_1 z_1$;
for $j = 1, ..., i-3$ set $a_{i-1,j}^{(k+1)} = 0$; $a_{j,i-1}^{(k+1)} = 0$;
set $a_{i-1,i-2}^{(k+1)} = a_{i-1,i-2}^{(k)} - v_{i-2} z_{i-1}$;
   $a_{i-2,i-1}^{(k+1)} = a_{i-1,i-2}^{(k+1)}$.
(*Note: The other elements of* $A^{(k+1)}$ *are the same as* $A^{(k)}$.)

Step 11   OUTPUT $(A^{(n-1)})$;
(*The process is complete.* $A^{(n-1)}$ *is symmetric, tridiagonal and similar to* A.)
STOP.

EXAMPLE 1   The 4×4 matrix

$$A = \begin{bmatrix} 4 & -2 & 1 & 2 \\ -2 & 3 & 0 & -2 \\ 1 & 0 & 2 & 1 \\ 2 & -2 & 1 & -1 \end{bmatrix}$$

is symmetric. To use Algorithm 8.9 to transform this matrix into a matrix that is tridiagonal and symmetric and has the same eigenvalues as A, we perform the following calculations:

Set $k = 1$.

   Set $i = 5$;
   $$q = \left(\sum_{j=1}^{3} [a_{4,j}^{(1)}]^2 \right)^{1/2} = 3;$$

$$s = 3a^{(1)}_{4,3}/|a^{(1)}_{4,3}| = 3;$$

$$RSQ = (q^2 + sa^{(1)}_{4,3}) = 12;$$

$$\mathbf{v} = (2, -2, 4, 0)^t;$$

$$\mathbf{u} = (1/RSQ)A\mathbf{v} = (\tfrac{4}{3}, -\tfrac{5}{6}, \tfrac{5}{6}, 1)^t;$$

$$PROD = \tfrac{23}{3};$$

$$\mathbf{z} = \mathbf{u} - (PROD/2RSQ)\mathbf{v} = (\tfrac{25}{36}, -\tfrac{7}{36}, -\tfrac{4}{9}, 1)^t;$$

$$A^{(2)} = \begin{bmatrix} \tfrac{11}{9} & -\tfrac{2}{9} & -\tfrac{8}{9} & 0 \\ -\tfrac{2}{9} & \tfrac{20}{9} & -\tfrac{1}{9} & 0 \\ -\tfrac{8}{9} & -\tfrac{1}{9} & \tfrac{50}{9} & -3 \\ 0 & 0 & -3 & -1 \end{bmatrix}.$$

Set $k = 2$.

Set $i = 4$;

$$q = \left(\sum_{j=1}^{2} [a^{(2)}_{3,j}]^2 \right)^{1/2} = 0.8958064164;$$

$$s = qa^{(2)}_{3,2}/|a^{(2)}_{3,2}| = -0.8958064164;$$

$$RSQ = (q^2 + sa^{(2)}_{3,2}) = 0.9020031820;$$

$$\mathbf{v} = (-0.8888888889, -1.006917528, 0, 0)^t;$$

$$\mathbf{u} = (1/RSQ)A^{(2)}\mathbf{v} = (-0.9563827706, -2.261703377, 1, 0)^t;$$

$$PROD = \mathbf{v}^t\mathbf{u} = 3.127466791;$$

$$\mathbf{z} = \mathbf{u} - (PROD/2RSQ)\mathbf{v} = (0.5846153854, -0.5160880660, 1, 0)^t;$$

$$A^{(3)} = \begin{bmatrix} 2.261538463 & -0.0923076911 & 0 & 0 \\ -0.0923076911 & 1.182905983 & 0.8958064169 & 0 \\ 0 & 0.8958064169 & 5.555555556 & -3 \\ 0 & 0 & -3 & -1 \end{bmatrix}.$$

The process is complete. The matrix $A^{(3)}$ is tridiagonal and has the same eigenvalues as A, except for round-off error.   ∎

In the remainder of this section it will be assumed that the symmetric matrix for which the eigenvalues are to be calculated is tridiagonal. When this is not the case, Householder's method will first be used. If we let A denote a matrix of this type, we can simplify the notation somewhat by labelling the entries of A as follows:

$$A = \begin{bmatrix} a_1 & b_2 & 0 & \cdots & & \cdots & 0 \\ b_2 & a_2 & b_3 & & & & \vdots \\ 0 & b_3 & a_3 & & & & \\ \vdots & & & & & & 0 \\ & & & & & & b_n \\ 0 & \cdots & & \cdots & 0 & b_n & a_n \end{bmatrix}.$$

The first observation that can be made is that, when $b_j = 0$ for some j, $2 < j < n$, the problem can be reduced to considering, instead of A, the smaller matrices

(8.60)

$$\begin{bmatrix} a_1 & b_2 & 0 & \cdots & & \cdots & 0 \\ b_2 & a_2 & & & & & \vdots \\ 0 & & & & & & \\ \vdots & & & & & & 0 \\ & & & & & & b_{j-1} \\ 0 & \cdots & & \cdots & 0 & b_{j-1} & a_{j-1} \end{bmatrix}$$

and

$$\begin{bmatrix} a_j & b_{j+1} & 0 & \cdots & & \cdots & 0 \\ b_{j+1} & a_{j+1} & & & & & \vdots \\ 0 & & & & & & \\ \vdots & & & & & & 0 \\ & & & & & & b_n \\ 0 & \cdots & & \cdots & 0 & b_n & a_n \end{bmatrix}.$$

If $b_2 = 0$ or $b_n = 0$, then the 1×1 matrix $[a_1]$ or $[a_n]$ immediately produces an eigenvalue a_1 or a_n of A.

If none of the b_j are zero, the QL Algorithm proceeds by forming a sequence of matrices $A^{(1)} = A$, $A^{(2)}$, $A^{(3)}$, ... as follows:

1) $A^{(1)} = A$ is factored as a product $A^{(1)} = Q^{(1)}L^{(1)}$, where $Q^{(1)}$ is orthogonal and $L^{(1)}$ is lower triangular;

2) $A^{(2)}$ is defined as $A^{(2)} = L^{(1)}Q^{(1)}$ and factored as a product $A^{(2)} = Q^{(2)}L^{(2)}$ where $Q^{(2)}$ is orthogonal and $L^{(2)}$ is lower triangular.

In general, $A^{(i)}$ is factored as a product $A^{(i)} = Q^{(i)}L^{(i)}$ of an orthogonal matrix $Q^{(i)}$ and a lower-triangular matrix $L^{(i)}$. Then $A^{(i+1)}$ is defined by the product of $L^{(i)}$ and $Q^{(i)}$ in the reverse direction $A^{(i+1)} = L^{(i)}Q^{(i)}$. Since $Q^{(i)}$ is orthogonal for each i,

(8.61) $$A^{(i+1)} = L^{(i)}Q^{(i)} = (Q^{(i)t}A^{(i)})Q^{(i)} = Q^{(i)t}A^{(i)}Q^{(i)}$$

and $A^{(i+1)}$ is symmetric and tridiagonal with the same eigenvalues as $A^{(i)}$. Continuing by induction, $A^{(i+1)}$ has the same eigenvalues as the original matrix A.

The success of the procedure is a result of the fact that $A^{(i+1)}$ tends to a diagonal matrix with the eigenvalues of A in increasing order of magnitude along the diagonal.

To describe the construction of the factoring matrices $Q^{(i)}$ and $L^{(i)}$, we must first introduce the notion of a **rotation matrix**.

DEFINITION 8.44   A **rotation matrix** R is a matrix that differs from the identity matrix in at most four elements. These four elements are of the form

$$r_{ii} = r_{jj} = \cos \theta \qquad \text{and} \qquad r_{ij} = -r_{ji} = \sin \theta$$

for some θ and some $i \neq j$.

It is easy to show (see Exercise 8) that, for any rotation matrix R, the matrix AR differs from A only in the ith and jth columns and the matrix RA differs from A only in the ith and jth rows. The angle θ can be chosen so that the product RA has a zero entry for $(RA)_{ij}$.

The factorization of $A^{(1)}$ into $A^{(1)} = Q^{(1)}L^{(1)}$ uses a product of $n-1$ rotation matrices of this type to construct

$$L^{(1)} = P_2 P_3 \cdots P_n A^{(1)}.$$

We choose the rotation matrix P_n first so that the matrix

$$A_2^{(1)} = P_n A^{(1)}$$

has the element in the $(n-1, n)$ position set to zero. Since the multiplication $P_n A^{(1)}$ effects both rows $n-1$ and n of $A^{(1)}$, the matrix $A_2^{(1)}$ does not necessarily retain zero entries in positions $(n, 1), (n, 2), \ldots, (n, n-2)$. However, since $A^{(1)}$ is tridiagonal, it is easily seen that the $(n, 1), (n, 2), \ldots, (n, n-3)$ entries of $A_2^{(1)}$ remain zero and only the $(n, n-2)$ entry can become nonzero.

In general, the matrix P_k is chosen so that the $(k-1, k)$ entry in $A_{n-k+2}^{(1)} = P_k A_{n-k+1}^{(1)}$ is zero, which results in the $(k, k-2)$ entry becoming nonzero. The matrix $A_{n-k+1}^{(1)}$ will have the form

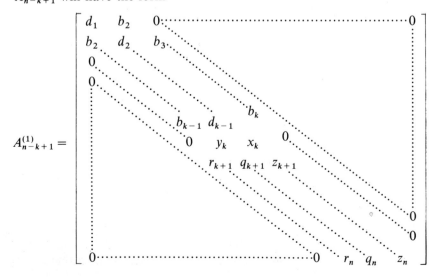

and P_k will have the form

$$P_k = \begin{bmatrix} I_{k-2} & O & O \\ O & \begin{matrix} c_{k-1} & -s_{k-1} \\ s_{k-1} & c_{k-1} \end{matrix} & O \\ O & O & I_{n-k} \end{bmatrix} \leftarrow \text{row } k.$$

$$\uparrow$$
$$\text{column } k$$

The constants $c_{k-1} = \cos\theta_{k-1}$ and $s_{k-1} = \sin\theta_{k-1}$ in P_k are chosen so that the $(k-1, k)$ entry in $A^{(1)}_{n-k+2}$ is zero. That is,

(8.62) $$c_{k-1}b_k - s_{k-1}x_k = 0.$$

Since $c^2_{k-1} + s^2_{k-1} = 1$, the solution to Eq. (8.62) is

$$s_{k-1} = \frac{b_k}{\sqrt{b_k^2 + x_k^2}} \qquad \text{and} \qquad c_{k-1} = \frac{x_k}{\sqrt{b_k^2 + x_k^2}}$$

and $A^{(1)}_{n-k+2}$ has the form

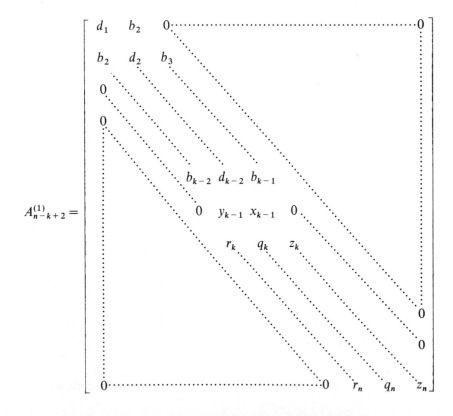

$$A^{(1)}_{n-k+2} =$$

Proceeding with this construction in sequence through $P_n, P_{n-1}, \ldots, P_2$ produces the lower-triangular matrix:

$$L^{(1)} \equiv A_{n-1}^{(1)} = \begin{bmatrix} x_1 & 0 & \cdots\cdots\cdots\cdots\cdots & 0 \\ q_2 & z_2 & & \\ r_3 & q_3 & z_3 & \\ 0 & & & \\ & & & 0 \\ 0 \cdots\cdots\cdots & 0 & r_n & q_n & z_n \end{bmatrix}$$

The orthogonal matrix $Q^{(1)}$ is then defined as

$$Q_1^{(1)} = P_n^t P_{n-1}^t \cdots P_2^t.$$

Consequently, the matrix $A^{(2)}$ is defined by

$$A^{(2)} = L^{(1)} Q^{(1)} = L^{(1)} P_n^t P_{n-1}^t \cdots P_2^t = P_2 P_3 \cdots P_n A P_n^t P_{n-1}^t \cdots P_2^t.$$

The matrix $A^{(2)}$ is tridiagonal, and, in general, the entries off the diagonal will be smaller in magnitude than the corresponding entries in $A^{(1)}$. The process is repeated to construct $A^{(3)}, A^{(4)}, \ldots$.

It can be shown that if the eigenvalues of A have distinct moduli, the rate of convergence of the entry $b_{j+1}^{(i+1)}$ to zero in the matrix $A^{(i+1)}$ is

$$O\left(\left| \frac{\lambda_j}{\lambda_{j+1}} \right|^{i+1} \right),$$

where the eigenvalues are ordered by $|\lambda_1| < |\lambda_2| < \cdots < |\lambda_n|$. The rate of convergence of $b_{j+1}^{(i+1)}$ to zero determines the rate at which the entry $a_j^{(i+1)}$ converges to the jth eigenvalue λ_j.

To accelerate this convergence a shifting technique is employed similar to that used with the inverse power method in Section 8.4. A constant s is selected close to an eigenvalue of A. This modifies the factorization in Eq. (8.61) to choosing $Q^{(i)}$ and $L^{(i)}$ so that

(8.63) $$A^{(i)} - sI = Q^{(i)} L^{(i)},$$

and correspondingly, the matrix $A^{(i+1)}$ is defined to be

(8.64) $$A^{(i+1)} = L^{(i)} Q^{(i)} + sI.$$

With this modification, the rate of convergence of $b_{j+1}^{(i+1)}$ to zero becomes

$$O\left(\left| \frac{\lambda_j - s}{\lambda_{j+1} - s} \right|^{i+1} \right),$$

which can result in a significant improvement over the original rate of convergence if s is close to λ_j but not close to λ_{j+1}.

In the listing of the QL Algorithm, we change s at each step so that when A has eigenvalues of distinct modulus, $b_2^{(i+1)}$ converges to zero faster than $b_{j+1}^{(i+1)}$ for any integer j greater than one. When $b_2^{(i+1)}$ is sufficiently small, we assume that $\lambda_1 \approx a_1^{(i+1)}$, delete the first row and column of the matrix, and proceed in the same manner to find an approximation to λ_2. The process is continued until approximations have been determined for each eigenvalue.

If A has eigenvalues of the same modulus, $b_j^{(i+1)}$ may tend to zero for some $j \neq 2$ at a faster rate than $b_2^{(i+1)}$. In this case, a matrix splitting technique of the type described in (8.60) can be employed to reduce the problem to one involving a pair of matrices of reduced order.

The following algorithm incorporates the shifting technique by choosing at the ith step the shifting constant s_i, where s_i is the eigenvalue of the matrix

$$E^{(i)} = \begin{bmatrix} a_1^{(i)} & b_2^{(i)} \\ b_2^{(i)} & a_2^{(i)} \end{bmatrix}$$

that is closest to $a_1^{(i)}$. This shift translates the eigenvalues of A by a factor s_i. The algorithm accumulates these shifts until $b_2^{(i+1)} \approx 0$ and then adds the shifts to $a_1^{(i+1)}$ to approximate the eigenvalue λ_1.

QL Algorithm 8.10

To obtain the eigenvalues of the symmetric, tridiagonal $n \times n$ matrix

$$A \equiv \hat{A}_1 \equiv \begin{bmatrix} a_1^{(1)} & b_2^{(1)} & 0 & \cdots & \cdots & 0 \\ b_2^{(1)} & a_2^{(1)} & & & & \vdots \\ 0 & & & & & \\ \vdots & & & & & b_n^{(1)} \\ 0 & \cdots & \cdots & 0 & b_n^{(1)} & a_n^{(1)} \end{bmatrix}$$

INPUT n; $a_1^{(1)}, \ldots, a_n^{(1)}$; $b_2^{(1)}, \ldots, b_n^{(1)}$; maximum number of iterations M.

OUTPUT eigenvalues of A or recommended splitting of A, or a message that the maximum number of iterations was exceeded.

Step 1 Set $k = 1$;

SHIFT $= 0$.   (*Accumulated shift.*)

Step 2 While $k \leq M$ do Steps 3–8.

Step 3 (*Test for success.*)

If $b_2^{(k)} \approx 0$ then set $\lambda = a_1^{(k)} + SHIFT$;

OUTPUT (λ);

set $n = n - 1$;

$a_1^{(k)} = a_2^{(k)}$;

$$\text{for } j = 2, \ldots, n$$
$$\text{set } a_j^{(k)} = a_{j+1}^{(k)};$$
$$b_j^{(k)} = b_{j+1}^{(k)}.$$

If $b_j^{(k)} \approx 0$ for $3 \leq j \leq n-1$ then
OUTPUT ('split into', $a_1^{(k)}, \ldots, a_{j-1}^{(k)}, b_2^{(k)}, \ldots, b_{j-1}^{(k)}$, 'and',
$a_j^{(k)}, \ldots, a_n^{(k)}, b_{j+1}^{(k)}, \ldots, b_n^{(k)}$, SHIFT);
STOP.

If $b_n^{(k)} \approx 0$ then set $\lambda = a_n^{(k)} + SHIFT$;
OUTPUT (λ);
set $n = n - 1$.

Step 4   (*Compute shift.*)
Set $b = -(a_2^{(k)} + a_1^{(k)})$;
$c = a_2^{(k)} a_1^{(k)} - [b_2^{(k)}]^2$;
$d = (b^2 - 4c)^{1/2}$.
If $b > 0$ then set $\mu_1 = -2c/(b + d)$;
$\mu_2 = -(b + d)/2$;
else set $\mu_1 = (d - b)/2$;
$\mu_2 = 2c/(d - b)$.
If $n = 2$ then set $\lambda_1 = \mu_1 + SHIFT$;
$\lambda_2 = \mu_2 + SHIFT$;
OUTPUT (λ_1, λ_2);
STOP.
Choose s so that $|s - a_1^{(k)}| = \min(|\mu_1 - a_1^{(k)}|, |\mu_2 - a_1^{(k)}|)$.

Step 5   (*Accumulate shift.*)
Set $SHIFT = SHIFT + s$.

Step 6   (*Perform shift.*)
For $j = 1, \ldots, n$ set $d_j = a_j^{(k)} - s$.

Step 7   (*Compute $L^{(k)}$.*)
Set $x_n = d_n$;
$y_n = b_n$.
For $j = n, n-1, \ldots, 2$
set $z_j = (x_j^2 + [b_j^{(k)}]^2)^{1/2}$;
$c_{j-1} = x_j/z_j$;
$s_{j-1} = b_j^{(k)}/z_j$;
$q_j = c_{j-1} y_j + s_{j-1} d_{j-1}$;
$x_{j-1} = c_{j-1} d_{j-1} - s_{j-1} y_j$.
If $j \neq 2$ then set $r_j = s_j = b_{j-1}$;
$y_{j-1} = c_{j-1} b_{j-1}$.
($A_{n-j+2}^{(k)} = P_j A_{n-j+1}^{(k)}$ *has just been computed where*

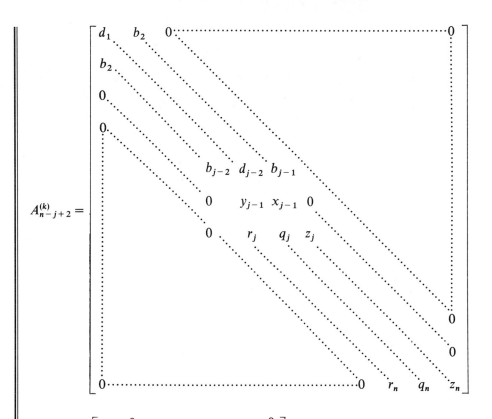

$$A^{(k)}_{n-j+2} = \begin{bmatrix} d_1 & b_2 & 0 & \cdots & & & & & 0 \\ b_2 & & & & & & & & \\ 0 & & & & & & & & \\ 0 & & & & & & & & \\ & & & b_{j-2} & d_{j-2} & b_{j-1} & & & \\ & & & 0 & y_{j-1} & x_{j-1} & 0 & & \\ & & & 0 & r_j & q_j & z_j & & \\ & & & & & & & 0 & \\ & & & & & & & & 0 \\ 0 & \cdots & & & & 0 & r_n & q_n & z_n \end{bmatrix}$$

$$L = A^{(k)}_n = \begin{bmatrix} x_1 & 0 & \cdots & & 0 \\ q_2 & z_2 & & & \\ r_3 & q_3 & z_3 & & \\ 0 & & & & \\ & & & & 0 \\ 0 & \cdots & 0 & r_n & q_n & z_n \end{bmatrix}$$ *has been computed.*)

Step 8   (*Compute $A^{(k+1)}$*).
Set $z_1 = x_1$;
$a^{(k+1)}_n = s_{n-1}q_n + c_{n-1}z_n$;
$b^{(k+1)}_n = s_{n-1}z_{n-1}$.
For $j = n-1, n-2, \ldots, 2$
set $a^{(k+1)}_j = s_{j-1}q_j + c_{j-1}c_j z_j$;
$b^{(k+1)}_j = s_{j-1}z_{j-1}$.
Set $a^{(k+1)}_1 = c_1 x_1$.
$k = k+1$.

Step 9   OUTPUT ('Maximum number of iterations exceeded');
STOP.

EXAMPLE 2   Let

$$A = A_1^{(1)} = \begin{bmatrix} 4 & 2 & 0 \\ 2 & 3 & 1 \\ 0 & 1 & 2 \end{bmatrix} = \begin{bmatrix} a_1^{(1)} & b_2^{(1)} & 0 \\ b_2^{(1)} & a_2^{(1)} & b_3^{(1)} \\ 0 & b_3^{(1)} & a_3^{(1)} \end{bmatrix}.$$

To find the acceleration parameter s requires finding the eigenvalues of

$$\begin{bmatrix} 4 & 2 \\ 2 & 3 \end{bmatrix}.$$

These can be easily found to be $\mu_1 = 5.561552$ and $\mu_2 = 1.438447$. The eigenvalue closest to $a_1^{(1)} = 4$ is $s = \mu_1 = 5.561552$. Thus, $SHIFT = 5.561552$ and

$$A_1^{(1)} - sI = \begin{bmatrix} -1.561552 & 2 & 0 \\ 2 & -2.561552 & 1 \\ 0 & 1 & -3.561552 \end{bmatrix} = \begin{bmatrix} d_1 & b_2^{(2)} & 0 \\ b_2^{(2)} & d_2 & b_3^{(2)} \\ 0 & y_3 & x_3 \end{bmatrix}.$$

The computation of L in Step 7 of Algorithm 8.10 is

$$P_3 = \begin{bmatrix} 1 & 0 & 0 \\ 0 & -0.9627697 & 0.2703230 \\ 0 & -0.2703230 & -0.9627697 \end{bmatrix},$$

$$A_2^{(1)} = P_3 A_1^{(1)} = \begin{bmatrix} -1.561552 & 2 & 0 \\ 1 & 2.195860 & 0 \\ 0.5406461 & -1.655216 & 3.699276 \end{bmatrix} = \begin{bmatrix} d_1 & b_2^{(1)} & 0 \\ y_2 & x_2 & 0 \\ r_3 & q_3 & z_3 \end{bmatrix},$$

$$P_2 = \begin{bmatrix} 0.7393092 & 0.6733661 & 0 \\ -0.6733661 & 0.7393092 & 0 \\ 0 & 0 & 1 \end{bmatrix},$$

$$L = A_3^{(1)} = P_2 A_2^{(1)} = \begin{bmatrix} 0.1421232 & 2 & 0 \\ -2.475064 & 2.970151 & 0 \\ 0.5406461 & -1.655216 & 3.699276 \end{bmatrix} = \begin{bmatrix} x_1 & 0 & 0 \\ q_2 & z_2 & 0 \\ r_3 & q_3 & z_3 \end{bmatrix}.$$

Computing $A^{(2)}$ gives

$$A^{(2)} = LP_3^t P_2^t = \begin{bmatrix} 0.1050729 & 0.09570091 & 0 \\ 0.09570091 & -3.780732 & 0.8029006 \\ 0 & 0.8029006 & -4.008994 \end{bmatrix}.$$

Iterating again using $A^{(2)}$ gives $s = 0.1074284$ and

$$A^{(3)} = \begin{bmatrix} 9.885922 \times 10^{-5} & 2.583158 \times 10^{-6} & 0 \\ 2.583158 \times 10^{-6} & -3.597252 & 0.7013983 \\ 0 & 0.7013983 & -4.409784 \end{bmatrix}.$$

Accepting $b_2^{(3)} = 2.583158 \times 10^{-6}$ as sufficiently small gives

$$\lambda \approx a_1^{(3)} + SHIFT = 9.885922 \times 10^{-5} + 5.668980 = 5.669077$$

as the first eigenvalue. The other two eigenvalues are computed in Step 4 as 2.476024 and 0.8548984. The actual eigenvalues are 5.669079087, 2.476023608, and 0.8548973088. ∎

A similar procedure can be employed to find approximations to the eigenvalues of a nonsymmetric $n \times n$ matrix A. The problem is reduced to finding the eigenvalues of a similar matrix with more easily determined eigenvalues. Then iterative matrix operations are performed so that the eigenvalues accumulate along the diagonal of a triangular matrix. Analogous to the symmetric case, we do not try to find the upper-triangular matrix directly. For a symmetric matrix, Theorem 8.39 on page 450 guarantees a diagonal matrix, but instead we constructed a tridiagonal matrix with the same eigenvalues. For the nonsymmetric case. Theorem 8.40 guarantees an upper-triangular matrix, but instead we find a matrix that can be considered as a combination of a tridiagonal and a triangular matrix. A matrix of this type is called **upper Hessenberg**, and contains only zero entries below the lower subdiagonal; that is, H is upper Hessenberg if $(H)_{ij} = 0$ for all $i \geq j + 2$. The reduction of an arbitrary nonsymmetric matrix to one that is in upper-Hessenberg form can be performed by a method similar to Householder's method or can be accomplished by a Gaussian-elimination type procedure that is modified to preserve eigenvalues.

After the upper-Hessenberg matrix H has been formed with the same eigenvalues as the matrix A, a factoring procedure is employed, which, at each step, factors a matrix into a product of an orthogonal matrix, denoted by Q, and an upper triangular matrix, denoted by R. The factoring process assumes the following form. First

(8.65) $$H \equiv H_1 = Q_1 R_1;$$

then H_2 is defined by

(8.66) $$H_2 = R_1 Q_1$$

and factored into

(8.67) $$H_2 = Q_2 R_2.$$

The method of factoring proceeds with the same aim as that of the QL factoring. The matrices are chosen to introduce zeros at appropriate entries of the matrix. The method is called the QR method. In practice, a shifting procedure is employed similar to that used in the QL method. However, the shifting is somewhat more complicated for nonsymmetric matrices since complex eigenvalues with the same modulus can occur. The shifting process for this procedure modifies the calculations in equations (8.65), (8.66), and (8.67) to obtain the double QR method $H_1 - s_1 I = Q_1 R_1$, $H_2 = R_1 Q_1 + s_1 I$, $H_2 - s_2 I = Q_2 R_2$, and $H_3 = R_2 Q_2 + s_2 I$, where s_1 and s_2 are complex conjugates and $H_1, H_2, \ldots$ are real upper-Hessenberg matrices.

A complete description of the QR method can be found in Wilkinson [142].

Detailed algorithms and ALGOL 60 programs for this method and most other commonly employed methods are given in Wilkinson and Reinsch [143]. We refer the reader to these works if the method we have discussed does not give satisfactory results.

Both the QL and QR methods can be performed in a manner that will produce the eigenvectors of a matrix as well as its eigenvalues. The QL Algorithm 8.10 has not been designed to accomplish this, however. If the eigenvectors of a symmetric matrix are needed as well as the eigenvalues, we suggest either using the inverse power method after Algorithms 8.9 and 8.10 have been employed or using a more powerful technique such as those listed in Wilkinson and Reinsch [143], methods that are designed expressly for this purpose.

The standard package from IMSL contains over a dozen subroutines for finding the eigenvalues and eigenvectors of a matrix. In addition, a collection of subroutines called EISPACK developed at Argonne National Laboratories is available from IMSL that contains many additional techniques. Details regarding these libraries can be found in Rice [108].

Exercise Set 8.5

1. Use Householder's method to place the following matrices in tridiagonal form.

a) $\begin{bmatrix} 12 & 10 & 4 \\ 10 & 8 & -5 \\ 4 & -5 & 3 \end{bmatrix}$

b) $\begin{bmatrix} 2 & -1 & -1 \\ -1 & 2 & -1 \\ -1 & -1 & 2 \end{bmatrix}$

c) $\begin{bmatrix} 2 & 0 & 1 \\ 0 & 3 & -2 \\ 1 & -2 & -1 \end{bmatrix}$

d) $\begin{bmatrix} 4.75 & 2.25 & -0.25 \\ 2.25 & 4.75 & 1.25 \\ -0.25 & 1.25 & 4.75 \end{bmatrix}$

e) $\begin{bmatrix} 4 & -1 & -1 & 0 \\ -1 & 4 & 0 & -1 \\ -1 & 0 & 4 & -1 \\ 0 & -1 & -1 & 4 \end{bmatrix}$

f) $\begin{bmatrix} 5 & -2 & -0.5 & 1.5 \\ -2 & 5 & 1.5 & -0.5 \\ -0.5 & 1.5 & 5 & -2 \\ 1.5 & -0.5 & -2 & 5 \end{bmatrix}$

g) $\begin{bmatrix} 8 & 0.25 & 0.5 & 2 & -1 \\ 0.25 & -4 & 0 & 1 & 2 \\ 0.5 & 0 & 5 & 0.75 & -1 \\ 2 & 1 & 0.75 & 5 & -0.5 \\ -1 & 2 & -1 & -0.5 & 6 \end{bmatrix}$

h) $\begin{bmatrix} 2 & -1 & -1 & 0 & 0 \\ -1 & 3 & 0 & -2 & 0 \\ -1 & 0 & 4 & 2 & 1 \\ 0 & -2 & 2 & 8 & 3 \\ 0 & 0 & 1 & 3 & 9 \end{bmatrix}$

2. Apply two iterations of the QL Algorithm to the following matrices.

a) $\begin{bmatrix} 2 & -1 & 0 \\ -1 & 2 & -1 \\ 0 & -1 & 2 \end{bmatrix}$

b) $\begin{bmatrix} 3 & 1 & 0 \\ 1 & 4 & 2 \\ 0 & 2 & 1 \end{bmatrix}$

c)
$$\begin{bmatrix} 0.5 & 0.25 & 0 & 0 \\ 0.25 & 0.8 & 0.4 & 0 \\ 0 & 0.4 & 0.6 & 0.1 \\ 0 & 0 & 0.1 & 1 \end{bmatrix}$$

d)
$$\begin{bmatrix} 4 & -3 & 0 & 0 \\ -3 & \frac{10}{3} & -\frac{5}{3} & 0 \\ 0 & -\frac{5}{3} & -\frac{99}{75} & \frac{68}{75} \\ 0 & 0 & \frac{68}{75} & \frac{149}{75} \end{bmatrix}$$

3. Use the QL Algorithm to determine, to within 10^{-5}, all the eigenvalues of the following matrices.

a)
$$\begin{bmatrix} 2 & -1 & 0 \\ -1 & -1 & -2 \\ 0 & -2 & 3 \end{bmatrix}$$

b)
$$\begin{bmatrix} 3 & 1 & 0 \\ 1 & 4 & 2 \\ 0 & 2 & 1 \end{bmatrix}$$

c)
$$\begin{bmatrix} 4 & 2 & 0 & 0 & 0 \\ 2 & 4 & 2 & 0 & 0 \\ 0 & 2 & 4 & 2 & 0 \\ 0 & 0 & 2 & 4 & 2 \\ 0 & 0 & 0 & 2 & 4 \end{bmatrix}$$

d)
$$\begin{bmatrix} 5 & -1 & 0 & 0 & 0 \\ -1 & 4.5 & 0.2 & 0 & 0 \\ 0 & 0.2 & 1 & -0.4 & 0 \\ 0 & 0 & -0.4 & 3 & 1 \\ 0 & 0 & 0 & 1 & 3 \end{bmatrix}$$

4. Use the QL Algorithm to determine, to within 10^{-5}, all the eigenvalues for the matrices given in Exercise 1.

5. Use the inverse power method to determine, to within 10^{-5}, the eigenvectors of the matrices in Exercise 1.

6. Let $\mathbf{w} \in R^n$ satisfy $\mathbf{w}^t\mathbf{w} = 1$. Show that $P^2 = I$ where $P = I - 2\mathbf{w}\mathbf{w}^t$.

7. a) Show that the rotation matrix $\begin{bmatrix} \cos\theta & -\sin\theta \\ \sin\theta & \cos\theta \end{bmatrix}$ applied to the vector $\mathbf{x} = (x_1, x_2)^t$ has the geometric effect of rotating $\mathbf{x}$ through the angle θ without changing its magnitude with respect to $\|\cdot\|_2$.

   b) Show that the magnitude of $\mathbf{x}$ with respect to $\|\cdot\|_\infty$ can be changed by a rotation matrix.

8. Let R be a rotation matrix with $r_{ii} = r_{jj} = \cos\theta$, $r_{ij} = -r_{ji} = \sin\theta$ for $j < i$. Show that for any $n \times n$ matrix A:

$$(AR)_{p,q} = \begin{cases} a_{p,q} & \text{if } q = i, j, \\ (\cos\theta)a_{p,j} + (\sin\theta)a_{p,i} & \text{if } q = j, \\ (\cos\theta)a_{p,i} - (\sin\theta)a_{p,j} & \text{if } q = i. \end{cases}$$

$$(RA)_{p,q} = \begin{cases} a_{p,q} & \text{if } p = i, j, \\ (\cos\theta)a_{j,q} - (\sin\theta)a_{i,q} & \text{if } p = j, \\ (\sin\theta)a_{j,q} + (\cos\theta)a_{i,q} & \text{if } p = i. \end{cases}$$

9. **Jacobi's method** for a symmetric matrix A is described by

$$A_1 = A,$$
$$A_2 = P_1 A_1 P_1^t,$$

and in general

$$A_{i+1} = P_i A_i P_i^t$$

tends to a diagonal matrix where P_i is a rotation matrix chosen to eliminate a large off-diagonal element in A_i. If $a_{j,k}$ and $a_{k,j}$ are to be set to zero where $j < k$, then either

$$(P_i)_{jj} = (P_i)_{kk} = \sqrt{\frac{1}{2}\left(1 + \frac{b}{\sqrt{c^2 + b^2}}\right)},$$

$$(P_i)_{k,j} = \frac{c}{2(P_i)_{jj}\sqrt{c^2 + b^2}} = -(P_i)_{j,k},$$

where

$$c = 2a_{jk}\,\text{sign}(a_{jj} - a_{kk}),$$

$$b = |a_{jj} - a_{kk}| \quad \text{if } a_{jj} \neq a_{kk},$$

or

$$(P_i)_{jj} = (P_i)_{kk} = \sqrt{2}/2,$$

$$(P_i)_{k,j} = -(P_i)_{j,k} = \sqrt{2}/2 \quad \text{if } a_{jj} = a_{kk}.$$

Develop an algorithm to implement Jacobi's method by setting $a_{21} = 0$, then set $a_{31}, a_{32}, a_{41}, a_{42}, a_{43}, \ldots, a_{n,1}, \ldots, a_{n,n-1}$ in turn to zero. This is repeated until matrix A_k is computed with

$$\sum_{i=1}^{n} \sum_{\substack{j=1 \\ j \neq i}}^{n} |a_{ij}^{(k)}|$$

sufficiently small. The eigenvalues of A can then be approximated by the diagonal entries of A_k.

10. Apply Jacobi's method to the matrices in Exercise 1.

11. Modify the QL Algorithm to find the eigenvalues of a nonsymmetric tridiagonal matrix

$$\begin{bmatrix} a_1 & b_2 & 0 & \cdots\cdots & 0 \\ c_2 & a_2 & b_3 & & \vdots \\ 0 & & & & 0 \\ \vdots & & & & b_n \\ 0 & \cdots\cdots & 0 & c_n & a_n \end{bmatrix}$$

12. Use the algorithm developed in Exercise 11 to find, within 10^{-4}, all the eigenvalues of

a) $\begin{bmatrix} 4 & 1 & 0 \\ -1 & 3 & 1 \\ 0 & 1 & 2 \end{bmatrix}$

b) $\begin{bmatrix} 5 & 2 & 0 \\ 1 & 4 & -1 \\ 0 & 3 & 7 \end{bmatrix}$

c) $\begin{bmatrix} 7 & 1 & 0 & 0 \\ 1 & 6 & 3 & 0 \\ 0 & -2 & 4 & 1 \\ 0 & 0 & 2 & 3 \end{bmatrix}$

d) $\begin{bmatrix} 6 & -2 & 0 & 0 \\ 3 & 4 & -1 & 0 \\ 0 & 0 & 5 & -1 \\ 0 & 0 & 2 & 3 \end{bmatrix}$

9

Numerical Solutions of Nonlinear Systems of Equations

The amount of pressure required to sink a large heavy object in a soft, homogeneous soil that lies above a hard base soil can be predicted by the amount of pressure required to sink smaller objects in the same soil. In particular, the amount of pressure p to sink a circular plate of radius r a distance d in the soft soil, where the hard base soil lies a distance $D > d$ below the surface, can be approximated by an equation of the form

$$p = k_1 e^{k_2 r} + k_3 r,$$

where k_1, k_2, and k_3 are constants depending on d and the consistency of the soil, but not on the radius of the plate.

To determine the minimal size plate required to sustain a large load, three small plates with differing radii are sunk to the same distance, and the loads required for this sinkage are measured. These calculations lead to three equations of the type given above in the three unknowns k_1, k_2, and k_3. Due to the nature of the equations, however, we cannot easily solve for the unknowns. Numerical methods are needed that can be used for solving systems of equations when the equations are not linear. See Exercise 7 of Section 9.2 for an application of this type.

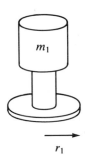

m_1

r_1

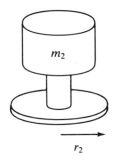

m_2

r_2

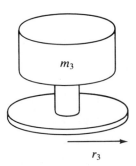

m_3

r_3

Fixed-point iteration methods dominated the study in Chapter 2 of numerical solutions to equations of the form $f(x) = 0$. The methods we will use for solving systems of nonlinear equations will basically be generalizations of Newton's method. Although other techniques are available, they are generally quite specialized with regard to the types of problem they will solve, and are probably inappropriate for inclusion in a first course in numerical analysis.

Most of the proofs of the theoretical results in this chapter will be omitted, since in general they involve methods that are usually studied in a course in advanced calculus. A good general reference for this material is Ortega's book entitled *Numerical Analysis—A Second Course* [94]. A more complete reference is Ortega and Rheinboldt [96].

9.1 Fixed Points for Functions of Several Variables

The general form of a system of nonlinear equations is:

(9.1)
$$
\begin{aligned}
f_1(x_1, x_2, \ldots, x_n) &= 0, \\
f_2(x_1, x_2, \ldots, x_n) &= 0, \\
&\vdots \qquad\qquad \vdots \\
f_n(x_1, x_2, \ldots, x_n) &= 0,
\end{aligned}
$$

where each function f_i can be thought of as mapping a vector $\mathbf{x} = (x_1, x_2, \ldots, x_n)^t$ of the n-dimensional space, R^n, into the real line R. The system can alternatively be represented by defining a function $\mathbf{F}$, mapping R^n into R^n by

$$\mathbf{F}(x_1, x_2, \ldots, x_n) = (f_1(x_1, x_2, \ldots, x_n), f_2(x_1, x_2, \ldots, x_n), \ldots, f_n(x_1, x_2, \ldots, x_n))^t.$$

Using vector notation to represent the variables $x_1, x_2, \ldots, x_n$, system (9.1) then assumes the form:

(9.2)
$$\mathbf{F}(\mathbf{x}) = \mathbf{0}.$$

The functions $f_1, f_2, \ldots, f_n$ are called the **coordinate functions of F**.

EXAMPLE 1   The three-by-three nonlinear system

$$
\begin{aligned}
3x_1 - \cos(x_2 x_3) - \tfrac{1}{2} &= 0, \\
x_1^2 - 81(x_2 + 0.1)^2 + \sin x_3 + 1.06 &= 0, \\
e^{-x_1 x_2} + 20x_3 + \frac{10\pi - 3}{3} &= 0,
\end{aligned}
$$

can be placed in the form (9.2) by defining the three functions f_1, f_2, and f_3 from R^3 to R as

$$f_1(x_1, x_2, x_3) = 3x_1 - \cos(x_2 x_3) - \tfrac{1}{2},$$

$$f_2(x_1, x_2, x_3) = x_1^2 - 81(x_2 + 0.1)^2 + \sin x_3 + 1.06,$$

$$f_3(x_1, x_2, x_3) = e^{-x_1 x_2} + 20x_3 + \frac{10\pi - 3}{3},$$

and **F** from $R^3 \to R^3$ by

$$
\begin{aligned}
\mathbf{F}(\mathbf{x}) &= \mathbf{F}(x_1, x_2, x_3) \\
&= (f_1(x_1, x_2, x_3), f_2(x_1, x_2, x_3), f_3(x_1, x_2, x_3))^t \\
&= \Big(3x_1 - \cos(x_2 x_3) - \tfrac{1}{2},\, x_1^2 - 81(x_2 + 0.1)^2 + \sin x_3 + 1.06, \\
&\qquad e^{-x_1 x_2} + 20x_3 + \frac{10\pi - 3}{3} \Big)^t.
\end{aligned}
$$

∎

Before discussing the solution of a system given in the form (9.1) or (9.2), we need to consider some results concerning continuity and differentiability of functions from R^n into R^n. Although this study could be presented directly (see Exercise 10), we prefer to use an alternative method that will allow us to present the more theoretically difficult concepts of **limits** and **continuity** in terms of functions from R^n into R.

DEFINITION 9.1   Let f be a function defined on a set $D \subset R^n$ and mapping into R. The function f is said to have the **limit** L at $\mathbf{x}_0$, written

$$\lim_{\mathbf{x} \to \mathbf{x}_0} f(\mathbf{x}) = L,$$

if, given any number $\varepsilon > 0$, a number $\delta > 0$ exists with the property that

$$|f(\mathbf{x}) - L| < \varepsilon$$

whenever $\mathbf{x} \in D$ and

$$0 < \|\mathbf{x} - \mathbf{x}_0\| < \delta.$$

Note that the existence of a limit is independent of the particular vector norm being used, because of the equivalence of vector norms in R^n (see Section 8.1).

DEFINITION 9.2   Let f be a function from a set $D \subset R^n$ into R. The function f is said to be **continuous** at $\mathbf{x}_0 \in D$ provided $\lim_{\mathbf{x} \to \mathbf{x}_0} f(\mathbf{x})$ exists and

$$\lim_{\mathbf{x} \to \mathbf{x}_0} f(\mathbf{x}) = f(\mathbf{x}_0).$$

Moreover, f is said to be **continuous** on a set D provided f is continuous at every point of D. This concept is expressed by writing $f \in C(D)$.

We can now define the limit and continuity concepts for functions from R^n into R^n by considering the coordinate functions from R^n into R.

DEFINITION 9.3   Let $\mathbf{F}$ be a function from $D \subset R^n$ into R^n and suppose $\mathbf{F}$ has the representation

$$\mathbf{F}(\mathbf{x}) = (f_1(\mathbf{x}), f_2(\mathbf{x}), ..., f_n(\mathbf{x}))^t,$$

where f_i for each i is a mapping from R^n into R. We define

$$\lim_{\mathbf{x} \to \mathbf{x}_0} \mathbf{F}(\mathbf{x}) = \mathbf{L} = (L_1, L_2, ..., L_n)^t$$

if and only if $\lim_{\mathbf{x} \to \mathbf{x}_0} f_i(\mathbf{x}) = L_i$ for each $i = 1, 2, ..., n$.

DEFINITION 9.4   Let $\mathbf{F}$ be a function from $D \subset R^n$ into R^n with the representation $\mathbf{F}(\mathbf{x}) = (f_1(\mathbf{x}), f_2(\mathbf{x}),$ $..., f_n(\mathbf{x}))^t$. The function $\mathbf{F}$ is said to be **continuous** at $\mathbf{x}_0 \in D$ provided $\lim_{\mathbf{x} \to \mathbf{x}_0} \mathbf{F}(\mathbf{x})$ exists and $\lim_{\mathbf{x} \to \mathbf{x}_0} \mathbf{F}(\mathbf{x}) = \mathbf{F}(\mathbf{x}_0)$. $\mathbf{F}$ is said to be **continuous** on the set D if $\mathbf{F}$ is continuous at each $\mathbf{x}$ in D. This concept is expressed by writing $\mathbf{F} \in C(D)$.

For functions from R into R, continuity can often be shown by showing that the function is differentiable (see Theorem 1.6, p. 3). Although this theorem generalizes to functions of several variables, the derivative (or total derivative) of a function of several variables is quite involved and will not be presented here. Instead we will state the following theorem, which relates the continuity of a function of n variables at a point to the partial derivatives of the function at the point.

THEOREM 9.5   Let f be a function from $D \subset R^n$ into R and $\mathbf{x}_0 \in D$. If constants $\delta > 0$ and $K > 0$ exist with

$$\left| \frac{\partial f(\mathbf{x})}{\partial x_j} \right| \leq K \qquad \text{for each } j = 1, 2, ..., n,$$

whenever $\|\mathbf{x} - \mathbf{x}_0\| < \delta$ and $\mathbf{x} \in D$, then f is continuous at $\mathbf{x}_0$.

In Chapter 2, an iterative process for solving an equation $f(x) = 0$ was developed by first transforming the equation into an equation of the form $x = g(x)$. The function g has fixed points precisely at solutions to the original equation. A similar procedure will be investigated here for functions from R^n into R^n.

DEFINITION 9.6   A function $\mathbf{G}$ from $D \subset R^n$ into R^n is said to have a **fixed point** at $\mathbf{p} \in D$ if $\mathbf{G}(\mathbf{p}) = \mathbf{p}$.

The following theorem extends the fixed-point Theorems 2.2 and 2.3 to the n-dimensional case. This theorem is a special case of the well-known **Contraction Mapping Theorem**, and its proof can be found in the previously mentioned book by Ortega [94], page 153.

THEOREM 9.7   Let $D = \{(x_1, x_2, \ldots, x_n)^t \mid a_i \le x_i \le b_i$ for each $i = 1, 2, \ldots, n\}$ for some collection of constants $a_1, a_2, \ldots, a_n$, and $b_1, b_2, \ldots, b_n$. Suppose **G** is a continuous function with continuous first partial derivatives from $D \subset R^n$ into R^n with the property that $\mathbf{G}(\mathbf{x}) \in D$ whenever $\mathbf{x} \in D$. Then **G** has a fixed point in D.

Moreover, suppose a constant $K < 1$ exists with

$$\left| \frac{\partial g_i(\mathbf{x})}{\partial x_j} \right| \le \frac{K}{n} \qquad \text{whenever } \mathbf{x} \in D,$$

for each $j = 1, 2, \ldots, n$ and each component function g_i. Then the sequence $\{\mathbf{x}^{(k)}\}_{k=0}^{\infty}$ defined by an arbitrarily selected $\mathbf{x}^{(0)}$ in D and generated by

$$\mathbf{x}^{(k)} = G(\mathbf{x}^{(k-1)}) \qquad \text{for each } k \ge 1$$

converges to the unique fixed point $\mathbf{p} \in D$ and

(9.3)
$$\|\mathbf{x}^{(k)} - \mathbf{p}\|_{\infty} \le \frac{K^k}{1 - K} \|\mathbf{x}^{(1)} - \mathbf{x}^{(0)}\|_{\infty}.$$

EXAMPLE 2   Consider the nonlinear system from Example 1 given by

(9.4)
$$\begin{aligned} 3x_1 - \cos(x_2 x_3) - \tfrac{1}{2} &= 0, \\ x_1^2 - 81(x_2 + 0.1)^2 + \sin x_3 + 1.06 &= 0, \\ e^{-x_1 x_2} + 20x_3 + \frac{10\pi - 3}{3} &= 0. \end{aligned}$$

If the ith equation is solved for x_i, the system can be changed into the fixed-point problem

(9.5)
$$\begin{aligned} x_1 &= \tfrac{1}{3}\cos(x_2 x_3) + \tfrac{1}{6}, \\ x_2 &= \tfrac{1}{9}\sqrt{x_1^2 + \sin x_3 + 1.06} - 0.1, \\ x_3 &= -\tfrac{1}{20}e^{-x_1 x_2} - \frac{10\pi - 3}{60}. \end{aligned}$$

Let $\mathbf{G}: R^3 \to R^3$ be defined by $\mathbf{G}(\mathbf{x}) = (g_1(\mathbf{x}), g_2(\mathbf{x}), g_3(\mathbf{x}))^t$ where

$$g_1(x_1, x_2, x_3) = \tfrac{1}{3}\cos(x_2 x_3) + \tfrac{1}{6},$$
$$g_2(x_1, x_2, x_3) = \tfrac{1}{9}\sqrt{x_1^2 + \sin x_3 + 1.06} - 0.1,$$
$$g_3(x_1, x_2, x_3) = -\tfrac{1}{20}e^{-x_1 x_2} - \frac{10\pi - 3}{60}.$$

Theorems 9.5 and 9.7 will be used to show that **G** has a unique fixed point in

$$D = \{(x_1, x_2, x_3)^t \mid -1 \le x_i \le 1 \qquad \text{for each } i = 1, 2, 3\}.$$

For $\mathbf{x} = (x_1, x_2, x_3)^t$ in D,

$$|g_1(x_1, x_2, x_3)| \leq \tfrac{1}{3}|\cos(x_2 x_3)| + \tfrac{1}{6} \leq 0.50,$$

$$|g_2(x_1, x_2, x_3)| = |\tfrac{1}{9}\sqrt{x_1^2 + \sin x_3 + 1.06} - 0.1|$$

$$\leq \tfrac{1}{9}\sqrt{1 + \sin 1 + 1.06} - 0.1 < 0.09,$$

and $$|g_3(x_1, x_2, x_3)| = \tfrac{1}{20}e^{-x_1 x_2} + \frac{10\pi - 3}{60} \leq \tfrac{1}{20}e + \frac{10\pi - 3}{60} < 0.61;$$

so $-1 \leq g_i(x_1, x_2, x_3) \leq 1$, for each $i = 1, 2, 3$. Thus, $\mathbf{G}(\mathbf{x}) \in D$ whenever $\mathbf{x} \in D$.
Finding bounds for the partial derivatives on D gives the following:

$$\left|\frac{\partial g_1}{\partial x_1}\right| = 0, \qquad \left|\frac{\partial g_2}{\partial x_2}\right| = 0, \qquad \text{and} \qquad \left|\frac{\partial g_3}{\partial x_3}\right| = 0,$$

while $$\left|\frac{\partial g_1}{\partial x_2}\right| \leq \tfrac{1}{3}|x_3||\sin x_2 x_3| \leq \tfrac{1}{3}\sin 1 < 0.281,$$

$$\left|\frac{\partial g_1}{\partial x_3}\right| = \tfrac{1}{3}|x_2||\sin x_2 x_3| \leq \tfrac{1}{3}\sin 1 < 0.281,$$

$$\left|\frac{\partial g_2}{\partial x_1}\right| = \frac{|x_1|}{9\sqrt{x_1^2 + \sin x_3 + 1.06}} < \frac{1}{9\sqrt{0.218}} < 0.238,$$

$$\left|\frac{\partial g_2}{\partial x_3}\right| = \frac{|\cos x_3|}{18\sqrt{x_1^2 + \sin x_3 + 1.06}} < \frac{1}{18\sqrt{0.218}} < 0.119,$$

$$\left|\frac{\partial g_3}{\partial x_1}\right| = \frac{|x_2|}{20}e^{-x_1 x_2} \leq \tfrac{1}{20}e < 0.14,$$

and $$\left|\frac{\partial g_3}{\partial x_2}\right| = \frac{|x_1|}{20}e^{-x_1 x_2} \leq \tfrac{1}{20}e < 0.14.$$

Since the partial derivatives of g_1, g_2, and g_3 are bounded on D, Theorem 9.5 implies that these functions are continuous on D. Consequently, G is continuous on D. Moreover, for every $\mathbf{x} \in D$

$$\left|\frac{\partial g_i(\mathbf{x})}{\partial x_j}\right| \leq 0.281 \quad \text{for each } i = 1, 2, 3 \text{ and } j = 1, 2, 3,$$

and the condition in the second part of Theorem 9.7 holds with $K = 0.843$.

In the same manner it can also be shown that $\partial g_i / \partial x_j$ is continuous on D for each $i = 1, 2, 3$, and $j = 1, 2, 3$. (This will be considered in Exercise 3.) Consequently, $\mathbf{G}$ has a unique fixed point in D and the nonlinear system (9.4) has a solution in D.

Note that G having a unique solution in D does not imply that the solution to the original system is unique on this domain. The solution for x_2 in (9.5) involved the choice of the principal square root. Exercise 7(d) examines the situation that occurs if instead the negative square root is chosen in this step.

To approximate the fixed point $\mathbf{p}$ we will choose $\mathbf{x}^{(0)} = (0.1, 0.1, -0.1)^t$. The sequence of vectors generated by

$$x_1^{(k)} = \tfrac{1}{3}\cos x_2^{(k-1)}x_3^{(k-1)} + \tfrac{1}{6},$$

$$x_2^{(k)} = \tfrac{1}{9}\sqrt{(x_1^{(k-1)})^2 + \sin x_3^{(k-1)} + 1.06} - 0.1,$$

$$x_3^{(k)} = -\tfrac{1}{20}e^{-x_1^{(k-1)}x_2^{(k-1)}} - \frac{10\pi - 3}{60},$$

will converge to the unique solution of (9.5). In this example the sequence was generated until k was found with

$$\|\mathbf{x}^{(k)} - \mathbf{x}^{(k-1)}\|_\infty < 10^{-5}.$$

The results are given in Table 9.1.

TABLE 9.1

k	$x_1^{(k)}$	$x_2^{(k)}$	$x_3^{(k)}$	$\|\mathbf{x}^{(k)} - \mathbf{x}^{(k-1)}\|_\infty$
0	0.10000000	0.10000000	−0.10000000	—
1	0.49998333	0.00944115	−0.52310127	0.423
2	0.49999593	0.00002557	−0.52336331	9.4×10^{-3}
3	0.50000000	0.00001234	−0.52359814	2.3×10^{-4}
4	0.50000000	0.00000003	−0.52359847	1.2×10^{-5}
5	0.50000000	0.00000002	−0.52359877	3.1×10^{-7}

Using the error bound (9.3) with $K = 0.843$ gives

$$\|\mathbf{x}^{(5)} - \mathbf{p}\|_\infty \le \frac{(0.843)^5}{1 - 0.843}(0.423) < 1.15,$$

which does not indicate the true accuracy of $\mathbf{x}^{(5)}$, because of the inaccurate initial approximation. The actual solution is

$$\mathbf{p} = \left(0.5, 0, -\frac{\pi}{6}\right)^t \approx (0.5, 0, -0.5235987757)^t,$$

so the true error is

$$\|\mathbf{x}^{(5)} - \mathbf{p}\|_\infty \le 2 \times 10^{-8}.$$

One way to generally accelerate convergence of the fixed-point iteration is to use the latest estimates $x_1^{(k)}, \ldots, x_{i-1}^{(k)}$, instead of $x_1^{(k-1)}, \ldots, x_{i-1}^{(k-1)}$, to compute $x_i^{(k)}$, as in the Gauss–Seidel method for linear systems (see Section 8.2). The component equations then become:

$$x_1^{(k)} = \tfrac{1}{3}\cos(x_2^{(k-1)}x_3^{(k-1)}) + \tfrac{1}{6},$$

$$x_2^{(k)} = \tfrac{1}{9}\sqrt{(x_1^{(k)})^2 + \sin x_3^{(k-1)} + 1.06} - 0.1,$$

$$x_3^{(k)} = -\tfrac{1}{20}e^{-x_1^{(k)}x_2^{(k)}} - \frac{10\pi - 3}{60}.$$

With $\mathbf{x}^{(0)} = (0.1, 0.1, -0.1)^t$, the results of these calculations are listed in Table 9.2.

The iterate $\mathbf{x}^{(4)}$ is actually accurate to within 10^{-7} in the l_∞ norm; so the convergence was indeed accelerated for this problem by using the Seidel method. It should be remarked, however, that the Seidel method does not *always* give an acceleration of the convergence.   ∎

TABLE 9.2

k	$x_1^{(k)}$	$x_2^{(k)}$	$x_3^{(k)}$	$\|\mathbf{x}^{(k)} - \mathbf{x}^{(k-1)}\|_\infty$
0	0.10000000	0.10000000	-0.10000000	—
1	0.49998333	0.02222979	-0.52304613	0.423
2	0.49997747	0.00002815	-0.52359807	2.2×10^{-2}
3	0.50000000	0.00000004	-0.52359877	2.8×10^{-5}
4	0.50000000	0.00000000	-0.52359877	3.8×10^{-8}

Exercise Set 9.1

1. Show that the function $\mathbf{F}: R^3 \to R^3$ defined by

$$\mathbf{F}(x_1, x_2, x_3) = (x_1 + 2x_3, x_1 \cos x_2, x_2^2 + x_3)^t$$

is continuous at each point of R^3.

2. Give an example of a function $\mathbf{F}: R^2 \to R^2$ that is continuous at each point of R^2 except at $(1, 0)$.

3. Show that the first partial derivatives in Example 2 are continuous on D.

4. Show that the nonlinear system

$$x_1^2 + x_2^2 - x_1 = 0,$$
$$x_1^2 - x_2^2 - x_2 = 0,$$

has a unique nontrivial solution. Approximate the solution graphically. Use the graphical solution as an initial approximation for an appropriate functional iteration. Determine the solution to within 10^{-3} in the l_2 norm.

5. The nonlinear system

$$x_1^2 - 10x_1 + x_2^2 + 8 = 0,$$
$$x_1 x_2^2 + x_1 - 10x_2 + 8 = 0,$$

can be transformed into the fixed-point problem

$$x_1 = g_1(x_1, x_2) = \frac{x_1^2 + x_2^2 + 8}{10},$$

$$x_2 = g_2(x_1, x_2) = \frac{x_1 x_2^2 + x_1 + 8}{10}.$$

a)   Use Theorem 9.7 to show that $\mathbf{G}=(g_1,g_2)^t\colon D\subset R^2\to R^2$ has a unique fixed point in

$$D=\{(x_1,x_2)^t|0\le x_1,\,x_2\le 1.5\}.$$

b)   Apply functional iteration to approximate the solution.

c)   Does the Seidel method accelerate convergence?

6.   The nonlinear system

$$3x_1^2-x_2^2=0,$$
$$3x_1x_2^2-x_1^3-1=0,$$

has a solution near $(\tfrac{1}{2},\tfrac{3}{4})^t$.

a)   Find a function $\mathbf{G}$ and a set $D\subset R^2$ such that $\mathbf{G}\colon D\to R^2$ and $\mathbf{G}$ has a unique fixed point in D.

b)   Apply functional iteration to approximate the solutions to within 10^{-3}, using $\|\cdot\|_\infty$.

c)   Does the Seidel method accelerate convergence?

7.   Use Theorem 9.7 to show that the following function $\mathbf{G}\colon D\subset R^3\to R^3$ has a unique fixed point in D. Apply functional iteration to approximate the solution to within 10^{-4}, using $\|\cdot\|_\infty$.

a) $\mathbf{G}(x_1,x_2,x_3)=\left(\dfrac{\cos{(x_2x_3)}+0.5}{3},\dfrac{1}{25}\sqrt{x_1^2+0.3125}-0.03,\,-\dfrac{1}{20}e^{-x_1x_2}-\dfrac{10\pi-3}{60}\right)^t;$

$D=\{(x_1,x_2,x_3)^t|-1\le x_i\le 1,i=1,2,3\}.$

b) $\mathbf{G}(x_1,x_2,x_3)=\left(\dfrac{7.17+3x_2^2+4x_3}{12},\dfrac{11.54+x_3-x_1^2}{10},\dfrac{7.631-x_2^3}{7}\right)^t;$

$D=\{(x_1,x_2,x_3)^t|0\le x_i\le 1.5,i=1,2,3\}.$

c) $\mathbf{G}(x_1,x_2,x_3)=(1-\cos{(x_1x_2x_3)},\,1-(1-x_1)^{1/4}-0.05x_3^2+0.15x_3,x_1^2+0.1x_2^2-0.01x_2+1)^t;$

$D=\{(x_1,x_2,x_3)^t|-0.1\le x_1\le 0.1,\,-0.1\le x_2\le 0.3,0.5\le x_3\le 1.1\}.$

d) $\mathbf{G}(x_1,x_2,x_3)=\left(\tfrac{1}{3}\cos{(x_2x_3)}+\tfrac{1}{6},\,-\tfrac{1}{9}\sqrt{x_1^2+\sin x_3+1.06}-0.1,\,-\dfrac{1}{20}e^{-x_1x_2}-\dfrac{10\pi-3}{60}\right)^t;$

$D=\{(x_1,x_2,x_3)^t|-1\le x_i\le 1,i=1,\,2,\,3\}.$

8.   Use Seidel's method to approximate the fixed points in Exercise 7 to within 10^{-4}, using $\|\cdot\|_\infty$.

9.   Use functional iteration to find solutions to the following nonlinear systems, accurate to within 10^{-4}, using $\|\cdot\|_\infty$.

a) $x_1^2+2x_2^2-x_2-2x_3=0,$
     $x_1^2-8x_2^2+10x_3=0,$
     $\dfrac{x_1^2}{7x_2x_3}-1=0.$

b) $2x_1+x_2+x_3-4=0,$
     $x_1+2x_2+x_3-4=0,$
     $x_1x_2x_3-1=0.$

c) $3x_1 - \cos(x_2 x_3) - \frac{1}{2} = 0,$ d) $x_1^2 + x_2 - 37 = 0,$

   $x_1^2 - 625x_2^2 = 0,$ $x_1 - x_2^2 - 5 = 0,$

   $e^{-x_1 x_2} + 20x_3 + \dfrac{10\pi - 3}{3} = 0.$ $x_1 + x_2 + x_3 - 3 = 0.$

10. Show that Definition 9.4 is equivalent to the following statement: A function **F**: $D \subset R^n \to R^n$ is said to be continuous at $x_0 \in D$ if, given any number $\varepsilon > 0$, a number $\delta > 0$ can be found with the property that

$$\| \mathbf{F}(\mathbf{x}) - \mathbf{F}(\mathbf{x}_0) \| < \varepsilon$$

whenever $\mathbf{x} \in D$ and $\| \mathbf{x} - \mathbf{x}_0 \| < \delta$.

11. In Exercise 6 of Section 5.9 we considered the problem of predicting the population of two species that compete for the same food supply. In that problem we made the assumption that the populations could be predicted by solving the system of equations:

$$\frac{dx_1(t)}{dt} = x_1(t)(4 - 0.0003x_1(t) - 0.0004x_2(t))$$

and

$$\frac{dx_2(t)}{dt} = x_2(t)(2 - 0.0002x_1(t) - 0.0001x_2(t)).$$

In this exercise we would like to consider the problem of determining equilibrium populations of the two species. The mathematical criteria that must be satisfied in order for the populations to be at equilibrium is that, simultaneously,

$$\frac{dx_1(t)}{dt} = 0 \quad \text{and} \quad \frac{dx_2(t)}{dt} = 0.$$

This clearly occurs when the first species is extinct and the second species has a population of 20,000, or when the second species is extinct and the first species has a population of 13,333. Can this equilibrium occur in any other situation?

9.2 Newton's Method

Although the problem presented in Example 2 of Section 9.1 can be transformed into a convergent fixed-point format quite easily, this is not often the situation. In this section we consider an algorithmic procedure that can be used to perform the transformation for a general problem.

To construct the algorithm that led to an appropriate fixed-point method in the one-dimensional case, we attempted to find a function ϕ with the property that

(9.6) $g(x) = x - \phi(x) f(x),$

gives quadratic convergence to the fixed point p of g. From this condition, Newton's method evolved, by choosing $\phi(x) = 1/f'(x)$.

Using a similar approach in the n-dimensional case involves a matrix

(9.7)
$$A(\mathbf{x}) = \begin{bmatrix} a_{11}(\mathbf{x}) & a_{12}(\mathbf{x}) & \cdots & a_{1n}(\mathbf{x}) \\ a_{21}(\mathbf{x}) & a_{22}(\mathbf{x}) & \cdots & a_{2n}(\mathbf{x}) \\ \vdots & \vdots & & \vdots \\ a_{n1}(\mathbf{x}) & a_{n2}(\mathbf{x}) & \cdots & a_{nn}(\mathbf{x}) \end{bmatrix}$$

where each of the entries $a_{ij}(\mathbf{x})$ is a function from R^n into R. The procedure requires that $A(\mathbf{x})$ be found so that

(9.8)
$$\mathbf{G}(\mathbf{x}) = \mathbf{x} - A(\mathbf{x})^{-1}\mathbf{F}(\mathbf{x})$$

gives quadratic convergence to the solution of $\mathbf{F}(\mathbf{x}) = \mathbf{0}$, provided, of course, that $A(\mathbf{x})$ is nonsingular at the fixed point of $\mathbf{G}$.

The following theorem verifies that this approach can be used to motivate the choice of A.

THEOREM 9.8 Suppose $\mathbf{p}$ is a solution of $\mathbf{G}(\mathbf{x}) = \mathbf{x}$ for some function $\mathbf{G} = (g_1, g_2, \ldots, g_n)^t$, mapping R^n into R^n. If a number $\delta > 0$ exists with the property that

i) $\partial g_i/\partial x_j$ is continuous on $N_\delta = \{\mathbf{x} \mid \|\mathbf{x} - \mathbf{p}\| < \delta\}$ for each $i = 1, 2, \ldots, n$ and $j = 1, 2, \ldots, n$,

ii) $\partial^2 g_i(\mathbf{x})/(\partial x_j \partial x_k)$ is continuous, and $|\partial^2 g_i(\mathbf{x})/(\partial x_j \partial x_k)| \le M$ for some constant M, whenever $\mathbf{x} \in N_\delta$ for each $i = 1, 2, \ldots, n, j = 1, 2, \ldots, n$, and $k = 1, 2, \ldots, n$,

iii) $\partial g_i(\mathbf{p})/\partial x_j = 0$ for each $i = 1, 2, \ldots, n$, and $j = 1, 2, \ldots, n$,

then a number $\hat{\delta} \le \delta$ exists such that the sequence generated by $\mathbf{x}^{(k)} = \mathbf{G}(\mathbf{x}^{(k-1)})$ converges quadratically to $\mathbf{p}$ for any choice of $\mathbf{x}^{(0)}$, provided that $\|\mathbf{x}^{(0)} - \mathbf{p}\| < \hat{\delta}$. Moreover

$$\|\mathbf{x}^{(k)} - \mathbf{p}\|_\infty \le \frac{n^2 M}{2} \|\mathbf{x}^{(k-1)} - \mathbf{p}\|_\infty^2 \qquad \text{for each } k \ge 1.$$

This theorem parallels Theorem 2.8 in Section 2.4, and its proof requires being able to express $\mathbf{G}$ in terms of its Taylor series in n variables about the point $\mathbf{p}$.

To use Theorem 9.8, suppose that $A(\mathbf{x})$ is an $n \times n$ matrix of functions from R^n into R in the form of Eq. (9.7), where the specific entries will be chosen later. Assume, moreover, that $A(\mathbf{x})$ is nonsingular near a solution $\mathbf{p}$ of $\mathbf{F}(\mathbf{x}) = \mathbf{0}$, and let $b_{ij}(\mathbf{x})$ denote the entry of $A(\mathbf{x})^{-1}$ in the ith row and jth column.

Since $\mathbf{G}(\mathbf{x}) = \mathbf{x} - A(\mathbf{x})^{-1}\mathbf{F}(\mathbf{x})$,

$$g_i(\mathbf{x}) = x_i - \sum_{j=1}^{n} b_{ij}(\mathbf{x}) f_j(\mathbf{x});$$

so
$$\frac{\partial g_i(\mathbf{x})}{\partial x_k} = \begin{cases} 1 - \sum_{j=1}^{n} \left(b_{ij}(\mathbf{x}) \dfrac{\partial f_j}{\partial x_k}(\mathbf{x}) + \dfrac{\partial b_{ij}}{\partial x_k}(\mathbf{x}) f_j(\mathbf{x}) \right), & \text{if } i = k, \\[2ex] - \sum_{j=1}^{n} \left(b_{ij}(\mathbf{x}) \dfrac{\partial f_j}{\partial x_k}(\mathbf{x}) + \dfrac{\partial b_{ij}}{\partial x_k}(\mathbf{x}) f_j(\mathbf{x}) \right), & \text{if } i \ne k. \end{cases}$$

Theorem 9.8 implies that we need to have $\partial g_i(\mathbf{p})/\partial x_k = 0$ for each $i = 1, 2, ..., n$ and $k = 1, 2, ..., n$. This means that for $i = k$,

$$0 = 1 - \sum_{j=1}^{n} b_{ij}(\mathbf{p}) \frac{\partial f_j}{\partial x_i}(\mathbf{p}),$$

so

(9.9)
$$\sum_{j=1}^{n} b_{ij}(\mathbf{p}) \frac{\partial f_j}{\partial x_i}(\mathbf{p}) = 1,$$

and, when $k \neq i$,

$$0 = - \sum_{j=1}^{n} b_{ij}(\mathbf{p}) \frac{\partial f_j}{\partial x_k}(\mathbf{p}),$$

so

(9.10)
$$\sum_{j=1}^{n} b_{ij}(\mathbf{p}) \frac{\partial f_j}{\partial x_k}(\mathbf{p}) = 0.$$

Defining the matrix $J(\mathbf{x})$ by

(9.11)
$$J(\mathbf{x}) = \begin{bmatrix} \dfrac{\partial f_1(\mathbf{x})}{\partial x_1} & \dfrac{\partial f_1(\mathbf{x})}{\partial x_2} & \cdots & \dfrac{\partial f_1(\mathbf{x})}{\partial x_n} \\[2mm] \dfrac{\partial f_2(\mathbf{x})}{\partial x_1} & \dfrac{\partial f_2(\mathbf{x})}{\partial x_2} & \cdots & \dfrac{\partial f_2(\mathbf{x})}{\partial x_n} \\[2mm] \vdots & \vdots & & \vdots \\[2mm] \dfrac{\partial f_n(\mathbf{x})}{\partial x_1} & \dfrac{\partial f_n(\mathbf{x})}{\partial x_2} & \cdots & \dfrac{\partial f_n(\mathbf{x})}{\partial x_n} \end{bmatrix},$$

we see that conditions (9.9) and (9.10) require

$$A(\mathbf{p})^{-1} J(\mathbf{p}) = I, \qquad \text{the identity matrix,}$$

so
$$A(\mathbf{p}) = J(\mathbf{p}).$$

An appropriate choice for $A(\mathbf{x})$ is consequently $A(\mathbf{x}) = J(\mathbf{x})$, since condition (iii) in Theorem 9.8 is satisfied with this choice.

The function $\mathbf{G}$ is defined by

$$\mathbf{G}(\mathbf{x}) = \mathbf{x} - J(\mathbf{x})^{-1} \mathbf{F}(\mathbf{x}),$$

and the functional iteration procedure evolves from selecting $\mathbf{x}^{(0)}$ and generating, for $k \geq 1$,

(9.12)
$$\mathbf{x}^{(k)} = \mathbf{G}(\mathbf{x}^{(k-1)}) = \mathbf{x}^{(k-1)} - J(\mathbf{x}^{(k-1)})^{-1} \mathbf{F}(\mathbf{x}^{(k-1)}).$$

This method is called, quite reasonably, **Newton's method for nonlinear systems** and is generally expected to give quadratic convergence, provided that a sufficiently accurate starting value is known and $J(\mathbf{p})^{-1}$ exists.

The matrix $J(\mathbf{x})$ is called the **Jacobian** matrix and has a number of applications in analysis. It might, in particular, be familiar to the reader due to its application in the multiple integration of a function of several variables over a region which requires a change of variables to be performed. (See, for example, Faires and Faires [43], p. 956.)

A definite weakness in the Newton's-method procedure arises from the necessity of inverting the matrix $J(\mathbf{x})$ at each step. In practice, the method is performed in a two-step manner. First, a vector $\mathbf{y}$ is found which will satisfy $J(\mathbf{x}^{(k)})\mathbf{y} = -\mathbf{F}(\mathbf{x}^{(k)})$. After this has been accomplished, the new approximation, $\mathbf{x}^{(k+1)}$, can be obtained by adding $\mathbf{y}$ to $\mathbf{x}^{(k)}$. The following algorithm uses this two-step procedure.

Newton's Method for Systems Algorithm 9.1

To approximate the solution of the nonlinear system $\mathbf{F}(\mathbf{x}) = \mathbf{0}$ given an initial approximation $\mathbf{x}$:

INPUT number n of equations and unknowns; initial approximation $\mathbf{x} = (x_1, \ldots, x_n)^t$, tolerance TOL; maximum number of iterations N.

OUTPUT approximate solution $\mathbf{x} = (x_1, \ldots, x_n)^t$ or a message that the number of iterations was exceeded.

Step 1   Set $k = 1$.

Step 2   While ($k \leq N$) do Steps 3–7.

    Step 3   Calculate $\mathbf{F}(\mathbf{x})$ and $J(\mathbf{x})$, where $J(\mathbf{x})_{i,j} = (\partial f_i(\mathbf{x})/\partial x_j)$ for $1 \leq i, j \leq n$.

    Step 4   Solve the $n \times n$ linear system $J(\mathbf{x})\mathbf{y} = -\mathbf{F}(\mathbf{x})$.

    Step 5   Set $\mathbf{x} = \mathbf{x} + \mathbf{y}$.

    Step 6   If $\|\mathbf{y}\| < TOL$ then OUTPUT $(\mathbf{x})$;
                    (*Procedure completed successfully.*)
                    STOP.

    Step 7   Set $k = k + 1$.

Step 8   OUTPUT ('Maximum number of iterations exceeded');
         STOP.

EXAMPLE 1   The nonlinear system

$$3x_1 - \cos(x_2 x_3) - \tfrac{1}{2} = 0,$$
$$x_1^2 - 81(x_2 + 0.1)^2 + \sin x_3 + 1.06 = 0,$$
$$e^{-x_1 x_2} + 20x_3 + \frac{10\pi - 3}{3} = 0$$

was shown, in Example 2 of Section 9.1, to have an approximate solution at $(0.5, 0, -0.52359877)^t$. Newton's method will be used to obtain this approximation when the initial approximation is $\mathbf{x}^{(0)} = (0.1, 0.1, -0.1)^t$.

The Jacobian matrix $J(\mathbf{x})$ for this system is given by

$$J(x_1, x_2, x_3) = \begin{bmatrix} 3 & x_3 \sin x_2 x_3 & x_2 \sin x_2 x_3 \\ 2x_1 & -162(x_2 + 0.1) & \cos x_3 \\ -x_2 e^{-x_1 x_2} & -x_1 e^{-x_1 x_2} & 20 \end{bmatrix}$$

and

$$\begin{bmatrix} x_1^{(k)} \\ x_2^{(k)} \\ x_3^{(k)} \end{bmatrix} = \begin{bmatrix} x_1^{(k-1)} \\ x_2^{(k-1)} \\ x_3^{(k-1)} \end{bmatrix} + \begin{bmatrix} y_1^{(k-1)} \\ y_2^{(k-1)} \\ y_3^{(k-1)} \end{bmatrix},$$

where
$$\begin{bmatrix} y_1^{(k-1)} \\ y_2^{(k-1)} \\ y_3^{(k-1)} \end{bmatrix} = -(J(x_1^{(k-1)}, x_2^{(k-1)}, x_3^{(k-1)}))^{-1} \mathbf{F}(x_1^{(k-1)}, x_2^{(k-1)}, x_3^{(k-1)}).$$

Thus, at the kth step, the linear system

$$\begin{bmatrix} 3 & x_3^{(k-1)} \sin x_2^{(k-1)} x_3^{(k-1)} & x_2^{(k-1)} \sin x_2^{(k-1)} x_3^{(k-1)} \\ 2x_1^{(k-1)} & -162(x_2^{(k-1)} + 0.1) & \cos x_3^{(k-1)} \\ -x_2^{(k-1)} e^{-x_1^{(k-1)} x_2^{(k-1)}} & -x_1^{(k-1)} e^{-x_1^{(k-1)} x_2^{(k-1)}} & 20 \end{bmatrix}$$

$$\times \begin{bmatrix} x_1^{(k)} - x_1^{(k-1)} \\ x_2^{(k)} - x_2^{(k-1)} \\ x_3^{(k)} - x_3^{(k-1)} \end{bmatrix} = - \begin{bmatrix} 3x_1^{(k-1)} - \cos x_2^{(k-1)} x_3^{(k-1)} - \frac{1}{2} \\ (x_1^{(k-1)})^2 - 81(x_2^{(k-1)} + 0.1)^2 + \sin x_3^{(k-1)} + 1.06 \\ e^{-x_1^{(k-1)} x_2^{(k-1)}} + 20x_3^{(k-1)} + \frac{10\pi - 3}{3} \end{bmatrix}$$

must be solved. The results obtained using this iterative procedure are shown in Table 9.3.

TABLE 9.3

k	$x_1^{(k)}$	$x_2^{(k)}$	$x_3^{(k)}$	$\|\mathbf{x}^{(k)} - \mathbf{x}^{(k-1)}\|_\infty$
0	0.10000000	0.10000000	-0.10000000	—
1	0.50003702	0.01946686	-0.52152047	0.422
2	0.50004593	0.00158859	-0.52355711	1.79×10^{-2}
3	0.50000034	0.00001244	-0.52359845	1.58×10^{-3}
4	0.50000000	0.00000000	-0.52359877	1.24×10^{-5}
5	0.50000000	0.00000000	-0.52359877	0

The convergence of Newton's method becomes very fast once an iterate is close to $\mathbf{p}$. This illustrates the quadratic convergence of the method near a solution. ∎

Exercise Set 9.2

1. Find a solution to the following nonlinear systems using Newton's method, and compute the error bound given in Theorem 9.8. Iterate until $\|\mathbf{x}^{(i)} - \mathbf{x}^{(i-1)}\|_\infty < 10^{-5}$.

   a) $x_1^2 - 10x_1 + x_2^2 + 8 = 0,$

   $x_1 x_2^2 + x_1 - 10x_2 + 8 = 0.$

   Compare the convergence to that of Exercise 5, Section 9.1.

   b) $x_1^2 + x_2^2 - x_1 = 0,$

   $x_1^2 - x_2^2 - x_2 = 0.$

   Compare the convergence to that of Exercise 4, Section 9.1.

   c) $3x_1^2 - x_2^2 = 0,$

   $3x_1 x_2^2 - x_1^3 - 1 = 0.$

   d) $\frac{1}{2}\sin(x_1 x_2) - \dfrac{x_2}{4\pi} - \dfrac{x_1}{2} = 0,$

   $\left(1 - \dfrac{1}{4\pi}\right)(e^{2x_1} - e) + \dfrac{e}{\pi}x_2 - 2ex_1 = 0.$

   [*Note*:   There is an infinite number of solutions.]

2. Find a solution to the following nonlinear systems using Newton's method and compute the error bound given in Theorem 9.8. Iterate until $\|\mathbf{x}^{(i)} - \mathbf{x}^{(i-1)}\|_\infty < 10^{-5}$.

   a) $x_1^2 + x_2 - 37 = 0,$

   $x_1 - x_2^2 - 5 = 0,$

   $x_1 + x_2 + x_3 - 3 = 0.$

   b) $x_1 + \cos(x_1 x_2 x_3) - 1 = 0,$

   $(1 - x_1)^{1/4} + x_2 + 0.05x_3^2 - 0.15x_3 - 1 = 0,$

   $-x_1^2 - 0.1x_2^2 + 0.01x_2 + x_3 - 1 = 0.$

   Compare the convergence to that of Exercise 7(c), Section 9.1.

   c) $12x_1 - 3x_2^2 - 4x_3 = 7.17,$

   $x_1^2 + 10x_2 - x_3 = 11.54,$

   $x_2^3 + 7x_3 = 7.631.$

   Compare the convergence to that of Exercise 7(b), Section 9.1.

   d) $4x_1 - x_2 + x_3 - 2 = 0,$

   $-x_1 + 6x_2 - 3x_3 - 11 = 0,$

   $x_1 - 3x_2 + 5x_3 + 5 = 0.$

3. Find a solution to the following nonlinear systems using Newton's method. Iterate until $\|\mathbf{x}^{(i)} - \mathbf{x}^{(i-1)}\|_\infty < 10^{-5}$.

   a) $6x_1 - 2\cos(x_2 x_3) - 1 = 0,$

   $9x_2 + \sqrt{x_1^2 + \sin x_3 + 1.06} + 0.9 = 0,$

   $60x_3 + 3e^{-x_1 x_2} + 10\pi - 3 = 0.$

b) $x_1^2 + 2x_2^2 - x_2 - 2x_3 = 0,$

   $x_1^2 - 8x_2^2 + 10x_3 = 0,$

   $\dfrac{x_1^2}{7x_2x_3} - 1 = 0.$

c) $2x_1 + x_2 + x_3 - 4 = 0,$

   $x_1 + 2x_2 + x_3 - 4 = 0,$

   $x_1x_2x_3 - 1 = 0.$

   (Find a solution other than $(1, 1, 1)^t$.)

d) $2x_1 + x_2 + x_3 + x_4 - 5 = 0,$

   $x_1 + 2x_2 + x_3 + x_4 - 5 = 0,$

   $x_1 + x_2 + 2x_3 + x_4 - 5 = 0,$

   $x_1x_2x_3x_4 - 1 = 0.$

   (Find a solution other than $(1, 1, 1, 1)^t$.)

4. The following nonlinear systems have singular Jacobian matrices at the solution. Can Newton's method still be used? How is the rate of convergence affected?

a) $3x_1 - \cos(x_2x_3) - 0.5 = 0,$ b) $x_1 - 10x_2 + 9 = 0,$

   $x_1^2 - 625x_2^2 = 0,$ $\sqrt{3}(x_3 - x_4) = 0,$

   $e^{-x_1x_2} + 20x_3 + \dfrac{10\pi - 3}{3} = 0.$ $(x_2 - 2x_3 + 1)^2 = 0,$

                                        $\sqrt{2}(x_1 - x_4)^2 = 0.$

5. Find the Jacobian matrix of the linear system:

$$c_{11}x_1 + c_{12}x_2 + \cdots + c_{1n}x_n - b_1 = 0,$$
$$c_{21}x_1 + c_{22}x_2 + \cdots + c_{2n}x_n - b_2 = 0,$$
$$\vdots \qquad\qquad\qquad\qquad \vdots$$
$$c_{n1}x_1 + c_{n2}x_2 + \cdots + c_{nn}x_n - b_n = 0.$$

When will this matrix have an inverse?

6. C. Chiarella, W. Charlton, and A. W. Roberts [27], in calculating the shape of a gravity-flow discharge chute that will minimize transit time of discharged granular particles, solve the following equations by Newton's method:

i) $f_n(\theta_1, \ldots, \theta_N) = \dfrac{\sin\theta_{n+1}}{v_{n+1}}(1 - \mu w_{n+1}) - \dfrac{\sin\theta_n}{v_n}(1 - \mu w_n) = 0,$

   for each $n = 1, 2, \ldots, N - 1.$

ii) $f_N(\theta_1, \ldots, \theta_N) = \Delta y \displaystyle\sum_{i=1}^{N} \tan\theta_i - X = 0,$

where

a) $v_n^2 = v_0^2 + 2gn\Delta y - 2\mu\Delta y \displaystyle\sum_{j=1}^{n} \dfrac{1}{\cos\theta_j}$ for each $n = 1, 2, \ldots, N,$ and

b) $w_n = -\Delta y v_n \sum\limits_{i=1}^{N} \dfrac{1}{v_i^3 \cos\theta_i}$ for each $n = 1, 2, \ldots, N$.

The constant v_0 is the initial velocity of the granular material, X is the x-coordinate of the end of the chute, μ is the friction force, N is the number of chute segments, and g is the gravitational constant. The variable θ_i is the angle of the ith chute segment from the vertical as shown in the following figure, and v_i is the particle velocity in the ith chute segment. Solve (i) and (ii) for $\boldsymbol{\theta} = (\theta_1, \ldots, \theta_N)^t$ using Newton's method with $\mu = 0$, $X = 2$, $\Delta y = 0.2$, $N = 20$, $v_0 = 0$, and $g = 32.2$ ft/sec^2, where the values for v_n and w_n can be obtained directly from (a) and (b). Iterate until $\|\boldsymbol{\theta}^{(i)} - \boldsymbol{\theta}^{(i-1)}\|_\infty < 10^{-2}$ radians.

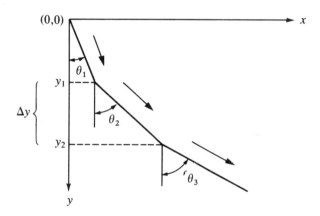

7. The amount of pressure required to sink a large heavy object in a soft homogeneous soil that lies above a hard base soil can be predicted by the amount of pressure required to sink smaller objects in the same soil. In particular, the amount of pressure p required to sink a circular plate of radius r a distance d in the soft soil, where the hard base soil lies a distance $D > d$ below the surface, can be approximated by an equation of the form

$$p = k_1 e^{k_2 r} + k_3 r,$$

where k_1, k_2, and k_3 are constants, with $k_2 > 0$, depending on d and the consistency of the soil but not on the radius of the plate. (See Bekker [7], pp. 89–94.)

a) Find the values of k_1, k_2, and k_3 if we assume that a plate of radius 1 inch requires a pressure of 10 lb/(in)^2 to sink 1 foot in a muddy field, a plate of radius 2 inches requires a pressure of 12 lb/(in)^2 to sink one foot, and a plate of radius 3 inches requires a pressure of 15 lb/(in)^2 to sink this distance (assuming that the mud is more than one foot deep).

b) Use your calculations from part (a) to predict the minimal size of circular plate that would be required to sustain a load of 500 lb on this field with sinkage of less than 1 foot.

8. Exercise 12 of Section 7.1 dealt with determining an exponential least-squares relationship of the form $R = bw^a$ to approximate a collection of data relating the weight and respiration rule of *Modest sphinx* moths. In that exercise, the problem was converted to a log-log relationship, and in part (c), a quadratic term was introduced in an attempt to improve the approximation. Instead of converting the problem, use Newton's method to determine the

constants a and b that minimize $\sum_{i=1}^{n} (R_i - bw_i^a)^2$ for the data listed in Exercise 12 of Section 7.1. Compute the error associated with this approximation, and compare this to the error of the previous approximations for this problem.

9. An interesting biological experiment, see Schroeder [118], concerns the determination of the maximum water temperature, X_M, at which various species of hydra can survive without shortened life expectancy. One approach to the solution of this problem uses a weighted least-squares fit of the form $f(x) = y = a/(x - b)^c$ to a collection of experimental data. The x values of the data refer to water temperatures above X_M, and the y values refer to the average life expectancy at that temperature. The constant b is the asymptote of the graph of f and as such is an approximation to X_M.

a) Show that choosing a, b, and c to minimize

$$\sum_{i=1}^{n} \left[w_i y_i - \frac{a}{(x_i - b)^c} \right]^2$$

reduces to solving the nonlinear system

$$a = \left(\sum_{i=1}^{n} \frac{y_i w_i}{(x_i - b)^c} \right) \Bigg/ \left(\sum_{i=1}^{n} \frac{w_i}{(x_i - b)^{2c}} \right)$$

$$0 = \sum_{i=1}^{n} \frac{y_i w_i}{(x_i - b)^c} \cdot \sum_{i=1}^{n} \frac{w_i}{(x_i - b)^{2c+1}} - \sum_{i=1}^{n} \frac{y_i w_i}{(x_i - b)^{c+1}} \cdot \sum_{i=1}^{n} \frac{w_i}{(x_i - b)^{2c}}$$

$$0 = \sum_{i=1}^{n} \frac{y_i w_i}{(x_i - b)^c} \cdot \sum_{i=1}^{n} \frac{w_i \ln (x_i - b)}{(x_i - b)^{2c}} - \sum_{i=1}^{n} \frac{w_i y_i \ln (x_i - b)}{(x_i - b)^c} \cdot \sum_{i=1}^{n} \frac{w_i}{(x_i - b)^{2c}}.$$

b) Solve the nonlinear system for the species with the following data. Use the weights $w_i = \ln y_i$.

i	1	2	3	4
y_i	2.40	3.80	4.75	21.60
x_i	31.8	31.5	31.2	30.2

9.3 Quasi-Newton Methods

A significant weakness of Newton's method for solving systems of nonlinear equations lies in the requirement that, at each iteration, a Jacobian matrix be computed and an $n \times n$ linear system solved that involves this matrix. To illustrate the magnitude of this weakness, let us consider the amount of calculation associated with one iteration of Newton's method. The Jacobian matrix associated with a system of n nonlinear equations written in the form $\mathbf{F}(\mathbf{x}) = \mathbf{0}$ requires that the n^2 partial derivatives of the n component functions of $\mathbf{F}$ be determined and evaluated. In most situations, the exact evaluation of the partial derivatives is inconvenient and in many applications impossible. This difficulty can be overcome by using finite difference approximations to the partial derivatives. For example,

(9.13)
$$\frac{\partial f_j}{\partial x_k}(\mathbf{x}^{(i)}) \approx \frac{f_j(\mathbf{x}^{(i)} + \mathbf{e}_k h) - f_j(\mathbf{x}^{(i)})}{h},$$

where h is small in absolute value and $\mathbf{e}_k$ is the vector whose only nonzero entry is a one in the kth coordinate. This approximation, however, still requires that at least n^2 scalar functional evaluations be performed to approximate the Jacobian and does not decrease the amount of calculation, in general $O(n^3)$, required for solving the linear system involving this approximate Jacobian. The total computational effort for just one iteration of Newton's method is consequently at least $n^2 + n$ scalar functional evaluations (n^2 for the evaluation of the Jacobian matrix and n for the evaluation of $\mathbf{F}$) together with $O(n^3)$ arithmetic operations to solve the linear system. This amount of computational effort is prohibitive except for relatively small values of n and easily evaluated scalar functions.

In this section we will consider a generalization of the Secant method to systems of nonlinear equations, in particular a technique known as **Broyden's method** (see Broyden [17]). The method requires only n scalar functional evaluations per iteration and also reduces the number of arithmetic calculations to $O(n^2)$. It is one of a class of methods known as *least-change secant updates* that produce algorithms called **quasi-Newton**. These methods replace the Jacobian matrix in Newton's method with an approximation matrix that is updated at each iteration. The disadvantage to the methods is that the quadratic convergence of Newton's method is lost, being replaced in general by a convergence called **superlinear**. Superlinear convergence was discussed in Section 2.5, Exercise 6. It implies here that

$$\lim_{i \to \infty} \frac{\|\mathbf{x}^{(i+1)} - \mathbf{p}\|}{\|\mathbf{x}^{(i)} - \mathbf{p}\|} = 0,$$

where $\mathbf{p}$ denotes the solution to $\mathbf{F}(\mathbf{x}) = \mathbf{0}$ and $\mathbf{x}^{(i)}$, $\mathbf{x}^{(i+1)}$ are consecutive approximations. In most applications, the reduction to superlinear convergence is a more than acceptable trade-off for the decrease in the amount of computation. An additional disadvantage of quasi-Newton methods is that, unlike Newton's method, they are not self correcting. Newton's method, for example, will generally correct for round-off error with successive iterations, Broyden's method will not.

Suppose that an initial approximation $\mathbf{x}^{(0)}$ is given to the solution $\mathbf{p}$ of $\mathbf{F}(\mathbf{x}) = \mathbf{0}$. We calculate the next approximation $\mathbf{x}^{(1)}$ in the same manner as Newton's method or, if it is inconvenient to determine $J(\mathbf{x}^{(0)})$ exactly, we use the difference equations given by (9.13) to approximate the partial derivatives. To compute $\mathbf{x}^{(2)}$, however, we depart from Newton's method and examine the Secant method for a single nonlinear equation. The Secant method uses the approximation

$$\frac{f(x_1) - f(x_0)}{x_1 - x_0}$$

as a replacement for $f'(x_1)$ in Newton's method. For nonlinear systems, $\mathbf{x}^{(1)} - \mathbf{x}^{(0)}$ is a vector, and the corresponding quotient is undefined. However, the method proceeds similarly in that we replace the matrix $J(\mathbf{x}^{(1)})$ in Newton's method by a matrix A_1 with the property that

(9.14) $$A_1(\mathbf{x}^{(1)} - \mathbf{x}^{(0)}) = \mathbf{F}(\mathbf{x}^{(1)}) - \mathbf{F}(\mathbf{x}^{(0)}).$$

Equation (9.14) does not define a unique matrix, because it does not describe how A_1 operates on vectors orthogonal to $\mathbf{x}^{(1)} - \mathbf{x}^{(0)}$. Since no information is available about the change in $\mathbf{F}$ in a direction orthogonal to $\mathbf{x}^{(1)} - \mathbf{x}^{(0)}$, we will require additionally of A_1 that

(9.15) $\qquad A_1\mathbf{z} = J(\mathbf{x}^{(0)})\mathbf{z} \qquad$ whenever $(\mathbf{x}^{(1)} - \mathbf{x}^{(0)})^t\mathbf{z} = 0.$

This condition specifies that any vector orthogonal to $\mathbf{x}^{(1)} - \mathbf{x}^{(0)}$ will be unaffected by the update from $J(\mathbf{x}^{(0)})$, which was used to compute $\mathbf{x}^{(1)}$, to A_1, which will be used in the determination of $\mathbf{x}^{(2)}$.

Conditions (9.14) and (9.15) uniquely define A_1 (see Dennis and Moré [35], page 54) as

$$A_1 = J(\mathbf{x}^{(0)}) + \frac{[\mathbf{F}(\mathbf{x}^{(1)}) - \mathbf{F}(\mathbf{x}^{(0)}) - J(\mathbf{x}^{(0)})(\mathbf{x}^{(1)} - \mathbf{x}^{(0)})](\mathbf{x}^{(1)} - \mathbf{x}^{(0)})^t}{\|\mathbf{x}^{(1)} - \mathbf{x}^{(0)}\|_2^2}.$$

It is this matrix that is used in place of $J(\mathbf{x}^{(1)})$ to determine $\mathbf{x}^{(2)}$:

$$\mathbf{x}^{(2)} = \mathbf{x}^{(1)} - A_1^{-1}\mathbf{F}(\mathbf{x}^{(1)}).$$

The method can then be repeated to determine $\mathbf{x}^{(3)}$ by using A_1 in place of $A_0 \equiv J(\mathbf{x}^{(0)})$ and with $\mathbf{x}^{(2)}$ and $\mathbf{x}^{(1)}$ in place of $\mathbf{x}^{(1)}$ and $\mathbf{x}^{(0)}$. In general, once $\mathbf{x}^{(i)}$ has been determined, $\mathbf{x}^{(i+1)}$ is computed by

(9.16) $\qquad A_i = A_{i-1} + \dfrac{(\mathbf{y}_i - A_{i-1}\mathbf{s}_i)}{\|\mathbf{s}_i\|_2^2}\mathbf{s}_i^t,$

(9.17) $\qquad \mathbf{x}^{(i+1)} = \mathbf{x}^{(i)} - A_i^{-1}\mathbf{F}(\mathbf{x}^{(i)}),$

where the notation $\mathbf{y}_i = \mathbf{F}(\mathbf{x}^{(i)}) - \mathbf{F}(\mathbf{x}^{(i-1)})$ and $\mathbf{s}_i = \mathbf{x}^{(i)} - \mathbf{x}^{(i-1)}$ is introduced into (9.16) to simplify the equation.

If the method is performed as outlined in Eqs. (9.16) and (9.17), the number of scalar functional evaluations is reduced from $n^2 + n$ to n (those required for evaluating $\mathbf{F}(\mathbf{x}^{(i)})$), but the method still requires $O(n^3)$ calculations to solve the associated $n \times n$ linear system (see Step 4 in Algorithm 9.1)

(9.18) $\qquad A_i\mathbf{y}_i = -\mathbf{F}(\mathbf{x}^{(i)}).$

Employing the method in this form would ordinarily not be justified because of the reduction to superlinear convergence from the quadratic convergence of Newton's method.

A considerable improvement can be incorporated, however, by employing a matrix inversion formula of Sherman and Morrison (see, for example, Dennis and Moré [35] page 55). This result states that if A is a nonsingular matrix and $\mathbf{x}$ and $\mathbf{y}$ are vectors, then $A + \mathbf{x}\mathbf{y}^t$ is nonsingular provided that $\mathbf{y}^t A^{-1}\mathbf{x} \neq -1$. Moreover, in this case,

(9.19) $\qquad (A + \mathbf{x}\mathbf{y}^t)^{-1} = A^{-1} - \dfrac{A^{-1}\mathbf{x}\mathbf{y}^t A^{-1}}{1 + \mathbf{y}^t A^{-1}\mathbf{x}}.$

This formula permits A_i^{-1} to be computed directly from A_{i-1}^{-1}, eliminating the need for a matrix inversion with each iteration. By letting $A = A_{i-1}$, $\mathbf{x} = (\mathbf{y}_i - A_{i-1}\mathbf{s}_i)/\|\mathbf{s}_i\|_2^2$, and $\mathbf{y} = \mathbf{s}_i$, formula (9.16) together with (9.19) implies that

$$A_i^{-1} = \left(A_{i-1} + \frac{(\mathbf{y}_i - A_{i-1}\mathbf{s}_i)}{\|\mathbf{s}_i\|_2^2}\mathbf{s}_i^t\right)^{-1}$$

$$= A_{i-1}^{-1} - \frac{A_{i-1}^{-1}\left(\frac{(\mathbf{y}_i - A_{i-1}\mathbf{s}_i)}{\|\mathbf{s}_i\|_2^2}\mathbf{s}_i^t\right)A_{i-1}^{-1}}{1 + \mathbf{s}_i^t A_{i-1}^{-1}\left(\frac{(\mathbf{y}_i - A_{i-1}\mathbf{s}_i)}{\|\mathbf{s}_i\|_2^2}\right)}$$

$$= A_{i-1}^{-1} - \frac{(A_{i-1}^{-1}\mathbf{y}_i - \mathbf{s}_i)\mathbf{s}_i^t A_{i-1}^{-1}}{\|\mathbf{s}_i\|_2^2 + \mathbf{s}_i^t A_{i-1}^{-1}\mathbf{y}_i - \|\mathbf{s}_i\|_2^2};$$

so

(9.20)
$$A_i^{-1} = A_{i-1}^{-1} + \frac{(\mathbf{s}_i - A_{i-1}^{-1}\mathbf{y}_i)\mathbf{s}_i^t A_{i-1}^{-1}}{\mathbf{s}_i^t A_{i-1}^{-1}\mathbf{y}_i}.$$

This computation involves only matrix multiplication at each step and as such requires $O(n^2)$ arithmetic calculations. The calculation of A_i is therefore bypassed, as is the necessity of solving the linear system (9.18). Algorithm 9.2 follows directly from this construction.

Broyden Algorithm 9.2

To approximate the solution of the nonlinear system $\mathbf{F}(\mathbf{x}) = \mathbf{0}$ given an initial approximation $\mathbf{x}$:

INPUT number n of equations and unknowns; initial approximation $\mathbf{x} = (x_1, \ldots, x_n)^t$; tolerance TOL; maximum number of iterations N.

OUTPUT approximate solution $\mathbf{x} = (x_1, \ldots, x_n)^t$ or a message that the number of iterations was exceeded.

Step 1   Set $A_0 = J(\mathbf{x})$ where $J(\mathbf{x})_{i,j} = \dfrac{\partial f_i(\mathbf{x})}{\partial x_j}$ for $1 \le i, j \le n$;

   $\mathbf{v} = \mathbf{F}(\mathbf{x})$.   (*Note:* $\mathbf{v} = \mathbf{F}(\mathbf{x}^{(0)})$.)

Step 2   Set $A = A_0^{-1}$.

Step 3   Set $k = 1$;

   $\mathbf{s} = -A\mathbf{v}$;   (*Note:* $\mathbf{s} = \mathbf{s}_1$.)

   $\mathbf{x} = \mathbf{x} + \mathbf{s}$.   (*Note:* $\mathbf{x} = \mathbf{x}^{(1)}$.)

Step 4   While ($k \le N$) do Steps 5–13.

   Step 5   Set $\mathbf{w} = \mathbf{v}$;   (*Save* $\mathbf{v}$.)

      $\mathbf{v} = \mathbf{F}(\mathbf{x})$;   (*Note:* $\mathbf{v} = \mathbf{F}(\mathbf{x}^{(k)})$.)

      $\mathbf{y} = \mathbf{v} - \mathbf{w}$.   (*Note:* $\mathbf{y} = \mathbf{y}_k$.)

   Step 6   Set $\mathbf{z} = -A\mathbf{y}$.   (*Note:* $\mathbf{z} = -A_{k-1}^{-1}\mathbf{y}_k$.)

   Step 7   Set $p = -\mathbf{s}^t\mathbf{z}$.   (*Note:* $p = \mathbf{s}_k^t A_{k-1}^{-1}\mathbf{y}_k$.)

Step 8   Set $C = pI + (\mathbf{s} + \mathbf{z})\mathbf{s}^t$.
          (*Note*: $C = \mathbf{s}_k^t A_{k-1}^{-1} \mathbf{y}_k I + (\mathbf{s}_k + A_{k-1}^{-1} \mathbf{y}_k)\mathbf{s}_k^t$.)

Step 9   Set $A = (1/p)CA$.   (*Note*: $A = A_k^{-1}$.)

Step 10   Set $\mathbf{s} = -A\mathbf{v}$.   (*Note*: $\mathbf{s} = -A_k^{-1}\mathbf{F}(\mathbf{x}^{(k)})$.)

Step 11   Set $\mathbf{x} = \mathbf{x} + \mathbf{s}$.   (*Note*: $\mathbf{x} = \mathbf{x}^{(k+1)}$.)

Step 12   If $\|\mathbf{s}\| < TOL$ then OUTPUT ($\mathbf{x}$);
                              (*Procedure completed successfully.*)
                              STOP.

Step 13   Set $k = k + 1$.

Step 14   OUTPUT ('Maximum number of iterations exceeded');
          STOP.

EXAMPLE 1   The nonlinear system

$$3x_1 - \cos(x_2 x_3) - \tfrac{1}{2} = 0,$$
$$x_1^2 - 81(x_2 + 0.1)^2 + \sin x_3 + 1.06 = 0,$$
$$e^{-x_1 x_2} + 20x_3 + \frac{10\pi - 3}{3} = 0,$$

was solved by Newton's method in Example 1 of Section 9.2. The Jacobian matrix for this system is

$$J(x_1, x_2, x_3) = \begin{bmatrix} 3 & x_3 \sin x_2 x_3 & x_2 \sin x_2 x_3 \\ 2x_1 & -162(x_2 + 0.1) & \cos x_3 \\ -x_2 e^{-x_1 x_2} & -x_1 e^{-x_1 x_2} & 20 \end{bmatrix}.$$

With $\mathbf{x}^{(0)} = (0.1, 0.1, -0.1)^t$,

$$\mathbf{F}(\mathbf{x}^{(0)}) = \begin{bmatrix} -1.199949 \\ -2.269832 \\ 8.462026 \end{bmatrix}.$$

$$A_0 = J(x_1^{(0)}, x_2^{(0)}, x_3^{(0)})$$
$$= \begin{bmatrix} 3 & 9.999836 \times 10^{-4} & -9.999836 \times 10^{-4} \\ 0.2 & -323.9999 & 0.9950041 \\ -9.900498 \times 10^{-2} & -9.900498 \times 10^{-2} & 20 \end{bmatrix}$$

$$A_0^{-1} = J(x_1^{(0)}, x_2^{(0)}, x_3^{(0)})^{-1}$$
$$= \begin{bmatrix} 0.3333331 & 1.023852 \times 10^{-5} & 1.615703 \times 10^{-5} \\ 2.108606 \times 10^{-3} & -3.086882 \times 10^{-2} & 1.535838 \times 10^{-3} \\ 1.660522 \times 10^{-3} & -1.527579 \times 10^{-4} & 5.000774 \times 10^{-2} \end{bmatrix},$$

$$\mathbf{x}^{(1)} = \mathbf{x}^{(0)} - A_0^{-1}\mathbf{F}(\mathbf{x}^{(0)}) = \begin{bmatrix} 0.4998693 \\ 1.946693 \times 10^{-2} \\ -0.5215209 \end{bmatrix},$$

$$\mathbf{F}(\mathbf{x}^{(1)}) = \begin{bmatrix} -3.404021 \times 10^{-4} \\ -0.3443899 \\ 3.18737 \times 10^{-2} \end{bmatrix},$$

$$\mathbf{y}_1 = \mathbf{F}(\mathbf{x}^{(1)}) - \mathbf{F}(\mathbf{x}^{(0)}) = \begin{bmatrix} 1.199608 \\ 1.925442 \\ -8.430152 \end{bmatrix},$$

$$\mathbf{s}_1 = \begin{bmatrix} 0.3998693 \\ -8.053307 \times 10^{-2} \\ -0.4215209 \end{bmatrix},$$

$$\mathbf{s}_1^t A_0^{-1} \mathbf{y}_1 = 0.3424604,$$

$$A_1^{-1} = A_0^{-1} + (1/0.3424604)[(\mathbf{s}_1 - A_0^{-1}\mathbf{y}_1)A_0^{-1}]$$

$$= \begin{bmatrix} 0.3333781 & 1.11077 \times 10^{-5} & 8.944584 \times 10^{-6} \\ -2.021271 \times 10^{-3} & -3.094847 \times 10^{-2} & 2.196909 \times 10^{-3} \\ 1.022381 \times 10^{-3} & -1.650679 \times 10^{-4} & 5.010987 \times 10^{-2} \end{bmatrix},$$

and

$$\mathbf{x}^{(2)} = \mathbf{x}^{(1)} - A_1^{-1}\mathbf{F}(\mathbf{x}^{(1)}) = \begin{bmatrix} 0.4999863 \\ 8.737888 \times 10^{-3} \\ -0.5231746 \end{bmatrix}.$$

Additional iterations are listed in Table 9.4. ∎

TABLE 9.4

k	$x_1^{(k)}$	$x_2^{(k)}$	$x_3^{(k)}$	$\|\mathbf{x}^{(k)} - \mathbf{x}^{(k-1)}\|_2$
3	0.5000066	8.672215×10^{-4}	-0.5236918	7.88×10^{-3}
4	0.5000005	6.087473×10^{-5}	-0.5235954	8.12×10^{-4}
5	0.5000002	-1.445223×10^{-6}	-0.5235989	6.24×10^{-5}

Procedures are also available that maintain quadratic convergence but significantly reduce the number of required functional evaluations. Methods of this type were originally proposed by Brown [16]. A survey and comparison of some commonly used methods of this type can be found in Moré and Cosnard [89]. In general, however, these methods are much more difficult to efficiently implement than Broyden's method.

Exercise Set 9.3

1. Use Algorithm 9.2 to approximate the solutions to the following systems of nonlinear equations. Iterate until $\|x^{(i)} - x^{(i-1)}\|_\infty < 10^{-5}$. Compare the number of iterations required for this accuracy to the number required for the corresponding approximation by Newton's method computed in Exercise 2 of Section 9.2.

a)
$$x_1^2 + x_2 - 37 = 0,$$
$$x_1 - x_2^2 - 5 = 0,$$
$$x_1 + x_2 + x_3 - 3 = 0.$$

b)
$$x_1 + \cos(x_1 x_2 x_3) - 1 = 0,$$
$$(1 - x_1)^{1/4} + x_2 + 0.05 x_3^2 - 0.15 x_3 - 1 = 0,$$
$$-x_1^2 - 0.1 x_2^2 + 0.01 x_2 + x_3 - 1 = 0.$$

c)
$$12 x_1 - 3 x_2^2 - 4 x_3 = 7.17,$$
$$x_1^2 + 10 x_2 - x_3 = 11.54,$$
$$x_2^3 + 7 x_3 = 7.631.$$

d)
$$4 x_1 - x_2 + x_3 - 2 = 0,$$
$$-x_1 + 6 x_2 - 3 x_3 - 11 = 0,$$
$$x_1 - 3 x_2 + 5 x_3 + 5 = 0.$$

2. Solve the following nonlinear systems using Algorithm 9.2. Iterate until $\|x^{(i)} - x^{(i-1)}\|_\infty < 10^{-5}$. Compare the number of iterations required for this accuracy to the number required for the corresponding approximation by Newton's method computed in Exercise 3 of Section 9.2.

a)
$$6 x_1 - 2 \cos(x_2 x_3) - 1 = 0,$$
$$9 x_2 + \sqrt{x_1^2 + \sin x_3 + 1.06} + 0.9 = 0,$$
$$60 x_3 + 3 e^{x_1 x_2} + 10\pi - 3 = 0.$$

b)
$$x_1^2 + 2 x_2^2 - x_2 - 2 x_3 = 0,$$
$$x_1^2 - 8 x_2^2 + 10 x_3 = 0,$$
$$\frac{x_1^2}{7 x_2 x_3} - 1 = 0.$$

c)
$$2 x_1 + x_2 + x_3 - 4 = 0,$$
$$x_1 + 2 x_2 + x_3 - 4 = 0,$$
$$x_1 x_2 x_3 - 1 = 0.$$

(Find a solution other than $(1, 1, 1)^t$.)

d)
$$2 x_1 + x_2 + x_3 + x_4 - 5 = 0,$$
$$x_1 + 2 x_2 + x_3 + x_4 - 5 = 0,$$
$$x_1 + x_2 + 2 x_3 + x_4 - 5 = 0,$$
$$x_1 x_2 x_3 x_4 - 1 = 0.$$

(Find a solution other than $(1, 1, 1, 1)^t$.)

3. Use both Algorithms 9.1 and 9.2 to attempt a solution to the system

$$x_1 + x_2^3 - 5 x_2^2 - 2 x_2 = 10$$
$$x_1 + x_2^3 + x_2^2 - 14 x_2 = 29$$

starting with $x_1 = 15$, $x_2 = -2$. Solve this problem instead by solving the first equation for x_1 and using this in the second to give a single nonlinear equation. Which method produces the best results?

4. The following nonlinear systems have singular Jacobian matrices at the solution. Can Broyden's method still be used? How is the rate of convergence affected?

a) $3x_1 - \cos(x_2 x_3) - 0.5 = 0$,

     $x_1^2 - 625x_2^2 = 0$,

     $e^{-x_1 x_2} + 20x_3 + \dfrac{10\pi - 3}{3} = 0$.

b) $x_1 - 10x_2 + 9 = 0$,

     $\sqrt{3}(x_3 - x_4) = 0$,

     $(x_2 - 2x_3 + 1)^2 = 0$,

     $\sqrt{2}(x_1 - x_4)^2 = 0$.

5. Exercise 6 of Section 9.2 requires that a system of 20 nonlinear equations be solved for $\boldsymbol{\theta} = (\theta_1, \theta_2, \ldots, \theta_{20})^t$ in order to determine the chute that will give the minimal transit time for granular particles. Use Broyden's method to approximate $\boldsymbol{\theta}$ by computing the iterations $\boldsymbol{\theta}^{(1)}$, $\boldsymbol{\theta}^{(2)}$, ..., until $\|\boldsymbol{\theta}^{(i)} - \boldsymbol{\theta}^{(i-1)}\|_\infty < 10^{-2}$ radians. Compare the amount of computation time required for this method to the amount required to solve the corresponding problem by Newton's method.

6. Exercise 12 of Section 7.1 dealt with determining an exponential least-squares relationship of the form $R = bw^a$ to approximate a collection of data relating the weight and respiration rule of *Modest sphinx* moths. In that exercise, the problem was converted to a log-log relationship, and in part (c), a quadratic term was introduced in an attempt to improve the approximation. Instead of converting the problem, use Broyden's method to determine the constants a and b that minimize $\sum_{i=1}^{n}(R_i - bw_i^a)^2$ for the data listed in Exercise 12 of Section 7.1. Compute the error associated with this approximation, and compare this to the error of the previous approximations for this problem.

7. Apply Broyden's method to the least-squares problem described in Exercise 9 of Section 9.2.

8. Verify that Eq. (9.19) is correct by showing that

$$\left[A^{-1} - \frac{A^{-1}\mathbf{x}\mathbf{y}^t A^{-1}}{1 + \mathbf{y}^t A^{-1}\mathbf{x}}\right](A + \mathbf{x}\mathbf{y}^t) = I.$$

9.4 Steepest Descent Techniques

The advantage of the Newton and quasi-Newton methods for solving systems of nonlinear equations is their speed of convergence once a sufficiently accurate approximation is known. The weakness of these methods is the usual requirement that an accurate initial approximation to the solution is needed to ensure convergence. The method considered in this section will generally converge only linearly to the solution, but is global in nature. As a consequence, the steepest descent method is used for solving nonlinear systems primarily to find sufficiently accurate starting approximations for the Newton techniques, in the same way that the bisection method is used for a single equation.

The method of steepest descent determines a local minimum for a multivariate function of the form $G\colon R^n \to R$. While the method is valuable quite apart from the application as a starting method for solving nonlinear systems, we will restrict our discussion to that situation. (Some other applications are considered in the exercises.)

The connection between the minimization of a function from R^n to R and the solution of a system of nonlinear equations is due to the fact that a system of the form

$$f_1(x_1, x_2, \ldots, x_n) = 0$$
$$f_2(x_1, x_2, \ldots, x_n) = 0$$
$$\vdots$$
$$f_n(x_1, x_2, \ldots, x_n) = 0$$

will have a solution at $\mathbf{x} = (x_1, x_2, \ldots, x_n)^t$ precisely when the function G defined by

$$G(x_1, x_2, \ldots, x_n) = \sum_{i=1}^{n} [f_i(x_1, x_2, \ldots, x_n)]^2$$

has the minimal value zero.

The method of steepest descent for finding a local minimum for a function G from R^n into R can be intuitively described as follows:

i) evaluate G at an initial approximation $\mathbf{x}^{(0)} = (x_1^{(0)}, x_2^{(0)}, \ldots, x_n^{(0)})^t$;
ii) determine a direction from $\mathbf{x}^{(0)}$ that results in a decrease in the value of G;
iii) decide the amount that should be moved in this direction and call the new value $\mathbf{x}^{(1)}$;
iv) repeat steps i) through iii) with $\mathbf{x}^{(0)}$ replaced by $\mathbf{x}^{(1)}$.

Before describing how to choose the correct direction and the appropriate distance to move in this direction, we need to review some results from calculus. The Extreme Value Theorem implies that a differentiable single variable function can have a minimum only when the derivative is zero. To extend this result to multivariate functions we need the following definition.

DEFINITION 9.9   If $G: R^n \to R$, the **gradient** of G at $\mathbf{x} = (x_1, x_2, \ldots, x_n)^t$ is denoted $\nabla G(\mathbf{x})$ and defined by

$$\nabla G(\mathbf{x}) = \left(\frac{\partial G}{\partial x_1}(\mathbf{x}), \frac{\partial G}{\partial x_2}(\mathbf{x}), \ldots, \frac{\partial G}{\partial x_n}(\mathbf{x}) \right)^t.$$

The gradient for a multivariate function is analogous to the derivative of a single variable function in the sense that a differentiable multivariate function can have a minimum at $\mathbf{x}$ only when the gradient is zero.

The gradient has another important property connected with the minimization of multivariate functions. Suppose that $\mathbf{v} = (v_1, v_2, \ldots, v_n)^t$ is a unit vector in R^n, that is,

$$\|\mathbf{v}\|_2^2 = \sum_{i=1}^{n} v_i^2 = 1.$$

The directional derivative of G in the direction of $\mathbf{v}$ is defined by

$$D_{\mathbf{v}}G(\mathbf{x}) = \lim_{h \to 0} \frac{1}{h}[G(\mathbf{x} + h\mathbf{v}) - G(\mathbf{x})].$$

The directional derivative of G at $\mathbf{x}$ in the direction of $\mathbf{v}$ measures the change in the value of the function G relative to the change in the variable in the direction of **v**. (*See Fig. 9.1* for an illustration when G is a function of two variables.)

FIGURE 9.1

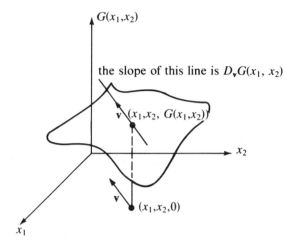

A standard result from the calculus of multivariate functions (see, for example, [43], page 823) states that the direction that produces the maximum value for the directional derivative occurs when **v** is chosen to be parallel to $\nabla G(\mathbf{x})$, provided that $\nabla G(\mathbf{x}) \neq \mathbf{0}$. As a consequence, the direction of greatest decrease in the value of G at **x** is the direction given by $-\nabla G(\mathbf{x})$. An appropriate choice for **x** is therefore

(9.21) $$\mathbf{x}^{(1)} = \mathbf{x}^{(0)} - \alpha \nabla G(\mathbf{x}^{(0)})$$

for some constant $\alpha > 0$.

The problem now reduces to choosing α so that $G(\mathbf{x}^{(1)})$ will be significantly less than $G(\mathbf{x}^{(0)})$. To determine an appropriate choice for the value α we consider the single variable function

$$h(\alpha) = G(\mathbf{x}^{(0)} - \alpha \nabla G(\mathbf{x}^{(0)})).$$

The value of α that minimizes h will be the value needed for Eq. (9.21).

Finding a minimal value for h directly would require differentiating h and determining the critical points, generally too costly a procedure. Instead we begin with three nonnegative approximations, α_1, α_2, and α_3 to $\hat{\alpha}$, the number that produces the minimal value of h. We then determine the quadratic polynomial that interpolates h at α_1, α_2 and α_3. Define α to be the number that minimizes this quadratic. This value is used to determine the new iterate for approximating the minimal value of G:

$$\mathbf{x}^{(1)} = \mathbf{x}^{(0)} - \alpha \nabla G(\mathbf{x}^{(0)}).$$

The following algorithm uses the method of steepest descent to determine an approximation to the minimal value of $G(\mathbf{x})$. To employ this method to approximate the solution to the system

$$f_1(x_1, x_2, \ldots, x_n) = 0$$
$$f_2(x_1, x_2, \ldots, x_n) = 0$$
$$\vdots$$
$$f_n(x_1, x_2, \ldots, x_n) = 0,$$

simply replace the function G with $\sum_{i=1}^{n} f_i^2$.

Steepest Descent Algorithm 9.3

To approximate a solution $\mathbf{p}$ to the minimization problem

$$G(\mathbf{p}) = \min_{\mathbf{x} \in R^n} G(\mathbf{x})$$

given an initial approximation $\mathbf{x}$:

INPUT number n of variables; initial approximation $\mathbf{x} = (x_1, \ldots, x_n)^t$; tolerance TOL; maximum number of iterations N.

OUTPUT approximate solution $\mathbf{x} = (x_1, \ldots, x_n)^t$ or a message of failure.

Step 1   Set $k = 0$.

Step 2   While ($k \leq N$) do Steps 3–15.

Step 3   Set $g_1 = G(x_1, \ldots, x_n)$;   (*Note:* $g_1 = G(\mathbf{x}^{(k)})$.)
$\quad\quad\quad$ $\mathbf{z} = \nabla G(x_1, \ldots, x_n)$;   (*Note:* $\mathbf{z} = \nabla G(\mathbf{x}^{(k)})$.)
$\quad\quad\quad$ $z_0 = \|\mathbf{z}\|_2$.

Step 4   If $z_0 = 0$ then OUTPUT ('ZERO GRADIENT');
$\quad\quad\quad\quad\quad\quad\quad\quad$ OUTPUT $(x_1, \ldots, x_n, g_1)$;
$\quad\quad\quad\quad\quad\quad\quad\quad$ (*Procedure completed, may have a minimum.*
$\quad\quad\quad\quad\quad\quad\quad\quad$ *Check further.*)
$\quad\quad\quad\quad\quad\quad\quad\quad$ STOP.

Step 5   Set $\mathbf{z} = \mathbf{z}/z_0$;   (*Make $\mathbf{z}$ a unit vector.*)
$\quad\quad\quad$ $\alpha_1 = 0$;
$\quad\quad\quad$ $\alpha_3 = 1$;
$\quad\quad\quad$ $g_3 = G(\mathbf{x} + \alpha_3 \mathbf{z})$.

Step 6   While ($|g_3| \geq |g_1|$) do Steps 7 and 8.

Step 7   Set $\alpha_3 = \alpha_3/2$;
$\quad\quad\quad$ $g_3 = G(\mathbf{x} - \alpha_3 \mathbf{z})$.

Step 8   If $\alpha_3 < TOL/2$ then
$\quad\quad\quad\quad\quad\quad$ OUTPUT ('NO LIKELY IMPROVEMENT');
$\quad\quad\quad\quad\quad\quad$ OUTPUT $(x_1, \ldots, x_n, g_1)$;
$\quad\quad\quad\quad\quad\quad$ (*Procedure completed, may have a minimum.*
$\quad\quad\quad\quad\quad\quad$ *Check further.*)
$\quad\quad\quad\quad\quad\quad$ STOP.

Step 9   Set $\alpha_2 = \alpha_3/2$;
$$g_2 = G(\mathbf{x} - \alpha_2 \mathbf{z}).$$

Step 10   Set $h_1 = (g_2 - g_1)/(\alpha_2 - \alpha_1)$;
$$h_2 = (g_3 - g_2)/(\alpha_3 - \alpha_2);$$
$$h_3 = (h_2 - h_1)/(\alpha_3 - \alpha_1).$$
(*Note: Quadratic* $P(\alpha) = g_1 + h_1\alpha + h_3\alpha(\alpha - \alpha_2)$ *interpolates* $h(\alpha)$ *at* $\alpha = 0$, $\alpha = \alpha_2$, $\alpha = \alpha_3$.)

Step 11   Set $\alpha_0 = 0.5(\alpha_1 + \alpha_2 - h_1/h_3)$; (*Critical point of P occurs at* α_0.)
$$g_0 = G(\mathbf{x} - \alpha_0 \mathbf{z}).$$

Step 12   Find α from $\{\alpha_0, \alpha_1, \alpha_2, \alpha_3\}$ so that
$$g = G(\mathbf{x} - \alpha\mathbf{z}) = \min \{g_0, g_1, g_2, g_3\}.$$

Step 13   Set $\mathbf{x} = \mathbf{x} + \alpha\mathbf{z}$.

Step 14   If $|g - g_1| < TOL$ then
OUTPUT $(x_1, \ldots, x_n, g)$;
(*Procedure completed successfully.*)
STOP.

Step 15   Set $k = k + 1$.

Step 16   OUTPUT ('MAXIMUM ITERATIONS EXCEEDED');
(*Procedure completed unsuccessfully.*)
STOP.

EXAMPLE 1   To find a reasonable initial approximation to the solution of the nonlinear system
$$f_1(x_1, x_2, x_3) = 3x_1 - \cos(x_2 x_3) - \tfrac{1}{2} = 0,$$
$$f_2(x_1, x_2, x_3) = x_1^2 - 81(x_2 + 0.1)^2 + \sin x_3 + 1.06 = 0,$$
$$f_3(x_1, x_2, x_3) = e^{-x_1 x_2} + 20x_3 + \frac{10\pi - 3}{3} = 0,$$

we use Algorithm 9.3 with $TOL = 0.005$, $N = 10$, and $\mathbf{x}^{(0)} = (0.5, 0.5, 0.5)^t$.
Let $G(x_1, x_2, x_3) = [f_1(x_1, x_2, x_3)]^2 + [f_2(x_1, x_2, x_3)]^2 + [f_3(x_1, x_2, x_3)]^2$; then

$$\nabla G(x_1, x_2, x_3) \equiv \nabla G(\mathbf{x}) = \left(2f_1(\mathbf{x})\frac{\partial f_1}{\partial x_1}(\mathbf{x}) + 2f_2(\mathbf{x})\frac{\partial f_2}{\partial x_1}(\mathbf{x}) + 2f_3(\mathbf{x})\frac{\partial f_3}{\partial x_1}(\mathbf{x}), \right.$$
$$2f_1(\mathbf{x})\frac{\partial f_1}{\partial x_2}(\mathbf{x}) + 2f_2(\mathbf{x})\frac{\partial f_2}{\partial x_2}(\mathbf{x}) + 2f_3(\mathbf{x})\frac{\partial f_3}{\partial x_2}(\mathbf{x}),$$
$$\left. 2f_1(\mathbf{x})\frac{\partial f_1}{\partial x_3}(\mathbf{x}) + 2f_2(\mathbf{x})\frac{\partial f_2}{\partial x_3}(\mathbf{x}) + 2f_3(\mathbf{x})\frac{\partial f_3}{\partial x_3}(\mathbf{x}) \right)^t$$
$$= 2\mathbf{J}(\mathbf{x})^t \mathbf{F}(\mathbf{x}).$$

For $\mathbf{x}^{(0)} = (0.5, 0.5, 0.5)^t$,

$$G(\mathbf{x}^{(0)}) = 1159.24, \quad z_0 = \|\mathbf{z}\|_2 = 5359.964,$$

and

$$
\begin{aligned}
\mathbf{z} &= (-0.0131206, 0.9897569, 0.1421648)^t, \\
\alpha_1 &= 0 \qquad g_1 = 1159.24 \\
\alpha_2 &= 0.25 \quad g_2 = 454.8059 \quad h_1 = -2817.738 \\
\alpha_3 &= 0.5 \quad\;\, g_3 = 363.5173 \quad h_2 = -365.1542 \quad h_3 = 4905.167
\end{aligned}
$$

so $P(\alpha) = 1159.24 - 2817.738\alpha + 4905.167\alpha(\alpha - 0.25)$. Hence,

$$\alpha_0 = 0.4122214, \; g_0 = 372.2808. \quad (\textit{Note } P(\alpha_0) = 325.723.)$$

Thus, $\alpha = 0.5$ and

$$\mathbf{x}^{(1)} = \mathbf{x}^{(0)} + 0.5\mathbf{z} = (0.5065602, 0.005121529, 0.4289176)^t.$$
$$G(\mathbf{x}^{(1)}) = 363.5173.$$

Table 9.5 contains the remainder of the results.

TABLE 9.5

k	$x_1^{(k)}$	$x_2^{(k)}$	$x_3^{(k)}$	$G(x_1^{(k)}, x_2^{(k)}, x_3^{(k)})$
0	0.5	0.5	0.5	1159.240
1	0.5065602	0.005121529	0.4289176	363.5173
2	0.5045557	0.06421983	−0.5155808	1.874157
3	0.5068673	0.001808132	−0.5179668	0.01322149
4	0.5067566	0.001208410	−0.5235992	0.0005774418

A solution to the nonlinear system is given in Example 2 of Section 9.1 as $(0.5, 0, -0.5235988)^t$, so the results here would certainly be adequate as initial approximations for Newton's and Broyden's methods. ∎

There are many variations of the method of steepest descent. Some of these involve more intricate methods for determining the value of α that will produce a minimum for the single variable function h defined in Eq. (9.21). Other techniques use a multidimensional Taylor polynomial to replace the original multivariate function G and minimize the polynomial instead of G. Although there are advantages to some of these methods over the procedure discussed here, all the steepest descent methods are, in general, linearly convergent and will converge independent of the starting approximation. In some instances, however, the methods may converge to other than the absolute minimum of the function G.

A more complete discussion of steepest descent methods can be found in Ortega and Rheinboldt [96] or Ralston and Rabinowitz [103]. A particular interesting topic in Chapter 8 of the latter reference concerns the Levenberg–Marquardt technique. The method incorporates features of both the steepest descent and Newton's method, emphasizing more heavily one or the other depending on the accuracy of the approximation at each step.

Exercise Set 9.4

1. Use the method of steepest descent with $TOL = 0.005$ to approximate the solutions of the following nonlinear systems.

   a) $\quad x_1^2 - 10x_1 + \quad x_2^2 + 8 = 0$

   $\quad\quad x_1 x_2^2 + \quad x_1 - 10x_2 + 8 = 0.$

   b) $\quad x_1^2 + x_2^2 - x_1 = 0$

   $\quad\quad x_1^2 - x_2^2 - x_2 = 0.$

   c) $\quad\quad 3x_1^2 - x_2^2 = 0$

   $\quad\quad 3x_1 x_2^2 - x_1^3 - 1 = 0.$

   d) $\quad \frac{1}{2} \sin(x_1 x_2) - \frac{x_2}{4\pi} - \frac{x_1}{2} = 0$

   $\quad \left(1 - \frac{1}{4\pi}\right)(e^{2x_1} - e) + \frac{e}{\pi} x_2 - 2ex_1 = 0.$

2. Use the results in Exercise 1 and Newton's method to approximate the solutions to within 10^{-5} for the nonlinear systems in Exercise 1. Do any of the results differ from those in Exercise 1, Section 9.2?

3. Use the method of steepest descent with $TOL = 0.005$ to approximate the solutions of the following nonlinear systems.

   a) $2x_1 + x_2 + x_3 - 4 = 0$

   $\quad x_1 + 2x_2 + x_3 - 4 = 0$

   $\quad\quad x_1 x_2 x_3 - 1 = 0.$

   b) $\quad\quad\quad\quad x_1 + \cos(x_1 x_2 x_3) - 1 = 0$

   $\quad (1 - x_1)^{1/4} + x_2 + 0.05x_3^2 - 0.15x_3 - 1 = 0$

   $\quad\quad -x_1^2 - 0.1x_2^2 + 0.01x_2 + x_3 - 1 = 0.$

   c) $12x_1 - 3x_2^2 - 4x_3 = \ 7.17$

   $\quad x_1^2 + 10x_2 - x_3 = 11.54$

   $\quad\quad x_2^3 + 7x_3 = \ 7.631.$

   d) $\quad 3x_1 - \cos(x_2 x_3) - \frac{1}{2} = 0$

   $\quad\quad x_1^2 - 625x_2^2 = 0$

   $\quad e^{-x_1 x_2} + 20x_3 + \frac{10\pi - 3}{3} = 0.$

4. Use the results of Exercise 3 and Newton's method to approximate the solutions to within 10^{-5} for the nonlinear system in Exercise 3.

5. Use the method of steepest descent to approximate minima to within 0.005 for the following functions.

   a) $G(x_1, x_2) = \cos(x_1 + x_2) + \sin x_1 + \cos x_2.$

   b) $G(x_1, x_2) = 100(x_1^2 - x_2)^2 + (1 - x_1)^2.$

   c) $G(x_1, x_2, x_3) = x_1^2 + 2x_2^2 + x_3^2 - 2x_1 x_2 + 2x_1 - 2.5x_2 - x_3 + 2.$

   d) $G(x_1, x_2, x_3) = x_1^4 + 2x_2^4 + 3x_3^4 + 1.01.$

10

Boundary-Value Problems for Ordinary Differential Equations

A common problem in civil engineering concerns the deflection of a beam of rectangular cross section subject to uniform loading, while the ends of the beam are supported so that they undergo no deflection.

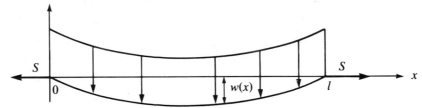

The differential equation approximating the physical situation is of the form

$$\frac{d^2w}{dx^2} = \frac{S}{EI}w + \frac{qx}{2EI}(x - l),$$

where $w = w(x)$ is the deflection a distance x from the left end of the beam, and l, q, E, S, and I represent, respectively, the length of the beam, the intensity of the uniform load, the modulus of elasticity, the stress at the endpoints, and the central moment of inertia. Associated with this differential equation are two conditions given by the assumption that no deflection occurs at the ends of the beam, $w(0) = w(l) = 0$.

When the beam is of uniform thickness, the product EI will be constant, and the exact solution can be easily obtained. In many applications, however, the thickness is not uniform, so the moment of inertia I is a function of x, and approximation techniques are required.

Although methods for finding approximate solutions to differential equations were studied in Chapter 5, those techniques required that all conditions imposed on the differential equation occur at an initial point. For a second order equation, we need to know both $w(0)$ and $w'(0)$, which is not the case in this problem. New techniques are required for handling problems when the conditions imposed are of a boundary-value rather than an initial-value type.

Physical problems that are position-dependent rather than time-dependent are often described in terms of differential equations with conditions imposed at more than one point. The general two-point boundary-value problems in this chapter involve a second-order differential equation of the form

(10.1) $$y'' = f(x, y, y'), \qquad a \le x \le b,$$

together with the boundary conditions

(10.2) $$y(a) = \alpha \quad \text{and} \quad y(b) = \beta.$$

Most of the material concerning second-order boundary-value problems can be extended to problems with boundary conditions of the form

(10.3) $$\alpha_1 y(a) - \beta_1 y'(a) = \alpha \quad \text{and} \quad \alpha_2 y(b) + \beta_2 y'(b) = \beta,$$

where $|\alpha_1| + |\beta_1| \ne 0$ and $|\alpha_2| + |\beta_2| \ne 0$, but some of the techniques become quite complicated. The reader who is interested in problems of this type is advised to consider a book specializing in boundary-value problems, such as Keller [75].

10.1 The Linear Shooting Method

The following theorem gives general conditions that ensure that the solution to a second-order boundary value problem will exist and be unique. The proof of this theorem in the general situation can be found in the book by Keller [75].

THEOREM 10.1   Suppose the function f in the boundary-value problem

$$y'' = f(x, y, y'), \qquad a \le x \le b, \quad y(a) = \alpha, \quad y(b) = \beta,$$

is continuous on the set

$$D = \{(x, y, y') | a \le x \le b, \ -\infty < y < \infty, \ -\infty < y' < \infty\},$$

and that $\partial f / \partial y$ and $\partial f / \partial y'$ are also continuous on D. If

i) $\dfrac{\partial f}{\partial y}(x, y, y') > 0$ for all $(x, y, y') \in D$, and

ii) a constant M exists, with

$$\left| \frac{\partial f}{\partial y'}(x, y, y') \right| \le M \qquad \text{for all } (x, y, y') \in D,$$

then the boundary-value problem has a unique solution.

EXAMPLE 1   The boundary-value problem

$$y'' + e^{-xy} + \sin y' = 0, \qquad 1 \le x \le 2, \quad y(1) = y(2) = 0,$$

has $$f(x, y, y') = -e^{-xy} - \sin y',$$

and since

$$\frac{\partial f}{\partial y}(x, y, y') = xe^{-xy} > 0 \quad \text{and} \quad \left| \frac{\partial f}{\partial y'}(x, y, y') \right| = |-\cos y'| \leq 1,$$

this problem has a unique solution.   ∎

When $f(x, y, y')$ can be expressed in the form

$$f(x, y, y') = p(x)y' + q(x)y + r(x),$$

the differential equation

$$y'' = f(x, y, y')$$

is called **linear**. Problems of this type occur quite often in practice, and this representation allows Theorem 10.1 to be simplified considerably.

COROLLARY 10.2   If the linear boundary-value problem

(10.4) $y'' = p(x)y' + q(x)y + r(x), \quad a \leq x \leq b, \quad y(a) = \alpha, \quad y(b) = \beta,$

satisfies:

i) $p(x)$, $q(x)$, and $r(x)$ are continuous on $[a, b]$,
ii) $q(x) > 0$ on $[a, b]$,

then the problem has a unique solution.

To approximate the unique solution guaranteed by the satisfaction of the hypotheses of Corollary 10.2, let us first consider the initial-value problems

(10.5) $y'' = p(x)y' + q(x)y + r(x), \quad a \leq x \leq b, \quad y(a) = \alpha, \quad y'(a) = 0,$

and

(10.6) $y'' = p(x)y' + q(x)y, \quad a \leq x \leq b, \quad y(a) = 0, \quad y'(a) = 1.$

Theorem 5.15 (p. 263) ensures that under the hypotheses in Corollary 10.2, both problems have a unique solution. If $y_1(x)$ denotes the solution to Eq. (10.5) and $y_2(x)$ denotes the solution to Eq. (10.6), it is not difficult to verify that

(10.7) $y(x) = y_1(x) + \dfrac{\beta - y_1(b)}{y_2(b)} y_2(x)$

is the unique solution to our boundary-value problem, provided, of course, that $y_2(b) \neq 0$. That $y_2(b) = 0$ is in conflict with the hypotheses of Corollary 10.2 is considered in Exercise 10.

The shooting method for linear equations is based on this replacement of the boundary-value problem (10.4) by the two initial-value problems (10.5) and (10.6). Numerous methods are available from Chapter 5 for approximating the solutions

$y_1(x)$ and $y_2(x)$, and once these approximations are available, the solution to the boundary-value problem can be approximated by using Eq. (10.7). Graphically, the method has the appearance shown in Fig. 10.1.

FIGURE 10.1

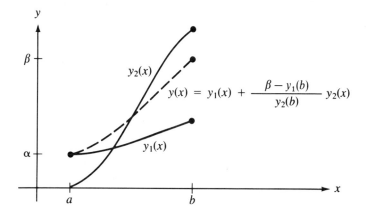

$$y(x) = y_1(x) + \frac{\beta - y_1(b)}{y_2(b)} y_2(x)$$

Algorithm 10.1 uses the fourth-order Runge–Kutta technique to find the approximations to $y_1(x)$ and $y_2(x)$, but any other technique for approximating the solutions to initial-value problems can be substituted into Step 4 (see Exercises 7 and 8). The algorithm has the additional feature of obtaining approximations for the *derivative* of the solution to the boundary-value problem as well as to the solution of the problem itself.

The use of the algorithm is not restricted to those problems for which the hypotheses of Corollary 10.2 can be verified and will, in fact, give satisfactory results for many problems that do not satisfy these hypotheses.

Linear Shooting Algorithm 10.1

To approximate the solution of the boundary-value problem

$$-y'' + p(x)y' + q(x)y + r(x) = 0, \qquad a \le x \le b, \quad y(a) = \alpha, \quad y(b) = \beta:$$

(*Note: Equations* (10.5), (10.6) *are written as first-order systems and solved.*)
INPUT endpoints a, b; boundary conditions α, β; number of subintervals N.
OUTPUT approximations $w_{1,i}$ to $y(x_i)$; $w_{2,i}$ to $y'(x_i)$ for each $i = 0, 1, \ldots, N$.

Step 1   Set $h = (b - a)/N$;
            $u_{1,0} = \alpha$;
            $u_{2,0} = 0$;
            $v_{1,0} = 0$;
            $v_{2,0} = 1$.

Step 2   For $i = 0, \ldots, N - 1$ do Steps 3 and 4.
            (*Runge–Kutta method for systems is used in Steps* 3 *and* 4.)

Step 3   Set $x = a + ih$.

Step 4   Set $k_{1,1} = hu_{2,i}$;

$k_{1,2} = h[p(x)u_{2,i} + q(x)u_{1,i} + r(x)]$;

$k_{2,1} = h[u_{2,i} + \frac{1}{2}k_{1,2}]$;

$k_{2,2} = h[p(x + h/2)(u_{2,i} + \frac{1}{2}k_{1,2})$
$\quad + q(x + h/2)(u_{1,i} + \frac{1}{2}k_{1,1}) + r(x + h/2)]$;

$k_{3,1} = h[u_{2,i} + \frac{1}{2}k_{2,2}]$;

$k_{3,2} = h[p(x + h/2)(u_{2,i} + \frac{1}{2}k_{2,2})$
$\quad + q(x + h/2)(u_{1,i} + \frac{1}{2}k_{2,1}) + r(x + h/2)]$;

$k_{4,1} = h[u_{2,i} + k_{3,2}]$;

$k_{4,2} = h[p(x + h)(u_{2,i} + k_{3,2}) + q(x + h)(u_{1,i} + k_{3,1}) + r(x + h)]$;

$u_{1,i+1} = u_{1,i} + \frac{1}{6}[k_{1,1} + 2k_{2,1} + 2k_{3,1} + k_{4,1}]$;

$u_{2,i+1} = u_{2,i} + \frac{1}{6}[k_{1,2} + 2k_{2,2} + 2k_{3,2} + k_{4,2}]$;

$k'_{1,1} = hv_{2,i}$;

$k'_{1,2} = h[p(x)v_{2,i} + q(x)v_{1,i}]$;

$k'_{2,1} = h[v_{2,i} + \frac{1}{2}k'_{1,2}]$;

$k'_{2,2} = h[p(x + h/2)(v_{2,i} + \frac{1}{2}k'_{1,2}) + q(x + h/2)(v_{1,i} + \frac{1}{2}k'_{1,1})]$;

$k'_{3,1} = h[v_{2,i} + \frac{1}{2}k'_{2,2}]$;

$k'_{3,2} = h[p(x + h/2)(v_{2,i} + \frac{1}{2}k'_{2,2}) + q(x + h/2)(v_{1,i} + \frac{1}{2}k'_{2,1})]$;

$k'_{4,1} = h[v_{2,i} + k'_{3,2}]$;

$k'_{4,2} = h[p(x + h)(v_{2,i} + k'_{3,2}) + q(x + h)(v_{1,i} + k'_{3,1})]$;

$v_{1,i+1} = v_{1,i} + \frac{1}{6}[k'_{1,1} + 2k'_{2,1} + 2k'_{3,1} + k'_{4,1}]$;

$v_{2,i+1} = v_{2,i} + \frac{1}{6}[k'_{1,2} + 2k'_{2,2} + 2k'_{3,2} + k'_{4,2}]$.

Step 5   Set $w_{1,0} = \alpha$;

$$w_{2,0} = \frac{\beta - u_{1,N}}{v_{1,N}};$$

OUTPUT $(a, w_{1,0}, w_{2,0})$.

Step 6   For $i = 1, \ldots, N$

set $W1 = u_{1,i} + w_{2,0}v_{1,i}$;

$W2 = u_{2,i} + w_{2,0}v_{2,i}$;

$x = a + ih$;

OUTPUT $(x, W1, W2)$.   (*Output is x_i, $w_{1,i}$, $w_{2,i}$.*)

Step 7   STOP.   (*Process is complete.*)

EXAMPLE 2   The boundary-value problem

(10.8) $y'' = -\frac{2}{x}y' + \frac{2}{x^2}y + \frac{\sin(\ln x)}{x^2}$, $1 \le x \le 2$, $y(1) = 1$, $y(2) = 2$,

has the exact solution

$$y = c_1 x + \frac{c_2}{x^2} - \tfrac{3}{10}\sin(\ln x) - \tfrac{1}{10}\cos(\ln x),$$

where $\qquad c_2 = \frac{1}{70}[8 - 12 \sin(\ln 2) - 4 \cos(\ln 2)] \approx -0.03920701320$

and $\qquad\qquad\qquad\qquad c_1 = \frac{11}{10} - c_2 \approx 1.1392070132.$

Applying Algorithm 10.1 to this problem requires approximating the solutions to the initial-value problems

$$y_1'' = -\frac{2}{x}y_1' + \frac{2}{x^2}y_1 + \frac{\sin(\ln x)}{x^2}, \qquad 1 \le x \le 2, \quad y_1(1) = 1, \quad y_1'(1) = 0,$$

and $\qquad y_2'' = -\frac{2}{x}y_2' + \frac{2}{x^2}y_2, \qquad 1 \le x \le 2, \quad y_2(1) = 0, \quad y_2'(1) = 1.$

The results of the calculations using Algorithm 10.1 with $N = 10$ and $h = 0.1$, are given in Table 10.1. The value listed as $u_{1,i}$ approximates $y_1(x_i)$, $v_{1,i}$ approximates $y_2(x_i)$, and w_i approximates $y(x_i)$. ∎

TABLE 10.1

x_i	$u_{1,i}$	$v_{1,i}$	w_i	$y(x_i)$	$\|y(x_i) - w_i\|$
1.0	1.00000000	0.00000000	1.00000000	1.00000000	—
1.1	1.00896058	0.09117986	1.09262917	1.09262930	1.43×10^{-7}
1.2	1.03245472	0.16851175	1.18708471	1.18708484	1.34×10^{-7}
1.3	1.06674375	0.23608704	1.28338227	1.28338236	9.78×10^{-8}
1.4	1.10928795	0.29659067	1.38144589	1.38144595	6.02×10^{-8}
1.5	1.15830000	0.35184379	1.48115939	1.48115942	3.06×10^{-8}
1.6	1.21248372	0.40311695	1.58239245	1.58239246	1.08×10^{-8}
1.7	1.27087454	0.45131840	1.68501396	1.68501396	5.43×10^{-10}
1.8	1.33273851	0.49711137	1.78889854	1.78889853	5.05×10^{-9}
1.9	1.39750618	0.54098928	1.89392951	1.89392951	4.41×10^{-9}
2.0	1.46472815	0.58332538	2.00000000	2.00000000	—

The accuracy exhibited in Example 2 is expected because the fourth-order Runge–Kutta method gives $O(h^4)$ accuracy to the solutions of the initial-value problems. Unfortunately there are problems hidden in this technique because of rounding errors. If $y_1(x)$ rapidly increases as x goes from a to b, then $u_{1,N} \approx y(b)$ will be large. Should β be small in magnitude compared to $u_{1,N}$, the term $w_{2,0} = (\beta - u_{1,N})/v_{1,N}$ will be approximately $-u_{1,N}/v_{1,N}$. The computations in Step 6 then would become

$$W1 = u_{1,i} + w_{2,0}v_{1,i} \approx u_{1,i} - \left(\frac{u_{1,N}}{v_{1,N}}\right)v_{1,i},$$

$$W2 = u_{2,i} + w_{2,0}v_{2,i} \approx u_{2,i} - \left(\frac{u_{1,N}}{v_{1,N}}\right)v_{2,i},$$

which allows a possibility of a loss of significant digits due to cancellation. However, since $u_{1,i}$ is an approximation to $y_1(x_i)$, the behavior of y_1 can easily be monitored, and if $u_{1,i}$ increases rapidly from a to b, the shooting technique can be employed in the other direction, that is, solving instead the initial-value problems

$$y'' = p(x)y' + q(x)y + r(x), \qquad a \le x \le b, \quad y(b) = \beta, \quad y'(b) = 0,$$

and $\qquad y'' = p(x)y' + q(x)y, \qquad a \le x \le b, \quad y(b) = 0, \quad y'(b) = 1.$

If the reverse shooting technique still gives cancellation of significant digits, and if increased precision does not yield greater accuracy, other techniques must be employed, techniques such as those presented later in this chapter. In general, however, if $u_{1,i}$ and $v_{1,i}$ are $O(h^n)$ approximations to $y_1(x_i)$ and $y_2(x_i)$, respectively, for each $i = 0, 1, ..., N$, then it can be shown that $w_{1,i}$ will be an $O(h^n)$ approximation to $y_{1,i}$. In particular,

$$|w_{1,i} - y(x_i)| \le Kh^n \left| 1 + \frac{v_{1,i}}{v_{1,N}} \right|,$$

for some constant K (see Isaacson and Keller [67], page 426).

Exercise Set 10.1

1. Use Algorithm 10.1 to approximate the solution to the following boundary-value problem.

    a) $y'' + y = 0$, $\quad 0 \le x \le \pi$, $\quad y(0) = 1$, $\quad y(\pi) = -1$;

    use $h = \pi/3$.

    b) Repeat part (a) using $h = \pi/4$.

2. Use Algorithm 10.1 to approximate the solution to

    $$y'' + 4y = \cos x, \quad 0 \le x \le \pi/4, \quad y(0) = 0, \quad y(\pi/4) = 0;$$

    use $h = \pi/12$.

3. Approximate the solution to the following boundary-value problems, using Algorithm 10.1:

    a) $y'' = -\dfrac{4}{x}y' + \dfrac{2}{x^2}y - \dfrac{2 \ln x}{x^2}$, $\quad 1 \le x \le 2$, $\quad y(1) = -\frac{1}{2}$, $\quad y(2) = \ln 2$;

    use $h = 0.05$.

    b) $y'' = -4y' + 4y$, $\quad 0 \le x \le 5$, $\quad y(0) = 1$, $\quad y(5) = 0$;

    use $h = 0.2$.

    c) $y'' = -3y' + 2y + 2x + 3$, $\quad 0 \le x \le 1$, $\quad y(0) = 2$, $\quad y(1) = 1$;

    use $h = 0.1$.

    d) $y'' = -(x + 1)y' + 2y + (1 - x^2)e^{-x}$, $\quad 0 \le x \le 1$, $\quad y(0) = y(1) = 0$;

    use $h = 0.1$.

4. Show that Corollary 10.2 does not apply to the following boundary-value problems, but that Algorithm 10.1 can still be used.

   a) $y'' = -\dfrac{4}{x}y' - \dfrac{2}{x^2}y + \dfrac{2}{x^2}\ln x,\quad 1 \le x \le 2,\quad y(1) = \tfrac{1}{2},\quad y(2) = \ln 2;$

   use $h = 0.05$.

   b) $y'' = -y,\quad 0 \le x \le \pi/4,\quad y(0) = 1,\quad y(\pi/4) = 1;$

   use $h = \pi/40$.

5. Which of the following boundary-value problems are guaranteed by Theorem 10.1 or Corollary 10.2 to have a unique solution?

   a) $y'' = -\dfrac{4}{x}y' + \dfrac{2}{x^2}y - \dfrac{2\ln x}{x^2},\quad 1 \le x \le 2,\quad y(1) = -\tfrac{1}{2},\quad y(2) = \ln 2.$

   b) $y'' = -y - x,\quad 0 \le x \le \pi,\quad y(0) = 1,\quad y(\pi) = 2.$

   c) $y'' = -\sin xy,\quad 0 \le x \le \pi,\quad y(0) = 0,\quad y(\pi) = 1.$

   d) $y'' = \tfrac{1}{2}y^3,\quad 1 \le x \le 2,\quad y(1) = -\tfrac{2}{3},\quad y(2) = -1.$

   e) $y'' = y^3 - yy',\quad 1 \le x \le 2,\quad y(1) = \tfrac{1}{2},\quad y(2) = \tfrac{1}{3}.$

   f) $y'' = -y,\quad 0 \le x \le \pi,\quad y(0) = 1,\quad y(\pi) = 1.$

   g) $y'' = -y,\quad 0 \le x \le \pi,\quad y(0) = 1,\quad y(\pi) = -1.$

   h) $y'' = -y,\quad 0 \le x \le \dfrac{\pi}{4},\quad y(0) = 1,\quad y\left(\dfrac{\pi}{4}\right) = 1.$

   i) $y'' = 2y^3 - 6y - 2x^3,\quad 1 \le x \le 2,\quad y(1) = 2,\quad y(2) = \tfrac{5}{2}.$

   j) $y'' = -(x+1)y' + 2y + (1-x^2)e^{-x},\quad 0 \le x \le 1,\quad y(0) = y(1) = 0.$

6. Write the second-order initial-value problems (10.5) and (10.6) as first-order systems, and derive the equations necessary to solve the systems, using the fourth-order Runge–Kutta method for systems.

7. Construct an algorithm similar to Algorithm 10.1, using the following initial-value methods.

   a) Adams–Bashforth and Adams–Moulton predictor–corrector method;

   b) Runge–Kutta–Fehlberg method.

8. Using the algorithms constructed in Exercise 7, repeat Exercises 3 and 4.

9. Use either Algorithm 10.1 or the algorithms constructed in Exercise 7 to approximate the solution $y = e^{-10x}$ to the boundary-value problem

   $$y'' = 100y,\qquad 0 \le x \le 1,\quad y(0) = 1,\quad y(1) = e^{-10}.$$

   Use $h = 0.1$ and 0.05. Can you explain the consequences?

10. Show that if y_2 is the solution to $y'' = p(x)y' + q(x)y$ and $y_2(a) = y_2(b) = 0$, then $y_2 \equiv 0$.

11. Devise a shooting method for the general boundary-value problem:

$$y'' = p(x)y' + q(x)y + r(x), \qquad a \le x \le b,$$
$$\alpha_1 y(a) - \beta_1 y'(a) = \alpha, \qquad |\alpha_1| + |\beta_1| \ne 0,$$
$$\alpha_2 y(b) + \beta_2 y'(b) = \beta, \qquad |\alpha_2| + |\beta_2| \ne 0.$$

12. Let u represent the electrostatic potential between two concentric metal spheres of radii R_1 and $R_2 (R_1 < R_2)$, such that the potential of the inner sphere is kept constant at V_1 volts and the potential of the outer sphere at 0 volts. The potential in the region between the two spheres is governed by Laplace's equation (see Eq. (11.2)), which in this particular application, reduces to

$$\frac{d^2 u}{dr^2} + \frac{2}{r}\frac{du}{dr} = 0, \qquad R_1 \le r \le R_2, \quad u(R_1) = V_1, \quad u(R_2) = 0.$$

Suppose $R_1 = 2$ in., $R_2 = 4$ in., and $V_1 = 110$ volts.

a) Approximate $u(3)$ using the Linear Shooting Algorithm 10.1 with $N = 20$.

b) Approximate $u(3)$ using the Linear Shooting Algorithm 10.1 with $N = 40$.

c) Compare the results of parts (a) and (b) with the actual potential $u(3)$ where

$$u(r) = \frac{V_1 R_1}{r}\left(\frac{R_2 - r}{R_2 - R_1}\right).$$

10.2 The Shooting Method for Nonlinear Problems

The shooting technique for the nonlinear second-order boundary-value problem

$$(10.9) \qquad y'' = f(x, y, y'), \qquad a \le x \le b, \quad y(a) = \alpha, \quad y(b) = \beta,$$

is similar to the linear case, except that the solution to a nonlinear problem cannot be simply expressed as a linear combination of the solutions to two initial-value problems. Instead, we need to use the solutions to a *sequence* of initial-value problems of the form

$$(10.10) \qquad y'' = f(x, y, y'), \qquad a \le x \le b, \quad y(a) = \alpha, \quad y'(a) = t,$$

involving a parameter t, to approximate the solution to our boundary-value problem. We do this by choosing the parameters $t = t_k$ in a manner that will ensure that

$$(10.11) \qquad \lim_{k \to \infty} y(b, t_k) = y(b) = \beta,$$

where $y(x, t_k)$ denotes the solution to the initial-value problem (10.10) with $t = t_k$ and $y(x)$ denotes the solution to the boundary-value problem (10.9).

This technique is called the "shooting" method by analogy to the procedure of firing objects at a stationary target. (*See Fig. 10.2.*) We start with a parameter t_0 that determines the initial elevation at which the object is fired from the point (a, α) and along the curve described by the solution to the initial-value problem:

(10.12) $y'' = f(x, y, y')$, $a \le x \le b$, $y(a) = \alpha$, $y'(a) = t_0$.

If $y(b, t_0)$ is not sufficiently close to β, we attempt to correct our approximation by choosing another elevation t_1 and so on, until $y(b, t_k)$ is sufficiently close to "hitting" β. (*See Fig. 10.3.*)

FIGURE 10.2

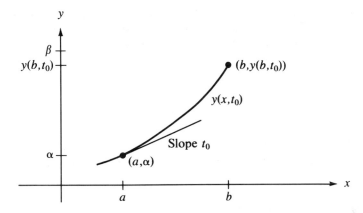

FIGURE 10.3

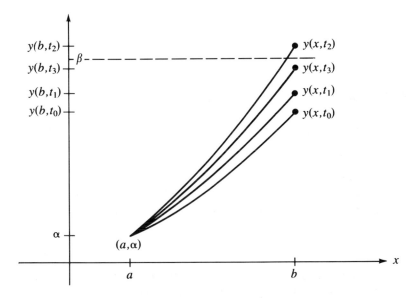

To determine how the parameters t_k can be chosen, suppose a boundary-value problem of the form (10.9) satisfies the hypotheses of Theorem 10.1. If $y(x, t)$ is used to denote the solution to the initial-value problem (10.10), the problem is to determine t so that

(10.13) $y(b, t) - \beta = 0$.

Since this is a nonlinear equation of the type considered in Chapter 2, a number of methods are available. If we wish to employ the Secant method (Algorithm 2.4 of

Section 2.3) to solve the problem, we need to choose initial approximations t_0 and t_1 and then generate the remaining terms of the sequence by

$$t_k = t_{k-1} - \frac{(y(b, t_{k-1}) - \beta)(t_{k-1} - t_{k-2})}{y(b, t_{k-1}) - y(b, t_{k-2})}, \qquad k = 2, 3, \ldots.$$

To use the more powerful Newton's method to generate the sequence $\{t_k\}$, only one initial value, t_0, is needed. However, the iteration has the form

$$(10.14) \qquad t_k = t_{k-1} - \frac{(y(b, t_{k-1}) - \beta)}{(dy/dt)(b, t_{k-1})} \qquad \text{where } (dy/dt)(b, t_{k-1}) \equiv \frac{dy}{dt}(b, t_{k-1}),$$

and requires the knowledge of $(dy/dt)(b, t_{k-1})$. This presents a difficulty, since an explicit representation for $y(b, t)$ is not known; we know only the values $y(b, t_0)$, $y(b, t_1), \ldots, y(b, t_{k-1})$.

Suppose we rewrite the initial-value problem (10.10), emphasizing that the solution depends on both x and t:

$$(10.15) \qquad y''(x, t) = f(x, y(x, t), y'(x, t)), \qquad a \leq x \leq b, \quad y(a, t) = \alpha, \quad y'(a, t) = t,$$

retaining the prime notation to indicate differentiation with respect to x. Since we are interested in determining $(dy/dt)(b, t)$ when $t = t_{k-1}$, we first take the partial derivative of (10.15) with respect to t. This implies that

$$\frac{\partial y''}{\partial t}(x, t) = \frac{\partial f}{\partial t}(x, y(x, t), y'(x, t))$$

$$= \frac{\partial f}{\partial x}(x, y(x, t), y'(x, t))\frac{\partial x}{\partial t} + \frac{\partial f}{\partial y}(x, y(x, t), y'(x, t))\frac{\partial y}{\partial t}(x, t)$$

$$+ \frac{\partial f}{\partial y'}(x, y(x, t), y'(x, t))\frac{\partial y'}{\partial t}(x, t)$$

or, since x and t are independent,

$$(10.16) \qquad \frac{\partial y''}{\partial t}(x, t) = \frac{\partial f}{\partial y}(x, y(x, t), y'(x, t))\frac{\partial y}{\partial t}(x, t) + \frac{\partial f}{\partial y'}(x, y(x, t), y'(x, t))\frac{\partial y'}{\partial t}(x, t)$$

for $a \leq x \leq b$. The initial conditions give

$$\frac{\partial y}{\partial t}(a, t) = 0 \qquad \text{and} \qquad \frac{\partial y'}{\partial t}(a, t) = 1.$$

If we simplify the notation by using $z(x, t)$ to denote $(\partial y/\partial t)(x, t)$ and assume that the order of differentiation of x and t can be reversed, Eq. (10.16) becomes the initial-value problem

$$(10.17) \qquad z'' = \frac{\partial f}{\partial y}(x, y, y')z + \frac{\partial f}{\partial y'}(x, y, y')z', \qquad a \leq x \leq b, \quad z(a) = 0, \quad z'(a) = 1.$$

Newton's method therefore requires that two initial-value problems be solved for each iteration, Eqs. (10.10) and (10.17). Then from Eq. (10.14),

$$(10.18) \qquad t_k = t_{k-1} - \frac{y(b, t_{k-1}) - \beta}{z(b, t_{k-1})}.$$

In practice, none of these initial-value problems is likely to be solved exactly; instead the solutions are approximated by one of the methods discussed in Chapter 5. Algorithm 10.2 uses the fourth-order Runge–Kutta method to approximate both solutions required by Newton's method. A similar procedure for the Secant method is considered in Exercise 4.

Nonlinear Shooting Algorithm 10.2

To approximate the solution of the nonlinear boundary-value problem

$$y'' = f(x, y, y'), \qquad a \le x \le b, \quad y(a) = \alpha, \quad y(b) = \beta:$$

(*Note: Equations (10.15), (10.17) are written as first-order systems and solved.*)

INPUT endpoints a, b; boundary conditions α, β; number of subintervals N; tolerance TOL; maximum number of iterations M.

OUTPUT approximations $w_{1,i}$ to $y(x_i)$; $w_{2,i}$ to $y'(x_i)$ for each $i = 0, 1, \ldots, N$ or a message that the maximum number of iterations was exceeded.

Step 1   Set $h = (b - a)/N$;
 $k = 1$;
 $TK = (\beta - \alpha)/(b - a)$.

Step 2   While $(k \le M)$ do Steps 3–10.

Step 3   Set $w_{1,0} = \alpha$;
 $w_{2,0} = TK$;
 $u_1 = 0$;
 $u_2 = 1$.

Step 4   For $i = 1, \ldots, N$ do Steps 5 and 6.
 (*Runge–Kutta method for systems is used in Steps 5 and 6.*)

Step 5   Set $x = a + (i - 1)h$.

Step 6 $k_{1,1} = hw_{2,i-1}$;
 $k_{1,2} = hf(x, w_{1,i-1}, w_{2,i-1})$;
 $k_{2,1} = h(w_{2,i-1} + \frac{1}{2}k_{1,2})$;
 $k_{2,2} = hf(x + h/2, w_{1,i-1} + \frac{1}{2}k_{1,1}, w_{2,i-1} + \frac{1}{2}k_{1,2})$;
 $k_{3,1} = h(w_{2,i-1} + \frac{1}{2}k_{2,2})$;
 $k_{3,2} = hf(x + h/2, w_{1,i-1} + \frac{1}{2}k_{2,1}, w_{2,i-1} + \frac{1}{2}k_{2,2})$;
 $k_{4,1} = h(w_{2,i-1} + k_{3,2})$;
 $k_{4,2} = hf(x + h, w_{1,i-1} + k_{3,1}, w_{2,i-1} + k_{3,2})$;
 $w_{1,i} = w_{1,i-1} + (k_{1,1} + 2k_{2,1} + 2k_{3,1} + k_{4,1})/6$;
 $w_{2,i} = w_{2,i-1} + (k_{1,2} + 2k_{2,2} + 2k_{3,2} + k_{4,2})/6$;
 $k'_{1,1} = hu_2$;

$$k'_{1,2} = h[f_y(x, w_{1,i-1}, w_{2,i-1})u_1$$
$$+ f_{y'}(x, w_{1,i-1}, w_{2,i-1})u_2];$$
$$k'_{2,1} = h[u_2 + \tfrac{1}{2}k'_{1,2}];$$
$$k'_{2,2} = h[f_y(x + h/2, w_{1,i-1}, w_{2,i-1})(u_1 + \tfrac{1}{2}k'_{1,1})$$
$$+ f_{y'}(x + h/2, w_{1,i-1}, w_{2,i-1})(u_2 + \tfrac{1}{2}k'_{2,1})];$$
$$k'_{3,1} = h(u_2 + \tfrac{1}{2}k'_{2,2});$$
$$k'_{3,2} = h[f_y(x + h/2, w_{1,i-1}, w_{2,i-1})(u_1 + \tfrac{1}{2}k'_{2,1})$$
$$+ f_{y'}(x + h/2, w_{1,i-1}, w_{2,i-1})(u_2 + \tfrac{1}{2}k'_{2,2})];$$
$$k'_{4,1} = h(u_2 + k'_{3,2});$$
$$k'_{4,2} = h[f_y(x + h, w_{1,i-1}, w_{2,i-1})(u_1 + k'_{3,1})$$
$$+ f_{y'}(x + h, w_{1,i-1}, w_{2,i-1})(u_2 + k'_{3,2})];$$
$$u_1 = u_1 + \tfrac{1}{6}[k'_{1,1} + 2k'_{2,1} + 2k'_{3,1} + k'_{4,1}];$$
$$u_2 = u_2 + \tfrac{1}{6}[k'_{1,2} + 2k'_{2,2} + 2k'_{3,2} + k'_{4,2}].$$

Step 7   If $|w_{1,N} - \beta| \le TOL$ then do Steps 8 and 9.

    Step 8   For $i = 0, 1, \ldots, N$
           set $x = a + ih$;
           OUTPUT $(x, w_{1,i}, w_{2,i})$.

    Step 9   (*Procedure is complete.*)
           STOP.

    Step 10   Set $TK = TK - \left(\dfrac{w_{1,N} - \beta}{u_1} \right);$ (*Newton's method is used to*

                                           compute TK.)

           $k = k + 1.$

Step 11   OUTPUT ('Maximum number of iterations exceeded');
           STOP.

In Step 7, the best approximation to β we can expect for $w_{1,N}(t_k)$ is $O(h^n)$ if the approximation method selected for Step 6 gives $O(h^n)$ rate of convergence.

The value t_0 selected in Step 1 is the slope of the straight line through (a, α) and (b, β). If the problem satisfies the hypotheses of Theorem 10.1, any choice of t_0 will give convergence; but in general the procedure will work for many problems for which these hypotheses are not satisfied, although a good choice of t_0 is necessary.

EXAMPLE 1   Consider the boundary-value problem

(10.19) $y'' = \tfrac{1}{8}(32 + 2x^3 - yy'),$ $1 \le x \le 3,$ $y(1) = 17,$ $y(3) = \tfrac{43}{3},$

which has the exact solution $y(x) = x^2 + 16/x$.

Applying the shooting method given in Algorithm 10.2 to this problem requires approximating the initial-value problems

$$y'' = \tfrac{1}{8}(32 + 2x^3 - yy'), \qquad 1 \le x \le 3, \quad y(1) = 17, \quad y'(1) = t_k,$$

and

$$z'' = \frac{\partial f}{\partial y} z + \frac{\partial f}{\partial y'} z' = -\tfrac{1}{8}(yz' + y'z), \qquad 1 \le x \le 3, \quad z(1) = 0, \quad z'(1) = 1,$$

at each step in the iteration. If the stopping technique

$$|w_{1,N}(t_k) - y(3)| \le 10^{-5}$$

is used, this problem requires four iterations and $t_4 = -14.000203$. The results obtained for this value of t are shown in Table 10.2. ■

TABLE 10.2

x_i	w_{1i}	$y(x_i)$	$|w_{1i} - y(x_i)|$
1.0	17.000000	17.000000	—
1.1	15.755495	15.755455	4.06×10^{-5}
1.2	14.773389	14.773333	5.60×10^{-5}
1.3	13.997752	13.997692	5.94×10^{-5}
1.4	13.388629	13.388571	5.71×10^{-5}
1.5	12.916719	12.916667	5.23×10^{-5}
1.6	12.560046	12.560000	4.64×10^{-5}
1.7	12.301805	12.301765	4.02×10^{-5}
1.8	12.128923	12.128889	3.41×10^{-5}
1.9	12.031081	12.031053	2.84×10^{-5}
2.0	12.000023	12.000000	2.32×10^{-5}
2.1	12.029066	12.029048	1.84×10^{-5}
2.2	12.112741	12.112727	1.40×10^{-5}
2.3	12.246532	12.246522	1.01×10^{-5}
2.4	12.426673	12.426667	6.68×10^{-6}
2.5	12.650004	12.650000	3.61×10^{-6}
2.6	12.913847	12.913846	9.17×10^{-7}
2.7	13.215924	13.215926	1.43×10^{-6}
2.8	13.554282	13.554286	3.47×10^{-6}
2.9	13.927236	13.927241	5.21×10^{-6}
3.0	14.333327	14.333333	6.69×10^{-6}

Although Newton's method used with the shooting technique requires the solution of an additional initial-value problem, it will generally be faster than the Secant method. Both methods are only locally convergent, since they require good initial approximations whenever the assumptions of Theorem 10.1 do not hold.

For a general discussion of the convergence of the shooting techniques for nonlinear problems, the reader is referred to the excellent text by Keller [75]. In this reference more general boundary conditions are discussed, and it is also noted that the shooting technique for nonlinear problems is sensitive to rounding errors, especially if the solutions $y(x)$ and $z(x)$ are rapidly increasing functions on $[a, b]$.

Exercise Set 10.2

1. Use Algorithm 10.2 with $h = 0.5$ to approximate the solution to the boundary-value problem

$$y'' = -(y')^2 - y + \ln x, \qquad 1 \le x \le 2, \quad y(1) - 0, \quad y(2) = \ln 2.$$

   Compare your answer to the actual solution $y = \ln x$.

2. Use Algorithm 10.2 with $h = 0.2$ to approximate the solution to the boundary-value problem

$$y'' = \frac{xy' - y}{x^2}, \qquad 1 \le x \le 1.6, \quad y(1) = -1, \quad y(1.6) = -0.847994.$$

3. Use the nonlinear shooting algorithm with Newton's method (Algorithm 10.2) and $TOL = 10^{-4}$ to approximate the solutions to the following boundary-value problems. The actual solution is given for comparison to your results. Do the hypotheses of Theorem 10.1 hold?

   a) $y'' = \frac{1}{2}y^3, \quad 1 \le x \le 2, \quad y(1) = -\frac{2}{3}, \quad y(2) = -1$;

      use $h = 0.05$ and compare the results to $y(x) = 2/(x - 4)$.

   b) $y'' = 2y^3, \quad 1 \le x \le 5, \quad y(1) = \frac{1}{4}, \quad y(5) = \frac{1}{8}$;

      use $h = 0.2$ and compare the results to $y(x) = 1/(x + 3)$.

   c) $y'' = y^3 - yy', \quad 1 \le x \le 2, \quad y(1) = \frac{1}{2}, \quad y(2) = \frac{1}{3}$;

      use $h = 0.1$ and compare the results to $y(x) = 1/(x + 1)$.

   d) $y'' = 2y^3 - 6y - 2x^3, \quad 1 \le x \le 2, \quad y(1) = 2, \quad y(2) = \frac{5}{2}$;

      use $h = 0.05$ and compare the results to $y(x) = x + (1/x)$.

4. Change Algorithm 10.2 to incorporate the Secant method instead of Newton's method. Use $t_0 = (\beta - \alpha)/(b - a)$ and $t_1 = t_0 + (\beta - y(b, t_0))/(b - a)$.

5. Repeat Exercise 3 using the Secant Algorithm derived in Exercise 4, and compare the number of iterations required for the two methods.

6. Change Algorithm 10.2 to incorporate, in place of the fourth-order Runge–Kutta method,

   a) Adams–Bashforth and Adams–Moulton predictor–corrector method;

   b) Runge–Kutta–Fehlberg method.

7. Use the algorithms developed in Exercise 6 to approximate the solutions to the problems given in Exercise 3. Determine which method is most accurate for each problem.

10.3 Finite-Difference Methods for Linear Problems

Although the shooting methods presented in the earlier part of this chapter can be used for both linear and nonlinear boundary-value problems, they often present problems of instability. The methods we present here have better stability characteristics, but generally require more work to obtain a specified accuracy.

Methods involving finite differences for solving boundary-value problems consist of replacing each of the derivatives in the differential equation by an appropriate difference-quotient approximation of the type considered in Section 4.1. The difference quotient is generally chosen so that a certain order of truncation error is maintained.

The linear second-order boundary-value problem,

(10.20) $y'' = p(x)y' + q(x)y + r(x)$, $a \le x \le b$, $y(a) = \alpha$, $y(b) = \beta$,

requires that difference-quotient approximations be used for approximating both y' and y''. To accomplish this, we select an integer $N > 0$ and divide the interval $[a, b]$ into $(N + 1)$ equal subintervals, whose endpoints are the meshpoints $x_i = a + ih$, for $i = 0, 1, ..., N + 1$, where $h = (b - a)/(N + 1)$. Choosing the constant h in this manner will facilitate the application of a matrix algorithm from Chapter 6, which in this form will require solving a linear system involving an $N \times N$ matrix.

At the interior meshpoints, x_i, $i = 1, 2, ..., N$, the differential equation to be approximated is

(10.21) $y''(x_i) = p(x_i)y'(x_i) + q(x_i)y(x_i) + r(x_i)$.

Expanding y in a third-degree Taylor polynomial about x_i evaluated at x_{i+1} and x_{i-1} we have:

(10.22) $y(x_{i+1}) = y(x_i + h) = y(x_i) + hy'(x_i) + \dfrac{h^2}{2} y''(x_i) + \dfrac{h^3}{6} y'''(x_i)$

$$+ \frac{h^4}{24} y^{(4)}(\xi_i^+),$$

for some ξ_i^+, $x_i < \xi_i^+ < x_{i+1}$, and

(10.23) $y(x_{i-1}) = y(x_i - h) = y(x_i) - hy'(x_i) + \dfrac{h^2}{2} y''(x_i) - \dfrac{h^3}{6} y'''(x_i)$

$$+ \frac{h^4}{24} y^{(4)}(\xi_i^-),$$

for some ξ_i^-, $x_{i-1} < \xi_i^- < x_i$, assuming $y \in C^4[x_{i-1}, x_{i+1}]$. If these equations are added, the terms involving $y'(x_i)$ and $y'''(x_i)$ are eliminated, and a simple algebraic manipulation gives

$$y''(x_i) = \frac{1}{h^2} [y(x_{i+1}) - 2y(x_i) + y(x_{i-1})] - \frac{h^2}{24} [y^{(4)}(\xi_i^+) + y^{(4)}(\xi_i^-)].$$

The Intermediate Value Theorem can be used to simplify this even further:

(10.24) $y''(x_i) = \dfrac{1}{h^2} [y(x_{i+1}) - 2y(x_i) + y(x_{i-1})] - \dfrac{h^2}{12} y^{(4)}(\xi_i)$,

for some point ξ_i, $x_{i-1} < \xi_i < x_{i+1}$. Equation (10.24) is called the **centered-difference formula** for $y''(x_i)$.

A centered-difference formula for $y'(x_i)$ can be obtained in a similar manner (the details are considered in Section 4.1) resulting in

(10.25)
$$y'(x_i) = \frac{1}{2h}[y(x_{i+1}) - y(x_{i-1})] - \frac{h^2}{6} y'''(\eta_i),$$

for some η_i where $x_{i-1} < \eta_i < x_{i+1}$.

The use of these centered-difference formulas in Eq. (10.21) results in the equation

$$\frac{y(x_{i+1}) - 2y(x_i) + y(x_{i-1})}{h^2} = p(x_i)\left[\frac{y(x_{i+1}) - y(x_{i-1})}{2h}\right] + q(x_i)y(x_i)$$

$$+ r(x_i) - \frac{h^2}{12}[2p(x_i)y'''(\eta_i) - y^{(4)}(\xi_i)].$$

A finite-difference method with truncation error of order $O(h^2)$ results by using this equation together with the boundary conditions $y(a) = \alpha$ and $y(b) = \beta$ to define

$$w_0 = \alpha, \qquad w_{N+1} = \beta,$$

and

(10.26)
$$\left(\frac{2w_i - w_{i+1} - w_{i-1}}{h^2}\right) + p(x_i)\left(\frac{w_{i+1} - w_{i-1}}{2h}\right) + q(x_i)w_i = -r(x_i)$$

for each $i = 1, 2, \ldots, N$.

In the form we will consider, Eq. (10.26) is rewritten as

$$-\left(1 + \frac{h}{2}p(x_i)\right)w_{i-1} + (2 + h^2 q(x_i))w_i - \left(1 - \frac{h}{2}p(x_i)\right)w_{i+1} = -h^2 r(x_i),$$

and the resulting system of equations is expressed in the tridiagonal $N \times N$-matrix form

(10.27)
$$A\mathbf{w} = \mathbf{b}, \quad \text{where}$$

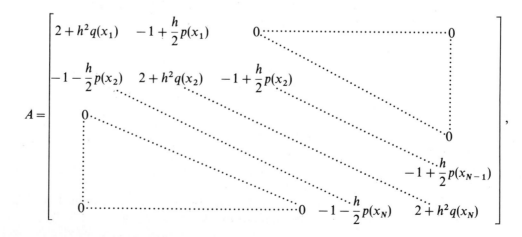

$$\mathbf{w} = \begin{bmatrix} w_1 \\ w_2 \\ \vdots \\ w_{N-1} \\ w_N \end{bmatrix} \quad \text{and } \mathbf{b} = \begin{bmatrix} -h^2 r(x_1) + \left(1 + \dfrac{h}{2} p(x_1)\right) w_0 \\ -h^2 r(x_2) \\ \vdots \\ -h^2 r(x_{N-1}) \\ -h^2 r(x_N) + \left(1 - \dfrac{h}{2} p(x_N)\right) w_{N+1} \end{bmatrix}$$

The following theorem gives conditions under which the tridiagonal linear system (10.27) has a unique solution. Its proof is a consequence of Theorem 6.24 (p. 356) and is considered in Exercise 6.

THEOREM 10.3   Suppose that p, q, and r are continuous on $[a, b]$. If $q(x) \geq 0$ on $[a, b]$, then the tridiagonal linear system (10.27) has a unique solution provided that $h < 2/L$ where $L = \max_{a \leq x \leq b} |p(x)|$.

It should be noted that the hypotheses of Theorem 10.3 guarantee a unique solution to the boundary-value problem (10.20), but they do not guarantee that $y \in C^4[a, b]$. It is necessary to establish that $y^{(4)}$ is continuous on $[a, b]$ in order to ensure that the truncation error has order $O(h^2)$.

Linear Finite-Difference Algorithm 10.3

To approximate the solution of the boundary-value problem

$$y'' = p(x) y' + q(x) y + r(x), \qquad a \leq x \leq b, \quad y(a) = \alpha, \quad y(b) = \beta:$$

INPUT endpoints a, b; boundary conditions α, β; integer N.

OUTPUT approximations w_i to $y(x_i)$ for each $i = 0, 1, \ldots, N + 1$.

Step 1   Set $h = (b - a)/(N + 1)$;
$x = a + h$;
$a_1 = 2 + h^2 q(x)$;
$b_1 = -1 + (h/2) p(x)$;
$d_1 = -h^2 r(x) + (1 + (h/2) p(x)) \alpha$.

Step 2   For $i = 2, \ldots, N - 1$
set $x = a + ih$;
$a_i = 2 + h^2 q(x)$;
$b_i = -1 + (h/2) p(x)$;
$c_i = -1 - (h/2) p(x)$;
$d_i = -h^2 r(x)$.

Step 3   Set $x = b - h$;
$$a_N = 2 + h^2 q(x);$$
$$c_N = -1 - (h/2)p(x);$$
$$d_N = -h^2 r(x) + (1 - (h/2)p(x))\beta.$$

Step 4   Set $l_1 = a_1$;   (Steps 4–10 *solve a tridiagonal linear system using*
Algorithm 6.7.)
$$u_1 = b_1/a_1.$$

Step 5   For $i = 2, \ldots, N - 1$ set $l_i = a_i - c_i u_{i-1}$;
$$u_i = b_i/l_i.$$

Step 6   Set $l_N = a_N - c_N u_{N-1}$.

Step 7   Set $z_1 = d_1/l_1$.

Step 8   For $i = 2, \ldots, N$ set $z_i = (d_i - c_i z_{i-1})/l_i$.

Step 9   Set $w_0 = \alpha$;
$$w_{N+1} = \beta;$$
$$w_N = z_N.$$

Step 10   For $i = N - 1, \ldots, 1$ set $w_i = z_i - u_i w_{i+1}$.

Step 11   For $i = 0, \ldots, N + 1$ set $x = a + ih$;
$$\text{OUTPUT } (x, w_i).$$

Step 12   STOP.   (*Procedure is complete.*)

EXAMPLE 1   Algorithm 10.3 will be used to approximate the solution to the linear boundary-value problem

$$y'' = -\frac{2}{x}y' + \frac{2}{x^2}y + \frac{\sin(\ln x)}{x^2}, \qquad 1 \le x \le 2, \quad y(1) = 1, \quad y(2) = 2,$$

which was also approximated by the shooting method in Example 2 of Section 10.1. For this example, we will use $N = 9$, so $h = 0.1$ and we have the same spacing as in Example 2 of Section 10.1. The results are listed in Table 10.3.

Note that these results are considerably less accurate than those obtained in Example 2 of Section 10.1. This is not surprising since the method used in that example involved a Runge–Kutta technique with truncation error of order $O(h^4)$, while the difference method used here has truncation error of order $O(h^2)$.   ∎

To obtain a difference method with greater accuracy, we could proceed in a number of ways. Using fifth-order Taylor series for approximating $y''(x_i)$ and $y'(x_i)$ results in a truncation error term involving h^4. However, this requires using multiples not only of $y(x_{i+1})$ and $y(x_{i-1})$, but also $y(x_{i+2})$ and $y(x_{i-2})$ in the approximation formulas for $y''(x_i)$ and $y'(x_i)$. This leads to difficulty at $i = 0$ and $i = N$. Moreover, the

TABLE 10.3

x_i	w_i	$y(x_i)$	$\lvert w_i - y(x_i)\rvert$
1.0	1.00000000	1.00000000	—
1.1	1.09260052	1.09262930	2.88×10^{-5}
1.2	1.18704313	1.18708484	4.17×10^{-5}
1.3	1.28333687	1.28338236	4.55×10^{-5}
1.4	1.38140205	1.38144595	4.39×10^{-5}
1.5	1.48112026	1.48115942	3.92×10^{-5}
1.6	1.58235990	1.58239246	3.26×10^{-5}
1.7	1.68498902	1.68501396	2.49×10^{-5}
1.8	1.78888175	1.78889853	1.68×10^{-5}
1.9	1.89392110	1.89392951	8.41×10^{-6}
2.0	2.00000000	2.00000000	—

resulting system of equations analogous to (10.27) would not be in tridiagonal form, and the solution to the system would require many more calculations.

Instead of attempting to obtain a difference method with a higher-order truncation error in this manner, it is generally more satisfactory to consider a reduction in step size. In addition, it can be shown (see, for example, Keller [75], page 81) that the Richardson's extrapolation technique can be used for this method, since the error term is expressed in even powers of h with coefficients independent of h, provided y is sufficiently differentiable.

EXAMPLE 2   If we use the Richardson extrapolation method discussed in Sections 4.2, 4.6, and 5.8, to approximate the solution to the boundary-value problem

$$y'' = -\frac{2}{x}y' + \frac{2}{x^2}y + \frac{\sin(\ln x)}{x^2}, \qquad 1 \le x \le 2, \quad y(1) = 1, \quad y(2) = 2,$$

with $h = 0.1$, 0.05, and 0.025, we obtain the results listed in Table 10.4. The first extrapolation is

$$\text{Ext}_{1i} = \frac{4w_i(h = 0.05) - w_i(h = 0.1)}{3};$$

the second extrapolation is

$$\text{Ext}_{2i} = \frac{4w_i(h = 0.025) - w_i(h = 0.05)}{3};$$

and the final extrapolation is

$$\text{Ext}_{3i} = \frac{16\text{Ext}_{2i} - \text{Ext}_{1i}}{15}.$$

All of the results of Ext_{3i} are correct to the decimal places listed. In fact, this approximation gives results that agree with the exact solution with maximum error of 6.3×10^{-11} at the mesh points.   ■

TABLE 10.4

x_i	$w_i(h=0.1)$	$w_i(h=0.05)$	$w_i(h=0.025)$	Ext_{1i}	Ext_{2i}	Ext_{3i}
1.0	1.00000000	1.00000000	1.00000000	1.00000000	1.00000000	1.00000000
1.1	1.09260052	1.09262207	1.09262749	1.09262925	1.09262930	1.09262930
1.2	1.18704313	1.18707436	1.18708222	1.18708477	1.18708484	1.18708484
1.3	1.28333687	1.28337094	1.28337950	1.28338230	1.28338236	1.28338236
1.4	1.38140205	1.38143493	1.38144319	1.38144589	1.38144595	1.38144595
1.5	1.48112026	1.48114959	1.48115696	1.48115937	1.48115941	1.48115942
1.6	1.58235990	1.58238429	1.58239042	1.58239242	1.58239246	1.58239246
1.7	1.68498902	1.68500770	1.68501240	1.68501393	1.68501396	1.68501396
1.8	1.78888175	1.78889432	1.78889748	1.78889852	1.78889853	1.78889853
1.9	1.89392110	1.89392740	1.89392898	1.89392950	1.89392951	1.89392951
2.0	2.00000000	2.00000000	2.00000000	2.00000000	2.00000000	2.00000000

Exercise Set 10.3

1. Use Algorithm 10.3 to approximate the solution to the following boundary-value problems.

   a) $y'' + y = 0$, $0 \le x \le \pi$, $y(0) = 1$, $y(\pi) = -1$;

      use $h = \pi/3$.

   b) $y'' + 4y = \cos x$, $0 \le x \le \pi/4$, $y(0) = 0$, $y(\pi/4) = 0$;

      use $h = \pi/12$.

2. Show that the following linear boundary-value problems satisfy the hypotheses of Theorem 10.3, and use Algorithm 10.3 to approximate their solutions.

   a) $y'' = -\dfrac{4}{x}y' + \dfrac{2}{x^2}y - \dfrac{2\ln x}{x^2}$, $1 \le x \le 2$, $y(1) = -\frac{1}{2}$, $y(2) = \ln 2$;

      use $h = 0.05$, and compare the results to the actual solution.

   b) $y'' = -4y' + 4y$, $0 \le x \le 5$, $y(0) = 1$, $y(5) = 0$;

      use $h = 0.2$, and compare the results to the actual solution.

   c) $y'' = -3y' + 2y + 2x + 3$, $0 \le x \le 1$, $y(0) = 2$, $y(1) = 1$;

      use $h = 0.1$, and compare the results to the actual solution.

   d) $y'' = -(x+1)y' + 2y + (1 - x^2)e^{-x}$, $0 \le x \le 1$, $y(0) = y(1) = 0$;

      use $h = 0.1$ and compare the results to the actual solution $y = (x-1)e^{-x}$.

3. Repeat Exercise 2 using the extrapolation discussed in Example 2.

4. Use Algorithm 10.3 to approximate the solution $y = e^{-10x}$ to the boundary value problem

$$y'' = 100y, \quad 0 \le x \le 1, \quad y(0) = 1, \quad y(1) = e^{-10}.$$

Use $h = 0.1$ and $h = 0.05$. Compare the results to the actual solution and to Exercise 9, Section 10.1.

5. Show that the following linear boundary-value problems do not satisfy the hypotheses of Theorem 10.3. Use Algorithm 10.3 to approximate their solutions.

a) $y'' = -\dfrac{4}{x} y' - \dfrac{2}{x^2} y + \dfrac{2}{x^2} \ln x$, $1 \le x \le 2$, $y(1) = \tfrac{1}{2}$, $y(2) = \ln 2$;

   use $h = 0.05$.

b) $y'' = -y$, $0 \le x \le \pi/4$, $y(0) = 1$, $y(\pi/4) = 1$;

   use $h = \pi/40$.

6. Prove Theorem 10.3.

7. Show that if $y \in C^6[a, b]$ and if $w_0, w_1, \ldots, w_{N+1}$ satisfy Eq. (10.26), then

$$w_i - y(x_i) = Ah^2 + O(h^4),$$

where A is independent of h, provided $q(x) \ge w > 0$ on $[a, b]$ for some w.

8. The deflection of a uniformly loaded, long rectangular plate under an axial tension force, for small deflections, is governed by a second-order differential equation. Let S represent the axial force and q the intensity of the uniform load. The deflection W along the elemental length is given by:

$$W''(x) - \frac{S}{D} W(x) = \frac{-ql}{2D} x + \frac{q}{2D} x^2, \quad 0 \le x \le l, \quad W(0) = W(l) = 0,$$

where l is the length of the plate, and D is the flexural rigidity of the plate. Let $q = 200$ lb/in.^{2}, $S = 100$ lb/in., $D = 8.8 \times 10^7$ lb in. and $l = 50$ in. Approximate the deflection at one inch intervals by the linear finite-difference Algorithm 10.3.

10.4 Finite-Difference Methods for Nonlinear Problems

For the general nonlinear boundary-value problem

(10.28) $\qquad y'' = f(x, y, y')$, $\quad a \le x \le b$, $\quad y(a) = \alpha$, $\quad y(b) = \beta$,

the difference method is similar to the method applied to linear problems in Section 10.3. Here, however, the system of equations that is derived will not be linear, so an iterative process is required to solve it.

For the development of the procedure, we will assume throughout that f satisfies the following conditions:

i) f and the partial derivatives $f_y \equiv \partial f / \partial y$ and $f_{y'} \equiv \partial f / \partial y'$ are all continuous on

$$D = \{(x, y, y') | a \le x \le b,\ -\infty < y < \infty,\ -\infty < y' < \infty\};$$

ii) $f_y(x, y, y') \ge \delta > 0$ on D for some $\delta > 0$;
iii) constants k and L exist, with

$$k = \max_{(x, y, y') \in D} |f_y(x, y, y')|, \qquad L = \max_{(x, y, y') \in D} |f_{y'}(x, y, y')|.$$

This will ensure, by Theorem 10.1, that a unique solution to Eq. (10.28) exists.

As in the linear case, we divide $[a, b]$ into $(N + 1)$ equal subintervals whose endpoints are at $x_i = a + ih$ for $i = 0, 1, \ldots, N + 1$. Assuming that the exact solution has a bounded fourth derivative allows us to replace $y''(x_i)$ and $y'(x_i)$ in each of the equations

$$(10.29) \qquad\qquad y''(x_i) = f(x_i, y(x_i), y'(x_i))$$

by the appropriate centered-difference formula given in Eqs. (10.24) and (10.25), to obtain, for each $i = 1, 2, \ldots, N$,

$$(10.30) \qquad \frac{y(x_{i+1}) - 2y(x_i) + y(x_{i-1})}{h^2} = f\left(x_i, y(x_i), \frac{y(x_{i+1}) - y(x_{i-1})}{2h} - \frac{h^2}{6} y'''(\eta_i)\right)$$
$$+ \frac{h^2}{12} y^{(4)}(\xi_i),$$

for some ξ_i and η_i in the interval (x_{i-1}, x_{i+1}).

As in the linear case, the difference method results when the error terms are deleted and the boundary conditions employed:

$$w_0 = \alpha, \qquad w_{N+1} = \beta,$$

and

$$-\frac{w_{i+1} - 2w_i + w_{i-1}}{h^2} + f\left(x_i, w_i, \frac{w_{i+1} - w_{i-1}}{2h}\right) = 0,$$

for each $i = 1, 2, \ldots, N$.

The $N \times N$ nonlinear system obtained from this method:

$$(10.31)$$

$$2w_1 - w_2 + h^2 f\left(x_1, w_1, \frac{w_2 - \alpha}{2h}\right) - \alpha = 0,$$

$$-w_1 + 2w_2 - w_3 + h^2 f\left(x_2, w_2, \frac{w_3 - w_1}{2h}\right) = 0,$$

$$\vdots \qquad\qquad\qquad \vdots$$

$$-w_{N-2} + 2w_{N-1} - w_N + h^2 f\left(x_{N-1}, w_{N-1}, \frac{w_N - w_{N-2}}{2h}\right) = 0,$$

$$-w_{N-1} + 2w_N + h^2 f\left(x_N, w_N, \frac{\beta - w_{N-1}}{2h}\right) - \beta = 0$$

will have a unique solution provided that $h < 2/L$, as shown in the previously referenced book by Keller [75], page 86.

To approximate the solution to this system, we will use Newton's method for nonlinear systems discussed in Section 9.2. A sequence of iterates $\{(w_1^{(k)}, w_2^{(k)}, \ldots, w_N^{(k)})^t\}$ is generated that will converge to the solution of system (10.31), provided that the initial approximation $(w_1^{(0)}, w_2^{(0)}, \ldots, w_N^{(0)})^t$ is sufficiently close to the solution, $(w_1, w_2, \ldots, w_N)^t$, and that the Jacobian matrix for the system, defined by Eq. (9.11), is nonsingular. However, for the system (10.31), the Jacobian matrix given in (10.32) is tridiagonal, and the assumptions presented at the beginning of this discussion ensure that J is a nonsingular matrix.

(10.32)

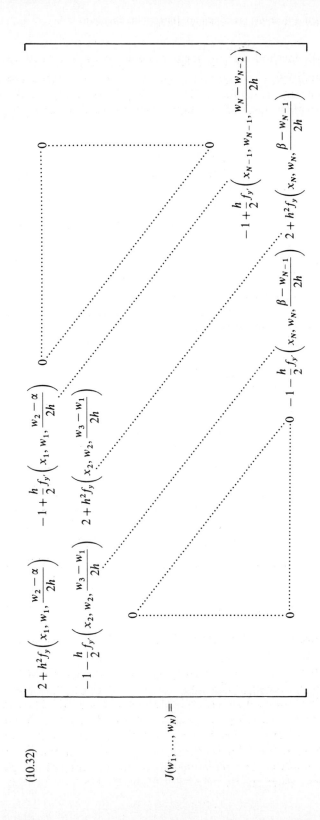

$$J(w_1, \ldots, w_N) =$$

Newton's method for nonlinear systems requires that at each iteration, the $N \times N$ linear system

$$J(w_1, \ldots, w_N)(v_1, \ldots, v_n)^t = -\left(2w_1 - w_2 - \alpha + h^2 f\left(x_1, w_1, \frac{w_2 - \alpha}{2h}\right),\right.$$

$$-w_1 + 2w_2 - w_3 + h^2 f\left(x_2, w_2, \frac{w_3 - w_1}{2h}\right), \ldots,$$

$$-w_{N-2} + 2w_{N-1} - w_N$$

$$+ h^2 f\left(x_{N-1}, w_{N-1}, \frac{w_N - w_{N-2}}{2h}\right),$$

$$\left.-w_{N-1} + 2w_N + h^2 f\left(x_N, w_N, \frac{\beta - w_{N-1}}{2h}\right) - \beta\right)^t$$

be solved for $v_1, v_2, \ldots, v_N$, since

$$w_i^{(k)} = w_i^{(k-1)} + v_i, \qquad \text{for each } i = 1, 2, \ldots, N.$$

Since J is tridiagonal, this is not as formidable a problem as it might at first appear; the Crout reduction algorithm for tridiagonal systems (Algorithm 6.7) can be easily applied. The entire process is detailed in the following algorithm.

Nonlinear Finite-Difference Algorithm 10.4

To approximate the solution to the nonlinear boundary-value problem

$$y'' = f(x, y, y'), \qquad a \le x \le b, \quad y(a) = \alpha, \quad y(b) = \beta:$$

INPUT endpoints a, b; boundary conditions α, β; integer N; tolerance TOL; maximum number of iterations M.

OUTPUT approximations w_i to $y(x_i)$ for each $i = 0, 1, \ldots, N + 1$ or a message that the maximum number of iterations was exceeded.

Step 1   Set $h = (b - a)/(N + 1)$;
$w_0 = \alpha$;
$w_{N+1} = \beta$.

Step 2   For $i = 1, \ldots, N$ set $w_i = \alpha + i\left(\frac{\beta - \alpha}{b - a}\right)h$.

Step 3   Set $k = 1$.

Step 4   While $k \le M$ do Steps 5–18.

Step 5   Set $x = a + h$;
$t = (w_2 - \alpha)/(2h)$;
$a_1 = 2 + h^2 f_y(x, w_1, t)$;
$b_1 = -1 + (h/2)f_{y'}(x, w_1, t)$;
$d_1 = -(2w_1 - w_2 - \alpha + h^2 f(x, w_1, t))$.

Step 6   For $i = 2, \ldots, N - 1$
    set $x = a + ih$;
     $t = (w_{i+1} - w_{i-1})/(2h)$;
     $a_i = 2 + h^2 f_y(x, w_i, t)$;
     $b_i = -1 + (h/2) f_{y'}(x, w_i, t)$;
     $c_i = -1 - (h/2) f_{y'}(x, w_i, t)$;
     $d_i = -(2w_i - w_{i+1} - w_{i-1} + h^2 f(x, w_i, t))$.

Step 7   Set $x = b - h$;
    $t = (\beta - w_{N-1})/(2h)$;
    $a_N = 2 + h^2 f_y(x, w_N, t)$;
    $c_N = -1 - (h/2) f_{y'}(x, w_N, t)$;
    $d_N = -(2w_N - w_{N-1} - \beta + h^2 f(x, w_N, t))$.

Step 8   Set $l_1 = a_1$; (*Steps 8–14 solve a tridiagonal linear system using Algorithm 6.7.*)
    $u_1 = b_1/a_1$.

Step 9   For $i = 2, \ldots, N - 1$ set $l_i = a_i - c_i u_{i-1}$;
          $u_i = b_i/l_i$.

Step 10   Set $l_N = a_N - c_N u_{N-1}$.

Step 11   Set $z_1 = d_1/l_1$.

Step 12   For $i = 2, \ldots, N$ set $z_i = (d_i - c_i z_{i-1})/l_i$.

Step 13   Set $v_N = z_N$;
     $w_N = w_N + v_N$.

Step 14   For $i = N - 1, \ldots, 1$ set $v_i = z_i - u_i v_{i+1}$;
          $w_i = w_i + v_i$.

Step 15   If $\|\mathbf{v}\| \leq TOL$ then do Steps 16 and 17.

  Step 16 For $i = 0, \ldots, N + 1$ set $x = a + ih$;
            OUTPUT (x, w_i).

  Step 17 STOP. (*Procedure completed successfully.*)

Step 18   Set $k = k + 1$.

Step 19   OUTPUT ('Maximum number of iterations exceeded');
    STOP.

The initial approximations $w_i^{(0)}$ to w_i for each $i = 1, 2, \ldots, N$, are obtained in Step 2 by passing a straight line through (a, α) and (b, β) and evaluating at x_i.

It can be shown (see Isaacson and Keller [67], page 433) that this nonlinear finite-difference method is of order $O(h^2)$ so the stopping criteria in Step 15 could be based on the condition that $|v_j| = O(h^2)$ for each $j = 1, 2, \ldots, N$.

Since a good initial approximation is required when the satisfaction of conditions (i), (ii), and (iii), given at the beginning of this presentation, cannot be verified, an upper bound for k should be specified and, if exceeded, a new initial approximation or a reduction in step size considered.

EXAMPLE 1   Applying Algorithm 10.4, with $h = 0.1$, to the nonlinear boundary-value problem

$$y'' = \tfrac{1}{8}(32 + 2x^3 - yy'), \qquad 1 \le x \le 3, \quad y(1) = 17, \quad y(3) = \tfrac{43}{3}$$

gives the results in Table 10.5. The stopping procedure used in this example was to iterate until all values of successive iterates differed by less than 10^{-8}. This was accomplished with four iterations. Note that the problem in this example is the same as that considered for the nonlinear shooting method, Example 1 of Section 10.2.   ∎

TABLE 10.5

x_i	w_i	$y(x_i)$	$|w_i - y(x_i)|$
1.0	17.000000	17.000000	—
1.1	15.754503	15.755455	9.520×10^{-4}
1.2	14.771740	14.773333	1.594×10^{-3}
1.3	13.995677	13.997692	2.015×10^{-3}
1.4	13.386297	13.388571	2.275×10^{-3}
1.5	12.914252	12.916667	2.414×10^{-3}
1.6	12.557538	12.560000	2.462×10^{-3}
1.7	12.299326	12.301765	2.438×10^{-3}
1.8	12.126529	12.128889	2.360×10^{-3}
1.9	12.028814	12.031053	2.239×10^{-3}
2.0	11.997915	12.000000	2.085×10^{-3}
2.1	12.027142	12.029048	1.905×10^{-3}
2.2	12.111020	12.112727	1.707×10^{-3}
2.3	12.245025	12.246522	1.497×10^{-3}
2.4	12.425388	12.426667	1.278×10^{-3}
2.5	12.648944	12.650000	1.056×10^{-3}
2.6	12.913013	12.913846	8.335×10^{-4}
2.7	13.215312	13.215926	6.142×10^{-4}
2.8	13.553885	13.554286	4.006×10^{-4}
2.9	13.927046	13.927241	1.953×10^{-4}
3.0	14.333333	14.333333	—

Richardson's extrapolation procedure can also be used for the nonlinear difference method. If this method is applied to our problem, using $h = 0.1$, 0.05, and 0.025, and four iterations in each case, the results in Table 10.6 are obtained. The notation is the same as in Example 2 of Section 10.3, and the values of EXT_{3i} are all accurate to the places listed, with an actual maximum error of 3.68×10^{-10}. The values of $w_i(h = 0.1)$ are omitted from the table since they were listed previously.

TABLE 10.6

x_i	$w_i(h = 0.05)$	$w_i(h = 0.025)$	Ext_{1i}	Ext_{2i}	Ext_{3i}
1.0	17.00000000	17.00000000	17.00000000	17.00000000	17.00000000
1.1	15.75521721	15.75539525	15.75545543	15.75545460	15.75545455
1.2	14.77293601	14.77323407	14.77333479	14.77333342	14.77333333
1.3	13.99718996	13.99756680	13.99769413	13.99769242	13.99769231
1.4	13.38800424	13.38842973	13.38857346	13.38857156	13.38857143
1.5	12.91606471	12.91651628	12.91666881	12.91666680	12.91666667
1.6	12.55938618	12.55984665	12.56000217	12.56000014	12.56000000
1.7	12.30115670	12.30161280	12.30176684	12.30176484	12.30176471
1.8	12.12830042	12.12874287	12.12889094	12.12888902	12.12888889
1.9	12.03049438	12.03091316	12.03105457	12.03105275	12.03105263
2.0	11.99948020	11.99987013	12.00000179	12.00000011	12.00000000
2.1	12.02857252	12.02892892	12.02904924	12.02904772	12.02904762
2.2	12.11230149	12.11262089	12.11272872	12.11272736	12.11272727
2.3	12.24614846	12.24642848	12.24652299	12.24652182	12.24652174
2.4	12.42634789	12.42658702	12.42666773	12.42666673	12.42666667
2.5	12.64973666	12.64993420	12.65000086	12.65000005	12.65000000
2.6	12.91363828	12.91379422	12.91384683	12.91384620	12.91384615
2.7	13.21577275	13.21588765	13.21592641	13.21592596	13.21592593
2.8	13.55418579	13.55426075	13.55428603	13.55428573	13.55428571
2.9	13.92719268	13.92722921	13.92724153	13.92724139	13.92724138
3.0	14.33333333	14.33333333	14.33333333	14.33333333	14.33333333

Exercise Set 10.4

1. Use Algorithm 10.4 to approximate the solution to the following boundary-value problems:

   a) $y'' = -(y')^2 - y + \ln x$, $1 \le x \le 2$, $y(1) = 0$, $y(2) = \ln 2$;

      use $h = 0.5$.

   b) $y'' = \dfrac{xy' - y}{x^2}$, $1 \le x \le 1.6$, $y(1) = -1$, $y(1.6) = -0.847994$;

      use $h = 0.2$.

2. Show that the following nonlinear boundary-value problems satisfy the hypotheses

   i) $f, f_y, f_{y'}$ are all continuous on

   $$D = \{(x, y, y') | a \le x \le b, -\infty < y, y' < \infty\};$$

   ii) $f_y(x, y, y') \ge \delta > 0$ on D for some $\delta > 0$;
   iii) constants k and L exist, with

   $$k = \max_{(x, y, y') \in D} |f_y(x, y, y')|, \qquad L = \max_{(x, y, y') \in D} |f_{y'}(x, y, y')|;$$

   and use Algorithm 10.4 to approximate their solutions.

a) $y'' = \frac{1}{2}y^3$, $1 \le x \le 2$, $y(1) = -\frac{2}{3}$, $y(2) = -1$;

use $h = 0.05$, and compare the results to Exercise 3(a), Section 10.2.

b) $y'' = 2y^3$, $1 \le x \le 5$, $y(1) = \frac{1}{4}$, $y(2) = \frac{1}{8}$;

use $h = 0.2$, and compare the results to Exercise 3(b), Section 10.2.

c) $y'' = y^3 - yy'$, $1 \le x \le 2$, $y(1) = \frac{1}{2}$, $y(2) = \frac{1}{3}$;

use $h = 0.1$, and compare the results to Exercise 3(c), Section 10.2.

d) $y'' = 2y^3 - 6y - 2x^3$, $1 \le x \le 2$, $y(1) = 2$, $y(2) = \frac{5}{2}$;

use $h = 0.05$, and compare the results to Exercise 3(d), Section 10.2.

3. Repeat Exercise 2 using extrapolation.

4. Show that the hypotheses listed in Exercise 2 ensure the nonsingularity of the Jacobian matrix J for $h < 2/L$.

10.5 Rayleigh–Ritz Method

An important linear two-point boundary-value problem in beam-stress analysis is given by the differential equation

(10.33)
$$-\frac{d}{dx}\left(p(x)\frac{dy}{dx}\right) + q(x)y = f(x) \qquad \text{for } 0 \le x \le 1,$$

with the boundary conditions

(10.34)
$$y(0) = y(1) = 0.$$

This differential equation describes the deflection $y(x)$ on a beam of length one with variable cross section represented by $q(x)$. The deflection is due to the added stresses $p(x)$ and $f(x)$.

In the discussion that follows we assume that $p \in C^1[0, 1]$ and $q, f \in C[0, 1]$. Further, we require that there exist a constant $\delta > 0$ such that

$$p(x) \ge \delta > 0 \qquad \text{for } 0 \le x \le 1$$

and that
$$q(x) \ge 0 \quad \text{for } 0 \le x \le 1.$$

These assumptions are sufficient to guarantee that the boundary-value problem (10.33) and (10.34) has a unique solution (see Bailey, Shampine, and Waltman [5]).

As is the case of many boundary-value problems that describe physical phenomena, the solution to the beam equation satisfies a **variational** property. The variational principle for the beam equation is fundamental to the development of the Rayleigh–Ritz method and characterizes the solution to the beam equation as the function that minimizes a certain integral over all functions in $C_0^2[0, 1]$, the set of those functions u in $C^2[0, 1]$ with the property that $u(0) = u(1) = 0$. The following theorem gives the characterization. The proof of this theorem, while not difficult, is lengthy; it can be found in Schultz [119], pp. 88–89.

THEOREM 10.4   Let $p \in C^1[0, 1]$, q, $f \in C[0, 1]$, and

$$(10.35) \qquad p(x) \geq \delta > 0, \quad q(x) \geq 0 \qquad \text{for } 0 \leq x \leq 1.$$

The function $y \in C_0^2[0, 1]$ is the unique solution to the differential equation

$$(10.36) \qquad -\frac{d}{dx}\left(p(x)\frac{dy}{dx}\right) + q(x)y = f(x), \qquad 0 \leq x \leq 1,$$

if and only if y is the unique function in $C_0^2[0, 1]$ that minimizes the integral

$$(10.37) \qquad I[u] = \int_0^1 \{p(x)[u'(x)]^2 + q(x)[u(x)]^2 - 2f(x)u(x)\}\, dx.$$

In the Rayleigh–Ritz procedure the integral I is minimized not over all the functions in $C_0^2[0, 1]$, but over a smaller set of functions consisting of linear combinations of certain basis functions $\phi_1, \phi_2, \ldots, \phi_n$. The basis functions $\{\phi_i\}_{i=1}^n$ must be linearly independent and satisfy

$$\phi_i(0) = \phi_i(1) = 0 \qquad \text{for each } i = 1, 2, \ldots, n.$$

An approximation $\phi(x) = \sum_{i=1}^n c_i\phi_i(x)$ to the solution $y(x)$ of Eq. (10.36) is obtained by finding constants $c_1, c_2, \ldots, c_n$ to minimize $I[\sum_{i=1}^n c_i\phi_i]$.

From Eq. (10.37),

$$(10.38) \quad \begin{aligned} I[\phi] &= I\left[\sum_{i=1}^n c_i\phi_i\right] \\ &= \int_0^1 \left\{p(x)\left[\sum_{i=1}^n c_i\phi_i'(x)\right]^2 + q(x)\left[\sum_{i=1}^n c_i\phi_i(x)\right]^2 - 2f(x)\sum_{i=1}^n c_i\phi_i(x)\right\} dx, \end{aligned}$$

and, for a minimum to occur it is necessary, when considering I as a function of c_1, $c_2, \ldots, c_n$, to have

$$(10.39) \qquad \frac{\partial I}{\partial c_j} = 0 \qquad \text{for each } j = 1, 2, \ldots, n.$$

Differentiating (10.38) gives

$$\frac{\partial I}{\partial c_j} = \int_0^1 \left\{2p(x)\sum_{i=1}^n c_i\phi_i'(x)\phi_j'(x) + 2q(x)\sum_{i=1}^n c_i\phi_i(x)\phi_j(x) - 2f(x)\phi_j(x)\right\} dx,$$

and substituting into Eq. (10.39) yields

$$(10.40) \quad 0 = \sum_{i=1}^n \left[\int_0^1 \{p(x)\phi_i'(x)\phi_j'(x) + q(x)\phi_i(x)\phi_j(x)\}\, dx\right]c_i - \int_0^1 f(x)\phi_j(x)\, dx,$$

for each $j = 1, 2, \ldots, n$.

The equations described in Eq. (10.40) can be considered as a linear system $Ac = b$, where the symmetric matrix A is given by

$$a_{ij} = \int_0^1 [p(x)\phi_i'(x)\phi_j'(x) + q(x)\phi_i(x)\phi_j(x)]\,dx$$

and b is defined by

$$b_i = \int_0^1 f(x)\phi_i(x)\,dx.$$

The first choice of basis functions we will discuss involves piecewise linear polynomials. The first step is to form a partition on $[0, 1]$ by choosing points $x_0, x_1, \ldots, x_{n+1}$ with

$$0 = x_0 < x_1 < \cdots < x_n < x_{n+1} = 1.$$

Letting $h_i = x_{i+1} - x_i$ for each $i = 0, 1, \ldots, n$, we define the basis functions $\phi_1(x), \phi_2(x), \ldots, \phi_n(x)$ by

$$(10.41) \qquad \phi_i(x) = \begin{cases} 0, & 0 \le x \le x_{i-1}, \\[2mm] \dfrac{(x - x_{i-1})}{h_{i-1}}, & x_{i-1} < x \le x_i, \\[2mm] \dfrac{(x_{i+1} - x)}{h_i}, & x_i < x \le x_{i+1}, \\[2mm] 0, & x_{i+1} < x \le 1, \end{cases}$$

for each $i = 1, 2, \ldots, n$. (*See Fig. 10.4.*)

FIGURE 10.4

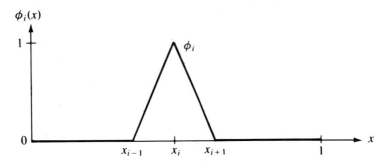

Since the functions ϕ_i are piecewise linear, the derivatives ϕ_i', while not continuous, are constant on the open subinterval (x_j, x_{j+1}) for each $j = 0, 1, \ldots, n$. Thus, we have

$$(10.42) \qquad \phi_i'(x) = \begin{cases} 0, & 0 < x < x_{i-1}, \\[2mm] \dfrac{1}{h_{i-1}}, & x_{i-1} < x < x_i, \\[2mm] -\dfrac{1}{h_i}, & x_i < x < x_{i+1}, \\[2mm] 0, & x_{i+1} < x < 1, \end{cases}$$

for each $i = 1, 2, \ldots, n$.

Because ϕ_i and ϕ_i' are nonzero only on (x_{i-1}, x_{i+1}),

$$\phi_i(x)\phi_j(x) \equiv 0 \qquad \text{and} \qquad \phi_i'(x)\phi_j'(x) \equiv 0,$$

except when j is $i-1$, i, or $i+1$. As a consequence the linear system (10.40) reduces to an $n \times n$ tridiagonal linear system. The nonzero entries in A are given by

$$a_{ii} = \int_0^1 \{p(x)[\phi_i'(x)]^2 + q(x)[\phi_i(x)]^2\}\,dx$$

$$= \int_{x_{i-1}}^{x_i} \left(\frac{1}{h_{i-1}}\right)^2 p(x)\,dx + \int_{x_i}^{x_{i+1}} \left(\frac{-1}{h_i}\right)^2 p(x)\,dx$$

$$+ \int_{x_{i-1}}^{x_i} \left(\frac{1}{h_{i-1}}\right)^2 (x - x_{i-1})^2 q(x)\,dx + \int_{x_i}^{x_{i+1}} \left(\frac{1}{h_i}\right)^2 (x_{i+1} - x)^2 q(x)\,dx,$$

for each $i = 1, 2, \ldots, n$;

$$a_{i,i+1} = \int_0^1 \{p(x)\phi_i'(x)\phi_{i+1}'(x) + q(x)\phi_i(x)\phi_{i+1}(x)\}\,dx$$

$$= \int_{x_i}^{x_{i+1}} -\left(\frac{1}{h_i}\right)^2 p(x)\,dx + \int_{x_i}^{x_{i+1}} \left(\frac{1}{h_i}\right)^2 (x_{i+1} - x)(x - x_i)q(x)\,dx,$$

for each $i = 1, 2, \ldots, n-1$; and

$$a_{i,i-1} = \int_0^1 \{p(x)\phi_i'(x)\phi_{i-1}'(x) + q(x)\phi_i(x)\phi_{i-1}(x)\}\,dx$$

$$= \int_{x_{i-1}}^{x_i} -\left(\frac{1}{h_{i-1}}\right)^2 p(x)\,dx + \int_{x_{i-1}}^{x_i} \left(\frac{1}{h_{i-1}}\right)^2 (x_i - x)(x - x_{i-1})q(x)\,dx,$$

for each $i = 2, \ldots, n$. The entries in **b** are given by

$$b_i = \int_0^1 f(x)\phi_i(x)\,dx$$

$$= \int_{x_{i-1}}^{x_i} \frac{1}{h_{i-1}}(x - x_{i-1})f(x)\,dx + \int_{x_i}^{x_{i+1}} \frac{1}{h_i}(x_{i+1} - x)f(x)\,dx,$$

for each $i = 1, 2, \ldots, n$. The entries in **c** are the unknown coefficients $c_1, c_2, \ldots, c_n$ from which the Rayleigh–Ritz approximation ϕ given by

$$\phi(x) = \sum_{i=1}^n c_i\phi_i(x)$$

is constructed.

The following algorithm sets up the tridiagonal linear system and incorporates the Tridiagonal Algorithm 6.7 to solve the system. In Steps 3, 4, and 5 integrations must be performed so a quadrature formula must be included. It is recommended that the quadrature be accomplished by interpolating p, q, and f with the piecewise polynomials $\phi_1, \ldots, \phi_n$, unless those integrals can be easily evaluated (see Section 3.6).

Piecewise Linear Rayleigh–Ritz Algorithm 10.5

To approximate the solution to the boundary-value problem

$$-\frac{d}{dx}\left(p(x)\frac{dy}{dx}\right) + q(x)y = f(x), \qquad 0 \le x \le 1, \quad y(0) = y(1) = 0,$$

with the piecewise linear function

$$\phi(x) = \sum_{i=1}^{n} c_i \phi_i(x):$$

INPUT integer n; points $x_0 = 0 < x_1 < \cdots < x_n < x_{n+1} = 1$.

OUTPUT coefficients $c_1, \ldots, c_n$.

Step 1 For $i = 0, \ldots, n$ set $h_i = x_{i+1} - x_i$.

Step 2 For $i = 1, \ldots, n$
  define the piecewise linear basis ϕ_i by

$$\phi_i(x) = \begin{cases} 0, & 0 \le x \le x_{i-1}, \\[1em] \dfrac{x - x_{i-1}}{h_{i-1}}, & x_{i-1} < x \le x_i, \\[1em] \dfrac{x_{i+1} - x}{h_i}, & x_i < x \le x_{i+1}, \\[1em] 0, & x_{i+1} < x \le 1. \end{cases}$$

Step 3 Set $IOLD1 = \left(\dfrac{1}{h_0}\right)^2 \displaystyle\int_{x_0}^{x_1} p(x)\,dx;$ (*Note: The integrals in Steps 3–5 can be evaluated using a numerical integration procedure.*)

$$IOLD2 = \left(\frac{1}{h_0}\right)^2 \int_{x_0}^{x_1} (x - x_0)^2 q(x)\,dx.$$

Step 4 For $i = 1, \ldots, n-1$ set $INEW1 = \left(\dfrac{1}{h_i}\right)^2 \displaystyle\int_{x_i}^{x_{i+1}} p(x)\,dx;$

$$INEW2 = \left(\frac{1}{h_i}\right)^2 \int_{x_i}^{x_{i+1}} (x - x_i)^2 q(x)\,dx;$$

$$I3 = \left(\frac{1}{h_i}\right)^2 \int_{x_i}^{x_{i+1}} (x_{i+1} - x)^2 q(x)\,dx;$$

$$I4 = \left(\frac{1}{h_i}\right)^2 \int_{x_i}^{x_{i+1}} (x_{i+1} - x)(x - x_i)q(x)\,dx;$$

$$I5 = \frac{1}{h_{i-1}} \int_{x_{i-1}}^{x_i} (x - x_{i-1})f(x)\,dx;$$

$$I6 = \frac{1}{h_i} \int_{x_i}^{x_{i+1}} (x_{i+1} - x)f(x)\,dx;$$

$$\alpha_i = IOLD1 + INEW1 + IOLD2 + I3;$$
$$\beta_i = -INEW1 + I4;$$
$$b_i = I5 + I6;$$
$$IOLD1 = INEW1;$$
$$IOLD2 = INEW2.$$

Step 5 Set $INEW1 = \left(\dfrac{1}{h_n}\right)^2 \displaystyle\int_{x_n}^{x_{n+1}} p(x)\,dx;$

$$I3 = \left(\dfrac{1}{h_n}\right)^2 \int_{x_n}^{x_{n+1}} (x_{n+1} - x)^2 q(x)\,dx;$$

$$I5 = \dfrac{1}{h_{n-1}} \int_{x_{n-1}}^{x_n} (x - x_{n-1}) f(x)\,dx;$$

$$I6 = \dfrac{1}{h_n} \int_{x_n}^{x_{n+1}} (x_{n+1} - x) f(x)\,dx;$$

$$\alpha_n = IOLD1 + INEW1 + IOLD2 + I3;$$
$$b_n = I5 + I6.$$

Step 6 Set $a_1 = \alpha_1;$ (*Steps 6–12 solve a symmetric tridiagonal linear system using Algorithm 6.7.*)
$$\zeta_1 = \beta_1/\alpha_1.$$

Step 7 For $i = 2, \ldots, n-1$ set $a_i = \alpha_i - \beta_{i-1}\zeta_{i-1};$
$$\zeta_i = \beta_i/a_i.$$

Step 8 Set $a_n = \alpha_n - \beta_{n-1}\zeta_{n-1}.$

Step 9 Set $z_1 = b_1/a_1.$

Step 10 For $i = 2, \ldots, n$ set $z_i = (b_i - \beta_{i-1}z_{i-1})/a_i.$

Step 11 Set $c_n = z_n;$
    OUTPUT $(c_n).$

Step 12 For $i = n-1, \ldots, 1$ set $c_i = z_i - \zeta_i c_{i+1};$
    OUTPUT $(c_i).$

Step 13 STOP. (*Procedure is complete.*)

The following example uses Algorithm 10.5. Because of the nature of this example, the integrations in Steps 3, 4, and 5 were found directly.

EXAMPLE 1   Consider the boundary-value problem

$$-y'' + \pi^2 y = 2\pi^2 \sin(\pi x), \qquad 0 \le x \le 1, \quad y(0) = y(1) = 0.$$

Let $h_i = h = 0.1$, so that $x_i = 0.1i$ for each $i = 0, 1, \ldots, 9$. Following the steps in Algorithm 10.5 gives:

Step 3 $IOLD1 = 100 \displaystyle\int_0^{0.1} dx = 10;$

$\qquad IOLD2 = 100 \displaystyle\int_0^{0.1} x^2\pi^2 \, dx = \dfrac{\pi^2}{30}.$

Step 4   For each $i = 1, \ldots, 8$

$\qquad\qquad INEW1 = 100 \displaystyle\int_{0.1i}^{0.1i+0.1} dx = 10;$

$\qquad\qquad INEW2 = 100 \displaystyle\int_{0.1i}^{0.1i+0.1} (x-0.1i)^2\pi^2 \, dx = \dfrac{\pi^2}{30};$

$\qquad\qquad I3 = 100 \displaystyle\int_{0.1i}^{0.1i+0.1} (0.1i+0.1-x)^2\pi^2 \, dx = \dfrac{\pi^2}{30};$

$\qquad\qquad I4 = 100 \displaystyle\int_{0.1i}^{0.1i+0.1} (0.1i+0.1-x)(x-0.1i)\pi^2 \, dx = \dfrac{\pi^2}{60};$

$\qquad\qquad I5 = 10 \displaystyle\int_{0.1i-0.1}^{0.1i} (x-0.1i+0.1)2\pi^2 \sin(\pi x) \, dx$

$\qquad\qquad\quad = -2\pi \cos 0.1\pi i + 20[\sin(0.1\pi i) - \sin((0.1i-0.1)\pi)];$

$\qquad\qquad I6 = 10 \displaystyle\int_{0.1i}^{0.1i+0.1} (0.1i+0.1-x)2\pi^2 \sin(\pi x) \, dx$

$\qquad\qquad\quad = 2\pi \cos 0.1\pi i - 20[\sin((0.1i+0.1)\pi) - \sin(0.1\pi i)];$

$\qquad\qquad \alpha_i = 20 + \dfrac{\pi^2}{15};$

$\qquad\qquad \beta_i = -10 + \dfrac{\pi^2}{60};$

$\qquad\qquad b_i = 40 \sin(0.1\pi i)(1 - \cos(0.1\pi));$

$\qquad\quad IOLD1 = 10;$

$\qquad\quad IOLD2 = \dfrac{\pi^2}{30}.$

Step 5 $INEW1 = 100 \displaystyle\int_{0.9}^{1} dx = 10;$

$\qquad\qquad I3 = 100 \displaystyle\int_{0.9}^{1} (1-x)^2\pi^2 \, dx = \dfrac{\pi^2}{30};$

$\qquad\qquad I5 = -2\pi \cos(0.9\pi) + 20[\sin(0.9\pi) - \sin(0.8\pi)];$

$\qquad\qquad I6 = 2\pi \cos(0.9\pi) + 20 \sin(0.9\pi);$

$\qquad\qquad \alpha_9 = 20 + \dfrac{\pi^2}{15};$

$\qquad\qquad b_9 = 40 \sin(0.9\pi) - 20 \sin(0.8\pi).$

Steps 6, 7, and 8   These give the following values of a_i and ζ_i:

$$
\begin{aligned}
a_1 &= 20.65797363, & \zeta_1 &= -0.4761118767, \\
a_2 &= 15.97517213, & \zeta_2 &= -0.6156745300, \\
a_3 &= 14.60250273, & \zeta_3 &= -0.6735493754, \\
a_4 &= 14.03327431, & \zeta_4 &= -0.7008704010, \\
a_5 &= 13.76455818, & \zeta_5 &= -0.7145530183, \\
a_6 &= 13.62998710, & \zeta_6 &= -0.7216081489, \\
a_7 &= 13.56059192, & \zeta_7 &= -0.7253006839, \\
a_8 &= 13.52427397, & \zeta_8 &= -0.7272483991, \\
a_9 &= 13.50511721.
\end{aligned}
$$

Steps 9 and 10   These generate the following values of z_i:

$$
\begin{aligned}
z_1 &= 0.0292852890, & z_6 &= 0.3687722609, \\
z_2 &= 0.0900626268, & z_7 &= 0.3842683589, \\
z_3 &= 0.1691255264, & z_8 &= 0.3645448400, \\
z_4 &= 0.2512140726, & z_9 &= 0.3102866742. \\
z_5 &= 0.3217362267,
\end{aligned}
$$

Steps 11 and 12   These give the coefficients c_i:

$$
\begin{aligned}
c_9 &= 0.3102866742, & c_4 &= 0.9549641893, \\
c_8 &= 0.5902003271, & c_3 &= 0.8123410598, \\
c_7 &= 0.8123410598, & c_2 &= 0.5902003271, \\
c_6 &= 0.9549641893, & c_1 &= 0.3102866742. \\
c_5 &= 1.004108771,
\end{aligned}
$$

The piecewise linear approximation is given by

$$
\phi(x) = \sum_{i=1}^{9} c_i \phi_i(x).
$$

The actual solution to the boundary-value problem is

$$
y(x) = \sin \pi x
$$

and Table 10.7 lists the error in the approximation at x_i for each $i = 1, \ldots, 9$.   ∎

It can be shown that the tridiagonal matrix A given by the piecewise linear basis functions is positive definite (see Exercise 8), so, by Theorem 6.22 on page 337 the linear system is stable. Under the hypotheses presented at the beginning of this section we have

$$
|\phi(x) - y(x)| = O(h^2), \qquad 0 \le x \le 1.
$$

A proof of this result can be found in Schultz [119], pages 103 and 104.

TABLE 10.7

i	x_i	$\phi(x_i)$	$y(x_i)$	$|\phi(x_i) - y(x_i)|$
1	0.1	0.3102866742	0.3090169943	0.00127
2	0.2	0.5902003271	0.5877852522	0.00242
3	0.3	0.8123410598	0.8090169943	0.00332
4	0.4	0.9549641896	0.9510565162	0.00391
5	0.5	1.004108771	1.0000000000	0.00411
6	0.6	0.9549641893	0.9510565162	0.00391
7	0.7	0.8123410598	0.8090169943	0.00332
8	0.8	0.5902003271	0.5877852522	0.00242
9	0.9	0.3102866742	0.3090169943	0.00127

The use of piecewise-linear basis functions results in an approximate solution to Eqs. (10.33) and (10.34) that is continuous but not differentiable on $[0, 1]$. A more complicated set of basis functions is required to construct an approximation that belongs to $C_0^2[0, 1]$. These basis functions are similar to the cubic interpolatory splines that were discussed in Section 3.6.

A cubic *interpolatory* spline S on the five nodes x_0, x_1, x_2, x_3, and x_4 for a function f is defined by:

a) S is a cubic polynomial, denoted by S_j, on $[x_j, x_{j+1}]$ for $j = 0, 1, 2, 3$. (This gives 16 selectable constants for S, 4 constants for each cubic.)

b) $S(x_j) = f(x_j)$ for $j = 0, 1, 2, 3, 4$. (5 specified conditions)

c) $S_{j+1}(x_{j+1}) = S_j(x_{j+1})$ for $j = 0, 1, 2$. (3 specified conditions)

d) $S'_{j+1}(x_{j+1}) = S'_j(x_{j+1})$ for $j = 0, 1, 2$. (3 specified conditions)

e) $S''_{j+1}(x_{j+1}) = S''_j(x_{j+1})$ for $j = 0, 1, 2$. (3 specified conditions)

f)   One of the following boundary conditions is satisfied:

    i)   Free: $S''(x_0) = S''(x_4) = 0$. (2 specified conditions)

    ii)   Clamped: $S'(x_0) = f'(x_0)$ and $S'(x_4) = f'(x_4)$. (2 specified conditions)

Since uniqueness of solution requires that the number of constants in (a), 16, must equal the number of conditions in (b) through (f), only one of the boundary conditions in (f) can be specified for the interpolatory cubic splines.

The cubic spline functions we will use for our basis functions are called **B-splines** or **bell-shaped splines**. These splines differ from interpolatory splines in that both sets of boundary conditions in (f) are satisfied. This requires the relaxation of two of the conditions in (b) through (e). Since the spline must have two continuous derivatives on $[x_0, x_4]$ the only reasonable choice is to modify (b) to

$$b')\quad S(x_j) = f(x_j), \text{ for } j = 0, 2, 4.$$

The basic B-spline S defined below uses the equally spaced nodes $x_0 = -2$, $x_1 = -1$, $x_2 = 0$, $x_3 = 1$, and $x_4 = 2$. It satisfies the interpolatory conditions

$$b')\quad S(x_0) = 0, \quad S(x_2) = 1, \quad S(x_4) = 0;$$

as well as both sets of conditions

    i) $S''(x_0) = S''(x_4) = 0$,

    ii) $S'(x_0) = S'(x_4) = 0$.

As a consequence, $S \in C_0^2(-\infty, \infty)$.

$$S(x) = \begin{cases} 0, & x \le -2; \\ \frac{1}{4}[(2-x)^3 - 4(1-x)^3 - 6x^3 + 4(1+x)^3], & -2 < x \le -1; \\ \frac{1}{4}[(2-x)^3 - 4(1-x)^3 - 6x^3], & -1 < x \le 0; \\ \frac{1}{4}[(2-x)^3 - 4(1-x)^3], & 0 < x \le 1; \\ \frac{1}{4}(2-x)^3, & 1 < x \le 2; \\ 0, & 2 < x. \end{cases}$$

    To construct the basis functions ϕ_i in $C_0^2[0, 1]$ we first partition $[0, 1]$ by choosing a positive integer n and defining $h = 1/(n+1)$. This produces the equally spaced nodes $x_i = ih$, for each $i = 0, 1, \ldots, n+1$. We then define S_i by $S_i(x) = S((x - x_i)/h)$ for each $i = 0, 1, \ldots, n+1$. The graph of a typical S_i is shown in Figure 10.5.

FIGURE 10.5

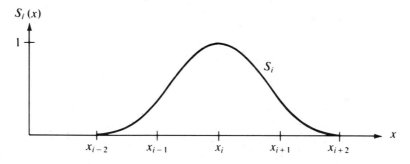

    It is easy to show that $\{S_i\}_{i=0}^{n+1}$ is a linearly independent set of cubic splines (see Exercise 7). For the set $\{S_i\}_{i=0}^{n+1}$ to satisfy the boundary conditions $\phi_i(0) = \phi_i(1) = 0$, it is necessary to modify $S_0, S_1, S_n,$ and S_{n+1}. The basis with this modification is defined by

$$\phi_i(x) = \begin{cases} S_0(x) - 4S\left(\dfrac{x+h}{h}\right), & i = 0, \\[2mm] S_1(x) - S\left(\dfrac{x+h}{h}\right), & i = 1, \\[2mm] S_i(x), & 2 \le i \le n-1, \\[2mm] S_n(x) - S\left(\dfrac{x-(n+2)h}{h}\right), & i = n, \\[2mm] S_{n+1}(x) - 4S\left(\dfrac{x-(n+2)h}{h}\right), & i = n+1. \end{cases}$$

(*See Fig. 10.6.*)

FIGURE 10.6

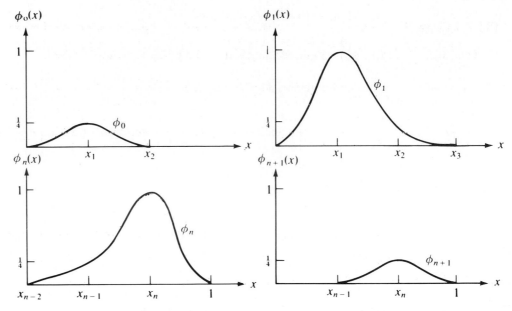

Since $\phi_i(x)$ and $\phi_i'(x)$ are nonzero only for $x_{i-2} \le x \le x_{i+2}$, the matrix in the Rayleigh–Ritz approximation will be a band matrix with band-width at most seven:

$$(10.43) \quad A = \begin{bmatrix} a_{00} & a_{01} & a_{02} & a_{03} & 0 & \cdots & & & & 0 \\ a_{10} & a_{11} & a_{12} & a_{13} & a_{14} & & & & & \\ a_{20} & a_{21} & a_{22} & a_{23} & a_{24} & a_{25} & & & & \\ a_{30} & a_{31} & a_{32} & a_{33} & a_{34} & a_{35} & a_{36} & & & \\ 0 & & & & & & & & & 0 \\ & & & & & & & & & a_{n-2,n+1} \\ & & & & & & & & & a_{n-1,n+1} \\ & & & & & & & & & a_{n,n+1} \\ 0 & \cdots & & 0 & a_{n+1,n-2} & a_{n+1,n-1} & a_{n+1,n} & a_{n+1,n+1} \end{bmatrix}$$

where

$$a_{ij} = \int_0^1 \{ p(x)\phi_i'(x)\phi_j'(x) + q(x)\phi_i(x)\phi_j(x) \}\, dx,$$

for each $i, j = 0, 1, \ldots, n + 1$. The matrix A is positive definite (see Exercise 12), so the linear system (10.43) can be easily solved by Choleski's Algorithm 6.6 or by Gaussian elimination. The following algorithm details the construction of the cubic spline approximation $\phi(x)$ by the Rayleigh–Ritz method for the boundary-value problem (10.33) and (10.34) given on page 546.

Cubic Spline Rayleigh–Ritz Algorithm 10.6

To approximate the solution to the boundary-value problem

$$-\frac{d}{dx}\left(p(x)\frac{dy}{dx}\right) + q(x)y = f(x), \qquad 0 \le x \le 1, \quad y(0) = y(1) = 0$$

with the cubic B-spline

$$\phi(x) = \sum_{i=0}^{n+1} c_i \phi_i(x):$$

INPUT integer n.

OUTPUT coefficients $c_0, \ldots, c_{n+1}$.

Step 1 Set $h = 1/(n+1)$.

Step 2 For $i = 0, \ldots, n+1$ set $x_i = ih$. Set $x_{-2} = x_{-1} = 0$; $x_{n+2} = x_{n+3} = 1$.

Step 3 Define the function S by

$$S(t) = \begin{cases} 0, & t \le -2, \\ \frac{1}{4}[(2-t)^3 - 4(1-t)^3 - 6t^3 + 4(1+t)^3], & -2 < t \le -1, \\ \frac{1}{4}[(2-t)^3 - 4(1-t)^3 - 6t^3], & -1 < t \le 0, \\ \frac{1}{4}[(2-t)^3 - 4(1-t)^3], & 0 < t \le 1, \\ \frac{1}{4}(2-t)^3, & 1 < t \le 2, \\ 0, & 2 < t. \end{cases}$$

Step 4 Define the cubic spline basis $\{\phi_i\}_{i=0}^{n+1}$ by

$$\phi_0(x) = S\left(\frac{x}{h}\right) - 4S\left(\frac{x+h}{h}\right),$$

$$\phi_1(x) = S\left(\frac{x-x_1}{h}\right) - S\left(\frac{x+h}{h}\right),$$

$$\text{for } i = 2, \ldots, n-1, \ \phi_i(x) = S\left(\frac{x-x_i}{h}\right),$$

$$\phi_n(x) = S\left(\frac{x-x_n}{h}\right) - S\left(\frac{x-(n+2)h}{h}\right),$$

$$\phi_{n+1}(x) = S\left(\frac{x-x_{n+1}}{h}\right) - 4S\left(\frac{x-(n+2)h}{h}\right).$$

Step 5 For $i = 0, \ldots, n+1$ do Steps 6–9.
(*Note: The integrals in Steps 6 and 9 can be evaluated using a numerical integration procedure.*)

Step 6 For $j = i, i+1, \ldots, min(i+3, n+1)$
$\qquad$ set $L = max(x_{j-2}, 0)$;
$\qquad\qquad U = min(x_{i+2}, 1)$;

$$a_{ij} = \int_L^U [p(x)\phi_i'(x)\phi_j'(x) + q(x)\phi_i(x)\phi_j(x)]\, dx;$$

$$a_{ji} = a_{ij}. \quad (\text{Since } A \text{ is symmetric.})$$

Step 7   If $i \geq 4$ then for $j = 0, \ldots, i - 4$ set $a_{ij} = 0$.

Step 8   If $i \leq n - 3$ then for $j = i + 4, \ldots, n + 1$ set $a_{ij} = 0$.

Step 9   Set $L = max(x_{i-2}, 0)$;
$$U = min(x_{i+2}, 1);$$
$$b_i = \int_L^U f(x)\phi_i(x)\, dx.$$

Step 10   Solve the linear system $A\mathbf{c} = \mathbf{b}$, where $A = (a_{ij})$, $\mathbf{b} = (b_0, \ldots, b_{n+1})^t$ and $\mathbf{c} = (c_0, \ldots, c_{n+1})^t$.

Step 11   For $i = 0, \ldots, n + 1$ OUTPUT (c_i).

Step 12   STOP.   (*Procedure is complete.*)

EXAMPLE 2   Consider the boundary-value problem

$$-y'' + \pi^2 y = 2\pi^2 \sin(\pi x), \qquad 0 \leq x \leq 1, \quad y(0) = y(1) = 0.$$

In Example 1 we let $h = 0.1$ and generated approximations using piecewise-linear basis functions. Table 10.8 lists the results obtained by applying the *B*-splines as detailed in Algorithm 10.6 with this same choice of nodes.   ■

TABLE 10.8

i	c_i	x_i	$\phi(x_i)$	$y(x_i)$	$|y(x_i) - \phi(x_i)|$
0	$0.50964361 \times 10^{-5}$	0	0.00000000	0.00000000	0.00000000
1	0.20942608	0.1	0.30901644	0.30901699	0.00000055
2	0.39835678	0.2	0.58778549	0.58778525	0.00000024
3	0.54828946	0.3	0.80901687	0.80901699	0.00000012
4	0.64455358	0.4	0.95105667	0.95105652	0.00000015
5	0.67772340	0.5	1.00000002	1.00000000	0.00000020
6	0.64455370	0.6	0.95105713	0.95105652	0.00000061
7	0.54828951	0.7	0.80901773	0.80901699	0.00000074
8	0.39835730	0.8	0.58778690	0.58778525	0.00000165
9	0.20942593	0.9	0.30901810	0.30901699	0.00000111
10	$0.74931285 \times 10^{-5}$	1.0	0.00000000	0.00000000	0.00000000

In practice it is recommended that the integrations in Steps 6 and 9 be performed by interpolating p, q, and f with the cubic spline interpolatory polynomial and then integrating. The hypotheses assumed at the beginning of this section are sufficient to guarantee that

$$\left\{ \int_0^1 |y(x) - \phi(x)|^2 \, dx \right\}^{1/2} = O(h^4), \qquad 0 \leq x \leq 1.$$

For a proof of this result, see Schultz [119], pages 107 and 108.

B-splines can also be defined for unequally spaced nodes, but the details are more complicated. A presentation of the technique can be found in Schultz [119], p. 73. Another basis used often in practice is the piecewise cubic Hermite polynomials. For an excellent presentation of this method see Schultz [119], p. 24ff.

Other methods that have received considerable attention recently are Galerkin or "weak form" methods. For the boundary-value problem we have been considering

$$-\frac{d}{dx}\left(p(x)\frac{dy}{dx}\right) + q(x)y = f(x), \qquad y(0) = y(1) = 0, \quad 0 \le x \le 1,$$

under the assumptions listed on page 546, the Galerkin and Rayleigh–Ritz methods are both determined by Eq. (10.40). This is not the case for an arbitrary boundary-value problem, however. A treatment of the similarities and differences in the two methods and a discussion of the wide application of the Galerkin method can be found in Schultz [119] and in Strang and Fix [132].

Another popular technique for solving boundary-value problems is called the method of collocation. This procedure begins by selecting a set of basis functions $\{\phi_1, ..., \phi_N\}$, a set of numbers $\{x_1, ..., x_n\}$ in $[0, 1]$, and requiring that an approximation

$$\sum_{i=1}^{N} c_i \phi_i(x)$$

satisfy the differential equation at each of the numbers x_j for $1 \le j \le n$. If, in addition, we require that $\phi_i(0) = \phi_i(1) = 0$ for $1 \le i \le N$, the boundary conditions are automatically satisfied. Much attention in the literature has been given to the choice of the numbers $\{x_j\}$ and the basis functions $\{\phi_i\}$. One of the more popular choices is to let the ϕ_i be the basis functions for cubic or quintic spline functions relative to a partition of $[0, 1]$, and to let the nodes $\{x_j\}$ be the Gaussian points or roots of certain orthogonal polynomials, transformed to the proper subinterval. A comparison of various collocation methods and finite difference methods is contained in Russell [111]. His conclusion is that the collocation methods using higher-degree splines are competitive with finite difference techniques using extrapolation. Other references for collocation methods are DeBoor and Swartz [34] and Lucas and Reddien [83].

Exercise Set 10.5

1. Use Algorithm 10.5 to approximate the solution to each of the following boundary-value problems.

   a) $y'' + \frac{\pi^2}{4}y = \frac{\pi^2}{16}\cos\frac{\pi}{4}x, \quad 0 \le x \le 1, \quad y(0) = 0, \quad y(1) = 0;$

   use $x_0 = 0$, $x_1 = 0.3$, $x_2 = 0.7$, and $x_3 = 1$.

   b) $-\frac{d}{dx}(xy') + 4y = 4x^2 - 8x + 1, \quad 0 \le x \le 1, \quad y(0) = 0, \quad y(1) = 0;$

   use $x_0 = 0$, $x_1 = 0.4$, $x_2 = 0.8$, and $x_3 = 1$.

2. Use Algorithm 10.6 with $n = 3$ to approximate the solution to each of the following boundary-value problems.

   a) $y'' + \dfrac{\pi^2}{4}y = \dfrac{\pi^2}{16}\cos\dfrac{\pi}{4}x, \quad 0 \le x \le 1, \quad y(0) = 0, \quad y(1) = 0.$

   b) $-\dfrac{d}{dx}(xy') + 4y = 4x^2 - 8x + 1, \quad 0 \le x \le 1, \quad y(0) = 0, \quad y(1) = 0.$

3. Show that the following boundary-value problems satisfy the hypotheses of Theorem 10.4 and approximate their solutions using Algorithm 10.5.

   a) $-y'' + y = x, \quad 0 \le x \le 1, \quad y(0) = 0, \quad y(1) = 0;$

   use $h = 0.1$, and compare the results to the actual solution

   $$y = x + \left(\dfrac{e}{e^2 - 1}\right)(e^{-x} - e^x).$$

   b) $-x^2y'' - 2xy' + 2y = -4x^2, \quad 0 \le x \le 1, \quad y(0) = 0, \quad y(1) = 0;$

   use $h = 0.05$, and compare the results to the actual solution $y = x^2 - x$.

   c) $-\dfrac{d}{dx}(e^x y') + e^x y = x + (2 - x)e^x, \quad 0 \le x \le 1, \quad y(0) = y(1) = 0;$

   use $h = 0.1$ and compare the results to the actual solution $y = (x - 1)(e^{-x} - 1)$.

   d) $-\dfrac{d}{dx}(e^{-x}y') + e^{-x}y = (x - 1) - (x + 1)e^{-(x-1)}, \quad 0 \le x \le 1, \quad y(0) = y(1) = 0;$

   use $h = 0.05$ and compare the results to the actual solution $y = x(e^x - e)$.

4. Repeat Exercise 3 using Algorithm 10.6.

5. Apply Algorithm 10.5 to the following boundary-value problems, which do not satisfy the hypotheses of Theorem 10.4.

   a) $-y'' - y = x, \quad 0 \le x \le 1, \quad y(0) = 0, \quad y(1) = 0;$

   use $h = 0.1$, and compare the results to the actual solution.

   b) $-y'' = 1, \quad 0 \le x \le 1, \quad y(0) = 0, \quad y(1) = 0;$

   use $h = 0.1$, and compare the results to the actual solution.

6. Show that $\{\phi_i\}_{i=1}^n$ is a linearly independent set of functions on $[0, 1]$ for the functions ϕ_i defined in Eq. (10.41).

7. Repeat Exercise 6 using the cubic spline basis $\{\phi_i\}_{i=0}^{n+1}$.

8. Show that the matrix given by the piecewise-linear basis functions is positive definite.

9. Using $\phi_j(x) = \sin(j\pi x)$ for each $j = 1, 2, \ldots, 5$, find a Rayleigh–Ritz approximation for the boundary-value problem

   $$-y'' + 4y = 4x(1 - e^{-2}), \qquad 0 \le x \le 1, \quad y(0) = 0, \quad y(1) = 0.$$

10. Show that the matrix A given by the piecewise-linear basis functions for the boundary-value problems (10.33) and (10.34) with $p(x) = 1$ and $q(x) = 0$ is the same as the matrix obtained in the finite-difference method.

11. Show that the boundary-value problem

$$-\frac{d}{dx}(p(x)y') + q(x)y = f(x), \qquad 0 \le x \le 1, \quad y(0) = \alpha, \quad y(1) = \beta$$

can be transformed by the change of variable

$$z = y - \beta x - (1 - x)\alpha$$

into the form

$$-\frac{d}{dx}(p(x)z') + q(x)z = F(x), \qquad 0 \le x \le 1, \quad z(0) = 0, \quad z(1) = 0.$$

12. Using Exercise 11 and Algorithm 10.5 with $n = 9$, approximate the solution to the boundary-value problem

$$-y'' + y = x, \qquad 0 \le x \le 1, \quad y(0) = 1, \quad y(1) = 1 + e^{-1}.$$

13. Show that the boundary-value problem

$$-\frac{d}{dx}(p(x)y') + q(x)y = f(x), \qquad a \le x \le b, \quad y(a) = \alpha, \quad y(b) = \beta,$$

can be transformed into the form of Eqs. (10.33) and (10.34) by a method similar to that given in Exercise 11.

14. Show that the matrix A in (10.43) is positive definite.

15. Repeat Exercise 12 using Algorithm 10.6.

11

Numerical Solutions to Partial-Differential Equations

A body is called *isotropic* if the thermal conductivity at each point in the body is independent of the direction of heat flow through the point. The temperature, $u \equiv u(x, y, z, t)$ in an isotropic body can be found by solving the partial-differential equation

$$\frac{\partial}{\partial x}\left(k\frac{\partial u}{\partial x}\right) + \frac{\partial}{\partial y}\left(k\frac{\partial u}{\partial y}\right) + \frac{\partial}{\partial z}\left(k\frac{\partial u}{\partial z}\right) = c\rho\frac{\partial u}{\partial t},$$

where k, c, and ρ are functions of (x, y, z) and represent, respectively, the thermal conductivity, specific heat, and density of the body at the point (x, y, z).

When k, c, and ρ are constants, this equation is known as the simple three-dimensional heat equation, and can be expressed as

$$\frac{\partial^2 u}{\partial x^2} + \frac{\partial^2 u}{\partial y^2} + \frac{\partial^2 u}{\partial z^2} = \frac{c\rho}{k}\frac{\partial u}{\partial t}.$$

The solution to this equation can be found by using Fourier series if the boundary of the body is relatively simple.

In most situations where k, c, and ρ are not constant, or when the boundary is irregular, the solution to the equation must be obtained by approximation techniques. An introduction to techniques of this type is presented in this chapter.

Physical situations involving more than one variable can often be expressed using equations involving partial derivatives. In this chapter we present a brief introduction to some of the techniques available for approximating the solution to partial-differential equations involving two variables by showing how these techniques can be applied to certain standard physical problems. We limit our treatment to problems of this type because most of the more advanced techniques require a stronger background in analysis than this book assumes.

11.1 Physical Problems Involving Partial-Differential Equations

In Section 11.2 we consider an **elliptic** partial-differential equation known as the **Poisson equation**:

(11.1) $$\frac{\partial^2 u}{\partial x^2}(x,\ y) + \frac{\partial^2 u}{\partial y^2}(x,\ y) = f(x,\ y).$$

We assume, in this equation, that the function f describes the input to the problem on a plane region R whose boundary we will denote by S. Equations of this type arise naturally in the study of various time-independent physical problems such as the steady-state distribution of heat in a plane region, the potential energy of a point in a plane acted on by gravitational forces in the plane, and two-dimensional steady-state problems involving incompressible fluids.

To obtain a unique solution to the Poisson equation, additional constraints must be placed on the solution. For example, the study of the steady-state distribution of heat in a plane region requires that $f(x, y) \equiv 0$, resulting in a simplification of Eq. (11.1) to:

(11.2) $$\frac{\partial^2 u}{\partial x^2}(x,\ y) + \frac{\partial^2 u}{\partial y^2}(x,\ y) = 0,$$

which is called **Laplace's equation**. If the temperature within the region is determined by the temperature distribution on the boundary of the region, the constraints are called the **Dirichlet boundary conditions**, given by

(11.3) $$u(x,\ y) = g(x,\ y)$$

for all $(x,\ y)$ on S, the boundary of the region R. (*See Fig. 11.1.*)

FIGURE 11.1

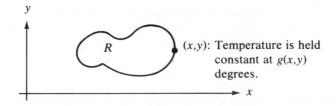

(x,y): Temperature is held constant at $g(x,y)$ degrees.

In Section 11.3 we consider the numerical solution to a problem involving a **parabolic** partial-differential equation of the form

(11.4)
$$\frac{\partial u}{\partial t}(x,\ t) - \alpha^2 \frac{\partial^2 u}{\partial x^2}(x,\ t) = 0.$$

The physical problem considered here concerns the flow of heat along a rod of length l (*see Fig. 11.2*), which is assumed to have a uniform temperature within each cross-sectional element. This condition requires the rod to be perfectly insulated on its lateral surface. The constant α in Eq. (11.4) is determined by the heat-conductive properties of the material of which the rod is composed and is assumed to be independent of the position in the rod.

FIGURE 11.2

0 l x

One of the typical sets of constraints for a heat-flow problem of this type is to specify the initial heat distribution in the rod

$$u(x,\ 0) = f(x),$$

and describe the behavior at the ends of the rod. For example, if we held the ends of the rod at constant temperatures U_1 and U_2, the boundary conditions would have the form

$$u(0,\ t) = U_1 \quad \text{and} \quad u(l,\ t) = U_2,$$

and the heat distribution in the rod would approach the limiting temperature distribution

$$\lim_{t \to \infty} u(x,\ t) = U_1 + \frac{U_2 - U_1}{l}\,x.$$

If instead we insulated the rod so that no heat would flow through the ends, the boundary conditions would be

$$\frac{\partial u}{\partial x}(0,\ t) = 0 \quad \text{and} \quad \frac{\partial u}{\partial x}(l,\ t) = 0,$$

and result in a constant temperature in the rod as the limiting case.

The parabolic partial-differential equation is also of importance in the study of gas diffusion and, in fact, Eq. (11.4) is known in some circles as the **diffusion equation**.

The problem studied in Section 11.4 is called the one-dimensional **wave equation** and is an example of a **hyperbolic** partial-differential equation.

Suppose that an elastic string of length l is stretched between two supports at the same horizontal level (*see Fig. 11.3*). If the string is set in motion so that it vibrates in a vertical plane, the vertical displacement $u(x, t)$ of a point x at time t satisfies the partial-differential equation

(11.5)
$$\alpha^2 \frac{\partial^2 u}{\partial x^2}(x,\ t) = \frac{\partial^2 u}{\partial t^2}(x,\ t), \qquad 0 < x < l, \quad 0 < t,$$

FIGURE 11.3

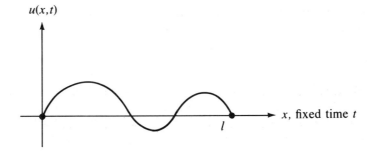

provided that damping effects are neglected and the amplitude is not too large. To impose constraints on this problem, assume that the initial position and velocity of the string are given by

$$u(x, 0) = f(x) \qquad \text{and} \qquad \frac{\partial u}{\partial t}(x, 0) = g(x), \qquad 0 \le x \le l;$$

and use the fact that the endpoints are fixed. This implies that $u(0, t) = 0$ and $u(l, t) = 0$.

Other physical problems involving the hyperbolic partial-differential equation (11.5) occur in the study of vibrating beams with one or both ends clamped, and in the transmission of electricity in a long transmission line in which there is some leakage of the current to the ground.

Exercise Set 11.1

The following problems summarize the Fourier-series solutions to the partial-differential equations of this section. It will be assumed that each series converges, and that the series obtained by termwise differentiations converge to the appropriate partial derivatives.

1. **Laplace's Equation on a Rectangle**   Assume that the functions $f_1(x)$, $f_2(x)$, $g_1(y)$, and $g_2(y)$ have the Fourier series:

$$f_1(x) = \sum_{n=1}^{\infty} A_n \sin\left(\frac{n\pi x}{a}\right), \qquad \text{where } A_n = \frac{2}{a}\int_0^a f_1(x)\sin\left(\frac{n\pi x}{a}\right)dx;$$

$$f_2(x) = \sum_{n=1}^{\infty} B_n \sin\left(\frac{n\pi x}{a}\right), \qquad \text{where } B_n = \frac{2}{a}\int_0^a f_2(x)\sin\left(\frac{n\pi x}{a}\right)dx;$$

$$g_1(y) = \sum_{n=1}^{\infty} C_n \sin\left(\frac{n\pi y}{b}\right), \qquad \text{where } C_n = \frac{2}{b}\int_0^b g_1(y)\sin\left(\frac{n\pi y}{b}\right)dy;$$

and

$$g_2(y) = \sum_{n=1}^{\infty} D_n \sin\left(\frac{n\pi y}{b}\right), \qquad \text{where } D_n = \frac{2}{b}\int_0^b g_2(y)\sin\left(\frac{n\pi y}{b}\right)dy.$$

Show that Laplace's equation on a rectangle.

$$\frac{\partial^2 u}{\partial x^2} + \frac{\partial^2 u}{\partial y^2} = 0, \qquad 0 < x < a, \quad 0 < y < b,$$

with boundary conditions

$$u(x, 0) = f_1(x), \quad u(x, b) = f_2(x), \qquad 0 \le x \le a,$$
$$u(0, y) = g_1(y), \quad u(a, y) = g_2(y), \qquad 0 \le y \le b;$$

has the solution

$$u(x, y) = \sum_{n=1}^{\infty} \left\{ \left[A'_n \sinh\left(\frac{n\pi(b - y)}{a}\right) + B'_n \sinh\left(\frac{n\pi y}{a}\right) \right] \sin\left(\frac{n\pi x}{a}\right) \right.$$
$$\left. + \left[C'_n \sinh\left(\frac{n\pi(a - x)}{b}\right) + D'_n \sinh\left(\frac{n\pi x}{b}\right) \right] \sin\left(\frac{n\pi y}{b}\right) \right\},$$

where

$$A'_n = \frac{A_n}{\sinh\left(\dfrac{n\pi b}{a}\right)}, \quad B'_n = \frac{B_n}{\sinh\left(\dfrac{n\pi b}{a}\right)}, \quad C'_n = \frac{C_n}{\sinh\left(\dfrac{n\pi a}{b}\right)}, \quad \text{and} \quad D'_n = \frac{D_n}{\sinh\left(\dfrac{n\pi a}{b}\right)}.$$

2. **Heat Equation**   Assume that the function f has the Fourier series

$$f(x) = \sum_{n=1}^{\infty} A_n \sin\frac{n\pi x}{l} \quad \text{where } A_n = \frac{2}{l} \int_0^l f(x) \sin\frac{n\pi x}{l}\, dx.$$

a)   Show that the solution to the heat equation

$$\frac{\partial u}{\partial t} - \alpha^2 \frac{\partial^2 u}{\partial x^2} = 0, \qquad 0 < x < l, \quad 0 < t$$

with boundary conditions

$$u(0, t) = u(l, t) = 0, \qquad 0 < t$$

and initial condition

$$u(x, 0) = f(x), \qquad 0 \le x \le l$$

is given by

$$u(x, t) = \sum_{n=1}^{\infty} A_n \sin\left(\frac{n\pi x}{l}\right) \exp\left(\frac{-\alpha^2 n^2 \pi^2}{l^2} t\right).$$

b)   Let $w(x) = U_1 + x(U_2 - U_1)/l$. Show that if the boundary conditions in part (a) are replaced by

$$u(0, t) = U_1, \qquad u(l, t) = U_2,$$

then the solution is given by

$$u(x, t) = w(x) + \sum_{n=1}^{\infty} B_n \sin\left(\frac{n\pi x}{l}\right) \exp\left(\frac{-\alpha^2 n^2 \pi^2}{l^2} t\right),$$

where

$$B_n = \frac{2}{l} \int_0^l [f(x) - w(x)] \sin\left(\frac{n\pi x}{l}\right) dx.$$

3. **Heat Equation**   Assume that the Fourier series for $f(x)$ is given by

$$f(x) = A_0 + \sum_{n=1}^{\infty} A_n \cos\frac{n\pi x}{l},$$

where $\qquad A_0 = \frac{1}{l}\int_0^l f(x)\,dx \qquad$ and $\qquad A_n = \frac{2}{l}\int_0^l f(x)\cos\frac{n\pi x}{l}\,dx,$

for each $n = 1, 2, \dots$. Show that the solution to the heat equation

$$\frac{\partial u}{\partial t} - \alpha^2 \frac{\partial^2 u}{\partial x^2} = 0, \qquad 0 < x < l, \quad 0 < t,$$

with boundary conditions

$$\frac{\partial u}{\partial x}(0, t) = \frac{\partial u}{\partial x}(l, t) = 0, \qquad 0 < t,$$

and initial condition

$$u(x, 0) = f(x), \qquad 0 \le x \le l,$$

is given by

$$u(x, t) = A_0 + \sum_{n=1}^{\infty} A_n \cos\left(\frac{n\pi x}{l}\right)\exp\left(\frac{-\alpha^2 n^2 \pi^2}{l^2} t\right).$$

4. **Wave Equation**   Assume that the functions $f(x)$ and $g(x)$ have Fourier series given by:

$$f(x) = \sum_{n=1}^{\infty} A_n \sin\left(\frac{n\pi x}{l}\right), \qquad \text{where } A_n = \frac{2}{l}\int_0^l f(x)\sin\left(\frac{n\pi x}{l}\right)dx,$$

and $\qquad g(x) = \sum_{n=1}^{\infty} B_n \sin\left(\frac{n\pi x}{l}\right), \qquad \text{where } B_n = \frac{2}{l}\int_0^l g(x)\sin\left(\frac{n\pi x}{l}\right)dx.$

Show that the solution to the wave problem,

$$\frac{\partial^2 u}{\partial t^2} - \alpha^2 \frac{\partial^2 u}{\partial x^2} = 0, \qquad 0 < x < l, \quad 0 < t,$$

$$u(0, t) = u(l, t) = 0, \qquad 0 < t,$$

$$u(x, 0) = f(x), \qquad 0 \le x \le l,$$

$$\frac{\partial u}{\partial t}(x, 0) = g(x), \qquad 0 \le x \le l,$$

is given by

$$u(x, t) = \sum_{n=1}^{\infty}\left[A_n \cos\left(\frac{\alpha n\pi}{l}t\right) + B'_n \sin\left(\frac{\alpha n\pi}{l}t\right)\right]\sin\left(\frac{n\pi x}{l}\right),$$

where $B'_n = (l/\alpha n\pi)B_n$.

11.2 Elliptic Partial-Differential Equations

The elliptic partial-differential equation we consider is the Poisson equation,

$$(11.6) \qquad \nabla^2 u(x, y) \equiv \frac{\partial^2 u}{\partial x^2}(x, u) + \frac{\partial^2 u}{\partial y^2}(x, y) = f(x, y)$$

for $(x, y) \in R$ and

$$u(x, y) = g(x, y) \qquad \text{for } (x, y) \in S,$$

where
$$R = \{(x, y) | a < x < b, \, c < y < d\},$$

and S denotes the boundary of R. For this discussion we assume that both f and g are continuous on their domains so a unique solution to Eq. (11.6) is ensured.

The method to be used is an adaptation of the finite-difference method for boundary-value problems, which was discussed in Section 10.3. The first step is to choose integers n and m, and define step sizes h and k by $h = (b - a)/n$ and $k = (d - c)/m$. Partitioning the interval $[a, b]$ into n equal parts of width h and the interval $[c, d]$ into m equal parts of width k (*see Fig. 11.4*) provides a means of placing a grid on the rectangle R by drawing vertical and horizontal lines through the points with coordinates (x_i, y_j), where

$$x_i = a + ih \qquad \text{for each } i = 0, 1, \ldots, n,$$

and
$$y_j = c + jk \qquad \text{for each } j = 0, 1, \ldots, m.$$

FIGURE 11.4

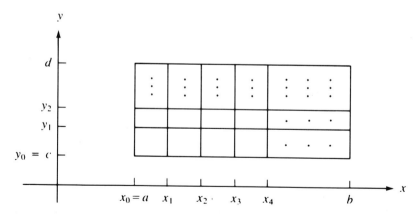

The lines $x = x_i$ and $y = y_j$ are called **grid lines** and their intersections are called the **mesh points** of the grid. For each mesh point in the interior of the grid, (x_i, y_j), $i = 1, 2, \ldots, n - 1$, and $j = 1, 2, \ldots, m - 1$, we use the Taylor series in the variable x about x_i to generate the central-difference formula

$$(11.7) \qquad \frac{\partial^2 u}{\partial x^2}(x_i, y_j) = \frac{u(x_{i+1}, y_j) - 2u(x_i, y_j) + u(x_{i-1}, y_j)}{h^2} - \frac{h^2}{12} \frac{\partial^4 u}{\partial x^4}(\xi_i, y_j),$$

where $\xi_i \in (x_{i-1}, x_{i+1})$; and the Taylor series in the variable y and y_j to generate the central-difference formula

(11.8) $\dfrac{\partial^2 u}{\partial y^2}(x_i, y_j) = \dfrac{u(x_i, y_{j+1}) - 2u(x_i, y_j) + u(x_i, y_{j-1})}{k^2} - \dfrac{k^2}{12}\dfrac{\partial^4 u}{\partial y^4}(x_i, \eta_j),$

where $\eta_j \in (y_{j-1}, y_{j+1})$.

Using these formulas in Eq. (11.6) allows us to express the Poisson equation at the points (x_i, y_j) as:

(11.9)

$$\dfrac{u(x_{i+1}, y_j) - 2u(x_i, y_j) + u(x_{i-1}, y_j)}{h^2} + \dfrac{u(x_i, y_{j+1}) - 2u(x_i, y_j) + u(x_i, y_{j-1})}{k^2}$$

$$= f(x_i, y_j) + \dfrac{h^2}{12}\dfrac{\partial^4 u}{\partial x^4}(\xi_i, y_j) + \dfrac{k^2}{12}\dfrac{\partial^4 u}{\partial y^4}(x_i, \eta_j),$$

for each $i = 1, 2, \ldots, (n-1)$ and $j = 1, 2, \ldots, (m-1)$, and the boundary conditions as

(11.10)

$$u(x_0, y_j) = g(x_0, y_j) \qquad \text{for each } j = 0, 1, \ldots, m,$$
$$u(x_n, y_j) = g(x_n, y_j) \qquad \text{for each } j = 0, 1, \ldots, m,$$
$$u(x_i, y_0) = g(x_i, y_0) \qquad \text{for each } i = 1, 2, \ldots, n-1,$$
$$u(x_i, y_m) = g(x_i, y_m) \qquad \text{for each } i = 1, 2, \ldots, n-1.$$

In difference-equation form, this results in a method, called the **central-difference** method, with local truncation error of order $O(h^2 + k^2)$, that can be written:

(11.11)

$$2\left[\left(\dfrac{h}{k}\right)^2 + 1\right]w_{i,j} - (w_{i+1,j} + w_{i-1,j}) - \left(\dfrac{h}{k}\right)^2(w_{i,j+1} + w_{i,j-1}) = -h^2 f(x_i, y_j),$$

for each $i = 1, 2, \ldots, n-1$ and $j = 1, 2, \ldots, m-1$, and

(11.12)

$$w_{0,j} = g(x_0, y_j) \qquad \text{for each } j = 0, 1, \ldots, m,$$
$$w_{n,j} = g(x_n, y_j) \qquad \text{for each } j = 0, 1, \ldots, m,$$
$$w_{i,0} = g(x_i, y_0) \qquad \text{for each } i = 1, 2, \ldots, n-1$$
$$w_{i,m} = g(x_i, y_m) \qquad \text{for each } i = 1, 2, \ldots, n-1,$$

where $w_{i,j}$ approximates $u(x_i, y_j)$.

The typical equation in (11.11) involves approximations to $u(x, y)$ at the points

$$(x_{i-1}, y_j), \quad (x_i, y_j), \quad (x_{i+1}, y_j), \quad (x_i, y_{j-1}), \quad \text{and} \quad (x_i, y_{j+1}).$$

Reproducing the portion of the grid where these points are located (Fig. 11.5) shows that each equation involves approximations in a star-shaped region about (x_i, y_j).

If we use the information from the boundary conditions (11.12) whenever appropriate in the system given by (11.11) — that is, at all points (x_i, y_j) that are adjacent to a boundary mesh point — we have an $(n-1)(m-1)$ system of linear equations in $(n-1)(m-1)$ unknowns, the unknowns being the approximations $w_{i,j}$ to $u(x_i, y_j)$ for the interior mesh points.

FIGURE 11.5

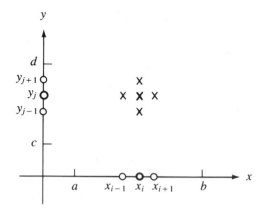

The linear system involving these unknowns can be expressed for matrix calculations more efficiently if a relabeling of the interior mesh points is introduced. A recommended labeling of these points (see Varga [135], page 187) is to let

$$P_l = (x_i, y_j) \qquad \text{and} \qquad w_l = w_{i,j},$$

where $l = i + (m - 1 - j)(n - 1)$ for each

$$i = 1, 2, ..., n - 1 \qquad \text{and} \qquad j = 1, 2, ..., m - 1.$$

This in effect labels the mesh points consecutively from left to right and top to bottom. For example, with $n = 4$ and $m = 5$, the relabeling would result in a grid whose points are shown in Fig. 11.6. Labeling the points in this manner ensures that the system needed to determine the $w_{i,j}$ will be a banded matrix with band width at most $2n - 1$.

FIGURE 11.6

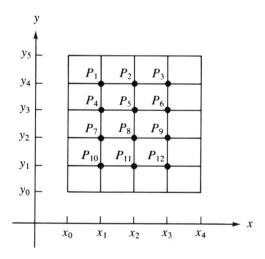

EXAMPLE 1   Consider the problem of determining the steady-state heat distribution in a thin metal plate in the shape of a square with dimensions 0.5 meters by 0.5 meters, which is held at 0° Celsius on two adjacent boundaries while the heat on the other boundaries

increases linearly from 0° Celsius at one corner to 100° Celsius where these sides meet. If we place the sides with zero boundary conditions along the x- and y-axes, the problem is expressed mathematically as:

(11.13)
$$\frac{\partial^2 u}{\partial x^2}(x, y) + \frac{\partial^2 u}{\partial y^2}(x, y) = 0,$$

for (x, y) in the set $R = \{(x, y) | 0 < x < 0.5, 0 < y < 0.5\}$, with the boundary conditions

(11.14) $u(0, y) = 0, \quad u(x, 0) = 0, \quad u(x, 0.5) = 200x, \quad u(0.5, y) = 200y.$

If $n = m = 4$, the problem has the grid given in Fig. 11.7, and the difference equation (11.11) is

$$4w_{i,j} - w_{i+1,j} - w_{i-1,j} - w_{i,j-1} - w_{i,j+1} = 0,$$

for each $i = 1, 2, 3$, and $j = 1, 2, 3$.

FIGURE 11.7

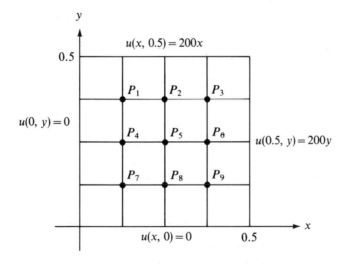

Expressing this in terms of the relabeled interior grid points $w_i = u(P_i)$ implies that the equations are:

$$4w_1 - w_2 - w_4 = w_{0,3} + w_{1,4},$$
$$4w_2 - w_3 - w_1 - w_5 = w_{2,4},$$
$$4w_3 - w_2 - w_6 = w_{4,3} + w_{3,4},$$
$$4w_4 - w_5 - w_1 - w_7 = w_{0,2},$$
$$4w_5 - w_6 - w_4 - w_2 - w_8 = 0,$$
$$4w_6 - w_5 - w_3 - w_9 = w_{4,2},$$
$$4w_7 - w_8 - w_4 = w_{0,1} + w_{1,0},$$
$$4w_8 - w_9 - w_7 - w_5 = w_{2,0},$$
$$4w_9 - w_8 - w_6 = w_{3,0} + w_{4,1},$$

where the right-hand sides of the equations are obtained from the boundary conditions. In fact, the conditions given in (11.14) imply that

$$w_{1,0} = w_{2,0} = w_{3,0} = w_{0,1} = w_{0,2} = w_{0,3} = 0,$$

$$w_{1,4} = w_{4,1} = 25, \quad w_{2,4} = w_{4,2} = 50, \quad \text{and} \quad w_{3,4} = w_{4,3} = 75.$$

The linear system associated with this problem has the form

$$
\begin{bmatrix}
4 & -1 & 0 & -1 & 0 & 0 & 0 & 0 & 0 \\
-1 & 4 & -1 & 0 & -1 & 0 & 0 & 0 & 0 \\
0 & -1 & 4 & 0 & 0 & -1 & 0 & 0 & 0 \\
-1 & 0 & 0 & 4 & -1 & 0 & -1 & 0 & 0 \\
0 & -1 & 0 & -1 & 4 & -1 & 0 & -1 & 0 \\
0 & 0 & -1 & 0 & -1 & 4 & 0 & 0 & -1 \\
0 & 0 & 0 & -1 & 0 & 0 & 4 & -1 & 0 \\
0 & 0 & 0 & 0 & -1 & 0 & -1 & 4 & -1 \\
0 & 0 & 0 & 0 & 0 & -1 & 0 & -1 & 4
\end{bmatrix}
\begin{bmatrix}
w_1 \\ w_2 \\ w_3 \\ w_4 \\ w_5 \\ w_6 \\ w_7 \\ w_8 \\ w_9
\end{bmatrix}
=
\begin{bmatrix}
25 \\ 50 \\ 150 \\ 0 \\ 0 \\ 50 \\ 0 \\ 0 \\ 25
\end{bmatrix}
$$

The values of $w_1, w_2, \ldots, w_9$, found by applying the Gauss–Seidel method to this matrix, are given as follows:

i	1	2	3	4	5	6	7	8	9
w_i	18.75	37.50	56.25	12.50	25.00	37.50	6.25	12.50	18.75

These answers are correct, since the solution $u(x, y) = 400xy$ has

$$\frac{\partial^4 u}{\partial x^4} = \frac{\partial^4 u}{\partial y^4} \equiv 0;$$

so the truncation error is zero at each step. ∎

The problem we considered in Example 1 has a mesh size 0.125 on each axis and requires solving only a 9×9 linear system. This simplifies the situation considerably, and does not introduce the computational problems that are present when the system is much larger. Algorithm 11.1 uses the Gauss–Seidel iterative method for solving the linear system that is produced and allows for unequal mesh sizes on the axes.

Poisson Equation Finite-Difference Algorithm 11.1

To approximate the solution to the Poisson equation

$$\frac{\partial^2 u}{\partial x^2}(x, t) + \frac{\partial^2 u}{\partial y^2}(x, t) = f(x, y), \qquad a \le x \le b, \quad c \le y \le d,$$

subject to the boundary conditions

$$u(x, y) = g(x, y) \qquad \text{if } x = a \text{ or } x = b \text{ and } c \leq y \leq d$$

and $\qquad u(x, y) = g(x, y) \qquad \text{if } y = c \text{ or } y = d \text{ and } a \leq x \leq b:$

INPUT endpoints a, b, c, d; integers m, n; tolerance TOL; maximum number of iterations $LBOUND$.

OUTPUT approximations $w_{i,j}$ to $u(x_i, y_j)$ for each $i = 1, \ldots, n-1$ and $j = 1, \ldots, m-1$ or a message that the maximum number of iterations was exceeded.

Step 1   Set $h = (b - a)/n$;
$\qquad\qquad k = (d - c)/m$.

Step 2   For $i = 1, \ldots, n-1$ set $x_i = a + ih$.   (*Steps 2 and 3 construct mesh points.*)

Step 3   For $j = 1, \ldots, m-1$ set $y_j = c + jk$.

Step 4   For $i = 1, \ldots, n-1$
$\qquad\qquad$ for $j = 1, \ldots, m-1$ set $w_{i,j} = 0$.

Step 5   Set $\lambda = h^2/k^2$;
$\qquad\qquad \mu = 2(1 + \lambda)$;
$\qquad\qquad l = 1$.

Step 6   While $l \leq LBOUND$ do Steps 7–20.   (*Steps 7–20 perform Gauss–Seidel iterations.*)

    Step 7   Set $z = (-h^2 f(x_1, y_{m-1}) + g(a, y_{m-1}) + \lambda g(x_1, d) + \lambda w_{1,m-2} + w_{2,m-1})/\mu$;
$\qquad\qquad\qquad NORM = |z - w_{1,m-1}|$
$\qquad\qquad\qquad w_{1,m-1} = z$.

    Step 8   For $i = 2, \ldots, n-2$
$\qquad\qquad\qquad$ set $z = (-h^2 f(x_i, y_{m-1}) + \lambda g(x_i, d) + w_{i-1,m-1}$
$\qquad\qquad\qquad\qquad + w_{i+1,m-1} + \lambda w_{i,m-2})/\mu$;
$\qquad\qquad\qquad$ if $|w_{i,m-1} - z| > NORM$ then set $NORM = |w_{i,m-1} - z|$;
$\qquad\qquad\qquad$ set $w_{i,m-1} = z$.

    Step 9   Set $z = (-h^2 f(x_{n-1}, y_{m-1}) + g(b, y_{m-1}) + \lambda g(x_{n-1}, d)$
$\qquad\qquad\qquad\qquad + w_{n-2,m-1} + \lambda w_{n-1,m-2})/\mu$;
$\qquad\qquad\qquad$ if $|w_{n-1,m-1} - z| > NORM$ then set $NORM = |w_{n-1,m-1} - z|$;
$\qquad\qquad\qquad$ set $w_{n-1,m-1} = z$.

    Step 10   For $j = m-2, \ldots, 2$ do Steps 11, 12, and 13.

        Step 11   Set $z = (-h^2 f(x_1, y_j) + g(a, y_j) + \lambda w_{1,j+1} + \lambda w_{1,j-1} + w_{2,j})/\mu$;
$\qquad\qquad\qquad\qquad$ if $|w_{1,j} - z| > NORM$ then set $NORM = |w_{1,j} - z|$;
$\qquad\qquad\qquad\qquad$ set $w_{1,j} = z$.

        Step 12   For $i = 2, \ldots, n-2$
$\qquad\qquad\qquad\qquad\qquad$ set $z = (-h^2 f(x_i, y_j) + w_{i-1,j} + \lambda w_{i,j+1}$
$\qquad\qquad\qquad\qquad\qquad\qquad + w_{i+1,j} + \lambda w_{i,j-1})/\mu$;
$\qquad\qquad\qquad\qquad\qquad$ if $|w_{i,j} - z| > NORM$ then set $NORM = |w_{i,j} - z|$;
$\qquad\qquad\qquad\qquad\qquad$ set $w_{i,j} = z$.

Step 13   Set $z = (-h^2 f(x_{n-1}, y_j) + g(b, y_j) + w_{n-2,j}$
$+ \lambda w_{n-1,j+1} + \lambda w_{n-1,j-1})/\mu;$
if $|w_{n-1,j} - z| > NORM$ then set $NORM = |w_{n-1,j} - z|;$
set $w_{n-1,j} = z.$

Step 14   Set $z = (-h^2 f(x_1, y_1) + g(a, y_1) + \lambda g(x_1, c) + \lambda w_{1,2} + w_{2,1})/\mu;$
if $|w_{1,1} - z| > NORM$ then set $NORM = |w_{1,1} - z|;$
set $w_{1,1} = z.$

Step 15   For $i = 2, \ldots, n-2$
set $z = (-h^2 f(x_i, y_1) + \lambda g(x_i, c) + w_{i-1,1} + \lambda w_{i,2} + w_{i+1,1})/\mu;$
if $|w_{i,1} - z| > NORM$ then set $NORM = |w_{i,1} - z|;$
set $w_{i,1} = z.$

Step 16   Set $z = (-h^2 f(x_{n-1}, y_1) + g(b, y_1) + \lambda g(x_{n-1}, c) + w_{n-2,1}$
$+ \lambda w_{n-1,2})/\mu;$
if $|w_{n-1,1} - z| > NORM$ then set $NORM = |w_{n-1,1} - z|;$
set $w_{n-1,1} = z.$

Step 17   If $NORM \le TOL$ then do Steps 18 and 19.

Step 18   For $i = 1, \ldots, n-1$
for $j = 1, \ldots, m-1$ OUTPUT $(x_i, y_j, w_{i,j}).$

Step 19   STOP.   (*Procedure completed successfully.*)

Step 20   Set $l = l + 1.$

Step 21   OUTPUT ('Maximum number of iterations exceeded');
STOP.

Although the Gauss–Seidel iterative procedure is incorporated into Algorithm 11.1 for simplicity, it is generally advisable to use a direct technique such as Gaussian elimination when the system is small, on the order of 100 or less, since the symmetry and positive definiteness will ensure stability with respect to rounding errors. In particular, the generalization of the Crout Reduction Algorithm 6.7, which is discussed in Exercise 13 of Section 6.7, is very efficient for solving this system, since the matrix is in symmetric-block tridiagonal form

$$
\begin{bmatrix}
A_1 & C_1 & 0 & \cdots & & & 0 \\
C_1 & A_2 & C_2 & & & & \\
0 & C_2 & & & & & \\
\vdots & & & & & & 0 \\
& & & & & & C_{m-1} \\
0 & \cdots & & 0 & & C_{m-1} & A_{m-1}
\end{bmatrix}
$$

with square blocks of size $(n-1)$ by $(n-1)$.

For very large systems it is recommended that an iterative method be used, specifically, the SOR method discussed in Algorithm 8.3. The choice of ω that is optimal in this situation comes from the fact that when A is decomposed into its diagonal D and upper- and lower-triangular parts U and L.

$$A = D - L - U,$$

and B is the Jacobi matrix,

$$B = D^{-1}(L + U),$$

then the spectral radius of B is (see Varga [135], page 203)

$$\rho(B) = \frac{1}{2}\left[\cos\left(\frac{\pi}{m}\right) + \cos\left(\frac{\pi}{n}\right)\right].$$

The value of ω to be used is consequently

$$\omega = \frac{2}{1 + \sqrt{1 - [\rho(B)]^2}} = \frac{4}{2 + \sqrt{4 - \left[\cos\left(\frac{\pi}{m}\right) + \cos\left(\frac{\pi}{n}\right)\right]^2}}.$$

For faster convergence of the SOR procedure, a block technique can be incorporated into the algorithm. For a presentation of the technique involved, see Varga [135], pp. 194–199.

EXAMPLE 2   Consider Poisson's equation

$$\frac{\partial^2 u}{\partial x^2}(x, y) + \frac{\partial^2 u}{\partial y^2}(x, y) = xe^y, \qquad 0 < x < 2, \quad 0 < y < 1,$$

with the boundary conditions

$$u(0, y) = 0, \qquad u(2, y) = 2e^y, \qquad 0 \le y \le 1,$$
$$u(x, 0) = x, \qquad u(x, 1) = ex, \qquad 0 \le x \le 2.$$

We will use Algorithm 11.1 to approximate the exact solution $u(x, y) = xe^y$ with $n = 6$ and $m = 5$. The stopping criterion for Step 17 required

$$|w_{i,j}^{(l)} - w_{i,j}^{(l-1)}| \le 10^{-10},$$

for each $i = 1, \ldots, 5$, and $j = 1, \ldots, 4$; so the solution to the difference equation was accurately obtained, and the procedure stopped at $l = 61$. The results, along with the correct values, are presented in Table 11.1.   ∎

TABLE 11.1

i	j	x_i	y_i	$w_{i,j}^{(61)}$	$u(x_i, y_i)$	$\lvert u(x_i, y_j) - w_{i,j}^{(61)} \rvert$
1	1	0.3333	0.2000	0.40726	0.40713	1.30×10^{-4}
1	2	0.3333	0.4000	0.49748	0.49727	2.08×10^{-4}
1	3	0.3333	0.6000	0.60760	0.60737	2.23×10^{-4}
1	4	0.3333	0.8000	0.74201	0.74185	1.60×10^{-4}
2	1	0.6667	0.2000	0.81452	0.81427	2.55×10^{-4}
2	2	0.6667	0.4000	0.99496	0.99455	4.08×10^{-4}
2	3	0.6667	0.6000	1.2152	1.2147	4.37×10^{-4}
2	4	0.6667	0.8000	1.4840	1.4837	3.15×10^{-4}
3	1	1.0000	0.2000	1.2218	1.2214	3.64×10^{-4}
3	2	1.0000	0.4000	1.4924	1.4918	5.80×10^{-4}
3	3	1.0000	0.6000	1.8227	1.8221	6.24×10^{-4}
3	4	1.0000	0.8000	2.2260	2.2255	4.51×10^{-4}
4	1	1.3333	0.2000	1.6290	1.6285	4.27×10^{-4}
4	2	1.3333	0.4000	1.9898	1.9891	6.79×10^{-4}
4	3	1.3333	0.6000	2.4302	2.4295	7.35×10^{-4}
4	4	1.3333	0.8000	2.9679	2.9674	5.40×10^{-4}
5	1	1.6667	0.2000	2.0360	2.0357	3.71×10^{-4}
5	2	1.6667	0.4000	2.4870	2.4864	5.84×10^{-4}
5	3	1.6667	0.6000	3.0375	3.0369	6.41×10^{-4}
5	4	1.6667	0.8000	3.7097	3.7092	4.89×10^{-4}

Exercise Set 11.2

1. Use Algorithm 11.1 to approximate the solution to the following elliptic partial-differential equation

$$\frac{\partial^2 u}{\partial x^2} + \frac{\partial^2 u}{\partial y^2} = 4, \qquad 0 < x < 1, \quad 0 < y < 2;$$

$$u(x, 0) = x^2, \quad u(x, 2) = (x-2)^2, \qquad 0 \le x \le 1;$$

$$u(0, y) = y^2, \quad u(1, y) = (y-1)^2, \qquad 0 \le y \le 2.$$

Use $h = k = 0.5$.

2. Use Algorithm 11.1 to approximate the solution to the following elliptic partial-differential equation

$$\frac{\partial^2 u}{\partial x^2} + \frac{\partial^2 u}{\partial y^2} = x, \qquad 0 < x, \quad y < 1;$$

$$u(x, 0) = \tfrac{1}{6}x^3, \quad u(x, 1) = \tfrac{1}{6}x^3, \qquad 0 \le x \le 1;$$

$$u(0, y) = 0, \quad u(1, y) = \tfrac{1}{6}, \qquad 0 \le y \le 1.$$

Use $h = k = \tfrac{1}{3}$.

3. Approximate the solutions to the following elliptic partial-differential equations, using Algorithm 1.1:

a) $\dfrac{\partial^2 u}{\partial x^2} + \dfrac{\partial^2 u}{\partial y^2} = 0,$ $0 < x, y < 1;$

$u(x, 0) = 0,$ $u(x, 1) = x,$ $0 \le x \le 1;$

$u(0, y) = 0,$ $u(1, y) = y,$ $0 \le y \le 1.$

Use $h = k = 0.1,$ and compare the results with the solution $u(x, y) = xy$ and with the first five terms of the Fourier-series solution.

b) $\dfrac{\partial^2 u}{\partial x^2} + \dfrac{\partial^2 u}{\partial y^2} = -2,$ $0 < x, y < 1;$

$u(0, y) = 0,$ $u(1, y) = \sinh(\pi)\sin \pi y,$ $0 \le y \le 1;$

$u(x, 0) = u(x, 1) = x(1 - x),$ $0 \le x \le 1.$

Use $h = k = 0.05,$ and compare the results with the solution $u(x, y) = \sinh(\pi x)\sin(\pi y) + x(1 - x).$

c) $\dfrac{\partial^2 u}{\partial x^2} + \dfrac{\partial^2 u}{\partial y^2} = (x^2 + y^2)e^{xy},$ $0 < x < 2,$ $0 < y < 1;$

$u(0, y) = 1,$ $u(2, y) = e^{2y},$ $0 \le y \le 1;$

$u(x, 0) = 1,$ $u(x, 1) = e^x,$ $0 \le x \le 2.$

Use $h = 0.2$ and $k = 0.1,$ and compare the results with the solution $u(x, y) = e^{xy}.$

d) $\dfrac{\partial^2 u}{\partial x^2} + \dfrac{\partial^2 u}{\partial y^2} = -(\cos (x + y) + \cos (x - y)),$ $0 < x < \pi,$ $0 < y < \dfrac{\pi}{2};$

$u(0, y) = \cos y,$ $u(\pi, y) = -\cos y,$ $0 \le y \le \dfrac{\pi}{2};$

$u(x, 0) = \cos x,$ $u\left(x, \dfrac{\pi}{2}\right) = 0,$ $0 \le x \le \pi.$

Use $h = \pi/5$ and $k = \pi/10$ and compare the results with the solution $u(x, y) = \cos x \cos y.$

4. Repeat Exercise 3, parts (a) and (b), using extrapolation with $h_0 = 0.2,$ $h_1 = h_0/2,$ and $h_2 = h_0/4.$ [*Hint:* Review the discussion at the end of Section 10.3.]

5. Construct an algorithm similar to Algorithm 11.1, except use Choleski's method instead of the Gauss–Seidel iterative method for solving the linear system.

6. Apply the algorithm from Exercise 5 to the problems in Exercise 3.

7. Construct an algorithm similar to Algorithm 11.1, except use the SOR method with optimal ω instead of the Gauss–Seidel method for solving the linear system.

8. Repeat Exercise 3, using the algorithm constructed in Exercise 7.

9. A coaxial cable is made of a 0.1-inch-square inner conductor and a 0.5-inch-square outer conductor. The potential at a point in the cross section of the cable is described by Laplace's equation. Suppose the inner conductor is kept at zero volts while the outer conductor is kept at 110 volts. Find the potential between the two conductors by placing a grid with horizontal mesh spacing $h = 0.1$ inch and vertical mesh spacing $k = 0.1$ inch on the region

$$D = \{(x, y) | 0 \leq x, y \leq 0.5\}.$$

Approximate the solution to Laplace's equation at each grid point, and use the two sets of boundary conditions to derive a linear system to be solved by the Gauss–Seidel method.

10. A 6-cm × 5-cm rectangular silver plate has heat being uniformly generated at each point at the rate $q = 1.5$ cal/cm$^3 \cdot$ sec. Let x represent the distance along the edge of the plate of length 6 cm and y the distance along the edge of the plate of length 5 cm. Suppose the temperature u along the edges is kept at the following temperatures:

$$u(x, 0) = x(6 - x), \quad u(x, 5) = 0, \qquad 0 \leq x \leq 6,$$

$$u(0, y) = y(5 - y), \quad u(6, y) = 0, \qquad 0 \leq y \leq 5,$$

where the origin lies at a corner of the plate with coordinates $(0, 0)$ and the edges lie along the positive x- and y-axes. The steady-state temperature $u = u(x, y)$ satisfies Poisson's equation:

$$\frac{\partial^2 u}{\partial x^2}(x, y) + \frac{\partial^2 u}{\partial y^2}(x, y) = \frac{-q}{K}, \qquad 0 < x < 6, \quad 0 < y < 5,$$

where K, the thermal conductivity, is 1.04 cal/cm $\cdot$ deg $\cdot$ sec. Approximate the temperature $u(x, y)$ using Algorithm 11.1 with $h = 0.4$ and $k = \frac{1}{3}$.

11.3 Parabolic Partial-Differential Equations

The parabolic partial-differential equation we will study is the heat or diffusion equation

(11.15)
$$\frac{\partial u}{\partial t}(x, t) = \alpha^2 \frac{\partial^2 u}{\partial x^2}(x, t), \qquad 0 < x < l, \quad t > 0,$$

subject to the conditions

$$u(0, t) = 0, \quad u(l, t) = 0, \qquad t > 0,$$

and
$$u(x, 0) = f(x), \qquad 0 \leq x \leq l.$$

The approach we use to approximate the solution to this problem involves finite differences and is similar to the method used in Section 11.2.

First select two mesh constants h and k, with the stipulation that $m = l/h$ is an integer. The grid points for this situation are (x_i, t_j), where $x_i = ih$ for $i = 0, 1, \ldots, m$, and $t_j = jk$, for $j = 0, 1, \ldots$

We obtain the difference method by using the Taylor series in t to form the difference quotient

(11.16)
$$\frac{\partial u}{\partial t}(x_i, t_j) = \frac{u(x_i, t_j + k) - u(x_i, t_j)}{k} - \frac{k}{2}\frac{\partial^2}{\partial t^2}u(x_i, \mu_j),$$

for some $\mu_j \in (t_j, t_{j+1})$, and the Taylor series in x to form the difference quotient

(11.17)
$$\frac{\partial^2 u}{\partial x^2}(x_i, t_j) = \frac{u(x_i + h, t_j) - 2u(x_i, t_j) + u(x_i - h, t_j)}{h^2}$$

$$-\frac{h^2}{12}\frac{\partial^4 u}{\partial x^4}(\xi_i, t_j), \text{ where } \xi_i \in (x_{i-1}, x_{i+1}).$$

The partial-differential equation (11.15) implies that at the interior gridpoint (x_i, t_j) for each $i = 1, 2, \ldots, m - 1$ and $j = 1, 2, \ldots$, we have

$$\frac{\partial u}{\partial t}(x_i, t_j) - \alpha^2 \frac{\partial^2 u}{\partial x^2}(x_i, t_j) = 0;$$

so the difference method using the difference quotients (11.16) and (11.17) is

(11.18)
$$\frac{w_{i,j+1} - w_{i,j}}{k} - \alpha^2 \frac{w_{i+1,j} - 2w_{i,j} + w_{i-1,j}}{h^2} = 0,$$

where w_{ij} approximates $u(x_i, t_j)$.

The local truncation error for this difference equation is

(11.19)
$$\tau_{i,j} = \frac{k}{2}\frac{\partial^2 u}{\partial t^2}(x_i, \mu_j) - \alpha^2 \frac{h^2}{12}\frac{\partial^4 u}{\partial x^4}(\xi_i, t_j).$$

If Eq. (11.18) is solved for $w_{i,j+1}$,

(11.20)
$$w_{i,j+1} = \left(1 - \frac{2\alpha^2 k}{h^2}\right)w_{i,j} + \alpha^2 \frac{k}{h^2}(w_{i+1,j} + w_{i-1,j}),$$

for each $i = 1, 2, \ldots, (m - 1)$ and $j = 1, 2, \ldots$. Since the initial condition $u(x, 0) = f(x)$, for each $0 \le x \le l$, implies that $w_{i,0} = f(x_i)$, for each $i = 0, 1, \ldots, m$, these values can be used in Eq. (11.20) to find the value of $w_{i,1}$ for each $i = 1, 2, \ldots, (m - 1)$. The additional conditions $u(0, t) = 0$ and $u(l, t) = 0$ imply that $w_{0,1} = w_{m,1} = 0$; so all the entries of the form $w_{i,1}$ can be determined. If the procedure is reapplied once all the approximations $w_{i,1}$, are known, the values of $w_{i,2}, w_{i,3}, \ldots, w_{i,m-1}$ can be obtained in a similar manner.

The explicit nature of the difference method expressed in Eq. (11.20) implies that the $(m - 1) \times (m - 1)$ matrix associated with this system can be written in the tridiagonal form

$$A = \begin{bmatrix} (1 - 2\lambda) & \lambda & 0 \cdots\cdots\cdots\cdots 0 \\ \lambda & (1 - 2\lambda) & \lambda \\ 0 & & & 0 \\ \vdots & & & & \lambda \\ 0 \cdots\cdots\cdots\cdots 0 & & \lambda & (1 - 2\lambda) \end{bmatrix},$$

where $\lambda = \alpha^2 (k/h^2)$. If we let

$$\mathbf{w}^{(0)} = (f(x_1), f(x_2), \ldots, f(x_{m-1}))^t$$

and

$$\mathbf{w}^{(j)} = (w_{1,j}, w_{2,j}, \ldots, w_{m-1,j})^t \qquad \text{for each } j = 1, 2, \ldots,$$

then the approximate solution is given by

$$\mathbf{w}^{(j)} = A\mathbf{w}^{(j-1)} \qquad \text{for each } j = 1, 2, \ldots.$$

This difference method is known as the **forward-difference** method. If the solution to the partial-differential equation has four continuous partial derivatives in x and two in t, then Eq. (11.19) implies that the method is of order $O(k + h^2)$.

EXAMPLE 1   Consider the heat equation

$$\frac{\partial u}{\partial t}(x, t) - \frac{\partial^2 u}{\partial x^2}(x, t) = 0, \quad 0 < x < 1, \quad 0 < t,$$

with boundary conditions

$$u(0, t) = u(1, t) = 0, \qquad 0 < t,$$

and initial conditions

$$u(x, 0) = \sin(\pi x), \qquad 0 \le x \le 1.$$

It is easily verified that the solution to this problem is

$$u(x, t) = e^{-\pi^2 t} \sin(\pi x).$$

The solution at $t = 0.5$ will be approximated using the forward-difference method first with $h = 0.1$, $k = 0.0005$, and $\lambda = 0.05$, and then with $h = 0.1$, $k = 0.01$, and $\lambda = 1$. The results are presented in Table 11.2.   ■

TABLE 11.2

x_i	$u(x_i, 0.5)$	$w_{i,1000}$ $k = 0.0005$	$\|u(x_i, 0.5) - w_{i,1000}\|$	$w_{i,50}$ $k = 0.01$	$\|u(x_i, 0.5) - w_{i,50}\|$
0.0	0	0	—	0	—
0.1	0.00222241	0.00228652	6.411×10^{-5}	8.19876×10^7	8.199×10^7
0.2	0.00422728	0.00434922	1.219×10^{-4}	-1.55719×10^8	1.557×10^8
0.3	0.00581836	0.00598619	1.678×10^{-4}	2.13833×10^8	2.138×10^8
0.4	0.00683989	0.00703719	1.973×10^{-4}	-2.50642×10^8	2.506×10^8
0.5	0.00719188	0.00739934	2.075×10^{-4}	2.62685×10^8	2.627×10^8
0.6	0.00683989	0.00703719	1.973×10^{-4}	-2.49015×10^8	2.490×10^8
0.7	0.00581836	0.00598619	1.678×10^{-4}	2.11200×10^8	2.112×10^8
0.8	0.00422728	0.00434922	1.219×10^{-4}	-1.53086×10^8	1.531×10^8
0.9	0.00222241	0.00228652	6.511×10^{-5}	8.03604×10^7	8.036×10^7
1.0	0	0	—	0	—

Although a local truncation error of order $O(k + h^2)$ is expected, this is certainly not the case in our example when $h = 0.1$ and $k = 0.01$. To explain the difficulty we must look at the round-off stability of the forward-difference method.

If the error $\mathbf{e}^{(0)} = (e_1^{(0)}, e_2^{(0)}, \ldots, e_{m-1}^{(0)})^t$ is made in representing the initial data $\mathbf{w}^{(0)} = (f(x_1), f(x_2), \ldots, f(x_{m-1}))^t$ or, for that matter, in any particular step (the choice of the initial step is simply for convenience), an error of $A\mathbf{e}^{(0)}$ would be propagated in $\mathbf{w}^{(1)}$ since

$$\mathbf{w}^{(1)} = A(\mathbf{w}^{(0)} + \mathbf{e}^{(0)}) = A\mathbf{w}^{(0)} + A\mathbf{e}^{(0)}.$$

This process would continue. At the nth time step, the error in $\mathbf{w}^{(n)}$ due to $\mathbf{e}^{(0)}$ would be $A^n\mathbf{e}^{(0)}$. The method is consequently stable if and only if these errors do not grow as n increases — that is, if and only if $\| A^n\mathbf{e}^{(0)} \| \le \| \mathbf{e}^{(0)} \|$ for all n. This implies that $\| A^n \| \le 1$, a condition which, by Theorem 8.16 (p. 419), requires that the spectral radius $\rho(A^n) = (\rho(A))^n \le 1$. The forward-difference method will therefore be stable only if $\rho(A) \le 1$.

The eigenvalues of A can be shown (See Exercise 7) to be

$$\mu_i = 1 - 4\lambda \left(\sin\left(\frac{i\pi}{2m}\right) \right)^2 \qquad \text{for each } i = 1, 2, \ldots, (m-1).$$

The condition for stability consequently reduces to determining whether:

$$\rho(A) = \max_{1 \le i \le m-1} \left| 1 - 4\lambda \left(\sin\left(\frac{i\pi}{2m}\right) \right)^2 \right| \le 1,$$

which simplifies to

$$0 \le \lambda \left(\sin\left(\frac{i\pi}{2m}\right) \right)^2 \le \tfrac{1}{2} \qquad \text{for each } i = 1, 2, \ldots, m-1.$$

Since stability requires that this inequality condition hold as $h \to 0$ or, equivalently, as $m \to \infty$, the fact that

$$\lim_{m \to \infty} \left[\sin\left(\frac{(m-1)\pi}{2m}\right) \right]^2 = 1$$

means that stability will occur only if $0 \le \lambda \le \tfrac{1}{2}$. Since $\lambda = \alpha^2(k/h^2)$, this inequality requires that h and k be chosen so that

$$\alpha^2 \frac{k}{h^2} \le \tfrac{1}{2}.$$

This condition was satisfied in our example when $h = 0.1$ and $k = 0.0005$; but, when k was increased to 0.01 with no corresponding increase in h, the ratio became

$$\frac{0.01}{(0.1)^2} = 1 > \tfrac{1}{2}$$

and stability problems became apparent. Consistent with the terminology of Chapter 5, we call this method **conditionally stable** and remark that the method converges to the solution of Eq. (11.15) with rate of convergence $O(k + h^2)$, provided

$$\alpha^2 \frac{k}{h^2} \leq \tfrac{1}{2}$$

and the required continuity conditions on the solution are met. (For a detailed proof of this fact, see Isaacson and Keller [67], pp. 502–505.)

To obtain a method that is **unconditionally stable**, we consider an implicit-difference method that results from using the backward-difference quotient for $(\partial u / \partial t)(x_i, t_j)$ in the form

$$\frac{\partial u}{\partial t}(x_i, t_j) = \frac{u(x_i, t_j) - u(x_i, t_{j-1})}{k} + \frac{k}{2}\frac{\partial^2 u}{\partial t^2}(x_i, \mu_j),$$

where $\mu_j \in (t_{j-1}, t_j)$. Substituting this equation, together with Eq. (11.17) for $\partial^2 u / \partial x^2$, into the partial-differential equation gives

$$\frac{u(x_i, t_j) - u(x_i, t_{j-1})}{k} - \alpha^2 \frac{u(x_{i+1}, t_j) - 2u(x_i, t_j) + u(x_{i-1}, t_j)}{h^2}$$
$$= -\frac{k}{2}\frac{\partial^2 u}{\partial t^2}(x_i, \mu_j) - \frac{h^2}{12}\frac{\partial^4 u}{\partial x^4}(\xi_i, t_j),$$

for some $\xi_i \in (x_{i-1}, x_{i+1})$. The **backward-difference method** that results is

$$(11.21) \qquad \frac{w_{i,j} - w_{i,j-1}}{k} - \alpha^2 \frac{w_{i+1,j} - 2w_{ij} + w_{i-1,j}}{h^2} = 0,$$

for each $i = 1, 2, \ldots, m - 1$, and $j = 1, 2, \ldots$.

This method involves, at a typical step, the mesh points

$$(x_i, t_j), \quad (x_i, t_{j-1}), \quad (x_{i-1}, t_j), \quad \text{and} \quad (x_{i+1}, t_j),$$

and, in grid form, involves approximations at the points marked with X's in Fig. 11.8.

FIGURE 11.8

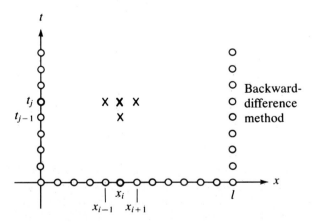

Since the boundary and initial conditions associated with the problem give information at the circled mesh points, it is clear from the figure that no explicit procedures can be used to solve Eq. (11.21). Recall that, in the forward-difference method (*see Fig. 11.9*), approximations at

$$(x_{i-1}, t_j), \quad (x_i, t_j), \quad (x_i, t_{j+1}), \quad \text{and} \quad (x_{i+1}, t_j)$$

FIGURE 11.9

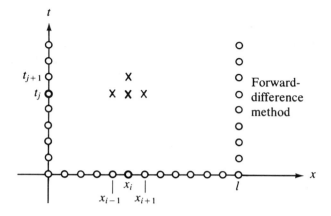

were used; so an explicit method for finding the approximations, based on the information from the initial and boundary conditions, was available.

If we again let λ denote the quantity $\alpha^2(k/h^2)$, the backward-difference method becomes

(11.22)
$$(1 + 2\lambda)w_{i,j} - \lambda w_{i+1,j} - \lambda w_{i-1,j} = w_{i,j-1},$$

for each $i = 1, 2, \ldots, m - 1$, and $j = 1, 2, \ldots$. Using the knowledge that $w_{i,0} = f(x_i)$ for each $i = 1, 2, \ldots, m - 1$, and $w_{m,j} = w_{0,j} = 0$ for each $j = 1, 2, \ldots$, this difference method has the matrix representation:

(11.23)

$$\begin{bmatrix} (1 + 2\lambda) & -\lambda & 0 & \cdots & \cdots & 0 \\ -\lambda & (1 + 2\lambda) & -\lambda & & & \vdots \\ 0 & & & & & 0 \\ \vdots & & & & & -\lambda \\ 0 & \cdots & \cdots & 0 & -\lambda & (1 + 2\lambda) \end{bmatrix} \begin{bmatrix} w_{1,j} \\ w_{2,j} \\ \vdots \\ w_{m-1,j} \end{bmatrix} = \begin{bmatrix} w_{1,j-1} \\ w_{2,j-1} \\ \vdots \\ w_{m-1,j-1} \end{bmatrix},$$

or $A\mathbf{w}^{(j)} = \mathbf{w}^{(j-1)}$ for each $j = 1, 2, \ldots$. Since $\lambda > 0$, the matrix A is positive definite and strictly diagonally dominant, as well as being tridiagonal and symmetric. We can use either the Crout Reduction for Tridiagonal Linear Systems (Algorithm 6.7) or the SOR method (Algorithm 8.3), for solving this system. The following algorithm solves (11.23) using Crout Reduction, which would be an acceptable method unless m is large. In this algorithm we assume, for stopping purposes, that a bound is given for t.

Heat Equation Backward-Difference Algorithm 11.2

To approximate the solution to the parabolic partial-differential equation

$$\frac{\partial u}{\partial t}(x,\ t) - \alpha^2 \frac{\partial^2 u}{\partial x^2}(x,\ t) = 0, \qquad 0 < x < l, \quad 0 < t < T,$$

subject to the boundary conditions

$$u(0,\ t) = u(l,\ t) = 0, \qquad 0 < t < T,$$

and the initial conditions

$$u(x,\ 0) = f(x), \qquad 0 \le x \le l:$$

INPUT endpoint l; maximum time T; constant α; integers m, N.

OUTPUT approximations $w_i(t_j)$ to $u(x_i,\ t_j)$ for each $i = 1, \ldots, m-1$ and $j = 1, \ldots, N$.

Step 1 Set $h = l/m$;
$\qquad k = T/N$;
$\qquad \lambda = \alpha^2 k/h^2$.

Step 2 For $i = 1, \ldots, m-1$ set $w_i = f(ih)$.   (*Initial values.*)

(*Steps 3–11 solve a tridiagonal linear system using Algorithm 6.7.*)

Step 3 Set $l_1 = 1 + 2\lambda$;
$\qquad u_1 = -\lambda/l_1$.

Step 4 For $i = 2,\ldots, m-2$ set $l_i = 1 + 2\lambda + \lambda u_{i-1}$;
$\qquad\qquad u_i = -\lambda/l_i$.

Step 5 Set $l_{m-1} = 1 + 2\lambda + \lambda u_{m-2}$.

Step 6 For $j = 1, \ldots, N$ do Steps 7–11.

Step 7 Set $t = jk$; (*Current t_j.*)
$\qquad z_1 = w_1/l_1$.

Step 8 For $i = 2, \ldots, m-1$ set $z_i = (w_i + \lambda z_{i-1})/l_i$.

Step 9 Set $w_{m-1} = z_{m-1}$.

Step 10 For $i = m-2, \ldots, 1$ set $w_i = z_i - u_i w_{i+1}$.

Step 11 OUTPUT (t);
$\qquad$ For $i = 1, \ldots, m-1$ set $x = ih$;
$\qquad\qquad\qquad\qquad$ OUTPUT $(x,\ w_i)$.

Step 12 STOP. (*Procedure is complete.*)

EXAMPLE 2 The backward-difference method (Algorithm 11.2) with $h = 0.1$ and $k = 0.01$ will be used to approximate the solution to the heat equation

$$\frac{\partial u}{\partial t}(x,\ t) - \frac{\partial^2 u}{\partial x^2}(x,\ t) = 0, \quad 0 < x < 1, \quad 0 < t,$$

subject to the constraints

$$u(0, t) = u(1, t) = 0, \quad 0 < t, \quad \text{and} \quad u(x, 0) = \sin \pi x, \quad 0 \le x \le 1,$$

which was considered in Example 1. To demonstrate the unconditional stability of the backward-difference method, we will again compare $w_{i, 50}$ to $u(x_i, 0.5)$ where $i = 0$, 1, ..., 10.

The results listed in Table 11.3 should be compared with the fifth and sixth columns of Table 11.2.   ■

TABLE 11.3

x_i	$w_{i, 50}$	$u(x_i, 0.5)$	$|w_{i, 50} - u(x_i, 0.5)|$
0.0	0	0	—
0.1	0.00289802	0.00222241	6.756×10^{-4}
0.2	0.00551236	0.00422728	1.285×10^{-3}
0.3	0.00758711	0.00581836	1.769×10^{-3}
0.4	0.00891918	0.00683989	2.079×10^{-3}
0.5	0.00937818	0.00719188	2.186×10^{-3}
0.6	0.00891918	0.00683989	2.079×10^{-3}
0.7	0.00758711	0.00581836	1.769×10^{-3}
0.8	0.00551236	0.00422728	1.285×10^{-3}
0.9	0.00289802	0.00222241	6.756×10^{-4}
1.0	0	0	—

The backward-difference method does not have the stability problems of the forward-difference method. This can be seen by analyzing the eigenvalues of the matrix A. In this case (see Exercise 8), the eigenvalues are of the form

$$\mu_i = 1 + 4\lambda \left[\sin \left(\frac{i\pi}{2m} \right) \right]^2 \quad \text{for each } i = 1, 2, \ldots, (m-1);$$

and since $\lambda > 0$, $\mu_i > 1$ for all $i = 1, 2, \ldots, (m-1)$. This implies that A^{-1} exists, since zero cannot be an eigenvalue of A. An error $\mathbf{e}^{(0)}$ in the initial data produces an error $(A^{-1})^n \mathbf{e}^{(0)}$ at the nth step. Since the eigenvalues of A^{-1} are the reciprocals of the eigenvalues of A, the spectral radius of A^{-1} is bounded above by 1 and the method is stable, independent of the choice of $\lambda = \alpha^2(k/h^2)$. In the terminology of Chapter 5, we call the backward-difference method an **unconditionally stable** method. The local truncation error for the method is of order $O(k + h^2)$, provided the solution of the differential equation satisfies the usual differentiability conditions. In this case the method converges to the solution of the partial-differential equation with this same rate of convergence (see Isaacson and Keller [67], page 508).

The weakness in the backward-difference method results from the fact that the local truncation error has a portion with order $O(k)$, requiring that time intervals be made much smaller that spatial intervals. It would clearly be desirable to devise a procedure with local truncation error of order $O(k^2 + h^2)$. The first step in this direction is to use a difference equation that has $O(k^2)$ error for $u_t(x, t)$ instead of those we have used previously, whose error was $O(k)$. This can be done by using the Taylor series in t for the function $u(x, t)$ at the point (x_i, t_j) and evaluating at (x_i, t_{j+1}) and (x_i, t_{j-1}) to obtain the central-difference formula

$$\frac{\partial u}{\partial t}(x_i, t_j) = \frac{u(x_i, t_{j-1}) - u(x_i, t_{j-1})}{2k} - \frac{k^2}{6}\frac{\partial^3 u}{\partial t^3}(x_i, \mu_j),$$

where $\mu_j \in (t_{j-1}, t_{j+1})$. The difference method that results from substituting this and the usual difference quotient for $(\partial^2 u/\partial x^2)$, Eq. (11.17), into the differential equation is called **Richardson's method**, and is given by

$$\frac{w_{i,j+1} - w_{i,j-1}}{2k} - \alpha^2 \frac{w_{i+1,j} - 2w_{i,j} + w_{i-1,j}}{h^2} = 0.$$

This method does have local truncation error of order $O(k^2 + h^2)$, but unfortunately also has serious stability problems (see Exercise 6).

A more rewarding method can be derived by averaging the forward-difference method at the jth step in t,

$$\frac{w_{i,j+1} - w_{i,j}}{k} - \alpha^2 \frac{w_{i+1,j} - 2w_{i,j} + w_{i-1,j}}{h^2} = 0,$$

which has local truncation error

$$\tau_F = \frac{k}{2}\frac{\partial^2 u}{\partial t^2}(x_i, \mu_j) + O(h^2),$$

and the backward-difference formula at the $(j+1)$st step in t,

$$\frac{w_{i,j+1} - w_{i,j}}{k} - \alpha^2 \frac{w_{i+1,j+1} - 2w_{i,j+1} + w_{i-1,j+1}}{h^2} = 0,$$

which has local truncation error

$$\tau_B = -\frac{k}{2}\frac{\partial^2 u}{\partial t^2}(x_i, \hat{\mu}_j) + O(h^2).$$

If we assume that $\mu_j \approx \hat{\mu}_j$, then the averaged difference method,

$$\frac{w_{i,j+1} - w_{i,j}}{k} - \frac{\alpha^2}{2}\left[\frac{w_{i+1,j} - 2w_{i,j} + w_{i-1,j}}{h^2} + \frac{w_{i+1,j+1} - 2w_{i,j+1} + w_{i-1,j+1}}{h^2}\right] = 0,$$

has local truncation error of order $O(k^2 + h^2)$, provided, of course, that the usual differentiability conditions are satisfied.

This method is known as the **Crank–Nicolson** method, and can be represented in the matrix form $A\mathbf{w}^{(j+1)} = B\mathbf{w}^{(j)}$ for each $j = 0, 1, 2, \ldots$, where

$$\lambda = \alpha^2 \frac{k}{h^2}, \qquad \mathbf{w}^{(j)} = (w_{1,j}, w_{2,j}, \ldots, w_{m-1,j})^t,$$

and the matrices A and B are given by:

$$
A = \begin{bmatrix}
(1+\lambda) & -\dfrac{\lambda}{2} & 0 & \cdots\cdots\cdots & 0 \\[1em]
-\dfrac{\lambda}{2} & (1+\lambda) & -\dfrac{\lambda}{2} & & \\[1em]
0 & -\dfrac{\lambda}{2} & & & 0 \\[1em]
 & & & & -\dfrac{\lambda}{2} \\[1em]
0 & \cdots\cdots\cdots & 0 & -\dfrac{\lambda}{2} & (1+\lambda)
\end{bmatrix}
$$

and

$$
B = \begin{bmatrix}
(1-\lambda) & \dfrac{\lambda}{2} & 0 & \cdots\cdots\cdots & 0 \\[1em]
\dfrac{\lambda}{2} & (1-\lambda) & \dfrac{\lambda}{2} & & \\[1em]
0 & \dfrac{\lambda}{2} & & & 0 \\[1em]
 & & & & \dfrac{\lambda}{2} \\[1em]
0 & \cdots\cdots\cdots & 0 & \dfrac{\lambda}{2} & (1-\lambda)
\end{bmatrix}
$$

Since A is a positive definite, symmetric, strictly diagonally dominant and tridiagonal matrix, it is nonsingular. Either the Crout Reduction for Tridiagonal Linear System (Algorithm 6.7) or the SOR method (Algorithm 8.3) can be used to obtain $\mathbf{w}^{(j+1)}$ from $\mathbf{w}^{(j)}$, for each $j = 0, 1, 2, \ldots$ The following algorithm incorporates Crout Reduction into the Crank–Nicolson technique. As in Algorithm 11.2, a finite length for the time interval must be specified to determine a stopping procedure.

Crank–Nicolson Algorithm 11.3

To approximate the solution to the parabolic partial-differential equation

$$\frac{\partial u}{\partial t}(x, t) - \alpha^2 \frac{\partial^2 u}{\partial x^2}(x, t) = 0, \qquad 0 < x < l, \quad 0 < t < T,$$

subject to the boundary conditions

$$u(0, t) = u(l, t) = 0, \qquad 0 < t < T,$$

and the initial conditions

$$u(x, 0) = f(x), \qquad 0 \le x \le l:$$

INPUT endpoint l; maximum time T; constant α; integers m, N.

OUTPUT approximations $w_i(t_j)$ to $u(x_i, t_j)$ for each $i = 1, \ldots, m - 1$ and $j = 1, \ldots, N$.

Step 1   Set $h = l/m$;
   $k = T/N$;
   $\lambda = \alpha^2 k/h^2$;
   $w_n = 0$.

Step 2   For $i = 1, \ldots, m - 1$ set $w_i = f(ih)$.   (*Initial values.*)
   (*Steps 3–11 solve a tridiagonal linear system using Algorithm 6.7.*)

Step 3   Set $l_1 = 1 + \lambda$;
   $u_1 = -\lambda/(2l_1)$.

Step 4   For $i = 2, \ldots, m - 2$ set $l_i = 1 + \lambda + \lambda u_{i-1}/2$;
   $u_i = -\lambda/(2l_i)$.

Step 5   Set $l_{m-1} = 1 + \lambda + \lambda u_{m-2}/2$.

Step 6   For $j = 1, \ldots, N$ do Steps 7–11.

Step 7   Set $t = jk$;   (*Current t_j.*)
$$z_1 = \left[(1 - \lambda)w_1 + \frac{\lambda}{2} w_2 \right] / l_1.$$

Step 8   For $i = 2, \ldots, m - 1$ set $z_i = \left[(1 - \lambda)w_i + \frac{\lambda}{2}(w_{i+1} + w_{i-1} + z_{i-1}) \right] / l_i.$

Step 9   Set $w_{m-1} = z_{m-1}$.

Step 10   For $i = m - 2, \ldots, 1$ set $w_i = z_i - u_i w_{i+1}$.

Step 11   OUTPUT (t);
   For $i = 1, \ldots, m - 1$ set $x = ih$;
   OUTPUT (x, w_i).

Step 12   STOP.   (*Procedure is complete.*)

The Crank–Nicolson method is unconditionally stable and has order of convergence $O(k^2 + h^2)$. The verification of these facts can be found in the previously mentioned book by Isaacson and Keller [67], pp. 508–512.

EXAMPLE 3   The Crank–Nicolson method can be used to approximate the solution to the problem in Examples 1 and 2, consisting of the equation

$$\frac{\partial u}{\partial t}(x, t) - \frac{\partial^2 u}{\partial x^2}(x, t) = 0, \qquad 0 < x < 1, \quad 0 < t,$$

subject to the conditions

$$u(0, t) = u(1, t) = 0, \qquad 0 < t,$$

and

$$u(x, 0) = \sin(\pi x), \qquad 0 \le x \le 1.$$

The choices $m = 10$, $h = 0.1$, $N = 50$, $k = 0.01$, and $\lambda = 1$ are used in Algorithm 11.3, as they were in the previous examples. The results in Table 11.4 indicate the increase in accuracy of the Crank–Nicolson method over the backward-difference method, the best of the two previously discussed techniques. ∎

TABLE 11.4

x_i	$w_{i, 50}$	$u(x_i, 0.5)$	$|w_{i, 50} - u(x_i, 0.5)|$
0.0	0	0	—
0.1	0.00230512	0.00222241	8.271×10^{-5}
0.2	0.00438461	0.00422728	1.573×10^{-4}
0.3	0.00603489	0.00581836	2.165×10^{-4}
0.4	0.00709444	0.00683989	2.546×10^{-4}
0.5	0.00745954	0.00719188	2.677×10^{-4}
0.6	0.00709444	0.00683989	2.546×10^{-4}
0.7	0.00603489	0.00581836	2.165×10^{-4}
0.8	0.00438461	0.00422728	1.573×10^{-4}
0.9	0.00230512	0.00222241	8.271×10^{-5}
1.0	0	0	—

Exercise Set 11.3

1. Approximate the solution to each of the following parabolic partial-differential equations using Algorithm 11.2.

a) $\dfrac{\partial u}{\partial t} - \dfrac{\partial^2 u}{\partial x^2} = 0, \qquad 0 < x < 2, \quad 0 < t;$

$u(0, t) = u(2, t) = 0, \qquad 0 < t;$

$u(x, 0) = \sin \dfrac{\pi}{2} x, \qquad 0 \le x \le 2.$

Use $m = 4$, $T = 0.1$, and $N = 2$.

b) $\dfrac{\partial u}{\partial t} - \dfrac{1}{16}\dfrac{\partial^2 u}{\partial x^2} = 0,$ $0 < x < 1,\ \ 0 < t;$

   $u(0, t) = u(1, t) = 0,$ $0 < t;$

   $u(x, 0) = 2 \sin 2\pi x,$ $0 \le x \le 1.$

   Use $m = 3,\ T = 0.1,$ and $N = 2.$

2. Repeat Exercise 1, using Algorithm 11.3.

3. Use the forward-difference method to approximate the solutions to the following parabolic partial-differential equations.

   a) $\dfrac{\partial u}{\partial t} - \dfrac{\partial^2 u}{\partial x^2} = 0,$ $0 < x < 2,\ \ 0 < t;$

      $u(0, t) = u(2, t) = 0,$ $0 < t;$

      $u(x, 0) = \sin 2\pi x,$ $0 \le x \le 2.$

      Use $h = 0.1$ and $k = 0.01,$ and compare your answers at $t = 0.5$ to the actual solution. Then use $h = 0.1$ and $k = 0.005,$ and compare the answers.

   b) $\dfrac{\partial u}{\partial t} - \dfrac{\partial^2 u}{\partial x^2} = 0,$ $0 < x < 2,\ \ 0 < t;$

      $u(0, t) = u(2, t) = 0,$ $0 < t;$

      $u(x, 0) = x(2 - x),$ $0 \le x \le 2.$

      Use $h = 0.1$ and choose an appropriate value for $k.$ Compare your answers to the first five terms of the Fourier-series solution at $t = 0.5.$

   c) $\dfrac{\partial u}{\partial t} - \dfrac{4}{\pi^2}\dfrac{\partial^2 u}{\partial x^2} = 0,$ $0 < x < 4,\ \ 0 < t;$

      $u(0, t) = u(4, y) = 0,$ $0 < t;$

      $u(x, 0) = \sin\dfrac{\pi}{4}x\left(1 + 2\cos\dfrac{\pi}{4}x\right),$ $0 \le x \le 4.$

      Use $h = 0.2$ and $k = 0.04.$ Compare your answers to the actual solution $u(x, t) = e^{-t}\sin\dfrac{\pi}{2}x + e^{-t/4}\sin\dfrac{\pi}{4}x$ at $t = 0.4.$

   d) $\dfrac{\partial u}{\partial t} - \dfrac{1}{\pi^2}\dfrac{\partial^2 u}{\partial x^2} = 0,$ $0 < x < 1,\ \ 0 < t;$

      $u(0, t) = u(1, t) = 0,$ $0 < t;$

      $u(x, 0) = \cos \pi(x - \tfrac{1}{2}),$ $0 \le x \le 1.$

      Use $h = 0.1$ and $k = 0.04.$ Compare your answers to the actual solution $u(x, t) = e^{-t}\cos \pi(x - \tfrac{1}{2})$ at $t = 0.4.$

4. Repeat Exercise 3, using the Backward-Difference Algorithm 11.2.

5. Repeat Exercise 3, using the Crank–Nicolson Algorithm 11.3.

6. Repeat Exercise 3, using Richardson's method.

7. Show that the eigenvalues for the $(m-1) \times (m-1)$ tridiagonal matrix A given by

$$a_{ij} = \begin{cases} \lambda, & j = i - 1 \quad \text{or} \quad j = i + 1, \\ (1 - 2\lambda), & j = i, \\ 0, & \text{otherwise} \end{cases}$$

are $\mu_i = 1 - 4\lambda \left[\sin\left(\dfrac{i\pi}{2m}\right) \right]^2$ for each $i = 1, 2, \ldots, m - 1$,

with corresponding eigenvectors $\mathbf{v}^{(i)}$, where $v_j^{(i)} = \sin \dfrac{ij\pi}{m}$.

8. Show that the $(m-1) \times (m-1)$ tridiagonal matrix A given by

$$a_{ij} = \begin{cases} -\lambda, & j = i - 1 \quad \text{or} \quad j = i + 1, \\ 1 + 2\lambda, & j = i, \\ 0, & \text{otherwise,} \end{cases}$$

where $\lambda > 0$, is positive definite and diagonally dominant and has eigenvalues

$$\mu_i = 1 + 4\lambda \left[\sin\left(\frac{i\pi}{2m}\right) \right]^2 \qquad \text{for each } i = 1, 2, \ldots, m - 1,$$

with corresponding eigenvectors $\mathbf{v}^{(i)}$, where $v_j^{(i)} = \sin \dfrac{ij\pi}{m}$.

9. Modify Algorithms 11.2 and 11.3 to include the parabolic partial-differential equation

$$\frac{\partial u}{\partial t} - \frac{\partial^2 u}{\partial x^2} = F(x), \qquad 0 < x < l, \quad 0 < t;$$

$$u(0, t) = u(l, t) = 0, \qquad 0 < t;$$

$$u(x, 0) = f(x), \qquad 0 \le x \le l.$$

10. Use the results of Exercise 9 to approximate the solution to

$$\frac{\partial u}{\partial t} - \frac{\partial^2 u}{\partial x^2} = 2, \qquad 0 < x < 1, \quad 0 < t;$$

$$u(0, t) = u(1, t) = 0, \qquad 0 < t;$$

$$u(x, 0) = \sin \pi x + x(1 - x),$$

with $h = 0.1$ and $k = 0.01$. Compare your answers to the actual solution $u(x, t) = e^{-\pi^2 t} \sin \pi x + x(1 - x)$ at $t = 0.25$.

11. Change Algorithms 11.2 and 11.3 to accommodate the partial-differential equation

$$\frac{\partial u}{\partial t} - \alpha^2 \frac{\partial^2 u}{\partial x^2} = 0, \qquad 0 < x < l, \quad 0 < t;$$

$$u(0, t) = \phi(t), \quad u(l, t) = \Psi(t), \qquad 0 < t;$$
$$u(x, 0) = f(x), \qquad 0 \le x \le l,$$

where $f(0) = \phi(0)$ and $f(l) = \Psi(l)$.

12. The temperature $u(x, t)$ in a long thin rod of constant cross section and homogeneous conducting material is governed by the one-dimensional heat equation. If heat is generated in the material, for example, by resistance to current or nuclear reaction, the heat equation becomes

$$\frac{\partial^2 u}{\partial x^2} + \frac{Kr}{\rho C} = K \frac{\partial u}{\partial t}, \qquad 0 < x < l, \quad 0 < t,$$

where l is the length, ρ the density, C the specific heat, and K the thermal diffusivity of the rod. The function $r = r(x, t, u)$ represents the heat generated per unit volume. Suppose that

$$l = 1.5 \text{ cm}, \qquad K = 1.04 \text{ cal/cm} \cdot \text{deg} \cdot \text{sec},$$
$$\rho = 10.6 \text{ g/cm}^3, \qquad C = 0.056 \text{ cal/g} \cdot \text{deg},$$

and
$$r(x, t, u) = 5.0 \text{ cal/sec} \cdot \text{cm}^3.$$

If the ends of the rod are kept at 0°C, then

$$u(0, t) = u(l, t) = 0, \qquad t > 0.$$

Suppose the initial temperature distribution is given by

$$u(x, 0) = \sin \frac{\pi x}{l}, \qquad 0 \le x \le l.$$

Use the results of Exercise 9 to approximate the temperature distribution with $h = 0.15$ and $k = 0.0225$. Also use a modified forward-difference method to approximate the distribution.

13. V. Sagar and D. J. Payne [113], in analyzing the stress–strain relationships and material properties of a cylinder alternately subjected to heating and cooling, consider the equation

$$\frac{\partial^2 T}{\partial r^2} + \frac{1}{r} \frac{\partial T}{\partial r} = \frac{1}{4K} \frac{\partial T}{\partial t}, \qquad \tfrac{1}{2} < r < 1, \quad 0 < T,$$

where $T = T(r, t)$ is the temperature, r is the radial distance from the center of the cylinder, t is time, and K is a diffusivity coefficient.

a) Find approximations to $T(r, 10)$ for a cylinder with outside radius one, given the initial and boundary conditions:

$$T(1, t) = 100 + 40t, \qquad 0 \le t \le 10,$$
$$T(\tfrac{1}{2}, t) = t, \qquad 0 \le t \le 10,$$
$$T(r, 0) = 200(r - 0.5), \qquad 0.5 \le r \le 1.$$

Use a modification of the backward-difference method with $K = 0.1$, $k = 0.5$, and $h = \Delta r = 0.1$.

b) Using the temperature distribution of part (a), calculate the strain by approximating the integral

$$I = \int_{0.5}^{1} \alpha T(r, t) r \, dr,$$

where $\alpha = 10.7$ and $t = 10$. Use the Composite Trapezoidal method given in Theorem 4.5 (p. 165) with $m = 5$.

14. The equation describing one-dimensional, single-phase, slightly compressible flow in a producing petroleum reservoir is given, for $0 < x < 1000$ and $0 < t$, by

$$\frac{\phi \mu C}{K} \frac{\partial p}{\partial t}(x, t) = \frac{\partial^2 p}{\partial x^2}(x, t) - \begin{cases} 0, & \text{if } x \neq 500, \\ 1000, & \text{if } x = 500, \end{cases}$$

where it has been assumed that the porous medium and the reservoir are homogeneous, that the liquid is ideal, and that gravitational effects are negligible. The symbols are defined as x representing distance (in feet), t the time (in days), p the pressure (in pounds per square inch), ϕ the dimensionless constant porosity of the medium, μ the viscosity (in centipoise), K the permeability of the medium (in millidarcies), and C the compressibility (in [pounds per square inch]$^{-1}$). Assume that $\alpha = \phi \mu C / K = 0.00004$ days/ft^2, and that the following conditions hold:

$$p(x, 0) = 2.5 \times 10^7, \qquad 0 \leq x \leq 1000,$$

$$\frac{1}{K} \frac{\partial p}{\partial x}(0, t) = \frac{\partial p}{\partial x}(1000, t) = 0, \qquad 0 < t.$$

Find the pressure p at $t = 5$, using the Crank–Nicolson method with $k = \Delta t = 0.5$ and $h = \Delta x = 100$.

11.4 Hyperbolic Partial-Differential Equations

In this section we consider the numerical solution to the wave equation, an example of a hyperbolic partial-differential equation. The wave equation is given by the differential equation

$$(11.24) \qquad \frac{\partial^2 u}{\partial t^2}(x, t) - \alpha^2 \frac{\partial^2 u}{\partial x^2}(x, t) = 0, \qquad 0 < x < l, \quad t > 0,$$

subject to the conditions

$$u(0, t) = u(l, t) = 0, \qquad t > 0,$$

$$u(x, 0) = f(x), \qquad 0 \leq x \leq l,$$

and

$$\frac{\partial u}{\partial t}(x, 0) = g(x), \qquad 0 \leq x \leq l,$$

where α is a constant. To set up the finite-difference method, select an integer $m > 0$ and time-step size $k > 0$. With $h = l/m$, the mesh points (x_i, t_j) are defined by

$$x_i = ih, \qquad \text{for each } i = 0, 1, \ldots, m,$$

and

$$t_j = jk, \qquad \text{for each } j = 0, 1, \ldots.$$

At any interior mesh point (x_i, t_j) the wave equation becomes

(11.25)
$$\frac{\partial^2 u}{\partial t^2}(x_i, t_j) - \alpha^2 \frac{\partial^2 u}{\partial x^2}(x_i, t_j) = 0.$$

The difference method is obtained by using the centered-difference quotient for the second partial derivatives given by

(11.26)
$$\frac{\partial^2 u}{\partial t^2}(x_i, t_j) = \frac{u(x_i, t_{j+1}) - 2u(x_i, t_j) + u(x_i, t_{j-1})}{k^2} - \frac{k^2}{12} \frac{\partial^4 u}{\partial t^4}(x_i, \mu_j),$$

where $t_{j-1} < \mu_j < t_{j+1}$ and

(11.27)
$$\frac{\partial^2 u}{\partial x^2}(x_i, t_j) = \frac{u(x_{i+1}, t_j) - 2u(x_i, t_j) + u(x_{i-1}, t_j)}{h^2} - \frac{h^2}{12} \frac{\partial^4 u}{\partial x^4}(\xi_i, t_j),$$

where $x_{i-1} < \xi_i < x_{i+1}$. Substituting these into Eq. (11.25) gives

(11.28)

$$\frac{u(x_i, t_{j+1}) - 2u(x_i, t_j) + u(x_i, t_{j-1})}{k^2} - \alpha^2 \frac{u(x_{i+1}, t_j) - 2u(x_i, t_j) + u(x_{i-1}, t_j)}{h^2}$$

$$= \frac{1}{12}\left[k^2 \frac{\partial^4 u}{\partial t^4}(x_i, \mu_j) - \alpha^2 h^2 \frac{\partial^4 u}{\partial x^4}(\xi_i, t_j) \right].$$

Neglecting the truncation error

(11.29)
$$\tau_{i,j} = \frac{1}{12}\left[k^2 \frac{\partial^4 u}{\partial t^4}(x_i, \mu_j) - \alpha^2 h^2 \frac{\partial^4 u}{\partial x^4}(\xi_i, t_j) \right]$$

leads to the difference equation

(11.30)
$$\frac{w_{i,j+1} - 2w_{i,j} + w_{i,j-1}}{k^2} - \alpha^2 \frac{w_{i+1,j} - 2w_{i,j} + w_{i-1,j}}{h^2} = 0.$$

If λ is used to denote $\alpha k/h$, we can write the difference equation as

$$w_{i,j+1} - 2w_{i,j} + w_{i,j-1} - \lambda^2 w_{i+1,j} + 2\lambda^2 w_{i,j} - \lambda^2 w_{i-1,j} = 0$$

and solve for $w_{i,j+1}$, the most advanced time-step approximation, to obtain

(11.31)
$$w_{i,j+1} = 2(1 - \lambda^2)w_{i,j} + \lambda^2(w_{i+1,j} + w_{i-1,j}) - w_{i,j-1}.$$

This equation holds for each $i = 1, 2, \ldots, (m-1)$, and $j = 1, 2, \ldots$. The boundary conditions give

(11.32)
$$w_{0,j} = w_{m,j} = 0 \qquad \text{for each } j = 1, 2, 3, \ldots,$$

and the initial condition implies that

(11.33) $\qquad\qquad w_{i,0} = f(x_i) \qquad$ for each $i = 1, 2, \ldots, m - 1$.

Writing this set of equations in matrix form implies that

(11.34)

$$
\begin{bmatrix} w_{1,j+1} \\ w_{2,j+1} \\ \vdots \\ w_{m-1,j+1} \end{bmatrix} = \begin{bmatrix} 2(1 - \lambda^2) & \lambda^2 & 0 \cdots\cdots 0 \\ \lambda^2 & 2(1 - \lambda^2) & \lambda^2 \\ 0 & & & 0 \\ \vdots & & & \\ \vdots & & & \lambda^2 \\ 0 \cdots\cdots 0 & \lambda^2 & 2(1 - \lambda^2) \end{bmatrix} \begin{bmatrix} w_{1,j} \\ w_{2,j} \\ \vdots \\ w_{m-1,j} \end{bmatrix} - \begin{bmatrix} w_{1,j-1} \\ w_{2,j-1} \\ \vdots \\ w_{m-1,j-1} \end{bmatrix}
$$

Equations (11.31) and (11.32) imply that the $(j + 1)$st time step requires values from the jth and $(j - 1)$st time steps. (*See Fig. 11.10.*) This produces a minor starting problem since values for $j = 0$ are given by Eq. (11.33), but values for $j = 1$, which are needed in Eq. (11.31) to compute $w_{i,2}$, must be obtained from the initial-velocity condition

$$\frac{\partial u}{\partial t}(x, 0) = g(x), \qquad 0 \le x \le l.$$

FIGURE 11.10

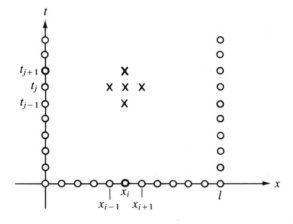

The first approach is to replace $(\partial u/\partial t)$ by a forward-difference approximation,

(11.35) $\qquad \dfrac{\partial u}{\partial t}(x_i, 0) = \dfrac{u(x_i, t_1) - u(x_i, 0)}{k} - \dfrac{k}{2} \dfrac{\partial^2 u}{\partial t^2}(x_1, \tilde{\mu}_i). \qquad 0 < \tilde{\mu}_i < t_1.$

Solving for $u(x_i, t_1)$ gives

$$u(x_i, t_1) = u(x_i, 0) + k \frac{\partial u}{\partial t}(x_i, 0) + \frac{k^2}{2} \frac{\partial^2 u}{\partial t^2}(x_i, \tilde{\mu}_i)$$

$$= u(x_i, 0) + kg(x_i) + \frac{k^2}{2} \frac{\partial^2 u}{\partial t^2}(x_i, \tilde{\mu}_i).$$

As a consequence

(11.36) $w_{i,1} = w_{i,0} + kg(x_i)$ for each $i = 1, \ldots, m - 1$.

From Eq. (11.35), however, we can see that Eq. (11.36) gives an approximation that has local truncation error of only $O(k)$. A better approximation to $u(x_i, 0)$ can be rather easily obtained, particularly when the second derivative of f at x_i can be determined. Consider

(11.37) $\dfrac{u(x_i, t_1) - u(x_i, 0)}{k} = \dfrac{\partial u}{\partial t}(x_i, 0) + \dfrac{k}{2}\dfrac{\partial^2 u}{\partial t^2}(x_i, 0) + \dfrac{k^2}{6}\dfrac{\partial^3 u}{\partial t^3}(x_i, \tilde{\mu}_i)$

for some $0 < \tilde{\mu}_i < t_1$ and suppose the wave equation also holds on the initial line; that is,

$$\frac{\partial^2 u}{\partial t^2}(x_i, 0) - \alpha^2 \frac{\partial^2 u}{\partial x^2}(x_i, 0) = 0 \qquad \text{for each } i = 0, 1, \ldots, m.$$

If f'' exists, then

$$\frac{\partial^2 u}{\partial t^2}(x_i, 0) = \alpha^2 \frac{\partial^2 u}{\partial x^2}(x_i, 0) = \alpha^2 \frac{d^2 f}{dx^2}(x_i) = \alpha^2 f''(x_i).$$

Substituting into Eq. (11.37) and solving for $u(x_i, t_1)$ gives

$$u(x_i, t_1) = u(x_i, 0) + kg(x_i) + \frac{\alpha^2 k^2}{2} f''(x_i) + \frac{k^3}{6}\frac{\partial^3 u}{\partial t^3}(x_i, \tilde{\mu}_i)$$

and

(11.38) $w_{i,1} = w_{i,0} + kg(x_i) + \dfrac{\alpha^2 k^2}{2} f''(x_i).$

This is an approximation with local truncation error $O(k^2)$ for each $i = 1, 2, \ldots, m - 1$.

If $f \in C^4[0, 1]$ but $f''(x_i)$ is not readily available we can use Eq. (4.20) on page 142 to write

$$f''(x_i) = \frac{f(x_{i+1}) - 2f(x_i) + f(x_{i-1})}{h^2} - \frac{h^2}{12} f^{(4)}(\tilde{\xi}_i),$$

for some $\tilde{\xi}_i$ in (x_{i-1}, x_{i+1}). This implies that the approximation becomes

$$\frac{u(x_i, t_1) - u(x_i, 0)}{k} = g(x_i) + \frac{k\alpha^2}{2h^2}[f(x_{i+1}) - 2f(x_i) + f(x_{i-1})]$$
$$+ O(k^2 + h^2 k)$$

or, letting $\lambda = \dfrac{k\alpha}{h}$,

$$u(x_i, t_1) = u(x_i, 0) + kg(x_i) + \frac{\lambda^2}{2}[f(x_{i+1}) - 2f(x_i) + f(x_{i-1})] + O(k^3 + h^2 k^2)$$

$$= (1 - \lambda^2)f(x_i) + \frac{\lambda^2}{2} f(x_{i+1}) + \frac{\lambda^2}{2} f(x_{i-1}) + kg(x_i) + O(k^3 + h^2 k^2).$$

Thus, the difference equation

(11.39) $$w_{i,1} = (1 - \lambda^2)f(x_i) + \frac{\lambda^2}{2}f(x_{i+1}) + \frac{\lambda^2}{2}f(x_{i-1}) + kg(x_i),$$

can be used to find $w_{i,1}$ for each $i = 1, 2, \ldots, m-1$.

The following algorithm uses Eq. (11.39) to approximate $w_{i,1}$, although Eq. (11.36) could also be used. It is assumed that there is an upper bound for the value of t, to be used in the stopping technique.

Wave Equation Finite-Difference Algorithm 11.4

To approximate the solution to the wave equation

$$\frac{\partial^2 u}{\partial t^2}(x, t) - \alpha^2 \frac{\partial^2 u}{\partial x^2}(x, t) = 0, \qquad 0 < x < l, \quad 0 < t < T,$$

subject to the boundary conditions

$$u(0, t) = u(l, t) = 0, \qquad 0 < t < T,$$

and the initial conditions

$$u(x, 0) = f(x), \qquad 0 \le x \le l,$$

$$\frac{\partial u}{\partial t}(x, 0) = g(x), \qquad 0 \le x \le l:$$

INPUT endpoint l; maximum time T; constant α; integers m, N.

OUTPUT approximations $w_{i,j}$ to $u(x_i, t_j)$ for each $i = 0, \ldots, m$ and $j = 0, \ldots, N$.

Step 1 Set $h = l/m$;
$k = T/N$;
$\lambda = k\alpha/h$.

Step 2 For $j = 1, \ldots, N$ set $w_{0,j} = 0$;
$w_{m,j} = 0$.

Step 3 Set $w_{0,0} = f(0)$;
$w_{m,0} = f(l)$.

Step 4 For $i = 1, \ldots, m-1$ (*Initialize for $t = 0$ and $t = k$.*)
set $w_{i,0} = f(ih)$;

$$w_{i,1} = (1 - \lambda^2)f(ih) + \frac{\lambda^2}{2}[f((i+1)h) + f((i-1)h)] + kg(ih).$$

Step 5 For $j = 1, \ldots, N-1$ (*Perform matrix multiplication.*)
for $i = 1, \ldots, m-1$
set $w_{i,j+1} = 2(1 - \lambda^2)w_{i,j} + \lambda^2(w_{i+1,j} + w_{i-1,j}) - w_{i,j-1}$.

Step 6   For $j = 0, \ldots, N$
    set $t = jk$;
    for $i = 0, \ldots, m$
        set $x = ih$;
        OUTPUT $(x, t, w_{i, j})$.

Step 7   STOP.   (*Procedure is complete.*)

EXAMPLE 1   Consider the hyperbolic problem

$$\frac{\partial^2 u}{\partial t^2}(x, t) - 4\frac{\partial^2 u}{\partial x^2}(x, t) = 0, \qquad 0 < x < 1, \quad 0 < t,$$

with boundary conditions $u(0, t) = u(1, t) = 0$, $0 < t$, and initial conditions

$$u(x, 0) = \sin(\pi x), \quad 0 \leq x \leq 1, \qquad \text{and} \qquad \frac{\partial u}{\partial t}(x, 0) = 0, \quad 0 \leq x \leq 1.$$

It is easily verified that the solution to this problem is

$$u(x, t) = \sin(\pi x)\cos(2\pi t).$$

The finite-difference method (Algorithm 11.4) is used in this example with $m = 10$, $T = 1$, and $N = 20$, which implies that $h = 0.1$, $k = 0.05$, and $\lambda = 1$. The following table lists the results of the approximation, $w_{i, N}$, for $i = 0, 1, \ldots, 10$. The values listed in Table 11.5 are correct to the places given.   ∎

TABLE 11.5

x_i	$w_{i, 20}$
0.0	0.0000000000
0.1	0.3090169944
0.2	0.5877852523
0.3	0.8090169944
0.4	0.9510565163
0.5	1.0000000000
0.6	0.9510565163
0.7	0.8090169944
0.8	0.5877852523
0.9	0.3090169944
1.0	0.0000000000

The results of the example were very accurate, more so than the truncation error $O(k^2 + h^2)$ would lead us to believe. The explanation for this phenomenon lies in the fact that the true solution to the equation is infinitely differentiable. When this is the case, it is easy to show, using Taylor series, that:

$$\frac{u(x_{i+1}, t_j) - 2u(x_i, t_j) + u(x_{i-1}, t_j)}{h^2}$$

$$= \frac{\partial^2 u}{\partial x^2}(x_i, t_j) + 2\left[\frac{h^2}{4!}\frac{\partial^4 u}{\partial x^4}(x_i, t_j) + \frac{h^4}{6!}\frac{\partial^6 u}{\partial x^6}(x_i, t_j) + \cdots\right]$$

and

$$\frac{u(x_i, t_{j+1}) - 2u(x_i, t_j) + u(x_i, t_{j-1})}{k^2}$$

$$= \frac{\partial^2 u}{\partial t^2}(x_i, t_j) + 2\left[\frac{k^2}{4!}\frac{\partial^4 u}{\partial t^4}(x_i, t_j) + \frac{k^4}{6!}\frac{\partial^6 u}{\partial t^6}(x_i, t_j) + \cdots\right].$$

Thus,

(11.40)

$$\frac{u(x_i, t_{j+1}) - 2u(x_i, t_j) + u(x_i, t_{j-1})}{k^2} - \alpha^2 \frac{u(x_{i+1}, t_j) - 2u(x_i, t_j) + u(x_{i-1}, t_j)}{h^2}$$

$$= 2\left[\frac{1}{4!}\left(k^2\frac{\partial^4 u}{\partial t^4}(x_i, t_j) - \alpha^2 h^2\frac{\partial^4 u}{\partial x^4}(x_i, t_j)\right)\right.$$

$$\left. + \frac{1}{6!}\left(k^4\frac{\partial^6 u}{\partial t^6}(x_i, t_j) - \alpha^2 h^4\frac{\partial^6 u}{\partial x^6}(x_i, t_j)\right) + \cdots\right].$$

However, by differentiating the wave equation,

$$k^2\frac{\partial^4 u}{\partial t^4}(x_i, t_j) = k^2\frac{\partial^2}{\partial t^2}\left[\alpha^2\frac{\partial^2 u}{\partial x^2}(x_i, t_j)\right] = \alpha^2 k^2\frac{\partial^2}{\partial x^2}\left[\frac{\partial^2 u}{\partial t^2}(x_i, t_j)\right]$$

$$= \alpha^2 k^2\frac{\partial^2}{\partial x^2}\left[\alpha^2\frac{\partial^2 u}{\partial x^2}(x_i, t_j)\right] = \alpha^4 k^2\frac{\partial^4 u}{\partial x^4}(x_i, t_j),$$

we see that

$$\frac{1}{4!}\left[k^2\frac{\partial^4 u}{\partial t^4}(x_i, t_j) - \alpha^2 h^2\frac{\partial^4 u}{\partial x^4}(x_i, t_j)\right] = \frac{\alpha^2}{4!}[\alpha^2 k^2 - h^2]\frac{\partial^4 u}{\partial x^4}(x_i, t_j) = 0$$

since

$$\lambda^2 = \frac{\alpha^2 k^2}{h^2} = 1.$$

Continuing in this manner, all the terms on the right-hand side of (11.40) are zero, implying a zero local truncation error. The only errors in Example 1 are those due to the approximation of $w_{i,1}$ and to rounding.

As in the case of the forward-difference method for the heat equation, the explicit finite-difference method for the wave equation has stability problems. In fact, it is necessary that $\lambda = \alpha k/h \leq 1$ for the method to be stable. (See Isaacson–Keller [67], page 489.) The explicit method given in Algorithm 11.4, with $\lambda \leq 1$ is $O(h^2 + k^2)$-convergent if f and g are sufficiently differentiable. For verification of this, see Isaacson–Keller [67], page 491.

Although we will not discuss them, there are implicit methods that are unconditionally stable. A discussion of these methods can be found in Ames [2], page 199 or in the books by Mitchell [88] or Smith [127].

Exercise Set 11.4

1. Approximate the solution to the wave equation

$$\frac{\partial^2 u}{\partial t^2} - \frac{\partial^2 u}{\partial x^2} = 0, \qquad 0 < x < 1, \quad 0 < t;$$

$$u(0, t) = u(1, t) = 0, \qquad 0 < t,$$

$$u(x, 0) = \sin 2\pi x, \qquad 0 \le x \le 1,$$

$$\frac{\partial u}{\partial t}(x, 0) = \sin \pi x, \qquad 0 \le x \le 1,$$

using Algorithm 11.4 with $m = 4$, $N = 2$, and $T = 0.5$.

2. Approximate the solution to the wave equation

$$\frac{\partial^2 u}{\partial t^2} - \frac{1}{4}\frac{\partial^2 u}{\partial x^2} = 0, \qquad 0 < x < 0.5, \quad 0 < t;$$

$$u(0, t) = u(0.5, t) = 0, \qquad 0 < t,$$

$$u(x, 0) = 0, \qquad 0 \le x \le 0.5,$$

$$\frac{\partial u}{\partial t}(x, 0) = \sin 4\pi x, \qquad 0 \le x \le 0.5,$$

using Algorithm 11.4 with $m = 2$, $N = 2$, and $T = 0.5$.

3. Approximate the solution $u(x, t) = \sin \pi x \cos \pi t$ to the wave equation

$$\frac{\partial^2 u}{\partial t^2} - \frac{\partial^2 u}{\partial x^2} = 0, \qquad 0 < x < 1, \quad 0 < t;$$

$$u(0, t) = u(1, t) = 0, \qquad 0 < t,$$

$$u(x, 0) = \sin \pi x, \qquad 0 \le x \le 1,$$

$$\frac{\partial u}{\partial t}(x, 0) = 0, \qquad 0 \le x \le 1,$$

using Algorithm 11.4 with $h = 0.1$ and $k = 0.05$, with $h = 0.05$ and $k = 0.1$, and then with $h = 0.05$ and $k = 0.05$. Compare your results to the exact solution at $t = 0.5$.

4. Repeat Exercise 3, using the approximation

$$w_{i,1} = w_{i,0} + kg(x_i) \qquad \text{for each } i = 0, 1, \ldots, m.$$

5. Approximate the solution to the wave equation

$$\frac{\partial^2 u}{\partial t^2} - \frac{\partial^2 u}{\partial x^2} = 0, \qquad 0 < x < 1, \quad 0 < t;$$

$$u(0, t) = u(1, t) = 0, \qquad 0 < t,$$

$$u(x, 0) = \sin 2\pi x, \qquad 0 \le x \le 1,$$

$$\frac{\partial u}{\partial t}(x, 0) = 2\pi \sin 2\pi x, \qquad 0 \le x \le 1,$$

using Algorithm 11.4 with $h = 0.1$ and $k = 0.1$. Compare your results to the actual solution $u(x, t) = \sin 2\pi x[\cos 2\pi t + \sin 2\pi t]$, at $t = 0.3$.

6. Approximate the solution to the wave equation

$$\frac{\partial^2 u}{\partial t^2} - \frac{\partial^2 u}{\partial x^2} = 0, \qquad 0 < x < 1, \quad 0 < t;$$

$$u(0, t) = u(1, t) = 0, \qquad 0 < t,$$

$$u(x, 0) = \begin{cases} 1, & 0 \le x \le \frac{1}{2}, \\ -1, & \frac{1}{2} < x \le 1, \end{cases}$$

$$\frac{\partial u}{\partial t}(x, 0) = 0, \qquad 0 \le x \le 1.$$

Use $h = 0.1$ and $k = 0.1$, and compare your answer to the Fourier series solution at $t = 0.5$.

7. The air pressure $p(x, t)$ in an organ pipe is governed by the wave equation

$$\frac{\partial^2 p}{\partial x^2} = \frac{1}{c^2} \frac{\partial^2 p}{\partial t^2}, \qquad 0 < x < l, \quad 0 < t,$$

where l is the length of the pipe and c is a physical constant. If the pipe is open, the boundary conditions are given by

$$p(0, t) = p_0 \qquad \text{and} \qquad p(l, t) = p_0.$$

If the pipe is closed at the end where $x = l$, the boundary conditions are

$$p(0, t) = p_0 \qquad \text{and} \qquad \frac{\partial p}{\partial x}(l, t) = 0.$$

Assume that $c = 1$, $l = 1$ and the initial conditions are

$$p(x, 0) = p_0 \cos 2\pi x, \qquad 0 \le x \le l,$$

$$\frac{\partial p}{\partial t}(x, 0) = 0, \qquad 0 \le x \le 1.$$

a) Approximate the pressure for an open pipe with $p_0 = 0.9$ at $x = \frac{1}{2}$ for $t = 0.5$ and $t = 1$, using Algorithm 11.4 with $h = k = 0.1$.

b) Modify Algorithm 11.4 for the closed pipe problem with $p_0 = 0.9$, and approximate $p(0.5, 0.5)$ and $p(0.5, 1)$, using $h = k = 0.1$.

8. In an electric transmission line of length l that carries alternating current of high frequency (called a "lossless" line), the voltage V and current i are described by

$$\frac{\partial^2 V}{\partial x^2} = LC \frac{\partial^2 V}{\partial t^2}, \qquad 0 < x < l, \quad 0 < t,$$

$$\frac{\partial^2 i}{\partial x^2} = LC \frac{\partial^2 i}{\partial t^2}, \qquad 0 < x < l, \quad 0 < t,$$

where L is the inductance per unit length and C is the capacitance per unit length. Suppose the line is 200 feet long and the constants C and L are given by

$$C = 0.1 \text{ farads/ft},$$
$$L = 0.3 \text{ henries/ft}.$$

Suppose the voltage and current also satisfy

$$V(0, t) = V(200, t) = 0, \qquad 0 < t,$$

$$V(x, 0) = 110 \sin \frac{\pi x}{200}, \qquad 0 \le x \le 200,$$

$$\frac{\partial V}{\partial t}(x, 0) = 0, \qquad 0 \le x \le 200,$$

$$i(0, t) = i(200, t) = 0, \qquad 0 < t,$$

$$i(x, 0) = 5.5 \cos \frac{\pi x}{200}, \qquad 0 \le x \le 200,$$

$$\frac{\partial i}{\partial t}(x, 0) = 0, \qquad 0 \le x \le 200.$$

a) Approximate the voltage and current at $t = 0.2$ and $t = 0.5$, using Algorithm 11.4 with $h = 10$ and $k = 0.1$.

b) Find the first three terms of the Fourier-series solution for V and i, and compare the values $V(50, 0.2)$, $V(50, 0.5)$, $i(50, 0.2)$, and $i(50, 0.5)$ to the approximations in part (a).

11.5 An Introduction to the Finite-Element Method

A method frequently used to solve partial-differential equations that occur in engineering applications is called the **finite-element method**. This method is similar to the Rayleigh–Ritz procedure discussed in Section 10.5, but generalized to higher dimensions.

One advantage of the finite-element method over finite-difference methods is the relative ease with which the boundary conditions of the problem are handled. Many physical problems have boundary conditions involving derivatives and, in general, the boundary of the region is irregularly shaped. Boundary conditions of this type are very difficult to handle using finite-difference techniques, since each boundary condition involving a derivative must be approximated by a difference quotient at the grid points, and irregular shaping of the boundary makes placing the grid points difficult. The finite-element method includes the boundary conditions as integrals in a functional that is being minimized, so the construction procedure is independent of the particular boundary conditions of the problem.

In our discussion, we consider the partial-differential equation

(11.41)
$$\frac{\partial}{\partial x}\left(p(x, y)\frac{\partial u}{\partial x}\right) + \frac{\partial}{\partial y}\left(q(x, y)\frac{\partial u}{\partial y}\right) + r(x, y)u(x, y) = f(x, y),$$

with $(x, y) \in \mathcal{D}$, where $\mathcal{D}$ is a plane region with boundary $\mathcal{S}$.

Boundary conditions of the form

(11.42)
$$u(x, y) = g(x, y)$$

are imposed on a portion, $\mathscr{S}_1$, of the boundary. On the remainder of the boundary, $\mathscr{S}_2$, $u(x, y)$ is required to satisfy

(11.43) $p(x, y)\dfrac{\partial u}{\partial x}\cos \theta_1 + q(x, y)\dfrac{\partial u}{\partial y}\cos \theta_2 + g_1(x, y)u(x, y) = g_2(x, y),$

where θ_1 and θ_2 are the direction angles of the outward normal to the boundary at the point (x, y). (*See Fig. 11.11.*)

FIGURE 11.11

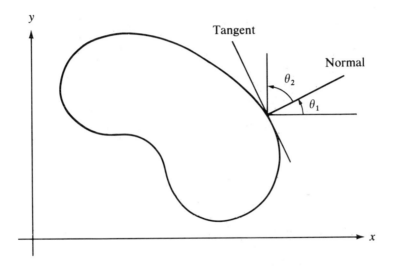

Physical problems in the areas of solid mechanics and elasticity have associated partial-differential equations similar to Eq. (11.41). The solution to a problem of this type is typically the minimization of a certain functional, involving integrals, over a class of functions determined by the problem.

Suppose p, q, r, and f are all continuous in $\mathscr{D} \cup \mathscr{S}$, p and q have continuous first partials, and g_1 and g_2 are continuous on $\mathscr{S}_2$. Suppose, in addition, that $p(x, y) > 0$, $q(x, y) > 0$, $r(x, y) \leq 0$, and $g_1(x, y) > 0$. Then a solution to Eq. (11.41) uniquely minimizes the functional

(11.44)

$$ I[w] = \iint\limits_{\mathscr{D}} \left\{ \frac{1}{2}\left[p(x, y)\left(\frac{\partial w}{\partial x}\right)^2 + q(x, y)\left(\frac{\partial w}{\partial y}\right)^2 - r(x, y)w^2 \right] + f(x, y)w \right\} dx\, dy $$

$$ + \int_{\mathscr{S}_2} \left\{ -g_2(x, y)w + \tfrac{1}{2}g_1(x, y)w^2 \right\} dS $$

over all functions w, satisfying Eq. (11.42) on $\mathscr{S}_1$, which are twice continuously differentiable. The finite-element method approximates this solution by minimizing the functional I over a smaller class of functions, just as the Raleigh–Ritz method did for the boundary-value problem considered in Section 10.5.

The first step in the procedure is to divide the region into a finite number of sections, or elements, of a regular shape, either rectangles or, more commonly, triangles. (*See Fig. 11.12.*)

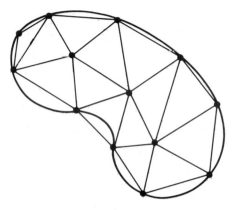

The set of functions used for approximation is generally a set of piecewise polynomials of fixed degree in x and y, and the approximation requires that the polynomials be pieced together in such a manner that the resulting function will be continuous with an integrable or continuous first or second derivative on the entire region. Polynomials of linear type in x and y,

$$\phi(x, y) = a + bx + cy,$$

are commonly used with triangular elements while polynomials of bilinear type in x and y,

$$\phi(x, y) = a + bx + cy + dxy,$$

are used with rectangular elements.

For our discussion, suppose that the region $\mathscr{D}$ has been subdivided into triangular elements. The collection of triangles will be denoted D, and the vertices of these triangles are called **nodes**. The method seeks an approximation of the form

$$\phi(x, y) = \sum_{i=1}^{m} \gamma_i \phi_i(x, y),$$

where $\phi_1, \phi_2, ..., \phi_m$ are linearly independent piecewise linear polynomials and $\gamma_1, \gamma_2, ..., \gamma_m$ are constants. Some of these constants, say, $\gamma_{n+1}, \gamma_{n+2}, ..., \gamma_m$, are used to ensure that the boundary condition

$$\phi(x, y) = g(x, y)$$

is satisfied on $\mathscr{S}_1$, while the remaining constants $\gamma_1, \gamma_2, ..., \gamma_n$ are used to minimize the functional $I[\sum_{i=1}^{m} \gamma_i \phi_i]$.

From Eq. (11.44) the functional is of the form

$$(11.45) \qquad I[\phi] = I\left[\sum_{i=1}^{m} \gamma_i \phi_i\right]$$

$$= \iint_{\mathscr{D}} \left(\frac{1}{2}\left\{p(x, y)\left[\sum_{i=1}^{m} \gamma_i \frac{\partial \phi_i}{\partial x}(x, y)\right]^2\right.\right.$$

$$\left. + q(x, y)\left[\sum_{i=1}^{m} \gamma_i \frac{\partial \phi_i}{\partial y}(x, y)\right]^2 - r(x, y)\left[\sum_{i=1}^{m} \gamma_i \phi_i(x, y)\right]^2\right\}$$

$$\left. + f(x, y) \sum_{i=1}^{m} \gamma_i \phi_i(x, y)\right) dy\, dx$$

$$+ \int_{\mathscr{S}_2} \left\{-g_2(x, y) \sum_{i=1}^{m} \gamma_i \phi_i(x, y) + \tfrac{1}{2}g_1(x, y)\left[\sum_{i=1}^{m} \gamma_i \phi_i(x, y)\right]^2\right\} dS.$$

For a minimum to occur, considering I as a function of $\gamma_1, \ldots, \gamma_n$, it is necessary to have

$$(11.46) \qquad \frac{\partial I}{\partial \gamma_j} = 0 \qquad \text{for each } j = 1, 2, \ldots, n.$$

Differentiating (11.45) gives:

$$\frac{\partial I}{\partial \gamma_j} = \iint_{\mathscr{D}} \left\{p(x, y) \sum_{i=1}^{m} \gamma_i \frac{\partial \phi_i}{\partial x}(x, y)\frac{\partial \phi_j}{\partial x}(x, y)\right.$$

$$+ q(x, y) \sum_{i=1}^{m} \gamma_i \frac{\partial \phi_i}{\partial y}(x, y)\frac{\partial \phi_j}{\partial y}(x, y)$$

$$\left. - r(x, y) \sum_{i=1}^{m} \gamma_i \phi_i(x, y)\phi_j(x, y) + f(x, y)\phi_j(x, y)\right\} dx\, dy$$

$$+ \int_{\mathscr{S}_2} \left\{-g_2(x, y)\phi_j(x, y) + g_1(x, y) \sum_{i=1}^{m} \gamma_i \phi_i(x, y)\phi_j(x, y)\right\} dS;$$

so

$$(11.47) \qquad 0 = \sum_{i=1}^{m} \left[\iint_{\mathscr{D}} \left\{p(x, y)\frac{\partial \phi_i}{\partial x}(x, y)\frac{\partial \phi_j}{\partial x}(x, y) + q(x, y)\frac{\partial \phi_i}{\partial y}(x, y)\frac{\partial \phi_j}{\partial y}(x, y)\right.\right.$$

$$\left. - r(x, y)\phi_i(x, y)\phi_j(x, y)\right\} dx\, dy$$

$$\left. + \int_{\mathscr{S}_2} g_1(x, y)\phi_i(x, y)\phi_j(x, y)\, dS\right] \gamma_i$$

$$+ \iint_{\mathscr{D}} f(x, y)\phi_j(x, y)\, dx\, dy - \int_{\mathscr{S}_2} g_2(x, y)\phi_j(x, y)\, dS,$$

for each $j = 1, 2, \ldots, n$. This set of equations can be written as a linear system:

$$(11.48) \qquad \qquad \qquad A\mathbf{c} = \mathbf{b},$$

where $A = (\alpha_{ij})$, $\mathbf{c} = (\gamma_1, \ldots, \gamma_n)^t$ and $\mathbf{b} = (\beta_1, \ldots, \beta_n)^t$ are defined by

(11.49) $\alpha_{ij} = \iint\limits_{\mathcal{D}} \left[p(x, y)\frac{\partial \phi_i}{\partial x}(x, y)\frac{\partial \phi_j}{\partial x}(x, y) + q(x, y)\frac{\partial \phi_i}{\partial y}(x, y)\frac{\partial \phi_j}{\partial y}(x, y) \right.$

$$\left. - r(x, y)\phi_i(x, y)\phi_j(x, y) \right] dx\, dy + \int_{\mathscr{S}_2} g_1(x, y)\phi_i(x, y)\phi_j(x, y)\, dS,$$

for each $i = 1, 2, \ldots, n$, $j = 1, 2, \ldots, m$, and

(11.50) $\beta_i = -\iint\limits_{\mathcal{D}} f(x, y)\phi_i(x, y)\, dx\, dy + \int_{\mathscr{S}_2} g_2(x, y)\phi_i(x, y)\, dS - \sum_{k=n+1}^{m} \alpha_{ik}\gamma_k,$

for each $i = 1, \ldots, n$.

The particular choice of basis functions is very important since the appropriate choice can often make the matrix A positive definite and banded. For our second-order problem, (11.41), we will assume that $\mathcal{D}$ is polygonal and that $\mathscr{S}_1$ is a contiguous set of straight lines so that $\mathcal{D} = D$. To begin the procedure we divide the region D into a collection of triangles $T_1, T_2, \ldots, T_M$ with the ith triangle having three vertices, or nodes, denoted

$$V_j^{(i)} = (x_j^{(i)}, y_j^{(i)}) \qquad \text{for } j = 1, 2, 3.$$

To simplify the notation, we write $V_j^{(i)}$ simply as $V_j = (x_j, y_j)$ when working with the fixed triangle T_i. With each vertex V_j we associate a linear polynomial

$$N_j^{(i)} \equiv N_j = a_j + b_j x + c_j y, \qquad \text{where} \quad N_j^{(i)}(x_k, y_k) = \begin{cases} 1, & \text{if } j = k, \\ 0, & \text{if } j \neq k. \end{cases}$$

This produces linear systems of the form.

$$\begin{bmatrix} 1 & x_1 & y_1 \\ 1 & x_2 & y_2 \\ 1 & x_3 & y_3 \end{bmatrix} \begin{bmatrix} a_j \\ b_j \\ c_j \end{bmatrix} = \begin{bmatrix} 0 \\ 1 \\ 0 \end{bmatrix},$$

with the element one occurring in the jth row in the vector on the right.

Let $E_1, \ldots, E_n$ be a labeling of the nodes lying in $D \cup \mathscr{S}$ in a left-to-right, top-to-bottom fashion. With each node E_k, we associate a function ϕ_k that is linear on each triangle, has the value one at E_k, and is zero at each of the other nodes. This choice makes ϕ_k identical to $N_j^{(i)}$ on triangle T_i when the node E_k is the vertex denoted $V_j^{(i)}$.

EXAMPLE 1   Suppose that a finite element problem contains the triangles T_1 and T_2 shown in Figure 11.13. The linear function $N_1^{(1)}(x, y)$ that assumes the value one at $(1, 1)$ and zero at both $(0, 0)$ and $(-1, 2)$ satisfies

$$a_1^{(1)} + b_1^{(1)}(1) + c_1^{(1)}(1) = 1,$$

$$a_1^{(1)} + b_1^{(1)}(-1) + c_1^{(1)}(2) = 0,$$

and $$a_1^{(1)} + b_1^{(1)}(0) + c_1^{(1)}(0) = 0.$$

FIGURE 11.13

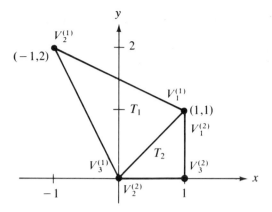

So $a_1^{(1)} = 0$, $b_1^{(1)} = \frac{2}{3}$, $c_1^{(1)} = \frac{1}{3}$, and

$$N_1^{(1)}(x, y) = \tfrac{2}{3}x + \tfrac{1}{3}y.$$

In a similar manner, $N_1^{(2)}(x, y)$, the linear function that assumes the value one at $(1, 1)$ and zero at both $(0, 0)$ and $(1, 0)$ must satisfy

$$a_1^{(2)} + b_1^{(2)}(1) + c_1^{(2)}(1) = 1,$$

$$a_1^{(2)} + b_1^{(2)}(0) + c_1^{(2)}(0) = 0,$$

and

$$a_1^{(2)} + b_1^{(2)}(1) + c_1^{(2)}(0) = 0,$$

so $a_1^{(2)} = 0$, $b_1^{(2)} = 0$, and $c_1^{(2)} = 1$. As a consequence $N_1^{(2)}(x, y) = y$. Note that on the common boundary of T_1 and T_2, $N_1^{(1)}(x, y) = N_1^{(2)}(x, y)$, since $y = x$.   ■

To generate the entries in the matrix A, consider as an example Fig. 11.14, the upper left portion of the region shown in Fig. 11.12.

FIGURE 11.14

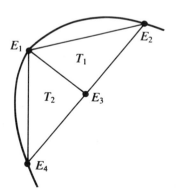

For simplicity we assume that E_1 is not one of the nodes on $\mathscr{S}_2$. The relationship between the nodes and the vertices of the triangles for this portion is

$$E_1 = V_3^{(1)} = V_1^{(2)}, \quad E_4 = V_2^{(2)}, \quad E_3 = V_2^{(1)} = V_3^{(2)}, \quad \text{and} \quad E_2 = V_1^{(1)}.$$

Since ϕ_1 and ϕ_3 are both nonzero on T_1 and T_2, the entries $\alpha_{1,3} = \alpha_{3,1}$ are computed by:

$$\alpha_{1,3} = \iint_D \left[p \frac{\partial \phi_1}{\partial x} \frac{\partial \phi_3}{\partial x} + q \frac{\partial \phi_1}{\partial y} \frac{\partial \phi_3}{\partial y} - r\phi_1 \phi_3 \right] dx\, dy$$

$$= \iint_{T_1} \left[p \frac{\partial \phi_1}{\partial x} \frac{\partial \phi_3}{\partial x} + q \frac{\partial \phi_1}{\partial y} \frac{\partial \phi_3}{\partial y} - r\phi_1 \phi_3 \right] dx\, dy$$

$$+ \iint_{T_2} \left[p \frac{\partial \phi_1}{\partial x} \frac{\partial \phi_3}{\partial x} + q \frac{\partial \phi_1}{\partial y} \frac{\partial \phi_3}{\partial y} - r\phi_1 \phi_3 \right] dx\, dy$$

$$= b_3^{(1)} b_2^{(1)} \iint_{T_1} p\, dx\, dy + c_3^{(1)} c_2^{(1)} \iint_{T_1} q\, dx\, dy$$

$$- \iint_{T_1} r(a_3^{(1)} + b_3^{(1)}x + c_3^{(1)}y)(a_2^{(1)} + b_2^{(1)}x + c_2^{(1)}y)\, dx\, dy$$

$$+ b_1^{(2)} b_3^{(2)} \iint_{T_2} p\, dx\, dy + c_1^{(2)} c_3^{(2)} \iint_{T_2} q\, dx\, dy$$

$$- \iint_{T_2} r(a_1^{(2)} + b_1^{(2)}x + c_1^{(2)}y)(a_3^{(2)} + b_3^{(2)}x + c_3^{(2)}y)\, dx\, dy.$$

Part of the entry β_1 is contributed by ϕ_1 restricted to T_1 and the remainder by ϕ_1 restricted to T_2. In general, an entry β_k has contributions from ϕ_k restricted to each of the triangles of which E_k is a vertex. In addition, nodes that lie on $\mathscr{S}_2$ will have line integrals added to their entries in A and $\mathbf{b}$.

The following algorithm performs the finite-element method on a second-order elliptic differential equation. The algorithm sets all values of the matrix A and vector $\mathbf{b}$ initially at zero and, after all the integrations have been performed on all the triangles, adds these values to the appropriate entries in A and $\mathbf{b}$.

Finite-Element Algorithm 11.5

To approximate the solution to the partial-differential equation

$$\frac{\partial}{\partial x}\left(p(x,y) \frac{\partial u}{\partial x} \right) + \frac{\partial}{\partial y}\left(q(x,y) \frac{\partial u}{\partial y} \right) + r(x,y)u = f(x,y), \qquad (x,y) \in D$$

subject to the boundary conditions

$$u(x,y) = g(x,y), \qquad (x,y) \in \mathscr{S}_1$$

and

$$p(x, y)\frac{\partial u}{\partial x}(x, y)\cos\theta_1 + q(x, y)\frac{\partial u}{\partial y}(x, y)\cos\theta_2 + g_1(x, y)u(x, y) = g_2(x, y), (x, y) \in \mathscr{S}_2,$$

where $\mathscr{S}_1 \cup \mathscr{S}_2$ is the boundary of D, and θ_1 and θ_2 are the direction angles of the normal to the boundary:

Step 0 Divide the region D into triangles $T_1, \ldots, T_M$ such that: $T_1, \ldots, T_K$ are the triangles with all vertices interior to D;
(*Note: $K = 0$ implies that no triangle is interior to D.*)
$T_{K+1}, \ldots, T_N$ are the triangles with at least one edge on $\mathscr{S}_2$; $T_{N+1}, \ldots, T_M$ are the remaining triangles.
(*Note: $M = N$ implies that all triangles have edges on $\mathscr{S}_2$.*)
Label the three vertices of the triangle T_i by
$(x_1^{(i)}, y_1^{(i)})$, $(x_2^{(i)}, y_2^{(i)})$, and $(x_3^{(i)}, y_3^{(i)})$.
Label the nodes (vertices) $E_1, \ldots, E_m$ where
$E_1, \ldots, E_n$ are in $D \cup \mathscr{S}_2$ and $E_{n+1}, \ldots, E_m$ are on $\mathscr{S}_1$.
(*Note: $n = m$ implies that $\mathscr{S}_1$ contains no nodes.*)

INPUT integers K, N, M, n, m; vertices $(x_1^{(i)}, y_1^{(i)}), (x_2^{(i)}, y_2^{(i)}), (x_3^{(i)}, y_3^{(i)})$ for each $i = 1, \ldots, M$; nodes E_j for each $j = 1, \ldots, m$.
(*Note: All that is needed is a means of corresponding a vertex $(x_k^{(i)}, y_k^{(i)})$ to a node $E_j = (x_j, y_j)$.*)
OUTPUT constants $\gamma_1, \ldots, \gamma_m$; $a_j^{(i)}, b_j^{(i)}, c_j^{(i)}$ for each $j = 1, 2, 3$ and $i = 1, \ldots, M$.

Step 1 For $l = n+1, \ldots, m$ set $\gamma_l = g(x_l, y_l)$,   (*Note: $E_l = (x_l, y_l)$*).

Step 2 For $i = 1, \ldots, n$
set $\beta_i = 0$;
for $j = 1, \ldots, n$ set $\alpha_{i,j} = 0$.

Step 3 For $i = 1, \ldots, M$

$$\text{set } \Delta_i = \det \begin{vmatrix} 1 & x_1^{(i)} & y_1^{(i)} \\ 1 & x_2^{(i)} & y_2^{(i)} \\ 1 & x_3^{(i)} & y_3^{(i)} \end{vmatrix};$$

$$a_1^{(i)} = \frac{x_2^{(i)}y_3^{(i)} - y_2^{(i)}x_3^{(i)}}{\Delta_i}; \quad b_1^{(i)} = \frac{y_2^{(i)} - y_3^{(i)}}{\Delta_i}; \quad c_1^{(i)} = \frac{x_3^{(i)} - x_2^{(i)}}{\Delta_i};$$

$$a_2^{(i)} = \frac{x_3^{(i)}y_1^{(i)} - y_3^{(i)}x_1^{(i)}}{\Delta_i}; \quad b_2^{(i)} = \frac{y_3^{(i)} - y_1^{(i)}}{\Delta_i}; \quad c_2^{(i)} = \frac{x_1^{(i)} - x_3^{(i)}}{\Delta_i};$$

$$a_3^{(i)} = \frac{x_1^{(i)}y_2^{(i)} - y_1^{(i)}x_2^{(i)}}{\Delta_i}; \quad b_3^{(i)} = \frac{y_1^{(i)} - y_2^{(i)}}{\Delta_i}; \quad c_3^{(i)} = \frac{x_2^{(i)} - x_1^{(i)}}{\Delta_i};$$

for $j = 1, 2, 3$
define $N_j^{(i)}(x, y) = a_j^{(i)} + b_j^{(i)}x + c_j^{(i)}y$.

Step 4 For $i = 1, \ldots, M$,   (*The integrals in Steps 4 and 5 can be evaluated using numerical integration procedures.*)
for $j = 1, 2, 3$,
for $k = 1, \ldots, j$ (*Compute all double integrals over the triangles.*)

$$\text{set } z_{j,k}^{(i)} = b_j^{(i)} b_k^{(i)} \iint\limits_{T_i} p(x, y)\, dx\, dy + c_j^{(i)} c_k^{(i)} \iint\limits_{T_i} q(x, y)\, dx\, dy$$

$$-\iint\limits_{T_i} r(x, y) N_j^{(i)}(x, y) N_k^{(i)}(x, y)\, dx\, dy;$$

$$\text{set } H_j^{(i)} = -\iint\limits_{T_i} f(x, y) N_j^{(i)}(x, y)\, dx\, dy.$$

Step 5   For $i = K + 1, \ldots, N$ *(Compute all line integrals.)*
         for $j = 1, 2, 3$
             for $k = 1, \ldots, j$

$$\text{set } J_{j,k}^{(i)} = \int_{\mathscr{S}_2} g_1(x, y) N_j^{(i)}(x, y) N_k^{(i)}(x, y)\, dS;$$

$$\text{set } I_j^{(i)} = \int_{\mathscr{S}_2} g_2(x, y) N_j^{(i)}(x, y)\, dS.$$

Step 6   For $i = 1, \ldots, M$ do Steps 7–12.   *(Assembling the integrals over each triangle into the linear system.)*

   Step 7   For $k = 1, 2, 3$ do Steps 8–12.

      Step 8   Find l so that $E_l = (x_k^{(i)}, y_k^{(i)})$.

      Step 9   If $k > 1$ then for $j = 1, \ldots, k - 1$ do Steps 10, 11.

         Step 10   Find t so that $E_t = (x_j^{(i)}, y_j^{(i)})$.

         Step 11   If $l \leq n$ then
                 if $t \leq n$ then set $\alpha_{lt} = \alpha_{lt} + z_{k,j}^{(i)}$;
                       $\alpha_{tl} = \alpha_{tl} + z_{k,j}^{(i)}$;
                       else set $\beta_l = \beta_l - \gamma_t z_{k,j}^{(i)}$;
                 else
                 if $t \leq n$ then set $\beta_t = \beta_t - \gamma_l z_{k,j}^{(i)}$.

      Step 12   If $l \leq n$ then set $a_{ll} = \alpha_{ll} + z_{k,k}^{(i)}$;
                 $\beta_l = \beta_l + H_k^{(i)}$.

Step 13   For $i = K + 1, \ldots, N$ do Steps 14–19.   *(Assembling the line integrals into the linear system.)*

   Step 14   For $k = 1, 2, 3$ do Steps 15–19.

      Step 15   Find l so that $E_l = (x_k^{(i)}, y_k^{(i)})$.

      Step 16   If $k > 1$ then for $j = 1, \ldots, k - 1$ do Steps 17, 18.

         Step 17   Find t so that $E_t = (x_j^{(i)}, y_j^{(i)})$.

Step 18   If $l \leq n$ then

if $t \leq n$ then set $\alpha_{lt} = \alpha_{lt} + J_{k,j}^{(i)}$;
$$\alpha_{tl} = \alpha_{tl} + J_{k,j}^{(i)};$$
else set $\beta_l = \beta_l - \gamma_t J_{k,j}^{(i)}$;
else
if $t \leq n$ then set $\beta_t = \beta_t - \gamma_l J_{k,j}^{(i)}$.

Step 19   If $l \leq n$ then set $\alpha_{ll} = \alpha_{ll} + J_{k,k}^{(i)}$;
$$\beta_l = \beta_l + I_k^{(i)}.$$

Step 20   Solve the linear system $A\mathbf{c} = \mathbf{b}$ where $A = (\alpha_{l,t})$, $\mathbf{b} = (\beta_l)$ and $\mathbf{c} = (\gamma_t)$ for $1 \leq l, t \leq n$.

Step 21   OUTPUT $(\gamma_1, \ldots, \gamma_m)$.
(*For each $k = 1, \ldots, m$ let $\phi_k = N_j^{(i)}$ on T_i if $E_k = (x_j^{(i)}, y_j^{(i)})$.*

Then $\phi(x, y) = \displaystyle\sum_{k=1}^{m} \gamma_k \phi_k(x, y)$ approximates $u(x, y)$ on $D \cup \mathscr{S}_1 \cup \mathscr{S}_2$.)

Step 22   For $i = 1, \ldots, M$
for $j = 1, 2, 3$ OUTPUT $(a_j^{(i)}, b_j^{(i)}, c_j^{(i)})$.

Step 23 STOP.   (*Procedure is complete.*)

EXAMPLE 2   The temperature, $u(x, y)$, in a two-dimensional region D satisfies Laplace's equation

$$\frac{\partial^2 u}{\partial x^2}(x, y) + \frac{\partial^2 u}{\partial y^2}(x, y) = 0 \qquad \text{on } D.$$

Consider the region D shown in Fig. 11.15 and suppose that the following boundary conditions are given:

$$u(x, y) = 4 \quad \text{for } (x, y) \in L_6 \text{ and } (x, y) \in L_7,$$

$$\frac{\partial u}{\partial n}(x, y) = x \qquad \text{for } (x, y) \in L_2 \text{ and } (x, y) \in L_4,$$

$$\frac{\partial u}{\partial n}(x, y) = y \qquad \text{for } (x, y) \in L_5,$$

$$\frac{\partial u}{\partial n}(x, y) = \frac{x + y}{\sqrt{2}} \quad \text{for } (x, y) \in L_1 \text{ and } (x, y) \in L_3,$$

where $\partial u / \partial n$ denotes the directional derivative in the direction of the normal to the boundary of the region D at the point (x, y).

We will first subdivide D into triangles with the labeling suggested in Step 0 of the algorithm. For this example, $\mathscr{S}_1 = L_6 \cup L_7$ and $\mathscr{S}_2 = L_1 \cup L_2 \cup L_3 \cup L_4 \cup L_5$. The labeling of triangles is shown in Fig. 11.16.

FIGURE 11.15

11-15

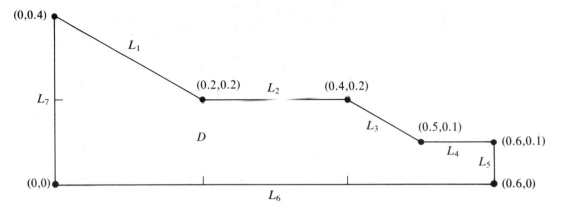

FIGURE 11.16

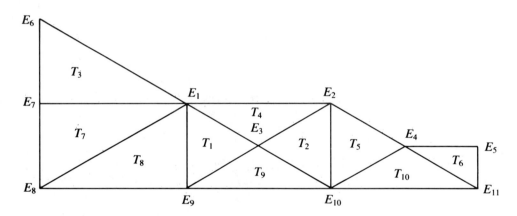

The boundary condition $u(x, y) = 4$ on L_6 and L_7 implies that $\gamma_l = 4$ when $l = 6, 7, ..., 11$. To determine the values of γ_l for $l = 1, 2, ..., 5$, apply the remaining steps of the algorithm and generate the matrix

$$A = \begin{bmatrix} 2.5 & 0 & -1 & 0 & 0 \\ 0 & 1.5 & -1 & -0.5 & 0 \\ -1 & -1 & 4 & 0 & 0 \\ 0 & -0.5 & 0 & 2.5 & -0.5 \\ 0 & 0 & 0 & -0.5 & 1 \end{bmatrix}$$

and the vector

$$\mathbf{b} = \begin{bmatrix} 6.066\overline{6} \\ 0.063\overline{3} \\ 8.0000 \\ 6.056\overline{6} \\ 2.031\overline{6} \end{bmatrix}.$$

The solution to the equation $A\mathbf{c} = \mathbf{b}$ is:

$$\mathbf{c} = \begin{bmatrix} \gamma_1 \\ \gamma_2 \\ \gamma_3 \\ \gamma_4 \\ \gamma_5 \end{bmatrix} = \begin{bmatrix} 4.0383 \\ 4.0782 \\ 4.0291 \\ 4.0496 \\ 4.0565 \end{bmatrix},$$

which gives the following approximation to the solution of Laplace's equation and the boundary conditions on the respective triangles:

T_1: $\phi(x, y) = 4.0383(1 - 5x + 5y) + 4.0291(-2 + 10x) + 4(2 - 5x - 5y),$

T_2: $\phi(x, y) = 4.0782(-2 + 5x + 5y) + 4.0291(4 - 10x) + 4(-1 + 5x - 5y),$

T_3: $\phi(x, y) = 4(-1 + 5y) + 4(2 - 5x - 5y) + 4.0383(5x),$

T_4: $\phi(x, y) = 4.0383(1 - 5x + 5y) + 4.0782(-2 + 5x + 5y) + 4.0291(2 - 10y),$

T_5: $\phi(x, y) = 4.0782(2 - 5x + 5y) + 4.0496(-4 + 10x) + 4(3 - 5x - 5y),$

T_6: $\phi(x, y) = 4.0496(6 - 10x) + 4.0565(-6 + 10x + 10y) + 4(1 - 10y).$

T_7: $\phi(x, y) = 4(-5x + 5y) + 4.0383(5x) + 4(1 - 5y),$

T_8: $\phi(x, y) = 4.0383(5y) + 4(1 - 5x) + 4(5x - 5y),$

T_9: $\phi(x, y) = 4.0291(10y) + 4(2 - 5x - 5y) + 4(-1 + 5x - 5y),$

T_{10}: $\phi(x, y) = 4.0496(10y) + 4(3 - 5x - 5y) + 4(-2 + 5x - 5y).$

The actual solution to the boundary-value problem is $u(x, y) = xy + 4$. Table 11.6 compares the value of u to the value of ϕ at E_i for each $i = 1, \ldots, 5$. ■

TABLE 11.6

x	y	$\phi(x, y)$	$u(x, y)$	$|\phi(x, y) - u(x, y)|$
0.2	0.2	4.0383	4.04	0.0017
0.4	0.2	4.0782	4.08	0.0018
0.3	0.1	4.0291	4.03	0.0009
0.5	0.1	4.0496	4.05	0.0004
0.6	0.1	4.0565	4.06	0.0035

Typically, the error for elliptic second-order problems of the type (11.41) with smooth coefficient functions is $O(h^2)$, where h is the maximum diameter of the triangular elements. Piecewise bilinear basis functions on rectangular elements are also expected to give $O(h^2)$ results, where h is the maximum length of the rectangular elements. Other classes of basis functions can be used to give $O(h^4)$ results, but the construction is more complex. Efficient error theorems for finite-element methods are difficult to state and apply because the accuracy of the approximation depends on the continuity properties of the solution and the regularity of the boundary.

The finite-element method can also be applied to parabolic and hyperbolic partial-differential equations, but the minimization procedure is more difficult and the finite-difference procedures are more competitive. A good survey on the advantages and techniques of the finite-element method applied to various physical problems can be found in a paper by Fix [47]. For a more extensive discussion, refer to the book by Strang and Fix [132] or Zienkiewicz [149].

Exercise Set 11.5

1. Use Algorithm 11.5 to approximate the solution to the following partial-differential equation (see the following figure).

$$\frac{\partial}{\partial x}\left(y^2 \frac{\partial u}{\partial x}(x, y)\right) + \frac{\partial}{\partial y}\left(y^2 \frac{\partial u}{\partial y}(x, y)\right) - yu(x, y) = -x, \qquad (x, y) \in D,$$

$$u(x, 0.5) = 2x, \qquad 0 \le x \le 0.5,$$

$$u(0, y) = 0, \qquad 0.5 \le y \le 1,$$

$$y^2 \frac{\partial u}{\partial x}(x, y) \cos \theta_1 + y^2 \frac{\partial u}{\partial y}(x, y) \cos \theta_2 = \frac{\sqrt{2}}{2}(y - x) \qquad \text{for } (x, y) \in \mathscr{S}_2.$$

EX 11-5

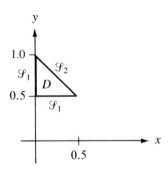

Let $M = 2$; T_1 have vertices $(0, 0.5)$, $(0.25, 0.75)$, $(0, 1)$; and T_2 have vertices $(0, 0.5)$, $(0.5, 0.5)$, and $(0.25, 0.75)$.

2. Repeat Exercise 1. using the triangles:

$$T_1: \quad (0, 0.75), (0, 1), (0.25, 0.75);$$
$$T_2: \quad (0.25, 0.5), (0.25, 0.75), (0.5, 0.5);$$
$$T_3: \quad (0, 0.5), (0, 0.75), (0.25, 0.75);$$
$$T_4: \quad (0, 0.5), (0.25, 0.05), (0.25, 0.75).$$

3. Approximate the solution to the partial-differential equation

$$\frac{\partial^2 u}{\partial x^2}(x, y) + \frac{\partial^2 u}{\partial y^2}(x, y) - 12.5\pi^2 u(x, y) = -25\pi^2 \sin\frac{5\pi}{2}x \sin\frac{5\pi}{2}y, \qquad 0 < x, \quad y < 0.4,$$

subject to the Dirichlet boundary condition

$$u(x, y) = 0,$$

using the Finite-Element Algorithm 11.5 with the elements given in the following figure. Compare the approximate solution to the actual solution

$$u(x, y) = \sin \frac{5\pi}{2} x \, \sin \frac{5\pi}{2} y$$

at the interior vertices and at the points (0.125, 0.125), (0.125, 0.25), (0.25, 0.125), and (0.25, 0.25).

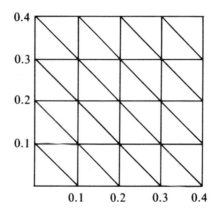

4. Repeat Exercise 3, with $f(x, y) = -25\pi^2 \cos \frac{5\pi}{2} x \cos \frac{5\pi}{2} y$, using the Neumann boundary condition

$$\frac{\partial u}{\partial n}(x, y) = 0.$$

The actual solution for this problem is

$$u(x, y) = \cos \frac{5\pi}{2} x \, \cos \frac{5\pi}{2} y.$$

5. Construct an algorithm for the finite element method using rectangular elements and basis functions of the form $a + bx + cy + dxy$.

6. Use the algorithm developed in Exercise 5 to solve the problem discussed in

   a) Exercise 3.   b) Exercise 4.

7. A silver plate in the shape of a trapezoid (see the following figure) has heat being uniformly generated at each point at the rate $q = 1.5$ cal/cm$^3 \cdot$sec. The steady-state temperature $u(x, y)$ of the plate satisfies Poissons equation

$$\frac{\partial^2 u}{\partial x^2}(x, y) + \frac{\partial^2 u}{\partial y^2}(x, y) = \frac{-q}{k},$$

where k, the thermal conductivity, is 1.04 cal/cm · deg · sec. Assume that the temperature is held at $15°C$ on L_2, that heat is lost on the slanted edges L_1 and L_3 according to the boundary condition $\partial u/\partial n = 4$, and that no heat is lost on L_4, that is, $\partial u/\partial n = 0$. Approximate the temperature of the plate at $(1, 0)$, $(4, 0)$, and $(\frac{5}{2}, \sqrt{3}/2)$ by using Algorithm 11.5.

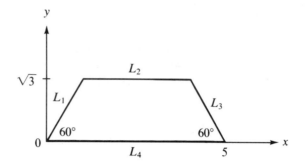

Bibliography

[1.] AHO, A. V., J. E. HOPCROFT, and J. D. ULLMAN (1974), *The design and analysis of computer algorithms.* Addison-Wesley, Reading, Mass.; 470 pp. QA76.6.A36.

[2.] AMES, W. F. (1977), *Numerical methods for partial differential equations* (Second edition). Academic Press, New York; 365 pp. QA374.A46.

[3.] BAILEY, N. T. J. (1967), *The mathematical approach to biology and medicine.* John Wiley & Sons, London; 296 pp. QH324.B28.

[4.] BAILEY, N. T. J. (1957), *The mathematical theory of epidemics.* C. Griffin, London; 194 pp. RA652.B3.

[5.] BAILEY, P. B., L. F. SHAMPINE, and P. E. WALTMAN (1968), *Nonlinear two-point boundary-value problems.* Academic Press, New York; 171 pp. QA372.B27.

[6.] BARTLE, R. G. (1976), *The elements of real analysis* (Second edition). John Wiley & Sons, New York; 480 pp. QA300.B29.

[7.] BEKKER, M. G. (1969), *Introduction to terrain vehicle systems.* University of Michigan Press, Ann Arbor, Mich.; 846 pp. TL243.B39.

[8.] BERNADELLI, H. (1941), "Population Waves." *Journal of the Burma Research Society,* **31**, 1–18.

[9.] BIRKHOFF, G., and C. DE BOOR (1964), "Error bounds for spline interpolation." *Journal of Mathematics and Mechanics,* **13**, 827–836.

[10.] BIRKHOFF, G., and R. E. LYNCH (1984), *Numerical solution of elliptic problems.* SIAM Publications, Philadelphia, Pa.; 320 pp. QA374.B57.

[11.] BIRKHOFF, G., and G. ROTA (1978), *Ordinary differential equations.* John Wiley & Sons, New York; 342 pp. QA372.B58.

[12.] BRACEWELL, R. (1978), *The Fourier transform and its application* (Second edition). McGraw-Hill, New York; 444 pp. QA403.5.B7.

[13.] BRENT, R. (1973), *Algorithms for minimization without derivatives.* Prentice-Hall, Englewood Cliffs, N.J.; 195 pp. QA402.5.B74.

[14.] BRIGHAM, E. O. (1974), *The fast Fourier transform.* Prentice-Hall, Englewood Cliffs, N.J.; 252 pp. QA403.B74.

[15.] BROGAN, W. L. (1982), *Modern control theory.* Prentice-Hall, Englewood Cliffs, N.J.; 393 pp. QA402.3.B76.

[16.] BROWN, K. M. (1969), "A quadratically convergent Newton-like method based upon Gaussian elimination." *SIAM Journal on Numerical Analysis,* **6**, No. 4, 560–569.

[17.] BROYDEN, C. G. (1965), "A class of methods for solving nonlinear simultaneous equations." *Mathematics of Computation,* **19**, 577–593.

[18.] BULIRSCH, R. (1964), "Bemerkungen zur Romberg-integration." *Numerische Mathematik,* **6**, 6–16.

[19.] BULIRSCH, R., and J. STOER (1966), "Numerical treatment of ordinary differential equations by extrapolation methods." *Numerische Mathematik,* **8**, 1–13.

[20.]   BULIRSCH, R., and J. STOER (1964), "Fehlerabschätzungen und extrapolation mit rationalen Funktionen bei Verfahren von Richardson-typus." *Numerische Mathematik*, **6**, 413–427.

[21.]   BULIRSCH, R., and J. STOER (1966), "Asymptotic upper and lower bounds for results of extrapolation methods." *Numerische Mathematik*, **8**, 93–104.

[22.]   BUNCH, J. R., and D. J. ROSE, editors (1976), *Sparse matrix computations.* Proceedings of a conference held at Argonne National Laboratories, September 9–11, 1975. Academic Press, New York; 453 pp. QA188.S9.

[23.]   BUTCHER, J. C. (1965), "On the attainable order of Runge–Kutta methods." *Mathematics of Computation*, **19**, 408–417.

[24.]   CANTONI, A., and P. BUTLER (1976), "Properties of the eigenvectors of persymmetric matrices with applications to communication theory." *IEEE Transactions on Communications*, Vol. Com-24, **8**, 804–809.

[25.]   CHAMBERS, J. M. (1977), *Computational methods for data analysis.* John Wiley & Sons, New York; 268 pp. QA276.4.C48.

[26.]   CHEN, B. H. (1976), "Holdup and axial mixing in bubble columns containing screen cylinders." *Industrial and Engineering Chemistry, Process Design and Development*, **15**, No. 1, 20–24.

[27.]   CHIARELLA, C., W. CHARLTON, and A. W. ROBERTS (1975), "Optimum chute profiles in gravity flow of granular materials: A discrete segment solution method." *Transactions of the ASME, Journal of Engineering for Industry*, Series B, **97**, 10–13.

[28.]   CODY, W. J., and W. WAITE (1980), *Software manual for the elementary functions.* Prentice-Hall, Englewood Cliffs, N.J.; 269 pp. QA331.C635.

[29.]   COOLEY, J. W., and J. W. TUKEY (1965), "An algorithm for the machine calculation of complex Fourier series." *Mathematics of Computation*, **19**, No. 90, 297–301.

[30.]   CUPPEN, J. J. M. (1981), "A divide and conquer method for the symmetric tridiagonal eigenproblem." *Numerische Mathematik*, **36**, 177–195.

[31.]   DAHLQUIST, G., and Å. BJÖRCK (Translated by N. Anderson) (1974), *Numerical methods.* Prentice-Hall, Englewood Cliffs, N.J.; 573 pp. QA297.D3313.

[32.]   DAVIS, P. J., and P. RABINOWITZ (1975), *Methods of numerical integration.* Academic Press, New York; 459 pp. QA299.3.D28.

[33.]   DE BOOR, C. (1978), *A practical guide to splines.* Springer-Verlag, New York; 392 pp. QA1.A647, Vol. 27.

[34.]   DE BOOR, C., and B. SWARTZ (1973), "Collocation at Gaussian points." *SIAM Journal on Numerical Analysis*, **10**, No. 4, 582–606.

[35.]   DENNIS, J. E. JR., and J. J. MORÉ (1977), "Quasi-Newton methods, motivation and theory." *SIAM Review*, **19**, No. 1, 46–89.

[36.]   DENNIS, J. E. JR., and R. B. SCHNABEL (1979), "Least change secant updates for quasi-Newton methods." *SIAM Review*, **21**, No. 4, 443–459.

[37.]   DEUFLHARD, P. (1983), "Order and step-size control in extrapolation methods." *Numerische Mathematik*, **41**, 399–422.

[38.]   DEUFLHARD, P., and G. BADER (1978), *A semi-implicit midpoint rule for stiff systems of ordinary differential equations.* Techniche Universität München, Munich; 48 pp.

[39.]   DORN, G. L., and A. B. BURDICK (1962), "On the recombinational structure of complementation relationships in the *m-dy* complex of the *Drosophila melanogaster.*" *Genetics*, **47**, 503–518.

[40.] ENRIGHT, W. H. (1974), "Optimal second derivative methods for stiff systems." *Stiff differential systems*, R. A. Willoughby, editor. Plenum Press, New York; 95–109. QA371.I56.

[41.] ENRIGHT, W. H., T. E. HULL, and B. LINDBERG (1975), "Comparing numerical methods for stiff systems of O.D.E.'s." *BIT*, **15**, 10–48.

[42.] ENRIGHT, W. H., and T. E. HULL (1976), "Test results on initial-value methods for nonstiff ordinary differential equations." *SIAM Journal on Numerical Analysis*, **13**, No. 6, 944–961.

[43.] FAIRES, J. D., and B. T. FAIRES (1983), *Calculus and analytic geometry*. PWS Publishers, Boston, Mass.; 1026 pp. QA303.F294.

[44.] FEHLBERG, E. (1964), "New high-order Runge–Kutta formulas with step-size control for systems of first- and second-order differential equations." *Zeitschrift für Angewandte Mathematik und Mechanik*, **44**, 17–29.

[45.] FEHLBERG, E. (1966), "New high-order Runge–Kutta formulas with an arbitrarily small truncation error." *Zeitschrift für Angewandte Mathematik und Mechanik*, **46**, 1–16.

[46.] FEHLBERG, E. (1970), "Klassische Runge–Kutta Formeln vierter und niedrigerer Ordnung mit Schrittweiten-Kontrolle und ihre Anwendung auf Wärmeleitungsprobleme." *Computing*, **6**, 61–71.

[47.] FIX, G. (1975), "A survey of numerical methods for selected problems in continuum mechanics." *Proceedings of a Conference on Numerical Methods of Ocean Circulation, National Academy of Sciences*, Durham, N. H., October 17–20, 1972, 268–283.

[48.] FORSYTHE, G. E., M. A. MALCOLM, and C. A. MOLER (1977), *Computer methods for mathematical computations*. Prentice-Hall, Englewood Cliffs, N.J.; 259 pp. QA297.F568.

[49.] FORSYTHE, G. E., and C. B. MOLER (1967), *Computer solution of linear algebraic systems*, Prentice-Hall, Englewood Cliffs, N.J.; 148 pp. QA297.F57.

[50.] FULKS, W. (1978), *Advanced calculus* (Third edition). John Wiley & Sons, New York; 731 pp. QA303.F954.

[51.] GARCIA, C. B., and F. J. GOULD (1980), "Relations between several path-following algorithms and local and global Newton methods." *SIAM Review*, **22**, No. 3, 263–274.

[52.] GEAR, C. W. (1971), *Numerical initial-value problems in ordinary differential equations*. Prentice-Hall, Englewood Cliffs, N.J.; 253 pp. QA372.G4.

[53.] GEAR, C. W. (1981), "Numerical solution of ordinary differential equations: Is there anything left to do?" *SIAM Review*, **23**, No. 1, 10–24.

[54.] GEORGE, J. A. (1973), "Nested dissection of a regular finite-element mesh." *SIAM Journal on Numerical Analysis*, **10**, No. 2, 345–362.

[55.] GEORGE, J. A., and J. W. H. LIU (1981), *Computer solution of large sparse positive definite systems*. Prentice-Hall, Englewood Cliffs, NJ.; 324 pp. QA188. G46.

[56.] GLADWELL, I., and R. WAIT (1979), *A survey of numerical methods for partial differential equations.* Oxford University Press; 424 pp. QA377.S96.

[57.]   GRAGG, W. B. (1965), "On extrapolation algorithms for ordinary initial-value problems," *SIAM Journal on Numerical Analysis*, **2**, 384–403.

[58.]   HAGEMAN, L. A., and D. M. YOUNG (1981), *Applied iterative methods*. Academic Press, New York; 386 pp. QA297.8.H34.

[59.]   HAMMING, R. W. (1973), *Numerical methods for scientists and engineers* (Second edition). McGraw-Hill, New York; 721 pp. QA297.H28.

[60.]   HATCHER, T. R. (1982), "An error bound for certain successive overrelaxation schemes." *SIAM Journal on Numerical Analysis*, **19**, No. 5, 930–941.

[61.]   HENRICI, P. (1962), *Discrete variable methods in ordinary differential equations*. John Wiley & Sons, New York; 407 pp. QA372.H48.

[62.]   HENRICI, P. (1964), *Elements of numerical analysis*. John Wiley & Sons, New York; 328 pp. QA297.H4.

[63.]   HENRICI, P. (1963), *Error propagation for difference methods*. John Wiley & Sons, New York; 73 pp. QA431.H44.

[64.]   HILDEBRAND, F. B. (1974), *Introduction to numerical analysis* (Second edition). McGraw-Hill, New York; 669 pp. QA297.H54.

[65.]   HOUSEHOLDER, A. S. (1970), *The numerical treatment of a single nonlinear equation*. McGraw-Hill, New York; 216 pp. QA218.H68.

[66.]   HULL, T. E., W. H. ENRIGHT, B. M. FELLEN, and A. E. SEDGEWICK (1972), "Comparing numerical methods for ordinary differential equations." *SIAM Journal on Numerical Analysis*, **9**, No. 4, 603–637.

[67.]   ISAACSON, E., and H. B. KELLER (1966), *Analysis of numerical methods*. John Wiley & Sons, New York; 541 pp. QA297.I8.

[68.]   JACKSON, K. R., W. H. ENRIGHT, and T. E. HULL (1978), "A theoretical criterion for comparing Runge–Kutta formulas." *SIAM Journal on Numerical Analysis*, **15**, No. 3, 618–641.

[69.]   JACOBS, D., editor (1977), *The state of the art in numerical analysis*. Academic Press, New York; 978 pp. QA297.C646.

[70.]   JENKINS, M. A., and J. F. TRAUB (1970), "A three-stage algorithm for real polynomials using quadratic iteration." *SIAM Journal on Numerical Analysis*, **7**, No. 4, 545–566.

[71.]   JOHNSON, G. W., and N. H. AUSTRIA (1983), "A quasi-Newton method employing direct secant updates of matrix factorizations." *SIAM Journal on Numerical Analysis*, **20**, No. 2, 315–325.

[72.]   JOHNSTON, R. L. (1982), *Numerical methods: A software approach*. John Wiley & Sons, New York; 276 pp. QA297.J64.

[73.]   JOYCE, D. C. (1971), "Survey of extrapolation processes in numerical analysis." *SIAM Review*, **13**, No. 4, 435–490.

[74.]   KAMMERER, W. J., G. W. REDDIEN, and R. S. VARGA (1974), "Quadratic splines." *Numerische Mathematik*, **22**, 241–259.

[75.]   KELLER, H. B. (1968), *Numerical methods for two-point boundary-value problems*. Blaisdell, Waltham, Mass.; 184 pp. QA372.K42.

[76.]   KELLER, J. B. (1984), "Probability of a shutout in racquetball." *SIAM Review*, **26**, No. 2, 267–268.

[77.]   KINCAID, D. R., and D. M. YOUNG (1979), "Survey of iterative methods."

Encyclopedia of Computer Science and Technology, J. Belzer, A. G. Holzman, and A. Kent, editors. M. Dekker, New York; QA76.15.E5.

[78.] LAMBERT, J. D. (1977), "The initial value problem for ordinary differential equations." *The state of the art in numerical analysis*, D. Jacobs, editor. Academic Press, New York; 451–501. QA297.C646.

[79.] LAPIDUS, L., and W. E. SCHIESSER (1976), *Numerical methods for differential systems*. Academic Press, New York; 291 pp. QA372.N85.

[80.] LARSON, H. J. (1982), *Introduction to probability theory and statistical inference* (Third edition), John Wiley & Sons, New York; 637 pp. QA273.L352.

[81.] LAUFER, H. B. (1984), *Discrete mathematics and applied modern algebra*. PWS Publishers, Boston, Mass.; 538 pp. QA161.L38.

[82.] LAWSON, C. L., and R. J. HANSON (1974), *Solving least squares problems*. Prentice-Hall, Englewood Cliffs, N.J.; 340 pp. QA275.L38.

[83.] LUCAS, T. R., and G. W. REDDIEN, JR. (1972), "Some collocation methods for nonlinear boundary value problems." *SIAM Journal on Numerical Analysis*, **9**, No. 2, 341–356.

[84.] LYNESS, J. N. (1983), "When not to use an automatic quadrature routine." *SIAM Review*, **25**, No. 1, 63–87.

[85.] LYNESS, J. N., and J. J. KAGANOVE (1976), "Comments on the nature of automatic quadrature routines." *ACM Transactions on Mathematical Software*, **6**, 65–81.

[86.] MACHURA, M., and R. SWEET (1980), "Survey of software for partial differential equations." *ACM Transactions on Mathematical Software*, **6**, 461–488.

[87.] MANO, M. M. (1982), *Computer system architecture*. Prentice-Hall, Englewood Cliffs, N.J.; 531 pp. QA76.9.A73M36.

[88.] MITCHELL, A. R. (1969), *Computational methods for partial-differential equations*. John Wiley & Sons, London; 255 pp. QA374.M68.

[89.] MORÉ, J. J., and M. Y. COSNARD (1979), "Numerical solution of nonlinear equations." *ACM Transactions on Mathematical Software*, **5**, No. 1, 64–85.

[90.] MÜLLER, D. E. (1956), "A method for solving algebraic equations using an automatic computer." *Mathematical Tables and Other Aids to Computation*, **10**, 208–215.

[91.] NA, T. Y., and G. M. KURAJIAN (1976), "Initial-curvature and lateral-load effects on thin struts with large elastic displacements." *Transactions of the ASME, Journal of Engineering for Industry*, Series B, **98**, 34–38.

[92.] NOBLE, B., and J. W. DANIEL (1977), *Applied linear algebra* (Second edition). Prentice-Hall, Englewood Cliffs, N.J.; 477 pp. QA184.N6.

[93.] OLVER, F. W. J. (1978), "A new approach to error arithmetic." *SIAM Journal on Numerical Analysis*, **15**, No. 2, 368–393.

[94.] ORTEGA, J. M. (1972), *Numerical analysis—A second course*. Academic Press, New York; 201 pp. QA297.O78.

[95.] ORTEGA, J. M., and W. G. POOLE, JR. (1981), *An introduction to numerical methods for differential equations*. Pitman, Marshfield, Mass.; 329 pp. QA371.O65.

[96.] ORTEGA, J. M., and W. C. RHEINBOLDT (1970), *Iterative solution of nonlinear equations in several variables*. Academic Press, New York; 572 pp. QA297.8.O77.

[97.] PANDIT, S. M., T. L. SUBRAMANIAN, and S. M. WU (1975), "Modeling machine-

622   Bibliography

tool chatter by time series." *Transactions of the ASME, Journal of Engineering for Industry*, Series B, **97**, 211–215.

[98.]   PARLETT, B. (1978), "Progress in numerical analysis." *SIAM Review*, **20**, No. 3, 443–456.

[99.]   PARLETT, B. (1980), *The symmetric eigenvalue problem*. Prentice-Hall, Englewood Cliffs, N.J., 348 pp. QA188.P37.

[100.]   PETERS, G., and J. H. WILKINSON (1979), "Inverse interpolation, ill-conditioned equations and Newton's method." *SIAM Review*, **21**, No. 3, 339–360.

[101.]   POWELL, M. J. D. (1981), *Approximation theory and methods*. Cambridge University Press, Cambridge; 339 pp. QA221.P65.

[102.]   PRYCE, J. D. (1984), "A new measure of relative error for vectors." *SIAM Journal on Numerical Analysis*, **27**, No. 1, 202–215.

[103.]   RALSTON, A., and P. RABINOWITZ (1978), *A first course in numerical analysis* (Second edition). McGraw-Hill, New York; 556 pp. QA297.R3.

[104.]   RALSTON, A., and H. S. WILF, editors (1960 and 1967), *Numerical methods for digital computers*. Vols. 1 and 2. John Wiley & Sons, New York; 293 + 287 pp. QA76.5.R3.

[105.]   RASHEVSKY, N. (1968), *Looking at history through mathematics*. Massachusetts Institute of Technology Press, Cambridge, Mass.; 199 pp. D16.25.R3.

[106.]   RHEINBOLDT, W. C. (1974), *Methods for solving systems of nonlinear equations. Regional Conference Series in Applied Mathematics*, No. 14. SIAM Publications, Philadelphia, Pa.; 104 pp. QA214.R44.

[107.]   RICE, J. R. (1981), *Matrix computations and mathematical software*. McGraw-Hill, New York; 248 pp. QA188.R52.

[108.]   RICE, J. R. (1983), *Numerical methods, software, and analysis: IMSL Reference Edition*. McGraw-Hill, New York; 661 pp. QA297.R49.

[109.]   RICHARDSON, L. F., and J. A. GAUNT (1927), "The deferred approach to the limit." *Philosophical Transactions of the Royal Society of London*, **226A**, 299–361.

[110.]   ROSE, D. J., and R. A. WILLOUGHBY, editors (1972), "Sparse matrices and their applications." Proceedings of a conference held at IBM Research Center, New York, September 9–10, 1971. Plenum Press, New York; 215 pp.

[111.]   RUSSELL, R. D. (1977), "A comparison of collocation and finite differences for two-point boundary value problems." *SIAM Journal on Numerical Analysis*, **14**, No. 1, 19–39.

[112.]   SAFF, E. B., and A. D. SNIDER (1976), *Fundamentals of complex analysis for mathematics, science and engineering*. Prentice-Hall, Englewood Cliffs, N.J.; 444 pp. QA300.S18.

[113.]   SAGAR, V., and D. J. PAYNE (1975), "Incremental collapse of thick-walled circular cylinders under steady axial tension and torsion loads and cyclic transient heating." *Journal of the Mechanics and Physics of Solids*, **21**, No. 1, 39–54.

[114.]   SALE, P. F., and R. DYBDAHL (1975). "Determinants of community structure for coral-reef fishes in experimental habitat," *Ecology*, **56**, 1343–1355.

[115.]   SCHNEIDER, C. (1975), "Vereinfachte Rekursionen zur Richardson-extrapolation in Spezialfällen." *Numerische Mathematik*, **24**, 177–184.

[116.]   SCHOENBERG, I. J. (1946), "Contributions to the problem of approximation of equidistant data by analytic functions." *Quarterly of Applied Mathematics*, **4**, Part A, 45–99; Part B, 112–141.

[117.]   SCHROEDER, L. A. (1973), "Energy budget of the larvae of the moth *Pachysphinx modesta*," *Oikos*, **24**, 278–281.

[118.]   SCHROEDER, L. A. (1981), "Thermal tolerances and acclimation of two species of hydras." *Limnology and Oceanography*, **26**, No. 4, 690–696.

[119.]   SCHULTZ, M. H. (1966), *Spline analysis*. Prentice-Hall, Englewood Cliffs, N.J.; 156 pp. QA221.S33.

[120.]   SEARLE, S. R. (1966), *Matrix algebra for the biological sciences*. John Wiley & Sons, New York; 296 pp. QH324.S439m.

[121.]   SECRIST, D. A., and R. W. HORNBECK (1976), "An Analysis of heat transfer and fade in disk brakes." *Transactions of the ASME, Journal of Engineering for Industry*, Series B, **98**, No. 2, 385–390.

[122.]   SHAMPINE, L. F., and C. W. GEAR (1979), "A user's view of solving stiff ordinary differential equations." *SIAM Review*, **21**, No. 1, 1–17.

[123.]   SHAMPINE, L. F., and M. K. GORDON (1975), *Computer solution of ordinary differential equations: The initial value problem*. W. H. Freeman, San Francisco; 318 pp. QA372.S416.

[124.]   SHAMPINE, L. F., H. A. WATTS, and S. M. DAVENPORT (1976), "Solving nonstiff ordinary differential equations—The state of the art." *SIAM Review*, **18**, No. 3, 376–411.

[125.]   SINGH, V. P. (1976), "Investigations of attentuation and internal friction of rocks by ultrasonics." *International Journal of Rock Mechanics and Mining Sciences*, 69–72.

[126.]   SKEEL, R. D. (1980), "Iterative refinement implies numerical stability for Gaussian elimination." *Mathematics of Computation*, **35**, No. 151, 817–832.

[127.]   SMITH, G. D. (1965), *Numerical solution of partial-differential equations*. Oxford University Press, Oxford; 179 pp. QA377.S59.

[128.]   STETTER, H. J. (1973), *Analysis of discretization methods for ordinary differential equations*. From Tracts in natural philosophy. Springer-Verlag, Berlin; 388 pp. QA372.S84.

[129.]   STEWART, G. W. (1973), *Introduction to matrix computations*. Academic Press, New York; 441 pp. QA188.S7.

[130.]   STOER, J. (1961), "Über zwei Algorithmen zur Interpolation mit rationalen Funktionen." *Numerische Mathematik*, **3**, 285–304.

[131.]   STRANG, W. G. (1980), *Linear algebra and its applications*. Academic Press, New York; 414 pp. QA184.S8.

[132.]   STRANG, W. G., and G. J. FIX (1973), *An analysis of the finite element method*. Prentice-Hall, Englewood Cliffs, N.J.; 306 pp. TA335.S77.

[133.]   STROUD, A. H. (1971), *Approximate calculation of multiple integrals*. Prentice-Hall, Englewood Cliffs, N.J.; 431 pp. QA311.S85.

[134.]   STROUD, A. H., and D. SECRIST (1966), *Gaussian quadrature formulas*. Prentice-Hall, Englewood Cliffs, N.J.; 374 pp. QA299.4.G3S7.

[135.]   VARGA, R. S. (1962), *Matrix iterative analysis*. Prentice-Hall, Englewood Cliffs, N.J.; 322 pp. QA263.V3.

[136.]   VEMURI, V., and W. J. KARPLUS (1981), *Digital computer treatment of partial differential equations*. Prentice-Hall, Englewood Cliffs, N.J.; 449 pp. QA374.V45.

[137.]   VICHNEVETSKY, R. (1981), *Computer methods for partial differential equations.* Volume 1: *Elliptic equations and the finite element method.* Prentice-Hall, Englewood Cliffs, N.J.; 400 pp. QA374.V53.

[138.]   WANG, H. T. (1975), "Determination of the accuracy of segmented representations of cable shape." *Transactions of the ASME, Journal of Engineering in Industry,* Series B, **97**, No. 2, 472–478.

[139.]   WATKINS, D. S. (1982), "Understanding the QR algorithm." *SIAM Review,* **24**, No. 4, 427–440.

[140.]   WENDROFF, B. (1966), *Theoretical numerical analysis.* Academic Press, New York; 239 pp. QA297.W43.

[141.]   WILKINSON, J. H. (1963), *Rounding errors in algebraic processes.* H.M. Stationery Office, London; 161 pp. QA76.5.W53.

[142.]   WILKINSON, J. H. (1965), *The algebraic eigenvalue problem.* Clarendon Press, Oxford; 662 pp. QA218.W5.

[143.]   WILKINSON, J. H., and C. REINSCH (1971), *Handbook for automatic computation.* Volume 2: *Linear algebra.* Springer-Verlag, Berlin; 439 pp. QA251.W67.

[144.]   WINOGRAD, S. (1978), "On computing the discrete Fourier transform." *Mathematics of Computation,* **32**, 175–199.

[145.]   YOUNG, D. M. (1971), *Iterative solution of large linear systems.* Academic Press, New York; 570 pp. QA195.Y68.

[146.]   YOUNG, D. M., and R. T. GREGORY (1972), *A survey of numerical mathematics.* Vol. 1. Addison-Wesley; Reading, Mass., 533 pp. QA297.Y63.

[147.]   YPMA, T. J. (1983), "Finding a multiple zero by transformation and Newton-like methods." *SIAM Review,* **25**, No. 3, 365–378.

[148.]   YPMA, T. J. (1983), "Local convergence difference Newton-like methods." *Mathematics of Computation,* **41**, No. 164, 527–536.

[149.]   ZIENKIEWICZ, O. (1977), *The finite-element method in engineering science.* McGraw-Hill, London; 787 pp. TA646.Z54.

Answers to Selected Exercises

CHAPTER 1

Exercise Set 1.1 (page 7)

1. Consider $f'(0)$ and $f'(1)$.

3. Let $f(x) = x - 3^{-x}$. Consider $f(0)$ and $f(1)$.

5. Use Theorem 1.7.

7. For $x < 0$, $f(x) < 2x + k < 0$, provided $x < -\frac{1}{2}k$. Similarly, for $x > 0$, $f(x) > 2x + k > 0$, provided $x > -\frac{1}{2}k$. By Theorem 1.12 there exists c with $f(c) = 0$. If $f(c) = 0$ and $f(c') = 0$ for some $c' \neq c$, then by Theorem 1.7 there would exist p between c and c' with $f'(p) = 0$. But $f'(x) = 3x^2 + 2 > 0$ for all x.

9. $P_4(x) = 1 + x - \dfrac{x^3}{3} - \dfrac{x^4}{6}$, $P_4\left(\dfrac{\pi}{16}\right) = 1.19357852$, $\left| R_4\left(\dfrac{\pi}{16}\right) \right| \leq 9.73 \times 10^{-6}$.

11. $P_3(x) = e^{-1}[1 - (x-1) + (x-1)^2/2 - (x-1)^3/6]$, $e^{-0.99} \approx P_3(0.99) = 0.3715766909$, $|R_3(0.99)| < 5 \times 10^{-10}$.

13. Since $42° = 7\pi/30$ radians, use $x_0 = \pi/4$.

$$\left| R_n\left(\frac{7\pi}{30}\right) \right| \leq \frac{\left(\dfrac{\pi}{4} - \dfrac{7\pi}{30}\right)^{n+1}}{(n+1)!} < \frac{(0.053)^{n+1}}{(n+1)!}$$

For $\left| R_n\left(\dfrac{7\pi}{30}\right) \right| < 10^{-6}$ it suffices to take $n = 3$. Since

$$P_3(x) = \cos\frac{\pi}{4} - \sin\frac{\pi}{4}\left(x - \frac{\pi}{4}\right) - \frac{1}{2}\cos\frac{\pi}{4}\left(x - \frac{\pi}{4}\right)^2$$
$$+ \frac{1}{6}\sin\frac{\pi}{4}\left(x - \frac{\pi}{4}\right)^3$$

$$P_3(42°) = P_3\left(\frac{7\pi}{30}\right) = 0.7431446.$$

To 7 digits, $\cos 42° = 0.7431448$, so the actual error is 2×10^{-7}.

15. $f^{(n)}(x) = (-1)^{n-1}(n-1)!/x^n$ so

$$|R_n(3)| \leq \frac{(3-e)^{n+1}}{e^{n+1}(n+1)}.$$

For $|R_n(3)| < 10^{-4}$, it suffices to take $n = 3$. Since

$$P_3(x) = 1 + \frac{1}{e}(x-e) - \frac{1}{2e^2}(x-e)^2 + \frac{1}{3e^3}(x-e)^3,$$

$P_3(3) = 1.09864$. To 5 decimal places, $\ln 3 = 1.09861$, so the actual error is 3×10^{-5}.

17. For $x < 0$,

$$f'(x) = \lim_{h \to 0} \frac{|x + h| - |x|}{h} = \lim_{h \to 0} \frac{-(x + h) + x}{h} = -1$$

and for $x > 0$,

$$f'(x) = \lim_{h \to 0} \frac{|x + h| - |x|}{h} = \lim_{h \to 0} \frac{x + h - x}{h} = 1.$$

But, when $x = 0$

$$\lim_{h \to 0^-} \frac{|h| - 0}{h} = \lim_{h \to 0^-} \frac{-h}{h} = -1 \quad \text{and} \quad \lim_{h \to 0^+} \frac{|h| - 0}{h} = \lim_{h \to 0^+} \frac{h}{h} = 1.$$

19. a) No, $f'(0)$ does not exist.   b) Yes, $c = 64/27$.
   c) No, $f'(0)$ does not exist.   d) Yes, c can be any number in $(0, 1)$.

21. a) Let $x_0 \in [a, b]$. Then $|f(x_0 + h) - f(x_0)| \le L|h|$ for small h. Given $\varepsilon > 0$ let δ be less than ε/L. For $|h| < \delta$, $|f(x_0 + h) - f(x_0)| \le L|h| < L\delta < \varepsilon$.
   b) Let x and y be in $[a, b]$. By Theorem 1.8 there is a number c in (a, b) with $|f(x) - f(y)| = |f'(c)(x - y)| \le L|x - y|$.
   c) Let $f(x) = x^{2/3}$ on $[0, 1]$. for $\varepsilon > 0$, $F \in C'[\varepsilon, 1]$ so that

$$|f(x) - f(y)| = |f'(c)||x - y| \quad \text{for} \quad \varepsilon \le x < c < y \le 1.$$

Thus, $|f(x) - f(y)| = \frac{2}{3}c^{-1/3}|y - x| > \frac{2}{3}y^{-1/3}|y - x|$. For any choice of $L > 0$, if we choose $x < y < \frac{8}{27}L^{-3}$ and $0 < x < y \le 1$, then

$$|f(x) - f(y)| > \frac{2}{3}y^{-1/3}|y - x| > \frac{2}{3}(\frac{8}{27}L^{-3})^{-1/3}|y - x| = L|y - x|.$$

Exercise Set
1.2 (page 17)

1. a) $3.14002 < x < 3.14316$

3. i) a) 4.16×10^{-2} b) 3.34×10^{-4} c) 4.16×10^{-5} d) 3.34×10^{-2}
   ii) a) 1.33×10^{-2} b) 1.00×10^{-3} c) 1.33×10^{-2} d) 1.00×10^{-3}

5. i) a) 14.1981   b) 3.9022
      c) 0.913   d) 0.913
   ii) a) 14.1 loss of 1   b) 3.90 no loss
      c) 0.000 loss of 3   d) 1.00 loss of 2
   iii) a) 14.2 no loss   b) 3.90 no loss
      c) 1.00 loss of 2   d) 0.900 loss of 1

7. a) 1.3780593×10^{-4} b) $-1.3780593 \times 10^{-4}$ c) 144.5   d) -144.5

9. $5.9604644775390625 \times 10^{-8}$

11. a) -1.827 b) 6.959×10^{-3}
    (b) is more accurate since subtraction is not involved.

13. 86.625, 39.375, 119.50, 71.50

Exercise Set
1.3 (page 25)

1. Using Theorem 1.13 with $f(x) = \sin x$, $x_0 = 0$ and $n = 2$ gives

$$|\sin x - x| = \frac{1}{6}|x|^3|\cos \xi| \text{ for some } \xi \text{ between } x \text{ and } 0.$$

Thus, $|\sin h - h| \le \frac{1}{6}|h|^3$ or

$$\frac{\left| \dfrac{\sin h}{h} - 1 \right|}{h^2} \le \frac{1}{6}.$$

3. a) 1.0000, 0.33333, 0.11111, 0.037037, 0.012346, 0.0041153, 0.0013718, 0.00045727, 0.00015242

b) 1.0000, 0.33333, 0.11112, 0.037050, 0.012365, 0.0041420, 0.0014080, 0.00050580, 0.00021725.

c) Stable   d) Unstable

5. INPUT n; $x_0, x_1, \ldots, x_n$; x.

OUTPUT P.

Step 1   Set $P = x - x_0$;
$\qquad i = 1$.

Step 2   While $P \neq 0$ and $i \leq n$ do
$\qquad$ set $P = P \cdot (x - x_i)$.

Step 3   OUTPUT (P);
$\qquad$ STOP.

7. INPUT number x, tolerance TOL, maximum number of iterations M.

OUTPUT number N of terms or a message of failure.

Step 1   Set $SUM = (1 - 2x)/(1 - x + x^2)$;
$\qquad S = (1 + 2x)/(1 + x + x^2)$;
$\qquad N = 2$.

Step 2   While $N \leq M$ do Steps 3–5.

Step 3   Set $j = 2^{N-1}$;
$\qquad y = x^j$;
$\qquad t_1 = j \dfrac{y}{x}(1 - 2y)$;
$\qquad t_2 = y(y - 1) + 1$;
$\qquad SUM = SUM + t_1/t_2$.

Step 4   If $|SUM - S| < TOL$ then
$\qquad$ OUTPUT (N);
$\qquad$ STOP.

Step 5   Set $N = N + 1$.

Step 6   OUTPUT ('Method failed');
$\qquad$ STOP.

$N = 4$, provided the precision of the calculating device is sufficient.

CHAPTER 2

Exercise Set
2.1 (page 32)

1. Consider $f'(x)$ on $[1, 2]$. 1.3203125   **3.** 4.4921875

5. a) 0.641181946   b) 1.82938385   c) 0.257530212

7. 2.924011   **9.** 14, 1.324768

11. Using $[1.2, 2.2]$ gives convergence to 1.75. Using $[1.5, 2.5]$ gives $p_1 = 2$ and $f(2)$ is undefined.

13. $p_n - p_{n-1} = 1/n$ so that $\lim_{n \to \infty} (p_n - p_{n-1}) = 0$. However, p_n is the nth partial sum of the divergent harmonic series.

Exercise Set 2.2 (page 40)

1. b)   i) $p_4 = 1.10782$ ii) $p_4 = 0.987506$
      iii) $p_4 = 1.12364$ iv) $p_4 = 1.12412$
   c)   (iv) gives best answer since $|p_3 - p_4|$ is smallest.

3. Using $p_0 = 1$ gives $p_{12} = 0.6412053$. Since $|g'(x)| = 2^{-x} \ln 2 \le 0.551$ on $[\frac{1}{3}, 1]$, with $k = 0.551$, $p_0 = 1$ Corollary 2.5 gives a bound of 15 iterations.

5. $p_3 = 1.3231$, $g(x) = \sqrt{1 + \dfrac{1}{x}}$, $p_0 = 1.5$

7. $p_{14} = 2.92399$, $g(x) = 5/\sqrt{x}$, $p_0 = 2.5$

9. a) $g(x) = \sqrt{\frac{1}{3} e^x}$, $[0, 1]$, $p_0 = 0.5$, $p_{14} = 0.91001$
   b) $g(x) = \cos x$, $[0, 1]$, $p_0 = 0.5$, $p_{28} = 0.7390817$

11. $g(x) = 1/\tan(x) - (1/x) + x$, $p_0 = 4$, $p_4 = 4.493409$

13. *Hint:*   First show the result when $x_0 > \sqrt{2}$. When $x_0 < \sqrt{2}$, consider x_1.

15. *Hint:*   Modify (2.7) in the proof.

Exercise Set 2.3 (page 48)

1. a) $p_0 = 2.5$, $p_4 = 2.6906475$ b) $p_0 = -1$, $p_3 = -0.65270365$
   c) $p_0 = 0.7854$, $p_3 = 0.7390851$ d) $p_0 = 0.7854$, $p_3 = 0.9643339$

3. Newton's method: $p_0 = 1$, $p_4 = 1.324718$
   Secant method: $p_0 = 1$, $p_1 = 2$, $p_7 = 1.324717$

5. a) $p_0 = 0$, $p_1 = 1$, $p_5 = 0.2575305$ b) $p_0 = 1.5$, $p_1 = 2.0$, $p_7 = 0.9100076$
   c) $p_0 = 1.5$, $p_1 = 2.0$, $p_6 = 1.829384$ d) $p_0 = 1.5$, $p_1 = 2.0$, $p_5 = 1.968873$

7. (1.8667604, 0.53568738)

11. $p_0 = \pi/2$, $p_{15} = 1.895486$; $p_0 = 5\pi$, $p_{19} = 1.895487$; $p_0 = 10\pi$, $p_{45} = 1.895486$

13. a) $p_5 = 1.75$ b) $p_5 = 1.75$ c) $p_1 = 2$, cannot continue
   d) $p_7 = 1.75$ e)   diverge   f)   diverge

15. Because of the subtraction of nearly equal numbers.

17. a) $p_0 = 0$, $p_1 = 1$, $p_5 = 0.2575305$ b) $p_0 = 0$, $p_1 = 1$, $p_7 = 0.9100077$
   c) $p_0 = 1$, $p_1 = 2$, $p_8 = 1.829382$ d) $p_0 = 1$, $p_1 = 2$, $p_7 = 1.968874$

19. $\lambda = 0.1010007$, $N(2) = 2,187,950$

21. 6.512849, 13.487151   **23.** 6.74%, 6.62%

25. $P_L = 259300$, $c = -0.720674$, $k = 0.047988$
   $P(1980) = 221,470,000$, $P(2000) = 243,379,000$

27. a)   2.10639, 4.51469, 7.22928, 10.12546   b) -0.69955

29. a)   1.52; 10^{-2} since that is the accuracy given for the data.   b)   1.52

Exercise Set 2.4 (page 60)

1. $p_3 = 1.895494$ **3.** $p_4 = -0.1831575$ **5.** a) $n \ge 20$ b) $n \ge 5$

7. $\dfrac{10^{-2^{n+1}}}{(10^{-2^n})^2} = 1$ **9.** $g(x) = x - \dfrac{m(x - p)q(x)}{mq(x) + (x - p)q'(x)}$
                                                        $g'(p) = 0$

11. $m = 2$; $p_4 = 1.895494$ **13.** *Hint:*   Consider Theorem 2.1.

Exercise Set
2.5 (page 65)

1. $g(x) = (1 + 1/x)^{1/2}$, $p_3 = 1.324718$ **3.** $p_3 = 1.7320508$

5. a) $\hat{p}_{10} = 0.0\overline{45}$ b) $\hat{p}_2 = 0.036324786$

Exercise Set
2.6 (page 75)

1. a) 2.69065   b) 0.532089, -0.652706, -2.87938
    c) 1.32472   d) 1.12412, -0.876053

3. a) $-0.345324 \pm 1.31873i$, -2.69065
    b) 0.532089, -0.652703, -2.87938
    c) 1.32472, $-0.662359 \pm 5.62280i$
    d) 1.12412, -0.876053, $-0.124035 \pm 1.74096i$

9. a) roots: 1.244, 8.847, -1.091; critical points: 0, 6.000
    b) roots: 0.5798, 1.521, 2.332, -2.432; critical points: 1.000, 2.001, -1.500

11. 16.2121

13. a) $P_2(x) = \frac{3}{2}x^2 - \frac{1}{2}$, $P_3(x) = \frac{5}{2}x^3 - \frac{3}{2}x$, $P_4(x) = \frac{35}{8}x^4 - \frac{15}{4}x^2 + \frac{3}{8}$
    $P_5(x) = \frac{63}{8}x^5 - \frac{35}{4}x^3 + \frac{15}{8}x$
    b) $P_6(x) = \frac{231}{16}x^6 - \frac{945}{48}x^4 + \frac{315}{48}x^2 - \frac{15}{48}$:
    roots: ± 0.932469514, ± 0.661209385, ± 0.238627205

15. a) $L_2(x) = x^2 - 4x + 2$, $L_3(x) = -x^3 + 9x^2 - 18x + 6$,
    $L_4(x) = x^4 - 16x^3 + 72x^2 - 96x + 24$,
    $L_5(x) = -x^5 + 25x^4 - 200x^3 + 600x^2 - 600x + 120$.

b)

n	Roots				
2	0.58578643,	3.4142136			
3	2.29431161,	0.415774557,	6.2899351		
4	0.32254769,	1.74576924,	4.5366176,	9.39507107	
5	0.26356032,	1.41340489,	3.5964294,	7.08580769,	12.6408013

CHAPTER 3

Exercise Set
3.1 (page 83)

1. a) $(x - 1)^2 + 2(x - 1) - 2$

3. 0.04761875. Error bound 3.1×10^{-7}, actual error 3.0×10^{-7}.

5. 8.86×10^{-7} **7.** 0.0953083   **11.** $n = 250,000$

Exercise Set
3.2 (page 91)

1. a)

Points used	Degree	Approximation
2.4, 2.6	1	0.4958727
2.4, 2.6, 2.2	2	0.4982120
2.4, 2.6, 2.2, 2.8	3	0.4980630
All	4	0.4980705

b)

Points used	Degree	Approximation
−0.1, 0.1	1	0.010070
−0.1, 0.1, −0.3	2	−0.00063250
−0.1, 0.1, −0.3, 0.3	3	−0.00063250
All	4	0.00010625

c)

Points used	Degree	Approximation
1.2, 1.3	1	1.22956
1.2, 1.3, 1.1	2	1.18451
1.2, 1.3, 1.1, 1.4	3	1.11778
All	4	1.13745

d)

Points used	Degree	Approximation
0.4, 0.6	1	0.8629029
0.4, 0.6, 0.2	2	0.8688582
0.4, 0.6, 0.2, 0.8	3	0.8696111
All	4	0.8693047

e)

Points used	Degree	Approximation
0.3, 0.1	1	1.5717608
0.3, 0.1, 0.4	2	1.5274061
0.3, 0.1, 0.4, 0.5	3	1.5325585
All	4	1.5316948

3. 0.33348. Error bound 1.2×10^{-9}, but the data is accurate to only 5×10^{-6}.

5. 0.2826. Error bound 7.4×10^{-6}, but the data is accurate to only 5×10^{-5}.

7. 1.75496, 2.4×10^{-4}

9. a) 1.32436   b) 2.18350   c) 1.15277, 2.01191
d) Parts (a) and (b) are better due to spacing of the nodes.

15. 81,045,000; 192,407,000; 571,329,000

Exercise Set
3.3 (page 97)

1. a) 0.4980705   b) 0.00010625   c) 1.13745
d) 0.8693047   e) 1.5316948

3. 0.198269   **5.** 1.75496   **7.** 0.8095
9. 1.708$\overline{3}$ **11.** 1.75496   **13.** 0.567142

Exercise Set
3.4 (page 107)

1. $-6 + 1.05170x + 0.57250x \, (x\text{–}0.1) + 0.21500x \, (x\text{–}0.1) \, (x\text{–}0.3)$
$+ 0.06301x \, (x\text{–}0.1) \, (x\text{–}0.3) \, (x\text{–}0.6)$

3. 1.05126   **5.** 1.53725

Exercise Set
3.5 (page 115)

1. a) 0.4980703   b) 1.657508   c) 2.65622×10^{-6}
    d) 1.15527   e) 0.8045975   f) 1.5318262

3. 0.80932362. Error bound 1.75×10^{-10}.

5. a) 1.2836451   b) 1.105×10^{-3}

7. a) 0.33350   b) Error bound 5.4×10^{-20}, actual error 0.00001 due to rounding.

9. $H_3(1.25) = 1.1690804$, 4.81×10^{-5}
    $H_5(1.25) = 1.1690161$, 4.43×10^{-5}

Exercise Set
3.6 (page 130)

1. a) 0.4976272   b) 1.637087   c) -0.00277301
    d) 1.09542   e) 0.8695049   f) 1.542323

3. $\int_0^1 S(x)\,dx = 0.000000$, $S'(0.5) = -3.24264$, $S''(0.5) = -0.000019$

5. $\int_0^1 s(x)\,dx = 0.000000$, $s'(0.5) = -3.13445$, $s''(0.5) = -0.000021$

7. a) 0.33348   b) 0.00001   c) 0.33349   d) 4×10^{-5}
    e) 0.94270   f) 0.94265   g) 0.015964   h) 0.015964

9. 0.8093241. Error bound 1.9×10^{-7}, actual error 4.5×10^{-7}.

11. On [0, 0.05], $F(x) = 20(e^{0.1} - 1)x + 1$ and on (0.05, 1],
    $F(x) = 20(e^{0.2} - e^{0.1})x + 2e^{0.1} - e^{0.2}$.

$$\int_0^{0.1} F(x)\,dx = 0.1107936.$$

15. $|f(x) - F(x)| \leq \dfrac{M}{8} \max_{0 \leq j \leq n-1} |x_{j+1} - x_j|^2$, where $M = \max_{a \leq x \leq b} |f''(x)|$.
    Error bounds for Exercise 11 are 1.53×10^{-3} and 1.53×10^{-4}.

17. e) On [0, 0.2], $s(x) = 0.9999999 + 1.000017x + 0.5169415x^2$.
        On [0.2, 0.6], $s(x) = 1.220678 + 1.206784(x - 0.2) + 0.7446967(x - 0.2)^2$
        On [0.6, 0.9], $s(x) = 1.822543 + 1.802541(x - 0.6) + 1.069964(x - 0.6)^2$
        $s(0.5) = 1.649736$
    f) On [1.00, 1.02], $s(x) = 0.7657899 + 1.53199(x - 1.00) - 2.664981(x - 1.00)^2$
        On [1.02, 1.04], $s(x) = 0.7953644 + 1.425388(x - 1.02) - 2.966(x - 1.02)^2$
        On [1.04, 1.06], $s(x) = 0.8226851 + 1.306751(x - 1.04) - 3.249785(x - 1.04)^2$
        $s(1.03) = 0.8093187$

23. The equation of the spline is

$$S(x) \equiv S_i(x) = f_i + b_i(x - x_i) + c_i(x - x_i)^2 + d_i(x - x_i)^3$$

on the interval $[x_i, x_{i+1}]$ where the results in Table A.1 are obtained.

TABLE A.1

	Sample 1				Sample 2			
x_i	f_i	b_i	c_i	d_i	f_i	b_i	c_i	d_i
0	6.67	−0.44687	0	0.06176	6.67	1.6629	0	−0.00249
6	17.33	6.2237	1.1118	−0.27099	16.11	1.3943	−0.04477	−0.03251
10	42.67	2.1104	−2.1401	0.28109	18.89	−0.52442	−0.43490	0.05916
13	37.33	−3.1406	0.38974	−0.01411	15.00	−1.5365	0.09756	0.00226
17	30.10	−0.70021	0.22036	−0.02491	10.56	−0.64732	0.12473	−0.01113
20	29.31	−0.05069	−0.00386	0.00016	9.44	−0.19955	0.02453	−0.00102

25. The free cubic spline predicts the populations to be 114,737,000; 191,844,000; 271,210,000.

CHAPTER 4

Exercise Set
4.1 (page 145)

1. a)

x	$f'(x)$
−0.3	0.35785
−0.1	0.78595
0.1	1.2141
0.3	1.6422

b)

x	$f'(x)$
1.1	1.9540
1.2	5.5574
1.3	14.500
1.4	28.781

3. a)

		$f'(0.4)$
(4.2)	$h = 0.6$	−0.8889958
	$h = 0.4$	−0.6979043
	$h = 0.2$	−0.5486810
	$h = -0.2$	−0.3104710
(4.12)	$h = 0.2$	−0.3994578
(4.13)	$h = 0.2$	−0.4295760

		$f''(0.4)$
(4.20)	$h = 0.2$	−1.191050

b)

		$f'(0.6)$
(4.2)	$h = 0.4$	−1.059153
	$h = 0.2$	−0.8471275
	$h = -0.2$	−0.5486810
	$h = -0.4$	−0.4295760
(4.12)	$h = 0.2$	−0.6351018
	$h = -0.2$	−0.6677860
(4.13)	$h = 0.4$	−0.7443646
	$h = 0.2$	−0.6979043
(4.14)	$h = 0.2$	−0.6824175

		$f''(0.6)$
(4.20)	$h = 0.4$	−1.573943
	$h = 0.2$	−1.492233

5. −3.10457, 3.98×10^{-2}

7. 2.27403, 2.27510

9. 0.0, $f(x)$ is symmetric about $x = 0.5$, 0.359

13. a) $f'(1.005) \approx 5.0$, $f'(1.015) \approx 6.0$ b) $f''(1.01) \approx 100$
    c) f' accurate to within 1.0, f'' accurate to within 200.

15.

Time	0	3	5	8	10	13
Speed	79	82.4	74.2	76.8	69.4	71.2

17.

t	1.00	1.01	1.02	1.03	1.04
$\mathscr{E}(t)$	2.400	2.403	3.386	5.352	7.320

19. a) Using $f'(x_0) = 1.5352695$ and $f'(x_3) = 1.179522$,
$$s(x) = 0.7657892 + 1.535267(x - 1.00) - 2.881544(x - 1.00)^2$$
$$+ 3.017642(x - 1.00)^3 \quad \text{on } [1.00, 1.02];$$
$$s(x) = 0.7953668 + 1.423640(x - 1.02) - 2.700481(x - 1.02)^2$$
$$- 8.823470(x - 1.02)^3 \quad \text{on } [1.02, 1.04];$$
$$s(x) = 0.8226882 + 1.305052(x - 1.04) - 3.229876(x - 1.04)^2$$
$$+ 3.055136(x - 1.04)^3 \quad \text{on } [1.04, 1.06].$$
$s(1.03) = 0.809324328.$

b) The equation of the spline is $s(x) = S_i(x) \equiv f_i + b_i(x - x_i) + c_i(x - x_i)^2 + d_i(x - x_i)^3$ on the interval $[x_i, x_{i+1}]$, the results in Table A.2 are obtained.

TABLE A.2

x_i	f_i	b_i	c_i	d_i
0.9	0.7	4.652147	−3.876282	−226.4519
0.8	0.3	2.308093	27.31677	−103.9769
0.7	0.2	1.115502	−15.39093	142.3590
0.6	0	2.229899	4.246964	−65.45966
0.5	−0.2	1.964892	−1.596888	19.47949
0.4	−0.4	1.910522	2.140592	−12.45826
0.3	−0.5	−0.6069813	23.03444	−69.64616
0.2	−0.3	−2.48258	−4.278495	91.04306
0.1	−0.1	−1.462688	−5.920434	5.473132
0.0	0	−0.6666667	−2.039782	−12.93550

Exercise Set
4.2 (page 152)

1. a) 1.0000109   b) 2.0000001   c) 2.2751459   d) −19.646799

3. a) 1.001   b) 1.999   c) 2.283   d) −19.61

5. 1.999999

Exercise Set
4.3 (page 161)

1.

	Trapezoidal	Error bound	Simpson's	Error bound	Actual
a)	0.34657	0.084	0.38583	2.1×10^{-3}	0.38629
b)	0.023208	none	0.032296	none	0.034812
c)	0.39270	0.192	0.30543	3.5×10^{-3}	0.30709
d)	0.39914	0.0114	0.40371	6.24×10^{-5}	0.40376
e)	0.39270	0.161	0.34778	8.31×10^{-3}	0.34657
f)	−0.39270	0.161	−0.34778	8.31×10^{-3}	−0.34657

3. a)   0.1024597, 0.1024596, 0.1024597, 0.1024661, 0.1024597, 0.1024596.
    b)   0.785397, 0.785397, 0.785397, 0.785398, 0.785397, 0.785397.
    c)   1.477534, 1.477526, 1.477519, 1.470979, 1.477508, 1.477513.
    d)   2.740906, 2.563390, 2.385700, 1.767856, 2.074892, 2.116376.
    e)   2.407900, 2.359771, 2.314754, 2.048634, 2.233249, 2.249001.
    f)   0.695800, 0.712603, 0.730634, 0.783471, 0.761114, 0.759357.

5. $n = 4$: 0.76680138, error bound 6.3×10^{-7}.
    $n = 3$: 0.76664406, error bound 2.6×10^{-4}.

7.   The errors in order are 1.6×10^{-6}, 5.3×10^{-8}, -6.7×10^{-7}, -7.2×10^{-7}, -1.3×10^{-6}.

Exercise Set
4.4 (page 167)

1. a)   1.1167, exact 1.09861   b)   4.25, exact 4
    c)   10.3122, exact 10.20759   d)   0.62201, exact 0.636620
    e) -5.9568, exact -6.28319 f)   0.72889, exact 0.718282

3. a)   1.098724   b)   4.000000   c)   10.20751
    d)   0.6366357   e) -6.284027 f)   0.7182836

5. a)   0.4215820   b)   0.4227162   c)   0.4241792

7. a)   0.3497582, 0.3473746   b)   0.0101, 0.0025
    c) $n \geq 4019$, $h \leq 1.96 \times 10^{-4}$

9. a)   0.3437928, 0.3455552   b)   0.00898, 0.00330
    c) $n \geq 5684$, $h < 0.0001382$

11. a) -0.3466741, -0.346580 b) 5.19×10^{-4}, 1.04×10^{-5}
    c) $h \leq 0.01301$, $n \geq 62$

13. a) $n \geq 16$, $h \leq 0.132$, 10.9501107
    b) $n \geq 516$, $h \leq 0.00388$, 10.9501443
    c) $n \geq 730$, $h \leq 0.00274$, 10.9501962

17. a)   2.61972   b)   2.62087   **19.**   58.47047

21. b)   2.924400   c)   1.075443   **23.**   6.945 miles/sec

Exercise Set
4.5 (page 176)

1. a)   1.098724   b)   4.000000   c)   10.20755
    d)   0.6367049   e) -6.283243 f)   0.7189079

3.   0.4227242   **5.**   58.46960   **7.**   0.7499954

9.   Adaptive quadrature: 0.13039×10^{-6};
    Simpson's composite method: -0.45253×10^{-6}.

Exercise Set
4.6 (page 182)

1. a)   1.09926   b)   4.00000   c)   10.2046
    d)   0.6361639   e) -7.01838 f)   0.718313

3.   0.4227250   **5.**   10.95017

7. a)   0.3465739   b) -0.3465738 **9.**   58.47047

Exercise Set
4.7 (page 188)

1. a)   1.09091   b)   4.00000   c)   10.2423
    d)   0.616191   e) -11.0616 f)   0.711942

3.   11.141495, 10.948403, 10.950140

7.	n	Roots	Coefficients
	2	0.5857864	0.8535534
		3.4142136	0.1464466
	3	0.4157746	0.7110930
		2.2942804	0.2785177
		6.2899451	0.0103893

When $n = 2$, the approximation is 0.432460. When $n = 3$, the approximation is 0.496023. The exact value is 0.5.

9. $n = 2$: 2.9865139; $n = 3$: 2.9958198

Exercise Set 4.8 (page 196)

1. a) 0.03920109   b) 1.577281×10^{-4} c) $-1.810904 \times 10^{-11}$
   d) 0.01000832   e) 16.50857   f) 0.1871799

3. 0.1479099   5. 0.6128008   7. 19.700

9. 1.469840   11. $\bar{x} = 0.380644$, $\bar{y} = 0.382249$

CHAPTER 5

Exercise Set 5.1 (page 204)

1. a)
$$f(t, y) = y \cos t, \qquad \frac{\partial f}{\partial y}(t, y) = \cos t$$

f satisfies Lipschitz condition in y on

$$D = \{(t, y) | 0 \leq t \leq 1, \, -\infty < y < \infty\} \quad \text{with } L = 1$$

f is continuous on D, so there exists a unique solution, which is $y(t) = e^{\sin t}$.

b)
$$f(t, y) = \frac{2}{t} y + t^2 e^t, \qquad \frac{\partial f}{\partial y} = \frac{2}{t}$$

f satisfies Lipschitz condition in y on

$$D = \{(t, y) | 1 \leq t \leq 2, \, -\infty < y < \infty\} \quad \text{with } L = 2$$

f is continuous on D, so there exists a unique solution, which is $y(t) = t^2(e^t - e)$.

c)
$$f(t, y) = -\frac{2}{t} y + t^2 e^t, \qquad \frac{\partial f}{\partial y} = -\frac{2}{t}$$

f satisfies Lipschitz condition in y on

$$D = \{(t, y) | 1 \leq t \leq 2, \, -\infty < y < \infty\} \quad \text{with } L = 2$$

f is continuous on D, so there exists a unique solution, which is

$$y(t) = (t^4 e^t - 4t^3 e^t + 12t^2 e^t - 24te^t + 24e^t + (\sqrt{2} - 9)e)/t^2.$$

d) $f(t, y) = \dfrac{4t^3}{y(1 + t^4)}$ does not satisfy a Lipschitz condition on

$$D = \{(t, y) | 0 \leq t \leq 1, \, -\infty < y < \infty\}$$

so Theorem 5.4 cannot be used. However, a unique solution

$$y(t) = (\ln(1 + t^4)^2 + 1)^{1/2}$$

does exist.

3. a) Differentiating $\dfrac{t}{y} - 1 - 2 \displaystyle\int_1^y \dfrac{\sin z}{z} dz = 0$, gives $\dfrac{1}{y} - \dfrac{t}{y^2} y' - \dfrac{2 \sin y}{y} y' = 0$. Solving for y' yields the original differential equation. Setting $t = 1$ and $y = 1$ verifies the initial condition. To approximate $y(2)$, solve the equation

$$\frac{2}{y} - 1 - 2 \int_1^y \frac{\sin z}{z} dz = 0$$

for y using Newton's method, where the integral $\displaystyle\int_1^y \dfrac{\sin z}{z} dz$ can be approximated using Simpson's composite rule. Requiring accuracy to within 10^{-4} in Newton's method and to within 10^{-5} in Simpson's composite rule gives $y(2) = 1.323791$.

b) Differentiating $y \sin t + t^2 e^y + 2y - 1 = 0$, gives $y' \sin t + y \cos t + 2te^y + t^2 e^y y' + 2y' = 0$. Solving for y' yields the original differential equation. To approximate $y(2)$, solve the equation

$$(2 + \sin 2)y + 4e^y - 1 = 0$$

using Newton's method. This gives $y(2) = -0.4946599$.

5. Given (t_1, y_1) and (t_2, y_2) in D, $a \leq t_1 \leq b$, $a \leq t_2 \leq b$, $-\infty < y_1 < \infty$ and $-\infty < y_2 < \infty$. Thus, for $0 \leq \lambda \leq 1$, $(1 - \lambda)a \leq (1 - \lambda)t_1 \leq (1 - \lambda)b$ and $\lambda a \leq \lambda t_2 \leq \lambda b$. Hence, $a = (1 - \lambda)a + \lambda a \leq (1 - \lambda)t_1 + \lambda t_2 \leq (1 - \lambda)b + \lambda b = b$. Clearly, $-\infty < (1 - \lambda)y_1 + \lambda y_2 < \infty$. Thus, D is convex.

7. a)

$$y(t) = 1 - e^{-t}$$
$$z(t) = 1 - e^{-t} + \delta(t - 1 + e^{-t}) + \varepsilon_0 e^{-t}$$
$$|y(t) - z(t)| \leq 2|\delta| + |\varepsilon_0| < 3\varepsilon$$

b)

$$y(t) = -t - 1$$
$$z(t) = -t - 1 + \delta(-t - 1 + e^t) + \varepsilon_0 e^t$$
$$|y(t) - z(t)| \leq 4.4|\delta| + 7.4|\varepsilon_0| < 11.8\varepsilon$$

c)

$$y(t) = t^2(e^t - e)$$
$$z(t) = t^2(e^t - e) + t^2(\varepsilon_0 + \delta \ln t)$$
$$|y(t) - z(t)| \leq 4(|\varepsilon_0| + \ln 2|\delta|) < 6.8\varepsilon$$

d)

$$y(t) = \frac{t^4 e^t - 4t^3 e^t + 12t^2 e^t - 24te^t + 24e^t}{t^2} + \frac{(\sqrt{2} - 9)e}{t^2}$$

$$z(t) = \frac{t^4 e^t - 4t^3 e^t + 12t^2 e^t - 24te^t + 24e^t}{t^2} + \frac{(\sqrt{2} - 9)e}{t^2}$$

$$+ \delta t + \frac{\varepsilon_0 - \delta}{t^2}$$

$$|y(t) - z(t)| \leq 2|\delta| + |\varepsilon_0 - \delta| < 4\varepsilon$$

**Exercise Set
5.2 (page 212)**

1. a)

i	t_i	w_i
1	1.1	1.2
2	1.2	1.4281

b)

i	t_i	w_i
1	0.5	0.5
2	1.0	1.04298

c)

i	t_i	w_i
1	1.5	-1.0
2	2.0	-1.0
3	2.5	-1.0
4	3.0	-1.0

d)

i	t_i	w_i
1	0.25	1.0000
2	0.50	1.1875
3	0.75	1.4601
4	1.00	1.7000

3. a)

i	t_i	w_i	$\lvert y(t_i) - w_i \rvert$
1	1.1	0.271828	0.07409
5	1.5	3.18744	0.7802
6	1.6	4.62080	1.100
9	1.9	11.7480	2.575
10	2.0	15.3982	3.285

b)

t	Approximation	$y(t)$	|Error|
1.04	0.108731	0.119986	0.01126
1.55	3.90412	4.78864	0.8845
1.97	14.3031	17.2793	2.976

c) $h < 0.00064$

5. a)

	$h = 0.2$	$h = 0.1$	$h = 0.05$
w_N	5.00377	5.00515	5.00591

$y(5) = 5.00674$

b) $\sqrt{2 \times 10^{-6}} = 0.0014142$

7

i	t_i	w_i
10	1.0	1.20319
20	2.0	2.14970

9. b) $w_{50} = 0.10430 \approx p(50)$

c) $p(t) = 1 - 0.99e^{-0.002t}$; $p(50) = 0.10421$

11. a)
$$\int_{t_i}^{t_{i+1}} y'(t)\,dt = \int_{t_i}^{t_{i+1}} f(t, y(t))\,dt$$

$$y(t_{i+1}) - y(t_i) = \int_{t_i}^{t_{i+1}} f(t, y(t))\,dt$$
$$\approx hf(t_i, y(t_i)),$$

using the quadrature formula $\int_{t_i}^{t_{i+1}} g(t)\,dt \approx (t_{i+1} - t_i)g(t_i)$. This leads to $w_{i+1} = w_i + hf(t_i, w_i)$.

b)
$$y(t_{i+1}) - y(t_i) = \int_{t_i}^{t_{i+1}} f(t, y(t))\,dt$$
$$\approx \frac{(t_{i+1} - t_i)}{2}[f(t_i, y(t_i)) + f(t_{i+1}, y(t_{i+1}))]$$
$$\approx \frac{h}{2}[f(t_i, y(t_i)) + f(t_{i+1}, y(t_{i+1}))]$$

This leads to

$$w_{i+1} = w_i + \frac{h}{2}[f(t_i, w_i) + f(t_{i+1}, w_{i+1})],$$

which defines w_{i+1} implicitly.

Exercise Set 5.3 (page 218)

1. a)

i	t_i	w_i
1	1.1	1.214999
2	1.2	1.465250

b)

i	t_i	w_i
1	0.5	0.5000000
2	1.0	1.076858

c)

i	t_i	w_i
1	1.5	-2.000000
2	2.0	-1.777776
3	2.5	-1.585732
4	3.0	-1.458882

d)

i	t_i	w_i
1	0.25	1.093750
2	0.50	1.312319
3	0.75	1.538468
4	1.0	1.720480

3. a)

i	t_i	Order 2	Order 4
5	0.5	-1.499997	-1.499997
10	1.0	-1.999990	-1.999994
15	1.5	-2.499980	-2.499987
20	2.0	-2.999963	-2.999973

b)

i	t_i	Order 2	Order 4
5	0.5	0.3929239	0.3934687
10	1.0	0.6314586	0.6321198
15	1.5	0.7762673	0.7768691
20	2.0	0.8641771	0.8646640

c)

i	t_i	Order 2	Order 4
10	1	1.368534	1.367873
20	2	2.135813	2.135326
30	3	3.050045	3.049776
40	4	4.018436	4.018302
50	5	5.006786	5.006725

d)

i	t_i	Order 2	Order 4
5	0.50	0.6474453	0.6487194
10	1.0	1.714074	1.718275
15	1.5	3.471287	3.481671
20	2.0	6.366201	6.389017

5. a)

i	t_i	w_i
1	1.1	0.3397848
5	1.5	3.910973
6	1.6	5.643064
9	1.9	14.15263
10	2.0	18.46992

b) $y(1.04) \approx 0.1359139$
$y(1.55) \approx 4.777019$
$y(1.97) \approx 17.17473$

c)

i	t_i	w_i
1	1.1	0.3459122
5	1.5	3.967590
6	1.6	5.720855
9	1.9	14.32284
10	2.0	18.68278

d) $y(1.04) \approx 0.1199702$
$y(1.55) \approx 4.788511$
$y(1.97) \approx 17.27896$

7. a)

Taylor method of order 2

i	t_i	w_i
2	0.2	5.86595
5	0.5	2.82145
7	0.7	0.84926
10	1.0	-2.09503

Taylor method of order 4

i	t_i	w_i
2	0.2	5.86433
5	0.5	2.81789
7	0.7	0.84455
10	1.0	-2.10154

b) 0.8 seconds

Exercise Set
5.4 (page 227)

1. a)

i	t_i	w_i
1	1.1	1.21405
2	1.2	1.46302

b)

i	t_i	w_i
1	0.5	0.521489
2	1.0	1.09531

c)

i	t_i	w_i
1	1.5	-1.5
2	2.0	-1.33594
3	2.5	-1.25246
4	3.0	-1.20209

d)

i	t_i	w_i
1	0.25	1.093750
2	0.50	1.294851
3	0.75	1.511425
4	1.0	1.692287

3. a)

i	t_i	w_i
1	1.1	1.21588
2	1.2	1.46755

b)

i	t_i	w_i
1	0.5	0.515898
2	1.0	1.09184

c)

i	t_i	w_i
1	1.5	-1.49541
2	2.0	-1.33056
3	2.5	-1.24804
4	3.0	-1.19850

d)

i	t_i	w_i
1	0.25	1.087168
2	0.50	1.289921
3	0.75	1.513531
4	1.0	1.701786

5. a)

i	t_i	w_i
1	1.1	0.3423771
5	1.5	3.936429
6	1.6	5.678886
9	1.9	14.23738
10	2.0	18.57879

b) $y(1.04) \approx 0.1369508$
$y(1.55) \approx 4.807658$
$y(1.97) \approx 17.27637$

c)

i	t_i	w_i
1	1.1	0.3459091
5	1.5	3.967585
6	1.6	5.720854
9	1.9	14.32286
10	2.0	18.68283

d) $y(1.04) \approx 0.1199692$
$y(1.55) \approx 4.788508$
$y(1.97) \approx 17.27900$

7. a)

t	Euler's method	Heun's method	Runge–Kutta 4
1.5	-0.3832803	-0.3967280	-0.3963564
2.0	-0.1395250	-0.1537860	-0.1534261

b)

t	Euler's method	Heun's method	Runge–Kutta 4
1	−1.999984	−1.999984	−1.999993
2	−2.999922	−2.999919	−2.999974

c)

t	Euler's method	Heun's method	Runge–Kutta 4
0.5	0.6004670	0.6190949	0.6178911
1.0	1.1176186	1.1365805	1.1353475

d)

t	Euler's method	Heun's method	Runge–Kutta 4
1	0.6367669	0.6319606	0.6321198
2	0.8680611	0.8645464	0.8646640

9. a)

i	t_i	w_i	Exact solution
5	0.5	1.62349375×10^5	0.02499998212
10	1.0	$7.858959155 \times 10^{10}$	0.9999992847

b)

i	t_i	w_i	Exact solution
10	0.25	0.06253916025	0.06250119209
20	0.50	0.02500364780	0.02499999404
30	0.75	0.05625362992	0.05624998808
40	1.0	1.000035286	0.9999998808

c)

i	t_i	w_i	Exact solution
25	0.25	0.06250184774	0.06250119209
50	0.50	0.02500003576	0.02499998808
75	0.75	0.05625001192	0.05624997020
100	1.0	1.000000000	0.9999995232

The only time a good approximation is achieved is when the step size h is small.

11. b) 7.9787 feet

Exercise Set
5.5 (page 235)

1. **a)**

i	t_i	w_i	h_i
1	1.05	1.103855	0.05
2	1.10	1.215881	0.05
3	1.15	1.336832	0.05
4	1.20	1.467560	0.05

b)

i	t_i	w_i	h_i
1	0.25	0.2522865	0.25
2	0.50	0.5158861	0.25
3	0.75	0.7959436	0.25
4	1.0	1.091815	0.25

c)

i	t_i	w_i	h_i
1	1.17783	-1.73767	0.17783
2	1.64199	-1.43898	0.46416
3	2.14199	-1.30525	0.5
4	2.64199	-1.23397	0.5

d)

i	t_i	w_i
1	0.5	0.0416666
2	1.0	0.333333
3	1.5	1.12500
4	2.0	2.66666

3. **a)**

i	t_i	w_i	h_i
4	0.8	0.2369365	0.2
8	1.6	0.8521823	0.2
12	2.4	1.173558	0.2
15	3.0	2.718405	0.2

b)

i	t_i	w_i	h_i
2	0.4	0.4064441	0.2
5	1.0	1.253376	0.2
8	1.563833	2.192151	0.1638336
10	1.893069	2.166748	0.1668379

c)

i	t_i	w_i	h_i
5	0.0534694	0.0258533	0.0111953
8	0.0969589	0.0120053	0.0165621
12	0.194914	0.0379899	0.0293748
19	0.414326	0.171637	0.0314950
25	0.603228	0.363855	0.0314687
31	0.792007	0.627246	0.0314658
37	0.980881	0.962098	0.0314338

d)

i	t_i	w_i	h_i
4	0.2140541	0.6975596	0.05535838
8	0.4428747	0.6085422	0.05838532
13	0.7450544	0.7804532	0.06182698
17	0.9914919	1.120710	0.06006820

5. a) 80,293   b) 80,296; $\lim\limits_{t\to\infty} y(t) = 100{,}000$

7. b) (1a)

i	t_i	w_i	h_i
1	1.05	1.099999	0.05
2	1.10	1.207255	0.05
3	1.15	1.322356	0.05
4	1.20	1.445960	0.05

(1b)

i	t_i	w_i	h_i
1	0.25	0.25	0.25
2	0.50	0.5065510	0.25
3	0.75	0.7780399	0.25
4	1.0	1.066540	0.25

(1c)

i	t_i	w_i	h_i
4	1.117748	-1.800182	0.03541044
7	1.273252	-1.628715	0.06274807
10	1.604489	-1.421488	0.1460301
13	2.673927	-1.179515	0.5000000

(1d)

i	t_i	w_i	h_i
5	0.9443970	0.2118899	0.1113544
10	1.374184	0.7538489	0.07467168
15	1.699229	1.492057	0.05986424
20	1.971521	2.383912	0.05137323

1. a)

i	t_i	w_i
1	1.05	1.10386
2	1.10	1.21588
3	1.15	1.33684
4	1.20	1.46757

b)

i	t_i	w_i
1	0.25	0.252287
2	0.50	0.515887
3	0.75	0.795945
4	1.0	1.09185

c)

i	t_i	w_i
1	1.5	-1.49541
2	2.0	-1.33056
3	2.5	-1.24804
4	3.0	-1.17988

d)

i	t_i	w_i
1	0.25	1.087168
2	0.50	1.289921
3	0.75	1.513531
4	1.0	1.703750

3. a)

i	t_i	w_i
2	0.2	0.1812691
5	0.5	0.3934668
7	0.7	0.5034131
10	1.0	0.6321199

b)

i	t_i	w_i
5	1.25	-0.7999989
10	1.50	-0.6666643
15	1.75	-0.5714253
20	2.00	-0.4999960

c)

i	t_i	w_i
2	1.2	0.8666181
5	1.5	3.967628
8	1.8	10.79362
10	2.0	18.68315

d) For $h = 0.1$:

i	t_i	w_i
1	0.1	4.60545
5	0.5	0.1993813×10^6
10	1.0	0.1655299×10^{12}

This is a stiff equation (see Section 5.11); these values do not accurately approximate the solution to the initial-value problem.

For $h = 0.025$:

i	t_i	w_i
4	0.1	-0.004602551
20	0.5	0.2456715
40	1.0	0.9968154

For $h = 0.01$:

i	t_i	w_i
10	0.1	0.01213566
50	0.5	0.2499997
100	1.0	0.9999993

5.

t_i	$w_i(h = 0.1)$	$w_i(h = 0.05)$	Actual
0.5	0.08091807	0.08203756	0.082085
1.0	0.006407015	0.006728194	0.0067379
1.5	0.0005059693	0.0005518035	0.00055308
2.0	0.0000399691	0.00004525537	0.000045400

9.

t_i	$w_i(h = 0.1)$	$w_i(h = 0.05)$
0.5	0.08212387	0.08205926
1.0	0.006474316	0.006720904
1.5	0.0005297423	0.0005358713
2.0	0.00005337294	0.00002521176

Exercise Set 5.7 (page 253)

1. a)

i	t_i	w_i	h
1	1.05	1.103856	0.05
2	1.10	1.215884	0.05
3	1.15	1.336835	0.05
4	1.20	1.467565	0.05

b)

i	t_i	w_i	h
1	0.2043788	0.2056590	0.2043788
2	0.4087576	0.4179090	0.2043788
3	0.6131365	0.6405029	0.2043788
4	0.8175153	0.8744465	0.2043788

c)

i	t_i	w_i	h
7	1.230612	-1.684344	0.03294554
14	1.529549	-1.485635	0.05571989
21	2.033099	-1.326123	0.08409792
28	2.896717	-1.208604	0.1299206

d)

i	t_i	w_i	h
4	0.2201681	1.068624	0.05504203
7	0.3852941	1.189073	0.05504203
10	0.5504202	1.335717	0.05504203
14	0.9230096	1.649312	0.09314733

3. a)

i	t_i	w_i	h
11	1.365286	0.6682541	0.1241170
23	1.996888	0.9997863	0.001399415
35	2.206795	1.043478	0.08956253
47	2.975817	2.590943	0.04714887

b)

i	t_i	w_i	h
8	0.9864113	1.226849	0.04660286
16	1.312483	1.889703	0.03491653
24	1.620551	2.216120	0.04210143
33	1.999459	2.097663	0.04210143

c)

i	t_i	w_i	h
13	0.05830010	0.02145894	0.004484623
26	0.1842373	0.03398216	0.01711229
39	0.5786425	0.3348414	0.02646209
53	0.9991178	0.9982529	0.02478282

d)

i	t_i	w_i	h
8	0.2408090	0.6757717	0.03010117
16	0.4816180	0.6136112	0.03010117
25	0.7525281	0.7883028	0.03010117
34	0.9933370	1.123867	0.03010117

Exercise Set 5.8 (page 260)

1. a)

i	t_i	w_i	h
1	1.05	1.103856	0.05
2	1.10	1.215883	0.05
3	1.15	1.336833	0.05
4	1.20	1.467561	0.05

b)

i	t_i	w_i	h
1	0.25	0.2522867	0.25
2	0.50	0.5158876	0.25
3	0.75	0.7959448	0.25
4	1.00	1.091816	0.25

c)

i	t_i	w_i	h
1	1.5	−1.500000	0.5
2	2.0	−1.333313	0.5
3	2.5	−1.249983	0.5
4	3.0	−1.199987	0.5

d)

i	t_i	w_i	h
1	0.25	1.087071	0.25
2	0.50	1.289778	0.25
3	0.75	1.513442	0.25
4	1.0	1.701845	0.25

3. a)

i	t_i	w_i
4	0.8	0.236758
8	1.6	0.851461
12	2.4	1.17255
16	3.0	2.71594

b)

i	t_i	w_i	h
1	0.20	0.2004003	0.2
4	0.80	0.9055567	0.2
7	1.4	2.031415	0.2
10	2.0	2.097315	0.2

c)

i	t_i	w_i
4	0.2	0.0407044
7	0.5	0.250004
10	0.8	0.640002
12	1.0	1.00000

d)

i	t_i	w_i
3	0.25	0.669028
5	0.45	0.609068
10	0.7	0.736592
17	0.975	1.09288

5. 56751

Exercise Set
5.9 (page 269)

1. a)

i	t_i	w_{1i}	w_{2i}
2	0.2	0.6328718	1.451603
4	0.4	2.237883	2.908204
6	0.6	6.505331	7.054143
8	0.8	18.02456	18.47389
10	1.0	49.26334	49.63122

b)

i	t_i	w_{1i}	w_{2i}
2	0.2	0.495482	-1.11554
4	0.4	0.831413	-1.11229
6	0.6	1.05990	-1.04403
8	0.8	1.12122	-0.944159
10	1.0	1.30654	-0.832917

c)

i	t_i	w_{1i}	w_{2i}	w_{3i}
2	0.2	2.997466	-0.03733361	0.7813973
5	0.5	2.963536	-0.2083733	0.3981578
7	0.7	2.905641	-0.3759597	0.1206261
10	1.0	2.749648	-0.6690454	-0.3011653

d)

i	t_i	w_{1i}	w_{2i}	w_{3i}
2	0.2	1.381651	1.007999	-0.6183311
5	0.5	1.907526	1.124998	-0.09090697
7	0.7	2.255029	1.342997	0.2634375
10	1.0	2.832112	1.999997	0.8821170

3. a)

i	t_i	w_{1i}	w_{2i}
2	0.2	0.632872	1.45160
5	0.5	3.85385	4.46038
7	0.7	10.8525	11.3490
10	1.0	49.2066	49.5744

b)

i	t_i	w_{1i}	w_{2i}
2	0.2	0.49548179	-1.1155433
4	0.4	0.83141276	-1.1122947
6	0.6	1.05990410	-1.0440282
8	0.8	1.12122424	-0.94415889
10	1.0	1.30656880	-0.83291735

c)

i	t_i	w_{1i}	w_{2i}	w_{3i}
2	0.2	2.997466	-0.03733361	0.7813973
5	0.5	2.963537	-0.2083736	0.3981572
7	0.7	2.905642	-0.3759606	0.1206247
10	1.0	2.749649	-0.6690474	-0.3011681

d)

i	t_i	w_{1i}	w_{2i}	w_{3i}
2	0.2	1.381651	1.007999	-0.6183311
5	0.5	1.907525	1.124998	-0.09090662
7	0.7	2.255025	1.342997	0.2634382
10	1.0	2.832104	1.999996	0.8821178

5. For $h = 0.5$:

i	t_i	w_{1i}	w_{2i}
2	1	8716	1435
4	2	7907	2120
6	3	6666	2813

A stable solution is $x_1 = 833$, $x_2 = 1500$.

Exercise Set 5.10 (page 280)

5. a) $\tau_{i+1} = \frac{1}{4}h^3 y^{(4)}(\xi_i)$, $t_{i-2} < \xi_i < t_{i+1}$
    b) Consistent but unstable and not convergent.

7. Unstable

9. b) If $h\lambda < -3$, there is a root $\beta_i < -1$.

Exercise Set
5.11 (page 288)

1. a)

i	t_i	Euler	Runge–Kutta	Trapezoidal
2	0.2	0.6795714	1.000793	0.9785822
5	0.5	0.08494663	0.2235730	0.2113739
7	0.7	0.02123668	0.08231317	0.07609468
10	1.0	0.00265459	0.01838843	0.01643646

c)

i	t_i	Euler	Runge–Kutta	Trapezoidal
3	0.15	0.01999998	0.04015712	0.03484569
4	0.20	0.03749998	0.04667348	0.04411522
7	0.50	0.2424998	0.2526543	0.2499998
10	0.80	0.6374997	0.6425052	0.6399995
12	1.0	0.9974995	1.002499	0.9999993

e)

i	t_i	Euler	Runge–Kutta	Trapezoidal
3	0.15	9.428946	20.13803	0.9747101
4	0.20	-13.87825	12.77384	1.011116
7	0.50	882.0559	32537.94	0.9967318
10	0.80	-56451.29	8.38184×10^7	1.000969
12	1.0	$-903219.$	1.575086×10^{12}	1.000430

5. b) (1a)

i	t_i	w_i
2	0.2	1.208124
5	0.5	0.3579629
7	0.7	0.1590946
10	1.0	0.04713917

(1c)

i	t_i	w_i
3	0.15	0.06635415
4	0.20	0.06317704
7	0.50	0.2556730
10	0.80	0.6450244
12	1.0	1.005002

(1e)

i	t_i	w_i
3	0.15	1.001982
4	0.20	1.000329
7	0.50	1.000000
10	0.80	1.000000
12	1.0	1.000000

7. $p(50) \simeq 0.1060$. If Euler's method was used for Exercise 9 of Section 5.2 with single precision, then 0.1061 would be obtained instead of 0.10430.

CHAPTER 6

Exercise Set
6.1 (page 295)

1. a) $x_1 = 11/5, x_2 = -2/5$
     b) $x_1 = 29/7, x_2 = 3/7$
     c) $x_1 = 1, x_2 = 2, x_3 = 0$
     d) $x_1 = -1, x_2 = -2, x_3 = -1$
     e) $x_1 = -4, x_2 = 2, x_3 = 0, x_4 = 3$
     f) $x_1 = 6.75, x_2 = 15.5, x_3 = 6, x_4 = 2.5$
     g) $x_1 = 1, x_2 = 1, x_3 = 1$
     h) $x_1 = -1579.342, x_2 = 4715.941, x_3 = -3283.854$

3. a) Infinite number of solutions   b) No solutions
     c) $x_1 = 2.75, x_2 = 1.125, x_3 = 4.5$ d) $x_1 = 9, x_2 = -13, x_3 = 0$
     e) No solutions   f) $x_1 = -1, x_2 = 2, x_3 = 0, x_4 = 1$

5. a) Yes
     b) Could add 200 of species 1, or 150 of species 2, or 100 of species 3, or 100 of species 4.
     c) Assuming none of the increases indicated in part (b) was selected, species 2 could be increased by 650, or species 3 could be increased by 150, or species 4 could be increased by 150.
     d) Assuming none of the increases indicated in parts (b) or (c) was selected, species 3 could be increased by 150, or species 4 could be increased by 150.

Exercise Set
6.2 (page 304)

1. a) $x_1 = 2.8, x_2 = -1.6, x_3 = 2.9$ b) $x_1 = 2.2, x_2 = -1.6, x_3 = 2.9$
     c) $x_1 = 1.1, x_2 = -1.1, x_3 = 2.9$ d) $x_1 = 1.0, x_2 = -1.0, x_3 = 3.0$

3. a) $x_1 = -227.0769, x_2 = 476.9231, x_3 = -177.6923.$
     b) $x_1 = x_2 = x_3 = 1.$
     c) $x_1 = x_2 = x_3 = x_4 = 1.$
     d) $x_1 = -0.03174600, x_2 = 0.5952377, x_3 = -2.380951, x_4 = 2.777777.$

5. a) $x_1 = 2.8, x_2 = -1.6, x_3 = 2.9$ b) $x_1 = 2.6, x_2 = -1.4, x_3 = 2.9$
     c) $x_1 = 1.1, x_2 = -1.0, x_3 = 2.9$ d) $x_1 = 1.0, x_2 = -0.70, x_3 = 3.0$

9. a) $x_1 = 2.8, x_2 = -1.6, x_3 = 2.9$ b) $x_1 = 1.9, x_2 = -1.4, x_3 = 2.9$
     c) $x_1 = 1.0, x_2 = -1.0, x_3 = 2.9$ d) $x_1 = 0.35, x_2 = -0.70, x_3 = 3.0$

Exercise Set
6.3 (page 316)

1. a) Singular   b) $\begin{bmatrix} -\frac{1}{4} & \frac{1}{4} & \frac{1}{4} \\ \frac{5}{8} & -\frac{1}{8} & -\frac{1}{8} \\ \frac{1}{8} & -\frac{5}{8} & \frac{3}{8} \end{bmatrix}$ c) Singular

     d) Singular   e) $\begin{bmatrix} \frac{1}{4} & 0 & 0 & 0 \\ -\frac{3}{14} & \frac{1}{7} & 0 & 0 \\ \frac{3}{28} & -\frac{11}{7} & 1 & 0 \\ -\frac{1}{2} & 1 & -1 & 1 \end{bmatrix}$ f) $\begin{bmatrix} 1 & 0 & 1 & -1 \\ -1 & \frac{5}{3} & \frac{5}{3} & -1 \\ -1 & \frac{2}{3} & \frac{2}{3} & 0 \\ 0 & -\frac{1}{3} & -\frac{4}{3} & 1 \end{bmatrix}$

3. Parts (a), (b), (c):

           first system $x_1 = 3, x_2 = -6, x_3 = -2, x_4 = -1$
           second system $x_1 = 1, x_2 = 1, x_3 = 1, x_4 = 1$

     d) Part (c) requires more work

7. b)
$$\begin{bmatrix} -0.25 & 0.25 & 0.25 \\ 0.625 & -0.125 & -0.125 \\ 0.125 & -0.625 & 0.375 \end{bmatrix}$$

e)
$$\begin{bmatrix} 0.25 & 0 & 0 & 0 \\ -0.214286 & 0.142857 & 0 & 0 \\ 0.107143 & -1.57143 & 1 & 0 \\ -0.5 & 1 & -1 & 1 \end{bmatrix}$$

f)
$$\begin{bmatrix} 1 & 0 & 1 & -1 \\ -1 & 1.66667 & 1.66667 & -1 \\ -1 & 0.666667 & 0.666667 & 0 \\ 0 & -0.333333 & -1.33333 & 1 \end{bmatrix}$$

Exercise Set 6.4 (page 323)

1. a) -8 b) 14 c) 0 d) 3

7. b) $(n+1)n! + n$ multiplications/divisions
$n! - 1$ additions/subtractions

Exercise Set 6.5 (page 330)

1. *i)* a) $x_1 = 0.0$, $x_2 = -7.0$, $x_3 = 5.0$ b) $x_1 = 1.0$, $x_2 = -1.0$, $x_3 = 3.0$
c) $x_1 = 2.2$, $x_2 = -1.6$, $x_3 = 2.9$ d) $x_1 = 1.8$, $x_2 = 0.64$, $x_3 = 1.9$

ii) a) $x_1 = 0.33$, $x_2 = -7.9$, $x_3 = 5.8$ b) $x_1 = 1.1$, $x_2 = -1.1$, $x_3 = 2.9$
c) $x_1 = 2.6$, $x_2 = -1.6$, $x_3 = 2.9$ d) $x_1 = 1.0$, $x_2 = 1.0$, $x_3 = 2.0$

iii) a) $x_1 = 0.0$, $x_2 = -7.0$, $x_3 = 5.0$
b) $x_1 = 1.1$, $x_2 = -1.1$, $x_3 = 2.9$
c) $x_1 = 2.6$, $x_2 = -1.6$, $x_3 = 2.9$
d) $x_1 = 1.8$, $x_2 = 0.64$, $x_3 = 1.9$

iv) a) $x_1 = 0$, $x_2 = -7$, $x_3 = 5$
b) $x_1 = 1$, $x_2 = -1$, $x_3 = 3$
c) $x_1 = 2.627119$, $x_2 = -1.610169$, $x_3 = 2.898305$
d) $x_1 = 1.827586$, $x_2 = 0.6551724$, $x_3 = 1.965517$

3. a) $x_1 = 30.0$, $x_2 = 0.990$; $x_1 = 10.0$, $x_2 = 1.00$; $x_1 = 10.0$, $x_2 = 1.00$
b) $x_1 = 1.00$, $x_2 = 9.98$; $x_1 = 1.00$, $x_2 = 9.98$; $x_1 = 1.00$, $x_2 = 9.98$
c) $x_1 = -10.0$, $x_2 = 45.4$, $x_3 = -36.3$; $x_1 = -9.93$, $x_2 = 46.1$, $x_3 = -37.0$; $x_1 = -10.3$, $x_2 = 47.7$, $x_3 = -38.5$
d) $x_1 = -1.50$, $x_2 = 4.00$, $x_3 = 2.42$; $x_1 = -0.102$, $x_2 = 1.38$, $x_3 = 2.42$; $x_1 = -0.103$, $x_2 = 1.38$, $x_3 = 2.42$

5. a) $x_1 = 1.001291$, $x_2 = 1.000000$, $x_3 = 1.00155$; $x_1 = 1.001291$, $x_2 = 1.000000$, $x_3 = 1.00155$; $x_1 = 1.000000$, $x_2 = 1.000000$, $x_3 = 0.9999997$
b) $x_1 = 53.99987$, $x_2 = -263.9992$, $x_3 = 239.9993$; $x_1 = 54.00071$, $x_2 = -264.0043$, $x_3 = 240.0044$; $x_1 = 54.00071$, $x_2 = -264.0043$, $x_3 = 240.0044$
c) $x_1 = 5.112152$, $x_2 = -122.0641$, $x_3 = 638.8149$, $x_4 = -1133.225$, $x_5 = 636.4389$; $x_1 = 5.170486$, $x_2 = -123.2355$, $x_3 = 644.0883$, $x_4 = -1141.421$, $x_5 = 640.5322$; $x_1 = 5.148010$, $x_2 = -122.7599$, $x_3 = 641.8918$, $x_4 = -1137.952$, $x_5 = 638.7805$
d) $x_1 = -0.08281284$, $x_2 = 0.8842493$, $x_3 = 0.2386317$, $x_4 = -1.920454$; $x_1 \equiv -0.08281332$, $x_2 = 0.8842485$, $x_3 = 0.2386294$, $x_4 = -1.920458$; $x_1 = -0.08281344$, $x_2 = 0.8842489$, $x_3 = 0.2386309$, $x_4 = -1.920455$

7. a) 1a) $x_1 = 0.0$, $x_2 = -7.7$, $x_3 = 5.5$
    1b) $x_1 = 0.94$, $x_2 = -1.0$, $x_3 = 2.9$
    1c) $x_1 = 3.3$, $x_2 = -1.8$, $x_3 = 2.9$
    1d) $x_1 = 1.8$, $x_2 = 0.67$, $x_3 = 2.0$
  e) 5a) $x_1 = 1.0000000$, $x_2 = 0.9999997$, $x_3 = 1.0000000$
    5b) $x_1 = 54.00022$, $x_2 = -264.0009$, $x_3 = 24.00008$
    5c) $x_1 = 5.164794$, $x_2 = -123.1002$, $x_3 = 643.4404$, $x_4 = -1140.374$, $x_5 = 639.9946$
    5d) $x_1 = -0.08281326$, $x_2 = 0.8842481$, $x_3 = 0.2386293$, $x_4 = -1.920458$

**Exercise Set
6.6 (page 338)**

1. *i*) Symmetric: (a), (b), (f)
  ii) Singular: (e), (h)
  iii) Strictly diagonally dominant: (a), (b), (c), (d)
  iv) Positive definite: (a), (f)

3. a) $\begin{bmatrix} 2 & 1 \\ 1 & 0 \end{bmatrix}$, $\begin{bmatrix} 1 & -1 \\ -1 & 1 \end{bmatrix}$
  b) True, since $(A^{-1})^t = (A^t)^{-1}$.
  c) Matrices in 3(a) provide a counterexample.

5. a) Yes
  b) Not necessarily. Consider

$$\begin{bmatrix} 2 & -1 \\ 3 & 4 \end{bmatrix}$$

  c) Not necessarily. Consider

$$\begin{bmatrix} 2 & 1 \\ 1 & 2 \end{bmatrix} \quad \text{and} \quad \begin{bmatrix} -2 & 1 \\ 1 & -2 \end{bmatrix}$$

  d) Not necessarily. Consider

$$\begin{bmatrix} 2 & -1 \\ 3 & 4 \end{bmatrix}$$

  e) Not necessarily. Consider

$$\begin{bmatrix} 2 & 1 \\ 1 & 2 \end{bmatrix} \quad \text{and} \quad \begin{bmatrix} 2 & -1 \\ -1 & 2 \end{bmatrix}$$

7. Yes
9. a) $\begin{bmatrix} 1 & 1 \\ 1 & 1 \end{bmatrix}$ b) No

17. a) Mating male i with female j, or male j with female i, yields offspring with same wing characteristics.
  b) No, $\mathbf{x} = (1, 0, -1)^t$

**Exercise Set
6.7 (page 356)**

1. a)

$$L = \begin{bmatrix} 1 & 0 & 0 \\ 1.5 & 1 & 0 \\ 1.5 & 1 & 1 \end{bmatrix}, \quad U = \begin{bmatrix} 2 & -1 & 1 \\ 0 & 4.5 & 7.5 \\ 0 & 0 & -4 \end{bmatrix}$$

b)

$$L = \begin{bmatrix} 1 & 0 & 0 \\ -0.5 & 1 & 0 \\ 5 & 2 & 1 \end{bmatrix}, \quad U = \begin{bmatrix} 2 & -1.5 & 3 \\ 0 & -0.75 & 3.5 \\ 0 & 0 & -8 \end{bmatrix}$$

c)

$$L = \begin{bmatrix} 1 & 0 & 0 \\ -2.106719 & 1 & 0 \\ 3.067193 & 1.19776 & 1 \end{bmatrix}$$

$$U = \begin{bmatrix} 1.012 & -2.132 & 3.104 \\ 0 & -0.3955249 & -0.4737443 \\ 0 & 0 & -8.939133 \end{bmatrix}$$

d) $L = I$, $U =$ original matrix

e)

$$L = \begin{bmatrix} 1 & 0 & 0 & 0 \\ 0.5 & 1 & 0 & 0 \\ 0 & -2 & 1 & 0 \\ 1 & -1.33333 & 2 & 1 \end{bmatrix}, \quad U = \begin{bmatrix} 2 & 0 & 0 & 0 \\ 0 & 1.5 & 0 & 0 \\ 0 & 0 & 0.5 & 0 \\ 0 & 0 & 0 & 1 \end{bmatrix}$$

f)

$$L = \begin{bmatrix} 1 & 0 & 0 & 0 \\ -1.849190 & 1 & 0 & 0 \\ -0.4596433 & -0.2501219 & 1 & 0 \\ 2.768661 & -0.3079435 & -5.35229 & 1 \end{bmatrix}$$

$$U = \begin{bmatrix} 2.175600 & 4.023099 & -2.173199 & 5.196700 \\ 0 & 13.43947 & -4.018660 & 10.80698 \\ 0 & 0 & -0.8929510 & 5.091692 \\ 0 & 0 & 0 & 12.03614 \end{bmatrix}$$

3. a)

$$L = \begin{bmatrix} 1.41423 & 0 & 0 \\ -0.7071069 & 1.224743 & 0 \\ 0 & -0.8164972 & 1.154699 \end{bmatrix}$$

b)

$$L = \begin{bmatrix} 2 & 0 & 0 & 0 \\ 0.5 & 1.658311 & 0 & 0 \\ 0.5 & -0.7537785 & 1.087113 & 0 \\ 0.5 & 0.4522671 & 0.08362442 & 1.240346 \end{bmatrix}$$

c)

$$L = \begin{bmatrix} 2 & 0 & 0 & 0 \\ 0.5 & 1.658311 & 0 & 0 \\ -0.5 & -0.4522671 & 2.132006 & 0 \\ 0 & 0 & 0.9380833 & 1.766351 \end{bmatrix}$$

d)
$$L = \begin{bmatrix} 2.449489 & 0 & 0 & 0 \\ 0.8164966 & 1.825741 & 0 & 0 \\ 0.4082483 & 0.3651483 & 1.923538 & 0 \\ -0.4082483 & 0.1825741 & -0.4678876 & 1.606574 \end{bmatrix}$$

5. a) $x_1 = 0.5$, $x_2 = 0.5$, $x_3 = 1$
   b) $x_1 = -0.9999995$, $x_2 = 1.999999$, $x_3 = 1$
   c) $x_1 = 1$, $x_2 = -1$, $x_3 = 0$
   d) $x_1 = -0.09357762$, $x_2 = 1.587155$, $x_3 = -1.16743$, $x_4 = 0.5412842$

11. $x_i = 1$, for each $i = 1, \ldots, 10$.   15. $x_i = 1$, for each $i = 1, \ldots, 9$.

17. n square roots, $\frac{1}{6}(n^3 + 9n^2 + 2n)$ multiplications/divisions and $\frac{1}{6}(n^3 + 6n^2 - 7n)$ additions/subtractions.

CHAPTER 7

Exercise Set
7.1 (page 370)

1. $1.70784x + 0.89968$ 3. $0.841679x^2 + 0.876603x + 1$

5. a) $72.0845x - 194.138$, 329
   b) $6.61822x^2 - 1.14357x + 1.23570$, 1.44×10^{-3}
   c) $-0.0137352x^3 + 6.84659x^2 - 2.38475x + 3.43896$, 5.27×10^{-4}
   d) $24.2588e^{0.372382x}$, 418
   e) $6.23903x^{2.01954}$, 0.00703

7. $0.22335x - 0.80283$. For minimal A, 406; for minimal D, 272. The A prediction is not reasonable.

9. $0.1795x + 8.2084$ 11. $3.87x + 25.70$

Exercise Set
7.2 (page 383)

1. a) $-x + 2.8333$ b) $3.6x - 2.6$
   c) $-0.2958x + 1.1410$ d) $e^{-1}[6(e - 3)x + 2(4 - e)]$
   e) $-2.4317x + 1.2159$ f) $0.68223x - 0.63706$

3. a) $-2x + \frac{10}{3}$, error $\frac{8}{45}$ b) $\frac{3}{5}x + 1$, error $\frac{8}{175}$

   d) $\frac{1}{e}\left[-3x + \frac{e^2 - 1}{2}\right]$, error 0.05265   e) 0, error 1

5. $S_2(x) = 2 \sin x$

7. a) $\phi_0(x) = 1$, $\phi_1(x) = x - 0.5$, $\phi_2(x) = x^2 - x + \frac{1}{6}$, $\phi_3(x) = x^3 - 1.5x^2 + 0.6x - 0.05$
      $P_3(x) = 0.841471\phi_0(x) - 0.467546\phi_1(x) - 0.431010\phi_2(x) + 0.078824\phi_3(x)$
   b) $\phi_0(x) = 1$, $\phi_1(x) = x - 1.5$, $\phi_2(x) = x^2 - 3x + \frac{13}{6}$, $\phi_3(x) = x^3 - 4.5x^2 + 6.6x - 3.15$
      $P_3(x) = 0.386294\phi_0(x) + 0.682234\phi_1(x) - 0.233508\phi_2(x) + 0.106686\phi_3(x)$
   c) $\phi_0(x) = 1$, $\phi_1(x) = x - 0.581977$, $\phi_2(x) = (x - 0.484843)\phi_1(x) - 0.0793264$,
      $\phi_3(x) = (x - 0.497483)\phi_2(x) - 0.0671167\phi_1(x)$
      $P_3(x) = 0.270349\phi_0(x) + 0.918895\phi_1(x) + 1.84493\phi_2(x) + 2.06350\phi_3(x)$
   d) $\phi_0(x) = 1$, $\phi_1(x) = x - 0.5$, $\phi_2(x) = x^2 - x + \frac{1}{6}$, $\phi_3(x) = x^3 - 1.5x^2 + 0.6x - 0.05$
      $P_3(x) = -0.951626\phi_0(x) - 0.0951467\phi_1(x) + 0.00475700\phi_2(x)$
      $- 1.58431 \times 10^{-4}\phi_3(x)$

9. a) $2 L_0(x) + 4 L_1(x) + L_2(x)$
    b) $\frac{1}{2}L_0(x) - \frac{1}{4}L_1(x) + \frac{1}{16}L_2(x) - \frac{1}{96}L_3(x)$
    c) $6 L_0(x) + 18L_1(x) + 9L_2(x) + L_3(x)$
    d) $\frac{1}{3}L_0(x) - \frac{2}{9}L_1(x) + \frac{2}{27}L_2(x) - \frac{4}{243}L_3(x)$

**Exercise Set
7.3 (page 391)**

1. a) $x_0 = 0.8660254, x_1 = 0, x_2 = -0.8660254;$
        $P_2(x) = 2.377443 + 1.590534(x - x_0) + 0.5320418x(x - x_0)$
    b) $x_0 = 2.931146, x_1 = 1.570796, x_2 = 0.2104468;$
        $P_2(x) = 0.2088969 - 0.5815441(x - x_0) - 0.4274961(x - x_0)(x - x_1)$
    c) $x_0 = 1.933013, x_1 = 1.5, x_2 = 1.066987;$
        $P_2(x) = 0.6590798 + 0.5856980(x - x_0) - 0.2320303(x - x_0)(x - x_1)$
    d) $x_0 = 0.9330127, x_1 = 0.5, x_2 = 0.0669873;$
        $P_2(x) = 1.9330127 + (x - x_0) = 1 + x$

3. $x_0 = 0.9238795, x_1 = 0.3826834, x_2 = -0.3826834, x_3 = -0.9238795;$
    $P_3(x) = 1.387040 + 0.3901806(x - x_0) - 0.09154931(x - x_0)(x - x_1)$
    $\quad + 0.1294703(x - x_0)(x - x_1)(x - x_2);$
    $P_3(0.1) = 1.058811$

5. $0.56510416T_0 + 1.4010416T_1 + 0.58723958T_2 + 0.13802083T_3 + 0.022395833T_4$
    approximates xe^x with error at most 0.00485.

**Exercise Set
7.4 (page 397)**

1. $r(x) = \dfrac{1 + \frac{2}{5}x + \frac{1}{20}x^2}{1 - \frac{3}{5}x + \frac{3}{20}x^2}$

3. $r(x) = \dfrac{x - \frac{7}{60}x^3}{1 + \frac{1}{20}x^2}$

5. a) $r(x) = \dfrac{1}{1 + x + \frac{1}{2}x^2 + \frac{1}{6}x^3 + \frac{1}{24}x^4 + \frac{1}{120}x^5}$

    b) $r(x) = \dfrac{1 - \frac{1}{5}x}{1 + \frac{4}{5}x + \frac{3}{10}x^2 + \frac{1}{15}x^3 + \frac{1}{120}x^4}$

    c) $r(x) = \dfrac{1 - \frac{3}{5}x + \frac{3}{20}x^2 - \frac{1}{60}x^3}{1 + \frac{2}{5}x + \frac{1}{20}x^2}$

    d) $r(x) = \dfrac{1 - \frac{4}{5}x + \frac{3}{10}x^2 - \frac{1}{15}x^3 + \frac{1}{120}x^4}{1 + \frac{1}{5}x}$

7. For 6(a): a) 5.63,   b) 5.63,   c) 5.60, exact value 5.61
    6(b): a) 1.03,   b) 1.02,   c) 0.800, exact value 0.887

9. $r_T(x) = \dfrac{0.88010T_1(x)}{T_0(x) + 0.044461T_2(x)}$

**Exercise Set
7.5 (page 406)**

1. $S_3(x) = -0.496899 + 0.239206 \cos x + 1.51539 \cos 2x + 0.239190 \cos 3x$
    $\quad - 1.15066 \sin x + 0.00001278 \sin 2x$

3. a) $S_4(x) = -4.62637 + 6.67949 \cos x - 3.701088 \cos 2x + 3.19008 \cos 3x$
        $\quad - 1.54212 \cos 4x + 5.95683 \sin x - 2.46740 \sin 2x$
        $\quad + 1.02203 \sin 3x$
    b) $S_4(x) = -4.62637 + 6.67950 \cos x - 3.70110 \cos 2x + 3.19008 \cos 3x$
        $\quad - 1.54212 \cos 4x + 5.95683 \sin x - 2.46740 \sin 2x$
        $\quad + 1.02204 \sin 3x$

5. a) 0.197672   b) 0.223244

CHAPTER 8

1. a) 4, 5.220153   b) 4, 5.477226
    c) $2^k, (1 + 4^k)^{1/2}$ d) $4/(k+1), (16/(k+1)^2 + 4/k^4 + k^4 e^{-2k})^{1/2}$

3. a) $(0, 0, 0)^t$ b) $(0, 1, 3)^t$ c) $(0, 0, \frac{1}{2})^t$ d) $(1, -1, 1)^t$

5. a) 3   b) 2   c) 25   d) 7

9. b) 5a) 2.645751, 5   5b) 2, 4
      5c) 18.05547, 26   5d) 6.244998, 11

15. a) $\lambda_1 = 3, \mathbf{x}_1 = (1, -1)^t; \lambda_2 = 1, \mathbf{x}_2 = (1, 1)^t$
    b) $\lambda_1 = \lambda_2 = 1, \mathbf{x} = (1, 0)^t$
    c) $\lambda_1 = \frac{1}{2}, \mathbf{x}_1 = (1, 1)^t; \lambda_2 = -\frac{1}{2}, \mathbf{x}_2 = (1, -1)^t$
    d) $\lambda_1 = 1, \mathbf{x}_1 = (-1, 1, 0)^t; \lambda_2 = 1, \mathbf{x}_2 = (-1, 0, 1)^t; \lambda_3 = 5, \mathbf{x}_3 = (1, 2, 1)^t$
    e) $\lambda_1 = 3, \mathbf{x}_1 = (0, 0, 1)^t; \lambda_2 = 3, \mathbf{x}_2 = (1, 1, 0)^t; \lambda_3 = 1, \mathbf{x}_3 = (1, -1, 0)^t$
    f) $\lambda_1 = 7, \mathbf{x}_1 = (1, 4, 4)^t; \lambda_2 = 3, \mathbf{x}_2 = (1, 2, 0)^t; \lambda_3 = -1, \mathbf{x}_3 = (1, 0, 0)^t$

17. a) 3   b) 1.618034   c) 0.5   d) 5.203527
    e) 3   f) 8.224257

21. $\rho(A_1) = 1, \rho(A_2) = \frac{1}{2}$

27. a) $\lambda = 1$ and $\mathbf{x} = (6, 3, 1)^t$ b) any multiple of $(6, 3, 1)^t$

1. a) $(-0.9, -1.9, 1.1)^t$
    b) Cannot apply method
    c) $(0.97, 0.91, 0.74)^t$
    d) $(1.5, 2, 4.8, 17)^t$
    e) $(-22.33332, -24.88887, 28.66665, 20.94442)^t$
    f) $(1.047271, 2.602274, -0.9927272, -2.364771)^t$
    g) $(0.1999999, 0.5500001, 1.183333, -1.166666)^t$
    h) $(0.6875, 1.125, 0.6875, 1.375, 0.5625, 1.375)^t$

3. a) $(-0.65, -1.75, 1.89)^t$
    b) Cannot apply method
    c) $(0.979, 0.9495, 0.7899)^t$
    d) $(1.5, 2, -1.2, 3)^t$
    e) $(-285.6665, -87.77769, 1414.443, 706.4433)^t$
    f) $(6.685455, 3.429051, -2.094185, -2.922667)^t$
    g) $(0.1999999, 0.6500001, 1.529166, -1.68611)^t$
    h) $(0.6875000, 1.546875, 0.7929687, 1.718750, 0.7226562, 1.878906)^t$

5. a) $(-0.5721604, -2.602366, 3.070938)^t$
    b) Cannot apply method
    c) $(0.9803522, 0.9923712, 0.7676284)^t$
    d) $(1.439999, 2.015999, -1.267203, 3.429122)^t$
    e) $(-251.2796, -124.7165, 1428.335, 1020.985)^t$
    f) $(1.190161, 3.179800, -0.9928426, -2.612629)^t$
    g) $(0.2879998, 0.7728000, 1.904158, -1.996286)^t$
    h) $(0.9899992, 1.748999, 1.050298, 1.853999, 1.018498, 2.010238)^t$

Exercise Set
8.3 (page 446)

1. a) $\tilde{\mathbf{x}}^{(4)} = (1.000, 1.000)^t$, $K_\infty(A) = 43$
   b) $\tilde{\mathbf{x}}^{(3)} = (1.41, 0.134)^t$, $K_\infty(A) = 44.2$
   c) $\tilde{\mathbf{x}}^{(4)} = (1.09, 0.488, 1.49)^t$, $K_\infty(A) = 667$
   d) $\tilde{\mathbf{x}}^{(3)} = (-227.0781, 476.9253, -177.6931)^t$, $K_\infty(A) = 476$

5. $\tilde{\mathbf{x}} = (-1.0000, 2.0000)^t$; yes.   **7.** $1.01(n^3 + 3n^2)$ is closer

Exercise Set
8.4 (page 466)

1. a) $\mu^{(3)} = 4.999999$, $\mathbf{x}^{(3)} = (-0.2599998, 1, -0.2399999)^t$
   b) $\mu^{(3)} = 4.999999$, $\mathbf{x}^{(3)} = (-0.26, 1, -0.28)^t$
   c) $\mu^{(3)} = 5.236839$, $\mathbf{x}^{(3)} = (1, 0.5678389, 0.1557790, 0.4874370)^t$
   d) $\mu^{(3)} = 4$, $\mathbf{x}^{(3)} = (1, 1, 1, 1)^t$

3. a) $\mu^{(3)} = 3.142854$, $\mathbf{x}^{(3)} = (0.6767157, -0.6767154, 0.2900211)^t$
   b) $\mu^{(3)} = 7.189565$, $\mathbf{x}^{(3)} = (0.5995309, 0.7367475, 0.3126760)^t$
   c) $\mu^{(3)} = 6.037039$, $\mathbf{x}^{(3)} = (0.5073714, 0.4878572, -0.6634859, -0.2536857)^t$
   d) $\mu^{(3)} = 5.142565$, $\mathbf{x}^{(3)} = (0.8373054, 0.3701771, 0.1939021, 0.3525497)^t$

5. a) $|\lambda| \leq 4$ b) $2 \leq |\lambda| \leq 6$
   c) $0 < \lambda \leq 6$ d) $1.25 \leq \lambda \leq 8.25$

7. a) $\lambda_1 = 5$, $(-0.2500799, 1, -0.2499199)^t$
      $\lambda_2 = 1$, $(-4, 0, 4)^t$
   b) $\lambda_1 = 5.124731$, $(-0.2424495, 1, -0.3199684)^t$
      $\lambda_2 = 1.636733$, $(1.783208, -1.135352, -3.124720)^t$
   c) $\lambda_1 = 5.23585$, $(1, 0.6178283, 0.1181719, 0.4999188)^t$
      $\lambda_2 = 3.618178$, $(0.7235748, -1.170626, 1.170681, -0.2762828)^t$
   d) $\lambda_1 = 8.998871$, $(1, -0.9999896, -0.9996932, 0.9997036)^t$
      $\lambda_2 = 5.000591$, $(1.999241, -2.000432, 1.998439, -1.999630)^t$

9. a) $\lambda_1 = 3.414204$, $(0.5009559, -0.7071061, 0.4990434)^t$
   b) $\lambda_1 = 7.223632$, $(0.6236797, 0.7205519, 0.3030332)^t$
   c) $\lambda_1 = 7.086096$, $(0.3340821, 0.2680588, -0.7587719, -0.4907129)^t$
   d) $\lambda_1 = 5.236042$, $(0.7806029, 0.4798963, 0.09384346, 0.3892983)^t$

11. For $q = 7$: $\mu^{(8)} = -1.000028$, $\mathbf{x}^{(8)} = (-1.000028, -0.7143087, 0.2500002)^t$;
      $q = 4$: $\mu^{(17)} = -1.000006$, $\mathbf{x}^{(17)} = (1.000013, 0.5000097, -1.000006)^t$;
      $q = 0$: $\mu^{(22)} = 0.4998214$, $\mathbf{x}^{(22)} = (-0.00035603, -0.00017801, 0.4998214)^t$

17. a) $\lambda_3 = 0.002534$, $(1, 0.9978877, 0.9957758)^t$
   b) $\lambda_3 = 0.240897$, $(1, 0.7596983, 0.4294162)^t$
   c) $\lambda_4 = 0.765082$, $(-0.4740131, 0.7610415, 1, -0.2287408)^t$
   d) $\lambda_4 = 2.003189$, $(1, 0.9999549, -0.9974821, -0.9975271)^t$

19. a) $\lambda_2 = 2$, $(0.7061744, 0.001948772, -0.7080360)^t$
   b) $\lambda_2 = 4.961711$, $(0.4837692, -0.04970267, -0.8737835)^t$
   c) $\lambda_2 = 4.428010$, $(0.7174588, 0.4232742, 0.1175253, 0.5406294)^t$
   d) $\lambda_2 = 3.618045$, $(0.3927337, -0.6424037, 0.6400135, -0.1531675)^t$

21. a) $|\lambda| \leq 6$ for all eigenvalues λ.
   b) $\lambda_1 = 0.6982681$, $\mathbf{x} = (1, 0.71606, 0.25638, 0.04602)^t$
   d) $P(\lambda) = \lambda^4 - \frac{1}{4}\lambda - \frac{1}{16}$;
      $\lambda_1 = 0.6976684972$,
      $\lambda_2 = -0.237313308$,
      $\lambda_3 = -0.2301775942 + 0.56965884i$,
      $\lambda_4 = -0.2301775942 - 0.56965884i$

658 Answers to Selected Exercises

e) The beetle population should approach zero since A is convergent.

23. a) $\lambda_1 = -6, \lambda_2 = -5, \lambda_3 = -2$; yes.
    b) $\lambda_1 = -2, \lambda_2 = -1.106711, \lambda_3 = -3.94664 + 0.82970i, \lambda_4 = -3.94664 - 0.82970i$; yes.

Exercise Set
8.5 (page 484)

1. a) $\begin{bmatrix} 20.19511 & -0.2439012 & 0 \\ -0.2439012 & -0.1951208 & 6.403123 \\ 0 & 6.403123 & 3.000000 \end{bmatrix}$
   b) $\begin{bmatrix} 3 & 0 & 0 \\ 0 & 1 & 1.41421 \\ 0 & 1.41421 & 2 \end{bmatrix}$

c) $\begin{bmatrix} 2.199999 & -0.3999996 & 0 \\ -0.3999996 & 2.799995 & 2.236067 \\ 0 & 2.236067 & -1.000000 \end{bmatrix}$

d) $\begin{bmatrix} 5.615384 & -2.076921 & 0 \\ -2.076921 & 3.884609 & -1.274754 \\ 0 & -1.274754 & 4.75 \end{bmatrix}$

e) $\begin{bmatrix} 4 & 0 & 0 & 0 \\ 0 & 4 & 0 & 0 \\ 0 & 0 & 4 & 1.414213 \\ 0 & 0 & 1.414213 & 4 \end{bmatrix}$

f) $\begin{bmatrix} 4.345381 & 0.6912802 & 0 & 0 \\ 0.6912802 & 4.269998 & 2.140755 & 0 \\ 0 & 2.140755 & 6.384615 & 2.549509 \\ 0 & 0 & 2.549509 & 5 \end{bmatrix}$

g) $\begin{bmatrix} 4.42618 & -0.639544 & 0 & 0 & 0 \\ -0.639544 & 5.15696 & 2.11891 & 0 & 0 \\ 0 & 2.11891 & 4.57685 & 5.69182 & 0 \\ 0 & 0 & 5.69182 & -0.16004 & 2.5 \\ 0 & 0 & 0 & 2.5 & 6 \end{bmatrix}$

h) $\begin{bmatrix} 2.442343 & -1.276837 & 0 & 0 & 0 \\ -1.276837 & 3.272677 & -0.9297242 & 0 & 0 \\ 0 & -0.9297242 & 2.484973 & 1.964689 & 0 \\ 0 & 0 & 1.964689 & 8.800008 & -3.162277 \\ 0 & 0 & 0 & -3.162277 & 9 \end{bmatrix}$

3. a) 2.12946, 3.91150, -2.04096
   b) 2.722244, 5.346457, -0.06870269
   c) 2, 6, 4, 7.46409, 0.535911
   d) 5.783995, 3.727556, 4.027436, 2.070721, 0.8903045

CHAPTER 9

Exercise Set
9.1 (page 494)

5. b) With $\mathbf{x}^{(0)} = (0, 0)^t$, $\mathbf{x}^{(6)} = (0.999328, 0.999329)^t$
   c) With $\mathbf{x}^{(0)} = (0, 0)^t$, $\mathbf{x}^{(6)} = (0.999552, 0.999630)^t$

7. a) With $\mathbf{x}^{(0)} = (1, 1, 1)^t$, $\mathbf{x}^{(4)} = (0.5000000, 0.0000000, -0.5235988)^t$
   b) With $\mathbf{x}^{(0)} = (0, 0, 0)^t$, $\mathbf{x}^{(9)} = (1.200425, 1.100663, 0.9000712)^t$
   c) With $\mathbf{x}^{(0)} = (0.05, 0.2, 0.8)^t$, $\mathbf{x}^{(3)} = (0.0000000, 0.1001078, 1.000007)^t$
   d) With $\mathbf{x}^{(0)} = (0, 0, 0)^t$, $\mathbf{x}^{(4)} = (0.4981453, -0.1996048, -0.5288248)^t$

9. a) With $\mathbf{G}(\mathbf{x}) = (\sqrt{2x_3 + x_2 - 2x_2^2}, \sqrt{\frac{1}{8}(10x_3 + x_1^2)}, x_1^2/7x_2)^t$ and $\mathbf{x}^{(0)} = (0.5, 0.3, 0.1)^t$,
    $\mathbf{x}^{(24)} = (0.5291938, 0.4000314, 0.09998558)^t$
   b) With $\mathbf{G}(\mathbf{x}) = (2 - 0.5x_2 - 0.5x_3, 2 - 0.5x_1 - 0.5x_3, \sqrt{x_3/x_1x_2})^t$ and
    $\mathbf{x}^{(0)} = (0.5, 0.6, 0.7)^t$, $\mathbf{x}^{(81)} = (1.000069, 1.000069, 0.9999725)^t$
   c) With $\mathbf{G}(\mathbf{x}) = (\frac{1}{3}\cos(x_2x_3) + \frac{1}{6}, \frac{1}{25}x_1, -\frac{1}{20}e^{-x_1x_2} - \pi/6 + 0.05)^t$ and $\mathbf{x}^{(0)} = (0, 0, 0)^t$,
    $\mathbf{x}^{(4)} = (0.4999818, 0.01999927, -0.5231013)^t$
   d) With $\mathbf{G}(\mathbf{x}) = (\sqrt{37 - x_2}, \sqrt{x_1 - 5}, 3 - x_2 - x_1)^t$ and $\mathbf{x}^{(0)} = (0.5, 0.5, 0.5)^t$,
    $\mathbf{x}^{(8)} = (5.999998, 0.9999985, -4.000018)^t$

11. Yes: $x_1 = 8000$, $x_2 = 40000$

Exercise Set
9.2 (page 501)

1. a) With $\mathbf{x}^{(0)} = (0, 0)^t$, $\mathbf{x}^{(4)} = (1, 1)^t$
   b) With $\mathbf{x}^{(0)} = (0, 0)^t$, $\mathbf{x}^{(5)} = (0.7718445, 0.4196434)^t$
   c) With $\mathbf{x}^{(0)} = (0.5, 0.5)^t$, $\mathbf{x}^{(5)} = (0.5, 0.8660253)^t$
   d) With $\mathbf{x}^{(0)} = (0, 0)^t$, $\mathbf{x}^{(4)} = (-0.3812285, 0)^t$

3. a) With $\mathbf{x}^{(0)} = (0.1, -0.1, 0.1)^t$, $\mathbf{x}^{(3)} = (0.4981446, -0.1996059, -0.5288259)^t$
   b) With $\mathbf{x}^{(0)} = (0.3, 0.3, 0.3)^t$, $\mathbf{x}^{(12)} = (0.5291502, 0.3999999, 0.09999996)^t$
   c) With $\mathbf{x}^{(0)} = (0.5, 0.5, 0.5)^t$, $\mathbf{x}^{(7)} = (-0.4342584, -0.4342588, 5.302775)^t$
   d) With $\mathbf{x}^{(0)} = (0.5, 0.5, 0.5, 0.5)^t$, $\mathbf{x}^{(14)} = (0.8688753, 0.8688760, 0.8688761, 1.524497)^t$

7. a) $k_1 = 8.77129$, $k_2 = 0.259695$, $k_3 = -1.37228$ b)   3.19 inches

Exercise Set
9.3 (page 510)

1. a) $\mathbf{x}^{(5)} = (6, 1, -4)^t$ b) $\mathbf{x}^{(5)} = (0.0000000, 0.1000013, 1.000000)^t$
   c) $\mathbf{x}^{(4)} = (1.2, 1.1, 0.9)^t$ d) $\mathbf{x}^{(2)} = (1, 2, 0)^t$

3. Both algorithms fail. Solution is $(15, -1)^t$.

5.

i	θ_i	i	θ_i
1	0.14062	11	0.48348
2	0.19954	12	0.50697
3	0.24522	13	0.52980
4	0.28413	14	0.55205
5	0.31878	15	0.57382
6	0.35045	16	0.59516
7	0.37990	17	0.61615
8	0.40763	18	0.63683
9	0.43397	19	0.65726
10	0.45920	20	0.67746

7. $a = 6.588$, $b = 29.840$, $c = 1.163$.

1. a) With $\mathbf{x}^{(0)} = (0, 0)^t$, $\mathbf{x}^{(5)} = (0.9960763, 0.9973286)^t$

b) With $\mathbf{x}^{(0)} = (1, 1)^t$, $\mathbf{x}^{(3)} = (0.7777234, 0.4242323)^t$

c) With $\mathbf{x}^{(0)} = (1, 1)^t$, $\mathbf{x}^{(2)} = (0.5026802, 0.8684244)^t$

d) With $\mathbf{x}^{(0)} = (0, 0)^t$, $\mathbf{x}^{(10)} = (-0.3781587, 0.01633489)^t$

3. a) With $\mathbf{x}^{(0)} = (0, 0, 0)^t$, $\mathbf{x}^{(4)} = (1.073144, 1.073144, 0.8044867)^t$

b) With $\mathbf{x}^{(0)} = (0, 0, 0)^t$, $\mathbf{x}^{(3)} = (-0.0003458932, 0.09484016, 1.000236)^t$

c) With $\mathbf{x}^{(0)} = (0, 0, 0)^t$, $\mathbf{x}^{(11)} = (1.198007, 1.100791, 0.8931340)^t$

d) With $\mathbf{x}^{(0)} = (0, 0, 0)^t$, $\mathbf{x}^{(7)} = (0.4730910, 0.01915232, -0.5231886)^t$

5. a) With $\mathbf{x}^{(0)} = (0.5, 0.5)^t$, $\mathbf{x}^{(2)} = (0.9999666, 0.1193127)^t$ and $G(\mathbf{x}^{(2)}) = 9.000202$

b) With $\mathbf{x}^{(0)} = (0, 0)^t$, $\mathbf{x}^{(2)} = (-2.839888, -0.8266732)^t$ and $G(\mathbf{x}^{(2)}) = -0.3208664$

c) With $\mathbf{x}^{(0)} = (0, 0, 0)^t$, $\mathbf{x}^{(5)} = (-0.6633785, 0.3145720, 0.5000740)^t$ and $G(\mathbf{x}^{(5)}) = 0.6921548$

d) With $\mathbf{x}^{(0)} = (1, 1, 1)^t$, $\mathbf{x}^{(4)} = (0.04022273, 0.01592477, 0.01594401)^t$ and $G(\mathbf{x}^{(4)}) = 1.010003$

CHAPTER 10

1. a)

i	x_i	w_{1i}
1	1.047197	-0.533308
2	2.094394	-1.51942

b)

i	x_i	w_{1i}
1	0.7853982	0.1631159
2	1.570796	-0.7662053
3	2.356194	-1.246696

3. a)

i	x_i	w_{1i}	w_{2i}
5	1.25	0.1676179	1.656001
10	1.50	0.4581901	0.8016986
15	1.75	0.6077718	0.4406008
20	2.00	0.6931460	0.2610475

b)

i	x_i	w_{1i}	w_{2i}
5	1.0	0.08655548	-0.04176616
10	2.0	0.00008010864	-0.0003566741
15	3.0	0.00002002716	0.00001239776
20	4.0	0.00004577636	0.00004577636

c)

i	x_i	w_{1i}	w_{2i}
3	0.3	0.7833204	-1.800761
6	0.6	0.6023521	0.2968196
9	0.9	0.8568906	1.305988

d)

i	x_i	w_{1i}	w_{2i}
3	0.3	-0.04603261	-0.02847886
6	0.6	-0.03460747	0.08129125
9	0.9	-0.007869184	0.08446383

5. (a), (d), (e), (i), (j)

9. For $h = 0.05$:

i	x_i	w_{1i}
5	0.25	0.08224765
10	0.50	0.006764673
15	0.75	0.000556349

**Exercise Set
10.2 (page 532)**

1. $w_1 = 0.405991 \approx \ln 1.5 = 0.405465$

3. a)

i	x_i	w_{1i}
5	1.25	-0.7272908
10	1.50	-0.8000371
15	1.75	-0.8889473

b)

i	w_{1i}	w_{2i}
2.6	0.1785917	-0.03187268
4.2	0.1389389	-0.01926722
5.0	0.1250701	-0.01559713

c)

i	x_i	w_{1i}
2	1.2	0.4545455
5	1.5	0.4000000
7	1.7	0.3703704

d)

x_i	w_{1i}	w_{2i}
1.25	2.050000	0.3600042
1.50	2.166665	0.5555525
1.75	2.321420	0.6734590
2.00	2.499969	0.7498404

5. a)

x_i	w_{1i}	w_{2i}
1.25	-0.7272728	-0.2644638
1.50	-0.8000001	-0.3200009
1.75	-0.8888891	-0.3950626
2.00	-1.0000000	-0.5000008

c)

x_i	w_{1i}	w_{2i}
1.50	0.4000011	-0.1599981
2.00	0.3333333	-0.1111111

Exercise Set 10.3 (page 538)

1. a)

i	x_i	w_i
1	1.04720	0.525382
2	2.09439	−0.525382

b)

x_i	w_i
$\dfrac{\pi}{12}$	−0.0877483
$\dfrac{\pi}{6}$	−0.0852364

2. d)

i	x_i	w_i
3	0.3	−0.04617565
6	0.6	−0.03471606
9	0.9	−0.007894191

3. a)

x_i	$w_i(h = 0.025)$	$w_i(h = 0.0125)$	EXT_{1i}	EXT_{2i}	EXT_{3i}
1.25	0.1677325	0.1677398	0.1676508	0.1677423	0.1677483
1.50	0.4582796	0.4583252	0.4582290	0.4583399	0.4583473
1.75	0.6078227	0.6078653	0.6078039	0.6078792	0.6078842

b)

x_i	$w_i(h = 0.1)$	$w_i(h = 0.05)$	EXT_{1i}	EXT_{2i}	EXT_{3i}
1.0	7.579267×10^{-3}	7.894285×10^{-3}	7.994696×10^{-3}	7.999289×10^{-3}	7.999591×10^{-3}
2.0	5.744566×10^{-5}	6.231994×10^{-5}	6.322536×10^{-5}	6.394466×10^{-5}	6.399261×10^{-5}
3.0	4.353907×10^{-7}	4.919569×10^{-7}	4.958567×10^{-7}	5.108119×10^{-7}	5.118088×10^{-7}
4.0	3.289033×10^{-9}	3.870230×10^{-9}	3.850686×10^{-9}	4.063959×10^{-9}	4.078177×10^{-9}
5.0	0.000000	0.000000	0.000000	0.000000	0.000000

c)

x_i	$w_i(h = 0.05)$	$w_i(h = 0.025)$	EXT_{1i}	EXT_{2i}	EXT_{3i}
0.2	1.021161	1.021937	1.022181	1.022195	1.022196
0.5	0.5964505	0.5970101	0.5971746	0.5971965	0.5971980
0.7	0.6525620	0.6528496	0.6529312	0.6529451	0.6529460

5. a)

i	x_i	w_i
5	1.25	0.643282
10	1.50	0.683328
15	1.75	0.692302

b)

x_i	w_i
0.2356193	1.069115
0.4712387	1.079113
0.7068580	1.029437

Exercise Set
10.4 (page 545)

1. a)

x_i	w_i
0.5	0.406800

b)

x_i	w_i
1.2	-0.980942
1.4	-0.928693

2. a)

i	x_i	w_i
5	1.25	-0.7272810
10	1.50	-0.8000128
15	1.75	-0.8889005

b)

x_i	w_i
2.0	0.2000223
3.0	0.1666879
4.0	0.1428695
5.0	0.1250000

c)

i	x_i	w_i
2	1.2	0.4545563
5	1.5	0.4000130
7	1.7	0.3703798

d)

i	x_i	w_i
6	1.25	2.050045
11	1.50	2.166697
16	1.75	2.321441

Exercise Set
10.5 (page 559)

1. a) $\phi(x) = -0.08057117\phi_1(x) - 0.07200348\phi_2(x)$
b) $\phi(x) = -0.125814\phi_1(x) - 0.240164\phi_2(x)$

2. a)

i	c_i
0	-4.38127×10^{-3}
1	-5.29927×10^{-2}
2	-6.86142×10^{-2}
3	-5.02453×10^{-2}
4	-3.10118×10^{-3}

b)

i	c_i
0	-1.38889×10^{-2}
1	-0.138889
2	-0.180556
3	-0.138889
4	-1.38889×10^{-2}

3. a)

i	c_i
1	1.4777309×10^{-2}
2	2.8700968×10^{-2}
3	4.0908778×10^{-2}
4	5.0521349×10^{-2}
5	5.6633300×10^{-2}
6	5.8304182×10^{-2}
7	5.4549063×10^{-2}
8	4.4328659×10^{-2}
9	2.6538926×10^{-2}

b)

i	c_i
1	-0.04829192
2	-0.09075051
3	-0.1282092
4	-0.1606679
5	-0.1881263
6	-0.2105847
7	-0.2280433
8	-0.2405021
9	-0.2479609

i	c_i
10	-0.2504198
11	-0.2478787
12	-0.2403376
13	-0.2277966
14	-0.2102548
15	-0.1877125
16	-0.1601701
17	-0.1276276
18	-0.09008508
19	-0.04754249

c)	i	c_i
	1	0.08569204
	2	0.1450892
	3	0.1815153
	4	0.1978995
	5	0.1968215
	6	0.1805512
	7	0.1510849
	8	0.1101763
	9	0.05936480

d)	i	c_i
	1	−0.08336102
	2	−0.1613321
	3	−0.2334986
	4	−0.2994176
	5	−0.3586155
	6	−0.4105871
	7	−0.4547930
	8	−0.4906576
	9	−0.5175666

i	c_i
10	−0.5348645
11	−0.5418520
12	−0.5377845
13	−0.5218675
14	−0.4932535
15	−0.4510389
16	−0.3942618
17	−0.3218967
18	−0.2328527
19	−0.1259662

4. a)

x_i	Approximation
0.0	0.00000000
0.1	0.014766286
0.2	0.028679521
0.3	0.040878126
0.4	0.050483351
0.5	0.056590497

x_i	Approximation
0.6	0.058259848
0.7	0.054507290
0.8	0.044294407
0.9	0.026518200
1.0	0.00000000

b)

x_i	Approximation
0.00	0.00000000
0.05	−0.047500000
0.10	−0.090000000
0.15	−0.12750000
0.20	−0.16000000
0.25	−0.18750000
0.30	−0.21000000
0.35	−0.22750000
0.40	−0.24000000
0.45	−0.24749999
0.50	−0.24999999

x_i	Approximation
0.55	−0.24749999
0.60	−0.23999999
0.65	−0.22749999
0.70	−0.20999999
0.75	−0.18749998
0.80	−0.15999998
0.85	−0.12749998
0.90	−0.089999975
0.95	−0.047499975
1.00	0.00000000

5. a)

i	c_i
1	0.01862440
2	0.03606449
3	0.05114782
4	0.06272542
5	0.06968343
6	0.07095396
7	0.06552600
8	0.05245540
9	0.03087436

b)

i	c_i
1	0.04499986
2	0.07999974
3	0.1049996
4	0.1199996
5	0.1249996
6	0.1199996
7	0.10499970
8	0.07999910
9	0.04499991

9. $0.045900355 \sin \pi x - 0.007321101 \sin 2\pi x + 0.002286058 \sin 3\pi x - 0.0009829617 \sin 4\pi x$
    $+ 0.0005077925 \sin 5\pi x$

15.

x_i	Approximation
0.0	0.00000000
0.1	-0.031970165
0.2	-0.054854433
0.3	-0.069553692
0.4	-0.076835058
0.5	-0.077408978
0.6	-0.071912530
0.7	-0.060922059
0.8	-0.044965103
0.9	-0.024502035
1.0	0.00000000

CHAPTER 11

Exercise Set
11.2 (page 576)

1. $w_3 = 1.00 = u(0.5, 1.5)$, $w_6 = 0.25 = u(0.5, 1)$, $w_9 = 0 = u(0.5, 0.5)$

3. a)

i	j	x_i	y_j	$w_{i,j}$
3	3	0.3	0.3	0.089941
3	7	0.3	0.7	0.209928
7	3	0.7	0.3	0.209951
7	7	0.7	0.7	0.489941

b)

i	j	x_i	x_j	$w_{i,j}$
6	6	0.3	0.3	1.068118
6	14	0.3	0.7	1.065449
14	6	0.7	0.3	3.795841
14	14	0.7	0.7	3.793414

c)

i	j	x_i	y_j	$w_{i,j}$
4	3	0.8	0.3	1.27136
4	7	0.8	0.7	1.75084
8	3	1.6	0.3	1.61675
8	7	1.6	0.7	3.06587

d)

i	j	x_i	y_j	$w_{i,j}$
2	1	1.256637	0.3141593	0.2951912
2	3	1.256637	0.9424778	0.1830968
4	1	2.513274	0.3141593	-0.7721915
4	3	2.513274	0.9424778	-0.4785097

9.

i	j	x_i	y_j	$w_{i,j}$
1	4	0.1	0.4	88
2	1	0.2	0.1	66
4	2	0.4	0.2	88

1. a)

i	j	x_i	t_j	w_{ij}
1	1	0.5	0.05	0.7709764
2	1	1.0	0.05	1.090324
3	1	1.5	0.05	0.7709770
1	2	0.5	0.1	0.8406147
2	2	1.0	0.1	1.188807
3	2	1.5	0.1	0.8406162

b)

i	j	x_i	t_j	$w_{i,j}$
1	1	1/3	0.05	1.59728
2	1	2/3	0.05	-1.59728
1	2	1/3	0.1	1.47300
2	2	2/3	0.1	-1.47300

3. a) For $h = 0.1$ and $k = 0.01$:

i	j	x_i	t_j	$w_{i,j}$
4	50	0.4	0.5	-9.3352×10^8
10	50	1.0	0.5	-9.1860×10^8
17	50	1.7	0.5	2.6047×10^8

b) For $h = 0.1$ and $k = 0.005$:

i	j	x_i	t_j	$w_{i,j}$
4	100	0.4	0.5	0.1757581
10	100	1.0	0.5	0.2990169
17	100	1.7	0.5	0.1357617

c)

i	j	x_i	t_j	$w_{i,j}$
4	10	0.8	0.4	1.166142
8	10	1.6	0.4	1.252404
12	10	2.4	0.4	0.4681804
16	10	3.2	0.4	-0.1027628

d)

i	j	x_i	t_j	$w_{i,j}$
2	10	0.2	0.4	0.3921147
4	10	0.4	0.4	0.6344550
6	10	0.6	0.4	0.6344550
8	10	0.8	0.4	0.3921148

5. a) For $h = 0.1$ and $k = 0.01$:

i	j	x_i	t_j	$w_{i,j}$
4	50	0.4	0.5	2.3541×10^{-9}
10	50	1.0	0.5	6.2567×10^{-17}
17	50	1.7	0.5	-3.8090×10^{-9}

b) For $h = 0.1$ and $k = 0.01$:

i	j	x_i	t_j	$w_{i,j}$
4	50	0.4	0.5	0.1770914
10	50	1.0	0.5	0.3012839
17	50	1.7	0.5	0.1367806

c)

i	j	x_i	t_j	$w_{i,j}$
5	10	1	0.4	1.312434
10	10	2	0.4	0.9050248
15	10	3	0.4	0.03253811

d)

i	j	x_i	t_j	$w_{i,j}$
3	10	0.3	0.4	0.5440574
5	10	0.5	0.4	0.6724913
7	10	0.7	0.4	0.5440568

13. a)

i	j	r_i	t_j	$w_{i,j}$
6	50	0.6	10	700.23
7	50	0.7	10	450.42
8	50	0.8	10	325.78
9	50	0.9	10	417.30

b) 1737.8

Exercise Set 11.4 (page 600)

1.

i	j	x_i	t_j	w_{ij}
1	1	0.25	0.25	0.17678
2	1	0.50	0.25	0.25000
3	1	0.75	0.25	-0.17678
1	2	0.25	0.50	-0.75000
2	2	0.50	0.50	0.00000
3	2	0.75	0.50	1.25000

3. For $h = 0.1$ and $k = 0.05$:

i	j	x_i	t_j	w_{ij}
2	10	0.2	0.5	0.00285202
5	10	0.5	0.5	0.00485216
7	10	0.7	0.5	0.00392548

For $h = 0.05$ and $k = 0.1$:

i	j	x_i	t_j	w_{ij}
4	5	0.2	0.5	-0.0028789
10	5	0.5	0.5	-0.0048977
14	5	0.7	0.5	-0.0039623

For $h = 0.05$ and $k = 0.05$:

i	j	x_i	t_j	w_{ij}
4	10	0.2	0.5	-1.6792×10^{-15}
10	10	0.5	0.5	-3.1780×10^{-15}
14	10	0.7	0.5	-2.8588×10^{-15}

5.

i	j	x_i	t_j	w_{ij}
2	3	0.2	0.3	0.6729902
5	3	0.5	0.3	6.317389×10^{-16}
7	3	0.7	0.3	-0.6729902

7. a) $p(0.5, 0.5) \approx 0.9$ and $p(0.5, 1.0) \approx 0.9$

Exercise Set 11.5 (page 614)

1. With $E_1 = (0.25, 0.75)$, $E_2 = (0, 1)$, $E_3 = (0.5, 0.5)$, and $E_4 = (0, 0.5)$, basis functions are

$$\phi_1(x) = \begin{cases} 4x & \text{on } T_1 \\ -2 + 4y & \text{on } T_2 \end{cases} \qquad \phi_2(x) = \begin{cases} -1 - 2x + 2y & \text{on } T_1 \\ 0 & \text{on } T_2 \end{cases}$$

$$\phi_3(x) = \begin{cases} 0 & \text{on } T_1 \\ 1 + 2x - 2y & \text{on } T_2 \end{cases} \qquad \phi_4(x) = \begin{cases} 2 - 2x - 2y & \text{on } T_1 \\ 2 - 2x - 2y & \text{on } T_2 \end{cases}$$

and $\gamma_1 = 0.323825$, $\gamma_2 = 0$, $\gamma_3 = 1.0000$, and $\gamma_4 = 0$.

3. (See diagram)

$K = 8$, $N = 8$, $M = 32$, $n = 9$, $m = 25$, $NL = 0$;

$\gamma_1 = 0.511023$ $\gamma_6 = 0.720476$
$\gamma_2 = 0.720476$ $\gamma_7 = 0.507897$
$\gamma_3 = 0.507898$ $\gamma_8 = 0.720476$
$\gamma_4 = 0.720475$ $\gamma_9 = 0.511023$
$\gamma_5 = 1.01885$ $\gamma_i = 0$, $10 \leq i \leq 25$

$u(0.125, 0.125) \approx 0.720475$
$u(0.125, 0.25) \approx 0.690343$
$u(0.25, 0.125) \approx 0.690343$
$u(0.25, 0.25) \approx 0.720475$

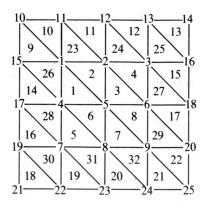

7. (See diagram)

$K = 0$, $N = 20$, $M = 32$, $n = 20$, $m = 27$, $NL = 14$;

$\gamma_1 = 19.25965$ $\gamma_8 = 17.17999$ $\gamma_{15} = 15.30939$ $\gamma_{22} = 15$
$\gamma_2 = 17.12355$ $\gamma_9 = 19.29031$ $\gamma_{16} = 15.54661$ $\gamma_{23} = 15$
$\gamma_3 = 15.94424$ $\gamma_{10} = 21.47760$ $\gamma_{17} = 16.26654$ $\gamma_{24} = 15$
$\gamma_4 = 15.34842$ $\gamma_{11} = 19.28071$ $\gamma_{18} = 17.54011$ $\gamma_{25} = 15$
$\gamma_5 = 15.17034$ $\gamma_{12} = 17.45838$ $\gamma_{19} = 19.41416$ $\gamma_{26} = 15$
$\gamma_6 = 15.36329$ $\gamma_{13} = 16.21978$ $\gamma_{20} = 21.61106$ $\gamma_{27} = 15$
$\gamma_7 = 15.97682$ $\gamma_{14} = 15.52535$ $\gamma_{21} = 15$

$u(1, 0) \approx 17.45838$

$u(4, 0) \approx 17.54011$

$u\left(\dfrac{5}{2}, \dfrac{\sqrt{3}}{2}\right) \approx 15.17034$

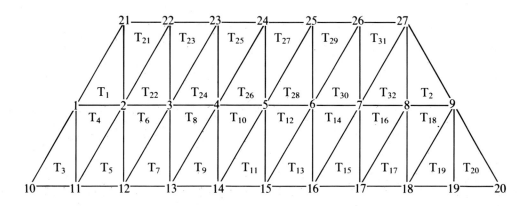

Index

Index of Algorithms